14th INTERNATIONAL SYMPOSIUM ON PROCESS SYSTEMS ENGINEERING

VOLUME 3

COMPUTER-AIDED CHEMICAL ENGINEERING, 49

14ᵗʰ INTERNATIONAL SYMPOSIUM ON PROCESS SYSTEMS ENGINEERING

VOLUME 3

Edited by

Yoshiyuki Yamashita
Professor and Chair of Department of Chemical Engineering
Tokyo University of Agriculture and Technology (TUAT), Tokyo, Japan
yama_pse@cc.tuat.ac.jp

Manabu Kano
Professor, Department of Systems Science
Kyoto University, Kyoto, Japan
manabu@human.sys.i.kyoto-u.ac.jp

ELSEVIER

Amsterdam – Boston – Heidelberg – London – New York – Oxford
Paris – San Diego – San Francisco – Singapore – Sydney – Tokyo

Elsevier
Radarweg 29, PO Box 211, 1000 AE Amsterdam, Netherlands
The Boulevard, Langford Lane, Kidlington, Oxford OX5 1GB, UK
50 Hampshire Street, 5th Floor, Cambridge, MA 02139, USA

British Library Cataloguing in Publication Data
A catalogue record for this book is available from the British Library

Library of Congress Cataloging-in-Publication Data
A catalog record for this book is available from the Library of Congress

ISBN (Volume 3):   978-0-443-18726-1
ISBN (Set) :   978-0-323-85159-6
ISSN: 1570-7946

For information on all Elsevier publications visit our
website at https://www.elsevier.com/

 Working together
to grow libraries in
developing countries

www.elsevier.com • www.bookaid.org

Publisher: Susan Dennis
Acquisition Editor: Anita Koch
Editorial Project Manager: Lena Sparks
Production Project Manager: Paul Prasad Chandramohan
Designer: Greg Harris

Typeset by STRAIVE

Contents

Contributed Papers: Energy, Food and Environmental Systems

Contributed Papers: Pharma and Healthcare Systems

Proceedings of the 14th International Symposium on Process Systems Engineering – PSE 2021+
June 19–23, 2022, Kyoto, Japan ©2022 Elsevier B. V. All rights reserved.
http://dx.doi.org/10.1016/B978-0-323-85159-6.50252-9

Equivalence Judgment of Equation Groups Representing Process Dynamics

Chunpu Zhang, Shota Kato, Manabu Kano*

Department of Systems Science, Kyoto University, Yoshida-honmachi, Sakyo-ku, Kyoto 606-8501, Japan

manabu@human.sys.i.kyoto-u.ac.jp

Abstract

Digital twins are expected to play a key role in digital transformation of the process industry. Although process informatics, i.e., process data analytics, has attracted a lot of attention, physical models are essential to realizing the digital twins. Building a physical model of a complex industrial process is toil. We aim to free the engineers from physical model building by developing an automated physical model builder, AutoPMoB. AutoP-MoB performs five tasks: 1) retrieving documents regarding a target process from literature databases, 2) converting the format of each document to HTML format, 3) extracting information necessary for building a physical model from the documents, 4) judging the equivalence of the information extracted from different documents, and 5) reorganizing the information to output a desired physical model. In this study, we propose a method of judging the equivalence of two equation groups to accomplish task 4. The proposed method first converts two equation groups in mathematical markup language (MathML) format into a format that can be manipulated in a computer algebra system (CAS). Then, the variables not shared between the groups are eliminated using the CAS. The method judges whether the two equation groups are equivalent. The results of several case studies demonstrated that the proposed method accurately judged the equivalence, including physical models of a continuous stirred-tank reactor. We also developed a web application that can easily judge the equivalence of MathML-formatted equation groups. This application is expected to reduce the effort required to find out the different models contained in multiple documents, and become an important part of AutoPMoB.

Keywords: Artificial intelligence, Equation groups equivalence, First principle model, Process modeling, Natural language processing

1. Introduction

The use of digital twins has become popular during the digitization of machinery and production systems in the manufacturing industry (El Saddik, 2018). Digital twins are virtual representations of physical entities and can be used to simulate the states inside plants under various conditions. Physical models are indispensable to enable digital twins to represent real-life phenomena accurately.

To build a physical model, experts need to survey the literature and build a physical model by trial and error until the model meets all requirements. The process of building the physical model is time-consuming.

This research aims to develop an automated physical model builder (AutoPMoB) that can automatically build a physical model. AutoPMoB first extracts information of variables, formulas, and experimental data from the literature and then combines the information to build a new physical model. Different documents may use different symbols to express the same variable and different ways to express the same formula. To combine information from multiple documents, AutoPMoB needs to recognize variables and formulas and judge their equivalence accurately.

Formulas are inherently hierarchical and can be represented as symbol layout trees (SLTs) or operator trees (OPTs) (Mansouri et al., 2019). SLTs capture the placement and nesting of symbols on writing lines, while OPTs capture the mathematical semantics of the application of operators to operands. The SLT and OPT of a formula $x - y^2 = 0$ are shown in Figure 1.

There has been no work on equivalence judgment of two equation groups. It seems possible to judge the equivalence by setting a threshold on the similarity of the two equation groups. Several similarity measures are available. Zhong and Zanibbi (2019) defined similarity between formulas using paths from OPTs. Mansouri et al. (2019) built an embedding model of formulas using SLTs and OPTs and defined the similarity as their cosine similarity. These similarities are based on appearance, and similar-looking formulas do not necessarily perform the same calculations; thus, the similarities do not properly work for the equivalence judgment.

In this work, we propose a rule-based method of judging the equivalence of two equation groups. In our proposed method, equation groups in mathematical markup language (MathML) format are converted into a format that can be manipulated using a computer algebra system (CAS). Then, the variables not shared between the groups are eliminated using the CAS. The equivalence of the two equation groups is judged by checking whether any equation in one equation group appears in the other equation group. We assume that variables having the same meaning are represented by the same symbol.

2. Proposed Method

This research aims to develop a system that can judge the equivalence of two equation groups not by their appearance but by the calculations they perform. Formulas in MathML format are just strings and cannot be manipulated following calculation rules. Thus, the

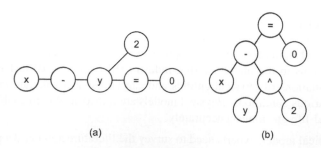

(a) (b)

Figure 1: (a) SLT and (b) OPT of the formula $x - y^2 = 0$.

proposed method converts formulas in MathML format into another format that can be manipulated mathematically using SymPy (Meurer et al., 2017), which is a CAS written in Python. Finally, the proposed method judges whether the equation groups are equivalent using the algorithms explained in sections 2.2. and 2.3.

2.1. Conversion of Equations

There are two types of MathML: presentation markup and content markup. Presentation markup captures notation structure while content markup captures mathematical structure. This research uses presentation markup since it is mainly used on websites. We convert a formula in MathML format into tangents-SLT by using tangent-s, a formula search system developed by Davila and Zanibbi (2017). Tangents-SLT is an SLT in which each symbol is assigned a certain type. A list holding the hierarchical information of the formula is created by parsing the tangents-SLT. The list is converted into a SymPy object. For example, the tangents-SLT of a formula $\frac{a+b}{c+d}$ is first converted into [tangents-SLT 1, tangents-SLT 2, 'frac'], where tangents-SLT 1 and tangents-SLT 2 refer to the tangents-SLT of the formula $a + b$ and the formula $c + d$, respectively. Tangents-SLT 1 and tangents-SLT 2 are then converted into a list of ['a', '+', 'b'] and ['c', '+', 'd']. Finally, the list [['a', '+', 'b'], ['c', '+', 'd'], 'frac'] is converted to a SymPy object.

2.2. Equivalence Judgment of Equations

In the proposed algorithm, two equations must have the same variables to be judged equivalent. The two equations are judged equivalent if the solutions for any variable of the equations are the same; otherwise, they are judged non-equivalent. As for equations that contain a derivative term such as $\frac{dV}{dt}$, the whole term is regarded as one variable. For example, the solutions of the equations $\frac{dV}{dt} = w_i - w$ and $\frac{dV}{dt} + w - w_i = 0$ for w_i are both $\frac{dV}{dt} + w$, which means they are judged equivalent. The solution for V is not calculated since the term $\frac{dV}{dt}$ is regarded as one variable. We use CAS to solve equations.

2.3. Equivalence Judgment of Equation Groups

In this research, we judge not only the equivalence between equations but also the equivalence between two equation groups consisting of multiple equations.

Two equation groups have the same degree of freedom (DOF) when they are equivalent; therefore, the DOF of the two equation groups are compared at first. If two equation groups have the same equations, they are obviously equivalent, and both groups have the same set of variables. Hence, we eliminate the variables not shared between the groups. If there remain variables not shared between the groups, the two equation groups are judged non-equivalent. Finally, for each equation in one equation group, our proposed method seeks the equation performing the same calculation in another equation group using the algorithm introduced in section 2.2. If such an equation does not exist, the two equation groups are judged non-equivalent; otherwise, they are judged equivalent.

3. Case Studies

We evaluated our proposed method through several case studies and confirmed that the proposed method could accurately judge the equivalence of equation groups. In this section, we introduce one case study.

In Figure 2, three equation groups are physical models of a continuous stirred-tank reactor (CSTR), where an exothermic and irreversible reaction (A→B) takes place. The two equation groups (a) and (b), which are used by Manzi and Carrazzoni (2008) and Nekoui et al. (2010), are equivalent but written in different ways. The equation group (c) is non-equivalent to the equation groups (a) and (b). The proposed method accurately judged that the equation groups (a) and (b) are equivalent while the equation groups (a) and (c) are non-equivalent.

We developed a web application for equivalence judgment. Figure 3 shows a screenshot to judge the equivalence of the two equation groups (a) and (b). To use this web application, we first choose two files, in which equation groups are written in MathML format (Part 1). Our web application then shows the selected equation groups (Part 2) and the equivalence judgment result (Part 3). Equation groups in MathML format can be easily

$$\frac{dC_A}{dt} = \frac{F}{V}(C_{A,in} - C_A) - k_0 \exp\left(-\frac{E}{RT}\right)C_A$$

$$\frac{dT}{dt} = \frac{F}{V}(T_{in} - T) + \frac{h_r}{\rho c_p}k_0 \exp\left(-\frac{E}{RT}\right)C_A - \frac{UA_r}{V\rho c_p}(T - T_j)$$

(a)

$$\frac{dC_A}{dt} = \frac{F}{V}(C_{A,in} - C_A) - k_0 \exp\left(-\frac{E}{RT}\right)C_A$$

$$\frac{dT}{dt} = \frac{F}{V}(T_{in} - T) + \frac{h_r}{\rho c_p}k_0 \exp\left(-\frac{E}{RT}\right)C_A - \frac{Q}{V\rho c_p}$$

$$Q = UA_r(T - T_j)$$

(b)

$$\frac{dC_A}{dt} = \frac{F}{V}(C_{A,in} - C_A) - k_0 \exp\left(-\frac{E}{RT}\right)$$

$$\frac{dT}{dt} = \frac{F}{V}(T_{in} - T) + \frac{h_r}{\rho c_p}k_0 \exp\left(-\frac{E}{RT}\right) - \frac{UA_r}{V\rho c_p}(T - T_j)$$

(c)

Figure 2: Physical models of a CSTR; (a) and (b) are equivalent while (c) is non-equivalent to (a) and (b).

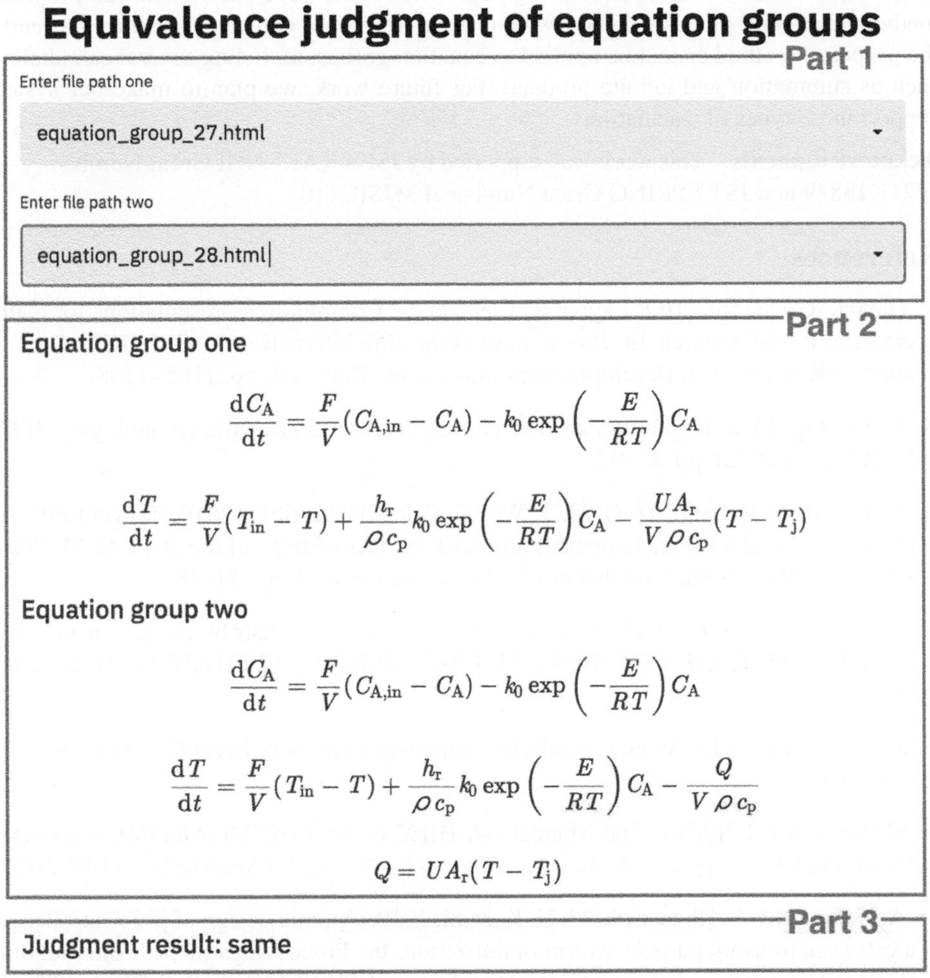

Part 3

Equivalence judgment of equation groups

Part 1

Enter file path one

equation_group_27.html

Enter file path two

equation_group_28.html

Part 2

Equation group one

$$\frac{\mathrm{d}C_\mathrm{A}}{\mathrm{d}t} = \frac{F}{V}(C_{\mathrm{A,in}} - C_\mathrm{A}) - k_0 \exp\left(-\frac{E}{RT}\right)C_\mathrm{A}$$

$$\frac{\mathrm{d}T}{\mathrm{d}t} = \frac{F}{V}(T_\mathrm{in} - T) + \frac{h_\mathrm{r}}{\rho\,c_\mathrm{p}}k_0 \exp\left(-\frac{E}{RT}\right)C_\mathrm{A} - \frac{U A_\mathrm{r}}{V\rho\,c_\mathrm{p}}(T - T_\mathrm{j})$$

Equation group two

$$\frac{\mathrm{d}C_\mathrm{A}}{\mathrm{d}t} = \frac{F}{V}(C_{\mathrm{A,in}} - C_\mathrm{A}) - k_0 \exp\left(-\frac{E}{RT}\right)C_\mathrm{A}$$

$$\frac{\mathrm{d}T}{\mathrm{d}t} = \frac{F}{V}(T_\mathrm{in} - T) + \frac{h_\mathrm{r}}{\rho\,c_\mathrm{p}}k_0 \exp\left(-\frac{E}{RT}\right)C_\mathrm{A} - \frac{Q}{V\,\rho\,c_\mathrm{p}}$$

$$Q = U A_\mathrm{r}(T - T_\mathrm{j})$$

Part 3

Judgment result: same

Figure 3: The screenshot of our web application; file selections for equivalence judgment (Part 1); selected equation groups (Part 2); the equivalence judgment result (Part 3).

obtained from PDF or TeX files using existing tools such as InftyReader (Suzuki et al., 2003) or L^AT_EXML (Miller, 2018). By using this web application, the equivalence of two equation groups can be easily judged.

4. Conclusion

In this work, we proposed a rule-based method for equation group equivalence judgment. In the proposed method, we convert formulas in MathML format into SymPy objects, which we can manipulate mathematically. We judge the equivalence of equation groups by checking whether any equation in one equation group performs the same calculation as the equation in the other equation group. The results have shown that the proposed method accurately judges whether two equation groups are equivalent or not. Currently, the proposed method cannot be applied to equation groups, including several calculations such as summation and infinite product. For future work, we plan to make our system support more types of calculations.

Acknowledgments This work was supported by JSPS KAKENHI Grant Number JP21K18849 and JST SPRING Grant Number JPMJSP2110.

References

K. Davila, R. Zanibbi, 2017. Layout and semantics: Combining representations for mathematical formula search. In: Proceedings of the 40th International ACM SIGIR Conference on Research and Development in Information Retrieval. pp. 1165–1168.

A. El Saddik, 2018. Digital twins: The convergence of multimedia technologies. IEEE MultiMedia 25 (2), pp. 87–92.

B. Mansouri, S. Rohatgi, D. Oard, J. Wu, C. Giles, R. Zanibbi, 2019. Tangent-cft: An embedding model for mathematical formulas. In: Proceedings of the 2019 ACM SIGIR international conference on theory of information retrieval. pp. 11–18.

J. Manzi, E. Carrazzoni, 2008. Analysis and optimization of a cstr by direct entropy minimization. JOURNAL OF CHEMICAL ENGINEERING OF JAPAN 41 (3), pp. 194–199.

A. Meurer, et al., 2017. Sympy: symbolic computing in python. PeerJ Computer Science 3, p. e103.

B. Miller, 2018. LaTeXML The Manual—A LaTeX to XML/HTML/MathML Converter, Version 0.8.3. https://dlmf.nist.gov/LaTeXML/, (Accessed on 11/03/2021).

M. A. Nekoui, M. A. Khameneh, M. H. Kazemi, 2010. Optimal design of pid controller for a cstr system using particle swarm optimization. In: Proceedings of 14th International Power Electronics and Motion Control Conference EPE-PEMC 2010. IEEE, pp. T7–63.

M. Suzuki, F. Tamari, R. Fukuda, S. Uchida, T. Kanahori, 2003. Infty: an integrated ocr system for mathematical documents. In: Proceedings of the 2003 ACM symposium on Document engineering. pp. 95–104.

W. Zhong, R. Zanibbi, 2019. Structural similarity search for formulas using leaf-root paths in operator subtrees. In: European Conference on Information Retrieval. Springer, pp. 116–129.

Proceedings of the 14th International Symposium on Process Systems Engineering – PSE 2021+
June 19-23, 2022, Kyoto, Japan © 2022 Elsevier B.V. All rights reserved.
http://dx.doi.org/10.1016/B978-0-323-85159-6.50253-0

Data-driven operation support for equipment deterioration detection in drug product manufacturing

Philipp Zürcher[a], Sara Badr[a*], Stephanie Knueppel[b], Hirokazu Sugiyama[a]

[a]*Department of Chemical System Engineering, The University of Tokyo, Tokyo 113-8656, JAPAN*
[b]*Engineering, Science & Technology, F. Hoffmann – La Roche Ltd., Wurmisweg, 4303 Kaiseraugst, SWITZERLAND*

badr@pse.t.u-tokyo.ac.jp

Abstract

This work presents a data-driven methodology for decision-support aiming at reliability experts in drug product manufacturing. The developed tool incorporates three consecutive stages. Firstly, equipment condition monitoring is performed through principal component analysis for dimensionality reduction on the process monitoring dataset. Equipment deteriorations are visualized by shifts in the monitored principal curve giving indication about deviating equipment condition. Secondly, a localization of the underlying physical source for the detected equipment deterioration is performed. Thereby, the impact from individual sensors to the observed shifts is investigated giving additional information to decision-makers on the underlying physical phenomena and location in the unit. In the last stage, prevention, the information from the two previous stages is combined in order to perform tailored maintenance actions during the production phase in order to minimize the occurrence of unplanned downtime. The developed methodology is demonstrated in the form of a case study. Industrial process data from a sterilization unit which is part of an aseptic filling line of F. Hoffmann – La Roche Ltd. located in Kaiseraugst, Switzerland is used.

Keywords: Predictive maintenance, decision-support, industrial application.

1. Introduction

Digitalization is the central pillar for the introduction of Industry 4.0. (Diez-Olivan *et al.* 2019) However, unlike other industry sectors, sophisticated data-driven applications are yet to be explored to fully take advantage of the abundance of recorded process data that is available in pharmaceutical manufacturing. The pharmaceutical manufacturing sector is highly regulated by government agencies, such as the FDA, in order to guarantee product safety. Thereby, regulators are requiring manufacturers to store process monitoring information over multiple years for backtracking purposes. (Casola *et al.*, 2019) A large historical data base is thus produced, which could potentially be used in order to gain data-driven insights into the current equipment condition.

Data-driven approaches including the application of principal component analysis (PCA) are well established in the field of process monitoring. However, few approaches have been proposed that focus on the long-time equipment related trends in the process data. (Reis *et al.*, 2017) In other industry sectors, such approaches have been presented under the term of predictive maintenance. (Bousdekis *et al.* 2019) However, in the pharmaceutical industry such approaches are scarce. One obstacle there is the costly

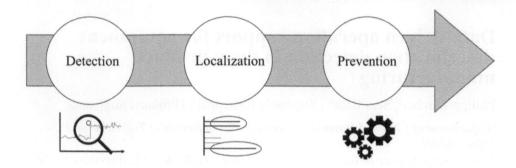

Figure 1: Overview of the individual steps of the developed methodology for equipment condition monitoring.

revalidation process associated with any changes in the production line including the installation of new sensors for predictive maintenance purposes. (Zürcher *et al.*, 2020)

Equipment reliability is of great importance in the manufacturing of highly valuable pharmaceutical products as any equipment malfunction or fault can potentially result in a loss of highly valuable drug product. Therefore, maintenance is performed on a regular basis within production facilities. It is based on a pure time-based maintenance scheme, which does not consider real-time equipment condition. Although having regular maintenance intervals, unexpected equipment faults induced by manual interventions or resulting from equipment deterioration are still frequent. Developing a method to determine the current condition of equipment is necessary, in order to minimize the occurrence of such unexpected events.

This work presents a decision-support tool for reliability experts to evaluate the current equipment condition. Thereby, aiming to provide support in early detection and localization of equipment deterioration to provide guidance for required maintenance interventions during production. Results are presented in the form of a case study for the aseptic filling process.

2. Methodology

In this work, real-time production data was used. Non-production time in the continuously collected process data was excluded. The production datasets form the basis for the developed methodology in this work. In comparison to traditional monitoring, the investigated time frame is on production campaigns and entire production phases, instead of individual batch runs.

The decision-support methodology is composed of three consecutive stages as summarized in Fig. 1. First, in the detection stage, equipment monitoring is performed, and data records are maintained. Multi-way PCA is performed for dimensionality reduction of process data. (Nomikos & MacGregor, 1994). The data is unfolded variable-wise and auto-scaled prior to the application of PCA. This is followed by the application of a noise-filter algorithm on the principal curves. (Savitzky & Golay, 1964) Long-term based deterioration becomes visible, expressed through fluctuations or shifts of the monitoring curve.

The ongoing production cycle dataset is investigated with regards to the observed shifts in any principal component which indicates a change in the equipment condition. The deterioration zone is defined based on historical analysis of the data where all shifts

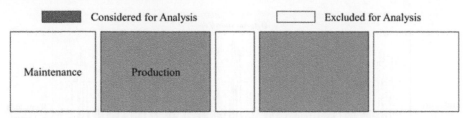

Figure 2: Overview over the production and maintenance schedule for the aseptic filling line of one year that is considered in this case study.

reaching the area ultimately led to an equipment fault within the near future (several days to weeks) after the detected shift and a restoration of the base line.

Secondly, the results from the principal component analysis are used for an in-depth loading analysis for the principal curve where a shift has been detected. Thereby, the individual contributions to the explained variance by the original variables representing measurements from installed sensors within the manufacturing unit are considered. Through the identification of the dominantly contributing sensor(s), the detected shift can be localized within the unit and the underlying physical phenomenon (e.g., leakage) is identified. Together with operator expert knowledge, this stage supports decision-makers to locate deteriorations and target maintenance actions.

Thirdly, by using the combined information from the detection and localization stage, maintenance actions can be proposed and windows where actions become necessary can be determined. Ultimately, a step towards more intelligent maintenance procedures in drug product manufacturing is achieved that supports experts in monitoring the actual equipment condition.

3. Case study

This work considers one year of industrial production data obtained from a sterilization unit that is part of the aseptic filling process. It represents a key process within drug product manufacturing.

Currently, maintenance is performed in a time-based manner in the facility twice a year. Consequently, two production phases over the course of a year are considered which are separated by maintenance actions and are combined in a data set within the scope of this work. The schematic overview is shown in Fig. 2.

4. Results & Discussion

4.1. Detection

In order to demonstrate the application of the methodology, the resulting principal curves for the sterilization unit were analysed for shifts. PC2 was selected as it showed long-term baseline shifts upwards into the deterioration zone, which persisted for several days to weeks as shown in Fig. 3.

Observed shifts having an intensity within the defined deterioration zone eventually led to an unexpected equipment fault. PC2 was identified to be suitable for long-term condition monitoring as shifts leading to an intensity within the deterioration zone always ended with an equipment fault requiring maintenance actions. The two production cycles – each representing about 3-4 months of commercial production - are combined, thereby visualizing the monitoring curve for an entire year of production. Steps in the principal curve can be observed which can be clearly separated from individual, short-

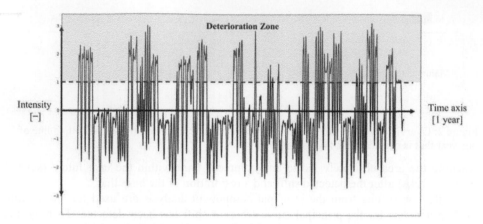

Figure 3: Representation of the principal curve (PC2) indicating a zone where increased equipment deterioration in observed which could eventually lead to an equipment fault.

term based peaks frequently occurring and resulting from intra and inter batch variance. Furthermore, observed shifts of PC2 last for up to several weeks until eventually a restoring of lower intensity is obtained. This restoration is connected to a performed maintenance action after an (unexpected) equipment fault has occurred. Therefore, an observed shift indicates that equipment deterioration is occurring which will eventually lead to a fault. Through the detection of such shifts, decision-makers can become aware of an underlying problem within the unit.

4.2. Localization

The loading plot for PC2 where shifts have been detected is shown in Fig.4. Highlighted bars represent sensors with the highest impact on the explained variance of the principal curve. Related sensors in the unit are the ventilator performance as well as the position

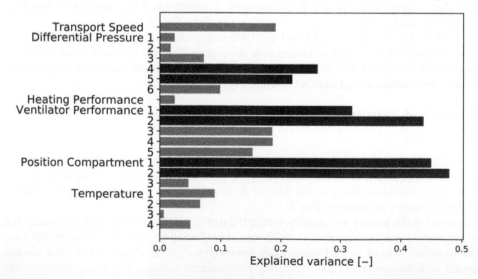

Figure 4: Overview of the loadings for principal component 2. Different numbers for a sensor type indicate different locations within the production unit. Highlighted loadings represent sensor values contributing most the monitoring curve in the detection stage.

setting of the compartments. Both are connected to the pre-heating zone (1) and cooling zone (2) respectively. Therefore, shifts observed can be located to belong to this area within the sterilization unit. Furthermore, the differential pressures between the pre-heating zone and sterilization zone (4) as well as cooling zone and sterilization zone (5), respectively, are dominantly expressed in PC2. Therefore, shifts observed for the principal curve are likely to be connected to pressure related equipment, such as valves which can suffer from deterioration eventually resulting in leakage and breakage of the sterile environment.

Results from the loading analysis show that the problem can be reduced to a small number of related sensors in specific areas of the unit. The localization stage therefore can aid decision-makers to identify the origin of detected equipment problems and increase maintenance actions through the offered possibility of better tailored actions.

4.3. Prevention

In the prevention step, the information gained from the detection and localization stage is combined. Thereby, tailored maintenance actions can be determined during phases with increased equipment deterioration represented by arrows in Fig. 6. Proposing such actions during a production cycle is an advantage for decision-makers in reliability engineering departments in comparison to the conventional practice where the condition is not monitored in real-time at all.

Phases with increased equipment deterioration eventually led to an equipment fault that is related to the identified location and underlying physical conditions that were determined in the localization stage. Therefore, the occurrence of unexpected equipment faults can be reduced by combining the information from detection and localization stages in the final prevention stage and tailored maintenance actions can be planned through the data-driven insights obtained by the proposed methodology.

5. Conclusion

In this work, a methodology for equipment condition monitoring as decision-support for reliability experts in drug product manufacturing is presented. A three-stage methodology is presented including a detection, localization and prevention stage. Thereby, equipment condition shifts due to deterioration are first identified in the detection stage

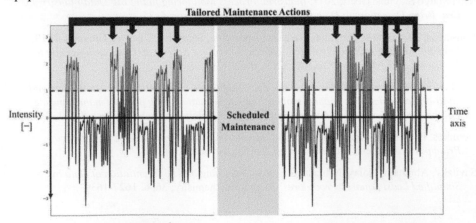

Figure 5: Schematic representation of the proposed type of maintenance after the implementation of the developed methodology, moving from scheduled to (partly) tailored maintenance.

by monitoring principal curves. Then, the deterioration is localized by loading analysis enabling both the identification of the underlying physical condition problem as well as the location within the unit.

Finally, in the prevention stage, through tailored maintenance actions, the occurrence of unexpected equipment faults from deterioration should be minimized. In phases where increased deterioration of the manufacturing equipment is detected, adequate maintenance actions have to be planned and production schedules altered accordingly. By using the information from the localization stage, tailored actions can be planned and executed thus reducing maintenance efforts.

The effective application of the methodology for monitoring the equipment condition of the sterilization unit as part of the aseptic filling process has been demonstrated. Its usefulness has been shown in both, the identification of phases with increased deterioration characterized by shifts in the monitoring curve, as well as the localization of the physical origin of the deterioration by a loading analysis. Consequently, decision-support can be provided for reliability experts on needed maintenance for special equipment. Thus, tailored maintenance actions can be performed leading to a decrease in maintenance effort, time, and resulting in a higher cost effectiveness. In order to further improve the applicability of data-driven monitoring, it is not enough to only predict changes in the underlying equipment condition. The timeframe until a failure would likely occur also needs to be addressed. Therefore, further research on how to predict the exact occurrence of such failures has to be conducted in order to further enhance more effective production and maintenance processes in drug product manufacturing.

References

Diez-Olivan, Alberto; Del Ser, Javier; Galar, Diego; Sierra, Basilio; 2019; *Data fusion and machine learning for industrial prognosis: Trends and perspectives towards Industry 4.0*; Information Fusion, 50, 92-111.

Casola, Gioele; Siegmund, Christian; Mattern., Markus; Sugiyama, Hirokazu; 2019; *Data mining algorithm for pre-processing of biopharmaceutical drug product manufacturing records*; Computers and Chemical Engineeering, 124, 253-269.

Reis, Marco S.; Gins, Geert; 2017; *Industrial Process Monitoring in the Big Data/Industry 4.0 Era: From Detection, to Diagnosis, to Prognosis;* Processes, 5, 35.

Bousdekis, Alexandros; Lepenioti, Katerina; Apostolou, Dimitris ; Mentzas, Gregoris; 2019; *Decision Making in Predictive Maintenance: Literature Review and Research Agenda for Industry 4.0*; IFAC-PapersOnLine; 52-13, 607-612

Zürcher, Philipp; Shirahata, Haruku; Badr, Sara; Sugiyama, Hirokazu; 2020; *Multi-stage and multi-objective decision-support tool for biopharmaceutical drug product manufacturing: Equipment technology evaluation*; Chemical Engineering Research and Design, 161, 240-252.

Nomikos, Paul; MacGregor, John F.; 1994; *Monitoring Batch Processes Using Multiway Principal Component Analysis;* AIChE Journal; 40; 8; 1361-1375.

Savitzky, Abraham; Golay, Marcel J.E.; 1964; *Smoothing and Differentiation of Data by Simplified Least Squares Procedures;* Analytical Chemistry; 36, 8, 1627-1639

Proceedings of the 14th International Symposium on Process Systems Engineering – PSE 2021+
June 19-23, 2022, Kyoto, Japan © 2022 Elsevier B.V. All rights reserved.
http://dx.doi.org/10.1016/B978-0-323-85159-6.50254-2

Pilot Plant 4.0: A Review of Digitalization Efforts of the Chemical and Biochemical Engineering Department at the Technical University of Denmark (DTU)

Mark Nicholas Jones, Mads Stevnsborg, Rasmus Fjordbak Nielsen, Deborah Carberry, Khosrow Bagherpour, Seyed Soheil Mansouri, Steen Larsen, Krist V. Gernaey, Jochen Dreyer, John Woodley, Jakob Kjøbsted Huusom, Kim Dam-Johansen*

Department of Chemical and Biochemical Engineering, Technical University of Denmark, Building 228A, 2800 Kgs. Lyngby, Denmark
kdj@kt.dtu.dk

Abstract

The pilot plant at the Chemical and Biochemical Engineering Department at DTU (DTU Kemiteknik, abbrv. DTU KT) serves as a facility for research & education with access to various process equipment, commonly employed in up-and down-stream processes. Among the available equipment are fermenters, membranes, distillation columns, absorbers, desorbers, extractors, crystallizers, chromatography columns and all kind of high-temperature reactors and process equipment for particulates. The equipment is supplemented by mobile demonstration units for use at industrial sites and a large-scale maritime test station. These units are perfectly suited in combination with laboratory facilities to perform scale up studies together with the capabilities of a modern digital infrastructure. Some of the units are only operated manually while other units can be operated through human machine interfaces (HMI). In line with DTU's strategic objectives, DTU KT focuses on the development and application of an Industry 4.0 framework for its research and educational activities. Therefore, the pilot plant and laboratory facilities are going through a digital transformation, creating a suitable infrastructure that provides remote accessibility to all research and operational data. These efforts are presented in this work.

Keywords: Digitalization, Database, Process Control, Digital Twin, Machine Learning

1. Introduction

The pilot plant facilities of DTU KT were established more than 50 years ago, and have been under constant development since then. A first big step after its establishment dates back to the 1990's with the construction of a large high-temperature facility followed by a general continuing update from 2005 to 2020, adding all kind of up- and down-stream equipment for the chemical, biochemical and energy related industries. Further, large pyrolysers and gazifiers for biomass and waste are located at the DTU Risø Campus and have recently been supplemented by advanced equipment for fermentations based biomanufacturing along with mobile units for industrial demonstrations of CO2 capture processes and a large-scale maritime test facility located at Hundested harbour.

The selection of units available in the pilot plant serves to recreate complete production lines in e.g. extensive scale-up studies. The available unit operations are however equipped with different degrees of automation and localized data storage. Some unit operations are only operated manually while other units are operated through a LabView interface and/or simple PID control loops. Thus, the current digital framework for the department's interaction with the raw process data generated in the pilot plant can be described as fragmented without connection to the internal network. This limits the ability of the department to test developed high-level control and optimization schemes such as digital twin implementations that embed various modelling and simulation environments with real-time data. A second limitation is the optimal use of the collected data. Until now, a minor subset of valuable simulation models, also referred to as digital twins, have been connected with the physical units available in the pilot plant. The framework presented in this contribution has the aim to develop a consistent guidance for researchers to easily integrate their simulation models at the department.

2.   Overall Framework and Infrastructure

In the present strategy period of DTU (2020-2025), the department will extend the capabilities of the pilot plant. The current efforts involve equipping the pilot plant with Industry 4.0 capabilities. In-house and industrial surveys were conducted and identified improved data handling as key needs for academia and industrial collaborators. To assurethe possibility of coupling developments with external partnership from industry, DTU KT has selected to modernize the data infrastructure (Udugama et al., 2021; Bähner et al., 2021; Lopez et al., 2021).

Currently, DTU KT is implementing the modernization efforts to overcome the previously described boundaries (Gargalo et al., 2021). The integration of the unit operations into automation platforms via SCADA, OPC-UA and API communication to digital twins are shown in Figure 1. The automation software is deployed on dedicated servers (Windows Server and Kubernetes cluster). Jupyter Hub, Kubeflow and git repositories will allow researchers to easily implement real-time optimization methods such as advanced process control and scheduling of operations (Ziaei-Halimejani et al., 2021). The developed framework establishes a digitalized research and education environment. By describing the framework from the bottom up, starting with the database and the connection to digital twin models, sensors and actuators, a prototype implementation has been established to evaluate and further improve the framework and will described in the fourth section.

Process scale-up and optimizationare complex, multi-faceted disciplines that engage several domains within chemical engineering. Knowledge and research within all of these areas must be combined to achieve a feasible large-scale design of an industrial process. Disciplines or aspects which are covered in scaling-up a process include: (I) design of experiments (DoE), (II) computational fluid dynamics (CFD), (III) kinetic studies, (IV) steady-state and dynamic simulations for mass and energy balances (V) quantification of disturbances (VI) thermodynamic modelling of multi-phase and multi-component systems (VII) control system design and implementation.

Process systems engineering (PSE) has advanced from previously solving scale-up problems by addressing each of the above-mentioned disciplines individually. Now, focus has shifted towardsadopting a hybrid multi-layer decomposition approach that makes extensive use of the operational data, to gain more insight and leverage the modelling capabilities with e.g. machine learning and data analysis during process characterization in laboratory and pilot scale experimentation.

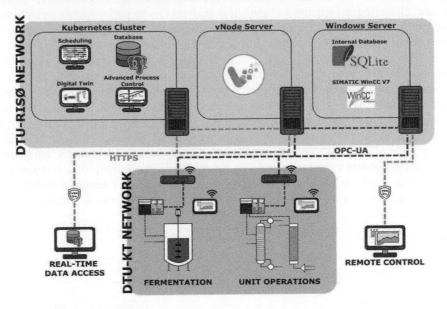

Figure 1: Overall framework for digitalizing the DTU KT pilot plant infrastructure

With the newly implemented framework, these information streams are being sent through various platforms and environments. Consolidating the whole infrastructure of such a multi-platform system has become a new task for engineers who must now spend considerable time to pipeline the data streams between process simulators, experimental equipment and various other computational tools.

3.   Database Design

This project is fundamentally about integrating data and data structures from multiple different sources. Accessible via well-defined interfaces, these new features and functions will provide for a diverse range of use cases and applications. As such, the digitalization project at DTU KT is heavily focusing on designing the database schema correctly to meet the needs of all stakeholders at the department. The database schema presented in this article will continue to be modified until a schema has been developed where all the research work of all research groups can be easily mapped to a relational database and updated accordingly.

While undertaking this bottomup approach of developing the infrastructure, the database is also being harmonized with the data delivered through the Siemens SCADA system and the vNode interface which stores the history data in a non-relational database (MongoDB).

The next step in the data pipeline is the treatment and cleaning step before the data is stored in a relational database (Postgres). As seen in Figure 1, a data processing step is performed when data is being transferred from the vNode system to the database. This allows to apply data clean up and data annotation methods on the raw data stored in the non-relational database (Perazzoli et al., 2018; Russo et al., 2021). Further, outlier treatment can be performed and other statistical metrics (e.g. means, frequencies, weights) generated to be stored in the relational database. Additionally, outlier treatment, clustering, annotation and statistical analysis can be performed to enhance the stored data. The relational mapping of the whole data allows for sophisticated data look up queries with SQL while machine learning research can more easily be performed with the treated data already stored. The database is harmonized with two other frameworks, namely with the Riffyn datalake software (Juergens et al.; 2018) which is being adopted by companies such as Novozymes and BASF. Here the Riffyn data agent allows to connect the DTU KT database to the Riffyn software. This will make it easier to join process data from ongoing experiments across different units to emulate a real process. The DTU KT database is also being harmonized with the AnIML data standard. This allows the captured data from a wide variety of laboratory equipment from different vendors to be integrated with the database. The AnIML data standard is a XML schema and is divided into AnIML Core and the AnIML Technique Definitions. For further information we refer to the reference literature (Schäfer et al.; 2004).

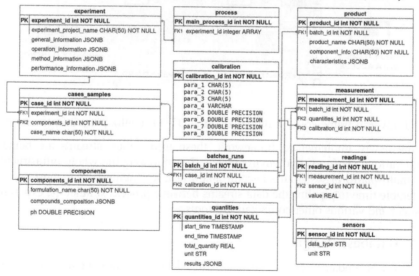

Figure 2: Current snapshot of the entity relationship (ER) diagram of the relational database under development

4.   Digital Twin Integration

To facilitate the development and evaluation of digital twins for each individual unit operation, a generic digital twin framework has been developed and has been named internally at the department DelphiTwin (Nielsen et al., 2021). This framework is designed to create and test digital twins for dynamic processes, and to control the equipment during operation. The setup of DelphiTwin, requires a total of four steps where the user defines the desired level of complexity and number of modules to include. After the initial setup, DelphiTwin can subsequently be used for a number of tasks. During process operation, the digital twin will automatically collect and store sensor and actuator data in an internal local database. It will furthermore handle and send scheduled set points to low-level controllers. Based on the process model, real-time predictions of future process states can furthermore be generated. Finally, the specified control algorithms can be used to calculate and schedule future set points of low level controllers. In between process operations, DelphiTwin can also be used to investigate various design spaces for different configurations.

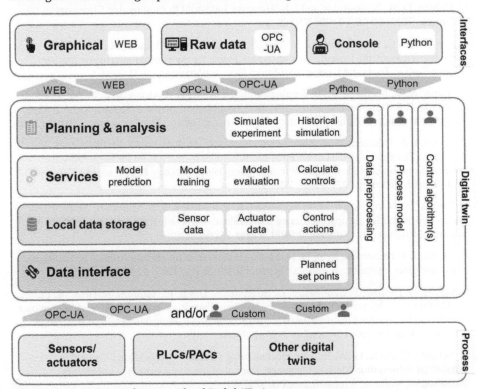

Figure 3: Digital twin framework of DelphiTwin

To allow for a high degree of flexibility when constructing a digital twin, DelphiTwin is implemented in Python with a modular architecture (as illustrated in Figure 3). Users can create custom modules to include system specific implementations of data interfaces, process models and control algorithms. DelphiTwin comes with a set of standard modules, including generic PID and MPC control algorithms and data-interfacing modules supporting OPC-UA, OPC-DA and TCP/IP. It is the plan to embed this framework in the digital infrastructure to better utilize pilot plant equipment in future plant scale-up, design and optimization research.

5.   Conclusions

This paper gives an overview of the current developments with respect to the digital transformation journey of an academic chemical and biochemical engineering department with a strong binding to industrial research. We present how a pilot plant can be retrofitted to accommodate the changes brought by the transition towards Industry 4.0. The efforts described in this article will provide a broad and foundational basis to perform advanced research tasks in a highly digitalized environment leveraging researchers' capabilities. From the educational point of view, the digital transformation of the department will allow to provide courses for students to equip them with skills for the digital era of chemical and biochemical engineering.

References

F. D. Bähner, O. A. Prado-Rubio and J. K. Huusom, Industrial & Engineering Chemistry Research, 2021, 60 (42), 14985-15003

C. L. Gargalo, S. C. de las Heras, M. N. Jones, I. Udugama, S. S. Mansouri, U. Krühne, K. V. Gernaey, Towards the Development of Digital Twins for the Bio-manufacturing Industry, Digital Twins: Tools and Concepts for Smart Biomanufacturing, 2021, 1-34.

H. Juergens, M. S. Niemeijer, L. D. Jennings-Antipov, R. Mans, J. Morel, A. J. A. van Maris, J. T. Pronk, T. S. Gardner, Evaluation of a novel cloud-based software platform for structured experiment design and linked data analytics, Scientific Data 5, Article number: 180195 (2018)

P. C. Lopez, I. A. Udugama, S. T. Thomsen, C. Roslander, H. Junicke, M. M. Iglesias, K. V. Gernaey., Transforming data to information: A parallel hybrid model for real-time state estimation in lignocellulosic ethanol fermentation, Biotechnology & Engineering, Volume 118, Issue 2, February 2021, Pages 579-591

R. F. Nielsen, N. Nazemzadeh, M. P. Andersson, K. V. Gernaey, S. S. Mansouri, An uncertainty-aware hybrid modelling approach using probabilistic machine learning, Computer Aided Chemical Engineering, Volume 50, 2021, Pages 591-597

S. Perazzoli, J. P. de Santana Neto, H. M. Soares, Prospects in bioelectrochemical technologies for wastewater treatment, Water Sci Technol (2018) 78 (6): 1237–1248.

S. Russo, M. D. Besmer, F. Blumensaat, D. Bouffard, A. Disch, F. Hammes, A. Hess, M. Lürig, B. Matthews, C. Minaudo, E. Morgenroth, V. Tran-Khac, K. Villez, The value of human data annotation for machine learning based anomaly detection in environmental systems, Water Research, Volume 206, 1 November 2021, 117695.

B. A. Schäfer, D. Poetz, G. W. Kramer, Documenting Laboratory Workflows Using the Analytical Information Markup Language, Journal of the Association for Laboratory Automation, Volume: 9 issue: 6, page(s): 375-381

I. A. Udugama, M. Öner, P. C. Lopez, C. Beenfeldt, C. Bayer, J. K. Huusom, K. V. Gernaey, G. Sin, Towards Digitalization in Bio-Manufacturing Operations: A Survey on Application of Big Data and Digital Twin Concepts in Denmark, Front. Chem. Eng., 16 September 2021.

H. Ziaei-Halimejani, N. Nazemzadeh, R. Zarghami, K. V. Gernaey, M. P. Andersson, S. S. Mansouri, N. Mostoufi, Fault diagnosis of chemical processes based on joint recurrence quantification analysis, Computers & Chemical Engineering, Volume 155, December 2021, 107549.

Proceedings of the 14th International Symposium on Process Systems Engineering – PSE 2021+
June 19-23, 2022, Kyoto, Japan © 2022 Elsevier B.V. All rights reserved.
http://dx.doi.org/10.1016/B978-0-323-85159-6.50255-4

Identification Method of Multiple Sequential Alarms that Occurred Simultaneously in Plant-operation Data

Ai Yanaga[a*], Masaru Noda[a]

[a]*Department of Chemical Engineering, Fukuoka University, 8-19-1, Nanakuma Jonan-ku, Fukuoka 814-0180, Japan*
mnoda@fukuoka-u.ac.jp

Abstract

Advances in distributed control systems in the chemical industry has made it possible to inexpensively and easily install numerous alarms in them. A poorly designed alarm system might cause nuisance alarms. One type of nuisance alarm is a sequential alarm, which reduces the capability of operators to cope with plant abnormalities because critical alarms are hidden in them. We propose an identification method of sequential alarms that occurred at the same time. With this method, similarities of all combinations of an alarm subsequence are compared using the Smith-Waterman algorithm (Smith *et al.*, 1981). We introduced a new objective function considering the time differences between alarms into this algorithm. We applied the proposed method to the simulation data of an extractive distillation column, and the simulation results indicate that the method can extract sequential alarms that occurred simultaneously in plant-operation data.

Keywords: Big data, Operation, Alarm system, Chemical plants

1. Introduction

1.1. Plant alarm systems

A plant-alarm system notifies an operator of plant-state deviations. An alarm is issued when the process variable deviates from the range set in consideration of safety. Advances in distributed control systems (DCS) in the chemical industry has made it possible to inexpensively and easily install numerous alarms in them. While most alarms help operators detect an abnormality and identify its cause, some are unnecessary. We called such unnecessary alarms nuisance alarms.

There are three typical types of nuisance alarms: sequential, repeating, and those without operations. Sequential alarms consist of numerous alarms in succession triggered by a single root cause. Repeating alarms occur routinely. Alarms without operations do not require corresponding operation. Nuisance alarms reduce the ability of operators to cope with plant abnormalities because critical alarms are hidden in many of them.

1.2. Previous research

Cheng *et al.* (2013) proposed a method of calculating the similarities of alarm-flood sequences in plant-operation data using the Smith-Waterman algorithm. The Smith-Waterman algorithm is a local-sequence-alignment tool for identifying common molecular subsequences (Smith *et al.*, 1981). Experts can conduct a thorough analysis, such as root cause, on the basis of the clustered patterns of alarm floods. However, it cannot be used to directly determine sequential alarms hidden in the plant-operation data. Wang and Noda (2017) proposed a mining method of sequential alarms in plant-operation data using a dot matrix method (Mount, 2004). The dot matrix method is a sequence-alignment method for identifying similar regions in deoxyribonucleic acid (DNA) or ribonucleic acid (RNA), which may be a consequence of functional, structural, or evolutionary relationships between the sequences. Proposed method can identify sequential alarms from the operation data of chemical plants but occasionally fails to detect sequential alarms between two related sequential alarms because the time information when alarms occurred is not used for evaluating the similarities between them. The above methods also occasionally fail to identify sequential alarms when two types of sequential alarms occurred at the same time.

1.3. Objective

A poorly designed alarm system triggers nuisance alarms, which might lead to oversight of critical alarms. Such oversight might cause plant accidents. We propose an identification method of sequential alarms. It takes into account multiple sequential alarms that occurred simultaneously and the time distances among alarms.

2. Proposed method

2.1. Plant-operation data

Plant-operation data consist of the alarm tag of alarm or operation and occurrence times. There are two types of alarm tags, process variable and operation variable. Table 1 shows an example of plant-operation data, where "A" indicates an alarm or operation, subscript values indicate their numbers of alarm or operation. When an alarm occurs, it is recorded in the plant database. Table 1 shows that A_1, A_2, A_3, and A_4 occurred in the order $A_1 \rightarrow A_2 \rightarrow A_4 \rightarrow A_2 \rightarrow A_3$ in about 20 minutes. A huge amount of data is accumulated every day, where nuisance alarms are hidden.

Table 1 Example of plant-operation data

Date	Time	Tag	Type
2021/1/1	0:08:53	A_1	Alarm
2021/1/1	0:09:36	A_2	Operation
2021/1/1	0:11:42	A_4	Alarm
2021/1/1	0:25:52	A_2	Operation
2021/1/1	0:30:34	A_3	Alarm

2.2. Problem formulation

Plant-operation data are characterized by these tags and the order of alarm occurrence. When an alarm sequence occurs repeatedly in plant-operation data, those alarms are considered to be sequential. The problem of identifying sequential alarms in such data is formulated as the problem of searching for a repeated subsequence of alarms in plant operation data.

2.3. Preparation

Our proposed method converts plant-operation data into an alarm subsequence by using a sliding window (Mount, 2004), which prevents the window from extracting a subsequence in which the time distances among alarms are critically different. The alarm sequence is converted into a set of windows consisting of occurrence alarms in a predetermined time window. By enabling overlapping and extracting of alarm subsequences, it is possible to deal with cases in which multiple types of sequential alarms are issued at the same time.

2.4. Similarity evaluation method

The proposed method uses a modified Smith-Waterman algorithm (Smith & Waterman, 1981) for searching sequential alarms. It finds a pair of segments, one from each of two long sequences, such that there is no other pair of segments with greater similarity. It is a dynamic algorithm for finding the highest scored local sequence alignment and is partially used with the method by Cheng *et al.* (2013).

Their method sums three types of scores, i.e., the match score when the corresponding alarms match the similarity between the two alarm columns, mismatch score when they do not match, and insert score of the gap when there is a gap.

The final similarity score is calculated by adding these scores. In this research, a mismatch score and gap score are considered constant, and the match score is calculated on the basis of the time distances among alarms.

For example, consider the following two sequences.

$$S_1 = a_1, a_2, \ldots a_m, \ldots, a_M, m = 1, 2, \ldots, M \tag{1}$$

$$S_2 = b_1, b_2, \ldots, b_n, \ldots, b_N, n = 1, 2, \ldots, N, \tag{2}$$

where a_m is the mth tag in aligned segment S_1 of length M, and b_n is the nth tag in aligned segment S_2 of length N. The similarity evaluation method is given a pair of contiguous subsequences, one from each of the two sequences by inserting gaps in one or both of them. The similarity score is positive for a match and negative for a mismatch. For a symbolic pair including a gap symbol, the similarity score is negative as a penalty of inserting a gap.

When the alarms do not match, the score is weighted by the time differences among the alarms, and the similarity between the sequences is evaluated. With this method, to consider the time differences among alarms, which is a problem in previous researches, the match score when alarms between two sequences are matched individually is calculated on the basis of the time differences among alarms.

It first calculates the weight of the time distances among matching alarms using the scaled Gaussian function Eq.(3).

$$w(D) = \exp(-D^2 / 2\sigma^2), \tag{3}$$

Where w is the weight based on the time distances among alarms, D, and σ is the standard deviation. The scaled Gaussian function is a normal distribution expressed from 0 to 1. The method then calculates the similarity score using Eq.(4), where s is the similarity score, μ is the mismatch score, and s_{max} is the maximum match score.

$$s(a_m, b_n) = \begin{cases} \mu(1 - w) + s_{max} \times w & \text{if } a_m = b_n \\ \mu & \text{if } a_m \neq b_n \end{cases}, \tag{4}$$

Eq.(4) indicates that a high match score means high similarity between those subsequences. The match score is close to the mismatch score when there is a long time distance and the maximum match score is a close range.

To obtain the alignment between S_1 and S_2, the score matrix H is calculated using the dynamic programming method. Eq.(5) expresses $H_{m,n}$ between a_m and b_n, where δ is the penalty of inserting a gap.

$$H_{m,n} = \max\{H_{m-1,n-1} + s(a_m, b_n), H_{m-1,n} + \delta, H_{m, n-1} + \delta, 0\}, \tag{5}$$

The Smith-Waterman algorithm finds the pair of segments with maximum similarity by first locating the maximum element because it indicates an optimal alignment. The Smith-Waterman algorithm searches for pairs of maximally similar subsequences on a mathematically rigorous basis. It can extract an optimum subsequence when multiple sequential alarms occurred simultaneously.

2.5. Grouping

The smith-Waterman algorithm cannot aggregate similar sequential alarm patterns when the same sequential alarm occurred several times in the plant-operation data. This algorithm is a local-sequence-alignment tool that searches for a pair of segments, one from each of two long sequences. The proposed method creates a color map on the basis of the similarity score of extracted sequential alarm patterns, and aggregates the patterns using the single-linkage method. The color map is used to identify sequential alarms if the similarity score is any given threshold or above. The proposed method can effectively aggregate similar sequential alarm groups when sequential alarms occurred several times.

2.6. Calculation procedure

The proposed method's calculation procedure is as follows.

(1) Converts plant-operation data into an alarm subsequence by using a pre-determined time window.
(2) Calculates the similarity score by using a modified Smith-Waterman algorithm.
(3) Identifies sequential alarms in accordance with a cluster of sequential alarm patterns by using the similarity-score color map.

3. Case study

3.1. Simulation Data

We applied the proposed method to the simulation data of the extractive distillation column shown in Fig.1. There was a total of 18 alarms in the DCS, and three types of malfunctions, low flow rate of coolant, low steam pressure, and valve stiction, were artificially induced in the process simulation. A defined operation for each malfunction was carried out after each malfunction occurred. Alarm occurrences were recorded in the plant-operation data. During a process simulation of 15 days, 265 alarms and operations were recorded in the plant-operation data. The grey area in Fig.2 is region of where two types of malfunctions occurred at the same time.

3.2. Results of identified sequential alarms

We set the window size to 120 minutes, maximum match score to 1, mismatch score to -0.6, gap score to -0.4, and delta to 42.8390. The method extracted a subsequence as a sequential alarm pattern when the similarity score was equal to or higher than 10. As a

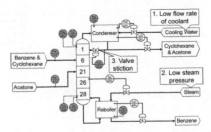

Fig.1 Process flow of extractive distillation column

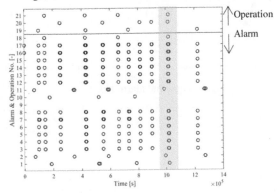

Fig.2 Simulated plant-operation data

result, it extracted 17 sequential alarm patterns and identified 3 types of sequential alarms in accordance with a cluster of patterns by using the similarity-score color map.

Fig.3 shows the identified sequential alarm No.1. X-axis is occurrence time of alarm and operation from the first alarm and y-axis is IDs of alarms and operations. There are 4 sequences identified as sequential alarm No.1. These alarms are considered to be sequential alarms issued due to a low flow rate of coolant based on the type and order of the alarms being issued. It is necessary to consider changing the alarm setting range so that this sequential alarm does not occur. All plot is overlapping, that the proposed method identifies sequential alarms correctly.

Fig.4 shows the identified sequential alarms in the grey area in Fig.2. The x-axis shows the occurrence times of alarms and operations, and the y-axis shows the IDs of alarms and operations. Because we wanted to verify if the proposed method can identify multiple sequential alarms that occurred at the same time, the grey area expanded. Between 1,018,000 to 1,019,000 seconds, 2 types of mulfanctions occurred simultaneously. Fig.4 shows that it has be identified as 2 types of sequential alarms as expected. Even though several sequential alarms occurred simultaneously, this method identified other type of sequential alarm.

4. Conclusions

We proposed a method for mining repeated similar sequential alarms. The method can identify multiple types of sequential alarms in plant-operation data even where the sequential alarms occurred simultaneously. This method can be used in more complex programs because it does not require process information.

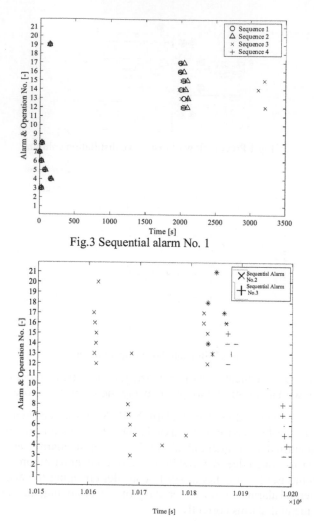

Fig.3 Sequential alarm No. 1

Fig.4 Results of identified sequential alarms of grey area in Fig.2

References

Cheng, Y., Izadi, I., and Chen, T., Pattern matching of alarm flood sequences by a modified Smith-Waterman algorithm, *Chemical Engineering Research and Design,* 91, 1085–1094 (2013)

Mount, D. W., Bioinformatics Sequence and Genome Analysis Second Edition, Cold Spring Harbor Laboratory Press, Cold Spring Harbor, New York (2004)

Nishiguchi, J., and Takai, T., IPL2 and 3 Performance Improvement Method for Process Safety using Event Correlation Analysis, *Computers & Chemical Engineering,* 34, 2007–20013 (2010)

Smith, T., Waterman, M., Identification of common molecular subsequences, *Journal of Molecular Biology,* 147, 195–197 (1981)

Wang, Z., and Noda, M., Identification of repeating Sequential Alarms in Noisy Plant Operation Data Using Dot Matrix Method with Sliding Window, *Journal of Chemical Engineering of Japan,* 50, 445–449 (2017)

Proceedings of the 14th International Symposium on Process Systems Engineering – PSE 2021+
June 19-23, 2022, Kyoto, Japan © 2022 Elsevier B.V. All rights reserved.
http://dx.doi.org/10.1016/B978-0-323-85159-6.50256-6

Understand how CNN diagnoses faults with Grad-CAM

Deyang Wu[a], Jinsong Zhao[a,b*]

[a]State Key Laboratory of Chemical Engineering, Department of Chemical Engineering, Tsinghua University, Beijing 100084, China
[b]Beijing Key Laboratory of Industrial Big Data System and Application, Tsinghua University, Beijing 100084, China
jinsongzhao@tsinghua.edu.cn

Abstract

CNN-based models for fault diagnosis have achieved high prediction accuracy, but the lack of explainability makes them hardly be understood by humans. In this paper, a technique used to produce visual explanations for CNN has been introduced to a CNN-based fault diagnosis model, DCNN, to make it more transparent and understandable. Experiments on Tennessee Eastman process showed variables that DCNN pays more attention to when diagnosing faults, which makes the decision making process of DCNN more explainable and understandable.

Keywords: fault diagnosis, explainable deep learning, convolutional neural network, chemical process safety

1. Introduction

Fault detection and diagnosis (FDD) is quite critical to safe operations of chemical processes to identity abnormal events that can hardly be controlled by distributed control systems (DCS). In decades, many researchers have been proposed different models of real-time FDD for processes. These models can be categories into quantitative model-based, qualitative model-based and process history-based models (Venkatasubramanian et al. 2003). Process history-based models are further classified into qualitative and quantitative models. The latter is also commonly termed as data-driven models. Data-driven models based on deep learning have drawn much attention of researchers these years, such as DBN-based models (Z. Zhang and Zhao 2017), CNN-based models (H. Wu and Zhao 2018), RNN-based models (S. Zhang, Bi, and Qiu 2020), autoencoder-based models (Cheng, He, and Zhao 2019) (Zheng and Zhao 2020) and GCN-based models (D. Wu and Zhao 2021). These models achieved high prediction accuracy, but the lack of explainablity makes them hard to understand by humans. The research of explainable deep learning models is critical to the application promotion for fault diagnosis.

In this paper, a technique used to produce visual explanation for CNN has been introduced to a CNN-based fault diagnosis model, DCNN, to make it more transparent and understandable. Grad-CAM was firstly introduced in Section 2, then Grad-CAM was applied to DCNN model with Tennessee Eastman process as the benchmark in Section 3, and the conclusio was drawn in Section 4 finally.

2. Explainable CNN and Grad-CAM

In recent years, deep learning had made great achievements in many research areas. Deep learning has many advantages such as low-cost modelling, self-directed learning from data and high prediction accuracy. However, the inference process of deep learning models cannot be fully understood by human beings. To understand how models make decisions, many researchers have been studying explainable deep learning recently.

CNN, as an important class of deep learning models, has made great success in the fields of computer vision and other engineering. Many models and methods have been proposed to make CNN more understandable. Grad-CAM (Selvaraju et al. 2020) is a technique for producing visual explanations for decisions from CNN based models, making them more transparent and explainable. As a gradient-based method, Grad-CAM uses the class-specific gradient information flowing into the final convolutional layer of a CNN in order to produce a coarse localization map of the important regions in the image when it comes to classification (Linardatos, Papastefanopoulos, and Kotsiantis 2021). Figure 1 is an example of Grad-CAM applied to an image classification task in (Selvaraju et al. 2020). It shows that Grad-CAM can highlight the key region on the image corresponding to different classification results. This localization map can be regarded as the basis for decision making of CNN, which is also reasonable and understandable to humans.

(a) Original Image   (b) Grad-CAM 'Cat'   (c) Grad-CAM 'Dog'

Figure 1: Example of Grad-CAM applied to an image classification task in (Selvaraju et al. 2020). (a) Original image with a cat and a dog. (b-c) Support for the cat or the dog category according to various visualizations for VGG-16.

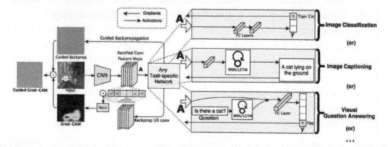

Figure 2: Grad-CAM overview

The workflow of computing the class-discriminative localization map is shown as Figure 2. Given a CNN that has been well trained for a classification task and a specific class c, the neuron importance weights α_k^c of feature maps in a convolutional layer can be calculated as Eq.(1).

$$\alpha_k^c = \frac{1}{Z} \sum_i \sum_j \frac{\partial y^c}{\partial A_{ij}^k} \tag{1}$$

y^c is the score for class c in the last fully connected layer before the softmax layer. A_{ij}^k is the activations of the kth feature map (indexed by i and j respectively over the width and height dimensions in the last convolutional layer. Eq.(1) means the gradients propagated from outputs back to feature maps, and α_k^c captures the importance of the kth feature map for a target class c. The localization map Grad-CAM can be calculated as Eq.(2).

$$L_{Grad-CAM}^c = ReLU \left(\sum_k \alpha_k^c A^k \right) \tag{2}$$

The localization map is finally obtained by weighted summing all the k feature map activations and applying the ReLU function.

3. Understand how CNN diagnoses faults

3.1. CNN-based fault diagnosis

Since CNN-based models have been applied to fault diagnosis successfully, Grad-CAM is also suitable for the interpretability of these models to study how CNNs diagnose faults for processes. In these models, fault diagnosis of a chemical process is generally regarded as a classification problem. Given an observation $X_t \in \mathbb{R}^{v \times w}$ at time t, the current operating state $y \in \mathbb{N}$ of the process should be identified out of a set of operating states consisting of normal state and different types of faults. The observation X_t is a matrix with the time window and process variable dimensions. If $w > 1$, it means the observation includes data before t in a time window w as shown in Eq.(3).

$$X_t = \begin{bmatrix} x_{1,t} & x_{1,(t-1)} & \cdots & x_{1,(t-w+1)} \\ x_{2,t} & x_{2,(t-1)} & \cdots & x_{2,(t-w+1)} \\ \vdots & \vdots & \ddots & \vdots \\ x_{v,t} & x_{v,(t-1)} & \cdots & x_{v,(t-w+1)} \end{bmatrix} \tag{3}$$

The operating state or class y is integer between 0 and N_c (the number of different fault states). $y = 0$ means the process is under normal operating state, and $y = c$ ($1 \le c \le N_c$) means the fault c has occurred. With enough pairs of observations and fault labels, a dataset $\left\{ X_t^{(i)}, y^{(i)} \right\}_{i=1}^N$ is obtained for training and testing a CNN-based fault diagnosis model.

In this paper, the DCNN model proposed by (H. Wu and Zhao 2018) is used to show how Grad-CAM can be applied to CNN-based fault diagnosis. The chosen architecture is 'Conv(64)-Conv(64)-Pool-Conv(128)-Pool(2×1)-FC(300)*-FC(21)' (model 7).

3.2. Tennessee Eastman process

Tennessee Eastman (TE) process (Downs and Vogel 1993) is a simulation process model modified based on an actual industrial process of Eastman Chemical Company in Tennessee, USA. In decades, TE process has been utilized for research in different fields such as process control, process monitoring and so on. TE process mainly consists of five unit operations and defines 20 different process disturbances. In the simulation program of TE process, 52 variables can be observed, and 20 different disturbances can be inserted at any time to make the process operate in a fault state.

3.3. Data preparation and model training

The process of data preparation and model training keeps the same with (D. Wu and Zhao 2021). To obtain enough data for training the DCNN model, the simulation program of TE process ran for 3000 h in normal state. Under every different fault, the simulation program ran for 20 h after the fault inserted, and 10 sets of parallel simulations were carried out. All the data have been normalized with the mean and standard deviation calculated from data in the normal state. The normalized data were then cut to slices with the time widow of 1 h, and the data are sampled every 3 min. Data from 8 sets of parallel simulations were used for training and the rest 2 sets were used for testing.

The DCNN model was trained for 50 epochs with Adam optimizer, the mini-batch size was set to 128 and the learning rate was set to 0.001. The trained DCNN model finally got an average classification rate of 0.93.

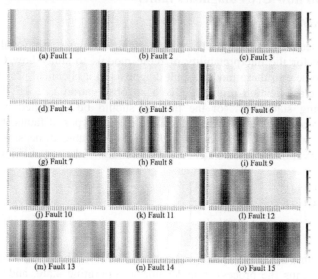

Figure 3: Average Grad-CAM localization map of 15 different faults of TE process

3.4. Explain fault diagnosis results with Grad-CAM

After the DCNN model was well trained, Grad-CAM was applied to the model following Eq.(1) and Eq.(2). The localization map of Grad-CAM is calculated based on a certain instance or observation i. In this paper, we care more about the variables but not time windows that DCNN focuses on when diagnosing faults. Thus, the average localization map for a certain type of fault c was calculated with all the maps obtained from all observations of the fault c as Eq.(4).

$$\bar{L}^c_{Grad-CAM} = \frac{1}{N} \sum_{i=1}^{N} L^c_{(i)} \tag{4}$$

$\bar{L}_{Grad-CAM}^c$ was then normalized using its maximum and minimum values to make every value falls into [0,1]. $\bar{L}_{Grad-CAM}^c$ of the first 15 faults with known causes in Table 1 is shown as Figure 3. When values changes from 0 to 1, the colour in Figure 3 changes from white to black. Figure 3 shows that what variables DCNN mainly focused on when diagnosing different faults. The deeper the colour is, the more important the variable is. Grad-CAM explains the basis of decision making of DCNN for fault diagnosis in a visual and understandable way.

Table 1: Comparison between variables affected by disturbances firstly and variables CNN focused on when diagnosing faults. T, P, L, Q, x, u mean temperature, pressure, level, flow rate, composition, control signal respectively. Subscript means unit operations or streams. Superscript means characteristics or components.

Number	Variables affected firstly	Variables CNN focused on
IDV (1)	$x_{feed}^A, x_{feed}^C, Q_{s1}, Q_{s4}, u_{s1}^Q, u_{s4}^Q$	$u_{cond}^{cwr}, u_{reac}^{cwr}, u_{strip}^{steam}, u_{comp}^{recy}, u_{purge}$
IDV (2)	$x_{feed}^A, \boldsymbol{x_{feed}^B}, \boldsymbol{x_{feed}^C}, Q_{s1}, Q_{s4}, u_{s1}^Q, u_{s4}^Q$	$x_{purge}^D, \boldsymbol{x_{feed}^C}, x_{purge}^E, x_{feed}^D, \boldsymbol{x_{feed}^B}$
IDV (3)	$T_{reac}, P_{reac}, L_{reac}, \boldsymbol{T_{reac}^{cwr}}, u_{reac}^{cwr}$	$\boldsymbol{T_{reac}^{cwr}}, W_{comp}, T_{cond}^{cwr}, Q_{strip}^{steam}, u_{cond}^{cwr}$
IDV (4)	$T_{reac}^{cwr}, \boldsymbol{u_{reac}^{cwr}}, T_{reac}, P_{reac}, L_{reac}$	$u_{cond}^{cwr}, \boldsymbol{u_{reac}^{cwr}}, u_{strip}^{steam}, Q_{s1}, Q_{s2}$
IDV (5)	$T_{cond}^{cwr}, \boldsymbol{u_{cond}^{cwr}}, T_{sep}, P_{sep}, L_{sep}$	$\boldsymbol{u_{cond}^{cwr}}, u_{reac}^{cwr}, u_{strip}^{steam}, u_{s2}^Q, x_{product}^H$
IDV (6)	$\boldsymbol{Q_{s1}}, u_{s1}^Q, Q_{feed}, x_{feed}^A$	$\boldsymbol{Q_{s1}}, Q_{s2}, Q_{s3}, u_{comp}^{recy}, u_{purge}$
IDV (7)	$Q_{s4}, u_{s4}^Q, T_{strip}, P_{strip}, L_{strip}$	$u_{purge}, u_{sep}^{under}, u_{comp}^{recy}, u_{strip}^{under}, u_{strip}^{steam}$
IDV (8)	$\boldsymbol{x_{feed}^A}, \boldsymbol{x_{feed}^B}, x_{feed}^C, Q_{s1}, Q_{s4}, u_{s1}^Q, u_{s4}^Q$	$\boldsymbol{x_{feed}^A}, \boldsymbol{x_{feed}^B}, T_{cond}^{cwr}, x_{purge}^D, x_{purge}^E$
IDV (9)	$T_{reac}, P_{reac}, L_{reac}, T_{reac}^{cwr}, u_{reac}^{cwr}$	$T_{strip}, u_{s1}^Q, u_{s3}^Q, u_{s4}^Q, Q_{strip}^{steam}$
IDV (10)	$T_{strip}, \boldsymbol{P_{strip}}, \boldsymbol{L_{strip}}, u_{strip}^{steam}, Q_{strip}^{steam}$	$\boldsymbol{P_{strip}}, T_{reac}^{cwr}, W_{comp}, \boldsymbol{L_{strip}}, Q_{strip}^{under}$
IDV (11)	$T_{reac}^{cwr}, \boldsymbol{u_{reac}^{cwr}}, T_{reac}, P_{reac}, L_{reac}$	$u_{cond}^{cwr}, \boldsymbol{u_{reac}^{cwr}}, Q_{s4}, Q_{s3}, Q_{recy}$
IDV (12)	$T_{cond}^{cwr}, u_{cond}^{cwr}, \boldsymbol{T_{sep}}, P_{sep}, \boldsymbol{L_{sep}}$	$T_{reac}, Q_{purge}, \boldsymbol{T_{sep}}, L_{reac}, \boldsymbol{L_{sep}}$
IDV (13)	$\boldsymbol{T_{reac}}, P_{reac}, L_{reac}, \boldsymbol{T_{reac}^{cwr}}, u_{reac}^{cwr}$	$\boldsymbol{T_{reac}}, W_{comp}, \boldsymbol{T_{reac}^{cwr}}, L_{reac}, P_{strip}$
IDV (14)	$T_{reac}^{cwr}, \boldsymbol{u_{reac}^{cwr}}, T_{reac}, P_{reac}, L_{reac}$	$Q_{s1}, Q_{s2}, P_{strip}, u_{cond}^{cwr}, \boldsymbol{u_{reac}^{cwr}}$
IDV (15)	$T_{cond}^{cwr}, u_{cond}^{cwr}, T_{sep}, P_{sep}, L_{sep}$	$x_{product}^G, x_{product}^H, x_{product}^F, u_{s2}^Q, u_{s3}^Q$

It should be noted that the important variables in Figure 3 are not necessarily the disturbance variables. Table 2 concludes the differences between the variables affected by disturbances firstly (observable root causes) and the first 5 variables that have highest values in Grad-CAM localization maps for every type of fault. For some faults, Grad-CAM localized some of the root causes or observable variables affected directly by root causes. For other faults, DCNN focused none of the root causes. This indicates that DCNN pays more attention to those sensitive variables that changes intensely with disturbances but not disturbance variables themselves. For those faults that are hard to identity (such as fault 3, 9 and 15), all variables only change slightly after the fault has occurred. Thus, DCNN can hardly localize some certain variables to determine the type of fault. This also reflects on the Grad-CAM localization maps that the variance of all the values in is small and they have little differences in colour.

4. Conclusions

In this paper, Grad-CAM, a technique for producing visual explanations for CNN, is introduced. Grad-CAM uses the gradient information flowing back into the last convolutional layer of the CNN to assign importance values to each neuron for a particular decision of interest (Selvaraju et al. 2020). Grad-CAM was applied to the DCNN model for fault diagnosis to make the inference process transparent. With a well trained DCNN model, Grad-CAM generates a localization map for an input matrix to show the importance of different neurons. This can help humans understand what variables DCNN pays more attention to when diagnosing the type of faults of the process. The results of experiments on TE process shows that the DCNN model pays more attention to those sensitive variables that changes intensely with disturbances but not disturbance variables themselves. When all variables only change slightly after faults happen, it'll be difficult for DCNN to identity the type of fault. In this case, the localization maps have only little differences in colour in different region. Future work will focus on construct more explainable deep learning model for fault diagnosis.

References

Cheng, Feifan, Q. Peter He, and Jinsong Zhao. 2019. "A Novel Process Monitoring Approach Based on Variational Recurrent Autoencoder." Computers & Chemical Engineering 129 (October): 106515. https://doi.org/10.1016/j.compchemeng.2019.106515.

Downs, James J, and Ernest F Vogel. 1993. "A Plant-Wide Industrial Process Control Problem." Computers & Chemical Engineering 17 (3): 245–55.

Linardatos, Pantelis, Vasilis Papastefanopoulos, and Sotiris Kotsiantis. 2021. "Explainable AI: A Review of Machine Learning Interpretability Methods." Entropy 23 (1): 18. https://doi.org/10.3390/e23010018.

Selvaraju, Ramprasaath R., Michael Cogswell, Abhishek Das, Ramakrishna Vedantam, Devi Parikh, and Dhruv Batra. 2020. "Grad-CAM: Visual Explanations from Deep Networks via Gradient-Based Localization." International Journal of Computer Vision 128 (2): 336–59. https://doi.org/10.1007/s11263-019-01228-7.

Venkatasubramanian, Venkat, Raghunathan Rengaswamy, Kewen Yin, and Surya N Ka. 2003. "A Review of Process Fault Detection and Diagnosis Part I: Quantitative Model-Based Methods." Computers and Chemical Engineering, 19.

Wu, Deyang, and Jinsong Zhao. 2021. "Process Topology Convolutional Network Model for Chemical Process Fault Diagnosis." Process Safety and Environmental Protection 150 (June): 93–109. https://doi.org/10.1016/j.psep.2021.03.052.

Wu, Hao, and Jinsong Zhao. 2018. "Deep Convolutional Neural Network Model Based Chemical Process Fault Diagnosis." Computers & Chemical Engineering 115 (July): 185–97. https://doi.org/10.1016/j.compchemeng.2018.04.009.

Zhang, Shuyuan, Kexin Bi, and Tong Qiu. 2020. "Bidirectional Recurrent Neural Network-Based Chemical Process Fault Diagnosis." Industrial & Engineering Chemistry Research 59 (2): 824–34. https://doi.org/10.1021/acs.iecr.9b05885.

Zhang, Zhanpeng, and Jinsong Zhao. 2017. "A Deep Belief Network Based Fault Diagnosis Model for Complex Chemical Processes." Computers & Chemical Engineering 107 (December): 395–407. https://doi.org/10.1016/j.compchemeng.2017.02.041.

Zheng, Shaodong, and Jinsong Zhao. 2020. "A New Unsupervised Data Mining Method Based on the Stacked Autoencoder for Chemical Process Fault Diagnosis." Computers & Chemical Engineering 135 (April): 106755. https://doi.org/10.1016/j.compchemeng.2020.106755.

Proceedings of the 14th International Symposium on Process Systems Engineering – PSE 2021+
June 19-23, 2022, Kyoto, Japan © 2022 Elsevier B.V. All rights reserved.
http://dx.doi.org/10.1016/B978-0-323-85159-6.50257-8

A Comprehensive Framework for the Modular Development of Condition Monitoring Systems for a Continuous Dry Granulation Line

Rexonni B. Lagare[a*], M. Ziyan Sheriff[a], Marcial Gonzalez[b], Zoltan Nagy[a], Gintaras V. Reklaitis[a]

[a]*Davidson School of Chemical Engineering, Purdue University, West Lafayette, IN 47907, USA*
[b]*School of Mechanical Engineering, Purdue University, West Lafayette, IN 47907, USA*
rlagare@purdue.edu

Abstract

The development of condition monitoring systems often follows a modular scheme where some systems are already embedded in certain equipment by their manufacturers, and some are distributed across various equipment and instruments. This work introduces a framework for guiding the modular development of monitoring systems and integrating them into a comprehensive model that can handle uncertainty of predictions from the constituent modules. Furthermore, this framework improves the robustness of the modular condition monitoring systems as it provides a methodology for maintaining quality assurance and preventing unnecessary shutdowns in the event of some modules going off-line due to condition-based maintenance interventions.

Keywords: Condition Monitoring, Probabilistic Programming, Modular, Machine Learning, Bayesian

1. Introduction

The challenges in modeling pharmaceutical powder processes, as outlined in Rogers and Ierapetritou (2014), has put an emphasis on the use of data-driven models as the basis for developing condition monitoring (CM) systems. While this approach offers a practical solution, as Webb and Romagnoli (2021) recently demonstrated for the Tennessee Eastman Process (TEP) case study, it ignores the modular nature of process control system development.

Since the data-driven models often require data spanning multiple unit operations in order to maximize the usage of process data, CM applications are likely to be on levels 1 and 2, which are distributed control systems in the process control implementation hierarchy introduced by Su *et al.* (2019). These modules are also likely to be focused on process faults, which differ from the CM modules at level 0, which are directly embedded into more advanced equipment. Because embedded modules are developed by the vendors, whose priority is on the safe and reliable operation of the equipment, they tend to focus on safety-related faults like electrical and mechanical faults.

All the aforementioned fault types need to be considered holistically, especially since they are likely correlated with each other. However, the varying levels at which these modules are installed in the process control implementation hierarchy, and the difference in their goals, create an integration challenge that needs to be addressed in order to have a safe and reliable operation of a continuous processing plant.

2. Condition Monitoring Framework Development

A natural framework for addressing this integration issue would be the probabilistic graphical modeling methodology, which is commonly used to implement hierarchical Bayesian models. As recently demonstrated by Radcliffe and Reklaitis (2021), this methodology is effective in systems where data is limited and there is significant uncertainty in the model parameters. Since this method is fundamentally based on modularity, where complex physical systems are constructed from simpler parts, it is sensible to utilize it for the CM module integration problem.

Under this methodology, the basic parts of the system are random variables with uncertain values, which are depicted as nodes in the graphical model. Conditional dependencies may be assigned based on expert knowledge on the system, and arcs can be drawn between one or more nodes to capture these relationships. Altogether, the nodes and the arcs comprising the probabilistic graphical model (PGM) form a compact representation of joint probability distributions where probability theory can be used to model the uncertainty in these variables and to make inferences on variables of interest.

2.1. A Probabilistic Condition Monitoring Model for Continuous Dry Granulation

The continuous dry granulation line of the Purdue University Pilot Plant comprises several unit operations that can blend pharmaceutical excipients and active pharmaceutical ingredients (API), granulate them, and then compress them into tablets using a rotary tablet press (see Figure 1). At the heart of this process is the granulation step, which takes place in an Alexanderwerk WP-120 roller compactor (RC). In the RC unit, the pharmaceutical blend is compacted into a ribbon, cut into flakes, and subsequently broken down into the desired granule size distribution in a classifier mill. For clarity in presenting basic concepts, the remaining discussion will focus on the RC.

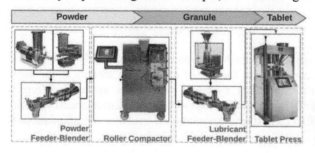

Figure 1. Dry granulation line at the Purdue University pilot plant.

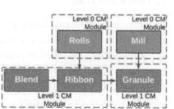

Figure 2. Roller compactor condition monitoring model.

A condition monitoring model can be constructed by considering two types of condition variables, the material condition and the equipment condition, and then forming appropriate relationships between them. For the RC, this model is shown in Figure 2. The roll and mill variables (green nodes) represent the condition of the respective roller compactor components. The WP-120 RC has a built-in condition monitoring system for each of these components, so these variables also represent "embedded" or "level 0" CM modules. The blend, ribbon, and granules variables do not have condition monitoring systems by default; so, these need to be developed as "distributed" or "level 1" CM modules that require the integration of additional PAT tools. Discussing the development of these modules is beyond the scope of the current paper, so they will be assumed to provide uncertain values for their corresponding variables, as is the case for the embedded modules.

At this stage, the model in Figure 2 serves as a useful guideline for the modular development of CM modules that supports the complete observability of variables pertinent to Quality-by-Design (QBD) principles. Under QBD, the Critical-Quality-Attributes (CQA) targets of a unit operation should be achieved by controlling the Control Process Parameters (CPP) and the CQA of the preceding unit operation. For the RC model, the ability to control the CPP is represented by the condition of the equipment (i.e., the rolls and the mill), and the CQAs are represented by the condition of the material (i.e., blend, ribbon, and granules).

Each random variable in this graph, whether it is a material condition (blue node) or equipment condition (green node), can have discrete states: normal or a faulty state. With multiple variations possible for each faulty state, each node represents a categorical distribution, which assigns a probability for each possible state. For clarity, these distributions are depicted as probability tables that are linked to its corresponding node via broken lines in Figure 3. Moreover, the variables are assumed to take only two possible states, whereas in reality, they can have up to "N" number of states, depending on the number of faulty conditions that are recognized for each node.

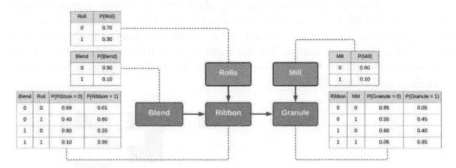

Figure 3. A probabilistic condition monitoring model for the roller compactor.

2.2. Probabilistic Programming and Inference

By the basic laws of probability, the entire graph represents the joint probability of all the condition monitoring variables per the following equation:

$$P(Blend, Rolls, Ribbon, Mill, Granule) = \tag{1}$$
$$P(Rolls)P(Mill)P(Blend)P(\ Ribbon \mid Blend, Rolls\ )P(\ Granule \mid Mill, Ribbon\ )$$

With this model, interesting analysis tasks such as probabilistic inference can be performed. For example, given observations on the condition of the roll and the ribbon, e.g., both are at normal state so their values equal 0, it is possible to directly compute the posterior distribution of the condition of the blend using Bayes' Rule.

$$P(\ Blend \mid Rolls = 0, Ribbon = 0\ ) = \frac{P(Blend, Rolls=0, Ribbon=0, Mill, Granule)}{P(Rolls=0, Ribbon=0)} \tag{2}$$

where: $P(Rolls = 0, Ribbon = 0) = \tag{3}$

$$\sum_{Blend, Mill, Granule} P(Blend, Rolls = 0, Ribbon = 0, Mill, Granule)$$

However, with modularity in mind, this model is expected to get bigger as adjacent unit operations along the manufacturing line are integrated. The increasing number of variables will slow down the exact inference computations to a point that makes it impractical for monitoring applications. To circumvent this, the graphical model can instead be encoded in a probabilistic programming framework like Infer.NET (Minka et al., 2018), where approximate inference tasks can be quickly performed via efficient message passing algorithms.

3. Results and Discussion

3.1. Parameter Learning

As demonstrated, a fully-defined model such as shown in Figure 3 can make useful predictions on variables based on observations from other variables. However, in practice, these parameters are not always initially available. Fortunately, the graphical modeling methodology can perform parameter learning by simply adding the parameters as variables in the graph, and then using the same approximate inference techniques to infer parameter values. Figure 4 shows the modified graph that addresses parameter learning; the yellow nodes represent the prior probabilities of the CM modules, and the block arrows depict message passing during the inference of the prior probabilities. In order for the message passing algorithms to remain computationally tractable as more modules are integrated, the probabilities of the parameter variables are assigned a Dirichlet distribution, which is a conjugate prior for a categorical distribution. This conjugacy ensures that the number of distribution parameters do not increase intractably during the implementation of the message passing algorithms. (Winn, Bishop, and Diethe 2015)

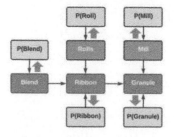

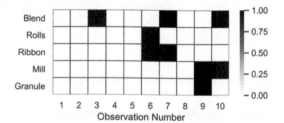

Figure 4. Parameter learning in a probabilistic graphical model.

Figure 5. Dataset required for parameter learning (where all condition variables are observed).

Initially, the prior variables can be assigned either non-informative or weakly informative priors. Then, parameter learning can be performed on data acquired when all the CM modules are functional (see Figure 5), and message passing algorithms can be used to infer the posterior distributions of the parameter variables. As more data is collected, the inferred distribution of the parameter variables would be more "informed" and have less variance.

This can be observed from the results in Figure 6, which shows the inferred probabilities of the blend and mill condition. After just 100 observations, the comprehensive model was able to correctly infer the "true" probabilities of the model shown in Figure 3, which is from where the dataset was randomly sampled. Beyond this

A Comprehensive Framework for the Modular Development of Condition
Monitoring Systems for a Continuous Dry Granulation Line

1547

number of observations, the mean of the inferred probability distributions barely changed, while the variance continued to decrease significantly.

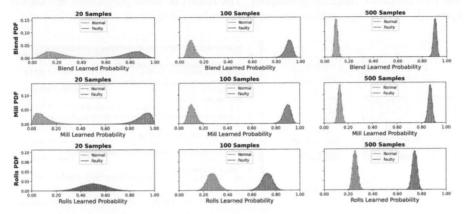

Figure 6. Learned probabilities for the blend, mill, and roll condition at varying sizes of training data.

3.2. Predictive Modeling

One of the main challenges in monitoring the CQA of the RC is the lack of current capability to measure the condition (e.g., flowability and tabletability) of the granules in real-time. Fortunately, the graphical modeling framework allows for the inference of the granule condition given observations from other CM modules. This scenario is shown in Figure 7, with the block arrows indicating message passing from observed variables that are not d-separated(Bishop 2006) from the granule variable. Results of this inference are shown in Figure 8, where based on observations from the surrounding CM modules, the probability of the granule condition changes correspondingly.

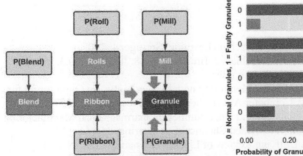

Figure 7. Granule Condition Inference Scheme   Figure 8. Inference Results on the Granule Condition

Throughout the operation of a continuous processing line, some CM modules might break down, rendering the condition of the actual material and equipment to be unobservable. For the RC example, the NIR sensor observing the ribbon could be undergoing maintenance because of fouling. While this temporary lack of observability could compromise quality assurance of the process, it should not be a reason for shutting down. Using the message passing scheme shown in Figure 9, the condition of the granule could still be inferred from available CM modules such as the blend, rolls, and the mill condition monitoring modules. The results of these predictions are shown in Figure 10, for varying conditions of the blend, rolls, and the mill.

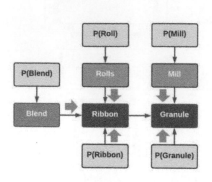

Figure 9. Inference Scheme with Multiple Unobserved Variables.

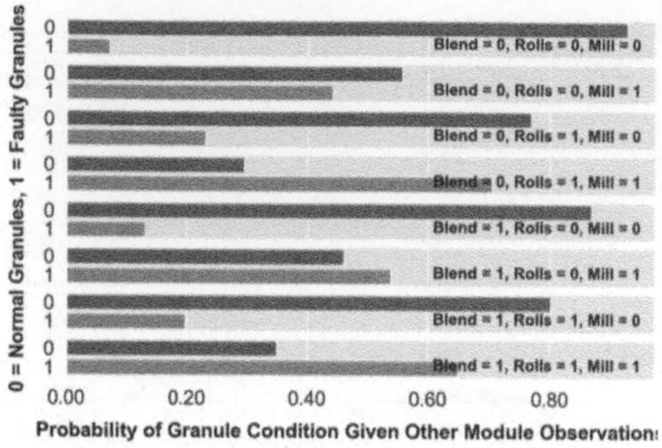

Figure 10. Inference Results on the Granule Condition with Unobserved Ribbon Condition.

4. Conclusions

A comprehensive condition monitoring model for a roller compactor was developed by first considering material and equipment condition variables that are involved, and then establishing logical relationships between them. The condition variables were assumed to be categorical variables with discrete states, and their relationships were encoded into a probabilistic programming framework. This framework was able to efficiently perform approximate inference to learn the parameters of the model, and most importantly, to infer the condition of other less-visible variables like the granule condition based on observations from other condition variables.

5. References

C.M. Bishop, 2006, Pattern Recognition and Machine Learning. Springer 1, 740.

T. Minka, J. Winn, J. Guiver, Y. Zaykov, D. Fabian, J. Bronskill, 2018. Infer.NET 0.3. Microsoft Research Cambridge. http://dotnet.github.io.infer.

A.J. Radcliffe, G. V. Reklaitis, 2021. Process Monitoring under Uncertainty: An Opportunity for Bayesian Multilevel Modelling. Computer Aided Chemical Engineering, 50, 1377-1382.

A. Rogers, M. Ierapetritou, 2014. Challenges and Opportunities in Pharmaceutical Manufacturing Modeling and Optimization. Computer Aided Chemical Engineering, 34,144–149.

S. Roweis, Z. Ghahramani, 1999. A Unifying Review of Linear Gaussian Models. Neural Computation, 11 (2), 305–45.

Q. Su, S. Ganesh, M. Moreno, Y. Bommireddy, M. Gonzalez, G.V. Reklaitis, Z. K. Nagy, 2019. A Perspective on Quality-by-Control (QbC) in Pharmaceutical Continuous Manufacturing. Computers & Chemical Engineering, 125, 216–31.

Z. Webb, J. Romagnoli, 2021. Real-Time Chemical Process Monitoring with UMAP. Computer Aided Chemical Engineering, 50, 2077–82.

J. Winn, C. Bishop, T. Diethe, 2015. Model-Based Machine Learning. Microsoft Research Cambridge. http://www.mbmlbook.com.

Proceedings of the 14th International Symposium on Process Systems Engineering – PSE 2021+
June 19–23, 2022, Kyoto, Japan ©2022 Elsevier B. V. All rights reserved.
http://dx.doi.org/10.1016/B978-0-323-85159-6.50258-X

Framework for Suppressing Transient Fault Alarms Online

Shu Xu[a]*, Mark Nixon[a]

[a] *Emerson Automation Solutions, 1100 W Louis Henna Blvd, Round Rock, TX, 78681, USA*

richard.xu@emerson.com

Abstract

While running process monitoring and fault detection software online, it is quite common to face alarms triggered by faults resulting from non-steady state transients. Usually soft sensor models (e.g., principal component analysis (PCA) models) are updated periodically to accommodate such process changes so that similar alarms will not be prompted in the future operation. During online operation, however, alarms might overwhelm operators before model updates and suppressing the alarms online temporarily after checking upon the process or equipment accordingly is deemed a better alternative. In this paper, a framework is presented on how to adjust standard deviations of parameters online to temporarily turn off T^2 and Q alarms in the PCA implementation for transient faults while keep the product's capability to detect other types of faults. Such a framework is tested and validated in a case study using Emerson's *Continuous Data Analytics* (CDA) product.

Keywords: fault detection, alarm suppression, online, transient state

1 Introduction

In modern chemical plant, numerous process variables must be kept within specific limits, and excursions of key variables beyond those limits are often bound to have ramifications in plant safety, the environment, product quality and plant profitability. Process monitoring and fault detection plays an increasingly important role in ensuring that the plant performance meets the operating objectives. Process monitoring takes its root from uni-variate approaches including limit checking, quality control charts and the Six Sigma Approach. However, such methods fail to provide satisfactory results as quality variables increase in number and become highly correlated. To overcome such challenges faced in process monitoring, multivariate methods such as principal component analysis (PCA), project to latent structure (PLS) have been widely used. Recently, uniform manifold approximation and projection (UMAP) from the machine learning community has also gained popularity among industrial practitioners thanks to its superior performance in dimensional reduction (Joswiak et al., 2019; Webb and Romgnoli, 2021). Indeed, several decent review papers on applying state-of-art statistical and machine learning techniques to process monitoring and fault detection have been published (Qin, 2014; Chiang et al., 2017; Ge, 2017; Qin et al., 2021). Yet in this paper we focus on solving problems related to online implementation of PCA in Emerson's *Continuous Data Analytics* (CDA) product used for process monitoring. During online implementation, the life span of PCA models is limited because most processes rarely stay at the same steady state and slow changes in equipment, feedstock and operating strategy may all compromise model performance. Although PCA models are updated periodically to accommodate process changes, operators are often overwhelmed by alarms generated by state-transient faults during online operation. In addition, simply acknowledging alarms associated with control limit violation is not desirable as the product might lose the capability to detect other faults that violate the control limits as well. After checking upon the process or equipment indicated by alarms, the operators request for tools to suppress them online temporarily before next model update. Industrial practitioners from Dow Chemical shared their experiences on online soft sensor maintenance where robust mean and variance estimators of the inputs and outputs are used (Chiang et al., 2017; Lu and Chiang, 2018) to overcome the long-term model degradation problem. Inspired by such work, a new framework is developed which automatically adjusts standard deviations of parameters online to temporarily turn off T^2 and Q alarms associated with transient faults and keeps the product's capability to detect other types of faults.

The remainder of this paper is organized as follows. Section 2 describes the proposed framework for suppressing transient fault alarms online, and it is validated by case study shown in Section 3. Section 4 draws conclusions based on results obtained in the study.

2 Algorithm for suppressing transient fault alarms online

Assuming each alarm suppressing step leads to the values of T^2 and Q falling around 0.8 of their control limits respectively, standard deviation of the ith parameter (σ_i) has to be adjusted to achieve such a goal.

2.1 T^2 alarms

If a T^2 alarm occurs at time t in a new steady state, since the standard deviation array of all parameters ($\sigma_{a,0}$) is available, by incorporating the contribution of T^2, the updated standard deviation ($\sigma_{b,0}$) is provided by:

$$\sigma_{b,j,0} = \left[\frac{|contT_{t,j}^2|}{\max_j \left\{ |contT_{t,j}^2| \right\}} \left(\frac{T_t^2}{0.8T_{UCL}^2} - 1 \right) + 1 \right] \sigma_{a,j,0} \qquad (1)$$

where T_{UCL}^2 is the control limit of *Hotelling T^2*, and $|contT_{t,j}^2|$ is the contribution of jth parameter. As shown in Eq. (1), if the jth parameter makes no contribution to the T^2, i.e., $|contT_{t,j}^2| = 0$, the corresponding jth standard deviation will not be changed. After obtaining $\sigma_{a,0}$ and $\sigma_{b,0}$, an interactive strategy based on the bisection method is used to obtain the standard deviation array, σ_c, which leads to updated T^2 values close to the target of $0.8T_{UCL}^2$.

2.2 Q alarms

After suppressing the T^2 alarms, there still might be Q alarms that require operators' attention. Although the Q values might have already decreased with increased standard deviations σ_c obtained in the T^2 alarm suppressing step, they need to be further reduced to quell the remaining Q alarms. An approach similar to the strategy used in T^2 alarm management is applied, shown as follows.

Assuming the standard deviation array of all parameters from the previous T^2 alarm suppressing step ($\sigma_{c,0}$) is available, by incorporating the contribution of Q, the updated standard deviation ($\sigma_{d,0}$) is provided by:

$$\sigma_{d,j,0} = \left[\frac{|contQ_{t,j}|}{\max_j \{|contQ_{t,j}|\}} \left(\frac{Q_t}{0.8 Q_{UCL}} - 1 \right) + 1 \right] \sigma_{c,j,0} \qquad (2)$$

where Q_{UCL} is the control limit of Q, and $|contQ_{t,j}|$ is the contribution of jth parameter.

Similar to Section 2.1, after obtaining $\sigma_{c,0}$ and $\sigma_{d,0}$, an iterative strategy based on the bisection method is used to obtain the standard deviation array, σ_e, which leads to updated Q values close to the target of $0.8 Q_{UCL}$.

3 Case study

In this section, the continuous process of mixing two ingredients (A and B) is studied, as shown in Figure 1. A process fault caused by changes in *Main Flow A* occurred during online operation, as indicated by soaring T^2 and Q values in Figure 2 which violate the control limits and trigger alarms. It is worth noting that in Emerson's product, T^2 and Q are normalized by their control limits T_{UCL}^2 and Q_{UCL} respectively so that they share an identical normalized control limit of 1.

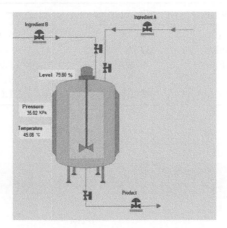

Figure 1: Mixing process of two ingredients

Variables with top contributions to the T^2 fault, together with associated changes in standard deviations are summarized in Table 1. Comparing the T^2-adjusted standard deviations σ_c with the original σ_a, it is observed that variables with large contributions show significant increases in their standard deviations, such as "Main Flow A "and "Main Flw Valve ".

After suppressing the T^2 alarm, results are obtained (see Figure 3) with T^2 values fall below the control limit. Q values have also been drastically reduced from 68.8 to 12.9, yet they are still above the control limit.

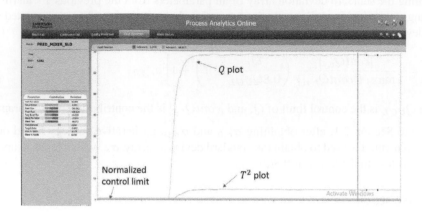

Figure 2: Process analytics online GUI with both T^2 and Q alarms

Table 1: Variables with top T^2 fault contribution and associated standard deviation changes

Variable Tag	T^2 contribution	σ_a	σ_c
Main Flow A	-18.74	45.08	81.74
Main Flw Valve	37.78	7.494	19.78
Targ Blend Flw	-14.89	21.95	36.13
Mixer Flow	-28.25	59.63	132.7
Blend Flow	-24.27	21.65	44.45

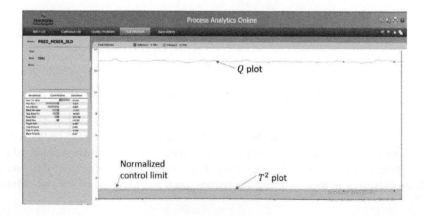

Figure 3: Process analytics online GUI with Q alarm only

In this case, further steps are taken to suppress the Q alarm. Variables with top contributions to the updated Q fault, together with updated standard deviations are summarized in Table 2. Similar to the results shown before, variables with the lion share of contribution enjoy drastic rises in standard deviations ($\sigma_c \to \sigma_e$), such as "Main Flow A "and "Main Flw Valve ". The resulting Q value drop (below the control limit) is demonstrate in Figure 4.

Table 2: Variables with top Q fault contribution and associated standard deviation changes

Variable Tag	Q contribution	σ_c	σ_e
Main Flow A	-1.602	81.74	382.6
Main Flw Valve	6.555	19.78	317.7
Actual Ration	-1.424	0.0622	0.2655
Targ Blend Flw	-0.596	36.13	85.60
Blend Flw Valve	-0.643	9.859	24.42

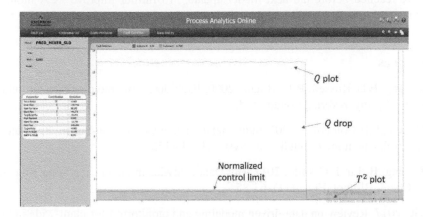

Figure 4: Process analytics online GUI with both T^2 and Q alarms suppressed

After suppressing the alarms caused by *Main Flow A* fault, the system is still able to sound alarms if a new fault occurs. As shown in Figure 5, later on the process is suffering a load disturbance where "Blend% Solid"decreases from 65 to 50 (the black line in the bottom figure), which leads to a quality deviation indicated by a jump of T^2.

4   Conclusions

In this paper, a new framework to suppress transient fault alarms online is proposed, and it is developed for Emerson's DeltaV *Continuous Data Analytics* product where PCA-based process monitoring approach is implemented. By adjusting standard deviations of parameters online based on variable contribution to faults, both T^2 and Q alarms can be turned off. Such an approach keeps the product's capability to detect other types of faults, as shown by the promising results in an industrial case study. It is worth pointing out that the new method only acts as a temporary solution and it works better when new steady

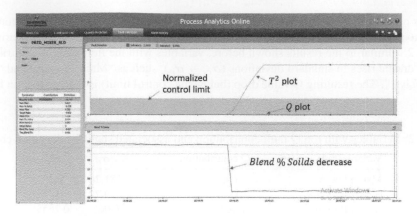

Figure 5: Process analytics online GUI with a new T^2 alarm

states are reached before the next model update. To further improve PCA monitoring performance, future work will be focused on incorporating filtering techniques to remove outliers.

References

L.H. Chiang, E.L. Russell, R.D. Braatz, 2000, Fault detection and diagnosis in industrial systems, Springer-Verlag London Ltd.

L.H. Chiang, L.F. Colegrove, 2007, Industrial implementation of on-line multivariate quality control, Chemometr. Intell. Lab. Syst., 88, 143-153

L.H. Chiang, B. Lu, I. Castillo, 2017, Big Data analytics in chemical engineering, Annu. Rev. Chem. Biomol. Eng., 8(1), 63-85

Z.Q. Ge, 2017, Review on data-driven modeling and monitoring for plant-wide industrial processes, Comput. Chem. Eng., 171, 16-25

M. Joswiak, Y. Peng, I. Castillo, L.H. Chiang, 2019, Dimensionality reduction for visualizing industrial chemical process data, Control Eng. Pract., 93, 104189

B. Lu, L.H. Chiang, 2018, Semi-supervised online soft sensor maintenance experiences in the chemical industry, J. Process Control, 67, 23-34

S.J. Qin, 2014, Process data analytics in the era of Big Data, AIChE J., 60(9), 3092-3100

S.J. Qin, L.H. Chiang, 2019, Advances and opportunities in machine learning for process data analytics,Comput. Chem. Eng., 126, 465-473

S.J. Qin, S. Guo, Z. Li, L.H. Chiang, I. Castillo, B. Braun, Z. Wang, 2021, Integration of process knowledge and statistical learning for the Dow data challenge problem, Comput. Chem. Eng., 153, 107451

Z. Webb, J. Romagnoli, 2021, Real-time chemical process monitoring with UMAP, Comput. Aided Chem. Eng., 50, 2077-2082

Proceedings of the 14th International Symposium on Process Systems Engineering – PSE 2021+
June 19-23, 2022, Kyoto, Japan © 2022 Elsevier B.V. All rights reserved.
http://dx.doi.org/10.1016/B978-0-323-85159-6.50259-1

Using Reinforcement Learning in a Game-like Setup for Automated Process Synthesis without Prior Process Knowledge

Quirin Göttl[a], Dominik G. Grimm[b,c,d] and Jakob Burger[a,*]

[a] Technical University of Munich, Campus Straubing for Biotechnology and
Sustainability, Laboratory of Chemical Process Engineering, Uferstraße 53, 94315
Straubing, Germany.
[b] Technical University of Munich, Campus Straubing for Biotechnology and
Sustainability, Bioinformatics, Schulgasse 22, 94315 Straubing, Germany.
[c] Weihenstephan-Triesdorf University of Applied Sciences, Petersgasse 18, 94315
Straubing, Germany.
[d] Technical University of Munich, Department of Informatics, Boltzmannstr. 3, 85748
Garching, Germany.
burger@tum.de

Abstract

The present work uses reinforcement learning (RL) for automated flowsheet synthesis. The task of synthesizing a flowsheet is reformulated into a two-player game, in which an agent learns by self-play without prior knowledge. The hierarchical RL scheme developed in our previous work (Göttl et al., 2021b) is coupled with an improved training process. The training process is analyzed in detail using the synthesis of ethyl tert-butyl ether (ETBE) as an example. This analysis uncovers how the agent's evolution is driven by the two-player setup.

Keywords: Automated Process Synthesis; Flowsheet Synthesis; Artificial Intelligence; Machine Learning; Reinforcement Learning.

1. Introduction

RL is a frequently used machine learning approach in the process engineering community. Besides many applications in process control, it is also employed for problems that require forward planning. For example, Wang et al. (2020) designed synthetic pathways for organic chemistry with an RL approach combined with Monte-Carlo tree search. Khan and Lapkin (2020) showed that RL can identify processing routes for hydrogen production.

We presented an approach called SynGameZero, which enables training an agent without prior knowledge to synthesize entire process flowsheets using RL (Göttl et al., 2021a). Thereby, flowsheet synthesis is transformed into a turn-based two-player game. Both players start with an empty flowsheet. In their turns, they add unit operations or recycles to their flowsheet while also seeing what their opponent does. After each turn, a flowsheet simulator generates stream tables and provides them to the players as a base for their next moves. The winner is determined by the net present value, calculated after flowsheet completion. In a tied game, the player who finishes the flowsheet first is the winner. The reward $r \in \{1, -1\}$ is a binary value indicating win or loss. The agent consists of an artificial neural network (ANN) combined with a tree search and is trained by playing this game against itself. The ANN's output contains a suggestion for the next move of the current player. It also includes an estimate of the chances of the current player to win the game. This output guides the tree search, which considers several moves in advance, imitating a typical human planning process. The search is adaptive regarding depth and explores only promising options selected based on the ANN's output. The tree consists of nodes (representing flowsheet states) connected by branches (possible moves of adding units or recycles). No prior knowledge is required to initialize the agent. During the training process, the parameters of the ANN are optimized to improve its ability to suggest good moves and estimating the chances to win the game. Recently, we have improved the two-player framework by structuring the agent's decision in three hierarchy levels and introducing hierarchical RL (Göttl et al., 2021b). Recasting the problem into this game framework enables powerful agent structures and training methods from the literature outside chemical engineering. Many authors proved that RL serves as a powerful tool to master complex problems like winning the board games of Go and Chess (Silver et al., 2018).

The present work builds upon the hierarchical RL scheme as presented in Göttl et al. (2021b) and modifies the agent's structure and the training process resulting in a slightly improved performance. For the first time, the agent's evolution during training is analyzed in detail. The analysis uncovers the importance of the two-player game setup for the success of the SynGameZero method.

2. Methodology

2.1. Example Process and Flowsheet simulation

For comparability, we adopt the process design problem of ETBE synthesis and the flowsheet simulator from Göttl et al. (2021b). Here, we give a brief summary; the reader is referred to the original paper for details. A quaternary system consisting of ethanol (EtOH), isobutene (IB), n-butane (nBut), and ETBE is considered. Two feed streams are sampled randomly at the start of the game (the first one containing EtOH and the second one a mixture of IB and nBut). The goal of the two-player game is to create the flowsheet that maximizes the net present value of the process. Thereby, the following idealized unit operations may be used:

Reactor (R): The following reversible reaction occurs in the reactor, which always reaches equilibrium.

$$EtOH + IB \leftrightarrow ETBE. \tag{1}$$

Distillation columns (D_L and D_H): Distillation columns are assumed with infinite height and total reflux for faster simulation. Due to binary azeotropes, a distillation boundary separates the quaternary system into two distillation regions. It is possible to choose between two different product splits. The split D_L obtains the light-boiler of the feed's distillation region with the highest possible yield, and the split D_H obtains the respective heavy-boiler with the highest possible yield.

Mixer / Recycle (M): any open stream can be admixed to another open stream or an already used stream. The latter results to a recycle in the process.

2.2. Agent structure and training procedure

The present work utilizes a similar framework as described by Göttl et al. (2021b), and we refer again for a detailed description. The agent's actions are structured hierarchically into three levels. At Level 1, the agent chooses an open stream in the flowsheet or the termination of the synthesis process. If an open stream was chosen at Level 1, the agent selects a unit operation from the above list as the destination of the open stream at Level 2. If the agent decides for Mixer / Recycle (M), then Level 3 is used. There, the agents select another already existing stream as the destination to admix/recycle the open stream chosen at Level 1. In contrast to Göttl et al. (2021b), the agent does no longer differentiate between a mixer and a recycle. They are conceptually the same. The flowsheet simulator determines automatically if the chosen action leads to a mixer or a recycle and simulates the respective option.

The ANN is structured according to the three hierarchy levels and explained using Figure 1 (for the specifications and the setup of the ANN, we refer to GitHub: https://github.com/grimmlab/SynGameZero/tree/ETBE_synthesis2.0). Its first part is a convolutional block, which processes the stream tables of both players. Each hierarchy level is represented by an actor-critic network, which receives data processed by the previous networks. At Level 2, information on the open stream chosen at Level 1 is also provided to the actor-critic network. Each actor-critic network i generates two kinds of outputs π_i and v_i. π_i represents a probability distribution, which is a suggestion for the decision at this level, while v_i is an estimate of the expected reward for the current player. Those outputs are used to guide the tree search by exploring promising flowsheet alternatives (promising is quantified by π_i and v_i). The agent's actions in the game are determined based on the results of the tree search.

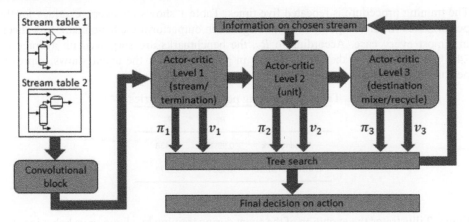

Figure 1. The agent's hierarchical structure. The arrows indicate flow of information. Artificial neural networks with trainable parameters are shown in rounded boxes.

During the training process, the agent plays $N_{steps} = 10,000$ games against itself with randomly sampled feed streams. In previous work (Göttl et al., 2021b), we did not consider feed streams with equimolar rates of EtOH and IB, although those configurations are particularly interesting. In the present work, equimolar feed streams are included in 20 % of all training games to enable the agent to deal with these configurations. After each game, the states (stream tables of both players), the results of the tree search (the basis for the decisions on actions), and the reward $r \in \{1, -1\}$ for each player are stored in a memory with size $N_{memory} = 160$. A batch of size $N_{batch} = 64$ is sampled out of the memory to perform a stochastic gradient descent step on the ANN with a learning rate of $\beta = 0.0001$.

2.3. Evaluation

After the training procedure, the agent is evaluated by playing 1,000 games against itself with randomly sampled feed streams. Due to combinatorics, it is not possible to store the optimal flowsheet for every conceivable feed stream combination. Therefore, the agent is evaluated by comparing its flowsheets with three benchmark flowsheets created by humans (Göttl et al., 2021b). They include the industrial Oxeno-process for ETBE synthesis (Ryll et al., 2014). The following metrics are used to quantify the performance. R_1 is the fraction of games where the agent proposed a flowsheet at least as good as the best benchmark. R_2 is the average relative gain in the net present value of the agent over the best benchmark. For a precise mathematical definition of the metrics, the reader is referred to (Göttl et al., 2021b).

To uncover the importance of the two-player game setup for the success of the SynGameZero method, the agent state (i.e., the ANN's parameters) is saved at various stages during the training process. The behavior of these agent versions is studied by letting it play the two-player game against itself for a fixed set of feed stream combinations: the molar flow rate of EtOH is varied between 15 and 95 kmol/hr. The molar flowrates of IB and nBut are set equal. They are also varied between 15 and 95 kmol/hr.

3. Results

The training procedure is repeated five times. Table 1 shows the average values for the performance metrics. R_1 shows that the agent can outperform the benchmark flowsheets in almost every case. According to R_2, the benchmarks are surpassed on average by 24.4 %. Compared to the results shown in previous work, the agent shows improved performance.

Table 1. Resulting performance metrics.

	R_1	R_2
Present work	0.9996	0.2438
Göttl et al. (2021b)	0.9864	0.2279

Figure 2 illustrates the evolution of the agent during training by showing its behavior after different numbers of training steps. The various feed stream combinations are depicted as cells in the matrix. The winning player of every combination is indicated with a color code. For one feed stream combination, the winning flowsheet (marked red) is displayed.

Without any training ($N_{steps} = 0$), the agent consists of a randomly initialized ANN and the tree search. The behavior of the agent is the same for all shown combinations of feed streams. In the role of Player 1 it terminates the synthesis right away. In the role of Player 2, it sets up the shown flowsheet and wins the game. After 100 training steps, Player 1 has copied this tactic (for all shown feeds) and wins all games (if both players do the same, the game is tied, and Player 1 wins because he/she finishes first).

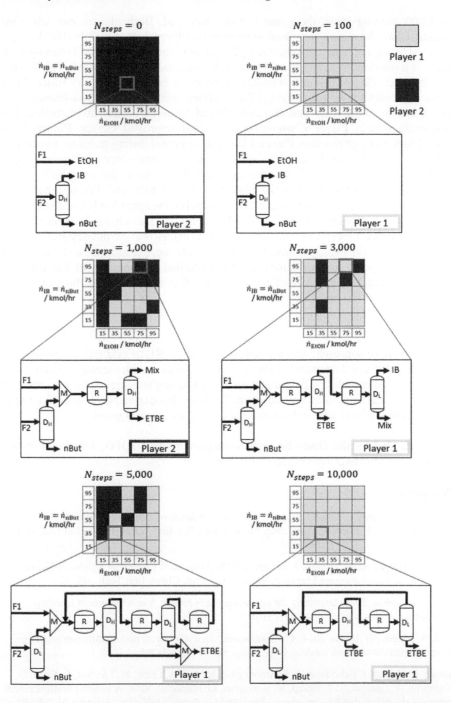

Figure 2. Illustration of the agent's evolution after different numbers N_{steps} of training steps. The matrix field represent different feed stream combinations. The color code marks the winning player. The red box shows the winning flowsheet for the respective feed streams.

After 1,000 training steps, the game is more balanced. Both players can win some situations. From the shown flowsheet generated by Player 2, it is visible that the agent has learned to use a reactor to synthesize ETBE, clear progress. After 3,000 training steps, Player 1 is more dominant, and the flowsheets become more sophisticated. This balance change between Players 1 and 2 winning in the game is observed many times during training. Typically, Player 1 is copying (if needed) and using the so-far best-known tactic. Player 2 has to avoid a tie and therefore forced to explore alternative tactics. It is consequently mainly Player 2 who uncovers novel improved tactics. Player 2 will afterward win more games than Player 1 for a short period during training. Eventually, Player 1 acquires the novel tactic and wins again. The bottom row in Figure 2 shows situations at the late stages of the training. After 5,000 steps, the complexity of the flowsheets further increases, while the game is still quite balanced. For equimolar feed rates of IB and EtOH (i.e., on the diagonal of the matrix), the agent has learned to generate flowsheets with complete conversion of IB and EtOH. The chemical equilibrium is overcome by using a recycle (cf. the shown flowsheet). However, the design is still not optimal. After 10,000 steps, the flowsheets are slightly more improved, and the training is completed. Player 1 wins all games. Even with further training, Player 2 is unable to find a better tactic. Such a constellation signifies that a local or maybe even global optimum for the performance has been reached.

4. Conclusions

The SynGameZero approach, which enables an agent via RL to synthesize flowsheets, is slightly improved and demonstrated for an example process with incomplete conversion in the reactor and recycles. Efficient and effective training is achieved by a hierarchical agent structure and a two-player game setup. The latter forces one player to explore novel tactics. The other player adopts them as soon as they are advantageous.

Funding: funded by the Deutsche Forschungsgemeinschaft (DFG, German Research Foundation) - Project-ID 466387255 - SPP 2331.

References

Q. Göttl, D. G. Grimm, J. Burger, 2021a, Automated synthesis of steady-state continuous processes using reinforcement learning, Front. Chem. Sci. Eng., DOI: 10.1007/s11705-021-2055-9

Q. Göttl, Y. Tönges, D. G. Grimm, J. Burger, 2021b, Automated Flowsheet Synthesis Using Hierarchical Reinforcement Learning: Proof of Concept, Chem. Ing. Tech., 93, 12

A. Khan, A. Lapkin, 2020, Searching for optimal process routes: A reinforcement learning approach, Comput. Chem. Eng., 141, 107027

O. Ryll, S. Blagov, H. Hasse, 2014, Thermodynamic analysis of reaction-distillation processes based on piecewise linear models, Chem. Eng. Sci., 109, 284-295

D. Silver, T. Hubert, J. Schrittwieser, I. Antonoglou, M. Lai, A. Guez, M. Lanctot, L. Sifre, D. Kumaran, T. Graepel, T. Lillicrap, K. Simonyan, D. Hassabis, 2018, A general reinforcement learning algorithm that masters chess, shogi, and Go through self-play, Science, 362, 1140-1144

X. Wang, Y. Qian, H. Gao, C. W. Coley, Y. Mo, R. Barzilayb, K. F. Jensen, 2020, Towards efficient discovery of green synthetic pathways with Monte Carlo tree search and reinforcement learning, Chem. Sci., 11, 10959-10972

Proceedings of the 14th International Symposium on Process Systems Engineering – PSE 2021+
June 19-23, 2022, Kyoto, Japan © 2022 Elsevier B.V. All rights reserved.
http://dx.doi.org/10.1016/B978-0-323-85159-6.50260-8

Generation and Benefit of Surrogate Models for Blackbox Chemical Flowsheet Optimization

Tim Janus[a*], Felix Riedl[a], Sebastian Engell[a]

[a]*TU Dortmund University, Emil-Figge-Straße 70, Dortmund 44137, Germany*
tim.janus@tu-dortmund.de

Abstract

Commercial process simulators are widely used in process design, due to their extensive library of models and ease of use. The results obtained from these simulators can be used for global flowsheet optimization but often gradient information is not provided so that derivative-free optimization methods must be used. The process simulator is called as a black box and this is computationally expensive, thus filtering out simulations that have a low probability of providing good results by machine learning is attractive to increase the efficiency of derivative-free optimization. The surrogate models used for filtering are initially based upon small data sets. We explore the generation of these initial data sets and we investigate two alternatives and suggest a heuristic for the choice of the decision function for inequality constraints.

Keywords: Aspen Plus, Process Optimization, Surrogate Models, Machine Learning, Evolutionary Algorithms

1. Introduction

Chemical process design usually is performed as an iterative process of comparing the performance and the costs of alternative process configurations and parameterizations by an interdisciplinary team of experts. In many cases, such design studies for chemical processes are performed interactively using commercial block-oriented flowsheet simulators e.g. Aspen Plus. Asprion et al. (2018) give an overview on process simulators. While this can lead to satisfactory solutions, optimal designs cannot be expected. On the other hand, the rigorous optimization of a chemical process is a very challenging task as the models that represent the process units are generally nonlinear and the problems are large and often nonconvex. Further, discrete decisions as e.g. the number of stages and the feed stage of a distillation column enlarge the complexity of the optimization problem. In research, usually mixed-integer non-linear program (MINLP) solvers and equation-based models are used. This approach has the disadvantage that specific model formulations and expert knowledge to set up the models are needed. In contrast, commercial process simulators offer a large model library, and different design alternatives can be simulated easily. It is therefore promising to combine the ease of modelling provided by commercial flowsheet simulators with the power of optimization. But the lack of interfaces to internal information like sensitivities is a big obstacle. An approach to combine the use of commercial simulators with the power of optimization is to use derivative-free methods that are not as efficient as the derivative-based optimization of equation-based models with mathematical programming. In our previous work, we have proposed to use evolutionary algorithms together with flowsheet simulators for optimization-based process design, for details see Urselmann et al. (2016). This however requires a large number of calls to the simulator of which a significant fraction do not converge or result

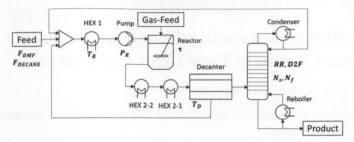

Figure 1 - Flowsheet of the case study of the hydroformylation of 1-dodecene in a thermomorphic solvent system with three major unit operations and two recycle streams. The degrees of freedom are highlighted in bold font.

in large violations of the design constraints. Due to the relatively long computation times for the single simulations, there is a strong interest to filter the design proposals and to steer the optimization to promising regions such that computation time with non-converging solutions and solutions that are far from the design specifications is not wasted. Surrogate-assisted optimization is a possible approach to enhance the efficiency of derivative-free blackbox optimization. It applies methods from Machine Learning (ML) to steer the optimization and to avoid the exploration of non-promising parts of the solution space, see Haftka et al. (2016) for a review. In Janus et al. (2021) it was shown that the required time for global flowsheet optimization could be halved by such methods. In this contribution, we analyze the use of surrogate models to decide on the execution of simulations in more detail, specifically the initial generation of the data for the training of the surrogate models and the choice of the decision rules.

2. Case Study

For the numerical studies, we use the case study of the homogenously catalyzed hydroformylation of 1-dodecene to n-tridecanal in a thermomorphic solvent system (TMS). The TMS process considered here has been investigated extensively experimentally in the Collaborative Research Center DFG Transregio SFB 63 "Integrated chemical processes in liquid multi-phase systems InPROMPT". The process is performed in a mixture of the solvents dimethylformamide (DMF) and decane in order to recover the expensive homogeneous rhodium catalyst from the product stream. The phase behavior of the mixture of DMF and decane is temperature-dependent and a change of the temperature is used to switch between a homogenous mixture in the reactor and a mixture with two liquid phases in the decanter, see Kiedorf et al. (2014). Figure 1 illustrates the flowsheet. The feed stream is heated and pressurized to create a homogenous mixture in the reactor and H_2 and CO are added as syngas. Beside the main reaction of 1-dodecene to n-tridecanal, four side reactions occur in the reactor, see Merchan et al. (2016). A cascade of two heat exchangers cools down the mixture. Then a liquid-liquid separator splits the two phases of the mixture. The polar DMF-rich phase contains the catalyst and is recycled back into the reactor. The decane-rich phase that contains the unconverted feed, the product and the byproducts is fed into the distillation column for purification. The top stream of the column is recycled back to the reactor and a product-rich liquid stream is obtained at the bottom of the column. The bottom stream must have a product purity of more than 99 mol %. The ten degrees of freedom (DOFs) of the process design considered here are the number of theoretical stages, the feed stage, the distillate to feed ratio and the reflux ratio of the column, the temperature, pressure and residence time of the reactor, the temperature of the decanter and the flow

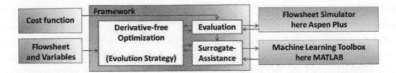

Figure 2 - Diagram of the flowsheet optimization framework that consists of three parts. The input for the framework, the optimization framework with the three modules: optimization, evaluation and surrogate-assistance, and the third party software. Here, Aspen Plus is used as a commercial flowsheet simulator and MATLAB is used as a machine learning toolbox.

rates of DMF and decane. The operating window is between 80 and 120 °C for the reactor and -5 to 25 °C for the decanter. The operating window of the reactor pressure is between 15 and 30 bar, the other vessels are operated at atmospheric pressure.

3. Framework for Flowsheet Optimization

Figure 2 shows the flowsheet optimization framework. The global derivative-free optimization is implemented by an evolution-strategy (ES) as described by Beyer and Schwefel (2002). An initial flowsheet, the degrees of freedom, and their bounds must be provided to the framework to start the design optimization. The framework calls the process simulator by setting the values of the degrees of freedom, and then uses the results, in particular the critical purities and the flow rates as well as equipment and operational parameters. If the simulation does converge, the framework calculates the cost function and stores the result and the internal process information. If a simulation does not converge, no internal process information, e.g. concentrations, is collected. The surrogate-assistance module in Figure 2 contains the candidate-rejection and the candidate-generation heuristics that guide the search of the optimizer. It generates and evaluates surrogate-models that estimate results of the flowsheet simulation, e.g. the product purity. The evaluation of a surrogate-model is several orders of magnitudes

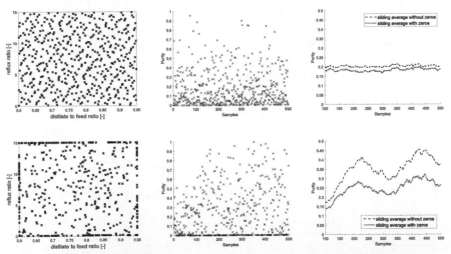

Figure 3 - Distribution plots. The first row contains plots with samples obtained from a 'Halton set' and the second row contains plots with samples from simulations that were triggered by the evolution strategy. The first column shows the distribution of the degrees of freedom 'distillate to feed ratio' and 'reflux ratio', the second column shows the purity distribution of the samples, and the third column shows a sliding average of the purity based on the last 100 samples.

faster than a flowsheet simulation. The generation of the surrogate-models is performed by an optimization that minimizes the difference between the estimates of the surrogate-models and the result of the simulator. If an initial data set is provided, the execution of the framework starts with the generation of data for the training of surrogate-models. After each evaluation of the process model, the data is stored and after a certain number of simulations, the machine learning toolbox (MLT) retrains the surrogate-models. The MLT supports multi-shot training and hyperparameter optimization.

4. Evaluation and generation of surrogate-models

ML methods generate surrogate-models based upon training data. We employ neural networks as surrogate-models. Beside the guidance of the optimization, the surrogate-assistance controls the iterative generation and evaluation of new surrogate-models.

In our approach, two surrogate-models are used. A classifier predicts if a simulation shall not be started because the process simulator will most likely not converge, and a regression model estimates the satisfaction of crucial constraints, in this example the product purity. The candidate rejection rule discards a simulation if either the classifier predicts non-convergence or the estimated product purity is equal to or below a threshold. For this, we use a dynamic boundary b that depends on the purity constraint c_p as a heuristic, see Eq. (1). The symbol mae here denotes the mean absolute error of the neural network and the index $feas$ refers to the feasible part of the training set that consists of the data that has been generated by the optimization before the training. Y_i and P_i represent the simulated and the estimated purity for data point i. If Eq (1) is not applicable, i.e. the number of feasible points in the test set N_{feas} is less than two, the boundary b is defined as $b = c_p - 3 \cdot mae_{all}$ whereby mae_{all} refers to the mean absolute error of the training set on the neural network.

$$b = c_p - \left(mae_{feas} + 1.96 \cdot \left(\sqrt{\frac{1}{N_{feas}} \sum \left(|Y_i - P_i| - mae_{feas} \right)^2} \right) \right) \tag{1}$$

Beside the configuration of the neural network, the distribution of the training data has a strong influence on the performance of the generated surrogate-models, e.g. it should be beneficial to have a high density of samples in the region of the purity constraint. A Halton set with RR2 scrambling and data from previous simulations that were triggered by the ES were compared for the generation of the initial samples. Figure 3 shows the

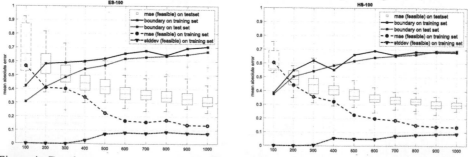

Figure 4 - Development of the mean absolute error of the neural networks that predict the product purity. On the left side, the first 100 samples origin from previous optimizations and on the right side the first 100 samples were generated as a Halton set. The test set consists of 12,000 process alternatives, 49 of them have a purity >= 99 %. The x-axis indicates the number of samples in the training set.

distribution of these two data sets for 500 samples. With samples from the Halton set, the space of the degrees of freedom is uniformly covered within the bounds. In the initial explorative phase, the evolution strategy tends to explore specifically the boundaries of the search space. After 100 samples, the average purity of the simulated process alternatives is around 20 %, but from there on the evolution strategy starts to find process alternatives with a higher purity. This is beneficial for the optimization as the goal of the surrogate-model is to predict high purities with a high accuracy, and therefore data points with higher purities should be part of the training data set.

When the surrogate models are updated, the training set contains the initial data (i.e. either the Halton set or the initial simulations generated by the ES) and all process alternatives that have been simulated during the optimization up to the start of the training procedure. The surrogate-assistance optimizes the neural networks on the entire training set, i.e. the training set equals the test set. As the heuristic boundary could be unreliable during the early stages of the optimization, we validate it with a large test set.

5. Results

We compare the use of an initial Halton set and the use of data from the initial simulations as training data for the surrogate models in the initial stage in Figure 4. Only candidate rejection was applied but not heuristic generation of promising candidates as in Janus et al. (2021). Ten runs were performed to reduce random effects. The training of the neural networks was repeated five times. The neural network architecture was chosen from previous studies as networks with two hidden layers with 45 nodes each. The neural networks were retrained every 100 simulations. When the training data consists of up to 1000 samples, the training requires approx. 20 s, while a single process simulation requires approx. 6 s, so the additional effort for the training of the surrogate models pays off quickly when later unpromising simulations are saved.

For surrogate models trained with 100 samples, the Halton set leads to more accurate models with a smaller variance than the data sequence from the calls of the optimization strategy, but this difference becomes smaller when more samples are available. Figure 4 shows the development of the mean absolute error (*mae*) of the networks. The x-axis shows the samples of the training set, where after the first training, the evolution strategy generated the training data. The box plots show the *mae* on the large test set for 50 trainings, i.e. ten optimizations with five repeats of the training. The solid line shows the values of the boundaries that were used in the purity-based candidate-rejection during the optimization which are based on the data seen so far. The dotted line shows the boundaries that were calculated based on the posteriori test set. Both lines are close to each other and have a similar behavior, which validates the proposed heuristic. The first dashed line shows the *mae* on the feasible part of the training set or three times the entire *mae* on the training set if the training data did not contain at least two feasible process alternatives. For the first 500 simulations, the median of the mean absolute error in the HS-100 variant is approximately 0.05 below the median *mae* of the ES-100 variant. From 500 simulations on, the difference shrinks but the variance remains lower for the HS-100 variant, i.e. the training of the neural network is more robust when it starts from a small uniformly distributed set of training data as a basis.

6. Conclusions

We discussed the data generation for and the evaluation of surrogate models that are used to guide global derivative-free flowsheet optimization, with the goal to avoid non-promising computationally expensive calls to the process simulator based on the evaluation of process constraints. We propose a heuristic for a dynamic boundary that is calculated based on the metrics of the surrogate-model, e.g. the mean absolute error, and can be applied whenever a process variable is restricted by an inequality constraint. The surrogate models and their metrics depend on training data that may not reflect the interesting regions of the search space, especially in the early stages of optimization. We confirmed the efficiency of the heuristic by an a posteriori evaluation of the surrogate models. We compared two methods to collect training data for the initial surrogate models, data from a Halton set and data generated by the initial simulations. The robustness of the training of the surrogate-models is improved if a small number of uniformly distributed samples are used, i.e. the application of the Halton set is beneficial.

In this work, we evaluated the performance of the surrogate-models after the optimization. However, an evaluation of past surrogate models during the optimization is advisable, e.g. a revision of decisions that were based on too aggressive surrogate-models. The training data consists of the data seen by the optimization and converges in the direction of regions of a high interest, e.g. around a constraint limit. In adaptive sampling methods, surrogate model predictions are used to generate promising new sample points. Ludl et al. (2021) propose an adaptive sampling approach that balances multiple goals, finding the border between convergent and divergent simulations and exploring sample points to reduce the uncertainty of the surrogate model. Winz et al. (2021) propose an upper confidence bound acquisition function to train surrogate models that provide more information regarding the target function of the optimization. To include such approaches is a promising direction for future research.

Acknowledgement

Gefördert durch die Deutsche Forschungsgemeinschaft (DFG) - TRR 63 "Integrierte chemische Prozesse in flüssigen Mehrphasensystemen" (Teilprojekt D1) - 56091768. Funded by the Deutsche Forschungsgemeinschaft (DFG, German Research Foundation) - TRR 63 "Integrated Chemical Processes in Liquid Multiphase Systems" (subprojects D1) – 56091768

References

N. Asprion, M. Bortz, 2018, Chemie-Ingenieur-Technik., 90, 11, 1727–1738.

H.-G. Beyer, H.-P. Schwefel, 2002, Nat. Comput., 1, 1, 3–52.

R. T. Haftka, D. Villanueva, A. Chaudhuri, 2016, Struct. Multidiscip. Optim., 54, 1, 3–13.

T. Janus, S. Engell, Chemie-Ingenieur-Technik. 2021, 93, 12, 2019-2028

G. Kiedorf, D. M. Hoang, A. Müller, A. Jörke, J. Markert, H. Arellano-garcia, A. Seidel-morgenstern, C. Hamel, 2014, Chem. Eng. Sci., 115, 31–48.

P. O. Ludl, R. Heese, J. Höller, N. Asprion, M. Bortz, 2021, Front. Chem. Sci. Eng.

V. A. Merchan, G. Wozny, 2016 Ind. Eng. Chem. Res., 55, 1, 293–310.

M. Urselmann, T. Janus, C. Foussette, S. Tlatlik, A. Gottschalk, M. T. M. M. Emmerich, T. Bäck, S. Engell, 2016, Comput. Aided Chem. Eng. 38, 187–192.

J. Winz, S. Engell, 2021, Comput. Aided Chem. Eng., 50, 2004, 953–958.

Proceedings of the 14th International Symposium on Process Systems Engineering – PSE 2021+
June 19–23, 2022, Kyoto, Japan ©2022 Elsevier B. V. All rights reserved.
http://dx.doi.org/10.1016/B978-0-323-85159-6.50261-X

Flowsheet Recognition using Deep Convolutional Neural Networks

Lukas Schulze Balhorn[a], Qinghe Gao[a], Dominik Goldstein[b], Artur M. Schweidtmann[a,*]

[a]*Department of Chemical Engineering, Delft University of Technology, Van der Maasweg 9, Delft 2629 HZ, The Netherlands*
[b]*Aachener Verfahrenstechnik - Process Systems Engineering, RWTH Aachen University, Aachen 52062, Germany*

a.schweidtmann@tudelft.nl

Abstract

Flowsheets are the most important building blocks to define and communicate the structure of chemical processes. Gaining access to large data sets of machine-readable chemical flowsheets could significantly enhance process synthesis through artificial intelligence. A large number of these flowsheets are publicly available in the scientific literature and patents but hidden among innumerable other figures. Therefore, an automatic program is needed to recognize flowsheets. In this paper, we present a deep convolutional neural network (CNN) that can identify flowsheets within images from literature. We use a transfer learning approach to initialize the CNN's parameter. The CNN reaches an accuracy of 97.9% on an independent test set. The presented algorithm can be combined with publication mining algorithms to enable an autonomous flowsheet mining. This will eventually result in big chemical process databases.

Keywords: Flowsheet, Data Mining, Image Classification, Deep Learning, Transfer Learning

1. Introduction

In recent years, machine learning (ML) has emerged as a popular method to solve complex problems in various domains. This popularity has predominantly been driven by (i) the increase of computational power, (ii) the improvement of ML algorithms, and (iii) the availability of big data (LeCun *et al.*, 2015). Chemical engineering has already seen many successful applications of ML (Schweidtmann *et al.*, 2021; Venkatasubramanian, 2018). However, literature on the structural synthesis of chemical processes through ML is scarce (c.f. (d'Anterroches & Gani, 2005; Zhang *et al.*, 2018; Oeing *et al.*, 2021)). While a variety of promising ML methods exist, big chemical process data is missing (Schweidtmann *et al.*, 2021; Weber *et al.*, 2021). We argue that this lack of structured chemical process data is hindering further progress of ML developments for chemical process synthesis.

The topological information about chemical processes is usually communicated through flowsheets. Flowsheets are technical drawings describing the unit operations connectivity of a process. There exists at least one flowsheet for every chemical process. Eventhough most flowsheets are only available in internal company reports, a large number of flow-sheets are also publicly available in scientific publications and patents. These flowsheets are mostly depicted on figures in PDF documents. However, searching for the flowsheet figures in scientific publications and patents can be as difficult as looking for a needle in a haystack. In particular, a manual search through the enormous amount of available literature would not only be a slow and labor-intense process, but it would also be prone to errors. Therefore, an algorithm is needed that autonomously recognizes flowsheet images.

In the previous literature, information extraction from scientific literature has mostly focused on text mining using natural language processing (Hong *et al.*, 2021; Nasar *et al.*, 2018). In the context of chemistry for example, Swain & Cole (2016) developed the Chem-DataExtractor which extracts chemical identifiers, spectroscopic attributes, and chemical property attributes from scientific literature. Furthermore, information extraction from scholarly images has been performed in the past. The majority of research on the classification of scientific images has been conducted on biomedical literature pushed by the yearly ImageCLEF challenge (c.f. (Pelka *et al.*, 2020)). Furthermore, a few works exist in chemistry on information extraction from images. This works mostly focus on the recognition and digitization of structural formulas (Tharatipyakul *et al.*, 2012; Beard & Cole, 2020). Another example for chemical image analysis is the ImageDataExtractor which mines microscopy images to extract information about the particle sizes and shapes (Mukaddem *et al.*, 2019). However, to the best of our knowledge, image classification has not been applied to chemical process design literature and there exists no previous algorithm that identifies chemical flowsheet images.

In this work, we propose an algorithm that recognizes flowsheet images from chemical engineering journal articles. The proposed algorithm will contribute to our long-term vision to build a database of chemical processes. In Section 2., we provide a brief background on Convolutional Neural Networks (CNNs). In Section 3., we present our methods, data set, and pre-processing. In Section 4., we evaluate the performance of the proposed flowsheet image classification model and discuss the results. Finally, we conclude our findings in Section 5.

2. Deep Convolutional Neural Networks

Inspired by the biological visual system (O'Shea & Nash, 2015), deep CNNs have been proposed as a computational method to bridge the gap between the capabilities of humans and machines for high-level tasks such as image classification, text recognition, and speech recognition (LeCun *et al.*, 2015). The powerful performance of deep CNNs in advanced tasks is achieved through the layout of the framework, which generally consists of three parts: Convolutional layers, pooling layers, and fully-connected layers. Convolutional layers contain a set of learnable filters that will convolve over the inputs to extract the underlying features. Intuitively, simple features such as edges, corners, and blotches will be detected in the early convolutional layers. Ultimately, more complex patterns such as 'unit operations' will appear with further layers. Pooling layers are usually periodically inserted between two convolutional layers to reduce the spatial dimension and the number of parameters. Average pooling and max pooling are the most common choices. Fully-connected neural network layers play the role of mapping the learned "distributed feature representation" to the sample label space, namely, making a classification. Additionally, to introduce nonlinearity into the output, activation functions such as sigmoid, ReLu, and hyperbolic tangent are usually included after convolutional or fully-connected layers. Furthermore, the size of the training data is an important factor for the performance of the deep CNNs and data-labeling is often expansive. Therefore, the concept of transfer learning emerged in recent years. In transfer learning, the CNN is first trained with a sufficiently big data set from one domain of interest. Afterward, the data set of the classification task from another domain of interest is used to fine-tune the CNN.

3. Method

The flowsheet recognition algorithm aims to identify flowsheets among a large number of images. We train a deep CNN for the recognition algorithm based on manually labeled images mainly from scientific journal articles.

3.1. Data Set

At present, no public data set of flowsheet images exists. To create a training data set, we automatically mine figures of scientific journal articles. First, we retrieve a list of all DOIs for a given journal ISSN from the crossref API. Then, the PDFs of the corresponding journal articles are downloaded through publisher APIs. Subsequently, all figures are extracted from the PDFs using the Python package PyMuPDF. The describe procedure is applied to the journals "Theoretical Foundations of Chemical Engineering" and "Frontiers of Chemical Science and Engineering" to generate an initial dataset. Subsequently, the extracted images are manually reviewed and labeled as being a flowsheet or not. In addition to the figures from scientific journal articles, we also add flowsheet images retrieved from a google search to our data set. In total, our data set contains about 1,000 flowsheet images and about 13,000 other images from scientific publications.

3.2. Data Augmentation and Oversampling

As a result of the data mining from journal articles, the data set is imbalanced. In particular, there exist far fewer flowsheet images than other images. This imbalance can cause the classifier to develop a bias towards the majority class. To overcome this issue, oversampling has been used in previous studies (Johnson & Khoshgoftaar, 2019). We oversample the flowsheet images by a factor of 13 to balance the data set. As this large oversampling factor can cause overfitting, we also employ a data augmentation technique (Shorten & Khoshgoftaar, 2019). Each copy of a flowsheet image is augmented by stretching it along the horizontal and vertical axis independently by a random factor between 0.7 and 1.2. Other common data augmentation techniques such as shifting, rotation, and shearing were dismissed because they are expected to destroy some key features of flowsheet images. For example, flowsheets usually include horizontal and vertical lines making image rotation pointless. The images of the negative "other" class are not augmented because of abundant data availability.

3.3. Model Training

The CNN architecture for the flowsheet recognition is based on the VGG16 network by Simonyan & Zisserman (2014). The network includes 13 convolutional layers, 5 max pooling layers, and 3 fully-connected layers. Since our data set is limited, we use a transfer learning approach. In particular, we use the publicly available VGG16 network that has been pre-trained on the ImageNet data set including tens of millions of images and 1,000 categories. To adapt the network to the use case of this work, we reduced the number of nodes in the output layer to two. The training is conducted using the PyTorch framework which is built on the Torch library. The model takes in images with a resolution of 224 × 224 pixels. We randomly divide our data set into training (70%), validation (15%), and test (15%) data set. The model is trained on the training data in batches of 150 images. The validation set was used to validate the training progress and tune the hyperparameters of the model. The independent test data set is used for the final performance evaluation. Notably, the test set is truly independent as it does not contain any augmented images from the training or validation sets.

4.   Results and Discussions

The most important performance metrics for classifiers is the accuracy as defined in Eq. 1. In the light of class imbalance, we also evaluate the precision (Eq. 2) and recall (Eq. 3):

$$Accuracy = \frac{TN + TP}{TP + FP + TN + FN},\tag{1}$$

$$Precision = \frac{TP}{TP + FP},\tag{2}$$

$$Recall = \frac{TP}{TP + FN},\tag{3}$$

where TP denotes true positive, TN denotes true negative, FP denotes false positive and FN denotes false negative. The training history is shown in Fig. 1. The classifier reaches a satisfying accuracy already after the first epoch. This good initial performance can be explained by the use of a pre-trained model. After the second epoch of training, the classifier shows a validation accuracy of over 98%. The training process was ended after 10 epochs. In training runs with more epochs no further improvement was experienced. The final training accuracy after 10 epochs is 98.1% while the validation accuracy is 98.2%. Notably, we do not observe any overfitting behavior in the training process.

Overall, the flowsheet recognition algorithm shows a promising performance on the independent test set. The confusion matrix on the test set is shown in Table 1. Of all predictions on the test set, 97.9% were correct. Furthermore, the precision is 80.7% and lower than the recall with 94.4%. The high recall shows that almost all flowsheet images are retrieved while the number of false negative flowsheets is very low. Furthermore, the fairly low precision could be explained by the class imbalance. The data set contains about thirteen times more images of the class "other". If only a small fraction of the class "other" is misclassified, these images already make up a great share of the flowsheet predictions.

Finally, the runtime of the image classification is investigated. The evaluation of an image by the trained CNN takes about 7 milliseconds on average on a personal computer. This short evaluation time allows for an online application that autonomously mines flowsheets from literature.

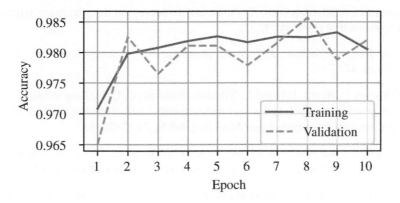

Figure 1: Training history of the CNN.

Table 1: Confusion matrix for the flowsheet recognition algorithm on the test set.

	Actual flowsheet	Actual other
Predicted flowsheet	151	36
Predicted other	9	1,976

5. Conclusions

We propose an image classification algorithm that can recognize flowsheet images. The algorithm consists of a deep CNN which classifies images with a high accuracy of 97.9%. In order to train the CNN, we mined about 1,000 flowsheet images from scientific literature and online search engines. Moreover, the transfer learning improved the prediction accuracy. The proposed tool can be used to automatically identify flowsheet images from scientific literature or other sources within a few milliseconds. In a preliminary study we applied our mining algorithm to the journal "Computers & Chemical Engineering" and identified more than 1500 flowsheets. Future work will digitize the flowsheet images to identify process topologies. This will eventually result in an open-source knowledge graph database providing chemical processes in a structured format. We believe that this database has a tremendous value for future process design because it allows the search and optimization over existing processes. In addition, our database will eventually serve as a training database for advanced ML algorithms able to design novel processes.

References

Beard, Edward J., & Cole, Jacqueline M. 2020. ChemSchematicResolver: A Toolkit to Decode 2D Chemical Diagrams with Labels and R-Groups into Annotated Chemical Named Entities. *Journal of Chemical Information and Modeling*, **60**(4), 2059–2072.

d'Anterroches, Loïc, & Gani, Rafiqul. 2005. Group contribution based process flowsheet synthesis, design and modelling. *Fluid Phase Equilibria*, **228-229**(Feb.), 141–146.

Hong, Zhi, Ward, Logan, Chard, Kyle, Blaiszik, Ben, & Foster, Ian. 2021. Challenges and Advances in Information Extraction from Scientific Literature: a Review. *JOM*, Oct.

Johnson, Justin M., & Khoshgoftaar, Taghi M. 2019. Survey on deep learning with class imbalance. *Journal of Big Data*, **6**(27).

LeCun, Yann, Bengio, Yoshua, & Hinton, Geoffrey. 2015. Deep learning. *Nature*, **521**(May), 436–444.

Mukaddem, Karim T., Beard, Edward J., Yildirim, Batuhan, & Cole, Jacqueline M. 2019. ImageDataExtractor: A Tool To Extract and Quantify Data from Microscopy Images. *Journal of Chemical Information and Modeling*, **60**(5), 2492–2509.

Nasar, Zara, Jaffry, Syed Waqar, & Malik, Muhammad Kamran. 2018. Information extraction from scientific articles: a survey. *Scientometrics*, **117**(3), 1931–1990.

Oeing, Jonas, Henke, Fabian, & Kockmann, Norbert. 2021. Machine Learning Based Suggestions of Separation Units for Process Synthesis in Process Simulation. *Chemie Ingenieur Technik*, Sept.

O'Shea, Keiron, & Nash, Ryan. 2015. An introduction to convolutional neural networks. *arXiv preprint arXiv:1511.08458*.

Pelka, Obioma, Friedrich, Christoph M, García Seco de Herrera, Alba, & Müller, Henning. 2020 (Sept.). Overview of the ImageCLEFmed 2020 Concept Prediction Task: Medical Image Understanding. *In: Proceedings of the CLEF 2020-Conference and labs of the evaluation forum.*

Schweidtmann, Artur M., Esche, Erik, Fischer, Asja, Kloft, Marius, Repke, Jens-Uwe, Sager, Sebastian, & Mitsos, Alexander. 2021. Machine Learning in Chemical Engineering: A Perspective. *Chemie Ingenieur Technik*, Oct.

Shorten, Connor, & Khoshgoftaar, Taghi M. 2019. A survey on Image Data Augmentation for Deep Learning. *Journal of Big Data*, **6**(60).

Simonyan, Karen, & Zisserman, Andrew. 2014. Very deep convolutional networks for large-scale image recognition. *arXiv preprint arXiv:1409.1556*.

Swain, Matthew C., & Cole, Jacqueline M. 2016. ChemDataExtractor: A Toolkit for Automated Extraction of Chemical Information from the Scientific Literature. *Journal of Chemical Information and Modeling*, **56**(10), 1894–1904.

Tharatipyakul, Atima, Numnark, Somrak, Wichadakul, Duangdao, & Ingsriswang, Supawadee. 2012. ChemEx: information extraction system for chemical data curation. *BMC Bioinformatics*, **13**(17).

Venkatasubramanian, Venkat. 2018. The promise of artificial intelligence in chemical engineering: Is it here, finally? *AIChE Journal*, **65**(2), 466–478.

Weber, Jana M., Guo, Zhen, Zhang, Chonghuan, Schweidtmann, Artur M., & Lapkin, Alexei A. 2021. Chemical data intelligence for sustainable chemistry. *Chemical Society Reviews*.

Zhang, Tong, Sahinidis, Nikolaos V., & Siirola, Jeffrey J. 2018. Pattern recognition in chemical process flowsheets. *AIChE Journal*, **65**(2), 592–603.

Proceedings of the 14th International Symposium on Process Systems Engineering – PSE 2021+
June 19-23, 2022, Kyoto, Japan © 2022 Elsevier B.V. All rights reserved.
http://dx.doi.org/10.1016/B978-0-323-85159-6.50262-1

Active learning for multi-objective optimization of processes and energy systems

Julia Granacher[a]* and François Maréchal[a]

aIndustrial Process and Energy Systems Engineering Group, Ecole Polytechnique Fédérale de Lausanne Valais Wallis, Rue de l'Industrie 17,Case Postale 440, 1951 Sion, Switzerland
julia.granacher@epfl.ch

Abstract

In superstructure optimization of processes and energy systems, the design space is defined as the combination of unit considerations, process conditions and model parameters that might be subjected to uncertainty. Most of the time, decision makers are not looking for a single best solution, but rather are interested in analyzing a set of Pareto-optimal superstructure designs. The generation of Pareto-optimal solutions is computationally expensive, especially if nonlinear process evaluations or simulation is required. In our approach, we address the question of how to efficiently generate Pareto-optimal sets of solutions by applying machine learning concepts. Using the criteria of Pareto-optimality to evaluate the performance of a set of design space variables and the corresponding solution, we train our algorithm on predicting if a solution is belonging to the Pareto frontier. Following the approach presented by Zuluaga et al., (2016) and applied to the design of materials by Jablonka et al., (2021), an adaptive learning concept is used to systematically identify the next best function evaluation to improve the confidence of the Pareto-frontier definition. Gaussian process surrogate models provide a prediction of the mean and the standard deviation of the relevant objectives. Design points with high probability to of being in the Pareto-optimal domain are evaluated by the original model, increasing the confidence with which the Pareto front is predicted. Simultaneously, the design space is continuously reduced by discarding the design points for which the probability of being in the set of Pareto-optimal solutions is low. The procedure is stopped when all points are labeled as Pareto-optimal or discarded. The algorithm is applied to the design of a utility superstructure for an industrial energy system. Our algorithm is compared and benchmarked with quasi random sampling of the design space.

Keywords: Multi-objective Optimization; Active Learning; Energy System Design; Utility Superstructure; Mathematical Programming; Machine Learning; Artificial Intelligence

1. Introduction

One of the most pressing challenges our society is facing is climate change, revealing the need for efficient and reliable design methods of energy systems that are sustainable in economic, environmental and social terms. Over the last decades, the methodologies on energy system and process design have evolved drastically. Process systems engineering (PSE) has developed as a conceptual element of chemical engineering, including the definition, design, planning and control of complex chemical processes (Mencarelli et al., 2020). PSE was initially dominated by the progress made in process simulation, where algebraic equations and flowsheeting methods are applied to describe a system's behavior. The emerging focus on quantitative descriptions of processes by means of simulation led to a more thorough analysis of the system performance (Rudd et al., 1973). Process synthesis addresses the development, simulation, and optimization of processes, where the unit operations are selected and interconnections are defined (Biegler et al., 1997; Douglas, 1988). In superstructure optimization, a network of all potential unit operations and connections is defined and translated into a mathematical programming model, which is then used for generating results by solving an optimization problem. Instead of generating one optimal design that may only be valid under certain external conditions, the generation of a set of feasible alternatives may be preferred.

Multi-objective optimization is widely applied for analyzing trade-offs between two or more objective functions. The set of optimal solutions obtained from a multi-objective optimization problem can be displayed in a Pareto-optimal curve, on which, for each point on the curve, none of the objectives can be improved without penalizing the others. However, most optimization techniques rely on the introduction of a total order in the search space, which biases the search and may introduce technical difficulties (Jablonka et al., 2021). Machine learning has gained interest recently for designing processes with complex design spaces, as they allow for the fast prediction of process performance. However, training large datasets makes the problem unnecessarily computationally expensive, especially when simulation-based approaches are included in the superstructure.

In this paper, we are addressing the question how Pareto fronts of large energy superstructure optimization problems with non-linear relations between the design space and the objectives can be generated efficiently with the assistance of machine learning algorithms, ensuring reliable predictions of the system performance along the Pareto front. Thereby, an active learning approach is applied, which iteratively improves the machine learning model where it is needed the most.

2. Methodology

A modified implementation of the of the ε-PAL algorithm introduced by Zuluaga et al., (2016) and implemented by Jablonka et al., (2021) is applied to the optimization of energy and process system superstructures. The ε-PAL algorithm iteratively reduces the effective design space by discarding those design points from which we know that they are Pareto-dominated by another design point. The design point with the highest dimensionless uncertainty from a set of possible design points predicted to be near the Pareto front is evaluated. When all points are either classified or discarded, the search ends. The approach offers the additional benefit of enabling the tuning between accuracy and efficiency, by tuning the granularity of the approximation to the Pareto front in every objective. In the following sections, the general superstructure problem as well as the ε-PAL algorithm is described.

2.1. Superstructure optimization

A process superstructure has the aim of describing the system's units and the way they can interact with others. By activating certain units and their connections, different system configurations are achieved. In this work, the methodology for superstructure modelling and optimization is adapted from Gassner and Maréchal, (2009). For each unit in the system, energy and mass flow balances are formulated, describing corresponding transformations. They are derived using either flowsheeeting or simplified black-box models. Binary decision variables describe whether a unit is installed and used in a certain period. Continuous decision variables describe the installed size of the unit and the level of usage at which it is operated in each period. Parameterized bounds constrain continuous and binary variables. In our approach, all process units are connected to a heat exchange system, which allows for the exchange between the process and the hot/cold utilities to close the energy balance. Operating and investment costs are derived based on equipment size. Environmental impacts of the system are estimated using the Life Cycle Inventory Ecoinvent database (Wernet et al., 2016). Pinch analysis is applied to model heat recovery opportunities and to investigate the integration of the utility system by introducing the heat cascade constraints explained in Maréchal and Kalitventzeff, (1998).

For generating a solution, the decision variables are fixed solving a mixed-integer linear programming (MILP) problem formulated in the AMPL optimization language (Fourer et al., 2002), using the CPLEX branch-and-bound algorithm (IBM, 2017). The superstructure model in the lower-level is integrated in an upper-level framework, in which optimization problems are formulated for exploring the impact of non-linear decision variables on the results, such as operating conditions of the utility system. For a problem communicated by the upper-level framework, the lower-level generates a solution and reports it to the upper-level. For generating and communicating optimization problems to the lower-level, different approaches could be envisioned. Besides the usage of random sampling, the application of evolutionary algorithms or the hereafter presented Pareto-active learning approach can be considered.

2.2. Pareto-active learning

For efficiently and reliably identifying the Pareto front, we apply the modified version of the ϵ-PAL algorithm presented in Jablonka et al., 2021. In the ϵ-PAL algorithm, the uncertainty estimation σ of a Gaussian process is used to construct hyperrectangles for a prediction (Figure 1A). The algorithm enables the classification of Pareto-optimal samples, as well as the proposition of the next best sample for evaluation.

The ϵ-PAL algorithm starts with a set of experiments, and the desired objectives are calculated by calling the original model for a subset of the generated samples. An initial machine learning model is trained on the obtained dataset of decision variables and objectives, and predictions for unlabeled datapoints are made. For each prediction, hyperrectangles around the prediction mean are constructed, the width being equivalent to the standard deviation of the posterior of the Gaussian process. The lower and upper limits are equivalent to the best/worst performance estimates (Figure 1A). The points to be discarded with confidence can be identified from the ϵ-Pareto dominance relation, as well as the ones with a high probability of being Pareto-optimal (Figure 1B). If the pessimistic estimate of a prediction is greater than a defined tolerance above the optimistic estimate of all other predictions, it will be part of the Pareto front. For estimating the accuracy of the Pareto front, one can connect the bottom left corners of hyperrectangles associated with the current estimate of the front, which gives the most pessimistic front (dashed blue line in Figure 1B). The optimistic front is then obtained by connecting the upper right corners.

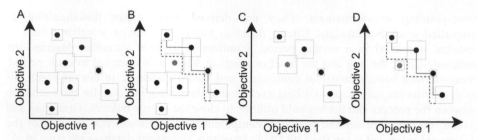

Figure 1: A: hyperrectangles based on predicted mean and variance. B: Pareto-optimal points identified. C: Relevant design space shrinks, red point identified as new point to label. D: Uncertainty in red point reduced after retraining the model with new sample, adapted from Jablonka et al., (2021).

Thus, a geometric construction is created that allows for classification whether a predicted solution is Pareto-optimal or whether it can be discarded. The next design point to be evaluated by the optimization is identified as the one that reduces the uncertainty in classifying points as Pareto-optimal. For this, it is assumed that the uncertainties are normalized by the predicted mean, so that the area of the hyperrectangles represent the relative error. The prediction model is then improved by reducing the uncertainty of the largest rectangle among the points presumed near the Pareto front (Jablonka et al., 2021). In Figure 1C, the red point is identified as assisting the model improvement most, and the more accurate estimate of the updated model is represented in Figure 1D. This procedure is repeated until the desired accuracy of the Pareto front is reached. For further information about the algorithm and its implementation, the reader may consult Jablonka et al., (2021); Zuluaga et al., (2016).

3. Application

The proposed methodology is applied to the design of a Pareto front for the optimal operation of a steam network integrated in a large-scale Kraft pulping process, producing 1000 air-dried tons of pulp per day. The pulp and paper industry is known as an energy-intensive industrial sector, consuming large quantities of water and energy. Integration techniques, including heat integration, water network optimization and steam cycle operation optimization as applied in Kermani et al., (2019) can significantly reduce water and energy consumption of the mill. In this contribution, we focus explicitly on the optimization of the operating conditions of a steam network integrated with the pulp mill by applying the ε-PAL algorithm. In the steam network, steam can be produced between 50 and 160 bar, while it can be consumed at 3 pressure levels. To meet the specification of combined steam and electricity production in industrial plants, steam production can only happen at the highest pressure level, and turbines are placed between the highest pressure level and the subsequent levels. The superstructure model is adapted from Kermani et al., (2019), which might be consulted for more information. Objectives selected for optimization and representation in a Pareto front are the annual operating cost (OPEX) and the annualized capital cost (CAPEX). The ε-PAL algorithm is applied on the upper-level, evaluating samples from the decision space for retrieving Pareto fronts. For the application of the ε-PAL algorithm, the decision variables in Table 1 are used. Two ε-PAL instances are created, one with 200 and one with 500 samples. In the first iteration, 50 samples are labeled, while in each following iteration, 10 new samples are labeled.

Table 1: Decision variables for designing the Pareto front, adapted from Kermani et al., (2019).

Decision variable	Range, Unit	Description
p_1^{st}	[50;160], bar	Boiler pressure
p_2^{st}	[9;14], bar	High-pressure steam header
p_3^{st}	[3;8], bar	Medium-pressure steam header
p_4^{st}	[0.5;2], bar	Low-pressure steam header
T_1^{sup}	[150;300], °C	Degree of superheating in the highest pressure level

Results

When running the ϵ-PAL algorithm on the small set of samples, five Pareto optimal points are identified. The learning curve in Figure 2A shows that after ten iterations, all design points are either discarded or identified as Pareto-optimal. Figure 2B shows the classification of all design points. The error bars indicate the obtained uncertainty which is used for calculating the hyperrectangles when qualifying whether a point might be Pareto-optimal or not. Overall, 140 design points are labeled, and the hypervolume obtained is 9614. Computation took 83 minutes, while labelling all 200 samples for obtaining the same Pareto front takes 118 minutes, indicating time savings of 30%.

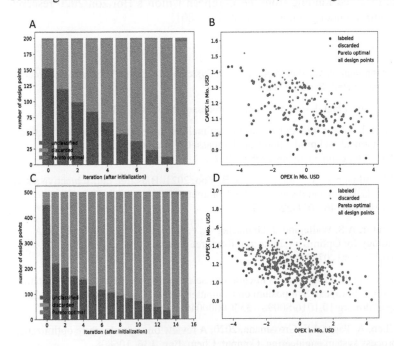

Figure 2: Results for applying ϵ-PAL algorithm. A: Learning curve for 200 samples, B: Pareto front for 200 samples, C: Learning curve for 500 samples, D: Pareto front for 500 samples.

Figure 2C shows the learning curve for 500 sampled design points. After 16 iterations, all design points are either discarded or classified as Pareto-optimal. A total of 11 Pareto-optimal points is identified, the rest is discarded. The hypervolume obtained from this Pareto front is 9651, so slightly higher than for the smaller design space. Over all iterations, a total of 200 samples were labeled by calling the original optimization model. The computational time was 122 minutes. Compared to random sampling and evaluation of 500 samples, this accounts for time savings of approximately 60%. It is worth noting that all points identified as Pareto-optimal are also labeled, meaning that the error in this region of the design space is minimized. This ensures reliable results in the Pareto-optimal domain.

4. Conclusion and outlook

In this contribution, we demonstrated that the ε-PAL algorithm can be used to identify Pareto-optimal design points for process and energy system superstructures. Compared to random sampling, significant time savings were obtained, the relative savings increasing with the size of the design space. The quality of the Pareto front, measured with the hypervolume error, increases with the design space size. The algorithm manages to identify relevant regions in the design space as Pareto-optimal or near Pareto-optimal, allowing for continuous improvement of the prediction quality where it is necessary. Future work will include benchmarking the applied methodology to a genetic algorithm, as well as obtaining results for larger design spaces.

Acknowledgements

This research has received funding from the European Union's Horizon 2020 research and innovation program under grant agreement No 81801.

References

L.T. Biegler, I.E. Grossmann, A.W. Westerberg, 1997. Systematic methods for chemical process design, 1st ed, Prentice Hall. Prentice Hall.

J.M., Douglas, 1988. Conceptual design of chemical processes. McGraw-Hill.

M. Gassner, F. Maréchal, 2009. Methodology for the optimal thermo-economic, multi-objective design of thermochemical fuel production from biomass. Comput. Chem. Eng., https://doi.org/10.1016/j.compchemeng.2008.09.017

K.M. Jablonka, G.M. Jothiappan, S. Wang, B. Smit, B. Yoo, 2021. Bias free multiobjective active learning for materials design and discovery. Nat. Commun. 12. https://doi.org/10.1038/s41467-021-22437-0

M. Kermani, I.D. Kantor, A.S. Wallerand, J. Granacher, A.V. Ensinas, F. Maréchal, 2019. A Holistic Methodology for Optimizing Industrial Resource Efficiency. Energies 12, 1315. https://doi.org/10.3390/en12071315

F. Maréchal, B. Kalitventzeff, 1998. Process integration: Selection of the optimal utility system. Comput. Chem. Eng., European Symposium on Computer Aided Process Engineering-8 22, S149–S156. https://doi.org/10.1016/S0098-1354(98)00049-0

L. Mencarelli, Q. Chen, A. Pagot, I.E. Grossmann, 2020. A review on superstructure optimization approaches in process system engineering. Comput. Chem. Eng. 136, 106808. https://doi.org/10.1016/j.compchemeng.2020.106808

D.F. Rudd, G.J. Powers, J.J. Siirola, 1973. Process Synthesis. Prentice-Hall.

G. Wernet, C. Bauer, B. Steubing, J. Reinhard, E. Moreno-Ruiz, B. Weidema, 2016. The ecoinvent database version 3 (part I): overview and methodology. Int. J. Life Cycle Assess. 21, 1218–1230. https://doi.org/10.1007/s11367-016-1087-8

M. Zuluaga, A. Krause, M. Puschel, 2016. e-PAL: An Active Learning Approach to the Multi-Objective Optimization Problem. Journal of Machine Learning Research 17

Proceedings of the 14th International Symposium on Process Systems Engineering – PSE 2021+
June 19-23, 2022, Kyoto, Japan © 2022 Elsevier B.V. All rights reserved.
http://dx.doi.org/10.1016/B978-0-323-85159-6.50263-3

Data-driven Stochastic Optimization of Numerically Infeasible Differential Algebraic Equations: An Application to the Steam Cracking Process

Burcu Beykal[a,b*], Zahir Aghayev[a,b], Onur Onel[c,d], Melis Onel[c,d], Efstratios N. Pistikopoulos[c,d]

[a]*Department of Chemical and Biomolecular Engineering, University of Connecticut, Storrs, CT 06269, USA*
[b]*Center for Clean Energy Engineering, University of Connecticut, Storrs, CT 06269, USA*
[c]*Texas A&M Energy Institute, Texas A&M University, College Station, TX 77843, USA*
[d]*Artie McFerrin Department of Chemical Engineering, Texas A&M University, College Station, TX 77843, USA*
burcu.beykal@uconn.edu

Abstract

In this work, we address the data-driven stochastic optimization of the numerically infeasible differential algebraic equations (DAEs) using Support Vector Machines (SVMs) and scenario analysis. Data-driven optimization is an attractive method for optimizing systems with highly complex first-principles models using iterative sampling, surrogate modeling, and optimization steps. Yet, the numerical stability and the unknown interconnections between the initial conditions of the DAEs determine the overall performance of the data-driven optimizer. Specifically, in the sampling step where the DAE system is initialized, varying samples of initial conditions can cause premature termination of the simulation due to numerical infeasibilities, without retrieving any viable output data that is essential for the surrogate modeling and grey-box optimization steps. These challenges are further amplified when there are stochastic elements present in the system which the numerically infeasible system of DAEs needs to handle to achieve robust solutions. Using the steam cracking process as our motivating example, the SVMs are used to accurately map the feasible region of the numerically infeasible system of DAEs representing the first-principles models of the cracking reactor, while the scenario analysis allows us to handle the uncertainty in the feed composition of the natural gas liquid (NGL). The resulting modeling framework is incorporated in a data-driven optimization solver and utilized to generate the guaranteed feasible solution for the design and operation of an NGL steam cracking reactor under uncertain feed compositions.

Keywords: Data-driven optimization, support vector machines, stochasticity, scenario analysis, steam cracking process.

1. Introduction

The first-principles representations of many unsteady-state or dynamic chemical engineering systems are composed of ordinary or partial differential equations (i.e., mass, energy, and momentum balances) and algebraic expressions (i.e., rate law), creating a system of differential algebraic equations (DAEs). The optimal decision-making with both differential and algebraic components is not straightforward since the deterministic optimization solvers cannot be directly implemented with such formulations. Typically, dynamic optimization of DAE systems is handled through: (1) Process simulation software, like gPROMS or Aspen Custom Modeler (Lang and Biegler, 2007); (2) Orthogonal collocation on finite elements to reduce a dynamic optimization problem to a constrained nonlinear problem (Biegler, 1984), or (3) Data-driven modeling and optimization (Beykal et al., 2020).

Optimization with dynamic programs can also be complicated by numerical infeasibilities, like stiffness, ill-conditioned algebraic equations, and undefined solutions. Orthogonal collocation on finite elements can handle such challenges easily with its discretization strategy, however, the resulting large-scale nonlinear program (NLP) is typically solved to local optimality (Caballero et al., 2015). Data-driven optimization enables the exploration of global solutions through sampling, surrogate modeling, and deterministic optimization steps (Beykal et al., 2018). Yet, such algorithms are not designed to handle numerical infeasibilities and implicit constraints. Especially at the sampling stage, when candidate sampling points for the decision variables (i.e., the initial conditions of the DAE system) are identified, sampling procedures like the Design of Experiments will assume that all decision variables are independent of each other. However, there might be interdependencies among variables that define the numerical stability of the solution and can only be represented in explicit mathematical forms if the analytical solution exists. As a result, without the explicit a priori knowledge on these interdependencies, data-driven optimization algorithms will return unrealistic solutions that cannot be validated by the first principles-based simulations. Recently, Beykal et al. (2020) developed a Support Vector Machine (SVM) based data-driven optimization framework to overcome numerical infeasibilities and implicit constraints by mapping the numerical feasibility boundary with a two-class classification model. This study showed that implicit constraints are accurately captured with SVMs, significant computational savings are achieved with the SVM classifier, and guaranteed feasible solutions are attained with data-driven optimization techniques for such difficult cases of DAE systems.

Despite recent efforts, the challenges in numerically infeasible DAE systems are further complicated with uncertain initial conditions. Even when the deterministic solution is retrieved by the aforementioned SVM approach, a slight deviation in the optimal input conditions due to system disturbances could still lead to failures and undesirable outcomes. For example, in a reaction system like steam cracking, the input natural gas liquid (NGL) feed composition can be uncertain, or the feed composition can be adjusted to maximize profit with changing market conditions. Hence, the steam cracker design and operation problem should consider these stochastic elements to be able to offer a solution that is flexible and robust against changing initial conditions.

Motivated by this, in this work, we further extend our previous analysis on handling numerically infeasible DAE systems using SVMs and introduce stochastic initial conditions in the problem formulation to study the effects of uncertainty on the data-driven optimization performance. We demonstrate the effectiveness of our approach on the optimal steam cracker design and operation problem which is subject to stochastic NGL feed compositions. Details on the problem formulation and the data-driven optimization results of the computational case study are provided in the following sections.

2. Methodology

2.1. Steam Cracking Optimization: Problem Formulation

The steam cracking process is modeled using a plug flow reactor (diameter = 0.108 m) with mass, energy, and momentum balances under coking effects. Steam and NGL streams are co-fed at the reactor entrance (Figure 1). The plug flow reactor is subject to constant external heat flux, Q, across the reactor length, L. The model is one-dimensional and dynamic along the spatial coordinate, z with spatial changes in the molar of species, reactor pressure, and temperature are provided as $F_j(z)$, $P(z)$, and $T(z)$, respectively.

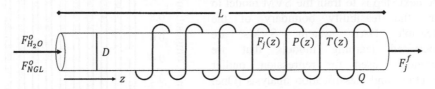

Figure 1 The plug flow reactor for modeling steam cracking of NGLs.

Our goal is to determine the optimal values of the reactor length, external heat flux, inlet pressure, inlet temperature, the inlet flowrates for the NGL and steam feeds to maximize the total stochastic profit from propylene and ethylene production subject to the stiff steam cracking model, the known constraint on the total initial flowrate, and output constraints on the reactor exit temperature and pressure (Eq.(1)). The objective function is calculated by multiplying the profit obtained from each scenario, n, and their corresponding probability of occurrence, ϕ^n. The profit at every scenario is obtained by numerically integrating the steam cracking model. Output reactor constraints are handled as grey-box constraints whereas the numerical feasibility of the mathematical models is handled through the SVM model. The detailed list of species, reaction mechanisms, and model equations considered in the formulation are available in Beykal et al., (2020).

$$\max \ Total\ Stochastic\ Profit = \sum_{n=1}^{N} \phi^n \cdot Profit^n$$

$$s.t. \quad Mass, Energy, Momentum\ Balances \quad \forall n$$

$$Rate\ Law\ \&\ Reaction\ Mechanism, Coking\ Effects \quad \forall n$$

$$F_{H_2O}^0 + F_{NGL}^o \leq 0.05\ kmol/s \quad \forall n$$

$$T^f \leq 1300\ K \quad \forall n, \qquad P^f \geq 80\ kPa \quad \forall n$$

$$T^{in} = [700\ K, 1100\ K], P^{in} = [290\ kPa, 500\ kPa], L = [5\ m, 100\ m]$$

$$Q = [10\ kW/m^2, 1000\ kW/m^2], F_{H_2O}^o, F_{NGL}^0 = [0.003\ kmol/s, 0.05\ kmol/s]$$

(1)

2.2. Scenario Analysis for the Stochastic Feed Compositions

Scenarios are created to represent the stochastic feed compositions in the steam cracking problem. We assume that the NGL feed is only composed of ethane and propane and create 11 scenarios with varying compositions of these compounds. The created scenarios aim to capture a variety of events that the steam cracker may encounter. For example, when ethylene demand increases, the reactor may be operated with a pure ethane feed. Likewise, when propylene demand increases, the reactor may be operated with pure propane feed. In less extreme cases, the NGL feed composition may vary depending on the supplier or due to system disturbances. Hence, by taking the pure ethane and propane compositions as the endpoints in our scenario analysis, we create 9 other scenarios with 0.1 increments in the propane composition (Figure 2). All scenarios are assumed to have an equal probability of occurrence with $\phi_n = 1/11$, and the steam cracker model is solved for every scenario to calculate the total stochastic cost.

2.3. Mapping the Feasible Region of Implicit Constraints with Support Vector Machines

The next step is to train the SVM model to map the feasibility boundary of the infeasible DAE system. SVMs are supervised learning models that are commonly used for regression, outlier detection, and classification analysis (Onel et al., 2019). Due to their highly flexible nature and ability to use nonlinear transformations, SVMs can learn highly nonlinear relationships within a dataset with high accuracy. To facilitate data-driven optimization and to ensure the validity of the final solution, the feasibility of the numerical

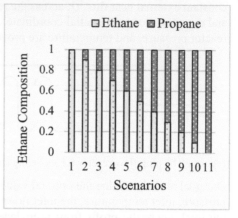

Figure 2 Representative scenarios for the stochastic NGL feed compositions.

integration of a DAE system is modeled using a two-class C-parametrized nonlinear SVM classification algorithm. As highlighted by Beykal et al. (2020), the main idea of this approach is two-fold: (1) to collect samples from the numerical integration and use their output information to train an SVM model of the feasible region for the DAE system (the offline phase); and (2) to incorporate this classifier within a grey-box optimization solver to eliminate infeasible solutions prior to sample collection (the online phase).

In the offline phase, we follow the recipe outlined in Beykal et al. (2020) and Onel et al. (2019) and construct a Latin Hypercube Design with 2000 points that satisfy the total maximum flowrate constraint in Eq.(1). Each sampling point is then numerically integrated across all feed composition scenarios and their discrete output information is collected as either "feasible - 0" or "infeasible - 1". The output value of a sample is deemed feasible if the sampling point is successfully integrated without any failures in simulation across all scenarios. Otherwise, the sample is deemed infeasible and an output tag of "1" is assigned. The input data is then min-max scaled within the bounds of the decision variables and randomly split into train, validation, and test sets. 90 % of the data set is reserved for training the SVM model with 5-fold cross-validation and the remaining 10 % is reserved for blind testing the model performance. Gaussian radial basis function is used as the nonlinear kernel for the SVM model and the respective hyperparameters are tuned through an exhaustive grid search. Finally, the predictive capability of the trained classifier is assessed using several performance metrics, including the accuracy, precision, recall, F_1 score, and area under the curve (AUC).

Once this offline phase is completed, the SVM model is ready to filter any numerically infeasible combinations of decision variables for data-driven optimization prior to the simulation call. In this study, we embed the SVM model in the ARGONAUT algorithm (Boukouvala and Floudas, 2017). The key findings for the SVM model performance and the stochastic steam cracking optimization problem are provided below.

3. Results

3.1. Performance Metrics for the SVM Classifier in the Offline Phase

The results summarizing the predictive performance of the SVM classifier are presented in Table 1. As ARGONAUT operates in sessions (i.e., the first session in the original variable bounds; the second session in the tightened variable bounds), two sets of performance metrics are reported. The results show that highly accurate SVM classifiers are trained using the numerical feasibility information obtained across all scenarios. When the SVM model is retrained within the tightened variable bounds, we observe that the model accuracy, precision, and F_1 score are improved.

Table 1 SVM model performance with the blind testing set.

SVM Model	Accuracy	Precision	Recall	F_1 score	AUC
Session 1	98.5 %	96.9 %	100 %	98.4 %	100 %
Session 2	99.5 %	100 %	99.1 %	99.5 %	100 %

Here, although we achieved highly accurate classifiers, one of the biggest drawbacks of the offline phase is the computational overhead required to collect samples for modeling. Especially with computationally expensive simulations, this step is very demanding. Our future work will focus on making the offline and online phases seamless to improve the computational efficiency of the overall framework.

3.2. Optimal Solution: Reactor Design Parameters and Key Results in the Online Phase

The highly accurate SVM model classifiers are then incorporated in the ARGONAUT algorithm and the stochastic steam cracking model is optimized over 10 random runs, each starting with a different set of Latin Hypercube Design. The optimal values for the decision variables in the best-found solution are: $T^{in} = 788.4$ K; $P^{in} = 341.4$ kPa; $Q = 350.9$ kW/m²; $L = 39.7$ m; $F_{NGL}^o = 0.0289$ kmol/s; and $F_{H_2O}^o = 0.018$ kmol/s. This set of optimal decision variables provides a guaranteed feasible solution for the 11 studied scenarios and achieves a total stochastic profit of $0.1677/s using the hybrid SVM and grey-box optimization approach. This total stochastic profit is almost 50 % less than the profit reported for pure ethane feed, but it is also almost 50 % greater than the profit reported for the pure propane feed in Beykal et al. (2020). This is an expected result as the previous study only considered pure feeds in the problem setup where the reactor parameters are fine tuned to maximize profit for the deterministic NGL feeds. However, in the current study, the optimal reactor parameters can handle a wide range of feed compositions which allows the reactor to be more flexible while achieving high profit values, even when the feed compositions change due to changing market conditions or other external factors.

In addition, Figure 3 shows the production of ethylene, propylene, and other key products under the provided optimal conditions for the two representative scenarios: (1) High ethane content (70 % ethane-30 % propane); and (2) high propane content (20 % ethane and 80 % propane). In both cases, we observe that the molar flowrate of the reactants is decreasing across the reactor length, whereas the molar flowrate of the desired products is increasing through carrying out the favorable reactions. The mean ethane conversion across all scenarios (except for the pure propane scenario) is 0.79 with a standard error of 0.015 across these scenarios. Likewise, the mean propane conversion is also high, 0.88, with a standard error of 0.006. These results indicate that guaranteed feasible solutions can be achieved for numerically infeasible stochastic problems using data-driven modeling and grey-box optimization.

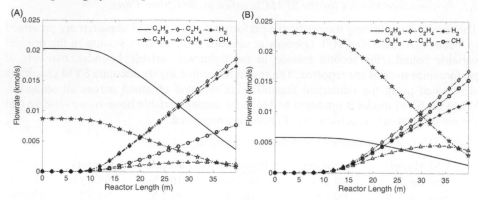

Figure 3 Molar flowrate of species for the favorable reactions in two scenarios with: (A) 70 % ethane – 30 % propane in the NGL feed; (B) 20 % ethane – 80 % propane in the NGL feed.

4. Conclusions

We present a framework to address the data-driven stochastic optimization of numerically infeasible DAE systems without the full discretization of the first-principles model. The stochastic elements in the formulation are handled via scenario analysis whereas the numerical infeasibilities are modeled using Support Vector Machines. By incorporating these two analysis methods in a grey-box optimization solver, we provide guaranteed feasible solutions to numerically infeasible stochastic problems using data-driven modeling. Results of the computational case study of the steam cracking of natural gas liquids with uncertain feed compositions show that high total stochastic profit is achieved using the proposed approach.

References

B. Beykal, F. Boukouvala, C.A. Floudas, N. Sorek, H. Zalavadia, E. Gildin, 2018, Global optimization of grey-box computational systems using surrogate functions and application to highly constrained oil-field operations, Computers & Chemical Engineering, 114, 99-110.

B. Beykal, M. Onel, O. Onel, E.N. Pistikopoulos, 2020, A data-driven optimization algorithm for differential algebraic equations with numerical infeasibilities, AIChE Journal, 66, 10, e16657.

L.T. Biegler, 1984, Solution of dynamic optimization problems by successive quadratic programming and orthogonal collocation, Computers & Chemical Engineering, 8, 3-4, 243-247.

F. Boukouvala, C.A. Floudas, 2017, ARGONAUT: AlgoRithms for Global Optimization of coNstrAined grey-box compUTational problems, Optimization Letters, 11, 5, 895-913.

D.Y. Caballero, L.T. Biegler, R.Guirardello, 2015, Simulation and optimization of the ethane cracking process to produce ethylene. Computer Aided Chemical Engineering, 37, 917-922.

Y.D. Lang, L.T. Biegler, 2007, A software environment for simultaneous dynamic optimization, Computers & Chemical Engineering, 31, 8, 931-942.

M. Onel, C.A. Kieslich, E.N. Pistikopoulos, 2019, A nonlinear support vector machine-based feature selection approach for fault detection and diagnosis: Application to the tennessee eastman process, AIChE Journal, 65, 3, 992-1005.

Proceedings of the 14th International Symposium on Process Systems Engineering – PSE 2021+
June 19–23, 2022, Kyoto, Japan ©2022 Elsevier B. V. All rights reserved.
http://dx.doi.org/10.1016/B978-0-323-85159-6.50264-5

Tensor-Based Autoencoder Models for Hyperspectral Produce Data

Charlotte Cronjaeger[a], Richard C. Pattison[b], Calvin Tsay[a]*

[a]*Department of Computing, Imperial College London, London SW7 2AZ, United Kingdom*
[b]*Apeel Sciences, Santa Barbara, CA 93117, United States of America*

c.tsay@imperial.ac.uk

Abstract

Effectively monitoring and controlling product quality is critical in produce supply chain management. Hyperspectral imaging has emerged as a promising technique for monitoring food products, but the size of hyperspectral datasets complicates storage and processing. This work develops a novel architecture for autoencoder models that is well-suited for nonlinear subspace learning on tensorial, hyperspectral data. In particular, separate submodels are used to (de)compress each mode of the data tensor, preserving spatial locality information and greatly reducing the number of autoencoder parameters. The approach enables memory-efficient training, nonlinear dimensionality reduction, and multi-task learning, as demonstrated by a real-world case study.

Keywords: Tensorial Data, Supply Chain Management, Hyperspectral Images, Dimensionality Reduction, Subspace Learning.

1. Introduction

Degradation of perishable products is highly dependent on storage/transport conditions and represents a considerable challenge to manage; approximately one-third of food produced each year is lost or wasted, at a cost of nearly \$1 trillion (USD) (World Food Program, 2020). Modeling and monitoring of product quality and degradation play a vital role in addressing these issues. For instance, degradation models can account for waste in supply chain optimization (Rong et al., 2011; Tsay and Baldea, 2019), while monitoring product quality can provide closed-loop "feedback" (Lejarza and Baldea, 2020). To this end, hyperspectral imaging is a promising technique for food products, bridging spectroscopy and computer vision (i.e., spectral and spatial information). Hyperspectral images can both reveal internal characteristics, such as firmness, dry matter, and sugar content, and detect external contaminants/defects. However, the size of hyperspectral data (often >100 MB/image) complicates storage and processing (Feng and Sun, 2012).

Owing to their size, many techniques can be applied to extract features from hyperspectral data (Huang et al., 2014). For example, linear subspace learning techniques such as principal component analysis (PCA) are widely applied for dimensionality reduction and pattern recognition. Analogous nonlinear techniques have since been proposed; autoencoders are often used as a form of nonlinear PCA, as they can be trained using methods tailored for large datasets (Kramer, 1991). However, hyperspectral data involve three *modes*: length, width, and spectral band. Such multi-modal data motivate the use of multilinear subspace learning (MSL) methods, e.g., Tucker decomposition, which are often more data-efficient, as they preserve/exploit spatial locality information among tensor entries (Lu et al., 2013).

In this work, we propose a novel nonlinear subspace learning technique for tensorial data based on autoencoders (AEs). Specifically, we avoid flattening (vectorizing) the tensors and instead use a separate sub-model to (de)compress each mode of the tensor. Exploiting the intrinsic structure in this manner greatly reduces the number of AE parameters that must be learned. Using a case study of real-world produce data, we show that our novel AE architecture with linear activations can closely match the compression ability of standard MSL approaches, while enabling memory-efficient training using semi-batch gradient descent. Nonlinear activations can further improve compression ability. Finally, as the model architectures are generic, we expand the AEs to include classification within the compressed, latent space (i.e., multi-task learning). In the context of providing feedback in supply chain management, this approach enables attributes to be predicted in online applications without requiring data to be reconstructed and re-processed.

2. Methodology

The goal of our proposed autoencoder (AE) architecture is to learn a low-dimensional, nonlinear manifold underlying tensorial data, while simultaneously sustaining the spatial locality of the data. For this, a residual tensor architecture consisting of N linear AEs and N non-linear AEs for an Nth-order tensor is trained, as shown in Figure 1. It can be shown that the linear AE architecture performs as well as common algebraic linear subspace learning techniques, such as Tucker decomposition and High-Order Singular Value Decomposition (HOSVD), if trained properly. Therefore, we do not simply train a nonlinear AE end-to-end, but rather a nonlinear AE that learns the residuals of a linear AE. This simplifies learning, enables stable training, and improves generalization performance.

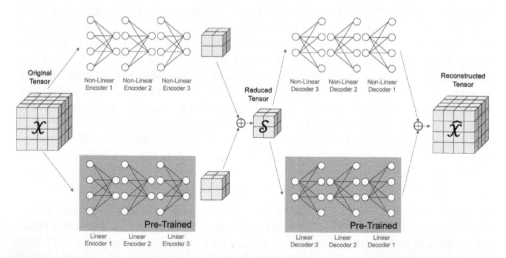

Figure 1: Residual tensor AE architecture for a third-order tensor with (fixed) pre-trained weights of the linear AE and trainable weights of the non-linear AE model.

2.1. Residual Tensor Autoencoder (AE) Architecture

Figure 1 depicts the architecture for a third-order tensor, but the proposed architecture can easily be generalized to Nth-order tensors. Both the linear and nonlinear encoders reduce the input tensor $\mathcal{X}$ to the desired subspace dimension. The reduced tensors are added together, resulting in a combined tensor $\mathcal{S}$, and subsequently fed into both the linear and non-linear decoders. The reconstructed tensor $\hat{\mathcal{X}}$ results from the addition of the outputs of the linear and non-linear decoders. We employ a mean squared error (MSE) loss function that minimizes the l_2-norm of the error between $\hat{\mathcal{X}}$ and $\mathcal{X}$, i.e., $\mathrm{MSE}(\mathcal{X}, \hat{\mathcal{X}}) := ||\mathcal{X} - \hat{\mathcal{X}}||_2^2$.

The weights of the linear AEs are pre-trained and fixed while training the nonlinear components. Our proposed architecture comprises N autoencoders: each reduces one dimension of an Nth-order tensor. The AE pairs encode "fibers" of the tensor independently. Each AE slices the tensor into its different fibers and feeds the fibers (i.e., vector data) into fully connected encoder and decoder models. Figure 2 depicts how an encoder model slices input tensor $\mathcal{X}$ into its component fibers to reduce dimensions sequentially. The decoder model works identically. Note that the number of data samples fed into the fully connected encoders and decoders decreases sequentially, as previous dimensions are already reduced, and fibers are processed independently by the AEs. This results in fewer data samples to be trained on by some AEs. However, we note that slicing the tensors creates more "samples" in the first place and hypothesize that, since these processed samples have fewer correlated dimensions, the effective dataset size remains similar.

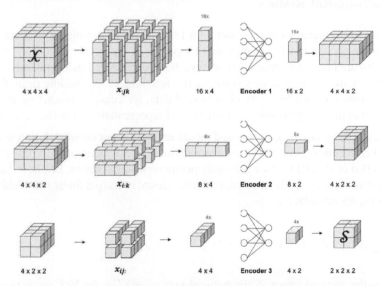

Figure 2: Encoder models for a third-order tensor which slice each tensor into its different fibers along each mode to reduce all three dimensions in sequence.

2.2. Non-Linear Architectures of the Fully Connected Autoencoder

For each of the N fully connected encoders and decoders, a simple nonlinear architecture is applied. In particular, each encoder and decoder comprises two fully connected layers.

After the first layer, a nonlinear activation function is applied, namely the Rectified Linear Unit (ReLU). While multiple extensions to this simple nonlinear architecture are possible, e.g., deeper architectures, batch normalization, or residual skip connections, none of these led to significantly better results in our experiments. This can be explained through the resulting depth—and hence number of trainable parameters—between input and output signals, which makes the non-convex loss function more difficult to train. For a third-order tensor with only two layers for each encoder and decoder, there are already $3 \times 2 \times 2 = 12$ nonlinear layers between input tensor $\mathcal{X}$ and reconstructed tensor $\hat{\mathcal{X}}$.

2.3. *Multi-Task Learning: Classification of Reduced Tensors*

One advantage of the proposed AE architecture, in addition to its trainability via gradient-descent-based optimization, is the potential to learn multiple tasks simultaneously and in an end-to-end fashion, i.e., multi-task learning. In this case, a classifier is learned in addition to minimizing the reconstruction loss when training the model. The loss function is a linear combination of the MSE loss (as described above) and the cross entropy loss for classification. The goal is to classify each tensor based on its compressed representation and, in particular, to learn a representation of tensors that yields good classification performance. For this, in addition to the existing architecture, a classification model on the compressed tensor is trained, which can be simultaneously learned using backpropagation.

3. Experimental Results

A real-world hyperspectral image dataset with 186 tensor data samples of avocados is used for the experiments. All data samples are cropped to an equal size of $236 \times 187 \times 224$. The avocados are imaged in three-by-four trays, but each is stored independently. Ambient lighting differs among the four tray columns (far left, middle left, middle right, far right). Therefore, given the confidentiality of industrial data, we consider predicting the column an avocado was in during imaging as a simple, yet representative, classification task.

We compare performance of the proposed model against Tucker decomposition as a benchmark. Tucker decomposition can be seen as a multilinear extension of PCA to higher order tensor data (Lu et al., 2013). As we do not operate on a single tensor, i.e., how Tucker decomposition is conventionally denoted, but rather desire to learn a linear manifold for a set of K tensors, we optimize:

$$\min_{U(n)} \frac{1}{K} \sum_{k=1}^{K} \|\mathcal{X}_k - \mathcal{S}_k \times_1 U_1 \times_2 U_2 \cdots \times_N U_N\|_F^2, \tag{1}$$

where $\mathcal{X}$ is the original tensor, $\mathcal{S}$ the reduced tensor, and U_N the N*th* projection matrix. A low-rank approximation can be found via High Order Orthogonal Iteration (HOOI) (Sheehan and Saad, 2007).

Residual Tensor Architecture. We compare the performance of the proposed AE architecture to Tucker decomposition for several different subspace dimensions. For simplicity of comparison, the tensors are always reduced from their original size of $236 \times 187 \times 224$ to cubes with subspace dimensions between $1 \times 1 \times 1$ up to $40 \times 40 \times 40$. As we observe in Table 1, the proposed model outperforms Tucker decomposition, especially for small

Table 1: MSE reconstruction loss for residual tensor learning for various cube subspace dimensions using two-layer nonlinear encoder and decoder models and training for 100 epochs. Training times reported for a single GPU NVIDIA GeForce GTX TITAN X.

Dim.	Train Loss / 10^{-3}	Test Loss / 10^{-3}	%-Change v Tucker Test Loss	Training Time	Reduction in Size
1	7.088	8.386	-44.03%	23 min	99.99%
5	1.034	1.427	-34.00%	24 min	99.99%
10	0.918	1.138	+18.54%	20 min	99.98%
15	0.588	0.602	-2.90%	19 min	99.96%
20	0.400	0.424	-2.08%	19 min	99.91%
25	0.300	0.320	+4.92%	19 min	99.84%
30	0.240	0.236	+10.28%	20 min	99.72%
35	0.175	0.192	+22.29%	22 min	99.56%
40	0.139	0.149	+24.17%	22 min	99.35%

Table 2: MSE Reconstruction Loss & Cross Entropy Classification loss for simultaneous residual tensor learning and classification tasks for various cube subspace dimensions using two-layer nonlinear encoder and decoder models and training for 100 Epochs. Training times reported for a single GPU NVIDIA GeForce GTX TITAN X.

Dim.	Test Loss / 10^{-3}	%-Change v Tucker Test Loss	Classifier Test Loss / 10^{-3}	# Correctly classified (Test)	Training Time
1	7.871	-47.47%	1333.622	29/36	16 min
5	1.441	-33.33%	50.055	35/36	17 min
10	1.824	+90.00%	5.757	36/36	17 min
20	0.449	+3.68%	0.355	36/36	17 min
30	0.241	+12.76%	0.324	36/36	18 min
40	0.144	+19.73%	0.916	36/36	21 min

subspace dimensions up to $20 \times 20 \times 20$. For subspace sizes below $5 \times 5 \times 5$ the model reduces the reconstruction error on the test dataset by over 30%. The fact that the proposed architecture performs better for smaller subspace dimensions may be due to having fewer trainable parameters to optimize, given the highly non-convex loss function.

Simultaneous Classification. When training a classifier on the compressed tensor simultaneously with learning the lower dimensional manifold, the model is able to classify over 97% data samples of the test dataset correctly for subspace sizes of $5 \times 5 \times 5$ and larger. Even when reducing each tensor to a $1 \times 1 \times 1$ scalar, the model classifies $> 80\%$ of test data correctly. The reconstruction loss remains comparable to the results for models when only learning the lower dimensional manifold. The results are presented in Table 2.

4. Conclusions

In this work, we introduced a novel AE architecture for hyperspectral produce data. We demonstrated that the proposed residual tensor architecture outperforms existing subspace learning techniques, especially for smaller subspace dimensions, where our model can reduce the reconstruction loss compared to Tucker decomposition by over 30%. Furthermore, we showed that multi-task learning, i.e., including a classification task on the reduced tensors, is possible and promising. Compared to existing subspace learning techniques our model is more scalable, as it can be efficiently trained using stochastic gradient descent and involves fewer parameters compared to other AE-based methods.

Future work can involve more effective uses of spatial locality in tensorial data. Our proposed method assumes the vector input of each AE (the fibers) to be drawn as independent samples of a distribution. However, in the case of hyperspectral images, nearby pixels are highly correlated, and the data exhibits strong spatial structure as a result. The sheer size of hyperspectral images, however, may prohibit directly employing standard approaches, such as convolutional layers. Given the current optimization difficulties for higher dimensional subspaces, future work can also explore alternative gradient-based algorithms for training, as well as gradient-free methods. A final interesting direction of future research could investigate the impact of the order in which the dimensions are reduced, as this impacts the numbers of samples each AE is trained on.

5. Acknowledgements

The authors gratefully acknowledge support from the Engineering & Physical Sciences Research Council (EPSRC) through fellowship grant EP/T001577/1, Apeel Sciences, and an Imperial College Research Fellowship to CT.

References

Feng, Y.Z., and Sun, D.W., Application of hyperspectral imaging in food safety inspection and control: a review. *Crit. Rev. Food Sci. Nutr.*, 52(11), 1039-1058, 2012.

Huang, H., Liu, L., and Ngadi, M.O., Recent developments in hyperspectral imaging for assessment of food quality and safety. *Sensors*, 14(4), 7248-7276, 2014.

Kakarla, S., Gangula, P., Singh, C., and Sarma, T. H., Dimensionality Reduction in Hyperspectral Images Using Auto-encoders. In *ICACCI 2019*, 101-107, 2019

Kramer, M.A., Nonlinear principal component analysis using autoassociative neural networks. *AIChE J.*, 37(2), 233-243, 1991).

Lejarza, F., and Baldea, M., Closed-loop optimal operational planning of supply chains with fast product quality dynamics. *Comput. Chem. Eng.*, 132, 106594, 2020.

Lu, H., Plataniotis, K.N., and Venetsanopoulos, A., *Multilinear subspace learning: dimensionality reduction of multidimensional data*, CRC Press, 2013.

Rong, A., Akkerman, R., and Grunow, M., An optimization approach for managing fresh food quality throughout the supply chain. *Int. J. Prod. Econ.*, 131(1), 421-429, 2011.

Sheehan, B.N., and Saad, Y., Higher order orthogonal iteration of tensors (HOOI) and its relation to PCA and GLRAM. *SIAM Int. Conf. Data Min.*, 355-365, 2007.

Tsay, C., and Baldea, M., 110th anniversary: Using data to bridge the time and length scales of process systems. *Ind. Eng. Chem. Res.*, 58(36), 16696-16708, 2019.

World Food Programme (WFP), Zero Hunger. https://www.wfp.org/zero-hunger, 2020.

Proceedings of the 14th International Symposium on Process Systems Engineering – PSE 2021+
June 19-23, 2022, Kyoto, Japan © 2022 Elsevier B.V. All rights reserved.
http://dx.doi.org/10.1016/B978-0-323-85159-6.50265-7

Molecular Representations in Deep-Learning Models for Chemical Property Prediction

Adem R.N. Aouichaoui, Fan Fan, Seyed Soheil Mansouri, Jens Abildskov
Gürkan Sin[*]

Process and Systems Engineering Center (PROSYS), Department of Chemical and Biochemical Engineering, Technical University of Denmark, Kgs.Lyngby, DK-2800, Denmark
gsi@kt.dtu.dk

Abstract

A molecular property prediction model is dependent on the interplay between the quality of data, and expressive representation (or descriptor), and a suitable algorithm to relate the descriptors to the target property. In this work, a deep neural network (DNN) is used to regress two types of descriptors: fixed descriptors (Group fragments and Morgan fingerprints) and learned descriptors (from a Graph Neural Network, GNN). Bayesian optimization was used for hyperparameter tuning and a set of 5 models were benchmarked and used to predict the enthalpy of formation of organic compounds. GNN based models provided the best overall results compared to descriptor-based models which the attentive fingerprint model that combines RNN and graph attention mechanism (AFP) achieved the best results of 5.9 kJ/mol mean absolute deviation and a coefficient of determination of 0.99 in the training, validation, and test set. Despite not achieving chemical accuracy of 4 kJ/mol, the model has shown great promise in distinguishing between isomers and provides a baseline for future improvements to achieve chemical accuracy.

Keywords: Deep-Learning, Molecular Property Prediction, Enthalpy of Formation

1. Introduction

The molecular properties of chemical compounds must be known *a priori* to execute many chemical engineering applications such as risk assessment, P-V-T calculation using equations of states as well as material selection. These properties are either the direct product or derived quantities from experimental measurements. Conducting such measurements on demand whenever the need arises is not a viable option due to time and expenses. Predictive models capable of describing these properties provide an attractive alternative to quickly evaluate the properties of a compound. Quantitative structure-property relation (QSPR) models are predictive models relating the chemical structure of a compound to a target property. The molecular structure is converted into a numerical representation that is then used as input to a mathematical model to produce the target property. The mathematical model is usually selected by observing the trend of the property with increasing carbon numbers for homologous series. Recently, deep neural networks (DNN) have gained popularity in many engineering applications and have also been used as part of the QSPR model(Aouichaoui *et al.*, 2021). This increasing attention is due to their ability to approximate any non-linear functions (universal approximation theory). Despite being an integral part of any QSPR, the selection of the mathematical model is less challenging than developing the molecular representation or descriptors used as input, which remains an issue and a detriment factor to the success of the model (good accuracy and ability to distinguish compounds) to a higher degree than the mathematical model. The molecular representation used can either be the product or a combination of domain knowledge, heuristics, or a data-driven approach. Group-contribution models (GC) employ both domain knowledge and heuristics to represent the molecule as an occurrence vector of a set of predefined groups(Hukkerikar *et al.*, 2012). The Morgan fingerprints and it's variation such as the extended connectivity fingerprints (ECFP) are another widely popular descriptor in the fields of cheminformatics and drug discovery (Rogers and Hahn, 2010). The ECFPs are circular fingerprints that represent the presence of a particular substructure that is encoded through a hashing function. The circular fingerprints can be generated for different diameters by combining features from the previous diameter length that are stored in a variable-length bit vector. Various QSPR models have used the ECFP to model a variety of molecular properties such as predicting water solubility (Xiong *et al.*, 2020). The group fragments and the ECFP are fixed descriptors that are proper to the molecule chosen and known before the modeling process. However, the molecular descriptors can also be learned so they become not only property to the molecule but also exclusive for the dataset through graph neural networks (GNNs) (Gilmer *et al.*, 2017). These models take a graph representation of the molecule where the nodes represent the atoms and the edges represent the bonds. Each node and edge is assigned a feature vector with information related to the atoms (type, valency, etc) and the bonds (type of bonds, etc.). These feature vectors are then updated based on the information contained in the feature vector of their neighboring nodes by applying graph convolutions or message passing layers. A readout function is then applied to the graph representation to produce a vector representation that is then supplied to a DNN to produce the target prediction. The feature update and the repression procedures both employ a series of algorithms that integrates learnable parameters that are adjusted using error backpropagation, which produces a representation that fits the compound and the target property at the same time. In this work, we benchmark the above-presented descriptors using the same property data, to compare their performance and highlight some of their advantages and drawbacks.

2. Methods

2.1. Models

We distinguish between two types of models, those that use a fixed representation in the form of the group fragments and ECFP and those that generate their representation from a molecular graph representation.

2.1.1. Fixed representation models

Models with a fixed molecular representation are used in conjunction with a DNN to correlate the descriptor to the target property. The GC-DNN uses the group fragmentation developed by (Hukkerikar *et al.*, 2012), where the molecule is described through 3 levels (orders) with increasing levels of complexity containing: 224 first-order groups, 134-second order groups, and 74 third-order groups. Although third-level groups are based on convenience and a more heuristic approach, they are included in this study to take advantage of the full predictive power of GC-based methods. The ECFP-DNN uses the extended circular fingerprints generated through the Morgan algorithm as described in (Rogers and Hahn, 2010). These descriptors are generated using a Python-based cheminformatics package RDKit (Landrum, 2020). The representation is hashed into a bit vector of length 1024 as used in previous studies (Xiong *et al.*, 2020). The hyper-parameters were optimized: dimensions of hidden neurons in the first layer [256, 1024], number of layers [2, 4], activation functions ['LeakyReLU', 'Sigmoid', 'Tanh', 'SELU'], L2 regularization [0, 0.05], initial learning rate [1e-5, 1e-1], learning rate reduce factor [0.2, 0.8]. Note that the size of hidden neurons in the following layers is designed to be half of the previous layer.

2.1.2. Adaptive/learned representation models

Three graph neural networks are used to evaluate the performance of models with an adaptive representation of the molecule. In the following, the main features of the models are highlighted. For a more in-depth explanation of the models and their hyperparameters, the reader is encouraged to inspect the references provided.

The Message Passing Neural Network (MPNN) by (Gilmer *et al.*, 2017) is a versatile model used for various property prediction purposes such as predicting water solubility. The model takes an undirected molecular graph with attributed nodes and edges. The operation of the models is described in two phases: a message-passing phase where the node and edge features are transmitted to the neighboring nodes and used to update its representation and a readout phase transforming the graph representation to a vector representation that is supplied to a DNN to regress the target property. More details on mathematics can be found in (Gilmer *et al.*, 2017). The main hyper-parameters were optimized: hidden dimensions [1, 128], number of layers [1, 4], L2 regularization [0, 0.05], learning rate [1e-5, 1e-1], learning rate reduce factor [0.2, 0.8].

Graph Isomorphism Network (GIN) by (Xu *et al.*, 2019) is a simple GNN that is intended to achieve a similar ability to the Weisfeiler-Lehman graph isomorphism test. The model only relies on node features to aggregate and update the node feature through a deep neural network and uses the sum function and concatenating the resulting representation from each iteration as the readout function. The main hyper-parameters for tuning together with their search domains are the same as MPNN.

Attentive Fingerprint (AFP) by (Xiong *et al.*, 2020) is considered the state-of-the-art GNN model that combines a series of deep-learning techniques to enhance its representative capabilities. A recursive neural network is used to agglomerate the messages from nearby and distant nodes in addition to the graph attention mechanism that allows the model to weigh the information and assign importance to it thus only focusing on the relevant structural information. Besides those hyper-parameters mentioned above, Attentive FP has one extra hyper-parameter to be optimized: the number of time steps [1, 4].

All previously described GNN models operate on a graph representation of the molecule. The node is attributed with the atom type (C, N, F, Br, Cl, S, I), the atomic mass, the atom degrees (nr. of covalent bonds), the type of hybridization (sp,sp^2,sp^3,sp^3d,sp^3d^2), whether the atom is part of an aromatic configuration, whether it is part of a ring structure as well as the number of hydrogen attached and whether it is a chiral center. The edges contain information on whether the bond is single, double, triple, or part of an aromatic structure. The feature vector also includes whether the bond is conjugated or part of a ring structure as well as a stereo-configuration it might be part of (E/Z, cis/trans).

2.2. Training & Optimization

The data are split into three folds: 90% for training, 5% for validation, and 5% for testing. Training has been prioritized since the dataset size is small compared to other deep-learning applications such as image or speech recognition. During training, the adaptive learning rate is used to adapt to the optimization surface as well as early stopping and L2 weight regularization to avoid overfitting the objective function used for model training is the mean-squared error. The hyperparameters of the various models have been tuned using a multi-objective Bayesian optimization (MOBO) toolbox (Galuzio *et al.*, 2020) The root-mean-squared error (RMSE) of the training and validation set are chosen to describe the performance of the model. The hyper-parameters of each model and their range have been previously described when presenting the models. The MOBO is done by constructing a posterior distribution function using a Gaussian process using the Matén covariance function as shown in Eq.(1), where "l" is the length scale, $\Gamma()$ is the gamma function, "Kv()" is the modified Bassel function and "r" is the distance between two arguments of Kernel and "v" is a positive parameter set to 1.5 in this study.

$$C_v(\vec{x}, \vec{x}') = \frac{2^{1-v}}{\Gamma(v)}\left(\frac{r\sqrt{2v}}{l}\right)^v K_v(\frac{r\sqrt{2v}}{l}) \tag{1}$$

3. Case Study: Predicting the enthalpy of formation of organic compounds

3.1. Property Data

The methodology model described previously is applied to predict the standard enthalpy of formation (HFOR) of organic compounds. The HFOR is defined as the change in enthalpy associated with the reaction forming the given chemical in its standard state from the elements in its standard state. The data were collected from the DIPPR database (Wilding *et al.*, 2017) containing a total of 741 compounds. Figure 1 shows the distribution and range of the data. Only experimental values are selected and only organic compounds with either of the following atoms: Oxygen (O), Bromide (Br), fluorine (F), chlorine (Cl), Iodine (I), Nitrogen (N), and Sulfur (S). An important challenge associated with the enthalpy of formation is to achieve " chemical accuracy" which is stated as 4 kJ/mol (Meier, 2021).

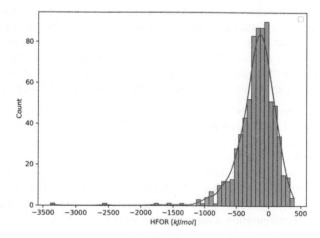

Figure 1: distribution of the heat of formation

3.2. Results

Table 1 provides the number of learned parameters (N), the coefficient of determination (R^2), the mean absolute error (MAE), and the mean absolute percentage error (MAPE) of the best-performing type of model. The best metrics are in bold.

Table 1: Perfoamnce metrics of the tuned models

Model	N	R^2	MAE(kJ/mol)	MAPE (%)
ECFP-DNN	1,494,510	0.977-0.916-0.944	26.0-74.9-60.5	27.4-80.9-56.7
GC-DNN	946,297	0.996-0.991-0.987	11.8-16.2-16.3	16.1-19.7-16.0
MPNN	557,665	0.997-0.973-0.981	6.7-28.7-31.1	10.5-33.6-26.8
GIN	123,481	0.994-0.992-0.991	12.9-34.8-16.2	23.3-29.8-43.5
AFP	97,995	**0.999-0.999-0.997**	**5.1-7.9-12.6**	**6.4-13.4-12.8**
GC+	428	0.999	1.75	-

3.3. Discussion

The AFP model outperforms all other GNN models as well as the models based on GC and ECFP across all metrics with an overall R^2 of 0.999, an MAE of 5.9 kJ/mol, and a MAPE of 8%. Compared to the model developed by (Hukkerikar *et al.*, 2013) (GC+) the AFP falls short of achieving the desired chemical accuracy of 2 kJ/mol. However, it is important to note that this model uses all data to perform the regression and even defined new additional higher order that is not funded in any chemical or property knowledge to reduce the error and accomplish the chemical accuracy. This raises a concern about the models' ability to extrapolate. The GC-DNN also falls short of the model by (Hukkerikar *et al.*, 2013) across all metrics. The reason is not all data are used for regression and the fact that it did not rely on all groups defined by (Hukkerikar *et al.*, 2013). Compared to classical QSPR models, GNN and in general deep-learning-based models are high parametric models with the number of parameters much higher than the number of data points available. Despite this, AFP is the GNN with the least model parameters and still outperforms the remaining models. This could be related to the attention mechanism and use of RNN to process node features.

Table 2: Prediction and experimental value for Methylpentenes. The absolute relative error compared to the experimental value in % are given in parenthesis

	2-METHYL-1-PENTENE	3-METHYL-1-PENTENE	2-METHYL-2-PENTENE
Experimental	-59.2 kJ/mol	-49.4 kJmol	-66.8 kJ/mol
GC-DNN	-57.0 kJ/mol (03.7)	-40.9 kJ/mol (17.2)	-59.4 kJ/mol (11.1)
ECFP-DNN	-48.6 kJ/mol (17.9)	-20.8 kJ/mol (57.9)	-87.5 kJ/mol (30.9)
MPNN	-57.8 kJ/mol (02.3)	-50.7 kJ/mol (02.6)	-62.6 kJ/mol (06.3)
GIN	-50.7 kJ/mol (14.4)	-46.2 kJ/mol (06.5)	-63.4 kJ/mol (05.1)
AFP	-57.9 kJ/mol (02.2)	-49.0 kJ/mol (00.8)	-65.2 kJ/mol (02.4)
GC+	-54.6 kJ/mol (07.8)	-45.7 kJ/mol (07.5)	-64.7 kJ/mol (03.1)

Table 2 contains the experimental value and corresponding predictions using various models of the methylpeneten isomers. The table clearly shows that the AFP model is superior in distinguishing between isomers compared to the rest of the models with the lowest absolute relative errors across all models. Although a rare occasion, the GC+ model can distinguish the isomers presented herein due to the presence of second-order groups.

Another interesting aspect is the fact that no prior knowledge is incorporated into the GNN models other than basic chemistry-related information, and despite this, the model achieves very promising results.

4. Conclusions

A successful QSPR model heavily relies on the interplay between data, representation, and model. The focus of this work was on some of the methodologies to represent the molecular structure in a machine-readable way. The data and the general prediction model were identical. GNN based models showed superior performance to descriptor-based models with the AFP model achieving the best results with 5.9 kJ/mol mean absolute error, and although it falls short of the target chemical accuracy (also referred to as the holy grail), the model has shown it is much better at distinguishing between isomers. Furthermore, the results suggest that deep-learning-based models such as GNN do provide a powerful tool to correlate molecular structure to the desired target property without time-consuming descriptor design or extensive domain knowledge.

References

Aouichaoui, A.R.N., Al, R., Abildskov, J. and Sin, G. (2021) 'Comparison of Group-Contribution and Machine Learning-based Property Prediction Models with Uncertainty Quantification', in *Computer Aided Chemical Engineering*. Elsevier Masson SAS, pp. 755–760.

Galuzio, P.P., de Vasconcelos Segundo, E.H., Coelho, L. dos S. and Mariani, V.C. (2020) 'MOBOpt — multi-objective Bayesian optimization', *SoftwareX*, 12.

Gilmer, J., Schoenholz, S.S., Riley, P.F., Vinyals, O. and Dahl, G.E. (2017) 'Neural Message Passing for Quantum Chemistry', *34th International Conference on Machine Learning, ICML 2017*, 3, pp. 2053–2070. Available at: http://arxiv.org/abs/1704.01212.

Hukkerikar, A.S., Meier, R.J., Sin, G. and Gani, R. (2013) 'A method to estimate the enthalpy of formation of organic compounds with chemical accuracy', *Fluid Phase Equilibria*, 348(1), pp. 23–32.

Hukkerikar, A.S., Sarup, B., Ten Kate, A., Abildskov, J., Sin, G. and Gani, R. (2012) 'Group-contribution + (GC +) based estimation of properties of pure components: Improved property estimation and uncertainty analysis', *Fluid Phase Equilibria*, 321, pp. 25–43.

Landrum, G. (no date) 'RDKit: Open-source cheminformatics'.

Meier, R.J. (2021) 'Group Contribution Revisited: The Enthalpy of Formation of Organic Compounds with "Chemical Accuracy"', *ChemEngineering*, 5(2).

Rogers, D. and Hahn, M. (2010) 'Extended-connectivity fingerprints', *Journal of Chemical Information and Modeling*, 50(5), pp. 742–754.

Wilding, W.V., Knotts, T.A., Giles, N.F. and Rowley, R.L. (2017) 'DIPPR® Data Compilation of Pure Chemical Properties', *Design Institute for Physical Properties, AIChE* [Preprint].

Xiong, Z., Wang, D., Liu, X., Zhong, F., Wan, X., Li, X., Li, Z., Luo, X., *et al.* (2020) 'Pushing the boundaries of molecular representation for drug discovery with the graph attention mechanism', *Journal of Medicinal Chemistry*, 63(16), pp. 8749–8760.

Xu, K., Hu, W., Leskovec, J. and Jegelka, S. (2019) 'How Powerful are GNN', *Int. Conf. on Learning Representations*, pp. 1–17.

Proceedings of the 14th International Symposium on Process Systems Engineering – PSE 2021+
June 19-23, 2022, Kyoto, Japan © 2022 Elsevier B.V. All rights reserved.
http://dx.doi.org/10.1016/B978-0-323-85159-6.50266-9

Deep Reinforcement Learning for Continuous Process Scheduling with Storage, Day-Ahead Pricing and Demand Uncertainty

Gustavo Campos[a], Simge Yildiz[a], Nael H. El-Farra[a,*], Ahmet Palazoglu[a]

[a]*Department of Chemical Engineering, University of California, Davis, CA 95616, USA*

nhelfarra@ucdavis.edu

Abstract

In this work, we evaluate the application of a Deep Reinforcement Learning (DRL) method for the scheduling of continuous process/energy systems under day-ahead electricity rate and demand forecast uncertainty. We employ the Soft Actor Critic (SAC) method, a stochastic, off-policy, actor-critic method with built-in entropy maximization that balances exploration and exploitation. We choose as a case study the dispatching of energy systems with storage, which can be posed as a continuous scheduling problem. Results from the computational case study demonstrate that the DRL agent is able to surpass a heuristic policy using very little data, and ultimately reaches a performance comparable to a model predictive control (MPC) solution. The effect of demand forecast uncertainty is further analysed and it is shown that, while the MPC performance degrades steadily as the forecast error and recalculation period increase, the DRL method exhibits a more robust performance.

Keywords: Deep Reinforcement Learning; Operation; Optimization; Energy Systems; Demand Response.

1.   Introduction

Data-based methods for optimization and control have been gaining traction over recent years due to advances in the field of Deep Reinforcement Learning (DRL), which merges the power of nonlinear approximators with strategies for online exploration, parameter estimation and optimization. Along with this trend, there has also been an increased effort to reduce carbon emissions and increase energy efficiency and renewable penetration. In this context, establishing an efficient operation of energy-intensive processes becomes a critical component for achieving these goals. The operation of these systems, particularly under time-varying electricity rates, has been typically approached using model-based optimization. For instance, in the case of district cooling plants that produce cooling utilities by running electricity-driven industrial chillers, previous works include Economic Model Predictive Control (Ma *et al.*, 2011) and closed-loop scheduling (Risbeck *et al.*, 2017; Campos *et al.*, 2021). While powerful, these methods require considerable modelling effort, which is exacerbated when attempting to model the effects of uncertainty (e.g., in stochastic programming or robust optimization). In this case, DRL emerges as a promising alternative for performing the operation of complex (e.g., nonlinear, stochastic, multiscale) systems while avoiding the high associated modelling and computational costs (Badgwell *et al.*, 2018).

In the Process Systems Engineering (PSE) literature, studies using classical data-based techniques such as Reinforcement Learning (e.g., Cassol *et al.*, 2018) or Approximate Dynamic Programming (e.g., Lee and Wong, 2010) have been regularly proposed over the past few decades. However, the recent developments in DRL and Deep Learning, more specifically over the last 5 years, have enabled more powerful applications with high dimensional state and action spaces, and policy complexity. DRL techniques have been recently employed in a few process systems applications, including chemical production scheduling using Advantage Actor-Critic (A2C) (Hubbs et al., 2020), control of liquid-liquid extraction columns in biopharmaceutical processes using Deep Q-Networks (DQN) (Hwangbo and Sin, 2020), and control of batch polymerization processes using Deep Deterministic Policy Gradient (DDPG) (Yoo et al., 2021).

While the performance of DRL methods for more traditional problems has been addressed, less attention has been paid to the demand responsive operation of processes. In addition, as opposed to the majority of works that employ deterministic agents (e.g., DDPG, TD3 and DQN), in this paper we employ a stochastic agent using the Soft Actor-Critic (SAC) method, which generates actions following a probability distribution and has been shown to be more robust and have lower brittleness to hyperparameters. Compared to other stochastic methods such as A2C (or A3C), SAC has the advantage of being an off-policy method, an important property for practical applications that allows the use of a replay memory buffer for reducing sample complexity. We provide insights into the applicability of SAC for demand response through a case-study that demonstrates how the agent's performance compares to heuristic and optimal policies with and without forecast uncertainty.

2.   Soft Actor-Critic (SAC) Method

The SAC formulation (Haarnoja et al., 2018a, b) aims to learn a policy that maximizes the expected sum of rewards (traditional RL objective), while simultaneously maximizing the policy's entropy or stochasticity. The entropy maximization encourages exploration of the state space and avoids local optima, in addition to other practical advantages (e.g., reducing hyperparameter sensitivity). The entropy augmented objective is posed as $\mathbb{E}_{(s_t,a_t)}[r(s_t,a_t) + \alpha H(\pi(\cdot|s_t))]$, where α is a temperature parameter that controls the relative importance of the policy's entropy function $H(\pi)$. By applying a soft policy iteration procedure using function approximators (for the Q-function and the policy) and stochastic gradient descent, one can derive equations for the loss functions (minimization objectives) of the actor L_π, critic L_Q and alpha L_α (its logarithm), which are given as follows.

$$L_\pi = -\left(Q_\theta(s_t,a_t) - \alpha \log \pi_\phi(a_t|s_t)\right) \tag{1}$$

$$L_Q = MSE\left(Q_\theta(s_t,a_t),\ r_t + \gamma\left(Q_{\underline{\theta}}(s_{t+1},a_{t+1}) - \alpha \log \pi_{\underline{\phi}}(a_{t+1}|s_{t+1})\right)\right) \tag{2}$$

$$L_{\log \alpha} = \log \alpha \left(\log \pi_\phi(a_t|s_t) + \hat{H}\right) \tag{3}$$

The policy update maximizes the Q-function, corrected by a term that steers the policy away from actions with high probability. The critic loss is a mean square error function between the current critic prediction and a target value, the latter calculated using target networks $Q_{\bar{\theta}}$ and $\pi_{\bar{\phi}}$, which are updated using a Polyak rule. Double estimation of the Q-function (i.e., two critic networks) is employed to mitigate positive bias. A condensed version of the algorithm is presented in Table 1, for a complete version the reader is referred to Haarnoja *et al.* (2018b).

Table 1 – Soft Actor Critic Algorithm (condensed).

1	*Initialize network parameters, hyperparameters, replay buffer and environment.*
2	**for** *each environment time step*, **do***:*
3	*Collect experience (choose a, observe r and s') and store in memory buffer.*
4	**for** *each training step*, **do***:*
5	*Sample minibatch of experiences.*
6	*Calculate losses: L_Q (Eq. (1)), L_π (Eq. (2)), L_α (Eq. (3)).*
7	*Calculate gradients (automated with PyTorch).*
8	*Update networks using the Adam stochastic optimizer.*
9	*Perform a Polyak update of the target actor and critics.*
10	*Update state $s_t \leftarrow s_{t+1}$*

The information flow through the actor and critic networks is shown in Figure 1. The yellow area represents the action calculation phase, while the blue area represents the training phase, in which the gradient of the parameterized Q-function with respect to the networks parameters is calculated. Reparameterization of the action sampling step explicitly with respect to the distribution parameters using a Gaussian white noise ϵ is employed to allow direct backpropagation of the gradient.

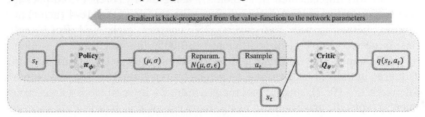

Figure 1. Information flow through networks for action generation and training phases.

3.   System Formulation and Description

Markov Decision Process Formulation. We formulate the problem of equipment and storage dispatching under day-ahead electricity prices as a Markov Decision Process (MDP). Our goal is to achieve a minimalistic description that is suitable for the application of DRL methods. We consider the classical MDP framework defined by the tuple (S, A, p, R), i.e., the state and action (continuous) spaces, the unknown state transition probability, and the bounded reward function. The state space includes the current storage value, future demands and future prices. The action space is defined as the current production level (i.e., equipment loads). The reward function is the negative of cost, including both operational cost (i.e., electricity consumption) and constraint violation penalties.

Scheduling of Energy Storage Systems. A typical on-line scheduling formulation is used to obtain the model-based solution. This problem type includes many practical applications such as the dispatching of distributed energy resources (e.g., solar PV and battery systems) and utility production facilities (e.g., district cooling and heating plants). A diagram of the system is presented in Figure 2. For the specific case-study, we consider the scheduling of district cooling plants, in which the production level represents the cooling load of the chillers, the storage represents a Thermal Energy Storage (TES) tank, the power consumption is a linear function of the production level, the demand corresponds to cooling demand from buildings, and the electricity price varies hourly according to a wholesale day-ahead program.

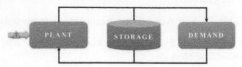

Figure 2. Energy/process system with production plant, storage and demand.

$$\min \ \sum_{t \in T} Power_t \ Price_t \tag{4}$$

$$\text{s.t.} \quad Storage_{t+1} = Storage_t + Production_t - Demand_t \tag{5}$$

$$Power_t = f(Production_t) \tag{6}$$

hard bounds for $Production_t$, soft bounds for $Storage_t$ $\qquad$ (7)

4.   Case Study Results

The algorithm was implemented in Python 3.7 using the PyTorch v1.6.0 package. The following settings were used: Adam optimizer with learning rates 1e-4 (actor) and 5e-4 (critic/alpha); networks with two layers, 256 neurons each, and ReLU activation functions; target network Polyak coefficient $\tau = 0.005$; discount factor $\gamma = 0.99$; buffer size = 1e6; samples per minibatch = 256; training steps/environment step = 3; weights/biases initialized uniformly $\in (-\mu, \mu)$, $\mu = $ (#weights)$^{-0.5}$; log_std output of the policy network $\in$ (-10, 2); single continuous action; 24 forecast steps (state vector with 49 elements); episodic environment using one week of real hourly data (non-episodic formulations were tested and work the same way); optimal solution obtained from a closed-loop MPC with long horizon and perfect information; heuristic policy operates equipment when electricity is cheaper on a weekly average; minimum-level (10%) constraint violation penalty = 30 \$/MWh (of storage load), proportional to the violation magnitude (penalty = 90 \$/MWh in the case of tank depletion).

4.1.   *Case Study 1: Perfect Information Scenario*

We first evaluate how the SAC method performs under a perfect information scenario and compare it to an optimal (upper bound), a heuristic (baseline), and a random policy (soft lower bound). Results are shown in Figure 3 (log-scale is used to amplify the data near the optimum), in which the plotted curve is an average of three runs (shaded area represents a single standard deviation). Within the first week of training, the SAC agent overcomes the performance of the random policy, meaning that the agent learns to avoid constraint violations (largest sources of penalties that dominate the random policy's performance). In the next 2-3 weeks, the agent overcomes the performance of the heuristic policy, which indicates a great potential for practical application. In the remaining weeks, the agent closes the gap towards the optimal solution. Figure 4 presents the scheduling variables for a representative week (action can be seen avoiding peak prices) and the network losses and alpha throughout the training period.

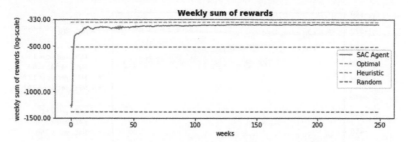

Figure 3. SAC training performance vs. optimal, heuristic, and random policy performances.

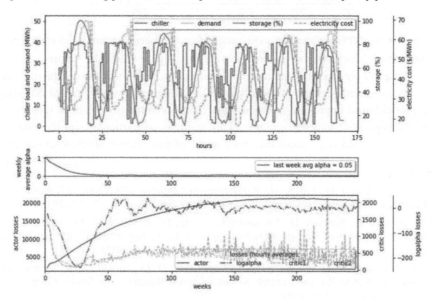

Figure 4. Weekly scheduling results, alpha and losses for the training phase.

4.2.   *Case Study 2: Demand Forecast Uncertainty*

We assume a forecast with additive Gaussian noise, which simulates the error increasing further into the future. Five error magnitudes are considered between 0 to 4% of an average demand (24.70 MWh) as the standard deviation of the additive Gaussian noise at each forecast step. Figure 5 presents the MPC solutions with varying recalculation periods (1, 6, 12, 24 h) averaged over 100 weeks to reduce sampling effects from the forecast distribution, and the SAC agent's performance at various training stages (50, 150, 250 weeks). The SAC solution shows good robustness against uncertainty, especially after 250 weeks of training, comparable to employing a fast recalculation frequency for the MPC. The SAC agent shows particular advantage when there is a combination of long recalculation period (or high computational cost) and considerable uncertainty.

5.   Conclusions

In this work, we evaluated the application of a DRL method for performing demand response of energy/process systems. The SAC method was employed, a stochastic, off-policy, actor-critic method with built-in entropy maximization that balances exploration and exploitation. Results demonstrate that the SAC agent quickly learns to avoid

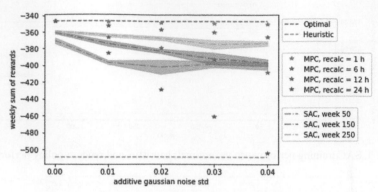

Figure 5. SAC vs. MPC performance under varying forecast uncertainty and recalculation period

constraint violations and continuously closes the gap towards the optimal solution, indicating a good potential for practical application. An analysis of the effects of forecast uncertainty indicated scenarios in which the DRL approach would present advantages over a model-based approach, i.e., when there is a combination of high recalculation period and high uncertainty magnitude. This result is of particular significance when dealing with complex (e.g., nonlinear, multi-scale, mixed-integer, stochastic) systems.

References

Badgwell, T. A., Lee, J. H., Liu, K.-H., Reinforcement Learning – Overview of Recent Progress and Implications for Process Control, *Comput. Aided Chem. Eng.*, 44, 71-85 (2018)

Campos, G., Liu, Y., Schmidt, D., Yonkoski, J., Colvin, D., Trombly, D. M., El-Farra, N. H., Palazoglu, A., Optimal Real-time Dispatching of Chillers and Thermal Storage Tank in a University Campus Central Plant, *Applied Energy*, 300, 117389, (2021)

Cassol, G. O., Campos, G., Thomaz, D. M., Capron, B. D. O., Secchi, A. R. Reinforcement Learning Applied to Process Control: A Van de Vusse Reactor Case Study. *Comput. Aided Chem. Eng.*, 44, 553–558 (2018)

Haarnoja, T., Zhou, A., Abbeel, P., and Levine, S., Soft Actor-Critic: Off-Policy Maximum Entropy Deep Reinforcement Learning with a Stochastic Actor, *Proc. of PMLR 80* (2018)

Haarnoja, T., Zhou, A., Hartikainen, K., Tucker, G., Ha, S., Tan, J., Kumar, V., Zhu, H., Gupta, A., Abbeel, P., and Levine, S., Soft Actor-Critic Algorithms and Applications. *CoRR* abs/1812.05905. URL http://arxiv.org/abs/1812.05905

Hubbs, C. D., Li, C., Sahinidis, N. V., Grossmann, I. E., and Wassick, J. M., A deep reinforcement learning approach for chemical production scheduling, *Comput. Chem. Eng.* 141, 106982 (2020)

Hwangbo S., Sin, G., Design of control framework based on deep reinforcement learning and monte-carlo sampling in downstream separation, *Comput. Chem. Eng.* 140, 106910 (2020)

Lee, J., Wong, W., Approximate dynamic programming approach for process control, *J. Process Control*, 20, 1038-1048, (2010)

Ma, Y., Borrelli, F., Hencey, B., Coffey, B., Bengea, S., Haves, P., Model predictive control for the operation of building cooling systems, *IEEE Transactions on control systems technology*, 20 (3), 796–803 (2011)

Risbeck, M. J., Maravelias, C. T., Rawlings, J. B., Turney, R. D., A mixed-integer linear programming model for real-time cost optimization of building heating, ventilation, and airconditioning equipment, *Energy Build*, 142, 220–235 (2017)

Yoo, H., Kim, B., Kim, J. W., Lee, J. H., "Reinforcement learning based optimal control of batch processes using monte-carlo deep deterministic policy gradient with phase segmentation, *Comput. Chem. Eng.*, 144, 107133 (2021)

Proceedings of the 14th International Symposium on Process Systems Engineering – PSE 2021+
June 19-23, 2022, Kyoto, Japan © 2022 Elsevier B.V. All rights reserved.
http://dx.doi.org/10.1016/B978-0-323-85159-6.50267-0

Convolutional Neural Network based Detection and Measurement for Microfluidic Droplets

Shuyuan Zhang[a,b], Xinye Huang[a,b], Kai Wang[a,c], Tong Qiu[a,b*]

[a]*Department of Chemical Engineering, Tsinghua University, Beijing, 100084, CHINA*
[b]*Beijing Key Laboratory of Industrial Big Data System and Application, Tsinghua University, Beijing, 100084, CHINA*
[c]*State Key Laboratory of Chemical Engineering, Department of Chemical Engineering, Tsinghua University, Beijing 100084, CHINA*

qiutong@tsinghua.edu.cn

Abstract

Modern microfluidic systems realize the envisioned idea to perform continuous process operations on a small scale using miniaturized devices and present superiorities in terms of plant modularization, reaction intensification and waste reduction. In microfluidic engineering, droplet size is central to desired function. Therefore, an effective droplet detection and size measurement method is highly-demand to quantitatively reveal the relationship between operation parameters and outcome droplet size. Herein, with recent impressive developments of computer vision, we propose a novel two-step convolutional neural network method to detect and measure droplets in microscopic images. The proposed model first locates droplets with bounding boxes and then calculate the droplet size with detailed coordinates. This convolutional neural network model not only exhibits outstanding performance for droplet size measurement, but also reveals the convenience of deep learning for digital, comprehensive and intelligent microfluidic researches.

Keywords: Microfluidics; Droplet detection; Size measurement; Deep learning; Convolutional neural network.

1. Introduction

Modern continuous-flow process engineering has been explored toward the small-scale device for the prominence in yield, selectivity, scale-up and controllability (Rossetti and Compagnoni, 2016). Among all small-scale reaction technologies, microfluidic engineering is appealing wide attentions and substantially growing in terms of pharmaceuticals, fine chemicals, green chemistry, catalytic reactions and material synthesis that is tough for traditional batch operations (Yan et al., 2021). Microfluidic engineering technology reveals the envisioned idea to perform continuous process operation on a small scale with miniaturized (lab-on-a-chip) devices. For reactions in microfluidic equipment, the process intensification is strengthened with increased mass/heat transfer rates, due to the relatively much larger mass/heat transfer area in the confined volume. Therefore, the size and size distribution of dispersed phase is central to the properties and functions of microfluids (Duraiswamy and Khan, 2009).

Currently, the microfluidic droplet size measurement methods are approximately divided into experimental and imaging methods. Experimental methods are based on probes, which can be intrusive or non-intrusive, focusing on relating some measurable parameters with the microdroplet size. In intrusive experiments, direct contact between probes and microfluids are inevitable, thus the flow state of microfluids would be disturbed and the measurement result uncertainty would be increased (Chen et al., 2004). Non-intrusive techniques avoid this drawback by adopting non-contacted experimental techniques. For instance, Lucas *et al.* used 3D printing technique to fabricate microfluidic devices with integrated electrodes based on contactless conductivity detection (Duarte et al., 2017). However, non-contacted techniques are still confronted with problems on expensive sensors, complicated installation and poor portability.

With the application of high-speed camera, optical imaging method is attracting wide attentions from microfluidic research communities. However, the post-processing of photographed microscopic images to measure microdroplet size is still rudimentary, mainly by manual measurement on the images at the present stage (Basu, 2013). Commonly, the mechanical image measurement would cost several weeks or months to obtain sufficient data to measure the droplet size. As a consequence, the low efficiency of imaging post-processing will severely harm the scale-up of microfluidic processes, as well as microfluidic device design.

Motivated by recent advancements of computer vision, especially deep learning, it is promising to realize an intelligent method to detect droplets and measure their sizes precisely (Cerqueira and Paladino, 2021; Zhang et al., 2022). Herein, we propose a two-step method based on convolutional neural network (CNN) for microdroplet size measurement. The proposed method first detects droplets in the image by locating them with bounding boxes (Bboxes), then calculates droplets' equivalent diameters with detailed coordinates of their Bboxes. In this way, the droplet size can be rapidly calculated and the size distribution curves can be easily acquired, with high droplet detection precision and low size measurement error. This method not only exhibits outstanding performance of CNN for droplet size measurement, but also reveals the convenience of deep learning for digital, comprehensive and intelligent microfluidic researches.

2. Experiment

We utilized a popular capillary-assembled microchannel as the research microfluidic system in this work, which was fabricated with polymethylmethacrylate (PMMA). The experimental setup is shown in Figure 1. The outer and inner diameters of the inserted capillary were 710 μm and 410 μm, respectively. A stepped T-junction was used as the micromixer to disperse one phase into another phase for producing microdroplets, due to its enhanced shearing effect (Wang et al., 2015). The narrow slit of the stepped T-junction is 200 μm high. After that, an observing chamber was attached with 2 mm wide window on both sides for microphotography. The channel size is ~3 times larger in the observing chamber, where the generated droplets are slowed down to acquire clear microscopic images. An optical microscope (XSP-63B, Shanghai Optical Instrument, China) is equipped with a CMOS camera (B742-F, PixeLink, Canada) to snap images of produced microfluidic droplets. The optical information was sent to computers and converted to digital images with a scale bar of 1.65 pix/μm for post-processing. A collector was set at the end to collect produced microdroplets.

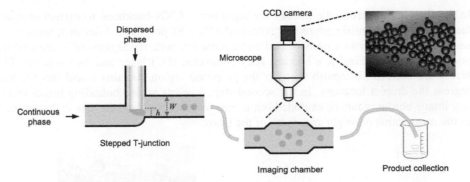

Figure 1. Microfluidic experimental setup.

In the experiment, styrene was selected as the dispersed phase, an aqueous solution of 3.0 wt% sodium dodecyl sulfonate (SDS) and 2.0 wt% polyvinylpyrrolidone K30 (PVP-K30) was selected as the continuous phase. SDS is a surfactant to decrease the interfacial tension, and PVP-K30 is the thickener of the solution for producing droplet swarm. The continuous and dispersed phases were continuously pumped into the T-junction with 50 mL and 10 mL gastight syringes (SEG, Australia), respectively, using commercial syringe pumps (Fusion 6000 for the continuous phase and Fusion 4000 for the dispersed phase). The width and height of photographed images are 1024 and 768 pixels, respectively. To realize a stable jetting flow regime, the flow rate of continuous phase is set between 3~8 mL/min and the flow rate of dispersed phase is set between 1~20 μL/min. This extreme phase ratio is as set to obtain tiny microdroplets. Besides, in some high flow rate ratio cases, the retracted neck of the dispersed phase may by stretched by the continuous phase, and break up into a smaller droplet, of which the size was smaller than dominant droplets.

After experiment, microfluidic images are collected to construct training and testing datasets, which are summarized in Table 1. For training neural networks, it is required to annotate droplets in these images (LeCun et al., 2015). Those droplets truncated by the image borders are also annotated with Bboxes but excluded out in size measurement, because the detailed ordinates of some corner points of the Bbox are missing. There are 2276 and 862 droplets in training and testing datasets, respectively. The mean, maximum and minimum diameters of the training dataset are close to those of the testing dataset, demonstrating the constructed datasets are qualified to evaluate the proposed CNN-based size measurement method.

Table 1. Annotated training and testing datasets of microdroplet image.

	The number of droplets	Mean diameter/μm	Min diameter/μm	Max diameter/μm
Train	2276	35.3	7.3	70.0
Test	862	38.2	8.8	69.4

3. Method

The proposed two-step CNN-based droplet size measurement method is exhibited in Figure 2. The first step is to detect droplets and regress their coordinates with Bboxes. The second step is to filter droplets not truncated by image borders and calculated the equivalent diameters.

In the first step, microfluidic images are input into a CNN backbone to extract abstract features. Then, potential regions that contain droplets are proposed. After that, features in each proposed regions are transformed into a same size with the regions of interest (RoI) pooling operation. Finally, a sharing fully-connected (FC) layer and two separate FC layers are used to distinguish whether the proposed region contains a real droplet and regress the droplet locations. In the second step, droplets whose bounding boxes reach any image border would be excluded out in size measurement and the size is determined as the mean value of height and width of the Bbox.

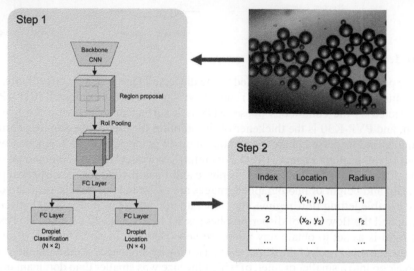

Figure 2. Schematic of the proposed CNN-based microdroplet size measurement method.

For model evaluation, the average precision (AP) under certain intersection-of-union (IoU) condition is utilized (Ren et al., 2016). For a pair of predicted and actual Bboxes, if their IoU is larger than a predetermined threshold, the prediction can be regarded as a matched one. AP is the averaged ratio of matched predictions to all predictions over all images, which can reflect the microdroplet detection accuracy. The IoU threshold is set as 0.75 in this work and the corresponding AP is denoted as AP75. Meanwhile, mean absolute error (MAE) of predicted diameter is also considered as the most straightforward criterion.

4. Results and discussions

4.1. Model tuning

During training, many reported CNN architectures can be implemented as the CNN backbone for effective feature extraction. In this work, an advanced CNN model series, ResNet, is selected as the backbone for feature extraction. ResNet is characterized by the skip identity mapping from shallower layers to deeper layers (Figure 3(a)), which realizes an incremental learning pattern in deep learning (He et al., 2016). Normally, ResNet that contains more CNN layers is inclined to perform better. Three ResNets with different number of layers are investigated, which are ResNet18, ResNet34 and ResNet50, respectively.

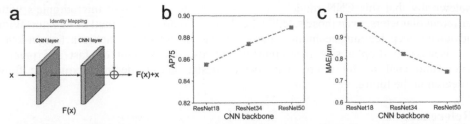

Figure 3. (a) Residual block of ResNet. (b) AP75 and (c) MAE results using ResNet backbones.

The AP75 and MAE results of ResNet50, ResNet75 and ResNet50 are exhibited in Figure 3(b) and (c), respectively. It can be observed that, as the ResNet backbone becomes deeper, AP75 is increasing and MAE is decreasing. AP75 reflects the droplet detection accuracy and MAE reflects the droplet measurement preciseness. Therefore, a deeper ResNet backbone can contribute a precise droplet detection and measurement, although the calculation cost is also increased because more CNN layers are included in the model. In application, ResNet with certain number of layers can be selected as the CNN backbone to meet the requirement of measurement preciseness or calculation speed (or both). Specifically, a high AP75 of 0.889 and a low MAE of 0.739 µm are achieved by ResNet50.

4.2. Measurement results

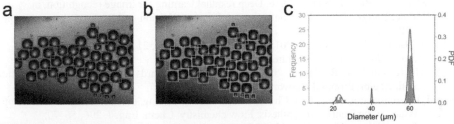

Figure 4. (a) All detected Bboxes, (b) non-truncated Bboxes, and (c) diameter frequency and probability density function (PDF) curve of a typical droplet image.

Figure 4 exhibits a typical photographed droplet image and corresponding inference result of the proposed CNN model. With the powerful feature extraction capability of CNN, almost all droplets can be detected and precisely located with a Bbox to depict their 2-dimensional existence limits (Figure 4(a)). After that, Bboxes that cover the image border are excluded out, as shown in Figure 4(b). The measured droplet diameter is determined as the average of the height and width of the Bbox. Finally, frequency can be easily calculated and PDF curve can be fitted with Gaussian mixture model. In Figure 4(c), the major peak can be ascribed to dominant droplets and the other two peaks can be ascribed to aforementioned satellite droplets.

5. Conclusions

In conclusion, this novel work proposes an intelligent and precise CNN-based model for detecting and measuring droplets in microfluids. This CNN model can comprehensively and precisely locate microdroplets with Bboxes. Based on that, microdroplet diameters can be easily measured and diameter distribution can be obtained for analysis. It is noteworthy that this CNN model can be implemented to analyse microdroplets with different diameters, even those satellite droplets with diameter as small as ~20 μm. With advanced deep learning technique, this CNN model not only reaches a human-level preciseness in droplet size measurement, but also sheds light on fast and precise microfluid analysis for microfluidic device design and in-depth microfluidic flow research in the future.

References

Basu, A.S., 2013. Droplet morphometry and velocimetry (DMV): a video processing software for time-resolved, label-free tracking of droplet parameters. Lab. Chip 13, 1892–1901. https://doi.org/10.1039/C3LC50074H

Cerqueira, R.F.L., Paladino, E.E., 2021. Development of a deep learning-based image processing technique for bubble pattern recognition and shape reconstruction in dense bubbly flows. Chem. Eng. Sci. 230, 116163. https://doi.org/10.1016/j.ces.2020.116163

Chen, J.Z., Darhuber, A.A., Troian, S.M., Wagner, S., 2004. Capacitive sensing of droplets for microfluidic devices based on thermocapillary actuation. Lab. Chip 4, 473. https://doi.org/10.1039/b315815b

Duarte, L.C., Chagas, C.L.S., Ribeiro, L.E.B., Coltro, W.K.T., 2017. 3D printing of microfluidic devices with embedded sensing electrodes for generating and measuring the size of microdroplets based on contactless conductivity detection. Sens. Actuators B Chem. 251, 427–432. https://doi.org/10.1016/j.snb.2017.05.011

Duraiswamy, S., Khan, S.A., 2009. Droplet-Based Microfluidic Synthesis of Anisotropic Metal Nanocrystals. Small 5, 2828–2834. https://doi.org/10.1002/smll.200901453

He, K., Zhang, X., Ren, S., Sun, J., 2016. Deep residual learning for image recognition, in: Proceedings of the IEEE Conference on Computer Vision and Pattern Recognition. pp. 770–778.

LeCun, Y., Bengio, Y., Hinton, G., 2015. Deep learning. Nature 521, 436–444. https://doi.org/10.1038/nature14539

Ren, S., He, K., Girshick, R., Sun, J., 2016. Faster R-CNN: Towards Real-Time Object Detection with Region Proposal Networks. arXiv:1506.01497.

Rossetti, I., Compagnoni, M., 2016. Chemical reaction engineering, process design and scale-up issues at the frontier of synthesis: Flow chemistry. Chem. Eng. J. 296, 56–70. https://doi.org/10.1016/j.cej.2016.02.119

Wang, K., Qin, K., Lu, Y., Luo, G., Wang, T., 2015. Gas/liquid/liquid three-phase flow patterns and bubble/droplet size laws in a double T-junction microchannel. AIChE J. 61, 1722–1734. https://doi.org/10.1002/aic.14758

Yan, Z., Tian, J., Wang, K., Nigam, K.D.P., Luo, G., 2021. Microreaction processes for synthesis and utilization of epoxides: A review. Chem. Eng. Sci. 229, 116071. https://doi.org/10.1016/j.ces.2020.116071

Zhang, S., Liang, X., Huang, X., Wang, K., Qiu, T., 2022. Precise and fast microdroplet size distribution measurement using deep learning. Chem. Eng. Sci. 247, 116926. https://doi.org/10.1016/j.ces.2021.116926

Proceedings of the 14th International Symposium on Process Systems Engineering – PSE 2021+
June 19-23, 2022, Kyoto, Japan © 2022 Elsevier B.V. All rights reserved.
http://dx.doi.org/10.1016/B978-0-323-85159-6.50268-2

Deep Reinforcement Learning Based Controller for Modified Claus Process

Jialin Liu[a*], Bing-Yen Tsai[a], Ding-Sou Chen[b]

[a]*Research Center for Smart Sustainable Circular Economy, Tunghai University, No. 1727, Sec.4, Taiwan Boulevard, Taichung, Taiwan*
[b]*Green Energy and System Integration Research and Development Department, China Steel Corporation, No. 1, Chung Kang Road, Hsiao Kang, Kaohsiung, Taiwan*
jialin@thu.edu.tw

Abstract

The modified Claus process is used to recover sulfur from acid gases containing high concentrations of H_2S. The downstream process, a tail gas treatment system, is required to reduce atmospheric emission of sulfur compounds to the level required by air pollution control regulations. In this study, a deep reinforcement learning (DRL) based controller was developed to minimize the concentration variations of H_2S and SO_2 in the tail gas from the modified Claus process. In addition, the sequence to sequence (Seq2Seq) networks were trained by the plant data to capture the dynamic information between the manipulated and controlled variables. Thereafter, the optimal operating policy can be found through that the advantage actor-critic (A2C) algorithm was implemented to the DRL agent by interacting with the environment constructed by Seq2Seq. The results show that the variations of H_2S and SO_2 can be reduced 40 % and 36 %, respectively, compared with that of applying the traditional control strategy.

Keywords: Deep Reinforcement Learning; Sequence to Sequence Networks; Modified Claus Process.

1. Introduction

The reinforcement learning (RL) framework consists of a learning agent interacting with a stochastic environment. The agent selects an action (a_t) at time t according to the probabilities that are generated by a learning policy $\pi(a|s)$ with the current state (s_t) observed from the environment. The selected action interacts with the environment to obtain the reward (r_t) and the next state (s_{t+1}). The deep reinforcement learning (DRL) is referred to approximate the learning policy and the cumulative reward, which is also known as the state-value function, by two deep neural networks (DNNs). Williams (1992) proposed a policy-gradient learning algorithm that generates an episode of actions, rewards and states following the policy. Thereafter, the DNN weightings of the learning policy is updated to maximize the cumulative rewards. Mnih et al. (2015) proposed a deep Q-network (DQN) to maximize the action-value function (also known as Q) and reported that the DQN can achieve a high level of performance on any of a collection of different problems by the same architecture of DNN. The actor-critic algorithm combines the features of the policy-gradient and value-based approaches (Sutton and Barto, 2018). The DNN weightings of the state-value function are updated by minimizing the errors of approximating the observed cumulative reward, whereas the probability function of the policy is updated to maximize the advantage between the reward and the estimated one by the state-value function.

Recently, the DRL approaches have been introduced into the model predictive control (MPC) community; for example, Spielberg et al. (2019) based on the actor-critic (AC) algorithm to develop a DRL-based controller that learned the control policy in real time by interacting with the simulation examples, in which setpoint tracking problem on single-input single-output, multiple-input multiple-output, and nonlinear systems were demonstrated. Petsagkourakis et al. (2020) applied the policy gradient (PG) method from batch-to-batch data to update a control policy to maximize the product concentration of a bioprocess. In their approach, a preliminary optimal control policy was obtained by interacting with the environment that simulated the real bioprocess. Subsequently, the policy was refined by implementing into the true system. Therefore, the number of evaluations with the true system was reduced, which may be costly and time consuming. Ma et al. (2019) designed a DRL controller based on the deep deterministic policy gradient (DDPG) method interacting with a simulated semi-batch polymerization system. More recently, Kang et al. (2021) proposed a two-stage training deep deterministic policy gradient (2S-DDPG) algorithm to control the boiler drum level, which was simulated by a set of transfer functions. The above-mentioned DRL approaches were interacted with the over-simplified mathematical models. The potential of applying DRL-based controllers to the real processes was not demonstrated.

Ma et al. (2020) applied a four-layer feedforward neural network to build a step-ahead prediction model using the experimental data from a bioreactor. In their approach, the DRL-based controller was developed by the asynchronous advantage actor-critic (A3C) algorithm. The experimental results showed that the A3C controller significantly improves the yield of the desired product compared to that of using a traditional control method. Adams et al. (2021) proposed a deep reinforcement learning optimization framework in which the environment was built by a 5-layer DNN from over 1.5 years of plant data with a 1 min sampling time, which interacted with an A2C agent. The objective of the framework was to maximize the power generation of a coal-fired plant while reducing the NO_x emission. In their approach, a static DNN was applied as the surrogate model to predict the power generation and the NO_x emission by the process variables. The actions of the manipulated variables were determined by the A2C agent with the predictions of the static DNN. Therefore, the process dynamic information was not incorporated into their proposed framework. In this study, the process model, which interacts with the DRL agent, is constructed by the sequence to sequence (Seq2Seq) networks (Sutskever et al., 2014). Chou et al. (2020) developed a physically consistent soft sensor by the Seq2Seq networks. They reported that the process dynamics can be fully extracted by the Seq2Seq model from the plant data, because the estimation of the process gains is consistent with the domain knowledge.

2. Seq2Seq and A2C Networks

In this study, the Seq2Seq networks are constructed by the gated recurrent units (GRUs, Cho et al., 2014), which was proposed to modify the drawback of RNNs. The original RNNs suffer from exploding or vanishing gradient problems through backpropagation on multiple time steps. The reset and update gates are added on the structure of RNNs to solve this issue. Figure 1 shows that the encoder extracts the dynamic information from the operating data with past window length w and capsules into a hidden state vector (z_0). In the encoder, the input layer of each node contains the previous hidden state vector (h_{i-1}) and the current data of input and output variables, which are the disturbance (d_i), manipulated (m_i) and controlled (c_i) variables, respectively. The decoder predicts the future f samples of the controlled variables by the corresponding data of the manipulated variables and the hidden state vector from the previous GRU, as shown in Figure 1.

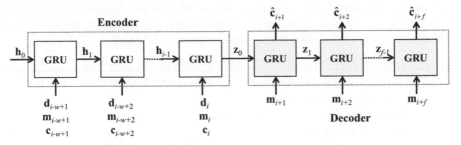

Figure 1. Seq2Seq networks for process modelling

The A2C algorithm uses two deep neural networks to approximate the learning policy and the state-value function, respectively. The probability function of learning policy, which is also called the actor network denoted by $\pi(A_t|S_t,\theta)$ with weightings θ, maps the current state (S_t) into an action (A_t), interacting with the environment to observe the reward (R_t) and the next state (S_{t+1}). The cumulative reward at time t can be expressed by $G_t = R_t + \gamma R_{t+1} + \gamma^2 R_{t+2} + ... = R_t + \gamma G_{t+1}$ with a discount rate γ. The state-value function, which is the critic network denoted by $\hat{v}(S_t,\mathbf{w}) \approx G_t$ with weightings $\mathbf{w}$, estimates the cumulative reward at the current state. The correlation of the cumulative reward can be applied into the state-value function, i.e., $\hat{v}(S_t,\mathbf{w}) = R_t + \gamma\hat{v}(S_{t+1},\mathbf{w})$. Therefore, the temporal difference (TD) error is defined as: $\delta \equiv R_t + \gamma\hat{v}(S_{t+1},\mathbf{w}) - \hat{v}(S_t,\mathbf{w})$. To improve the accuracy of the cumulative reward approximated by the critic network, the weightings of state-value function $(\mathbf{w})$ are updated by minimizing the square of TD error, i.e., $\mathbf{w} \leftarrow \mathbf{w} + \alpha^w \delta \nabla \delta$ where α^w is a learning rate. In addition, the TD error can be expressed as an advantage function, i.e., $\delta = R_t - \left[\hat{v}(S_t,\mathbf{w}) - \gamma\hat{v}(S_{t+1},\mathbf{w})\right]$ in which the square bracket term is the current reward estimated by the critic network. A proper policy function should be designed to maximize the advantage; therefore, the weightings of policy network (θ) are updated by $\theta \leftarrow \theta - \alpha^\theta \delta \nabla \ln \pi(A_t|S_t,\theta)$ where α^θ is a learning rate. The details of A2C algorithm can be found in Sutton and Barto (2018).

In this study, the A2C agent interacting with the Seq2Seq model is proposed. The pseudocode of the proposed approach is listed in Table 1. Two GRU networks, which were initialized with parameters θ and $\mathbf{w}$, are used as the actor and critic networks, respectively. Each network is deployed with two layers and 30 hidden nodes to map the current state into actions and state values. As listed in Table 1, the initial state at time t $(S_{t,0})$ is defined by the disturbance data $(\mathbf{d}_t)$, the measurements of manipulated variables $(\mathbf{m}_t)$ and the corresponding predictions $(\hat{\mathbf{c}}_t)$ by the decoder of the Seq2Seq networks at Line 3. Starting from Line 4, the weighting $\mathbf{w}$ and θ are updated 20 times $(T = 20)$ for each sample. The future actions $(\tilde{\mathbf{m}}_{t,i})$ with f prediction horizon and the corresponding state value $(v_{t,i})$ are generated by the actor-critic networks using the current state at Line 5. Thereafter, the predictions are made by the decoder incorporating with the future actions to form the next state with the same disturbance data at Line 6. Line 7 describes that the next state value ($v_{t,i+1}$) is estimated by the critic network with the new state and the previous state is replaced by the new one. The variance of predictions is calculated and compared with the benchmark (var_B) at Line 8. If the current variance is less than the benchmark, the reward is set to 10 and the benchmark is replaced. Otherwise, the reward is given by the negative variance. Consequently, the TD error can be calculated and used to update the network parameters at Line 9.

Table 1. Pseudocode for integrating A2C agent with Seq2Seq networks

1.	Input: a policy $\pi(a\|s,\theta)$ and a state-value function $\hat{v}(s,\mathbf{w})$ with parameters θ and $\mathbf{w}$
2.	For each sample in training dataset:
3.	Initialize the first state: $S_{t,0} = [\mathbf{d}_t \quad \mathbf{m}_t \quad \hat{\mathbf{c}}_t]$
4.	For $i = 0, 1, ..., T-1$:
5.	$\tilde{\mathbf{m}}_{t,i} = \pi(a\|S_{t,i},\theta), \quad v_{t,i} = \hat{v}(S_{t,i},\mathbf{w})$
6.	$\hat{\mathbf{c}}_{t,i} = Decoder(\tilde{\mathbf{m}}_{t,i}), \quad S_{t,i+1} = [\mathbf{d}_t \quad \tilde{\mathbf{m}}_{t,i} \quad \hat{\mathbf{c}}_{t,i}]$
7.	$v_{t,i+1} = \hat{v}(S_{t,i+1},\mathbf{w}), \quad S_{t,i} = S_{t,i+1}$
8.	$var_i = Var(\hat{\mathbf{c}}_{t,i})$; If $var_i \le var_B$: $R_t = 10$; $var_B = var_i$ else: $R_t = -var_i$
9.	$\delta \leftarrow R_t + \gamma v_{t,i+1} - v_{t,i}$; $\mathbf{w} \leftarrow \mathbf{w} + \alpha_w \delta \nabla \delta$; $\theta \leftarrow \theta - \alpha_\theta \delta \nabla \ln \pi(A_t\|S_t,\theta)$

3. Industrial Example

In this study, the DRL-based controller is applied to minimize the concentration variations of SO_2 and H_2S in the tail gas from the modified Claus process. The process flow diagram is shown in Figure 2. The sour gas is fed into the burner reactor, in which H_2S is burned with air to form SO_2 and H_2O, i.e., $2H_2S+3O_2 \rightarrow 2H_2O+2SO_2$. Thereafter, the effluent gas from the burner reactor is cooled and fed into the converter reactors, which are labeled as R1 and R2 in Figure 2, for catalytic conversion of H_2S and SO_2 to elemental sulfur and water ($2H_2S+SO_2 \rightarrow 2H_2O+3S$). In addition to maintain the reactor temperatures, two control loops are used to stabilize the process operations by adjusting the air flowrates. According to the operational guidelines, the primary air flowrate is determined by the sour gas flowrate with a ratio controller whose setpoint of the air to sour gas ratio is recommended as 1.1 by volume. Furthermore, the secondary air flowrate is manipulated to maintain the molar ratio of H_2S to SO_2 whose setpoint ought to be 2. However, the historical data show that the variation of the air to sour gas ratio ranges between 1.2 and 1.6. On the other hand, the molar ratio of H_2S to SO_2 spreads from 2 to 16. That indicates the air flowrate controllers might not work properly; thereafter, the downstream process, the tail gas treating unit, suffers from the large variations of the H_2S and SO_2 concentrations in the tail gas.

The Seq2Seq model was built by five-month operating data, which were collected once per minute around 180,000 samples. The encoder contained all variables listed in Figure 2. On the other hand, the inputs for the decoder were the primary and secondary air flowrates, and the outputs were the H_2S and S_2O concentrations in the tail gas. One layer of GRU with 30 hidden nodes was applied to the encoder and decoder, respectively. The optimal window length (w) of the encoder was determined as 40 samples by the mean absolute percentage errors (MAPEs) of the test dataset in which

5600 samples after modelling data were applied. The Seq2Seq networks predicted the future sixty samples of the controlled variables once per minute using the corresponding manipulated variables. The MAPEs of the test data, for which the predictions were made by the different time periods in 10, 30, and 60 minutes, are listed in Table 2. The MAPEs of predicted H_2S and SO_2 are around 4% and 8%, respectively. The accuracy of predictions show that the Seq2Seq networks capture the process dynamic behavior, properly. Therefore, the Seq2Seq model can be used as the environment interacting with the DRL agent.

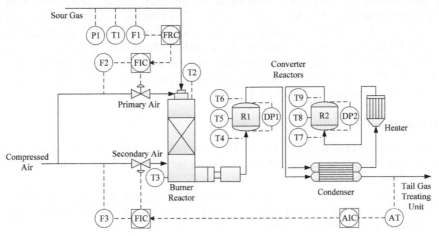

Figure 2. The process flow diagram of the modified Claus process

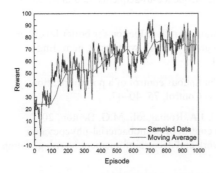

Figure 3. Reward evolution for training the A2C agent

Table 2. Result summary

		10 min	30 min	60 min
		MAPE		
	H_2S	3.7%	4.0%	4.1%
	SO_2	8.1%	8.0%	8.1%
Standard Deviation (H_2S: 0.045, SO_2: 0.013)				
S2S	H_2S	0.040	0.039	0.039
	SO_2	0.011	0.011	0.011
A2C	H_2S	0.024	0.026	0.028
	SO_2	0.007	0.008	0.009

The A2C agent was trained during 1000 episodes where the weightings of actor-critic networks were updated 20 times in each episode. For each update, the future sixty actions were drawn from the actor network; thereafter, the corresponding H_2S and SO_2 concentrations were predicted by the Seq2Seq model incorporating with the future actions. Consequently, the reward was evaluated by the variances of the predictions. As shown in Figure 3, the reward was initially around 20 that could be improved to 70 at the end of training. The standard deviations of the H_2S and SO_2 concentrations were 0.045 and 0.013 for the raw data, as listed in Table 2. The standard deviations of the predictions by the Seq2Seq model are comparable with those of the raw data in the different time periods, implementing with the original measurements of the primary and secondary air flowrates. On the other hand, the A2C agent was implemented to determine the future actions of primary and secondary air flowrates every minute in the periods of 10, 30, and 60 minutes, respectively. Compared with the results of the standard deviations by the Seq2Seq and A2C networks listed in Table 2, the variations were reduced around 28%–40% for the H_2S and 18%–36% for the SO_2, respectively.

4. Conclusions

The DRL-based controller was developed by the A2C agent interacting with the environment constructed by the Seq2Seq model. The process dynamic feature could be captured by the encoder; meanwhile, the correlation between manipulated and controlled variables was extracted by the decoder. Thereafter, the reward of future actions generated by the A2C networks was evaluated by the multistep-ahead predictions. The results showed that the proposed approach can reduce the variations of H_2S and SO_2 concentrations, effectively.

References

D. Adams, D-H Oh, D-W Kim, C-H Lee, M. Oh, 2021, Deep reinforcement learning optimization framework for a power generation plant considering performance and environmental issues, J. Clean. Prod., 291, 125915.

K. Cho, B. van Merrienboer, C. Gulcehre, D. Bahdanau, F. Bougares, H. Schwenk, Y. Bengio, 2014, Learning phrase representations using RNN encoder-decoder for statistical machine translation, in EMNLP, https://arxiv.org/abs/1406.1078.

C.H. Chou, H. Wu, J.L. Kang, D.S.H. Wong, Y. Yao, Y.C. Chuang, S.S. Jang, J. D.Y. Ou, 2020, Physically consistent soft-sensor development using sequence-to-sequence neural networks, IEEE T. Ind. Inform., 16, 2829–2838.

J-L Kang, S. Mirzaei, J-A Zhou, Robust control and training risk reduction for boiler level control using two-stage training deep deterministic policy gradient, J. Taiwan Inst. Chem. Eng., (2021), https://doi.org/10.1016/j.jtice.2021.06.050.

Y. Ma, W. Zhu, M.G. Benton, J. Romagnoli, 2019, Continuous control of a polymerization system with deep reinforcement learning, J. Process Control, 75, 40–47.

Y. Ma, D.A. Noreña-Caro, A.J. Adams, T.B. Brentzel, J.A. Romagnoli, M.G. Benton, 2020, Machine-learning-based simulation and fed-batch control of cyanobacterial-phycocyanin production in Plectonema by artificial neural network and deep reinforcement learning, Comp. Chem. Eng., 142, 107016.

V. Mnih, K. Kavukcuoglu, D. Silver, A.A. Rusu, J. Veness, M.G. Bellemare, ..., Ostrovski, G., 2015, Human-level control through deep reinforcement learning, Nature, 518, 529–533.

P. Petsagkourakis, I.O. Sandoval, E. Bradford, D. Zhang, E.A. del Rio-Chanona, 2020, Reinforcement learning for batch bioprocess optimization, Comp. Chem. Eng., 133, 106649.

S. Spielberg, A. Tulsyan, N.P. Lawrence, P.D. Loewen, R.B. Gopaluni, 2019, Toward self-driving processes: a deep reinforcement learning approach to control, AIChE, 65, e16689.

I. Sutskever, V. Oriol, Q.V. Le, Sequence to sequence learning with neural networks, 2014, Adv. Neural Inf. Process. Syst., 3104–3112.

R.S. Sutton, A.G. Barto, 2018, Reinforcement learning: An introduction (2nd ed.) MIT Press.

R.J. Williams, 1992, Simple statistical gradient-following algorithms for connectionist reinforcement learning. Machine Learning, 8, 229–256.

Proceedings of the 14th International Symposium on Process Systems Engineering – PSE 2021+
June 19-23, 2022, Kyoto, Japan © 2022 Elsevier B.V. All rights reserved.
http://dx.doi.org/10.1016/B978-0-323-85159-6.50269-4

Process performance prediction based on spatial and temporal feature extraction through bidirectional LSTM

Changrui Xie[a], Runjie Yao[b], Zhengbang Liu[b], Lingyu Zhu[b], Xi Chen[a*]

[a]: *State Key Laboratory of Industrial Control Technology, College of Control Science and Engineering, Zhejiang University, Hangzhou, Zhejiang 310027 P.R. China*
[b]: College of Chemical Engineering, Zhejiang University of Technology, Hangzhou, Zhejiang, 310014, China
Corresponding author：xi_chen@zju.edu.cn

Abstract

With the development of deep learning, it has been a trend to build data driven soft sensors in process industries with neural networks. There are a number of networks proposed to deal with time series prediction, such as Recurrent Neural Network (RNN) and Long Short-Term Memory Network (LSTM). However, it is a critical part to extract nonlinear and dynamic characteristics hiding in process data collected from industrial production. This paper proposes a novel approach for performance prediction based on the spatial and temporal feature extraction through bidirectional LSTM networks (BiLSTM) for a reactor network. Due to the superiority of processing sequences from both directions, BiLSTM are utilized to simulate the physical structure of the reactor network. With both spatial and temporal feature extraction, the deep learning model through BiLSTM achieves nice prediction performance.

Keywords: Bidirectional Long Short-Term Memory Network, Soft sensing, Feature extraction, Deep learning.

1. Introduction

In comparison to the traditional offline analysis in laboratory, soft sensing provides a more fast and economical way to predict critical quality variables, which has been widely used in plenty of industrial plants. With the rapid development of machine learning and statistics, great progress has been made in the field of data-driven model based soft sensing. Different from first-principle models, data-driven models are developed with available data collected during industrial productions even without exact mechanism. However, data-driven models have to attach more emphasis on how to extract as many relevant nonlinear and dynamic features as possible to capture the valid characteristics of the complex chemical processes, because the validity of features will determine the performance of soft sensor directly (Ma et al. 2018). Kaneko et al. (2009) developed a new soft sensor combining independent component analysis (ICA) and partial least squares (PLS) together, where independent components can be seen as features sensitive to the outliers, then a PLS model can be updated with normal samples. Corrigan et al. (2021) proposed a soft sensor model based on dynamic kernel slow feature analysis, which was utilized to extract slowly varying features. Sun et al. employed multi-layer perceptron (MLP) to model the complex desulfurization process, based on which a soft sensor was built for SO$_2$ emission. Although many approaches have been proposed for feature extraction for the purpose of soft sensing, such as principal component analysis (PCA), ICA and deep learning methods, there is still a lot of room for improvement to extract the hidden nonlinear and dynamic features from vast process data especially in certain industrial scenarios. In this paper, a deep learning model with spatial and temporal feature extraction is proposed for a reactor network (Dorgo et al. 2019). The BiLSTM networks are utilized to simulate the physical structure of the reactor network to extract spatial features, then a unidirectional LSTM network is followed to process the feature maps output from bidirectional LSTM networks to extract the temporal features at multiple time points. Thus, the nonlinear and dynamic characteristics of the reactor network can be well captured through with both spatial and temporal feature extraction.

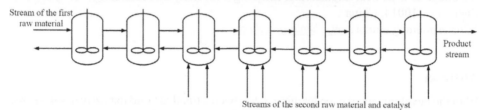

Figure 1. Structure diagram of the reactor network

2. Reactor network process

The reactor network considered in this paper is a system consisting of seven successively connected reactors, as is shown in Figure 1. There are mainly two raw materials entering the system, one from the first reactor directly and the other from the third to the seventh reactors. Overall, the reactor network keeps a counter-flow structure. In each reactor, two liquid phases exist and two outflow streams are kept via a separator inside. On account of the complex structure of the reactor network, it is natural that the dynamic and nonlinear characteristics of each reactor will propagate among the reactor network. During the industrial production, the outlet concentration of components from the fourth reactor are of the most significance, because they are seemed as a flag to reveal the real-time state of reactions in the whole reactor network. Therefore, they are usually selected as the critical quality variables that determines the process performance and operators usually adjust the feed flow of raw materials to maintain the production stability based on this observation. In this task, we select the outlet concentration of three components from the fourth reactors as the predicting target.

3. Feature extraction by bidirectional Long short-term memory networks

Sequence problems is considered as one of the hardest problems in many industrial cases. RNN is widely used to deal with this kind of problem. But RNN works well only towards short sequences, it may suffer from carrying information from earlier time steps to later ones when sequences are long enough due to gradient vanishing. LSTM, as a variant of RNN, improves the architecture by introducing a mechanism named cell state, by which it can preserve the relevant information to the later units even when the sequence is very long. As shown in Figure 2, LSTM propagates the information with cell state and hidden state produced by three gates inside, named forget gate, input gate and output gate. These three gates determine the information needed to be remembered or forgot.

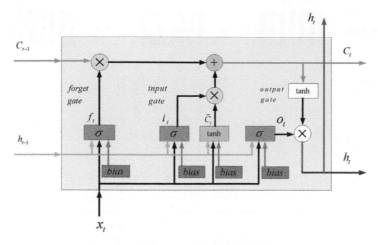

Figure 2. Data flow in the cell in LSTM network

With the help of mechanism of LSTM, information in the past can be preserved and passed to the future then an excellent prediction result can be obtained. However, the propagation of information in sequences may be in both directions in some cases, such as natural language processing. In terms of this issue, bidirectional long short-term memory network is proposed to construct two independent LSTMs at the same time, one of which processes the input sequences from past to future as traditional LSTM, and the other one processes from future to past inversely. By combining the outputs of both independent LSTMs together in some ways, the output of BiLSTM at each time point has the ability to preserve the information in both directions.

Considering the structural similarity between the reactor network and the BiLSTM, we propose using BiLSTM to build the soft sensor for the reactor network. Instead of extracting the temporal features from past to future and future to past in natural language processing, BiLSTM here is used to extract the spatial features at one time point. The input of BiLSTM layer is a feature matrix in shape with the complete information acquired from the whole reactor network. In this way, each column of feature maps output from BiLSTM layer denotes the spatial features for an individual reactor; thus, the fourth of them is selected to predict the outlet concentrations of the fourth reactor. In order to extract as many spatial characteristics as possible, two BiLSTM layers are set in order at the top of networks. Afterwards, these feature vectors are concatenated together and transferred into a unidirectional LSTM layer to extract the temporal features. Then some fully connected layers are followed and an output layer is added at last. Figure 3 illustrates the basic framework of the proposed system for the spatial and temporal feature extraction. Corresponding neuron number and activation function for each layer are listed in the Table 1. Besides, L2 regularization terms are added to the layers to avoid overfitting. The Root Mean Squared Propagation (RMSProp) is selected as the optimization algorithm to conduct gradient descent during network training with the learning rate set at 0.001.

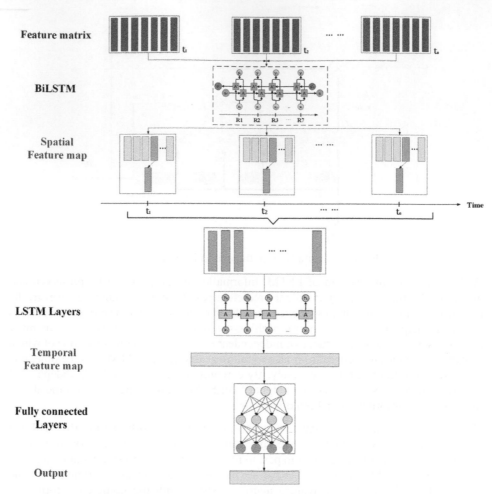

Figure 3. Spatial and temporal feature extraction based soft sensing network

Table 1. Number of neurons and activation function for each layer

Layers	Activation Function	Model with spatial and Temporal feature extraction
Input layer	-	(24,7,7)
BiLSTM1	sigmoid	64
BiLSTM2	sigmoid	32
LSTM	sigmoid	48
Fully connected layer1	tanh	48
Fully connected layer2	tanh	32
Output layer	sigmoid	3

4. Results

In the real industrial operation, the concentration of the target stream is collected and offline analysed for every 4 hours. Although what we aim to predict is the outlet concentration from the fourth reactor, the influence from other reactors is not negligible due to the non-decoupling property of the whole system. Therefore, it is necessary to take variables such as the feed flow of raw materials of other reactors into consideration. Eventually, 49 online measurable process variables are collected from the reactor network system with a sampling rate at 5 minutes. Considering the fact that the real process residence time of the whole system is about 2 hours, the input includes all the information from the past 2 hours. In the end, 1272 data pairs are prepared and are randomly divided as training set and testing set for the neural network model. To ensure consistent distribution between the training set and testing set, 70% of dataset are randomly split as the training set, and the remaining samples as the testing set.

To quantitatively evaluate the discrepancy, the root-mean-square error (RMSE) is used as:

$$RMSE = \sqrt{\frac{1}{N}\sum_{i=1}^{N}(\hat{y}_i - y_i)^2} \tag{1}$$

where i is the sample index and N is the number of samples in the testing set.

The RMSE for all samples in the testing set are presented in Table 2. The comparison for the first component of the target stream is presented in Figure 4. The soft sensing model with both spatial and temporal feature extraction achieves a good prediction performance. Instead of just inputting all relevant information as a flattened feature vector at each time point, the proposed soft sensing model has the ability to eliminate the useless features or the existence of information redundancy included in the feature matrices.

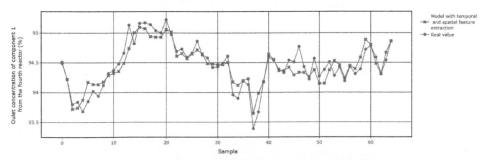

Figure 4. Prediction results of the testing samples

Table 2. RMSE of prediction results by soft sensing model

RMSE	Component 1	Component 2	Component 3
Model with spatial and Temporal feature extraction	0.189	0.152	0.147

5. Conclusions

Based on BiLSTM networks, a novel spatial and temporal features extraction method is proposed for reactor network system in this paper. The BiLSTM networks are utilized to simulate the physical structure of the equipment deployment, with the purpose of extracting spatial information of the whole reactor network at one time point as much as possible; then a unidirectional LSTM layer is followed to extract the temporal features from multiple timesteps. The prediction results indicate that feature extraction by deep learning is beneficial to capture nonlinear and dynamic characteristics of complex reactor systems, which contributes to more accurate predictions for soft sensing. This study provides a new point of view to take advantage of neural networks, that is simulating the real deployment of reactor network with similar structures designed by neural networks, especially for those non-decoupling complex systems.

Acknowledgement
Financial support of the Zhejiang Provincial Natural Science Foundation of China (No. LZ21B060001) is gratefully acknowledged.

References

Y. Ma, B. Huang, 2018, Extracting dynamic features with switching models for process data analytics and application in soft sensing, AIChE Journal, 64, 2037–2051.

H. Kaneko, M. Arakawa, K. Funatsu, 2009, Development of a new soft sensor method using independent component analysis and partial least squares, AIChE Journal, 55, 87–98.

J. Corrigan, J. Zhang, 2021, Developing accurate data-driven soft-sensors through integrating dynamic kernel slow feature analysis with neural networks. Journal of Process Control, 106, 208–220.

K. Sun, X. Wu, J. Xue, F. Ma, 2019, Development of a new multi-layer perceptron based soft sensor for SO2 emissions in power plant, Journal of Process Control, 84, 182-91.

G. Dorgo, J. Abonyi, 2019, Learning and predicting operation strategies by sequence mining and deep learning, Computers & Chemical Engineering, 128, 174-187.

Proceedings of the 14th International Symposium on Process Systems Engineering – PSE 2021+
June 19-23, 2022, Kyoto, Japan © 2022 Elsevier B.V. All rights reserved.
http://dx.doi.org/10.1016/B978-0-323-85159-6.50270-0

Exploring the Potential of Fully Convolutional Neural Networks for FDD of a Chemical Process

Ana Cláudia O. e Souza[a*], Maurício B. de Souza Jr.[b], Flávio V. da Silva[a]

[a] *Chemical Systems Engineering Department, School of Chemical Engineering, University of Campinas, 500 Albert Einstein Avenue, Campinas 13083-872, Brazil*
[b] *School of Chemistry, Federal University of Rio de Janeiro, 149 E-207 Athos da Silveira Ramos Avenue, Ilha do Fundão, Rio de Janeiro 20000-000, Brazil*
a192507@dac.unicamp.br

Abstract

This study investigated three distinct variations of convolutional neural network (CNN) topologies to model a fault detection and diagnosis system. The primary goal was to determine if fully convolutional networks, which do not present fully connected layers or apply the pooling operation, could outperform the well-known traditional convolutional topology. To explore this issue in a chemical process context, the Tennessee Eastman Process was the study case used. Data corresponding to four years of operation was simulated to mimic the big data scenario faced by many industries nowadays. The fully convolutional model provided better average precision and recall results. On top of it, there was a reduction of 80% of the time (elapsed real time) demanded in the training stage when compared with the traditional CNN model evaluated.

Keywords: Fault detection and diagnosis; Deep Learning; Big Data; Fully convolutional neural networks.

1. Introduction

Chemical process safety is one of the biggest concerns of engineers and process operators. It is crucial to ensure the protection of employees and facilities during the large-scale manufacturing of chemical products. Besides that, it is desirable to extend the useful life of equipment as much as possible to reduce operating costs and avoid compromising equipment availability. Therefore, the development of fault detection and diagnosis systems (FDD) is essential for any industrial process. In the last few years, the application of deep learning techniques to modeling FDD frameworks has achieved outstanding results revealing how promising artificial intelligence is to solve complex problems.

Convolutional neural networks (CNNs) are among the most known deep learning neural architectures. Its application for modeling FDD frameworks has recently been addressed in the literature (Wu and Zhao, 2018; Ge et al., 2021). Still, there is much more to explore regarding fully convolutional neural networks (FCNs). Fully convolutional neural networks consist of an end-to-end convolutional network. They do not have fully connected layers (FC) because the classification stage is also performed by convolutional layers (Conv). One of the advantages of FCNs is to demand simpler structures since only Conv layers are necessary. FCNs were successfully used to develop FDD models for navigation systems (Xu and Lian, 2018), insulators of power lines (Chen et al., 2019), and continental sandstone reservoirs (Wu et al., 2021). An excellent benchmark for evaluating FDD techniques in the process systems engineering field is the Tennessee Eastman Process (TEP) case study (Downs and Vogel, 1993). Therefore, TEP will be used for investigating if FCNs are a good choice for modeling FDD systems in chemical processes. Since the TEP data generation is done by using its available simulation models, it is possible to create a very realistic data distribution domain, similar to the big data scenario that is a reality for many processes and chemical industries nowadays.

This work aimed to model an FDD system for the TEP benchmark comparing the performance of a fully convolutional neural network and a traditional convolutional network with FC layers. In the following Section (2), the study case is described as well as the methodology to collect and preprocess the data. Section 3 is dedicated to presenting and discussing the FDD framework modeled. Finally, Section 5 summarizes the discussion, and some possible future developments are pointed out.

2. Methods

2.1. Tennessee Eastman Process Benchmark

Figure 1 presents the Tennessee Eastman Process (TEP) diagram. This benchmark was proposed by Downs and Vogel (1993) as an attempt to provide a complex study case well-suited for addressing topics like plant-wide control strategies, optimization, multivariable and predictive control, process monitoring and diagnostics, among others. The TEP represents an industrial chemical process composed of a reactor, a product condenser, a vapor-liquid separator, a recycle compressor, and a product stripper.

There are 11 manipulated variables, 22 continuous process measurements, and 19 sampled measurements, totalizing 52 variables that can be used as the input data for the model's training stage. There are also 20 process disturbances implemented, which will be used as process faults for the present work. Heat and material balance data, the list of process operating constraints, as well as the detailed description of the process operating modes can be found in Downs and Vogel (1993).

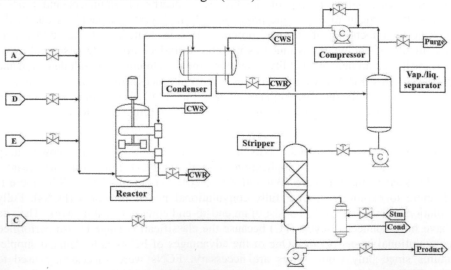

Figure 1. Tennessee Eastman Process diagram.

2.2. Data collection and preprocessing

Two different datasets were generated, one for training with data corresponding to three years of operation (525.600 samples) and the other for testing with one year of data (175.200 samples). Continuous simulations were performed with a sampling time equal to 3 minutes with the occurrence of only one fault at a time (which was randomly chosen among the 20 process faults implemented in the TEP simulator). The duration of each fault was also randomly selected between the established range of 24 and 48 hours.

Regarding data pretreatment, no feature selection technique was applied since the convolutional layers themselves can identify and isolate the most relevant attributes present in the input data. The only preprocessing consisted of normalizing each of the 52 input variables and transforming the data into 4-dimensional tensors. Therefore, $m \times n$ matrices were generated from the simulated data frames, where m represents the time span of each matrix and n is the number of input variables. To work with square matrices, which enhance the efficiency of the CNN training (Aggarwal, 2018), matrices with a shape of 52x52 were obtained. Thus, each matrix corresponds to a period of about two and a half hours of data points.

2.3. CNN based FDD framework

Figure 2 shows the proposed framework. After data acquisition, a validation dataset was separated from the training set for the application of early stopping to avoid overfitting. Then, the training, validation, and test data frames were normalized and transformed into 52x52 matrices. The best hyperparameters values for the traditional convolutional neural network (denoted by TCN) were determined by trial and error. The hyperparameters investigated were the number of convolutional layers, number of filters, optimizer method, the number of fully connected layers and their neurons, learning rate, and batch size. The size of the convolutional and pooling kernels was kept constant and equal to (3,3) and (2,2), respectively. Also, for every trial tested, max pooling was the pooling method, categorical cross-entropy was the loss function, ReLu was the activation for the intermediate layers, a SoftMax function was applied in the output layer, and the strides of convolution and pooling were 1 and 2, respectively.

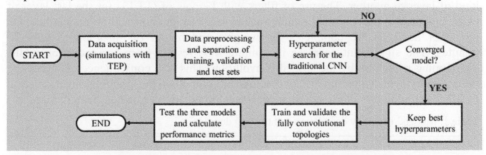

Figure 2. CNN-based fault detection and diagnosis system developing framework.

Once the best hyperparameters for TCN were found, the topology of the network was altered to generate two new networks, denoted by FCN1 and FCN2. The first fully convolutional network, FCN1, does not have fully connected layers (FC) to perform the final classification of the process state. Instead, the target classification is performed by convolutional layers (Conv) with adequate kernel sizes. In the second fully convolutional network, FCN2, not only the FC layers are replaced but also the pooling operation is omitted. Therefore, the downsampling operation is achieved by increasing the stride of the preceding convolutional layer. The idea of FCN2 corresponds to the "Strided-CNN" presented by Springenberg et al. (2015). No hyperparameter search was conducted for FCN1 and FCN2 to allow the straight comparison of the training and testing stages performances of the traditional CNN topology and the altered ones.

Precision and recall were the performance criteria applied. The simulations for data acquisition, and the training, validation, and testing stages of the models were conducted on a computer with Intel i7-9700 CPU (9th gen) 3.00 GHZ 12MB, 32 GB RAM, and Ubuntu 20.04.1 LTS. The programming language used was Python 3.7.4 with the open-source libraries Keras and Tensorflow.

3. Results and Discussion

For the traditional convolutional network, TCN, with pooling and FC layers, from all the topologies tested during the hyperparameter search, the best one was Conv(20)-Conv(30)-Pool-Conv(40)-Pool-FC(21) with a learning rate of 0.001, batch size of 500 and Adam as the optimizer. Once defined the TCN topology, FCN1 and FCN2 were created. For FCN1, the last FC layer was replaced by a Conv layer with stride 1 and kernel size equal to (13,13). On the other hand, for FCN2 the max pooling layers were removed, the stride of the two preceding Conv layers [Conv(30)-Conv(40)] was increased from 1 to 2, and the last FC layer was also replaced by a Conv layer with stride 1 but kernel size equal to (13,13). Finally, FCN1 and FCN2 were trained using the same hyperparameters of TCN. Table 1 summarizes the described topologies and presents the total number of trainable parameters as well as the time spent to update and optimize the weights and biases of each model (elapsed real time).

Despite the topology differences in the three models, the number of total trainable parameters remains the same, allowing a straightforward comparison between them. The TCN model took one hour and 39 minutes to complete the training stage. The FCN1, which does not use a fully connected layer of neurons for the final classification, was 43 minutes faster to complete the training step. In this case, the number of total parameters does not change because the only thing happening is the conversion of one kind of layer to another. The use of filters with kernels of (13,13) for the last layer of FCN1 guarantees that the total number of learnable parameters will be maintained. On the other hand, FCN2 completes the weights and biases optimization in only 20 minutes. The significant decrease in training time results from the reduction in the overlapping presented by the strided convolution. This is a consequence of increasing the stride by which the filters move across the output of the previous layer since in FCN2 the convolutions themselves are downsampling the intermediate feature maps.

Table 1. Models' topology description and training duration.

Model	Topology	Total training time (min)*	Total trainable parameters
TCN	Conv(20)-Conv(30)-Pool-Conv(40)-Pool-FC(21)	99	196,651
FCN1	Conv(20)-Conv(30)-Pool-Conv(40)-Pool-Conv(21)	56	196,651
FCN2	Conv(20)-Conv(30)-Conv(40)-Conv(21)	20	196,651

* Elapsed real time.

Since the reduction in the training time is not the primary goal here (despite being a useful outcome), the models were tested on a separate dataset never seen before during the training and validation stages to evaluate their performances. Precision and recall for the three models were calculated and are presented for each process state (normal operation and 20 different faults) in Table 2. The KPIs of FCN1 and FCN2 with equal or better performances than the traditional topology (TCN) are highlighted.

Among the three trained models, FCN2 is the one that better detects and diagnosis the fault occurrences of the test set with 80.8 % and 80.1 % of precision and recall, respectively. Compared with TCN, FCN2 shows an expressive improvement in detecting faults considered insipient and difficult to learn in the literature (Zhang and Zhao, 2017; Wu and Zhao), like Faults 3, 16, and 18. Also, FCN2 was able to provide a better separation between the faulty instances and the periods of normal operation; this is represented by the observed increase in the precision and recall of the normal status. Besides that, the maintenance of detection of faults with particular dynamic signatures, like Faults 1, 2, 5, 6, 7, and 14, proves that the conversion of FC layers into Conv layers and the removal of max pooling operations do not harm the performance of the FDD system in general.

Table 2. Detailed results of the three models on the test set.

Process status	TCN		FCN1		FCN2	
	Precision (%)	Recall (%)	Precision (%)	Recall (%)	Precision (%)	Recall (%)
Normal	84.5	82.0	84.5	70.4	86.8	88.2
Fault 1	99.8	99.9	99.9	100.0	99.8	99.9
Fault 2	100.0	100.0	100.0	100.0	99.2	100.0
Fault 3	11.6	13.1	8.4	6.3	29.3	44.9
Fault 4	99.7	98.7	99.2	98.8	99.1	99.9
Fault 5	100.0	100.0	100.0	100.0	100.0	100.0
Fault 6	100.0	92.0	100.0	100.0	100.0	98.0
Fault 7	100.0	100.0	100.0	100.0	100.0	100.0
Fault 8	91.7	94.9	92.5	91.5	93.8	91.6
Fault 9	6.4	2.3	7.3	21.5	8.4	4.2
Fault 10	96.6	85.1	79.0	93.4	97.0	88.6
Fault 11	99.8	99.5	100.0	98.6	99.8	98.5
Fault 12	100.0	100.0	89.7	100.0	100.0	96.2
Fault 13	93.4	89.2	89.7	93.4	92.3	92.7
Fault 14	100.0	100.0	100.0	100.0	100.0	100.0
Fault 15	4.2	13.7	2.9	4.6	4.8	5.8
Fault 16	36.7	48.2	34.8	37.7	58.0	53.9
Fault 17	97.8	97.6	95.8	99.1	97.5	97.9
Fault 18	44.4	30.8	48.8	51.3	62.1	46.2
Fault 19	99.8	96.9	99.8	96.5	91.5	98.8
Fault 20	97.4	98.0	98.2	97.7	98.5	98.6
Average	79.2	78.2	77.6	79.1	80.8	80.1

Therefore, the superior performance of the FCN2 model is clear. The fully convolutional network demanded a simpler topology with only convolution layers and outperformed the TCN model in all the metrics evaluated. Besides that, it is essential to emphasize that the better results were observed for the FCN2 with a reduction of 80% of the time demanded by the TCN to complete the training stage. The performance of FCN1 was not uniform. Despite the increase in the recall, due to improvements in the detection of some faults (like 9, 10, and 18), the model was not accurate regarding the normal operation, which led to the observation of some false alarms. The ideal scenario is the one where the rates of false alarms and false negatives are both low. So, FCN2 remains the best option between the models investigated in this work.

4. Conclusions

In this work, the potential of fully convolutional neural networks to model a fault detection and diagnosis system for a chemical process was explored. The outstanding performance of a convolutional topology – that does not possess fully connected layers nor max pooling operations – was proved using the Tennessee Eastman Process benchmark. The conversion of FC layers into Conv layers, and the increase of the convolutions stride to perform the downsampling of the internal feature maps, allowed the development of a model cheaper to train and with an improved ability of generalization when facing new data. The parsimonious nature of the fully convolutional neural networks appears to be a promising paradigm for designing adaptive FDD systems applied to processes subject to novel faults or new operational conditions. Given these promising outcomes, some other techniques can be tested to improve even more the achieved results. In future work, an automated hyperparameter tuning will be used to enlarge and optimize the search for the best FCN model settings. Also, the application of transfer learning to improve the detection of incipient faults by the FCN2 will be further investigated.

References

C. C. Aggarwal, 2018, Neural networks and deep learning – A Textbook. 1 ed., Switzerland: Springer Nature.

J. Chen, X. Xu, H. Dang, 2019, Fault detection of insulators using second-order fully convolutional network model, Mathematical Problems in Engineering, 1–10.

J. J. Downs, E. F. Vogel, 1993, A plant-wide industrial process control problem, Computers & Chemical Engineering, 17, 245–255.

X. Ge, B. Wang, X. Yang, Y. Pan, B. Liu, B. Liu, 2021, Fault detection for reactive distillation based on convolutional neural network, Computers & Chemical Engineering, 145, 1–14.

J. T. Springenberg, A. Dosovitskiy, T. Brox, M. Riedmiller, 2015, Striving for simplicity: the all convolutional net, International Conference on Learning Representations (ICLR), 1-14.

J. Wu, B. Liu, H. Zhang, S. He, Q. Yang, 2021, Fault detection based on fully convolutional networks (FCN), Journal of Marine Science and Engineering, 9, 1–13.

H. Wu, J. Zhao, 2018, Deep convolutional neural network model based chemical process fault diagnosis, Computers & Chemical Engineering, 115, 185–197.

H. Xu, B. Lian, 2018, Fault detection for multi-source integrated navigation system using fully convolutional neural network, IET Radar, Sonar & Navigation, 1, 774–782.

Z. Zhang, J. Zhao, 2017, A deep belief network based fault diagnosis model for complex chemical processes, Computers & Chemical Engineering, 107, 395-407.

Proceedings of the 14th International Symposium on Process Systems Engineering – PSE 2021+
June 19–23, 2022, Kyoto, Japan ©2022 Elsevier B. V. All rights reserved.
http://dx.doi.org/10.1016/B978-0-323-85159-6.50271-2

Data-driven online scenario selection for multistage NMPC

Zawadi Mdoe, Mandar Thombre, Johannes Jäschke*

Department of Chemical Engineering, Norwegian University of Science & Technology, Sem Sælands vei 4, 7491 Trondheim, Norway.

johannes.jaschke@ntnu.no

Abstract

This paper aims at reducing the conservativeness of the robust and computationally efficient sensitivity assisted multistage nonlinear model predictive controller. The approach uses a hyperbox over-approximation for the parametric uncertainty set that often results into conservativeness. We propose the use of principal component analysis (PCA) on available process data to extract more information to tighten the approximation of the parametric uncertainty set. It is approximated by a polytope whose vertices lie on the principal components. Then we define the multistage nonlinear problem with a linear transformation of the uncertain parameters. This transformation ensures consistency with the required conditions for sensitivity assisted multistage MPC algorithm used for scenario tree pruning. Finally, the method was implemented on a case study of a system of four tanks and the controller exhibited reduced conservativeness and fast computational performance.

Keywords: Robust MPC, Dynamic optimization, Parametric uncertainty, Data-driven

1. Introduction

Model predictive control (MPC) is a model based control strategy that reoptimizes a nonlinear process system with respect to a control objective subject to constraints at each sampling time. MPC includes constraints for online decision making, and has good control performance even when the system is disturbed away from the desired reference trajectory (Rawlings & Mayne, 2009). Although MPC has inherent robustness against uncertainty, the property may break when there are significant disturbances, causing infeasibilities. As a result, robust MPC approaches have been developed. One of them was proposed by Lucia et al. (2013) and is known as the multistage MPC.

1.1. Multistage MPC

Multistage MPC explicitly considers a selection of possible future scenarios along a prediction horizon to formulate its optimization problem. The scenarios are determined by propagating from the current state to the end of the prediction horizon, a finite number of uncertain parameter realizations using a scenario tree. When the prediction horizon is long the number of scenarios in the scenario tree increases exponentially resulting into an intractable problem. Lucia et al. (2013) proposed a robust horizon where the scenario tree branching is stopped before the end of the horizon, and the uncertain parameters are kept constant until the end of the prediction horizon. The robust horizon makes the problem practically feasible to solve but can still be expensive, especially for nonlinear problems, leading to a significant computational delay. In order to reduce the computational cost and computational delay of the multistage MPC, Thombre et al. (2020) proposed the sensitivity assisted multistage MPC. It has an algorithm to prune irrelevant scenarios from the scenario tree using NLP sensitivities in order to speed up computations. The sensitivity assisted multistage MPC is discussed further in Section 2.

1.2. Motivation

Even though multistage MPC is robust against constraint violations, it is rather conservative, resulting into performance loss. The conservativeness is highly dependent on how uncertainty set is represented. So far, its implementation has mainly been done using a hyperbox over-approximation of the uncertainty set. The over-approximation is often very poor if the true uncertainty set is ellipsoidal. Although the computational delay of multistage MPC can be reduced by the sensitivity assisted algorithms (Thombre et al., 2020), it has been implemented with an over-approximation of the uncertainty set leading to conservative control performance. However, in combination with statistical data analysis methods used for uncertainty identification, one can significantly reduce the conservativeness. Krishnamoorthy et al. (2018) suggested that detailed information on process uncertainty could be extracted via statistical data analysis to obtain more representative scenarios. Moreover, Shang & You (2019) rigorously present on calibration of approximate uncertainty sets for a scenario-based stochastic MPC in linear systems using support vector clustering with stability guarantees based on some mild assumptions. The contribution of this paper is to demonstrate how principal component analysis can be specifically applied to the sensitivity assisted multistage MPC framework in order to reduce conservativeness and retain its computational efficiency.

1.3. Notation

We assume a nonlinear system model $z_{i+1} = f(z_i, \nu_i, d_i)$ that predicts the evolution of the states z_i from time t_{k+i} with control actions ν_i and uncertain parameters d_i. Let us define the notation used in this manuscript. The time index $k \geq 0$ corresponds to sampling time t_k. A perfect state measurement is always assumed, and the state at time t_k is denoted by x_k. The time index of a model prediction is denoted by $i \in \mathbb{Z}_+$ which corresponds to sample time t_{k+i}. The nominal parameters are denoted as d_i^0 such that the nominal model becomes $z_{i+1}^0 = f(z_i^0, \nu_i^0, d_i^0)$. For a nonlinear system we obtain a nonlinear optimization problem (NLP) resuling inot a class of MPC known as nonlinear MPC (NMPC).

2. Sensitivity assisted multistage NMPC

The algorithm for the sensitivity assisted multistage NMPC (samNMPC) that performs online critical scenario selection based on NLP sensitivities was first developed by Holtorf et al. (2019). This selection is done by solving the NMPC problem for the nominal scenario together with a lower level optimization problem (LLP) that gives the parametric realizations that maximize the inequality constraints. This gives the constraints that are most likely violated. However, when the inequalities are interval bounds there exists a trivial solution to the LLP that lies on the vertices of the uncertainty hyperbox. Assume that the constraints are monotonically increasing or decreasing in the uncertain parameter space. The multistage MPC problem is parametric in the disturbances thus the online critical scenarios selection is based on parametric NLP sensitivities from the nominal scenario. This algorithm determines the realization most likely to violate a constraint using the sign of the parametric sensitivity. It formulates a pruned scenario tree with only the critical scenarios and the nominal, leading to a smaller NMPC problem that is cheaper to solve. The stability and recursive feasibility properties of the samNMPC were established by Thombre et al. (2020). A sensitivity assisted multistage NMPC problem at time t_k is written as follows:

$$
V_N^{sam}(x_k) = \min_{\substack{z_i^c, \nu_i^c \\ c \in \widehat{\mathbb{C}} \cup \{0\}}} \sum_{c \in \widehat{\mathbb{C}} \cup \{0\}} \omega_c \Big(\psi(z_N^c, d_{N-1}^c) + \sum_{i=0}^{N-1} \ell(z_i^c, \nu_i^c, d_i^c) \Big) +
$$

$$
\sum_{c \in \bar{\mathbb{C}}} \omega_c \Big(\psi(z_N^0 + \Delta z_N^c, d_{N-1}^c) + \sum_{i=0}^{N-1} \ell(z_i^0 + \Delta z_i^c, \nu_i^0 + \Delta \nu_i^c, d_i^c) \Big)
$$

$$
\text{s.t. } z_{i+1}^c = f(z_i^c, \nu_i^c, d_i^c), \quad i = 0, \dots, N-1 \tag{1b}
$$

$$
z_0^c = x_k, \; z_N^c \in \mathbb{X}_f, \tag{1c}
$$

$$
\nu_i^c = \nu_i^{c'}, \quad \{(c, c') \mid z_i^c = z_i^{c'}\} \tag{1d}
$$

$$
d_{i-1}^c = d_i^c, \quad i = N_R, \dots, N-1 \tag{1e}
$$

$$
z_i^c \in \mathbb{X}, \; \nu_i^c \in \mathbb{U}, \; d_i^c \in \mathbb{D}, \forall c, c' \in \widehat{\mathbb{C}} \cup \{0\} \tag{1f}
$$

where the sets $\widehat{\mathbb{C}}$ and $\bar{\mathbb{C}}$ are the critical and noncritical scenario index sets, respectively and $\{0\}$ repesents the nominal scenario. $\mathbb{D} \in \mathbb{R}^{n_d}$ is the uncertain parameter set containing a finite number of realizations, $\mathbb{X} \in \mathbb{R}^{n_x}, \mathbb{U} \in \mathbb{R}^{n_u}$ are the feasible sets for states and inputs, respectively and $\mathbb{X}_f$ represents the terminal set. N is the prediction horizon length and N_R is the robust horizon. z_i^c and ν_i^c are the predicted state and control variable vectors for scenario c at time t_{k+i}, respectively. The stage cost function is given by ℓ, terminal cost is denoted by ψ, and ω_c represents the weights on scenario c to the objective function. The variables and constraints in problem (1) are only those associated with critical constraints, thus making the problem smaller than that of the ideal multistage NMPC with a robust horizon.

3. Data driven sensitivity assisted multistage NMPC

This section presents the main idea which is to integrate principal component analysis (PCA) and samNMPC in order to reduce its conservativeness, hence enhancing its performance. The goal is to achieve that while retaining the computational speed of samNMPC.

3.1. Principal component analysis

Principal component analysis (PCA) is a multivariate data analysis tool that reveals hidden information from data. This method evaluates the variability in the data set and identifies principal components (PC) which are the unit directions that explain the total variation in the data. As a result, PCA fits a hyperellipsoid to the data with the principal components corresponding to the ellipsoids axes. The principal components are listed in order of decreasing component variance.

Assume we have a data set with n_s samples for each uncertain parameter and the data set is a represented by a matrix $\mathbf{D} \in \mathbb{R}^{n_s \times n_d}$. Before decomposition, the data set must be mean centered and scaled because PCA is sensitive to scale differences. Let the scaled and mean centered data corresponding to $\mathbf{D}$ be denoted as $\mathbf{D_0} \in \mathbb{R}^{n_s \times n_d}$. PCA on $\mathbf{D_0}$ results in the linear model $\mathbf{D_0} = \mathbf{\Lambda C}^\top$ where $\mathbf{\Lambda} \in \mathbb{R}^{n_s \times n_p}$ is a matrix with the scores corresponding to each data sample. The scores are a projection of the data points onto the principal components directions. The matrix $\mathbf{C} \in \mathbb{R}^{n_p \times n_p}$ is made up of the weights on the original samples required to obtain the component score.

3.2. Algorithm for scenario selection using both data and NLP sensitivities

This algorithm combines PCA that determines the maximum and minimum scores in the principal component directions with the samNMPC algorithm presented by Thombre et al. (2020). In order to use the samNMPC algorithm with data, we make a linear transformation of the uncertain parameters in the optimization problem using the PCA matrix. The algorithm has the following steps

(a) Scale or normalize and mean-center the data set $\mathbf{D}$ to obtain $\mathbf{D_0}$.
(b) Perform PCA on $\mathbf{D_0}$ to determine the principal component scores $\mathbf{\Lambda}$ and the corresponding principal component matrix $\mathbf{C}$.
(c) Transform the uncertain parameter vectors d_i^c into the new orthogonal space using the matrix $\mathbf{C}$, such that, $d_i^c = \mathbf{C}\bar{d}_i^c + d_i^0$ where $\bar{d}_i^c$ are the transformed parameters.
(d) Substitute the transformation from step (c) above in problem (1) to obtain an NLP in terms of the transformed parameters.
(e) At the current time t_k, determine critical scenarios $\hat{\mathbb{C}}$ and non-critical scenarios $\bar{\mathbb{C}}$ with respect to the transformed parameters using the samNMPC algorithm.
(f) Generate a pruned scenario tree with only the critical scenarios and the nominal scenario and then solve the transformed problem (1).

4. Case study

Consider the quadtank problem with a four tank configuration from Raff et al. (2006). The levels of water in the four tanks are described by the following set of differential equations:

$$\dot{x}_1 = -\frac{a_1}{A_1}\sqrt{2gx_1} + \frac{a_3}{A_1}\sqrt{2gx_3} + \frac{\gamma_1}{A_1}u_1 \qquad \dot{x}_3 = -\frac{a_3}{A_3}\sqrt{2gx_3} + \frac{1-\gamma_2}{A_3}u_2$$

$$\dot{x}_2 = -\frac{a_2}{A_2}\sqrt{2gx_2} + \frac{a_4}{A_2}\sqrt{2gx_4} + \frac{\gamma_2}{A_2}u_2 \qquad \dot{x}_4 = -\frac{a_4}{A_4}\sqrt{2gx_4} + \frac{1-\gamma_1}{A_4}u_1$$

where the states x_i are the tank levels, the inputs u_i are pump flow rates, and the uncertain parameters are the valve coefficients γ_1 and γ_2. The controller tracks setpoint levels x_1 and x_2 with minimum input usage such that the objective is $\ell = (x_1 - x_1^*)^2 + (x_2 - x_2^*)^2 + r_1 u_1^2 + r_2 u_2^2$. There are constraints on x_3 and x_4 and the system experiences predefined pulses in x_1 as described by Thombre et al. (2020).

4.1. Data analysis

The uncertain parameters have a process data cloud shown in the left plot of Figure 1. PCA on the data gives $\mathbf{C} = [0.6571, -0.7538; 0.7538, 0.6571]$. The red circled points

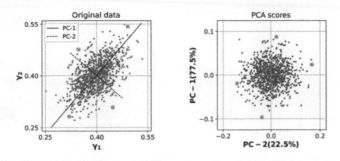

Figure 1: PCA on process data. Left shows original data, right shows the PCA scores.

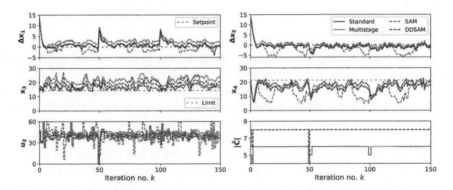

Figure 2: Comparing the control performance of the data-driven samNMPC with standard (nominal) NMPC, multistage, sensitivity-assisted multistage NMPC in the quadtank problem.

are the data points corresponding to the extreme scores on each principal component. The scores in the principal components are shown in plot on the right of Figure 1.

4.2. Results

The uncertain parameters γ_1 and γ_2 are random values generated from the multivariate distribution of the process. Then simulations were performed for both standard NMPC, multistage NMPC, samNMPC and the data-driven samNMPC. It was done for 150 iterations and the results for robust horizon $N_R = 2$ are shown in Figure 2. The tracking performance of the samNMPC is improved by the data transformation. Data-driven samNMPC tracks closer to the set point hence it is less conservative than original samNMPC and multistage NMPC. It is also robust against constraint violations for x_3 and x_4. To show the improvement of the tracking performance, we computed the accumulated cost in the simulation as shown in the bar chart on the right of Figure 3. For robust horizons lengths 1 to 3, data-driven samNMPC shows a slightly better setpoint tracking performance than the standard NMPC. It also shows a significant improvement from the original samNMPC tracking performance. In terms of computational efficiency, Figure 3 shows that the data-driven samNMPC is as fast as the original samNMPC.

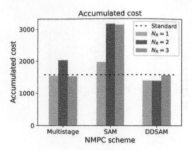
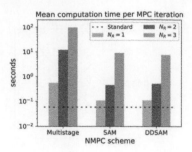

Figure 3: Comparing tracking costs (left - absolute scale) and the average computation time (right - logarithmic scale) for different robust horizons (N_R).

5.   Conclusions

We have demonstrated how analysis on process data can extract more information on the uncertainty set used to formulate the sensitivity assisted multistage MPC problem. The integration of data with samNMPC requires transforming the uncertain parameters into new variables corresponding to the principal components. As a result, the samNMPC becomes less conservative while still being computationally efficient. A caveat to the method is that the uncertainty set representation is an under-approximation using a polytope whose vertices are the maximum and minimum PCA scores. There is still a chance that the process may be outside the polytope especially when a dominant principal component does not exist. However, we expect such cases to be rare and we include soft constraints on the state bounds with penalties to avoid the infeasibilities. Future work would be investigating how scaling up to a higher-order system with more uncertain parameters affects the method's performance.

References

F. Holtorf, A. Mitsos, & L. T. Biegler (2019). Multistage NMPC with on-line generated scenario trees: Application to a semi-batch polymerization process. Journ. Proc. Control.

D. Krishnamoorthy, M. Thombre, S. Skogestad, & J. Jäschke (2018). Data-driven scenario selection for multistage robust model predictive control. IFACPapersOnLine.

S. Lucia, T. Finkler, S. Engell (2013). Multi-stage nonlinear model predictive control applied to a semi-batch polymerization reactor under uncertainty, Journ. Proc. Control.

T. Raff, S. Huber, Z. K. Nagy, & F. Allgower (2006). Nonlinear model predictive control of a four tank system: An experimental stability study. Int. Symp. on Intelligent Control. IEEE.

J. B. Rawlings, & D. Q. Mayne (2009). Model Predictive Control: Theory and Design.

C. Shang, & F. You (2019). A data-driven robust optimization approach to scenario-based stochastic model predictive control, Journ. Proc. Control.

M. Thombre, Y. Zhou, J. Jäschke, & L. T. Biegler (2020). Sensitivity-assisted multistage nonlinear model predictive control: Robustness, stability and computational efficiency, Comp. & Chem. Eng.

Proceedings of the 14th International Symposium on Process Systems Engineering – PSE 2021+
June 19-23, 2022, Kyoto, Japan © 2022 Elsevier B.V. All rights reserved.
http://dx.doi.org/10.1016/B978-0-323-85159-6.50272-4

Data-driven Robust Model Predictive Control with Disjunctive Uncertainty for Building Control

Guoqing Hu[a,*], Fengqi You[a]

[a]*Cornell University, Ithaca, New York 14853, USA*
gh429@cornell.edu

Abstract

Model Predictive Control (MPC) has gained popularity in recent years and is widely adopted in building control. This study proposes a novel data-driven robust MPC to make the optimal heating plan, specifically for the multi-zone single-floor building. In this study, the room temperature and relative humidity (RH) will be highly valued in the optimization decision. To better incorporate RH into the state-space model (SSM), the linear relations between RH and other room temperature parameters in the thermal zones are formulated, ensuring the better linear fitting of SSM to the original nonlinear model. Afterward, k-means clustered, principal component analysis (PCA), and kernel density estimation (KDE) based data-driven uncertainty set is constructed and applied to MPC. The other three kinds of MPC's are compared to our proposed data-driven robust MPC (RMPC), including conventional RMPC, k-means clustered, data-driven RMPC, PCA and KDE based data-driven RMPC. The results demonstrate that the optimality of our proposed k-means clustered, PCA and KDE based data-driven RMPC, which consumes 9.8 % to 17.9 % less energy in controlling both temperature and RH, compared to other data-driven robust MPC's, and essentially follow the constraints which certainty equivalent MPC and conventional RMPC cannot conform.

Keywords: model predictive control, disjunctive uncertainty, multi-zone building control

1. Introduction

According to the EIA report in 2019, heating and humidity control dominate energy usage, contributing 30 % of total power consumption. Controlling temperature is essential to the building control since overheating is another problem that consumes significant energy and deteriorates the living condition.

Among all possible control methods, model predictive control (MPC) provides the new scope for controlling the building temperature, saving a tremendous amount of energy usage compared to the rule-based control strategies (Prívara et al., 2011). However, the conventional MPC does not possess the capability of hedging against the uncertainty (Shang et al., 2019), i.e. being applied under stochastic conditions (Ning and You, 2019). In building control, weather information can never be perfectly predicted, and thus can be treated as the sources for uncertainties (Shang et al., 2020). Consequently, it remains a knowledge gap needs to be filled with the new designed MPC which is not only robust to the disturbances from uncertainty, but also can avoid the "over-conservative" problem proposed by Chen et al. (2021). Therefore, we focus on developing the better control strategy to multi-zone building's room temperature and RH under realistic condition, k-mean clustered, principal component analysis (PCA) and kernel density estimation (KDE) based data-driven RMPC (KM-PKDDRMPC). We apply this model to the multizone building's SSM, which incorporates both room temperature and RH. In this work, the SSM of the building is generated from based on both building element construction and the study of the dynamic airflow within the building. Afterward, the uncertainty set is constructed based on the historical forecast error to the weather information, i.e., the differences between forecast and real-measured values. This uncertainty set can be further clustered by the k-means algorithm, and PCA combined with KDE can return the polyhedral-shaped applied to the RMPC. The optimization problem at each control horizon is solved using affine disturbance policy (ADF). The contributions of this paper are summarized below:

- A novel data-driven robust model predictive control framework with disjunctive uncertainty to control the multi-zone building's room temperature and RH;
- A simulation of multi-zone building's temperature and RH control based on actual weather data demonstrates better control performance of KM-PKDDRMPC comparing to other MPC's

2. Model formulation

2.1. Complete state-space model

The BRCM MATLAB toolbox is adopted for finding the state space matrix (SSM). BRCM can generate the linear resistance-capacitance models from self-designed building geometry construction. The dynamic multi-input multi-output system is given by:

$$x_{t+1} = Ax_t + B_u u_t + B_v v_t + B_w w_t \tag{1}$$

Where A is the state matrix that correlates state variables x_t to SSM. The state variables returned from BRCM are room temperature, wall temperature, floor temperature and ceil temperature. B_u, B_v, B_w are control input matrix, disturbance matrix, and uncertainty matrix, respectively, corresponding to u_t, v_t, w_t, which are control input, disturbances, and uncertainty. The control inputs include heater, radiator, humidifiers and dehumidifiers; the disturbances are from ambient temperature and ambient RH condition. Uncertainties are the forecasted temperature and RH errors. Meanwhile, RH within each room is calculated based on the air dynamic within the building (Cengel, 1997). The mass of airflow is initially found as:

$$m_{air,t-1} = \frac{Q_{t-1}}{c_p \Delta T} \tag{2}$$

ΔT is calculated as follows:

$$\Delta T = \max \left(\left(T_{room,t} + \delta T - T_{air,t} \right), 0 \right) \tag{3}$$

where δT is temperature difference of room and air heating unit (AHU). Unlike in previous research, $m_{air,t-1}$ is not regarded as a constant because the simulation process is conducted in the winter season. The constant intake airflow rate indicates that the room is constantly exchanging the air with a colder ambient environment. The heater, most of the time, is active to maintain room temperature within the thermal comfort standard. Alternatively, we assume the difference between the room temperature and heated air from the AHU is constant. Subsequently, the heating airflow can be turned off when heating is not necessary. When the mass of airflow is calculated, the mass of water vapor brought by airflow can be found by the following equation:

$$m_{AC,in,t-1} = \rho_{AC,in,t-1} \cdot RH_{out,t-1} \cdot \frac{m_{air,t-1}}{\rho_{air}} \tag{4}$$

And so can be found the mass of water vapor taken away by airflow:

$$m_{AC,out,t-1} = \rho_{water,sat,t-1} \cdot RH_{t-1} \cdot \frac{m_{air,t-1}}{\rho_{air}} \tag{5}$$

Where SVD values are found through equation f, which is a linear equation of SVD values over temperature (T) expressed as follows:

$$\rho_{water} = f(T) = 1.0272T - 1.8959 \tag{6}$$

Afterward, the mass of water vapor stored in each room can be found as:

$$m_{water,t} = \rho_{water,sat,t-1} \cdot RH_{t-1} \cdot V_{room} + m_{hum} + m_{AC,in,t-1} - m_{dehum} - m_{AC,out,t-1} \tag{7}$$

Eventually, RH values within each room at t can be found, which is the ratio of absolute and SVD:

$$\rho_{abs,t} = \frac{m_{water,t}}{V_{room}}, \ \rho_{sat,t} = f(T_{room,t}), \ RH_t = \frac{\rho_{abs,t}}{\rho_{sat,t}} \tag{8}$$

At this point, the RH values within each room can be found based on the room temperature, control input and room volume. The next step is to add system identification toolbox found in MATLAB to obtain the SSM required for the MPC. The testing data, instead of training data, is used to ensure the feasibility of SSM to be applied in simulation within the real condition. The average value of mean absolute percentage error (MAPE) for RH in all rooms is 4.65 % and the average MAPE for temperature in all rooms is 0.95 %, indicating this model is acceptable for the MPC problem.

2.2. PCA and KDE based data-driven uncertainty sets clustered by K-means algorithm
Disjunctive uncertainty sets are constructed to better learn the trend of the uncertainty data (Ning and You, 2017). Therefore, the K-means clustering method is adopted in this work to cluster the uncertainty into multiple groups. The groups are identified by minimizing the sum of intracluster variances, i.e., squared Euclidean distance:

$$D^* = \arg\min\left(\sum_{i=1}^{k} \sum_{w \in D_i} \|w - \mu_i\|^2 \right) \tag{9}$$

Despite multiple groups of uncertainty data, the traditional norm-based uncertainty set cannot be applied directly to deal with the uncertainty data with varied structure and complexity (Ning and You, 2021). Therefore, PCA and KDE are adopted for coping with the data with polyhedral shapes. PCA can then maximise the variance of the uncertainty under the same scale. The covariance matrix can be approximated as:

$$S_i = \frac{1}{N-1} w_i^T w_i \tag{10}$$

As the covariance matrix S_i can be further decomposed as $S_i = Q_i \Lambda_i Q_i^T$, where Q_i's column contains all the eigenvectors, corresponding to the eigenvalues stored in the diagonal matrix Λ_i. The individual eigenvalue will represent the variance of this axis if data is projected on this eigenvector.

Finally, it can be further studied the distributional information of the uncertainty dataset within each component j within the cluster k via the KDE approach:

$$f_{j,k} = \frac{1}{N} \sum_{n=1}^{N} K(\xi_{j,k}, p_{j,k}^{(n)}) \tag{11}$$

With probability density function, the cumulative density function will be written as follows:

$$F_{j,k}^{-1}(\alpha) = \min\{\xi_{j,k} \mid F_{j,k}(\xi_{j,k} \geq \alpha)\} \tag{12}$$

where α is the pre-specified small quantile parameter, ranging from 0 to 0.5, and ξ is the inferred latent variable. The uncertainty set W_k within cluster k can be formulated by introducing forward and backward deviation variable z^+ and z^- (Ning and You, 2018):

$$\mathbb{W}^k = \left\{ \mathbf{w}^k \in R^H \left| \begin{array}{l} \mathbf{w}^k = \hat{\mu}_k + Q_k \xi_k, \ \xi_k = \underline{\xi}_k z^- + \overline{\xi}_k z^+ \\[4pt] 0 \leq z^+, z^- \leq 1, \ z^+ + z^- \leq 1, \ \mathbf{1}^T\left(z^+ + z^-\right) \leq \Gamma \\[4pt] \underline{\xi} = \left[\hat{F}_{1,k}^{-1}(\alpha), ..., \hat{F}_{1,k}^{-1}(\alpha)\right]^T \\[4pt] \overline{\xi} = \left[\hat{F}_{1,k}^{-1}(1-\alpha), ..., \hat{F}_{1,k}^{-1}(1-\alpha)\right]^T \end{array} \right. \right\} \tag{13}$$

3. Control strategy

The next step is to develop the optimization problem to get the control strategy to the multi-zone building. To ensure the tractability of the RMPC optimization problem, ADF is adopted to get control input u_t based on past disturbances. The equation is expressed as following (Goulart et al., 2006):

$$u_i = h_i + \sum_{j=0}^{i-1} M_{i,j} w_j, \ \forall w \in \mathbb{W}^k \tag{14}$$

where M is regulated as follows:

$$M = \begin{bmatrix} 0 & \cdots & \cdots & 0 \\ M_{1,0} & 0 & \cdots & 0 \\ \vdots & \vdots & \ddots & \vdots \\ M_{H,0} & M_{H,1} & \cdots & 0 \end{bmatrix} \tag{15}$$

Only the first u_0 will be applied for the control to the model and the rest will be discarded. The optimization problem with ADF can be formulated as follows (Shang et al., 2017):

$$\min \sum_{i \in B_u} c_i u_i + \lambda^T L \lambda$$

$$\text{s.t. } F_u[Mw + h] \leq f_u, \qquad\qquad \forall w \in \mathbb{W}^k \tag{16}$$

$$F_x[Ax_0 + B_u h + B_v v + (B_w + B_u M)w] \leq f_x + \lambda, \quad \forall w \in \mathbb{W}^k$$

where F_x, F_u, f_x, f_u represent the state variable constraints matrix, control input constraints matrix, constraints for state variables, and constraints for the input. L is the weighted cost matrix that penalizes the violation to the constraints. Λ is the slack variable that allows some extent of violation to the hard constraints (Jia et al., 2020).

4. Case study

In this study, the single-floor multi-zone building located in Ithaca, New York, USA is selected for the simulation of close-loop data-driven RMPC to control the temperature and RH in each individual room. The constraints for the control conditions are: For the room temperature should be within 15 °C to 25 °C, and RH should sit in between 30 % to 60 %, according to ASHRAE Standard 62-2001.

The model was simulated from 0:00 AM, November 1st, 2016 to 0:00 AM, on November 8th, 2016, ranging from precisely one week. The initial conditions for temperature values in all rooms are 21 °C and RH values are 40 %. One of rooms' results are selected for demonstration, as shown in Fig 1. Based on the result, both certainty equivalence MPC (CEMPC) and RMPC violate the constraints more severely. CEMPC which only considers the deterministic conditions, fails to compose the strategy against the prediction error from ambient temperature and RH. Meanwhile, the RMPC fails to obey the RH constraints, indicating an irregular shape of the uncertainty data of RH. On the other hand, the rest three control strategies can be more conservative in maintaining both temperature and RH within the constraints. KM-DDRMPC will be the most conservative one since there is nearly no violation at all, but, it will have the highest power consumption across all control methods. On the other hand, though there are slightly more violation cases and more computation time, KM-PKDDRMPC will draw significantly less power in controlling the temperature and RH compared to KM-PKDDRMPC and PCA coupled with KDE based data-driven RMPC (PKDDRMPC).

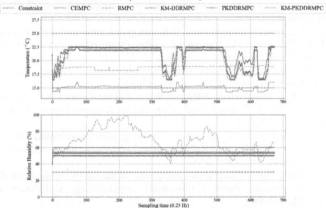

Figure 1. Control profile in Ithaca, New York, in the first week of November 2016

5. Conclusions

In this work, we develop a KM-PKDDRMPC framework for the multi-zone building SSM, which includes indoor temperature and RH control. In order to maintain temperature and RH within the comfortable range, KM-PKDDRMPC is capable of handling the uncertainty sets from temperature and RH forecast. The steady-state system with RH is constructed with the help of system identification. Then the optimization problem can be further developed with the SSM and disjunctive uncertainty sets. The proposed KM-PKDDRMPC was compared with the CEMPC and other MPC strategies, including RMPC, KM-DDRMPC, PKDDRMPC. The result demonstrated that the proposed KM-PKDDRMPC has outperformed the rest from the overall perspective, using 17.9 % less power consumption than KMDDRMPC and 9.8 % fewer compared to PKDDRMPC. Though CEMPC and RMPC have used less power, the high violation rate will exclude them from the final consideration to the practical application.

6. Nomenclature

SVD – saturated vapor density
VD –vapor density
$m_{air,t-1}$ – mass of airflow at t-1, kg
Q_{t-1} – heat input at t-1, J
c_p – specific heat of air, kJ/(kg-K)
ΔT – temperature change, °C
$T_{room,t}$ – room temperature at t, °C
$T_{air,t}$ – ambient temperature at t, °C
ρ_{air} – air density, kg/m³
RH_{t-1} – relative humidity in room at t-1
$RH_{out,t-1}$ – ambient relative humidity at t-1
$\rho_{water,sat,t-1}$ – SVD of T_{room} at t-1, g/m³

$\rho_{AC,sat,t-1}$ – SVD of T_{air} at t-1, g/m³
V_{room} – room volume, m³
$m_{water,t}$ – mass of VD at t, kg
m_{hum} –mass of VD from humidifier, kg
m_{dehum} – mass of VD taken by dehumidifier, kg
$m_{AC,in,t-1}$ – mass of VD from air circulation at t-1, kg
$m_{AC,out,t-1}$ – mass of VD taken by air circulation at t-1, kg
$\rho_{abs,t}$ – absolute VD density at t, kg/m³
$\rho_{sat,t}$ – SVD at t, kg/m³

References

ISO, 2001, Standard 62-2001, Ventilation for acceptable indoor air quality. American Society of Heating, Refrigerating and Air Conditioning Engineers.

Y. A. Cengel, 1997, Introduction to thermodynamics and heat transfer, McGraw-Hill New York.

W.-H. Chen, C. Shang, S. Zhu, et al., 2021, Data-driven robust model predictive control framework for stem water potential regulation and irrigation in water management. Control Engineering Practice, 113, 104841.

EIA, 2019, Use of energy explained[Online]. U.S. Energy Information Administration.

P. J. Goulart, E. C. Kkerrigan, J. M. Maciejowski, 2006, Optimization over state feedback policies for robust control with constraints. Automatica, 42, 523-533

R. Jia, F. You, 2020, Multi-stage economic model predictive control for a gold cyanidation leaching process under uncertainty. AIChE Journal, 67, e17043

C. Ning, F. You, 2017, Data-Driven Adaptive Nested Robust Optimization: General Modeling Framework and Efficient Computational Algorithm for Decision Making Under Uncertainty. AIChE Journal, 63, 3790-3817.

C. Ning, F. You, 2018, Data-driven decision making under uncertainty integrating robust optimization with principal component analysis and kernel smoothing methods, Computers & Chemical Engineering, 112, 190 – 210.

C. Ning, F. You, 2019, Optimization under uncertainty in the era of big data and deep learning: When machine learning meets mathematical programming. Computers & Chemical Engineering, 125, 434-448.

C. Ning, F. You, 2021, Online learning based risk-averse stochastic MPC of constrained linear uncertain systems. Automatica, 125, 109402.

C. Shang, W.-H. Chen, A.D. Stroock, et al., 2020, Robust Model Predictive Control of Irrigation Systems With Active Uncertainty Learning and Data Analytics. IEEE Transactions on Control Systems Technology, 28, 1493-1504.

C. Shang, X. Huang, F. You, 2017, Data-driven robust optimization based on kernel learning. Computers & Chemical Engineering, 106, 464-479.

C. Shang, F. You, 2019, A data-driven robust optimization approach to scenario-based stochastic model predictive control. Journal of Process Control, 75, 24-39.

C. Shang, F. You, 2019, Data Analytics and Machine Learning for Smart Process Manufacturing: Recent Advances and Perspectives in the Big Data Era. Engineering, 5, 1010-1016.

S. Privara, J. Siroky, L. Ferkl, J. Cigler, 2011, Model predictive control of a building heating system: The first experience. Energy and Buildings, 43, 564-572

Proceedings of the 14th International Symposium on Process Systems Engineering – PSE 2021+
June 19-23, 2022, Kyoto, Japan © 2022 Elsevier B.V. All rights reserved.
http://dx.doi.org/10.1016/B978-0-323-85159-6.50273-6

Low-Dimensional Input and High-Dimensional Output Modelling Using Gaussian Process

Jiawei Tang, Xiaowen Lin, Fei Zhao, Xi Chen*

State Key Laboratory of Industrial Control Technology, College of Control Science and Engineering, Zhejiang University, Hangzhou 310027, China
xi_chen@zju.edu.cn

Abstract

In this paper, a unified low-dimensional input and high-dimensional output modelling method is proposed to deal with complex molecular simulation and design problems. First, a convex optimization framework is constructed to decompose vertically stacked molecular weight distribution (MWD) matrix into low-rank and sparse parts, while the intrinsic structure can be explored, and abnormal points can be eliminated. Then, considering the correlations between independent output channels, an effective coregionalization kernel is adopted in Gaussian Process (GP) to implement the low-dimensional multi-output tasks. The whole procedure consists of data filtering, feature compressing and multi-output GP, which is named by DF-MGP. Case study of an ethylene homo-polymerization with the Ziegler-Natta catalyst system shows the effectiveness of the proposed DF-MGP strategy.

Keywords: Multi-output regression; Gaussian Process; coregionalization kernel; molecular weight distribution

1. Introduction

Machine learning (ML) has been widely applied in almost all areas of science. They are great at problems when inputs lie in high-dimension space and outputs lie in low-dimension space. However, this situation is inverse in molecular simulation, which macro manipulation space is much smaller than micro molecular space. Because of the complex and time-consuming features of molecular simulation, some scholars have utilized ML to improve such phenomenon. Elton et al. (2018) proved that ML techniques can be used to predict CNOHF energetic molecules from their molecular structures. Afzal et al. (2019) applied ML to develop a data-driven prediction model in the study of 1.5 million organic molecules. Moreover, main challenges of the practical applications are missing data, especially when some feature values cannot be observed, presence of noise, and coupling interactions between multiple target variables. To tackle these issues, multi-output regression methods are presented, which are capable to yielding better predictive performance than single-output methods. The multi-output regression methods aim to simultaneously predict multiple real-valued outputs. Kocev et al. (2009) applied ML methods to predict multiple targets describing conditions or quality of vegetation. Tuia et al. (2011) estimated different biophysical parameters from remote sensing images simultaneously.

Furthermore, problems with low-dimensional input and high-dimensional output feature make multi-output regression extremely difficult. For example, to represent a molecular weight distribution of a polymer usually needs a chain length as large as 10^5. Even after discretization, normally 100 grids are required to represent such a curve as measured by Gel Permeation Chromatograp (GPC). This kind of problems requires an elaborate and elegant technique to process data. In this paper, we propose to apply compressed sensing (CS) method, which is well known in machine vision, to decompose the MWD matrices into low-rank and sparse parts, while a novel strategy is developed to address data missing and noise. An ethylene homo-polymerization with the Ziegler-Natta catalyst system (Lin et al. 2021) is studied to show the effectiveness of the proposed method.

2. Related works

2.1. Decomposing Sparse and Low-Rank Matrices

Compressed sensing is widely studied in computer science, which has been proved to be a complete technique for signal treatment and analysis. It can recover the signal from few of samples. In addition, the compressed sensing technique can also be applied to matrices when signals are stacked vertically in sequence. The general formula is shown as follows.

$$\text{Minimize } \|L\|_* + \lambda\|S\|_1 \quad \text{s.t. } P_Q[L + S] = P_Q[Y] \tag{1}$$

where $\lambda > 0$ is a positive weight parameter; the subscripts, * and 1, denote nuclear norm and L1 norm, respectively; Y is the unprocessed data, P_Q is the projection operator. L and S represent the optimized low-rank and sparse matrices. Because the linear constraint and objective function are convex, the convex problem shown in Eq. (1) can be solved by alternating directions method of multipliers (ADMM).

2.2. Gaussian Process

Gaussian process (GP) is a kind of non-parametric Bayesian approaches (Schulz, E et al. 2009). Theoretically, it can capture a variety of relations between inputs and outputs by using an infinite number of parameters, while determining the level of complexity by means of Bayesian inference. Generally, GP can be formulated in traditional parametric weight space or non-parametric Bayesian function space. A univariate linear Bayesian regression formula is presented as follows:

$$p(f_*|x_*, X, \boldsymbol{y}) = N(\frac{1}{\sigma_\epsilon^2} x_*^T A^{-1} X \boldsymbol{y}, x_*^T A^{-1} x_*) \tag{2}$$

where f is an unknown function, which maps inputs x to outputs y: $f: X \rightarrow Y$. For the sake of simplicity, the dimension of Y is set to one. x_* is a test case; f_* is latent variable output; σ_ϵ^2 is variance of Gaussian noise. $A = \sigma_\epsilon^2 X X^T + \Sigma_p^{-1}$; X and $\boldsymbol{y}$ represent the training data.

GP is a probabilistic ML, which can predict uncertainties of f_*. We can specify different probability density and construct kinds of likelihood probability density. Moreover, many approximation inference methods can be used for modelling and Bayesian optimization.

3. High-dimensional-output modelling method

In this section, we use an ethylene homo-polymerization with Ziegler-Natta catalyst system to introduce the framework of proposed DF–MGP strategy. As shown in Fig. 1, it mainly contains four steps, i.e., data-collection, feature extraction, kernel design and inference chosen.

Step 1: Data collection

The manipulated variables (MV) involved in this case are hydrogen feed(H2), monomer feed(M), hexane feed (C6H14), catalyst feed (Cp), cocatalyst feed (A) and temperature (T). Latin hypercube sampling (LHS) is applied to sample points in six-dimensional space.

Step 2: Decomposing low-rank and sparse parts

Due to the numerical calculation error, some MVs can lead to a non-convergent MWD in the simulation process. A few outliers can degrade the performance of ML. To tackle this issue, the convex optimization model described in section 2.1 is constructed to minimize the L1 and nuclear norm of MWD matrix to explore the intrinsic structure and cut off the abnormal points. Considering the high dimensionality, the singular value decomposition (SVD) method is performed to recover the best low-rank approximation matrix of MWD and complete feature compressing.

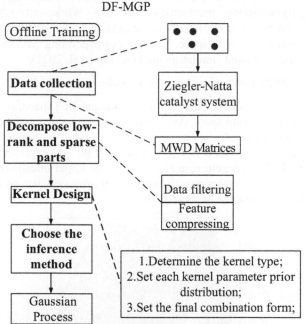

Fig. 1. Framework of the proposed DF–MGP strategy

Step 3: Kernel design

As a nonlinear regression method, the GP models involve many kernel functions. The kernel function defines a covariance function, which can describe the similarity of random variables. There are many kinds of kernel functions, such as linear, Matern32, periodic, polynomial, exponent and radial basis functions. Considering the requirements of multi-output analysis, the coregionalization kernel function is utilized to describe correlation of the outputs. Once the type of kernel is specified, the hyper-parameters involved can be found by Bayesian inference. Then, the corresponding prior of hyper-parameters can be determined. In addition, different kernel functions can be combined, such as sum, product, vertical scaling, warping.

Step 4: Inference chosen

Based on the Gaussian prior and Gaussian likelihood functions, the conditional posterior of f_* can be evaluated by Eq. (2). However, if the kernel function and likelihood function are changed, the conditional posterior will be rather intractable. Thus, different inference methods are proposed, such as Laplace approximation, expectation propagation, Markov Chain Monte Carlo (MCMC). Different methods can lead to different generalization performance; thus, the step of inference chosen is crucial. Note that the continuous and categorical hyper-parameters involved in above steps are chosen by trial and error, and there also exist some empirical ways to assist in hyper-parameters tuning.

4. Case studies

4.1. Homo-polymerization Ziegler-Natta Catalyst System

Molecular weight distribution (MWD) is a critical index of the optimization and control for industrial polymerization processes, which indicates the processability and properties of polymers. MWD is a probability distribution function of chain length, which can be predicted by polymerization mechanisms. In this work, an ethylene homo-polymerization with the Ziegler-Natta catalyst system is demonstrated. 1549 sample points are generated in the MV space, and the corresponding MWDs are generated through a kinetic modeling and simulation method (Lin et al. 2021).

4.2. Projection of MWD Matrices on Intrinsic-Low-dimensional Matrices

Since the MWD data is originally defined in a 132-dimensional space, we need to project it onto the low-dimensional subspace. As shown in Fig. 4(a), the raw data have abnormal points because of numerical calculation error. In Fig. 2(a), the low-rank part has extracted most features of the curve. Non-zero parts shown in Fig. 2(b) represent the noise and abnormal points of MWD. Therefore, in Fig. 3, we set an upper threshold 51 to eliminate those parts. Fig. 4(b) demonstrates effectiveness of our proposed data processing method through which the valid data can be identified.

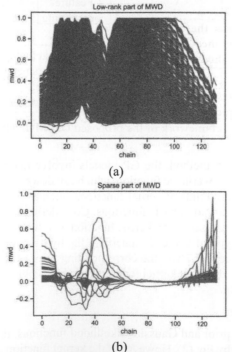

(a)

(b)

Fig. 2. Low-rank part and sparse part visualization.

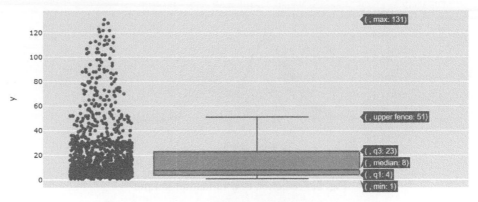

Fig. 3. Box plot of non-zero positions in sparse part

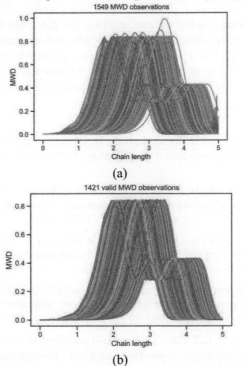

Fig. 4. Visualization results of MWD. (a) Raw data (b) Valid data

4.3. LS-MGP Prediction

MCMC inference is implemented to optimize the parameters of GP. 80% of the valid data are used to train the offline model, and the rest are used for validation. Fig. 5 shows one of the visualization results. The shaded part is the uncertainty evaluation of prediction bounded by 0.05 and 0.95 quantiles. The average root mean square error is 0.01. The first solid line in the legends represents the mean value of GP, and the second marker line with left triangle indicates the result of validation point. The rest two dash dot lines show the region of 0.05 and 0.95 quantiles

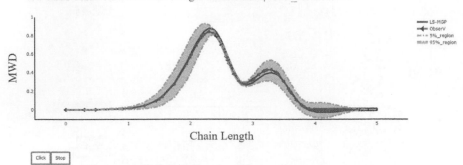

Fig. 5. Prediction of a validation sample

5. Conclusions

Through data processing and feature compressing, the proposed DF-MGP strategy can accurately predict the micro-structure of polymers with different manipulating conditions. Due to the flexibility of GP and powerful performance of MCMC inference, the proposed method is also applicable to other high-dimensional multi-output problems. The current work is a beginning of the future research for molecular design and optimization.

References

Elton, D. C., Boukouvalas, Z., Butrico, M. S., Fuge, M. D., & Chung, P. W. (2018). Applying machine learning techniques to predict the properties of energetic materials. Scientific reports, 8(1), 1-12.

Afzal, M. A. F., Sonpal, A., Haghighatlari, M., Schultz, A. J., & Hachmann, J. (2019). A deep neural network model for packing density predictions and its application in the study of 1.5 million organic molecules. Chemical science, 10(36), 8374-8383.

Kocev, D., Džeroski, S., White, M. D., Newell, G. R., & Griffioen, P. (2009). Using single-and multi-target regression trees and ensembles to model a compound index of vegetation condition. Ecological Modelling, 220(8), 1159-1168.

Tuia, D., Verrelst, J., Alonso, L., Pérez-Cruz, F., & Camps-Valls, G. (2011). Multioutput support vector regression for remote sensing biophysical parameter estimation. IEEE Geoscience and Remote Sensing Letters, 8(4), 804-808.

Lin, X., Chen, X., & Biegler, L. T. (2021). Generalized initialization for the dynamic simulation and optimization of grade transition processes using two‐dimensional collocation. AIChE Journal, 67(1), e17053.

Lin, X., Chen, X., Biegler, L. T., & Feng, L. F. (2021). A modified collocation modeling framework for dynamic evolution of molecular weight distributions in general polymer kinetic systems. Chemical Engineering Science, 237, 116519.

Schulz, E., Speekenbrink, M., & Krause, A. (2018). A tutorial on Gaussian process regression: Modelling, exploring, and exploiting functions. Journal of Mathematical Psychology, 85, 1-16.

Proceedings of the 14th International Symposium on Process Systems Engineering – PSE 2021+
June 19-23, 2022, Kyoto, Japan © 2022 Elsevier B.V. All rights reserved.
http://dx.doi.org/10.1016/B978-0-323-85159-6.50274-8

Piecewise Smooth Hybrid System Identification for Model Predictive Control

Ilya Stolyarov[c], Ilya Orson Sandoval[a], Panagiotis Petsagkourakis[b], Ehecatl Antonio del Rio-Chanona[a*]

[a]*Department of Chemical Engineering, Imperial College London, London - U.K.*
[b]*Centre for Process Systems Engineering, University College London, London - U.K.*
[c]*Department of Computing, Imperial College London, London - U.K.*
[*]*a.del-rio-chanona@imperial.ac.uk*

Abstract

Complex systems which exhibit different dynamics based on their operating region pose challenges for data driven control because a single global model may not capture the varying dynamics of the system. One solution is to use hybrid system identification to learn the location of operating regions and dynamics within each region from data, yielding a more accurate multi-model of the system. This article proposes a novel method of hybrid system identification through spectral clustering with a custom similarity function. A case study of a chemical process illustrates benefits of this approach for Model Predictive Control.

Keywords: Hybrid System Identification, Model Predictive Control, Data Driven Models

1. Introduction

Systems which exhibit different dynamics based on their operating region, termed hybrid systems, are prevalent within all areas of engineering, ranging from a four-stroke cycle of a combustion engine to chemical processes controlled by a thermostat (Lauer 2019). These systems pose challenges for data driven control, since a single model is often inadequate in capturing the varying dynamics of the system. However, the location or even a number of operating regions may be unknown *a priori*, and hence hybrid system identification is concerned with learning the location and local model of each operating region. This can lead to a more accurate piecewise system model and improve control of hybrid systems.

Due to the presence of discrete operating regions and continuous dynamics, hybrid system identification is an NP-hard mixed integer optimization problem (Lauer 2019), and as such, local methods or relaxation approaches must be employed. Most of the currently available methods further constrain the problem by assuming that the underlying system is linear, leading to piecewise affine (PWA) system identification (Lauer 2019, Ohlsson 2013). On the other hand, only a few algorithms tackle identification of general piecewise smooth (PWS) systems (Lauer 2019, Lauer 2014, Lee 2017) that can be used within the emergent field of nonlinear control. This article proposes a novel PWS system identification approach based on spectral clustering, with ramifications for control illustrated by a case study of a Continuously Stirred Tank Reactor (CSRT).

2. Hybrid System Identification

2.1. Problem Definition

For a hybrid system with states $x \in \mathbb{R}^{d_x}$, controls $u \in \mathbb{R}^{d_u}$ and p operating regions, a latent function $q(\cdot,\cdot) : \mathbb{R}^{d_x} \times \mathbb{R}^{d_u} \rightarrow \{1, \dots, p\}$ defines the operating region for a given set of states and controls. Assuming a full state feedback, the discrete state space model of the system then exhibits piecewise behaviour as described in Eq. (1), where $\varepsilon \sim \mathcal{N}(0, \Sigma)$ is some Gaussian noise.

$$x_{k+1} = f_{q(x_k, u_k)}(x_k, u_k) + \varepsilon \equiv f_{q(\zeta_k)}(\zeta_k) + \varepsilon \tag{1}$$

Given a training dataset $\mathcal{D} = \left\{ x_k^{(i)}, u_k^{(i)}, x_{k+1}^{(i)} \right\}_{i=1,\dots,N-1}$, the goal of hybrid system identification is to learn the latent function $q(\cdot,\cdot)$ as well as each local model $\{f_j(\cdot,\cdot)\}_{j=1,\dots,p}$. Notably, the number of regions p may be unknown and hence must also be learned. To simplify the notation, let's further aggregate the state and control vectors into a single vector $\zeta_i = \left[x_k^{(i)}, u_k^{(i)} \right]^T$ and rename $x_{k+1}^{(i)}$ as y_i, which allows us to reformulate Eq. (1) and rewrite the training dataset in a more familiar input-output form $\mathcal{D} = \{\zeta_i, y_i\}_{i=1,\dots,N}$

2.2. Challenges of Hybrid System Identification

The main difficulty of hybrid system identification comes from the dual nature of the problem: simultaneously assigning datapoints to regions and learning the model within each region (Lauer 2014). Moreover, the problem is naturally ill defined if the number of regions is unknown, which can be observed by a trivial solution of assigning each datapoint to a different region. The problem can be simplified by assuming that each local model $f_j(\cdot,\cdot)$ is linear, hence deriving a PWA system model. This is attractive because nonlinear systems can be approximated with sufficiently many local affine models and classical linear control can then be employed (Lauer 2019). However, this limits the use of hybrid system identification for nonlinear system control strategies such as nonlinear MPC.

In PWS system identification, on the other hand, the only assumption made about the nature of the underlying local models is that they are smooth. Hence, if the complexity of these models is not limited, a single flexible model can overfit the training dataset. Based on this intuition, (Ohlsson 2013) derives a regularization approach to PWA identification, which is then extended to PWS systems by considering functions in a Reproducing Kernel Hilbert Space (RKHS) in (Lauer 2014). These approaches work by estimating a regularized parametric local model for each datapoint, hence projecting it into some parameter space, where k-means clustering can be used to identify operating regions. Although, this works well for linear models where the dimensionality of the parameter space is small, extension to PWS systems faces the difficulty of the large dimensionality of the parameter space associated with functions in RKHS (i.e. dimensionality is equal to the number of training samples).

2.3. Spectral Clustering for Hybrid System Identification

An alternative to clustering points based on compactness in some parameter space, is to employ connectedness based clustering algorithms such as spectral clustering (von Luxburg 2007), which aims to identify clusters such that the similarity of points within a cluster is maximized and the similarity of points between clusters is minimized. This is done by using a positive similarity function $s(\cdot,\cdot) : \mathbb{R}^d \times \mathbb{R}^d \to \mathbb{R}^+$ to measure the similarity between any two points in the dataset. Notably, custom similarity function can be used to suit the application needs.

This approach is especially attractive for hybrid system identification because the similarity function can be viewed as a surrogate for the latent region function $q(\cdot)$, such that $s(\zeta_i, \zeta_j)$ is large if $q(\zeta_i) = q(\zeta_j)$, and small otherwise. Hence, the problem is reformulated as finding such positive similarity function. A thorough background on spectral clustering can be found in (von Luxburg 2007).

3. Local Predictive Clustering

3.1. Similarity Function

If a group of points is generated by the same local model, then they will contain information about each other. Hence, if a set of points is used to construct a local model, which can then accurately predict the value at another point, it is likely that the prediction and at least some training points belong to the same region. This intuition can be encoded into a similarity function by first assuming that the global training dataset $\mathcal{D} = \{\zeta_i, y_i\}$ is sampled uniformly (i.e. at regular intervals) along its input dimensions and defining a standardized distance metric $d(\cdot,\cdot) : \mathbb{R}^d \times \mathbb{R}^d \to \mathbb{R}^+$ shown in Eq. (2), where $\zeta_i[n]$ refers to the nth component of the vector ζ_i.

$$d(\zeta_i, \zeta_j) = \sqrt{\sum_{n=1}^{d} \frac{(\zeta_i[n] - \zeta_j[n])^2}{V_\zeta[n]}} \tag{2}$$

where $V_\zeta[n] = Variance(\{\zeta_i[n]\}_{i=1,\dots,N})$

Then for each point ζ_i, a local dataset $\mathcal{D}_i^{(l)}$ containing its k_l neighbours can be used to learn a local model $h_i(\cdot) : \mathbb{R}^d \to \mathbb{R}^{d_x}$. The neighbourhood of the model is then defined as complimentary dataset $\mathcal{D}_i^{(p)}$ such that it includes any point among k_p neighbours of any point in $\mathcal{D}_i^{(l)}$ (but excluding any point in $\mathcal{D}_i^{(l)}$). The local model $h_i(\cdot)$ is then used to make predictions for all points in $\mathcal{D}_i^{(p)}$ giving rise to some prediction RMSE for each point in $\mathcal{D}_i^{(p)}$. Intuitively, if that error is small, then it is likely that point in $\mathcal{D}_i^{(p)}$ is from the same region as some points in $\mathcal{D}_i^{(l)}$.

However, since local models are constructed around each dataset in $\mathcal{D}$, each datapoint will be used for prediction multiple times. Given two close points ζ_i and ζ_j, let's assemble all predictive errors where ζ_i was used for model construction and ζ_j for prediction, and vice versa, into a single array of error $E_{i,j}$. Then, the similarity value between ζ_i and ζ_j is taken to be the inverse of the minimum error in $E_{i,j}$, as defined by Eq. (3).

$$s(\zeta_i, \zeta_j) = \frac{1}{\min E_{i,j}} \tag{3}$$

3.2. Graph Construction and Clustering

The similarity function defined in Eq. (3) can then be used to derive a square similarity matrix $S \in \mathbb{R}^{N \times N}$, such that $S_{i,j} = s(\zeta_i, \zeta_j)$. Following the methodology of spectral clustering, an adjacency matrix $A \in \mathbb{R}^{N \times N}$ is constructed according to Eq. (4).

$$A_{i,j} = \begin{cases} S_{i,j} & \text{if } \zeta_i \in k_n \text{ neighbours of } \zeta_j \text{ or } \zeta_j \in k_n \text{ neighbours of } \zeta_i \\ 0 & \text{otherwise} \end{cases} \quad (4)$$

Let's now assume that the number of regions is known to be k_c. Then spectral clustering can be performed in a straight forward manner on the graph described by A to yield k_c clusters containing training points within each region. This approach for hybrid system identification, which we term Local Predictive Clustering (LPC), is described in Algorithm 1.

Algorithm 1: Calculating the Similarity Between Points for Local Predictive Clustering

Input: dataset $\mathcal{D} = (\zeta_i, y_i)_{i=1,\dots,N}$ where $\zeta_i \in \mathbb{R}^d$ and $y_i \in \mathbb{R}$, $k_l \in \mathbb{Z}^+$, $k_p \in \mathbb{Z}^+$, trainable model $h(\zeta_i | \{\zeta_j, y_j\}) \mapsto \mathbb{R}$

Do:

1. Initiate an empty list E of size $N \times N$, where each element contains an empty array
2. For each ζ_i in $\mathcal{D}$:
    2.1. Find a set $\mathcal{D}_l$ of k_l nearest neighbours of ζ_i using Eq. (3).
    2.2. Fit a local model h_i on the dataset $\mathcal{D}_l$
    2.3. Find a set $\mathcal{D}_p$ of k_p nearest neighbours of all points in $\mathcal{D}_l$
    2.4. Eliminate any point from $\mathcal{D}_p$ which belongs to $\mathcal{D}_l$ (i.e. $\mathcal{D}_p \cap \mathcal{D}_l = \emptyset$)
    2.5. Use the local model h_i to make prediction $\bar{y}_j$ for all points ζ_j in $\mathcal{D}_p$
    2.6. For each point ζ_j in $\mathcal{D}_p$, calculate Root Mean Square Error e_j between y_j and $\bar{y}_j$
    2.7. For each point x_k in $\mathcal{D}_l$, append e_j to $E[k,j]$ and $E[j,k]$ where k and j are original indices of x_k and x_j respectively.
3. Initiate an array $S \in \mathbb{R}^{N \times N}$ with zeros
4. For $i = 1, \dots, N$:
    4.1. For $j = 1, \dots, N$:
        4.1.1. $S[i,j] = \min(E[i,j])$

Return: S

3.3. Local Models

One of the advantages of LPC, is that any regression model can be used to approximate local models $h_i(\cdot)$. One preferred model is a Gaussian Process (GP) (Rasmussen 2006) due to its performance in low data settings and ability to encode additional prior assumptions about the underlying function through the kernel function. As such, LPC can be used for PWA system identification by simply using GPs with a linear kernel. On the other hand, Square Exponential (SE) kernel can be used for PWS system identification.

3.4. Parameter Tuning

The prediction neighborhood size k_p can be fixed to $N - k_l - 1$, i.e. making predictions at all points not used for local model construction, although smaller values can slightly reduce the computational cost without affecting performance. On the other hand, optimal

k_l and k_n are extremely sensitive to the dataset and even similar values can produce widely different clusters. These can be optimized by Bayesian Optimization within the range of 5-50% of the dataset size N.

Hierarchical clustering (Wang 2007) can be used to identify the number of clusters k_c, by iteratively splitting the dataset into two clusters while the resulting cross validation error decreases. An alternative method, enabled through the spectral clustering methodology, is to use the eigengap heuristic (von Luxburg 2007). One drawback of this method is that it is highly sensitive to the derived graph and hence k_l and k_n values, although this can be mitigated by summing the spectra of graphs produced by multiple candidate parameters.

3.5. From Local Predictive Clustering to Model Predictive Control

Once LPC divides the training dataset $\mathcal{D}$ into k_c clusters, the latent function $q(\cdot)$ can be approximated by any supervised classification model, such as an SVM. One caveat of hybrid system identification, is that the target value (i.e. $\boldsymbol{y}_i \equiv \boldsymbol{x}_{k+1}^{(i)} \in \mathbb{R}^{d_x}$) is a multidimensional vector, while LPC is aimed at scalar valued functions. A simple solution is then to treat each dimension of the target value separately, effectively resulting in d_x datasets and identified clusterings, where the optimal clustering is chosen through cross validation.

Once the latent function $q(\cdot)$ is learned, local functions $f_i(\cdot)$ for $i = 1, \dots, k_c$ can be readily approximated by a regression model of choice, for example a GP, hence learning a full piecewise multi-model of the hybrid system. This can then be leveraged in MPC by identifying the currently active local model at each iteration using the learned $q(\cdot)$ and only optimizing the controls over the finite horizon using the identified local model. At the next iteration, the active local model is re-identified.

4. Case Study

To illustrate LPC and advantages of proper hybrid system identification for MPC, let's consider an ideal continuous-stirred tank reactor (CSRT) described in (Kazantis 2000) where concentration and temperature of the reactant, $\boldsymbol{x}_k = [C_A, T] \in \mathbb{R}^2$, is controlled by a dilution rate and inlet temperature, $\boldsymbol{u}_k = [u_1, u_2] \in \mathbb{R}^2$. Hybrid behavior is introduced by modifying the dynamics described in [12], with the piecewise dynamics of the system described in Eq. (5). The system is then controlled from $\boldsymbol{x}_0 = [0.116, 368.5]$ to $\boldsymbol{x}_{target} = [0.666, 308.5]$.

$$\frac{dC_A}{dt} = \begin{cases} \left(\frac{F}{V} + u_1\right)(C_{A,in} - C_A) - 2k(T)C_A^2 & \text{if } C_A < 0.6 \\ \left(\frac{F}{V} + u_1\right)(C_{A,in} - C_A) - 2k(T)C_A & \text{otherwise} \end{cases}$$

$$\frac{dT}{dt} = \begin{cases} \left(\frac{F}{V} + u_1\right)(T_{in} + u_2 - T) + 2\frac{(-\Delta H)_R}{\rho c_p}k(T)C_A^2 - \frac{UA}{V\rho c_p}(T - T_j) & \text{if } C_A < 0.6 \\ \left(\frac{F}{V} + u_1\right)(T_{in} + u_2 - T) + 2\frac{(-\Delta H)_R}{\rho c_p}k(T)C_A - \frac{UA}{V\rho c_p}(T - T_j) & \text{otherwise} \end{cases}$$

$$(5)$$

The discretized system model is uniformly sampled to yield 1120 training points used to learn the location of operating regions through LPC. Another training dataset of 160 points is sampled randomly and used to train a single, global GP. MPC is then run using the global GP as a system model, failing to control the plant to the target states. Then, the training dataset is split into regions identified through LPC and local GPs are built for each region. A multi-model MPC is then run using local GPs by identifying an active local GP and using its mean prediction at each iteration, successfully controlling the plant.

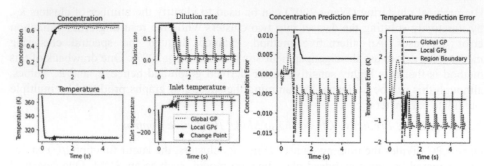

Figure 1: System behavior and model mismatch for MPC with a global and local models.

Figure 1 shows the system behavior as well as the model mismatch for both MPC with a single global GP and two local GPs. While local GPs exhibit large model mismatch close to the boundary ($C_A \approx 0.6$), it quickly decays. On the other hand, a single global GP results in significant model mismatch throughout, illustrating benefits of LPC.

5. Conclusions

Accurate hybrid system identification can significantly improve data driven control, and provide insight into the system itself. For MPC with GP system model, local models can also improve the speed of optimization. Although LPC is shown to be effective for general PWS system identification, it is limited to uniformly sampled datasets and is sensitive to noise and hyperparameters. Importantly, though, it illustrates the potential of spectral clustering in hybrid system identification through the proposed similarity function.

References

F. Lauer, G. Bloch, 2019, Hybrid System Identification: Theory and Algorithms for Learning Switching Methods

H. Ohlsson, L. Ljung, 2013, Identification of switched linear regression models using sum-of-norms regularization, Automatica 49, 1045-1050

B.J. Park, Y. Kim, J.M. Lee, 2021, Design of switching multilinear model predictive control using gap metric, Computers and Chemical Engineering 150

F. Lauer, G. Bloch, 2014, Piecewise smooth system identification in reproducing kernel Hilbert space, IEEE Conference on Decision and Control 53

G. Lee, Z. Marihno, A.M. Johnson, G.J. Gordon, S.S. Srinivasa, M.T. Mason, 2017, Unsupervised Learning for Nonlinear PieceWise Smooth Hybrid Systems

U. von Luxburg, 2007, A Tutorial on Spectral Clustering, Statistics and Computing 17

C.E. Rasmusen, K.I. Williams, Gaussian Processes for Machine Learning, the MIT Press, 2006

X. Wang, V.L. Syrmos, 2007, Identification of Nonlinear Dynamical System using Hierarchical Clustering Analysis and Local Linear Models, European Control Conference

N. Kazantzis, C. Kravaris, 2000, Synthesis of state feedback regulators for nonlinear processes, Chemical Engineering Science 55

Proceedings of the 14th International Symposium on Process Systems Engineering – PSE 2021+
June 19-23, 2022, Kyoto, Japan © 2022 Elsevier B.V. All rights reserved.
http://dx.doi.org/10.1016/B978-0-323-85159-6.50275-X

Distillation Column Temperature Prediction Based on Machine-Learning Model Using Wavelet Transform

Hyukwon Kwon[a], Yeongryeol Choi[a,b], Hyundo Park[a,b], Kwang Cheol Oh[c], Hyungtae Cho[a], Il Moon[b], Junghwan Kim[a*]

[a]*Green Materials and Processes R&D Group, Korea Institute of Industrial Technology, Ulsan 44413, Republic of Korea*
[b]*Department of Chemical and Biomolecular Engineering, Yonsei University, Seoul 03722, Republic of Korea*
[c]*Department of Biosystems Engineering, Agriculture and Life Sciences Research Institute, Kangwon National University, Gangwon-do 24341, Republic of Korea*
kjh31@kitech.re.kr

Abstract

This study presents a machine-learning-based prediction model for distillation process operation data using wavelet transform. The process operation data collected from a distillation column contain noise due to sensor errors. Developing a machine-learning model using noisy data reduces the accuracy of the model; therefore, the data should be denoised. Denoising was achieved using wavelet transform, and a long short-term memory (LSTM) machine-learning model was developed. Wavelet transforms generally decompose data into high- and low-frequency components using wavelet functions with various frequencies. The high-frequency components are the details comprising noisy data, and the low-frequency components correspond to the approximations of the original data. The approximations were used to develop the LSTM model. Depending on the type of wavelet function used for decomposition, the denoised values varied and affected the model accuracy. Case studies were conducted using various wavelet functions to develop models with optimum prediction performances. By applying the optimal wavelet transform to the LSTM model, the prediction performance improved by 10%.

Keywords: distillation column temperature, machine learning, wavelet transform, long short-term memory.

1. Introduction

The distillation process is a representative process for improving the purity of a product in the chemical industry. The main types of data in the distillation process are related to temperature, pressure, flow rate, and liquid level, which are measured in real time. In general, additional devices are required to measure the purity of the distillation process as real-time measurements are difficult. In addition, time delays are inevitable, which pose challenges to controlling real-time purity data (das Neves et al., 2018). Product purity is often controlled indirectly by estimating other related measurable data. Product purity is mainly controlled by temperature in the distillation process; therefore, accurate data collection and temperature prediction methods are required.

The distillation process data have time-series characteristics that accumulate over time. The artificial neural network (ANN)-based prediction model performs well in mapping and utilizing the input and output values of a complex nonlinear relationship (Joo et al., 2021). Additionally, predictions and control are easy in real time because there are slight time delays of the order of seconds to output values using the developed model (Himmelblau, 2008). However, time-series data have performance limitations with predictions that use traditional ANN-based models. In time-series data, the past information affects the future because the traditional ANNs cannot reflect this. Recurrent neural network (RNN)-based long short-term memory (LSTM) is therefore designed to remember and convey past information in the future, making it suitable for analyzing and predicting time-series data.

The performances of ANNs or linear and nonlinear regression models are affected by the data quality and require appropriate data preprocessing methods. Time-series data often include abnormal characteristics, such as tendency and periodicity; these characteristics are analyzed in several ways such as the autocorrelation and spectrum are mainly used. Spectral analysis is generally used to determine the periodicity of time-series data and is based on Fourier transform of the time-domain data to the frequency domain. Although the frequencies present in the signal can be analyzed after the Fourier transform, there is a disadvantage that the existence of each frequency in time is unknown because the time information is removed. Candidate (2019) reported that the wavelet transform (WT) is an approach to solving this problem. Khandelwal et al. (2015) conducted a study using ANN models and discrete wavelet transform (DWT) to predict time-series data. Kwon et al. (2021) reported that data from the distillation process could be predicted using the LSTM algorithm. The statistical technique called autoregressive integrated moving average (ARIMA) can be combined with ANNs to predict time-series data. The problem with ANNs as well as the linear and nonlinear regression models is that the predictive performances decrease for abnormal data that have not been appropriately preprocessed. The DWT can help analyze abnormal time-series data and decompose them into normal and noisy data.

In the present study, an LSTM-based prediction model with WT was developed to predict the temperature of the distillation process accurately. To improve the noise removal performance, a study was conducted on the basis function of the WT, and the most appropriate wavelet basis function was selected by measuring noise removal performance. The denoised data were then used as the input data to the LSTM model to develop the WT-LSTM model, and performance comparisons were made with the general LSTM model.

2.   Methods

2.1.   Description of the distillation process

The object of this study is a commercial distillation process that produces normal butane from mixed butane. The distillation column consists of 78 stages, and the raw material, i.e., mixed butane, is designed to flow into 35 stages. Mixed butane is composed of normal butane, isobutane, and pentane and produces 99% pure normal butane in 64 stages. Figure 1 shows a schematic of the distillation process and the sensor positions for data collected during the process.

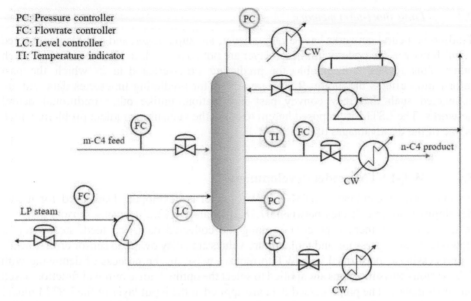

PC: Pressure controller
FC: Flowrate controller
LC: Level controller
TI: Temperature indicator

2.2.   Figure 1. Mixed butane splitter process diagram. Wavelet transforms and wavelet basis functions

WTs are divided into discrete and continuous types. The DWT consists of highpass and lowpass filters that separate the original signal into high- and low-frequency bands, respectively. In general, the approximation data from the lowpass filters determine the signal characteristics, and the detail data passed through the highpass filters are treated as noise. In decomposing the original signal into approximation and detail data, particular functions are used to determine the forms of the approximation data and are called as the wavelet basis functions.

The WT is defined as an extension of the wavelet basis function, which must satisfy the following two conditions:

$$\int_{-\infty}^{\infty}|\psi(t)|^2 dt < \infty \tag{1}$$

$$c_\psi = 2\pi \int_{-\infty}^{\infty}\frac{|\Psi(\omega)|^2}{|\omega|}d\omega < \infty \tag{2}$$

In Eq. (1) and (2), ψ is the wavelet basis function, and Ψ is the Fourier transform of ψ. Eq. (1) indicates that the function ψ has a finite value, and Eq. (2) indicates that $\Psi(0) = 0$ when Ψ is smoothed as an acceptance condition. Because the wavelet basis function is a simple function that needs to satisfy only the above two conditions, the types of wavelet basis functions used are so diverse that it is necessary to select suitable functions based on the data characteristics. The data collected during this process have different tendencies depending on the variable. Therefore, it is necessary to select the most suitable wavelet basis function for each variable. In this study, five wavelet basis functions are used: biorthogonal (bior), coiflet (coif), Daubechies (db), reverse biorthogonal (rbior), and symlet (sym). These five functions are the most commonly used in discrete wavelet transformations and have been useful in other signal processing studies. The degree of noise removal was evaluated for each wavelet function to select the most appropriate function for each data variable. In the process of selecting the basis function, all decomposition levels were set to five.

2.3. Long short-term memory

Traditional neural networks have an input layer, an output layer, and a hidden connected layer; however, the nodes within each layer are not connected, so they do not affect each other. This makes it unsuitable for predicting time-series data in which the past information affects the future. RNNs are ANNs for predicting timeseries data and are connected such that they convey past information, unlike other traditional neural networks. The LSTM is proposed herein to solve the vanishing gradient problem, which is one of the disadvantages of the RNN.

3. WT-LSTM model development

In this study, a combined LSTM and WT model is developed. Data used for model development were collected between 07.18.2019 and 07.23.2019 using sensors installed in the actual commercial process. Among the collected data, the feed, steam, reflux flowrate, bottom pressure, and temperature values are noisy owing to sensor errors; hence, preprocessing is performed with WT to remove noise. In the process of denoising with WT, various basis functions are applied to select the optimal noise removal function based on performance. The preprocessed data are applied at the input layer of the LSTM model for learning, and the temperature is predicted at the output layer. The number of hidden nodes in the LSTM was set to 20 with 128 batch sizes. The other parameters are summarized in Table 1.

Table 1. Parameters of LSTM model.

Parameter	Values
Optimizer	Adam
Learning rate	0.001
Activation function	ReLU
Training/test dataset ratio	70/30
Loss function	MSE
Wavelet basis functions	bior, coif, db, rbior, sym
Wavelet decomposition levels	5

The root mean-squared error (RMSE) is used to evaluate model performance and is calculated as follows:

$$RMSE = \sqrt{\frac{1}{N}\sum_{i=1}^{N}(x_i - y_i)^2} \tag{3}$$

Here, x_i and y_i are the ith values of the actual and predicted datasets respectively, and N is the total number of datasets. The RMSE is a measure of the error between the raw and predicted data, indicating that the smaller the error value, the more accurate is the performance of the prediction model.

4. Results and discussion

In this study, an LSTM model combined with a WT was developed to predict the temperature of the distillation process. Various basis functions were applied and evaluated to improve the noise removal performance of the WT. Performance comparisons with basic LSTM models were performed to verify the performance of the proposed model.

4.1. Wavelet basis functions

The selection of the wavelet basis function is important for removing noisy data using the DWT. In general, when denoising a signal using WT, greater similarity of the shape of the wavelet function to the signal indicates better noise removal. However, it is difficult to create useful wavelet functions, hence, an appropriate wavelet function is selected and used from among the existing wavelets. Daubechies, Symlets, Coiflets, biorthogonal, and other wavelets are included for decomposing signals and denoising with DWT. In this study, five wavelet basis functions were applied to each distillation process dataset for denoising and performance evaluation for optimal function selection. Table 2 shows the denoising results and RMSEs for bior, coif, db, rbior, and sym basis functions.

Table 2. Results of denoising evaluations for each variable with RMSE.

	Feed flowrate	Steam flowrate	Reflux flowrate	Bottom pressure	Temperature
bior	652.941	264.516	1828.90	0.089	0.390
coif	652.412	264.009	1823.67	0.089	0.390
db	652.457	264.859	1825.079	0.089	0.390
rbior	685.477	274.663	1914.24	0.092	0.403
sym	652.449	264.368	1824.54	0.089	0.390

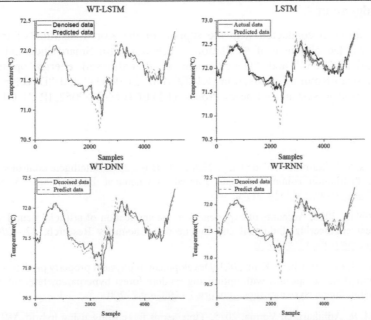

Figure 2. The comparisons between denoised and predicted temperature of WT-LSTM, LSTM, WT-DNN and WT-RNN model.

In this study, we confirmed that the coif function was most suitable for denoising data. However, the bior and db functions showed no significant differences compared to coif. Thus, we found that other functions were also suitable for denoising.

4.2. Performance evaluation

After removing data noise using the selected coif function, learning was conducted with the LSTM model to develop the WT-LSTM model. The performance of the proposed model was compared with the basic LSTM model, WT- DNN, and WT-RNN. The learning conditions of WT-LSTM, LSTM model, and WT-RNN were set the same, and the predictive performance was evaluated with RMSE. Figure 3 shows the prediction of test set data using WT-LSTM, LSTM, WT-DNN and WT-RNN model. The RMSE of each model was 0.0966, 0.144, 0.0976 and 0.117. we confirmed that a more accurate prediction could be performed by reducing RMSE by 33# than the basic-LSTM model by removing noise with WT.

5.   Conclusion

A machine-learning model combined with WT was developed herein to predict distillation process temperatures. The process data were denoised effectively using WT, and we confirmed that the coif function was most appropriate for optimal denoising performance. It was shown that the WT-LSTM model using the coif function could achieve more accurate predictions than other basis functions, with a 33% reduction in RMSE than the basic LSTM model. When using WT, the LSTM model achieved more accurate predictions. The temperature is thus calculated to be lower by approximately 0.4 °C. Hence, the WT approach needs to be improved, and future studies will focus on models that can predict the process more accurately through supplemental corrections.

Acknowledgement

This study has been conducted with the support of the Korea Institute of Industrial Technology as "Development of AI Platform Technology for Smart Chemical Process (KITECH JH-21-0005)" and "Development of digital-based energy optimization platform for manufacturing innovation (KITECH IZ-21-0063)" and "Development of Global Optimization System for Energy Process (KITECH IZ-21-0052, IR-21-0029, EM-21-0022)".

References

T.G. das Neves, W.B. Ramos, G.W. de Farias Neto, R.P. Brito, 2018, Intelligent control system for extractive distillation columns. Korean Journal of Chemical Engineering, 35, 826–834. https://doi.org/10.1007/s11814-017-0346-0

D.M. Himmelblau, 2008, Accounts of experiences in the application of artificial neural networks in chemical engineering. Industrial and Engineering Chemistry Research, 47, 5782–5796. https://doi.org/10.1021/ie800076s

C. Joo, H. Park, J. Lim, H. Cho, J. Kim, 2021, Development of physical property prediction models for polypropylene composites with optimizing random forest hyperparameters. International Journal of Intelligent Systems, https://doi.org/10.1002/int.22700

I. Khandelwal, R. Adhikari, G. Verma, 2015, Time series forecasting using hybrid ARIMA and ANN models based on DWT decomposition, Procedia Computer Science, 48, 173–179. https://doi.org/10.1016/j.procs.2015.04.167

H. Kwon, K.C. Oh, Y. Choi, Y.G. Chung, J. Kim, 2021, Development and application of machine learning-based prediction model for distillation column, International Journal of Intelligent Systems, 36, 1970–1997. https://doi.org/10.1002/int.22368

H. Lee, J. Lee, C. Yoo, 2019, Selecting a mother wavelet for univariate wavelet analysis of time series data, Journal of Korea Water Resources Association, 52, 575–587. https://doi.org/10.3741/JKWRA.2019.52.8.575

Proceedings of the 14th International Symposium on Process Systems Engineering – PSE 2021+
June 19-23, 2022, Kyoto, Japan © 2022 Elsevier B.V. All rights reserved.
http://dx.doi.org/10.1016/B978-0-323-85159-6.50276-1

Moisture Estimation in Woodchips Using IIoT Wi-Fi and Machine Learning Techniques

Kerul Suthar[a], Q. Peter He[a*]

[a]*Department of Chemical Engineering, Auburn University, Auburn, AL 36849, USA*

qhe@auburn.edu

Abstract

For the pulping process in a pulp & paper plant that uses woodchips as raw material, the moisture content (MC) of the woodchips is a major process disturbance that affects product quality and consumption of energy, water, and chemicals. Existing woodchip MC sensing technologies have not been widely adopted by the industry due to unreliable performance and/or high maintenance requirements that can hardly be met in a manufacturing environment. To address these limitations, we propose a non-destructive, economic, and robust woodchip MC sensing approach utilizing channel state information (CSI) from industrial Internet-of-Things (IIoT) based Wi-Fi. While these IIoT devices are small, low-cost, and rugged to stand for harsh environment, they do have their limitations such as the raw CSI data are often very noisy and sensitive to woodchip packing. Thus, direct application of machine learning (ML) algorithms leads to poor performance. To address this, statistics pattern analysis (SPA) is utilized to extract physically and statistically meaningful features from the raw CSI data, which are sensitive to woodchip MC but not to packing. The SPA features are then used for developing multiclass classification models as well as regression models using various linear and nonlinear ML techniques to provide potential solutions to woodchip MC estimation for the pulp and paper industry.

Keywords: systems engineering, machine learning, feature engineering, channel state information, IIoT sensors.

1. Introduction

The US pulp and paper industry ranks the third in energy consumption among US industries. The pulping process, which converts woodchips into pulp by displacing lignin from cellulose fibers, is one of the most energy intensive processes and has been identified as a major opportunity to improve energy productivity and efficiency of the industry (Brueske et al., 2015). Currently, vast majority of the US pulp is produced by chemical pulping processes and most of them utilize continuous Kamyr digesters. For Kamyr digesters, the incoming woodchip moisture content (MC) is a major disturbance that affects the cooking performance.

Currently, the woodchip MC is not measured in real-time due to the lack of affordable, reliable, and easy-to-maintain sensors. As a result, the performance of existing control solutions is often unsatisfactory and process engineers often overcook the woodchips to ensure pulp quality, which results in significant loss of pulp yield, overuse of heat/energy and chemicals. Chemical overuse also adds burdens to the downstream processes, such as washing and evaporation, and results in increased energy and chemical usages for downstream processes as well. To address this need, this work proposes a non-destructive, economic, and robust approach based on 5 GHz IIoT short-range Wi-Fi and use channel state information (CSI) to estimate MC in woodchips. Both classification and regression techniques are studied for MC estimation. For classification, we investigate linear discriminant analysis (LDA), support vector machine (SVM), artificial neural network (ANN), bagging with LDA, and ensemble boosting XGBoost. For regression, we study ANN, k-nearest neighbor regression (KNNR), Gaussian process regression (GPR), and support vector regression (SVR) with radial basis function (RBF) kernel.

The remainder of this work is organized as follows: Section 2 describes the experimental setup and software tools used   in this study, as well as the features proposed and the modeling techniques utilized in this work. Section 3 presents results and discussions of this work, and Section 4 draws conclusions.

2. Data collection and feature engineering

2.1. Channel state information for moisture estimation

Using Wi-Fi cards such as IWL5300, it is convenient to collect CSI measurements that record the channel variation during propagation of wireless signals. After being transmitted from a source, the wireless signal is expected to experience impairments caused by obstacles before the signal reaches the receiver. CSI can reflect indoor channel characteristics such as multipath effect, shadowing, fading, and delay. In this work, we collect CSI using CSItool, which is built on IWL5300 NIC using a custom modified firmware and open-source Linux wireless drivers. The channel response of the ith subcarrier can be represented as:

$$CSI_i = |CSI_i| \exp\{\angle CSI_i\} \tag{1}$$

where $|CSI_i|$ is the amplitude and $\angle CSI_i$ is the phase response of the ith subcarrier.

2.2. Data description

In this work, data are collected for 20 different MC classes or levels ranging from 53.39% to 11.81% on the wet basis (see Eqn (2)). A single antenna is used on the transmitter side which is configured in injection mode to send CSI and 3 antennas are used on the receiving side to take advantage of diversity. Woodchips are places in an airtight container between the transmitter and receiver to collect data. 10,000 packets are sent from the transmitter to the receivers for each sample collection. Total mass (m_T) is measured during each experiment and oven drying method was performed after all experiments were conducted to determine the oven dry weight (m_D). m_T and m_D are then used to determine the mass of water (m_W) and MC as the following.

$$MC = \frac{m_W}{m_T} \times 100\% = \frac{m_W}{m_W + m_D} \times 100\% \tag{2}$$

The 20 different MC levels are plotted in Figure 1(a), which shows that MC levels are narrowly separated at the high MC region and even more so at the low MC region. The minimum difference between MC levels is 0.05%, which is more than sufficient for pulping process optimization and control.

2.3. Methodology and feature engineering

To address the shortcoming of raw CSI features that lead to poor classification and prediction performance, in this work, statistics pattern analysis (SPA) is utilized to generate more robust and predictive features. In SPA, the statistics of the process variables, instead of process variables themselves, are used for modeling. This is based on the hypothesis that these statistics are sufficient and even better in capturing process characteristics than original process variables. This hypothesis has been supported in various applications, including fault detection (He et al., 2019; He & Wang, 2011, 2018; Wang & He, 2010), fault diagnosis (He & Wang, 2018), and virtual metrology or soft sensor (Shah et al., 2019, 2020; Suthar et al., 2019). SPA is selected in this work to extract robust and predictive features from raw CSI data. It is worth noting that SPA does not require preprocessing of the CSI data (e.g., outlier detection and handling, noise removal/reduction) that has been required in previous studies (Hu et al., 2019; Yang et al., 2018). A schematic for SPA based feature engineering is shown in Figure 1 (b). After a deeper exploration of candidate features and statistics, mean difference of consecutive subcarrier in CSI amplitude are chosen which leads to 87 features considering all 3 antennas on the receiving side.

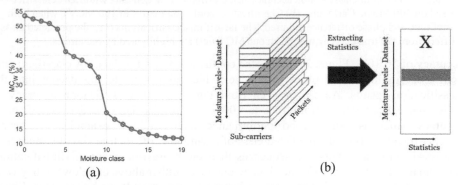

(a)   (b)

Figure 1 (a) 20 different moisture levels tested in this work; (b) SPA based feature engineering for MC estimation

3. Results and discussion

In this work, we conduct investigations from three perspectives: (1) comparing raw CSI data vs engineered features; (2) comparing the performance of different classification approaches; and (3) comparing the performance of different regression approaches. For each model, 9 samples are randomly selected as training samples from 10 shuffled samples at the same MC level for each of the 20 MC levels, which results in 180 training samples. The remaining shuffled sample for each of the MC levels is used for testing. In this work we use Monte Carlo validation and testing (MCVT) procedure 100 times for performance comparison. To assess various classification approaches, the mean and standard deviation of classification accuracy of the 100 MCVT simulations are reported. For regression approaches, the mean and standard deviation of root mean square error (RMSE) for the 100 MCVT simulations are reported.

First, raw CSI data are used for MC level classification. The results are similar across different classification techniques. Due to limited space, only results from LDA are discussed here. Figure 2 (a) shows the overall classification accuracy of all classes when the raw CSI data were used. The comparison indicates that LDA classifier using both amplitude and phase difference performs the best with 86.15% classification accuracy, followed by LDA classifier using phase difference with 83.85% classification accuracy, while the LDA classifier using amplitude alone results in the lowest classification

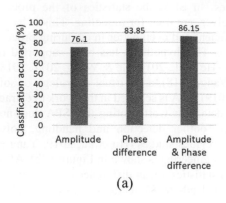

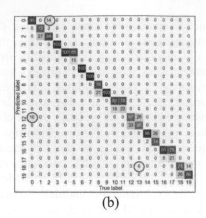

(a)   (b)

Figure 2 (a) Overall classification accuracy using different raw CSI data with LDA classifier based on 100 Monte Carlo runs. (b) Classification confusion matrix of 100 MCVT when both amplitude and phase difference are used. The far-off misclassifications (*i.e.*, the predicted class differs from the true class by more than one MC level) are highlighted by red circles.

accuracy of 76.10%. Figure 2 (b) plots the confusion matrix for the LDA classifier using both CSI amplitude and phase difference, which allows us to dig deeper into the classification results. As can be seen from Figure 2 (b), classification accuracy of individual classes ranges from 15% to 100%. It can also be seen that classification accuracy alone is not a good performance indicator. For example, the far-off misclassifications (i.e., the predicted class of a sample is off its true class by more than one level) will have worse consequences than the nearest-neighbor misclassifications (i.e., the predicted class is off true class by one level, either above or below) if they were used to control the white liquor usage or digester temperature. It can be seen from Figure 2 (b) that the classification results using raw CSI data are poor as there are samples misclassified far off their true classes. There are totally 478 misclassified samples, of which 30 are far-off misclassifications (highlighted by red circles in Figure 2 (b)). Also, the overall classification accuracy is not satisfactory.

Next the 87 rationally engineered features (i.e., the mean difference of consecutive subcarrier in CSI amplitude) are used for MC level classification and the results are summarized in Table 1. The classification accuracies shown in Table 1 indicate that all methods perform well with higher than 95% classification accuracy. The significantly improved performance compared to that of the raw CSI data demonstrates that the engineered features are more informative and characterize the MC in woodchips far better than the raw CSI data. Among all classification methods studied in this work, the bagging LDA performs the best with 98.75% average classification accuracy. The standard deviation of its classification accuracy is the lowest of 2.29%, indicating the

Table 1 Classification accuracy using engineered features

Method	Classification Accuracy	
	Mean	Std. dev.
SVM	95.50	3.79
ANN	95.85	4.15
XGBoost	96.40	3.70
LDA	97.55	2.89
Bagging (LDA)	98.75	2.29

bagging LDA is also the most robust or consistent classifier among all methods studied in this work.

Finally, we study different regression methods for MC estimation. When raw CSI data are used, all regression methods perform poorly, similar to the classification results when the raw CSI data are used. Due to limited space, they are not shown here. When the same 87 engineered features are used for regression-based MC estimation, a well-tuned ANN with two hidden layers outperforms other regression-based approaches as shown in Table 2. KNNR performs comparable to ANN while GPR and SVR with RBF kernel have relatively higher average RMSE's for 100 MCVT simulations.

Table 2 Regression for MC estimation using engineered features

Method	RMSE	
	Mean	Std. dev.
ANN	0.51	0.3921
KNNR	0.6573	0.5055
GPR	1.9223	0.5714
SVR(RBF)	2.0179	0.523

Figure 3 shows the measured vs predicted MC values for ANN and SVR(RBF). It can be seen from Figure 3(a) that the ANN predicted MC values agree very well with the actual or measured MC values. In comparison, while SVR captures the MC trend, its predictions have much higher standard deviation compared to ANN. It is worth noting for all the above-mentioned results, the models and their hyperparameters were tuned using random search followed by Bayesian optimization (Bergstra & Bengio, 2012).

4. Conclusions

In this work, we investigate the potential of an IIoT short-range Wi-Fi based woodchip MC sensing technology to overcome some limitations of the existing technologies. The proposed technology takes the advantages of IIoT devices (e.g., toughness, connectivity,

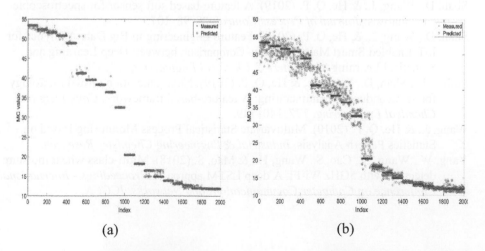

(a)   (b)

Figure 3 Measured vs predicted MC by (a) ANN and (b) SVR(RBF)

low-cost, small-size, etc.), while overcoming their shortcomings (e.g., the machine learning challenges of messy big data) through SPA-based feature engineering. We investigate the use various classification and regression approaches for the estimation of 20 different moisture levels. We demonstrate that with SPA-based features, all classification approaches studied in this work can successfully classify 20 different MC levels, some of which are separated by small margins. We also investigate the use of different regression approaches for continuous MC estimation. While SVR and GPR capture the trend of measured MC values but with relatively high RMSE's, methods including ANN and KNNR predict the moisture levels accurately. The relationship between the CSI and woodchip MC is very complex, which requires further work to get a better understanding of this relationship for further improvement of this work.

References

Bergstra, J., & Bengio, Y. (2012). Random search for hyper-parameter optimization. *Journal of Machine Learning Research*.

Brueske, S., Kramer, C., & Fisher, A. (2015). *Bandwidth Study on Energy Use and Potential Energy Saving Opportunities in US Pulp and Paper Manufacturing*. Energetics.

He, Q. P. P., & Wang, J. (2018). Statistical process monitoring as a big data analytics tool for smart manufacturing. *Journal of Process Control, 67*, 35–43.

He, Q. P., & Wang, J. (2011). Statistics pattern analysis: A new process monitoring framework and its application to semiconductor batch processes. *AIChE Journal, 57*(1), 107–121.

He, Q. P., & Wang, J. (2018). Statistical process monitoring as a big data analytics tool for smart manufacturing. *Journal of Process Control, 67*, 35–43.

He, Q. P., Wang, J., & Shah, D. (2019). Feature Space Monitoring for Smart Manufacturing via Statistics Pattern Analysis. *Computers & Chemical Engineering, 126*, 321–331.

Hu, P., Yang, W., Wang, X., & Mao, S. (2019). MiFi: Device-free wheat mildew detection using off-the-shelf wifi devices. *2019 IEEE Global Communications Conference, GLOBECOM 2019 - Proceedings*.

Shah, D., Wang, J., & He, Q. P. (2019). A feature-based soft sensor for spectroscopic data analysis. *Journal of Process Control, 78*, 98–107.

Shah, D., Wang, J., & He, Q. P. (2020). Feature Engineering in Big Data Analytics for IoT-Enabled Smart Manufacturing–Comparison between Deep Learning and Statistical Learning. *Computers & Chemical Engineering*, 106970.

Suthar, K., Shah, D., Wang, J., & He, Q. P. (2019). Next-generation virtual metrology for semiconductor manufacturing: A feature-based framework. *Computers and Chemical Engineering, 127*, 140–149.

Wang, J., & He, Q. P. (2010). Multivariate Statistical Process Monitoring Based on Statistics Pattern Analysis. *Industrial & Engineering Chemistry Research*.

Yang, W., Wang, X., Cao, S., Wang, H., & Mao, S. (2018). Multi-class wheat moisture detection with 5GHz Wi-Fi: A deep LSTM approach. *Proceedings - International Conference on Computer Communications and Networks, ICCCN*.

Proceedings of the 14th International Symposium on Process Systems Engineering – PSE 2021+
June 19-23, 2022, Kyoto, Japan © 2022 Elsevier B.V. All rights reserved.
http://dx.doi.org/10.1016/B978-0-323-85159-6.50277-3

Transfer Learning for Quality Prediction in a Chemical Toner Manufacturing Process

Shohta KOBAYASHI[a*], Masashi MIYAKAWA[a], Susumu TAKEMASA[a],

Naoki TAKAHASHI[a], Yoshio WATANABE[a], Toshiaki SATOH[b],

Manabu KANO[c]

[a] Ricoh Co., Ltd., Shizuoka 410-0004, JAPAN
[b] MEIWA e-TEC Co., LTD., Aichi 471-0047, JAPAN
[c] Kyoto University, Kyoto 606-8501, JAPAN
shohta.kobayashi@jp.rioch.com

Abstract

In chemical toner manufacturing plants, equipment and raw materials are frequently changed to improve the toner quality and productivity. These changes require reconstruction of the prediction model, which plays a key role in the automatic quality control system, and cause downtime. To reduce the downtime, we developed an efficient modelling method based on transfer learning, which can build an accurate model from small-size data obtained just after the changes. By extending Frustratingly Easy Domain Adaptation, a new heterogeneous domain adaptation technique was proposed. In addition, gaussian process regression (GPR) was adopted with bagging to improve the robustness and accuracy of the model. The proposed method showed superior performance to partial least squares regression, random forest, and GPR. Finally, the proposed prediction method was applied to a toner mass-production plant; the prediction accuracy target was satisfied for all toner qualities. As a result, a 75% reduction in plant control person-hours of the toner quality manager was achieved.

Keywords: Quality prediction, Transfer learning, Frustratingly Easy Domain Adaptation, Gaussian Process Regression, Chemical Toner Production.

1. Introduction

In recent years, automatic quality control has been used for stabilizing chemical toner quality and determining efficient operating conditions in toner plants (Khorami et al., 2017; Takahashi et al., 2020). In these plants, equipment and raw materials are often changed. As shown in Fig. 1, such a change alters the dimensions and distributions of input variables, makes it difficult to use existing prediction models, and makes it necessary to reconstruct the models. During the re-accumulation of training data, automatic quality control is forced to stop functioning, and manual quality control is required. This manual control requires many person-hours, therefore, it has been desired to construct an accurate prediction model using as short-term data as possible.

A promising approach to solve this problem is to use transfer learning. We expanded Frustratingly Easy Domain Adaptation (FEDA), which is a simple homogeneous domain adaptation method, to cope with a heterogeneous domain adaptation (HDA) problem without complex parameter tuning. The proposed method is referred to as Frustratingly Easy Heterogeneous Domain Adaptation (FEHDA). Moreover, we utilized a combination of Gaussian Process Regression (GPR) and bagging, a type of ensemble learning, for predicting the toner quality.

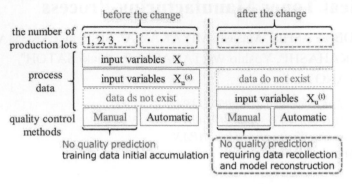

Fig. 1. The influence of changes in equipment and raw materials.

2. Chemical toner manufacturing process

The chemical toner manufacturing process treats one lot per day, and it takes seven days from raw material to final product, as shown in Fig. 2. The IoT-based manufacturing process data collection system handles several thousand variables (items), including raw material properties, equipment operation conditions, and toner quality, and stores data of several hundred lots or more.

Before the introduction of automatic quality control, toner quality was controlled manually by the toner quality manager, who determined the optimum operating condition for lot N based on the quality measurements of the lots whose manufacturing was finished (lot N-2 and older). The manual quality control consumes many person-hours and increases the risk of out-of-specification due to variations in toner quality.

The automatic quality control system currently in operation consists of a quality prediction module that predicts future toner quality and an operating condition optimization module that determines the operation amount (Takahashi et al., 2020).

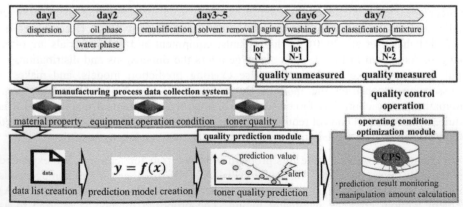

Fig. 2. Chemical toner manufacturing process and automatic quality control system (Takahashi et al., 2020).

This automatic quality control is feed-forward inferential control, which simulates the manual operation, as shown in Fig. 2. However, as mentioned in the previous section, the change of equipment and raw materials requires the data re-accumulation of at least 40 lots (days) so that the model reconstructed satisfies the accuracy target.

3. Prediction using transfer learning

3.1. Frustratingly Easy Heterogeneous Domain Adaptation

FEDA is a method of transfer learning that is easy to implement with simple feature space expansion (Daumé III, 2007). Assuming that the input variables $x^{(s)}$ in the source domain (hereinafter referred to as "SD") and $x^{(t)}$ in the target domain (hereinafter referred to as "TD") are K-dimensional, the input variables in both domains are expanded into $3K$-dimensional features as follows:

$$D_s = \left(x^{(s)}, x^{(s)}, 0 \right) \tag{1}$$
$$D_t = \left(x^{(t)}, 0 \ \ , x^{(t)}\right) \tag{2}$$

The expanded feature space consists of a space with features common to both domains, a space with features unique to SD, and a space with features unique to TD. Also, $0 = (0, 0, 0, \dots, 0) \in \mathcal{R}^K$ in Eqs. (1) and (2) is the zero vector.

In the manufacturing process, due to changes in equipment and raw materials, the configuration of the manufacturing equipment differs in both domains, which makes the location and number of installed sensors also different. Hence, when heterogeneous domain adaptation is required, FEDA cannot be used as it is. To make FEDA applicable to heterogeneous domain adaptation (HDA), heterogeneous feature augmentation (HFA) (Duan et al., 2012) was proposed. This method needs much computational time because to solve an optimization problem for finding the optimal latent space.

We propose frustratingly easy heterogeneous domain adaption (FEHDA), which is a direct and simple extension of FEDA and applicable to HDA. The proposed method does not require solving the optimization problem. We divide input variables $x^{(s)} \in \mathcal{R}^P$ in SD into $x_c^{(s)} \in \mathcal{R}^K$ that is common to SD and TD and $x_u^{(s)} \in \mathcal{R}^{P-K}$ that is unique to SD. Similarly, input variables $x^{(t)} \in \mathcal{R}^Q$ in TD is divided into the common input variables $x_c^{(t)} \in \mathcal{R}^K$ and the unique input variables $x_u^{(t)} \in \mathcal{R}^{Q-K}$. As shown in Fig. 3, $x_c^{(s)}$ and $x_c^{(t)}$ are expanded as in Eqs. (1) and (2), respectively, while $x_u^{(s)}$ and $x_u^{(t)}$ are placed in the space with unique features in each domain as follows:

$$D_s = \left(x_c^{(s)}, x_c^{(s)}, x_u^{(s)}, 0 \ \ , 0 \right) \tag{3}$$
$$D_t = \left(x_c^{(t)}, 0 \ , 0 \ \ , x_c^{(t)}, x_u^{(t)}\right) \tag{4}$$

3.2. Prediction Model

To build a prediction model, we propose a method that combines Gaussian process regression (GPR) and bagging. The input variables are the expanded ones in Eqs. (3) and (4). GPR can predict not only the expected values but also the standard deviations of output variables and provide the reliability of the prediction. Bagging is a form of ensemble learning that uses bootstrap sampling to construct many independent weak learners and then integrates the results of the weak learners into a prediction.

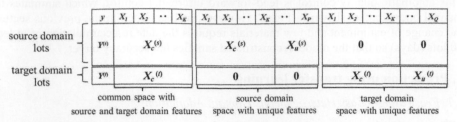

| y | X_1 | X_2 | $\cdot\cdot$ | X_K | X_1 | X_2 | $\cdot\cdot$ | X_K | $\cdot\cdot$ | X_P | X_1 | X_2 | $\cdot\cdot$ | X_K | $\cdot\cdot$ | X_Q |

source domain lots — $Y^{(s)}$ — $X_c^{(s)}$ — $X_c^{(s)}$ — $X_u^{(s)}$ — 0 — 0

target domain lots — $Y^{(t)}$ — $X_c^{(t)}$ — 0 — 0 — $X_c^{(t)}$ — $X_u^{(t)}$

common space with source and target domain features · source domain space with unique features · target domain space with unique features

Fig. 3. Feature space expansion in Frustratingly Easy Heterogeneous Domain

Kamishima et al. (2009) proposed TrBagg, which uses bagging for transfer learning. TrBagg builds weak learners using data sampled from SD and TD. The weak learners are adopted based on the classification errors for TD. The method may cause over-fitting or require the separation of TD for validation.

The chemical toner manufacturing process produces only one lot per day. To reduce the downtime of the automatic quality control system, the number of samples after each change, which are used for reconstructing the prediction model, needs to be limited. That means the number of TD lots must be small, i.e., about 10 lots. Since TrBagg does not work well in such a situation, we did not adopt it. In the proposed method, bagging is modified by selecting only weak learners with small standard deviations of the output variables when integrating the results of the weak learners. The weak learners with small standard deviations are expected to give a more reliable prediction because it is considered to use data with high similarity to the target lot preferentially. We use sequential updating of the prediction model for each lot.

4. Comparison of prediction methods

The proposed modeling method, i.e., GPR and bagging, was compared with the typical regression methods, partial least squares regression (PLSR), random forest (RF), and GPR in two cases: 1) change of coloring materials, representing material improvement, and 2) change of production scale, representing equipment improvement. FEHDA was used in both cases, and the two most important qualities were investigated. The dimensions of the input variables are shown in Table 1.

In case 1, black and magenta toners, which were made from almost the same materials except for the coloring one, were targeted, regarding black toner as SD and magenta toner as TD. In case 2, the same color toner manufactured by equipment with different scales was targeted, regarding the large scale plant as SD and the small scale one as TD.

The prediction accuracy was evaluated using Root Mean Squared Error (RMSE). In

Table 1

The dimensions of the input variables and the number of lots in two cases: 1) change of coloring materials and 2) change of production scale.

	input variables	case (1)	case (2)	Application to a mass production plant
the number of variables	$X^{(s)}$	2821	2826	2914
	$X^{(t)}$	2831	2266	2721
	$X_c^{(s)}$, $X_c^{(t)}$	2789	1803	2598
	$X_u^{(s)}$	32	1023	316
	$X_u^{(t)}$	42	463	123
the number of lots	$X^{(s)}$, $X_c^{(s)}$, $X_u^{(s)}$	450	450	333
	$X^{(t)}$, $X_c^{(t)}$, $X_u^{(t)}$	10-109	10-109	10-39

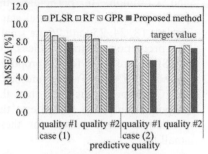

Fig. 4. Comparison of RMSE's of
prediction methods.

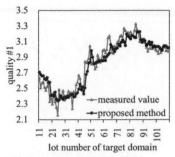

Fig. 5. Quality #1 prediction results
of the proposed method for case 1.

defining the target for the prediction accuracy, the following conditions were set; first, the center of the predicted distribution of the qualities is within 50% of the process specification width Δ, and second, the probability of out-of-specification is less than 0.3% when the quality prediction value is at the upper or lower limit of the process specification width Δ. The variability was assumed to be normally distributed (Takahashi et al., 2020). Based on these conditions, the target value for prediction accuracy became $0.5\Delta \geq 6$ RMSE, i.e., RMSE/$\Delta \leq 8.3$[%].

Fig. 4 shows the evaluation results for the 11th to 110th lots in TD. The proposed method outperformed the other methods in both cases and also satisfied the prediction accuracy target. In particular, a more significant improvement was obtained in case 1. Fig. 5 shows the predicted and measured values for each lot of quality #1. It was confirmed that the predicted values followed the trend of the actual measured values, and there were no large errors in all lots. On the other hand, the improvement achieved by the proposed method in case 2 was smaller than that in case 1. This can be attributed to the large proportion of intrinsic variables that account for 45% of the input variables in each domain, which implies that SD contains less valid information for the transfer.

5. Application to a mass-production plant

The proposed method was applied to a mass-production plant in RICOH. There are 12 quality items to be predicted, including particle size distribution, particle shape, and charging characteristics. The numbers of variables and lots are shown in Table 1. The proposed prediction method was compared with two different methods using only TD (hereinafter, referred to as Target) and using only common input variables in SD and TD (referred to as Common). In these two methods, we used random forest, which has been used in the existing automatic quality control (Takahashi et al., 2020).

We conducted the prediction of the 12 qualities from the 11th lot to the 40th lot in TD. While Target and Common failed to achieve the prediction accuracy target for two and three quality items, respectively, the proposed method achieved the prediction accuracy target for all quality items. Besides, the proposed method outperformed Target and Common in all qualities. The prediction accuracy in RMSE of the proposed method was 11.4% higher than Target on average, and particularly 17.4% for quality #10. Compared to Common, the average improvement was 15.4%, and the best was 25.4% in quality #4.

Fig. 6 shows the predicted and measured values for each lot of quality #2. The predicted values of the proposed method follow the measured values better than those of Target

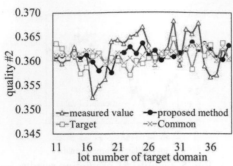

Fig.6. The prediction results of the proposed method, Target, and Common for quality #2.

and Common. In particular, the prediction errors in the initial stage for lots 18, 24, 25, and 26 are small. The prediction accuracy indices, i.e., RMSE/Δ, are 8.1% for the proposed method, 9.5% for Target, and 8.6% for Common, indicating that the proposed method satisfies the prediction target values for these small lots.

The toner qualities predicted by the transfer learning were used in the automatic quality control system based on feed-forward inference control (Takahashi et al., 2020) described in Section 2. Before applying prediction by the transfer learning, 40 lots (days) of data had to be accumulated to achieve the required accuracy target. With the proposed method, the data accumulation was reduced to 10 lots (days), and the person-hours required for monitoring and control by quality managers immediately after a change in equipment or raw materials were reduced by 75%.

6. Conclusions and future tasks

We first proposed a new transfer learning method that can cope with heterogeneous domain adaptation, i.e., FEHDA, which is simple extension of FEDA. Second, we proposed a new prediction method that combines Gaussian process regression (GPR) and bagging. Finally, the proposed method was adopted in the automatic control system of RICOH's chemical toner plant. The downtime of the automatic quality control system decreased from 40 lots (days) to 10 lots (days), and the person-hours required for manual quality control by toner quality managers have been reduced by 75%.

References

Daumé III, H., 2007. Frustratingly Easy Domain Adaptation. Proceedings of the 45th Annual Meeting of ACL 256–263.

Duan, L., Xu, D., Tsang, I.W., 2012. Learning with Augmented Features for Heterogeneous Domain Adaptation. Proceedings of the 29th International Conference on Machine Learning, ICML 2012 1, 711–718.

Kamishima, T., Hamasaki, M., Akaho, S., 2009. TrBagg: A simple transfer learning method and its application to personalization in collaborative tagging, in: Proceedings - IEEE International Conference on Data Mining, ICDM. pp. 219–228.

Khorami, H., Fgaier, H., Elkamel, A., Biglari, M., Chen, B., 2017. Multivariate Modeling of a Chemical Toner Manufacturing Process. Chemical Engineering and Technology 40, 459–469.

Takahashi, N., Miyakawa, M., Satoh, N., Watanabe, Y., Satoh, T., Kano, M., 2020. Automatic Quality Control for Chemical Toner Using Machine Learning Prediction Technology. Society of Instrument and Control Engineers 7th Multi Symposium on Control Systems 2A2-3, in Japanese.

Proceedings of the 14th International Symposium on Process Systems Engineering – PSE 2021+
June 19–23, 2022, Kyoto, Japan ©2022 Elsevier B. V. All rights reserved.
http://dx.doi.org/10.1016/B978-0-323-85159-6.50278-5

Towards An Automated Physical Model Builder: CSTR Case Study

Shota Kato and Manabu Kano*

Department of Systems Science, Kyoto University, Yoshida-honmachi, Sakyo-ku, Kyoto 606-8501, Japan

manabu@human.sys.i.kyoto-u.ac.jp

Abstract

Physical models are indispensable for the realization of digital twins, which are expected to enhance process design, operation, and optimization. The conventional physical model building relies entirely on experts and takes much time and effort. This arduousness has hampered the widespread use of digital twins. To overcome the difficulty and enable non-experts to build a practical physical model, we aim to develop an automated physical model builder, AutoPMoB. AutoPMoB conducts five tasks: 1) searching literature databases for documents relating to a target process, 2) converting the format of each document to HTML format, 3) extracting information required to build a physical model, 4) judging whether the extracted information is equivalent in different documents to unify the expressions, and 5) reorganizing the information to output a desired physical model. In the present study, we proposed an architecture of AutoPMoB and developed its prototype. By building a physical model of a continuous stirred-tank reactor, we have demonstrated that the prototype can automatically build a model that meets all requirements. AutoPMoB is expected to facilitate physical model building and foster the realization of digital twins.

Keywords: Artificial intelligence, First principle model, Process modeling, Natural language processing, Information extraction

1. Introduction

A digital twin is a core technology for realizing a cyber-physical system and has attracted much attention in recent years. It uses a model to reproduce the behavior of an actual plant and explore and predict unknown phenomena (Wang, 2020). Physical models based on the principles of chemistry, physics, and biology are applicable over a wide range of conditions, while statistical models should not be used outside the range of training data. In order to realize a digital twin, it is necessary to build an accurate physical model.

Conventionally, researchers and engineers with in-depth plant knowledge have surveyed the literature and built a process model that meets their demands. There are multiple pieces of literature to be investigated, and it is not easy to immediately find the equations that the desired model requires. Furthermore, when the accuracy of the model is inadequate, researchers and engineers need to improve its accuracy by trial and error. Thus, the conventional physical model building takes much toil.

To facilitate physical model building, we aim to develop an automated physical model builder, AutoPMoB. AutoPMoB extracts information of variables, formulas, experimental data, and prerequisites from documents and then integrates the information to build a desired physical model. AutoPMoB frees the engineers from laborious tasks of physical model building and provides access to information overlooked when manually processed.

In the present paper, we first propose an architecture of AutoPMoB and describe the fundamental technologies required for realizing it in section 2. We then develop a prototype of AutoPMoB and apply it for building a physical model of a continuous stirred-tank reactor (CSTR) in section 3.

2.   Automated Physical Model Builder

AutoPMoB first retrieves documents concerning a target process from literature databases. Next, AutoPMoB extracts information necessary to build a physical model. AutoPMoB then unifies the notations so that the information with the same meaning is not written differently. AutoPMoB finally integrates the information to build a physical model that meets all requirements.

We can obtain the documents regarding the target process by using the existing search engine. Here, it is assumed that such documents are collected in advance. This section describes the fundamental technologies required for realizing AutoPMoB.

2.1.   Document Format Conversion

The most widespread format of scientific digital documents is PDF. Extracting information directly from documents in PDF format using computers is difficult because PDF is designed for human viewing. On the other hand, documents in HTML format are relatively easy to extract information from because the information is tagged. In order to extract mathematical formulas, it is effective to represent the formulas in mathematical markup language (MathML) format (Ausbrooks et al., 2014). MathML is a standard format of web pages and is mainly used for information extraction and retrieval of mathematical expressions. Thus, AutoPMoB first converts the format of each document to HTML format. Documents in PDF format can be converted to those in HTML format with high accuracy using existing tools, such as InftyReader (Suzuki et al., 2003) and LaTeXML (Miller, 2018).

2.2.   Information Extraction

Various types of information are required to build physical models, such as variables, formulas, experimental data, and prerequisites. The essential pieces of information are variables and equations.

AutoPMoB must identify the variable definitions. All variable symbols can be extracted based on their tags in MathML format, but the challenge is to extract the definition of each symbol accurately. Some scientific documents have a table describing symbols and corresponding definitions. For such documents, AutoPMoB can accurately extract the definitions using noun phrases next to the symbols in the table. On the other hand, some scientific documents do not have such a table, and AutoPMoB needs to extract variable definitions from sentences. Although some studies (Schubotz et al., 2016, 2017) proposed variable definition extraction methods, the accuracy of the existing methods is insufficient for practical use. We have been currently developing a method to extract the variable definitions accurately (Kato and Kano, 2020).

2.3. Equivalence Judgment and Unification

AutoPMoB judges whether the variable definitions and equations extracted from different documents are equivalent and unifies their expressions.

We can determine whether two noun phrases represent the same variable by calculating their similarity and checking whether it exceeds a threshold. A language model trained with a large corpus, such as BERT (Devlin et al., 2019), is known to achieve high performance on natural language processing (NLP) tasks. The model's performance varies depending on the corpus used for training. For example, Beltagy et al. (2019) released SciBERT, a BERT-based language model trained on a large corpus of scientific texts, and achieved higher performance on a range of NLP tasks in the scientific domain than BERT. To judge the equivalence of the variable definitions, a language model trained on a corpus consisting of documents relating to physical models of processes would be useful.

In order to judge the equivalence of equations, it is necessary to consider the calculation they perform. Existing computer algebra systems can judge the equivalence of two polynomials; for example, in Wolfram Language, $(x+1)^2$ and $x^2 + 2x + 1$ are judged equivalent by expanding the former and then comparing the two polynomials (Wolfram Research, 2007). Physical models are mainly described by differential equations, algebraic equations, and partial differential equations. To our best knowledge, there has been no research focusing on the equivalence of equation groups consisting of multiple equations. We have been developing a method to judge the equivalence of two equation groups.

After judging the equivalence of variable definitions and equations, AutoPMoB unifies their expressions.

2.4. Integration & Scoring

AutoPMoB builds model candidates by integrating the unified equations. The desired model among the candidates varies depending on the purpose of model building. A user of AutoPMoB gives input variables of the model, and then AutoPMoB builds models whose numbers of degrees of freedom match the number of the input variables. Then, AutoPMoB scores and ranks the models because it takes time and effort to determine which one to choose when multiple models are built. The score is calculated based on the information, such as the number of documents containing the equations in the model and their citation information.

3. Case Study

3.1. Implementation

We developed a prototype of AutoPMoB using Streamlit, which is a Python library. Figure 1 shows a screenshot of the prototype. The prototype first takes HTML files as the input, extracts variables and equations from the files, and displays them. In this prototype, the variable definitions are extracted from a table in each file. A user selects variables required for the model and input variables. The prototype builds all models satisfying the requirements and shows them.

3.2.   Dataset

Based on the documents (Uppal et al., 1974; Marlin, 2000), we created two files, 01.html and 02.html, including physical models of an ideal jacketed CSTR, where a first-order, exothermic, irreversible reaction (A→B) takes place. The equations consisting the models are as follows:

$$\frac{dC_A}{dt} = \frac{F}{V}\left(C_{A0} - C_A\right) + r_A \tag{1}$$

$$\frac{dT}{dt} = \frac{F}{V}\left(T_0 - T\right) + \frac{h_r}{\rho C_p} r_A - \frac{Q}{V\rho C_p} \tag{2}$$

$$-r_A = k_0 \exp\left(-\frac{E}{RT}\right) C_A \tag{3}$$

$$Q = \frac{aF_c^{b+1}}{F_c + \left(\frac{aF_c^b}{2\rho_c C_{pc}}\right)} \left(T - T_{cin}\right) \tag{4}$$

$$Q = UA\left(T - T_c\right) \tag{5}$$

where the nomenclature is shown in Table 1. The model in 01.html consists of Eqs. (1)–(4), and the model in 02.html consists of Eqs. (1), (2) and (5).

3.3.   Results & Discussion

Figure 1 shows the result when we selected five variables as the required ones and one variable as the input variable. This prototype built an accurate model that met our requirements. However, the prototype has several limitations. To realize AutoPMoB, we need to develop methods for 1) improving the accuracy of extracting variable definitions from sentences, 2) accurately judging the equivalence of variable definitions, and 3) scoring and ranking the models.

Table 1: Nomenclature.

Symbol	Definition	Symbol	Definition
A	heat transfer area	T	reactor temperature
C_p	specific heat of the reacting meterial	T_c	coolant temperature
C_{pc}	specific heat of the coolant	T_{cin}	inlet temperature of the coolant
C_A	reactor concentration of A	T_0	feed temperature
C_{A0}	inlet concentration of A	U	heat transfer coefficient
E	activation energy	V	reactor volume
F	feed flow rate	h_r	heat of reaction
F_c	coolant flow rate	ρ	density of the reacting material
k_0	reation rate constant	ρ_c	density of the coolant
R	universal gas constant	a	parameter
t	time	b	parameter

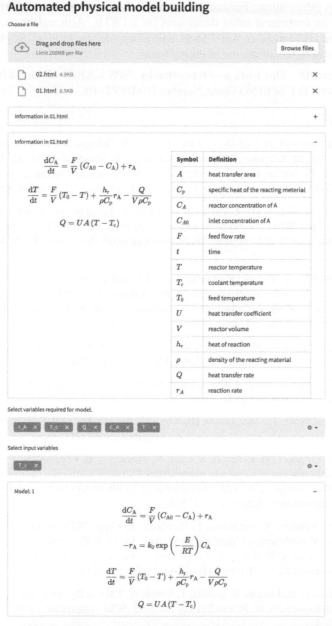

Figure 1: Screenshot of a prototype of an automated physical model builder, AutoPMoB.

4. Conclusions

To facilitate physical model building, we proposed an architecture for an automated physical model builder, AutoPMoB. AutoPMoB retrieves documents regarding a target process, extracts information such as variables and equations, and builds a physical model by integrating the information. Furthermore, we developed a prototype of AutoPMoB and demonstrated its usefulness using documents on a CSTR. Although there remain some challenges, the realization of AutoPMoB will eliminate the barriers to physical model building and lead to the utilization of digital twins.

Acknowledgments   This work was supported by JSPS KAKENHI Grant Number JP21K18849 and JST SPRING Grant Number JPMJSP2110.

References

R. Ausbrooks, S. Buswell, D. Carlisle, G. Chavchanidze, S. Dalmas, S. Devitt, A. Diaz, S. Dooley, R. Hunter, P. Ion, M. Kohlhase, A. Lazrek, P. Libbrecht, B. Miller, R. Miner, C. Rowley, M. Sargent, B. Smith, N. Soiffer, R. Sutor, S. Watt, 2014. Mathematical Markup Language (MathML) Version 3.0 2nd Edition. `https://www.w3.org/TR/MathML3`, (Accessed on 2021/11/06).

I. Beltagy, K. Lo, A. Cohan, 2019. SciBERT: A pretrained language model for scientific text. In: Proceedings of the 2019 Conference on Empirical Methods in Natural Language Processing and the 9th International Joint Conference on Natural Language Processing (EMNLP-IJCNLP). pp. 3615–3620.

J. Devlin, M.-W. Chang, K. Lee, K. Toutanova, 2019. BERT: Pre-training of deep bidirectional transformers for language understanding. In: Proceedings of the 2019 Conference of the North American Chapter of the Association for Computational Linguistics: Human Language Technologies, Volume 1 (Long and Short Papers). pp. 4171–4186.

S. Kato, M. Kano, 2020. Identifier information based variable extraction method from scientific papers for automatic physical model building. PSE Asia, Paper No. 210043.

T. E. Marlin, 2000. Process Control: Designing Processes and Control Systems for Dynamic Performance, 2nd Edition. McGraw-Hill.

B. Miller, 2018. LaTeXML The Manual—A LaTeX to XML/HTML/MathML Converter, Version 0.8.3. `https://dlmf.nist.gov/LaTeXML/`, (Accessed on 2021/11/06).

M. Schubotz, A. Grigorev, M. Leich, H. S. Cohl, N. Meuschke, B. Gipp, A. S. Youssef, V. Markl, 2016. Semantification of identifiers in mathematics for better math information retrieval. In: SIGIR 2016 - Proceedings of the 39th International ACM SIGIR Conference on Research and Development in Information Retrieval. pp. 135–144.

M. Schubotz, L. Krämer, N. Meuschke, F. Hamborg, B. Gipp, 2017. Evaluating and improving the extraction of mathematical identifier definitions. In: G. J. F. Jones, S. Lawless, J. Gonzalo, L. Kelly, L. Goeuriot, T. Mandl, L. Cappellato, N. Ferro (Eds.), Conference and Labs of the Evaluation Forum (CLEF), Dublin, Ireland. Vol. 10456. pp. 82–94.

M. Suzuki, F. Tamari, R. Fukuda, S. Uchida, T. Kanahori, 2003. Infty: an integrated ocr system for mathematical documents. In: Proceedings of the 2003 ACM symposium on Document engineering. pp. 95–104.

A. Uppal, W. Ray, A. Poore, 1974. On the dynamic behavior of continuous stirred tank reactors. Chemical Engineering Science 29 (4), pp. 967–985.

Z. Wang, 2020. Digital Twin Technology. Ch. 7, In: Industry 4.0 - Impact on Intelligent Logistics and Manufacturing.

Wolfram Research, 2007. Equal. `https://reference.wolfram.com/language/ref/Equal.html`, (Accessed on 2021/11/06).

Proceedings of the 14th International Symposium on Process Systems Engineering – PSE 2021+
June 19-23, 2022, Kyoto, Japan © 2022 Elsevier B.V. All rights reserved.
http://dx.doi.org/10.1016/B978-0-323-85159-6.50279-7

Forward physics-informed neural networks for catalytic CO_2 methanation via isothermal fixed-bed reactor

Son Ich Ngo[a], and Young-Il Lim[a,*]

[a]Center of Sustainable Process Engineering (CoSPE), Department of Chemical Engineering, Hankyong National University, Gyeonggi-do, Anseong-si, Jungang-ro 327, 17579 Korea
limyi@hknu.ac.kr

Abstract

We developed physics-informed neural networks (PINNs) to solve an isothermal fixed-bed (IFB) model for catalytic CO_2 methanation. The PINN is composed of a feed-forward artificial neural network (FF-ANN) with two inputs and physics-informed constraints for governing equations, boundary conditions, initial conditions, and nonlinear reaction kinetics. The forward PINN showed excellent extrapolation performance for the IFB model. The calculation speed of the PINN surrogate model is faster significantly than a stiff ODE numerical solver. These results suggest that forward PINNs can be used as a surrogate model for chemical reaction kinetics.

Keywords: Catalytic CO_2 methanation; Fixed-bed reactor; Reaction kinetics; System identification; Machine learning; Physics-informed neural network.

1. Introduction

CO_2 methanation (Ngo et al., 2021) combining captured CO_2 with H_2 produced via water electrolysis (Kim et al., 2021) is an alternative to existing energy systems that could be integrated with renewable electricity sources. CO_2 methanation technologies could considerably reduce carbon emissions by encouraging industrial symbiosis from industries with large CO_2 footprints such as thermal power plants (Kim et al., 2021). Because CH_4 is easier to store and transport than H_2 (Ngo et al., 2021), the synergistic integration of renewable electricity with a natural gas grid is expected via CO_2 methanation (Miguel et al., 2018).

Despite of advances in first principles and empirical elucidations, artificial neural network (ANN) models in the category of data-driven models, black-box models, or surrogate models (SMs), have become an alternative functional mapping between input and output data because of their prompt predictions, automated knowledge extraction, and high inference accuracy (Abiodun et al., 2018, Gusmão et al., 2020).

Recently, ANNs and conservation equations coupled with automatic differentiation (AD) that solve ordinary differential equations (ODEs) and partial differential equations (PDEs), called physics-informed neural networks (PINNs), have been reported (Raissi et al., 2019). Because PINNs are constrained to respect any symmetries, invariances, or first-principle laws (Raissi et al., 2019), they present a potential for solving chemical engineering problems, which usually deal with complex geometries and physical phenomena. In contrast to common ANNs, PINNs do not depend on empirical data because the initial and boundary conditions are directly used to adjust the network parameters such as weights and biases (Raissi et al., 2019). In addition, the extrapolation capability of PINNs is enhanced owing to physical constraints (Kim et al., 2020, Ngo and Lim, 2021). Nevertheless, there are few applications of PINNs in process modeling and chemical reactor design.

In this study, forward PINNs coupled with AD were developed for the solution and parameter identification of a highly nonlinear reaction rate model for catalytic CO_2 methanation in an IFB reactor. The results obtained from the PINNs were compared with those obtained using a common numerical solver of ODEs (ode15s in MATLAB). The extrapolation capability was analyzed by narrowing the collocation training domain and detaching the collocation training domain from the boundary. It was deomonstrated that the forward PINN solved fixed-bed models with highly nonlinear chemical reaction kinetics.

2. Isothermal fixed-bed reactor for CO₂ methanation

The single-tube IFB was assumed to be equipped with a heat exchanger that was able to transfer immediately the heat generated in the exothermic reactions to the coolant. The catalytic CO_2 methanation reaction, known as the Sabatier reaction, is (Ngo et al., 2020, Ngo and Lim, 2021)

$$CO_2 + 4H_2 \rightleftarrows CH_4 + 2H_2O, \Delta H_r^{298K} = -165 \text{ kJ·mol-1} \tag{1}$$

The operating conditions were set as a temperature (T) of 450 °C, a pressure (P) of 5 bar, and a volumetric flow rate (Q) of 10 Nm³/s. The pure gas reactants were fed to the inlet at a CO_2/H_2 molar ratio of 1/4.

The mass balances for the ith species (i = CO_2, H_2, CH_4, and H_2O) participating in the CO_2 methanation reaction in Eq. (1) are formulated as follows:

$$\frac{1}{A_t} \frac{dF_i}{dz} = \eta v_i r \tag{2}$$

where z (m) is the reactor tube axial position, F_i (mol/s) is the molar flow rate of a species i at position z, A_t (m²) is the tube cross-sectional area, v_i is the stoichiometric coefficient of species i, and r (mol/m³/s) is the volumetric reaction rate. η is the effectiveness factor of the chemical reaction, which is defined as the volume-averaged reaction rate with diffusion within catalyst particles divided by the area-averaged reaction rate at the catalyst particle surface (Ngo et al., 2020).

The boundary conditions for the molar flow rate (F_i) of the species at the inlet ($z = 0$) are as follows:

$$F_i|_{z=0} = x_{i,0} F_0 \tag{3}$$

where $x_{i,0}$ and F_0 (mol/s) are the inlet mole fraction of gas species i and the total molar flow rate of the inlet gas mixture, respectivel. A reaction kinetics model proposed by (Koschany et al., 2016) for catalytic CO_2 methanation, which was tested within a wide range of Ni contents and industrial operating conditions, was adopted in this study.

$$r = \rho_{cat}(1 - \varepsilon)k \cdot \frac{p_{H_2}^{0.31} p_{CO_2}^{0.16}}{1 + K_{ad}\frac{p_{H_2O}}{p_{H_2}^{0.5}}} \left(1 - \frac{p_{CH_4} p_{H_2O}^2}{p_{H_2}^4 p_{CO_2} K_{eq}}\right) \tag{4}$$

$$k = 6.41 \times 10^{-5} \exp\left(\frac{93.6}{R}\left(\frac{1}{555}-\frac{1}{T}\right)\right) \tag{5}$$

$$K_{ad} = 0.62 \times 10^{-5} \exp\left(\frac{64.3}{R}\left(\frac{1}{555}-\frac{1}{T}\right)\right) \tag{6}$$

$$K_{eq} = 137 \cdot T^{-3.998} \exp\left(\frac{158.7}{RT}\right) \tag{7}$$

where R (=8.314×10^{-3} kJ/mol/K) is the gas constant, T (K) is the temperature, p_i (bar) is the partial pressure of species i, k ($mol/g_{cat}/s$) is the reaction rate constant, K_{ad} (1/bar$^{0.5}$) is the adsorption constant, and K_{eq} is the thermodynamic equilibrium constant. The catalyst density (ρ_{cat}) was set to 2300×10^3 g_{cat}/m^3_{cat} (Koschany et al., 2016).

3. Forward PINN structure

The architecture of the forward PINN problem is shown in Fig. 1. The objective of the forward PINN problem is to solve the given governing equation with initial, boundary, and operating conditions. The 30,000 collocation points were used to train the governing equations over the reactor length ($0 < z \leq L$) except $z = 0$. The Dirichlet's boundary conditions fixed the value of $F_{i,0} = [97.74\ 378.9\ 0\ 0]$ mol/s at the reactor inlet ($z = 0$). The lower and upper bounds of η were 0 and 1, respectively.

The FF-ANN structure contained two inputs (z and η), four outputs (F_i), five hidden layers, and 128 neurons for each layer. The activation function (f_a) of hyperbolic tangent (*tanh*), was applied for each neuron. The weights ($w_{j,k}$) and biases ($b_{j,k}$) for the jth hidden layer and the kth neuron are adjusted to minimize the loss function (*Loss*). The AD for spatial derivatives ($\frac{dF_i}{dz}$) was calculated via the reverse accumulation mode which propagates derivatives backward from a given output (Güneş Baydin et al., 2018). The governing equations as the physics-informed part of the ANN included the reaction kinetic rate (r) in Eq. (4), the four ODEs in Eqs. (2), and the boundary conditions.

The optimized weights and biases (w^* and b^*) were obtained from the following optimization problem:

$$\{w^*, b^*\} = \underset{w,b}{\mathrm{argmin}}\{Loss = MSE_g(w,b) + MSE_b(w,b)\} \tag{8}$$

$$MSE_g(w,b) = \frac{1}{N_{train}}\sum_{j=1}^{N_{train}}\sum_{i=1}^{N_{comp}}\left|\frac{1}{A_t}\left(\frac{dF_i}{dz}\right)_j - \eta v_i r_j\right|^2 \tag{9}$$

$$MSE_b(w,b) = \frac{1}{N_{bnd}}\sum_{k=1}^{N_{bnd}}\sum_{i=1}^{N_{comp}}\left|F_{i,k}|_{z=0} - x_{i,0}F_0\right|^2 \tag{10}$$

where MSE_g and MSE_b are the mean squared errors for the governing equation and boundary condition, respectively. N_{train}, N_{comp}, and N_{bnd} are the number of training data sets, species (or components), and boundary condition sampling points, respectively. The loss function (*Loss*) sums MSE_g and MSE_b.

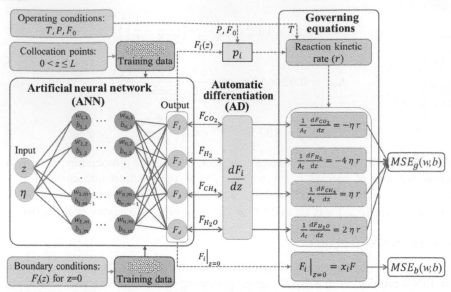

Figure 1. Structure of the physics-informed neural network (PINN) forward problem for CO_2 methanation in an isothermal fixed-bed (IFB) reactor.

An Adam optimizer (Kingma and Lei Ba, 2015) was used to solve Eq. (8), which combines a stochastic gradient descent with adaptive momentum, because of its good convergence speed (Rao et al., 2020). A mini-batch size of 128, which had a minor effect on the PINN training results, was used. The number of training epochs was set to 1,000. In the FF-ANN, the biases (b) were initialized to zeros and the weights (w) were initialized by the commonly used heuristic called Xavier's method (Xavier and Yoshua, 2010).

4. Results and discussion

Fig. 2 shows the performance of the forward PINN for 30,000 training data points in a limited range of z and 1,000 test data points in a full range of z ($0 \leq z \leq 2$) while η was fixed at 1. The collocation range of the training data starts from $z = 0$ and ends at $z = 0.5$ (Fig. 2a) and 1.0 (Fig. 2b). Even though the PINN was trained within one-sixth ($0 \leq z \leq 0.5$) of the full range, the PINN output ($F_{i,PINN}$) for the test data of the full range ($0 \leq z \leq 2$) agrees well the ODE solution ($F_{i,ODE}$) outside the training range (Fig. 2a).

Fig. 3 plots the performance of forward PINN for 30,000 training data points within the full range of $0 \leq z \leq 2$ and $0 \leq \eta \leq 1$. Fig. 3a and 3b show the interpolation for $\eta = 0.5$ and 1.0, respectively, whereas Fig. 3c demonstrats the extrapolation for $\eta = 1.5$. With the extrapolation of 150% higher than the trained bound, the PINN captures F_i with a prediction accuracy of 97.3%.

The extrapolation capability of the PINN is remarkable, unlike that of common ANNs (Abiodun et al., 2018). The accuracy of the PINN solution is closely related to the range and distribution of the training data (Jagtap et al., 2020). The forward PINN model is appropriate for solving governing equations with complex geometries or moving boundary conditions (Sun et al., 2020). In addition, numerical diffusion and round-off errors are minimized in PINNs with the aid of AD.

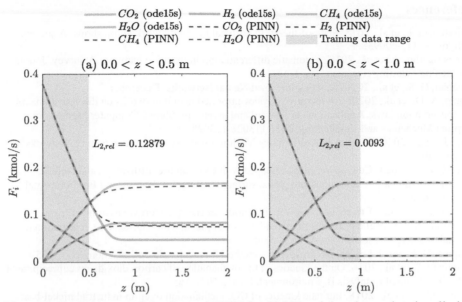

Figure 2. Performance of the forward PINN for 10,000 training data points in a limited range of the reactor length (z) and 1,000 test data points in the full range of z and $\eta = 1$.

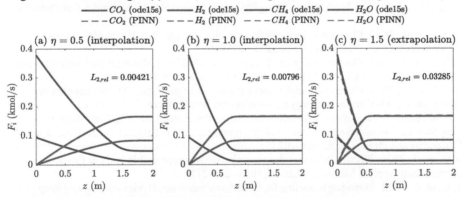

Figure 3. Performance of forward PINN for two inputs as η and z.

5. Conclusions

The physics-informed neural network (PINN) was developed for an isothermal fixed-bed (IFB) reactor model for catalytic CO_2 methanation. The PINN was composed of a feed-forward artificial neural network (FF-ANN), automatic differentiation (AD) for derivatives, and governing equations with a stiff reaction kinetic rate. The loss function of the PINN included two mean squared errors (MSEs) for the governing equations and boundary conditions. The one-dimensional reactor was initialized at a molar flow rate that was the same as the boundary condition at the reactor inlet.

The forward PINN model exhibited an excellent extrapolation performance because the PINN provided a solution satisfying physics-informed constraints. The current approach is useful for building a surrogate model for CO_2 methanation process design and optimization.

References

Abiodun, O. I., et al., 2018. State-of-the-art in artificial neural network applications: A survey. Heliyon, 4, 11, e00938.

Güneş Baydin, A., et al., 2018. Automatic differentiation in machine learning: A survey. Journal of Machine Learning Research, 18, 1-43.

Gusmão, G. S., et al., 2020. Kinetics-Informed Neural Networks. December.

Jagtap, A. D., et al., 2020. Conservative physics-informed neural networks on discrete domains for conservation laws: Applications to forward and inverse problems. Computer Methods in Applied Mechanics and Engineering, 365, 113028-113028.

Kim, J., et al., 2020. DPM: A Novel Training Method for Physics-Informed Neural Networks in Extrapolation.

Kim, S., et al., 2021. Coal power plant equipped with CO_2 capture, utilization, and storage: Implications for carbon emissions and global surface temperature. Energy and Environmental Science, in review.

Kim, S., et al., 2021. Effects of flue gas recirculation on energy, exergy, environment, and economics in oxy-coal circulating fluidized-bed power plants with CO_2 capture. International Journal of Energy Research, 45, 4, 5852-5865.

Kingma, D. P. and J. Lei Ba, 2015. Adam: A method for stochastic optimization. Iclr, 1-15.

Koschany, F., et al., 2016. On the kinetics of the methanation of carbon dioxide on coprecipitated $NiAl(O)_x$. Applied Catalysis B: Environmental, 181, 504-516.

Miguel, C. V., et al., 2018. Intrinsic kinetics of CO_2 methanation over an industrial nickel-based catalyst. Journal of CO2 Utilization, 25, 128-136.

Ngo, S. I. and Y.-I. Lim, 2021. Solution and Parameter Identification of a Fixed-Bed Reactor Model for Catalytic CO2 Methanation Using Physics-Informed Neural Networks. Catalysts, 11, 11.

Ngo, S. I., et al., 2020. Flow behaviors, reaction kinetics, and optimal design of fixed-and fluidized-beds for CO_2 methanation. Fuel, 275, 117886.

Ngo, S. I., et al., 2021. Flow behavior and heat transfer in bubbling fluidized-bed with immersed heat exchange tubes for CO_2 methanation. Powder Technology, 380, 462-474.

Ngo, S. I., et al., 2021. Experiment and numerical analysis of catalytic CO_2 methanation in bubbling fluidized bed reactor. Energy Conversion and Management, 233, 113863.

Raissi, M., et al., 2019. Physics-informed neural networks: A deep learning framework for solving forward and inverse problems involving nonlinear partial differential equations. Journal of Computational Physics, 378, 686-707.

Rao, C., et al., 2020. Physics-informed deep learning for incompressible laminar flows. Theoretical and Applied Mechanics Letters, 10, 3, 207-212.

Sun, L., et al., 2020. Surrogate modeling for fluid flows based on physics-constrained deep learning without simulation data. Computer Methods in Applied Mechanics and Engineering, 361, 112732-112732.

Xavier, G. and B. Yoshua (2010). Understanding the difficulty of training deep feedforward neural networks, PMLR. 9: 249-256.

Proceedings of the 14th International Symposium on Process Systems Engineering – PSE 2021+
June 19-23, 2022, Kyoto, Japan © 2022 Elsevier B.V. All rights reserved.
http://dx.doi.org/10.1016/B978-0-323-85159-6.50280-3

Hashing-based just-in-time learning for big data quality prediction

Xinmin Zhang[a]*, Jiang Zhai[a], Zhihuan Song[a], Yuan Li[b]

[a]*State Key Laboratory of Industrial Control Technology, College of Control Science and Engineering, Zhejiang University, Hangzhou 310027, China*
[b]*Department of Information Engineering, Shenyang University of Chemical Technology, Shenyang 110142, China*
xinminzhang@zju.edu.cn

Abstract

In recent years, the just-in-time (JIT) predictive models have attracted considerable attention due to their ability to prevent degradation of prediction accuracy. However, one of their practical limitations is expensive computation, which becomes a major factor that prevents them from being used for big data quality prediction. This is because the JIT modeling methods need to update the local regression model using the relevant samples that are searched through the lineal scan of the database during online operation. To solve this issue, the present work proposes a novel hashing-based JIT (HbJIT) modeling method that is suitable for big data quality prediction. In HbJIT, a family of locality-sensitive hash functions is firstly used to hash big data into a set of buckets, in which similar samples are grouped on themselves. During online prediction, HbJIT looks up multiple buckets that have a high probability of containing similar samples of a query object through the intelligent probing scheme, uses the data objects in the buckets as the candidate set of the results, and then filters the candidate objects using a linear scan. After filtering, the most relevant samples are used to construct the local regression model to yield the prediction of the query object. By integrating the multi-probe hashing strategy into the JIT learning framework, HbJIT can not only deal with process nonlinearity and time-varying characteristics but also is applicable to large-scale industrial processes. Experimental results on real-world dataset have demonstrated that the proposed HbJIT is time-efficient in processing large-scale datasets, and greatly reduces the online prediction time without compromising on the prediction accuracy.

Keywords: Virtual sensor, soft-sensor, big data quality prediction, hashing-based just-in-time modeling.

1. Introduction

In the modern process industry, with the widespread utilization of distributed control systems and the Internet of Things, large amounts of process data have been collected. Data-driven soft-sensors are important tools in process industries for online prediction of some quality variables that generally cannot be automatically measured at all, or can only be measured sporadically, with high delay, or at high cost (Kadlec et al., 2009; Kano and Ogawa, 2010). In recent years, a wide variety of data-driven soft-sensors ranging from linear models to nonlinear models have been developed (Zhang et al., 2019; Zhang et al., 2020).

In practical applications, there is a challenging problem with soft-sensors, that is, the predictive performance of soft-sensors might deteriorate due to the time-varying characteristics of industrial processes. As reported in (Kano and Ogawa, 2010), model maintenance is thought to be one of the most important issues related to soft-sensors. A simple and effective solution to solve the problem of predictive model degradation is to use the just-in-time (JIT) learning methods. Since the JIT modeling methods have the advantages of handling process nonlinearity and time-varying characteristics, they have been extensively used in various fields (Liu et al., 2012; Jin et al., 2019). Despite the JIT methods have some successful applications, they suffer from the problem of high computational cost when used for big data quality prediction. The reason is that the JIT methods need to update the local regression model using the relevant samples that are searched through the lineal scan of the database during online operation. The calculation of relevant samples through the linear scan of the database is very time-consuming. However, the size of the data collected in the process industry is increasing vastly, and the process industry has entered the era of big data (Qin, 2014). Thus, it is important to design a new JIT modeling algorithm that is suitable for big data quality prediction.

To solve this issue, the present work proposes a novel hashing-based JIT (HbJIT) modeling method that is suitable for big data quality prediction. HbJIT is designed based on the multi-probe hashing scheme, which first hashes big data into multiple buckets, in which similar samples are grouped on themselves. To perform a quality prediction, HbJIT looks up multiple buckets that have a high probability of containing the similar samples of a query object through the intelligent probing scheme, uses the data objects in the buckets as the candidate set of the results, and then filters the candidate objects using a linear scan. After filtering, the most relevant samples are used to construct the local regression model to yield the prediction of the query object. As a fast adaptive soft-sensor, HbJIT can not only deal with process nonlinearity and time-varying characteristics but also is suitable for large-scale industrial processes. The effectiveness of the proposed HbJIT is evaluated on real-world large-scale dataset. The results demonstrate that HbJIT can significantly reduce the online prediction time without sacrificing much in terms of accuracy.

The remainder of this paper is organized as follows. Section 2 gives a brief description of the JIT modeling method and Gaussian process regression. The proposed HbJIT modeling method is presented in Section 3. Section 4 provides the experimental results on real-world large-scale dataset. Conclusion is given in Section 5.

2. Preliminaries

2.1. Just-in-time (JIT) modelling method

Generally, the prediction accuracy of soft-sensors may be degraded due to changes in process characteristics (Kano and Ogawa, 2010). The JIT modeling method can deal with changes in process characteristics as well as nonlinearity, and thus it can prevent degradation of prediction accuracy.

Different from traditional soft-sensor which builds a global model of the process in an offline manner, the JIT modeling method constructs a query-driven local model. More specifically, given a historical dataset, the JIT modeling method consists of three steps: (1) When an output estimate is required for a new query, it searches for relevant samples to the query in the reference dataset based on some similarity measures. The most popular similarity measure is the Euclidean distance. (2) A local model is built using the relevant samples. (3) An output estimate is produced by the constructed local model, and then the constructed local model is discarded. When the next query sample arrives, one needs to follow the same steps as above to build a new local model.

2.2. Gaussian process regression (GPR)

In JIT modeling, the local model should be built using some regression methods. In this work, GPR is adopted to construct the local regression model. Given a dataset $S = \{X, y\} = \{(x_1, y_1), \cdots, (x_N, y_N)\}$, where $x \in \mathbb{R}^M$ and $y \in \mathbb{R}$ denote any input-output pair. Let $f(x)$ denote a latent function which maps input x to output y. $f = (f(x_1), f(x_2), \cdots, f(x_N))$ represents the function values of the input vectors. In GPR, the function $f(x)$ is regarded as a random variable following a Gaussian process (GP) prior distribution (Williams and Rasmussen, 2006). A GP is defined in terms of a positive definite kernel (or covariance) function $k(x, x')$ as follows

$$f(x) \sim GP\big(m(x), k(x, x')\big) \tag{1}$$

$$m(x) = \mathrm{E}[f(x)] \tag{2}$$

$$k(x, x') = \mathrm{E}\big[\big(f(x) - m(x)\big)\big(f(x') - m(x')\big)\big] \tag{3}$$

Generally, $m(x)$ is zero-mean. According to the principles of GP, the distribution of f follows the zero-mean Gaussian distributions with covariance matrix $K = \big(k(x_i, x_j)\big)_{ij}$:

$$p(f|X) = \mathrm{N}(0, K) \tag{4}$$

where $k(x_i, x_j)$ denotes the (i, j)-element of K. Given a query object x_q, the prediction of the function value $f(x_q)$ (denoted as f_q) is given by

$$\bar{f}_q = k_q [K + \delta_\epsilon^2 I]^{-1} y \tag{5}$$

where I is an identity matrix and δ_ϵ^2 is the variance of the Gaussian noise term.

3. Hashing-based just-in-time (HbJIT) modeling method

Notice that a major drawback of the JIT modeling methods is that a high computational cost is required in order to search for the relevant samples from the database to construct the local regression model, when a query object is provided. Especially, when the database provided is large scale, the computational cost of the JIT modeling will be very high, leading to a large delay in online quality prediction. However, the amount of data collected in the process industry is exploding, and the process industry has entered the era of big data. In such a scenario, if the JIT modeling method is used for big data quality prediction, the online prediction time will be very long. To handle the computational cost challenge of the JIT modeling methods, the present work proposes a novel hashing-based JIT (HbJIT) modeling method that is applicable to big data quality prediction. HbJIT is designed based on the multi-probe LSH scheme, which first uses the hashing functions to hash the big data into a set of buckets, in which similar samples are grouped on themselves. To perform a quality prediction, HbJIT looks up multiple buckets that have a high probability of containing the nearest neighbours of a query object through the intelligent probing scheme, uses the data objects in the buckets as the candidate set of the results, and then filters the candidate objects using a linear scan. After filtering, the most relevant samples are used to construct the local GPR model to yield the prediction of the query object. Different from the conventional JIT modeling method which calculates similar samples to the query in linear time with respect to the data size, the proposed HbJIT calculates similar samples in sub-linear time, and significantly speeds up the online quality prediction.

The proposed method is implemented in two steps. The first step is to construct the index data structure using the locality sensitive hashing (LSH) functions, and the second step is to construct the local GPR model based on the multi-probe scheme. More specifically, let S be the historical database, and D be the distance measure between two objects. A family of hash functions $H = \{h: S \rightarrow U\}$ is called (r, cr, p_1, p_2)-sensitive if the following conditions are satisfied for any two data objects $x_p, x_q \in S$ (Datar et al., 2004):

$$if\ D(x_p, x_q) \leq r, \quad then\ Pr_H[h(x_q) = h(x_p)] \geq p_1 \tag{6}$$

$$if\ D(x_p, x_q) > cr, \quad then\ Pr_H[h(x_q) = h(x_p)] \leq p_2 \tag{7}$$

where $c > 1$ is an approximation factor, $p_1, p_2 \in (0,1)$ represent two probability thresholds and satisfy $p_1 > p_2$. The characteristic of the LSH function is that similar objects have a higher probability of being hashed to the same bucket than distant ones. The family of LSH functions based on p-stable distributions is considered as follows

$$h(x) = \left\lfloor \frac{a^T x + b}{r} \right\rfloor \tag{8}$$

where a is a d-dimensional random vector whose entries are drawn independently from a p-stable distribution, b is a real number drawn uniformly from the range $[0, r]$. In practical applications, in order to construct the index data structure with high search precision, multiple hash tables need to be built, and each hash table contains multiple LSH functions. Let L and M denote the number of hash tables and LSH functions, respectively. A concatenation of M LSH functions in a hash table can be represented as

$$g(x_p) = \left(h_1(x_p), \cdots, h_M(x_p) \right) \tag{9}$$

For L hash tables, we can construct L independent copies of $g_1(x), \cdots, g_L(x)$. During the stage of constructing the index data structure, each data object in the database is hashed into one of the hash buckets of $g_1(x), \cdots, g_L(x)$ and stored.

During the online prediction phase, when a query x_q is coming, we need to find the most k relevant samples of x_q, which is used for local GPR modeling. Generally, the basic LSH algorithm will compute $g_1(x_q), \cdots, g_L(x_q)$, and search all these L buckets to get a set of candidates. Then, this candidate set is pruned through the linear scan algorithm to obtain the most k relevant samples of x_q. It is worth noting that in practical applicaitons, a large number of hash tables need to be built in order to obtain higher search precision. A large number of hash tables results in a memory footprint. When the space requirement of the hash tables exceeds the size of the main memory, disk I/O may be required to look up the hash bucket, which causes a large amount of delay in the query process. One of the solutions to solve this issue is to use a multi-probe scheme (Lv et al., 2007), which can significantly reduce the memory footprint of the LSH data structure. The intuition of multiprobe LSH is that it probes multiple buckets in each hash table instead of building many different hash tables. This is realized by using an intelligently generated probing sequence which probes multiple buckets that are likely to contain the nearest neighbours of the query object in each hash table. According to the characteristics of LSH, if a data object is close to x_q, but is not hashed into the same bucket as x_q, then it is likely to be in a nearby bucket. Based on this principle, multiprobe LSH defines a probeing sequence $(\Delta_1, \Delta_2, \cdots \Delta_J)$ where $\Delta_1 = (\delta_m, \cdots, \delta_M)$ is a hash perturbation vector with $\delta_m \in \{-1, 0, 1\}$, $m = 1, \cdots, M$, and $j = 1, \cdots, J$. Given x_q, the basic LSH probes the hash bucket $g_l(x_q)$, while multiprobe LSH probes not only the bucket $g_l(x_q)$ but also the buckets $g_l(x_q) + \Delta_1, \cdots, g_l(x_q) + \Delta_J$. These buckets are ordered according to the score of the perturbation vector, which is defined as

$$score(\Delta_j) = \sum_{m=1}^{M} x_m(\delta_m)^2 \tag{10}$$

where $x_m(\delta_m)$ denotes the distance from x_q to the bundary of the bucket $h_m(x_q) + \delta_m$. For the perturbation vectors that have smaller scores, they will have higher probability of yielding data objects near to the query object x_q.

4. Case studies

This section describes the experimental results of the proposed HbJIT-GPR method on a large-scale real-world blast furnace ironmaking dataset, and compared it against the GPR and JIT-GPR(brute) methods. JIT-GPR(brute) denotes the brute-force linear scan that is used to select similar samples. The experimental results are evaluated from two aspects: prediction accuracy and speed. The prediction accuracy is measured in terms of the root mean squared error (RMSE) criterion. The prediction speed is measure by the online prediction time.

The experiments were carried out in the blast furnace ironmaking process, which is a typical nonlinear time-varying process (Geerdes et al., 2020). It is difficult to accurately control the blast furnace to produce hot metal with consistent quality, because the harsh operating circumstances prevent the inside chemical heat from being directly detected. silicon content of hot metal is an important index indicating the chemical heat of molten iron. To carry on a steady operation of the blast furnace and produce hot metal with consistent quality, it is important to predict silicon content in real time.

To construct the prediction model, the collected dataset is randomly separated into a training dataset (80k samples), a validation dataset (10k samples) and a test dataset (10k samples). The dataset contains 110 process variables. The key parameters of the proposed HbJIT-GPR are selected using the grid search method in terms of the RMSE criterion on the validation set. Table 1 summarizes the prediction accuracy and prediction time for the silicon content by all methods on the testing dataset. GPR obtained the worst prediction accuracy although the online prediction time is the shortest. Compared with GPR, JIT-GPR(brute) has higher prediction accuracy and longer online prediction time. In comparison, the proposed HbJIT-GPR achieved the similar prediction accuracy as JIT-GPR$_{(brute)}$, but provided shorter online prediction time. As a result, the prediction accuracy couple with the reduction in online prediction time clearly demonstrated the success of the proposed HbJIT-GPR in handling large-scale dataset.

Table 1. Prediction results for the silicon content on the testing dataset.

Methods	RMSE	Time (ms)
GPR	0.0211	121
JIT-GPR$_{(brute)}$	0.0126	460
HbJIT-GPR	0.0127	350

5. Conclusions

In this paper, a novel hashing-based JIT (HbJIT) modeling method was proposed for big data quality prediction. HbJIT is a fast adaptive soft-sensor that can not only deal with process nonlinearity and time-varying characteristics but also be applicable to large-scale industrial processes. The usefulness of the proposed method was verified through an industrial blast furnace ironmaking process. The experimental results show that the proposed method can reduce the online prediction time by a huge amount without sacrificing much in terms of accuracy.

References

Kadlec, P., Gabrys, B. and Strandt, S., 2009, Data-driven soft sensors in the process industry, Computers & Chemical Engineering, 33(4), 795-814.

Kano, M., Ogawa, M., 2010. The state of the art in chemical process control in japan: Good practice and questionnaire survey. Journal of Process Control, 20, 969–982.

Zhang X., Kano M., Matsuzaki S., 2019. A comparative study of deep and shallow predictive techniques for hot metal temperature prediction in blast furnace ironmaking. Computers & Chemical Engineering, 130, 106575.

Zhang X., Kano M., Tani M., et al., 2020, Prediction and causal analysis of defects in steel products: Handling nonnegative and highly overdispersed count data, Journal of chemical engineering of Japan, 95, 104258.

Liu, Y., Gao, Z., Li, P., Wang, H., 2012. Just-in-time kernel learning with adaptive parameter selection for soft sensor modeling of batch processes. Industrial & Engineering Chemistry Research 51, 4313–4327.

Jin, H., Pan, B., Chen, X., Qian, B., 2019. Ensemble just-in-time learning framework through evolutionary multiobjective optimization for soft sensor development of nonlinear industrial processes. Chemometrics and Intelligent Laboratory Systems 184, 153–166.

Qin, S.J., 2014. Process data analytics in the era of big data. AIChE Journal 60, 3092–3100.

Williams, C.K. , Rasmussen, C.E. , 2006. Gaussian processes for machine learning. The MIT Press 2 (3), 4 .

Buhler, J., 2001. Efficient large-scale sequence comparison by locality-sensitive hashing. Bioinformatics 17, 419–428.

Datar, M., Immorlica, N., Indyk, P., Mirrokni, V.S., 2004. Locality-sensitive hashing scheme based on p-stable distributions, in: Proceedings of the twentieth annual symposium on Computational geometry, pp. 253–262.

Geerdes, M., Chaigneau, R., Lingiardi, O., 2020. Modern Blast Furnace Ironmaking: An Introduction (2020). Ios Press.

Proceedings of the 14th International Symposium on Process Systems Engineering – PSE 2021+
June 19-23, 2022, Kyoto, Japan © 2022 Elsevier B.V. All rights reserved.
http://dx.doi.org/10.1016/B978-0-323-85159-6.50281-5

Physics-Constrained Autoencoder Neural Network for the Prediction of Key Granule Properties in a Twin-Screw Granulation Process

Chaitanya Sampat[a] and Rohit Ramachandran[a,*]

[a]*Chemical and Biochemical Engineering, Rutgers, the State University of New Jersey, Piscataway, NJ 08823, USA*
rohitrr@soe.rutgers.edu

Abstract

With the advancement of digitization of industrial manufacturing, there has been an increase in the application of machine learning methods to model these processes. These data-driven models are multivariate in nature and on occasion may not deliver the accuracy that can be obtained from first-principle models. The statistical approach in data-driven models is completely data-dependent and may give erroneous or undesired results due to noisy and incomplete database. Though accurate, first-principle models are often slow to simulate and lack the ability to predict data in real-time (Chen et al., 2020). Thus, to obtain real-time process predictions with accuracy similar to first-principle models, there is a need to develop data-driven models with first principle-based process constraints within their framework. In this study, several experimental datasets for twin-screw granulators (TSG) were considered. The data for 13 different TSGs was collected from previously published studies. The collected data was sorted for process parameters, material properties and geometric conditions of the study. An autoencoder neural network was developed to model these processes. The output from this model not only predicted the data well but also showed granule growth characteristics with the output properties obeying first-principle laws. The encoding section of the neural network helped find correlated inputs creating a reduced order model and captured information about the underlying physics of the process.

Keywords: Physics constrained neural network; autoencoders; twin screw granulation; Physics informed neural networks; PINN; PCNN

1. Introduction

Wet granulation is the process of agglomeration of fine powder into larger granules by adding a liquid binder. These granules help achieve desired quality attributes which can aid in improved flow, better dissolution rates, and better compression characteristics (Iveson et al., 2001). Wet granulation find application in various powder processing industries like mineral processing, agricultural products, detergents, food, and pharmaceuticals. It is an important unit operation in downstream oral dosage manufacturing in the pharmaceutical industry to more uniform distribution and dissolution characteristics. Previously wet granulation has been performed in a batch manufacturing scenario where, powder was mixed using an impeller and a liquid binder was sprayed using a liquid binder. This high-shear granulation can produce less compressible granules and operate in a very narrow range (Kumar et al., 2013). These challenges were overcome by converting this batch process to continuous manufacturing process.

Twin-screw granulation (TSG) is a widely used continuous wet granulation process. This equipment consists of a barrel which contain 2 co-rotating screws along parallel axes helping in the transfer of material along its length (Seem et al., 2015). These screws are made up of smaller several screw elements which can help alter the flow of the material along the axis and can aid the mixing, breakage and other mechanisms which can affect the CQAs of the final granules. TSG can also support higher production volume compared to a batch granulator. TSGs have a larger design space due to the large number of independent operating parameters. This results in a large design of experiments which needs to be performed for optimization of the process to obtain the desired granule critical quality attributes (CQAs). Performing large number of experiments in early-stage process development when large amounts of active pharmaceutical ingredient (API) may not be available. Thus, there is a need for development of generalized models that can predict the outcome of the TSG. This model would need to be trained on a large data set of experiments which incorporates the effects of various independent operating parameters on the final granule CQAs and process outputs.

Neural networks with their dense structure have proven to be able to capture complicated relationships between inputs and outputs accurately. These neural networks can also be used to create reduce order models for faster prediction of these processes. Recently, to improve the prediction of neural networks for more complex physical processes, physical information about the process has been added to supplement its training (Mao et al., 2020; Raissi et al., 2019). Other studies have focused on constraining outputs of the neural networks with physics-based boundaries to make better informed models which, have the ability to accurately predict process outcomes(Zhu et al., 2019). These physics-based boundaries when incorporated into the loss these neural networks, help the model learn the underlying physics of the process leading to accurate predictions and more reliability under uncertain process conditions (Sampat and Ramachandran, 2021). These physics-based boundaries can be added to both the representation loss as well as the supervised loss, which leads to the addition of an extra loss function to the training. They have also resulted in neural networks requiring less data to train, which is especially useful with TSGs as this would reduce the amount of experimental data required.

In this work, a physics-constrained autoencoder (PCSAE) network was developed to create a reduced order model to represent a complete TSG process. Experimental data from 13 previously published literature was collected for various operating process parameters, process outputs and granule CQAs. The boundary conditions for each output were determined and were added the loss function of the developed PCSAE network. Sensitivity analysis was also performed on the PCSAE to determine whether it was able to capture the physical information about the process.

2. Methods

2.1. Data collection and completion methods

Twin-screw granulators with a wide design space have a large number of process parameters and geometry which can be varied. These variations when combined with changes in formulation can lead to an almost infinite combinations which can make the development of a general model for TSG very complicated. To incorporate all these effects a detailed data collection is required. In this study, data was collected from 9 different previously published experiments encompassing changes in formulation, process parameters and geometry(Dhenge et al., 2013, 2012; Kumar et al., 2016; Meier et al., 2017; Meng et al., 2019; Mundozah et al., 2020; Shirazian et al., 2017; Vercruysse et al., 2012). A total of 227 data points were collected for the creation of the model. Granule growth within a granulation process can be inferred from the process outcomes and critical quality attributes (CQAs) of the granules obtained. Some of the process outcomes of the TSG process commonly studied are residence time distribution (RTD), mixing, torque inside the system, while granule size distribution (GSD) and granule density/porosity are the commonly studied granule quality attributes (Seem et al., 2015). Table 1 lists all the input parameters and outputs collected from each of the sources to develop the PCSAE model for a TSG process. In some literature, outputs had been reported in figures and each figure was processed individually for relevant data. The data from each plot was extracted using WebPlotDigitizer (Rohatgi, 2021). The data was split in the ratio of 3:1 for training and validation.

Table 1: Inputs and output monitored for the development of the PCSAE model

Input Parameters			Output Parameters
Geometry	**Process**	**Material**	
Number of CE and KE (nCE,nKE)	L/S ratio	Initial PSD	Granule size distribution
Staggering angle (SA) of KE	Screw Speed	Binder viscosity	Torque
L/D Ratio	Feed Rate	% API in powder	Mean Residence Time
Granulator diameter	Temperature		
Liquid addition position			

In this study, a multivariate linear regression was used since the torque and MRT values for a TSG are dependent on several inputs instead of only a single input. The regression model ($Y = BX + X_i$) was trained using the *sklearn* (Pedregosa et al., 2011) package in Python. The regression model used the existing data for torque and MRT for training. Here, Y is the response matrix of size $n \times p$, X is the matrix containing all predictors with size of $n \times (q + 1)$. B is a $(q + 1) \times p$ matrix of fixed parameters, X_i is the intercept matrix of size $n \times p$. Here n represents the number of observations, q are the number of inputs or predictors and p represents the number of responses or outputs. This model is often referred to as deterministic regression imputation. Such an imputation can add a bias to the predictions. To remove such biases, uncertainty can be added back to these models.

2.2. Development of Physics-constrained supervised auto-encoders (PCSAE)
An auto-encoder (AE) is a neural network which output are the same as the inputs and during its training for reconstruction, they can extract underlying attributes which can enable accurate predictions (Le et al., 2018). Single-layer AEs with linear activation functions are equivalent to principal component analysis, moreover non-linear auto-encoders have found to extract key attributes (Vincent et al., 2010). A supervised auto-encoder (SAE) is an AE with the addition of a supervised loss on the representation layer. A single linear layer SAE would perform like a partial least square method. The addition of a supervise loss to the AE better directs the representation learning.

The PCSAE model for this study was developed in Python v3.7.6 using Keras (Chollet et al., 2015). Keras is a wrapper used for machine learning package Tensorflow (Martin Abadi et al., 2015) developed by Google. The network had 12 input nodes which were divided into three separate groups as shown in Table 1. This helped create the 3 different reduced dimensional bottleneck layers representing each of the group individually. This bottleneck layer was then used for both reconstruction of the inputs as well as training the outputs with the physical constraints. The output physical constraints were obtained using physics-based boundaries. Maximum value boundaries

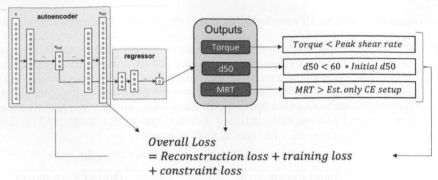

Figure 1: Physics-constrained supervised auto-encoder (PCSAE) model

for the median granule diameter and torque were determined using an empirical correlation and peak shear rate respectively, while a minimum boundary value was determined for the mean residence time (MRT). The minimum value was based on a screw configuration consisting only of conveying elements, which are known to aid conveying of material with little to no back-mixing. No physical constraints were introduced for the reconstruction. Figure 1 contains a schematic detailing the PCSAE model used. A single encoding layer with four nodes was used for the three individual inputs layers, a single decoding layer was used for reconstruction with eight nodes. Four layers were used for prediction of the outputs of the TSG with four nodes in each layer. All layers used the *'tanh'* activation function. The 'Adam' optimizer was used for optimization of the PCSAE with a learning rate of 0.008.

3. Results

3.1. Performance of the PCSAE

The total loss for the system was calculated as the summation of the reconstruction loss, training loss and the error due to the physical boundary constraints. These losses help aid the training of the system and prevent over-fitting of the model. For the PCSAE trained model no over-fitting was observed. The coefficient of determination (R^2) for prediction and reconstruction of outputs and inputs were 0.64 and 0.86 respectively. These values indicate that PCSAE was accurate to reconstruct the inputs to the model while the prediction of the outputs may not always be accurate. Figure 2(a) represents a parity plot for the predicted values of the outputs v/s the actual experimental values and it can be observed that some of the points are away from the $x = y$ line, indicating low accuracy. Figure 2(b) illustrates the parity plot for the reconstruction of the inputs and with an even spread across the $x = y$ line. The accuracy for the output prediction can be

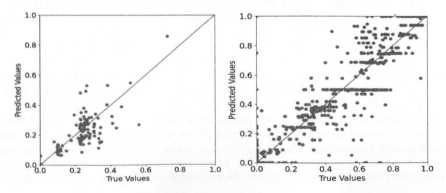

Figure 2: Parity plots for (a) Output prediction (b) reconstruction.

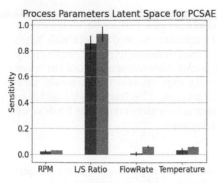

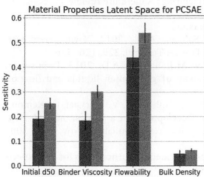

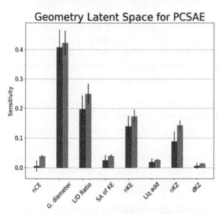

Figure 3: Sensitivity analysis of each input on the reduced dimension layer. The blue bars represent direct effects while the red bars indicate the total effect of the inputs.

improved in several ways including a deeper neural network for regression of the outputs, optimizing the hyperparameters of the neural network, improving the boundaries conditions.

3.2. Sensitivity analysis of PCSAE

To understand the effect of individual inputs on the outputs it is necessary to study their effects on the individual reduced dimension nodes. These nodes in-turn used to predict the outputs as well as reconstruct the inputs. It is vital for a model to capture the physics of the system well such that it considers the effect of each input. In this study, a Sobol sensitivity analysis was performed by varying the inputs across the range of values found in the literature. This sensitivity was compared against a supervised autoencoder without physical boundary constraints, and it was found that the PCSAE's sensitivity captured more physical information about the process than the non-physics constrained autoencoder. Figure 3 shows the sensitivity of the inputs on the reduced dimensions nodes. The effect of L/S ratio and RPM seems to the highest from the process parameters, while the contribution of different material properties seems to be almost equal, and the effect of staggering angle is the most prominent from the geometry parameters. These effects have been studied in literature and are in close accordance with the observed results. The effects of the inputs in the normal autoencoder system were observed to be skewed and did not align with physical observations.

4. Conclusions

Twin-screw granulation is a complicated process with an infinite number of combinations possible for its operation. With the help of developed physics-constrained autoencoder model, we were able to incorporate all the effects into a single model. This robust modelling framework is required could reduce the number of dimensions of the inputs to 3 segregated latent spaces for better process understanding. This framework which has been trained on several experimental datasets was able to capture the underlying physics of the system with accuracy of ~65%. The model was able to identify key inputs affecting the outputs which may not be captured using a regular autoencoder. The overall performance of the model can further be increased by optimizing the neural network structure and including more datasets with larger variations in the inputs. This model can further be used to reduce experimentation by supplementing the design of experiments. Prediction of the latent spaces could be used to assess the granule growth regimes and identify experiments which would yield desired granule CQAs resulting in material and cost saving. This framework could also be adapted to different unit operations with changes in boundary conditions for desired outputs for better cost saving during process development.

References

Chen, Y., Yang, O., Sampat, C., Bhalode, P., Ramachandran, R., Ierapetritou, M.G., 2020. Digital Twins in Pharmaceutical and Biopharmaceutical Manufacturing: A literature review. Processes 8, 1–33.

Chollet, F., others, 2015. Keras. Available at: https://keras.io

Dhenge, R.M., Cartwright, J.J., Hounslow, M.J., Salman, A.D., 2012. Twin screw wet granulation: Effects of properties of granulation liquid. Powder Technol. 229, 126–136.

Dhenge, R.M., Washino, K., Cartwright, J.J., Hounslow, M.J., Salman, A.D., 2013. Twin screw granulation using conveying screws: Effects of viscosity of granulation liquids and flow of powders. Powder Technol. 238, 77–90.

Kumar, A., Dhondt, J., Vercruysse, J., De Leersnyder, F., Vanhoorne, V., Vervaet, C., Remon, J.P., Gernaey, K. V., De Beer, T., Nopens, I., 2016. Development of a process map: A step towards a regime map for steady-state high shear wet twin screw granulation. Pow. Tech. 300, 73–82.

Mao, Z., Jagtap, A.D., Karniadakis, G.E., 2020. Physics-informed neural networks for high-speed flows. Comput. Methods Appl. Mech. Eng. 360, 112789.

Martin Abadi, and others, 2015. TensorFlow: Large-Scale Machine Learning on Heterogeneous Systems.

Meier, R., Moll, K.P., Krumme, M., Kleinebudde, P., 2017. Impact of fill-level in twin-screw granulation on critical quality attributes of granules and tablets. Eur. J. Phar. Biop. 115, 102–112.

Meng, W., Román-Ospino, A.D., Panikar, S.S., O'Callaghan, C., Gilliam, S.J., Ramachandran, R., Muzzio, F.J., 2019. Advanced process design and understanding of continuous twin-screw granulation via implementation of in-line process analytical technologies. Ad. Pow. Tech. 30, 879–894.

Mundozah, A.L., Yang, J., Omar, C., Mahmah, O., Salman, A.D., 2020. Twin screw granulation: A simpler re-derivation of quantifying fill level. Int. J. Pharm. 591, 119959.

Pedregosa, F. and others, 2011. Scikit-learn: Machine Learning in Python. J. Mach. Learn. Res. 12, 2835--2830.

Raissi, M., Perdikaris, P., Karniadakis, G.E., 2019. Physics-informed neural networks: A deep learning framework for solving forward and inverse problems involving nonlinear partial differential equations. J. Comput. Phys. 378, 686–707.

Rohatgi, A., 2021. Webplotdigitizer: Version 4.5.

Sampat, C., Ramachandran, R., 2021. Identification of granule growth regimes in high shear wet granulation processes using a physics-constrained neural network. Processes 9.

Seem, T.C., Rowson, N.A., Ingram, A., Huang, Z., Yu, S., de Matas, M., Gabbott, I., Reynolds, G.K., 2015. Twin screw granulation - A literature review. Powder Technol. 276, 89–102.

Shirazian, S., Kuhs, M., Darwish, S., Croker, D., Walker, G.M., 2017. Artificial neural network modelling of continuous wet granulation using a twin-screw extruder. Int. J. Pharm. 521, 102–109.

Vercruysse, J., Córdoba Díaz, D., Peeters, E., Fonteyne, M., Delaet, U., Van Assche, I., De Beer, T., Remon, J.P., Vervaet, C., 2012. Continuous twin screw granulation: Influence of process variables on granule and tablet quality. Eur. J. Pharm. Biopharm. 82, 205–211.

Zhu, Y., Zabaras, N., Koutsourelakis, P.S., Perdikaris, P., 2019. Physics-constrained deep learning for high-dimensional surrogate modeling and uncertainty quantification without labeled data. J. Comput. Phys. 394, 56–81.

Proceedings of the 14th International Symposium on Process Systems Engineering – PSE 2021+
June 19-23, 2022, Kyoto, Japan © 2022 Elsevier B.V. All rights reserved.
http://dx.doi.org/10.1016/B978-0-323-85159-6.50282-7

CSTR control with deep reinforcement learning

Borja Martínez[a], Manuel Rodríguez[a*] and Ismael Díaz[a]

[a]*Department of Chemical Engineering, Universidasd Politécnica de Madrid, José Gutierrez Abascal 2, Madrid 28006, Spain*
manuel.rodriguezh@upm.es

Abstract

In this paper we have applied the use of the deep reinforcement learning (DRL) for process control, to explore its applicability. The main objective is to develop a controller based on the deep reinforcement learning methodology in order to keep the level and composition of a continuous stirred-tank reactor under control.

Keywords: Deep Reinforcement learning; process control.

1. Introduction

The Oil&Gas and the process industry use mostly decentralized PID control and in some units Model Predictive Control. Both are mature technologies and well established in industry practices. But still, many PIDs are badly tuned and the costs of developing a model based predictive control are high. This classic control is difficult to implement and requires a lot of resources in complex processes that are not easy to model.

Today, machine learning has had a new outburst and specifically reinforcement learning. The use of neural networks with reinforcement learning, in what is called deep reinforcement learning has shown astounding results in some domains, such as games, (where the machine called AlphaGo created by DeepMind defeated the world champion of the game Go), self-driving cars, medicine o process control.

This paper is focused in the last one, specifically for the case of a continuous stirred-tank reactor with the aim of improving its classical process control, since this could have implications, such as replacing existing process control technology, mitigating the limitations of Model Predictive Control or helping to manage controller settings.

The structure of this paper has two distinct parts: the first one presents a brief explanation of the deep reinforcement learning methodology and its application in process control (section 2), and the second one shows the case study in which the work has been carried out, with the corresponding results and conclusions obtained (sections 3 to 5).

2. Reinforcement Learning and Deep Reinforcement Learning

Reinforcement learning is one of the three methodologies, together with supervised learning and unsupervised learning, that make up Machine Learning (Bishop, 2006). Unlike the other two, reinforcement learning does not use a set of labelled data, but rather the agent learns a task by interacting with the surrounding environment and evaluating the actions performed (Sutton and Barto, 2018). Schematically, the process carried out in reinforcement learning is illustrated in Fig 1.

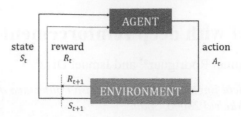

Fig. 1. Reinforcement Learning loop

The figure represents a cycle between the agent and the environment that starts with the observation of an action at time instant t, performed by the agent on the environment. This produces a series of changes at future instants (t+1) for both the reward and the new state of the environment, so that the cycle would continue with a new action by the agent (A_{t+1}).

The aim of all this is for the agent to be able to learn the relationship between the actions executed and the states obtained, known as policy π (A_t=a | S_t = s), and to maximise the long-term value of the rewards according to the value function $v_\pi(s)$:

$$v_\pi(s) = E\{R_{t+1} + \gamma R_{t+2} + \gamma^2 R_{t+3} \dots | S_t = s\} \tag{1}$$

The parameter $\gamma \in [0, 1]$, denoted as discount factor, determines the behaviour of $v_\pi(s)$ by prioritising long-term rewards.

Through this value function, the algorithms update their parameters iteratively with the intention of improving the policy associated with this function. In such a way that if the agent performs good actions, so that its policy is good and improves, it means that the values of the associated function will be greater at each iteration.

It is therefore conceivable that this function may reach a maximum value when the resulting policy is optimal. Optimal Control Theory supports this premise by means of Bellman's equation:

$$v_*(s) = {}^{max}_{a} \Sigma_{s',r} p(s',r \,|s,a) \,[r + \gamma v_*(s')] \tag{2}$$

Here the term v_* refers to the optimal policy value function, and *p(s', r |s, a)* is the transition probability. This factor indicates the probability that the environment transitions to a new state *s'* and offers a reward *r*, when the environment is in the previous state *s* and the action *a* has been executed.

Once the optimal value function is known, the associated optimal policy $\pi_*(s)$ can be found using the transition model (Shin, J. et al., 2019). The learning procedure is based on the Actor-Critic methodology, whereby according to the existing policy, an action is chosen to be performed on the environment (Actor), and subsequently evaluated, based on the reward issued by the environment (Critic). After this evaluation, the parameters governing the Actor policy are readjusted for immediate future actions.

This process can be used with neural networks (deep reinforcement learning, DRL) to eliminate the need to store all state and value pairs and allow the agent to estimate state values using an approximation function. Within DRL there are numerous algorithms that differ mainly in their architecture to optimise the Actor policy. Examples are: Proximal Policy Optimization (PPO), Deep Deterministic Policy Gradient (DDPG), Soft-Actor Critic (SAC) or Twin Delayed DDPG (TD3).

2.1. Twin Delayed DDPG: TD3 algorithm

The TD3 algorithm is a model-free, online, off-policy reinforcement learning method, in which the agent follows the Actor-Critic methodology to achieve an optimal policy that maximizes the value of the expected cumulative long-term reward (MathWorks, 2021).

As the name indicates, TD3 is an extension of the DDPG algorithm. This approach assumes that, for environments with continuous action spaces, the Bellman optimisation function with which the policy is learned is differentiable with respect to the action argument. Therefore, a gradient-based learning rule is established for a policy $\mu(s)$ that exploits this fact (Spinning Up, 2021).

Since this estimation of the value function can imply an erroneous learning by the agent, the TD3 algorithm aims to solve this by means of three adjustments: (1) learning two Critic ("twin") functions to form the targets in the Bellman error loss functions, (2) updating the Actor parameters less frequently than Critic ("delayed") and (3) adding noise to the action chosen by Actor with the intention of preventing the policy from exploiting errors made in the value function estimates.

2.2. Deep Reinforcement Learning for process control

The control of industrial processes is a rather complex task, largely due to the non-linearity of the processes and the fact that in many cases there is more than one control loop. This has motivated the development of new control techniques that adapt more efficiently to the process in question (Robayo, F., et al., 2015).

The use of neural networks stands out among these techniques because they are non-linear models that can represent systems based only on the input and output data of the system, and they are highly adaptive by adjusting their parameters to changing operating conditions (Morcego, B., 2000).

This technique has been applied in cases such as the gasoline blending process (Yu, W., et al., 2004) where the use of recurrent neural networks is proposed to model the process, without the need to know the equations that define it, the system of interconnected tanks (Robayo, F., et al., 2015) where a neural controller based on an inverse model is developed or the case of a predictive controller based on neural networks for the control of the water level in a steam generator (Parlos, A., 2001).

The main objective pursued with the application of deep reinforcement learning in process control is to ensure that the value of a desired variable is the one established by its set point, while at the same time complying with the constraints of the process.

It is important to emphasize that the action space to be performed by the agent is continuous, rather than discrete, as in other areas where reinforcement learning has been successful.

In this framework, the agent represents the controller and the environment represents the process, so that the interaction between the agent and the environment is achieved through the actions (control actions) that the agent performs depending on the state of the environment it receives. In addition, the reward system is added to evaluate the quality of the chosen action, according to the variation that the state of the environment has undergone (Spielberg, S.P.K., et al., 2017).

3. Case study

This section presents the continuous stirred-tank reactor (CSTR) unit that has been controlled using the TD3 algorithm.

In this reactor, the first order reaction (A → B) takes place isothermally, while it receives the input of two flows with different concentrations of reactant, and the product is obtained by gravity.

This work focuses on the control of two variables: the level of content inside the reactor and the concentration of the product. Moreover, it is intended to be able to work with different setpoints defined for each variable.

3.1. Implementation of the process

An agent has been designed to be responsible for performing two simultaneous actions on two other variables of the process, which have a direct impact on the variables to be controlled. These manipulated variables are the opening of the valve located in one of the inlet flows and the reactor temperature.

The environment, which as shown in Fig.1 is the other major player in the reinforcement learning loop, has been defined following the correct mass balance of the CSTR. The state of this environment, which contains the information that the agent receives from it, is defined by the variables that define each change of state: the instantaneous measurements of the variables to be controlled (level and product concentration), the absolute errors made in each control variable with respect to their setpoints, and the integrals and derivatives of these errors.

The most important hyperparameters of the reinforcement learning algorithm used by the agent in this work are shown in the following Table 1.

Table 1. Hyperparametros del agente para la unidad CSTR

Hyperparameter	Value	Note
Batch size n	64	
Replay memory size	10^6	Older transitions are replaced
Policy and Critic learning rate	$3 \cdot 10^{-4}$	Step size for ADAM
Policy hidden layers	2	1st: 400 neurons; 2nd: 200 neurons
Critic hidden layers	2	1st: 800 neurons; 2nd: 400 neurons
Hidden activation function	Relu	
Output activation function	Tanh	Only for policy
Loss function	MSE	
Target update rate (τ)	0.003	
Discount factor (γ)	0.99	
Exploration noise	0.2	Ornstein-Uhlenbeck process
Warm-up time	1,000	Timesteps until training starts

4. Simulation results

The resulting learning curve after application or the algorithm is shown in Fig.2 This curve represents the evolution of the reward values obtained throughout the episodes used to simulate the TD3 algorithm.

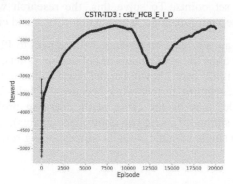

Fig. 2. Learning curve

The curve has been obtained for a simulation of 20,000 episodes. It starts with low reward values, as expected, and then grows progressively to high reward values as the episodes progress. It should be note that once a maximum reward value has been obtained, the learning curve declines from episode 10,000 onwards, which can be justified by a process of exploration by the agent seeking other possible solutions since the rewards already obtained do not continue to improve. This explanation is plausible because in episode 12,500, the curve is back on track and grows until it reaches the maximum values already obtained.

To complete the analysis of the learning curve, the TD3 algorithm is validated. This algorithm has been trained by achieving different set points for each control variable that are randomly generated episode after episode.

The Fig.3 show the response of the agent when the set point for each control variable is specified. In addition, the evolution of the reactant inside the CSTR is represented.

It can be seen that the evolution over time of the variables to be controlled is quite satisfactory, as the agent manages to achieve the objective of maintaining the control variables at their set points, when these are quite far apart.

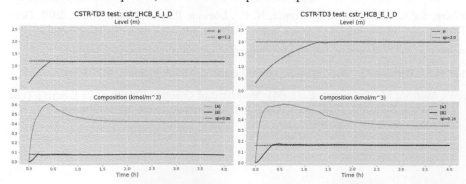

Fig. 3. Validation of the TD3 algorithm.

5. Conclusions and further work

The application of the TD3 algorithm to control the level and composition of a continuous stirred-tank reactor has generated favourable results, although the learning process is complex due to the difficulty of creating a multivariable and adaptive controller to be applicable to several set points. To solve this, the research will continue with the application of other deep reinforcement learning algorithms and configurations.

In addition, this work aims to give an advance with respect to PID controllers, since in this case the controller aims to cover a range of defined set points, without the need to be retuned. Furthermore, it has been shown that the reward function is important to obtain different results and it requires further work because a single function to achieve the control of two variables may be the cause of the learning problem, and it could be improved with the possibility of using two controllers (agents), so that each agent is in charge of controlling a single variable, with its corresponding reward function.

References

Bishop, C.M. , 2006. Pattern Recognition and Machine Learning (Information Science and Statistics). Springer-Verlag New York. Inc. Secaucus, NJ, USA.

MathWorks (2021) Retrieved from https://es.mathworks.com/help//reinforcement-learning/ug/create-agents-for-reinforcement-learning.html

Morcego, B. , 2000. Estudio de redes neuronales modulares para el modelado de sistemas dinámicos no lineales. Universitat Politècnica de Catalunya.

Parlos, A., Parthasarathy, S., & Atiya, A. (2001). Neuro-Predictive Process Control Using On-Line Controller Adaptation. IEEE TRANSACTIONS ON CONTROL SYSTEMS TECHNOLOGY, VOL. 9, NO. 5, 741-755.

Robayo, F., Barrera, A., & Polanco, L. , 2015. Desarrollo de un controlador basado en redes neuronales para un sistema multivariable de nivel y caudal. *Revista Ingeniería y Región*, 14(2), 43-54.

Shin, J. , Badgwell, T. A. , Liu, K., Lee, J. H., 2019. Reinforcement Learning –Overview of recent progress and implications for process control. Computers and Chemical Engineering. 127, 282–294.

Spielberg, S.P.K. , Gopaluni, R.B. , Loewen, P.D., 2017. Deep Reinforcement Learning Approaches for Process Control. 6th International Symposium on Advanced Control of Industrial Processes 201 - 206.

Spielberg, S.P.K. , Tulsyan, A. ,Lawrence, N. P. , Loewen, P.D., Gopaluni, R.B. , 2019. Deep Reinforcement Learning for Process Control : A Primer for Beginners. AICHE Journal 2019.

Spinning Up (2021). Retrieved from https://spinningup.openai.com/en/latest/index.html

Sutton, R.S. , Barto, A.G. , 2018. Reinforcement Learning: An Introduction. MIT Press.

Yu, W., Moreno, M., & Gómez, E. (2004). Modelling of Gasoline Blending via Discrete-Time Neural Networks. IEEE International Joint Conference, (págs. 1291-1296 vol 2).

Proceedings of the 14th International Symposium on Process Systems Engineering – PSE 2021+
June 19-23, 2022, Kyoto, Japan © 2022 Elsevier B.V. All rights reserved.
http://dx.doi.org/10.1016/B978-0-323-85159-6.50283-9

Application of machine learning and big data for smart energy management in manufacturing

Manu Suvarna[a], Pravin P.S[a], Ken Shaun Yap[b], Xiaonan Wang[a,c*]

[a] Department of Chemical and Biomolecular Engineering, National University of Singapore, 4 Engineering Drive 4, Singapore, 117585.
[b] Singapore Institute of Manufacturing Technology, 08-04, Innovis, Singapore, 138634.
[c] Department of Chemical Engineering, Tsinghua University, Beijing 100084, China
chewxia@nus.edu.sg

Abstract

Within the *Industry 4.0* context, platforms such as cyber-physical production system (CPPS) offer numerous opportunities for smart energy management in manufacturing. In this study, we demonstrate the application of big data and machine learning (ML) to foster such practices for real manufacturing environments by taking the Model Factory (MF) in Singapore as a testbed. We first used supervised learning algorithms to predict machine-specific load profiles via energy disaggregation at the MF shop floor. Here, the light gradient boosting machines had the best predictive performance with a mean absolute error and root mean squared error of 0.035 and 0.106 (units in Watts). We then coupled unsupervised learning with mathematical optimization to devise an optimal energy scheduling plan for facility management at the MF. When applied for day-ahead scheduling, the data-driven optimizer showed cost benefits of 14% in comparison to the current existing conditions. The study successfully demonstrated the application of big data and ML in the drive towards smart manufacturing practices.

Keywords: smart manufacturing, energy disaggregation, light gradient boost, k-mean clustering, data-driven optimization

1. Introduction

The emergence of the fourth industrial revolution in recent years, commonly referred to as *Industry 4.0* has challenged and disrupted conventional manufacturing norms. Platforms such as cyber physical production systems (CPPS) and technologies such as Internet of Things (IoT), Artificial Intelligence (AI), digital twins etc. within the *Industry 4.0* framework, are transforming global manufacturing practices (Suvarna et al., 2021; Tao et al., 2018). The growing popularity of these technologies have sparked an interest on their potential application to reduce the energy consumption of manufacturing industries.

The energy consumed in a typical manufacturing setting could stem from the power required by machines in the shop floor, process equipment, office buildings and facility management. Some of the common strategies employed to minimize the energy consumption in manufacturing include lean management principles (six-sigma) for machine performances, optimization of production planning and scheduling, and eluding power peaks during prolonged production (Suvarna et al., 2020; Tan et al., 2021). While recent works have proposed conceptual means to apply CPPS and AI for the above mentioned, there is dearth in literature on application of these technologies to real-world case studies. To this aim, we show how data-driven analytics, can be applied to real-world manufacturing practices via two case studies, 1) machine learning based energy disaggregation of individual machines in a shop floor, and 2) day-ahead energy scheduling via data-driven optimization.

Methodology

1.1. Machine specific energy disaggregation

The Model Factory (MF) at the Singapore Institute of Manufacturing Technology (SIMTech) which is an actual production environment was used as a test bed for this study (Tan et al., 2021). Four machines from the MF, namely, laser welder (LW), laser trimmer (LT), oven 1 and oven 2 were selected. For each of the machines, the following information were logged: timestamp, individual electrical load profile (Watts) and operational states (1 = *off*, 2 = *production* and 3 = *idle*) at frequency of 1 minute for a duration of 15 months spanning from October 2017 to December 2018. This resulted in the extraction of 600,000 data points an ideal representation of industrial big data (IBD).

The primary objective was to disaggregate the central power supply to the machine-specific load-profiles. As such, the *total load* was defined as the input feature whereas the *individual load profiles* for each of the four machines were defined as the target labels (Tan et al., 2021). Tree-based supervised algorithms including extreme gradient boost (XGBoost), light gradient boosting machines (LightGBM), and deep learning algorithms including ensemble regular and bidirectional long short-term memory (EnLSTM) and (EnBLSTM) were evaluated for their predictive performance for the task.

During the modelling process, the entire data was first split into training and test data with temporal specifications. The 12-month period from Oct 2017 - Sept 2018 was labelled as training set while the last 3 months from Oct - Dec 2018 was labelled as the test set. The training data was subjected to hyperparameter tuning to find the optimum combination of hyperparameters for the various algorithms evaluated, using the Bayesian optimization strategy followed by k-fold cross validation (where k = 3). Once the best hyperparameters were identified on the training set, they were also used on the test data for each of the algorithms. The best performing algorithm was identified based on the terms of lowest MAE and RMSE scores on the test dataset. These are calculated as described in equations (1) and (2).

$$MAE = \frac{1}{n}\sum_{i=1}^{n}(y_{act,i} - y_{pred,i}) \tag{1}$$

$$RMSE = \sqrt{\frac{\sum_{i=1}^{n}(y_{act,i} - y_{pred,i})^2}{n}} \tag{2}$$

where, $y_{act,i}$ and $y_{pred,i}$ are the actual and predicted values of the target variables and n is the total number of data points.

1.2. Day-ahead energy scheduling via data-driven optimization

Currently the MF sources all its energy requirements for the operations (shop floor and facility management) from the central power grid. To this cause, we proposed the implementation of a *hypothetical hybrid grid* – comprising of solar panels and waste to energy (WTE) plant, in addition to the central power grid. For this case study, we focused only on the energy consumption of the refrigeration system, which is part of the technical building services in the MF. The energy consumed by the refrigeration system E_{com}, subjected to uncertainty in ambient temperature was modelled as:

$$E_{com} = \dot{m}_{com}(h_{com,disc} - h_{com,suct})/\eta_{com,mec} \tag{3}$$

where, $\dot{m}_{com}$ is the mass flow rate of the working fluid (kg/s), $h_{com,disc}$ is the specific enthalpy of the discharge fluid (J/kg), $h_{com,suct}$ is the specific enthalpy of the suction fluid (J/kg), $\eta_{com,mec}$ is the mechanical efficiency of the compressor (-).

The energy consumption of the refrigeration system is simulated with the operating conditions presented in **Table 1** as per (ASHRAE, 2020). The T_{amb} is obtained from the scenario generation method (discussed below).

Table 1. Relevant operational data for refrigeration system

Parameters	Values
Working fluid	R123
Degree of superheat	27.8°C
Degree of subcooling	0°C
Saturated temperature of evaporator	7.2°C
Mechanical efficiency of the compressor	85%
Cooling capacity	40 kW
Compressor speed	50 Hz

The hybrid grid proposed in this study is subjected to uncertainty on both the supply and demand side. At the demand side, it has to meet the energy requirements by the refrigeration system which is significantly influenced by ambient temperature. On the supply side, it is subjected to the solar power availability (which is a function of the solar irradiance) and constantly changing price of the main power grid. Thus, the operation of the hybrid grid (comprising of solar panels, WTE and central grid) was formulated as a stochastic optimization for day-ahead energy scheduling; with the objective to minimize the total operating cost given as:

$$\min \sum_{t=1}^{NT}(C_S^t\, E_S^{t,s} + C_{WTE}^t\, E_{WTE}^t + C_M^{t,s}\, E_M^t) \qquad (4)$$

The optimization problem is devised to ensure that the energy delivered by the hybrid grid should satisfy the energy requirements to the air refrigeration system at all times. This constraint is formulated as follows:

$$ED^{t,s} \le \sum_e^E E_S^{t,s} + E_{WTE}^t + E_M^t \qquad (5)$$

The formulation also ensures that the amount of energy delivered by each energy sources should be always between the lower and upper bounds of the designed capacity of the energy sources. This constraint is formulated as follows:

$$E_{S,min}^t \le E_S^{t,s} \le E_{S,max}^t \qquad (6)$$
$$E_{WTE,min}^t \le E_{WTE}^t \le E_{WTE,max}^t \qquad (7)$$
$$E_{M,min}^t \le E_M^t \le E_{M,max}^t \qquad (8)$$

where, NT is the total number of time slots in the horizon, t is the time period, s denotes stochastic parameters, $e \in E$ denotes energy sources, C_S^t and $E_S^{t,s}$ is the cost associated and energy availability from solar at time period t, C_{WTE}^t and E_{WTE}^t is the cost associated and energy availability WTE at time period t, $C_M^{t,s}$ and E_M^t is the cost associated and energy availability from power grid at time period t, $ED^{t,s}$ is the energy demand by the refrigeration system at time period t, $E_{e,min}^t$ is the lower bound/capacity of the eth energy source and $E_{e,max}^t$ is the upper bound/capacity of the eth energy source.

The decision variables are the energy contribution from solar ($E_S^{t,s}$), WTE (E_{WTE}^t) and the power grid (E_M^t) respectively, while all the other variables in the problem formulation described from equations (1)-(8) are the given parameters.

Conventional mathematical optimization approaches assume that the uncertainty set/scenarios are perfectly given *a priori*, which is then modelled via probability distribution function (PDF) or user defined uncertainty via sample average approximations or Monte Carlo simulation (Ning and You, 2019). In contrast, we use a data-driven approach in the form of unsupervised learning to create the scenarios. The application of unsupervised learning on big data for variable of interest (on which the scenarios are created), presents benefits such as reduced scenarios with greater confidence, thereby resulting in enhanced optimization performance and faster computation (Ning and You, 2019; Tao et al., 2018). Thus, in this study, the hourly historical data (2015-2018) of solar radiance, ambient temperature and electricity pricing was sourced from relevant weather and electricity board in Singapore. The details of this historical data in the form of upper and lower bounds is presented in **Table 2**.

Table 2. The upper and lower bounds for relevant parameters in the stochastic optimization which determines the range uncertainty values

Parameters	Bounds	Solar	WTE	Mains
Price ($/kWh)	Lower	0.00	0.05	0.07
	Upper	0.07	0.11	0.11
Hourly energy availability (kWh)	Lower	0.00	0.00	0.00
	Upper	3.64	3.98	6.81

This data was then subjected to K-means clustering to create the scenarios. K-means partitions an N-dimensional population into k sets on a basis of a sample (Mehar et al., 2013). It is an unsupervised classification method to solve problems when no labels are available. In the dataset $D = \{x_t\}_{t=1}^n$, the number of clusters is K, the natural goal is to seek a partition of the dataset $D_1 \cup ... \cup D_K$, as well an associated set of cluster centroids $\mu = (\mu_1, ..., \mu_K)$, such that the sum of Euclidian distances between features and each centroid is as small as possible. Initial μ are randomly selected. While finding the global minimum of distance, there are two related steps: (1). Minimize the distance with respect to regulating the partitioned dataset D_K for fixed centroid μ_K; (2). Minimize the distance with respect to regulating the centroid μ_K for fixed partitioned dataset D_K. The K-means algorithm is optimized by alternating between these two steps until converges. Eventually, all points in the dataset locate around one of the K cluster centroids that achieve stable state. When centroids do not change, clusters are fixed, which means K scenarios are produced. Each probability of corresponding scene is the percentage of the number of labels in this cluster among the total count of labels in the dataset

The stochastic optimization was solved using the CPLEX solver. Its performance was compared with a base case analysis i.e., the actual energy consumed by the refrigeration system on 5th of September 2019. The base case is the actual working condition of the refrigeration system, and it uses only the central power grid as the current available energy source.

2. Results

2.1. Energy Disaggregation

In the energy disaggregation studies, it was observed that the tree-based algorithms (LightGBM and XGB) had lower MAE while the deep learning algorithms (EnLSTM and EnBLSTM) had lower RMSE to the tree-based counterparts. However, the performance of all the models were very comparable and the prediction accuracies were in close proximities to each other. Given the ease in tuning tree-based models, and the combined model balance in terms of MAE and RMSE, it was revealed that performance of order was LightGBM > EnBLSTM > XGBoost. Here it is worth a mention that, in the case of structured datasets as used in this study, tree-based models have very comparable performances to that of deep learning algorithms as they can fit the hyperparameters to the input features, which is a natural extension to their workflow. Moreover, both LightGBM and XGB are ensemble tree models and as such sum the predictions of many decision trees into a final one and thus closely compete to deep learning algorithms where the latter also use a multitude of neurons for their prediction performance. The average predictive performance of all the algorithms for the 4 machines is shown in **Table 3**. The actual (Act) v/s predicted (Pred) plot of energy disaggregation for LW and LT is show in **Figure 1**.

Table 3. Comparative evaluation of algorithms for energy disaggregation study

Ranking	Algorithm	MAE (Watts)	Algorithm	RMSE (Watts)
1	LightGBM	0.035	EnBLSTM	0.100
2	XGBoost	0.036	EnLSTM	0.103
3	EnBLSTM	0.039	LightGBM	0.106
4	EnLSTM	0.041	XGBoost	0.106

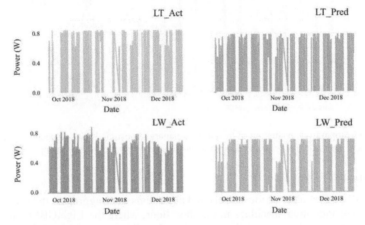

Figure 1. Actual v/s predicted plots of energy disaggregation for LW and LT for the best performing LightGBM model. Sourced with permission from (Tan et al., 2021)

2.2. Data-driven stochastic optimization for day-ahead energy scheduling

The energy demand for day ahead scheduling under uncertainty was first compared with the base case, where it was seen that the MAE and RMSE of energy demands between the two was 0.25 and 0.27 kWh respectively. With an average energy demand throughout the day between 6.5 and 6.8 kWh, the energy demand projected by the day-ahead scheduling deviated from the base case by approximately 4%.

On subjecting the day-ahead energy demand to stochastic optimization, it was identified that that the hybrid grid suitably met the demand by optimally distributing the three energy sources with the objective to minimize daily operating cost. Specifically, it was observed that the solar panel contributed significantly to the hybrid grid during the morning and afternoon hours, i.e., almost 100% of its available energy to meet energy demand thereby minimizing the overall cost. On the other end, the early morning and the late-night hours were purely met by the optimal combination of WTE, and power mains based on their hourly cost distribution. In addition to the distribution of the individual energy sources in the hybrid grid, its performance was also compared to the base case condition for cost saving evaluation. Here it was realized that the actual energy consumed by the refrigeration system using power mains only for the entire day operation was 14.18 SGD. In contrast, the overall cost incurred by the hybrid grid inclusive of all the uncertainties was found to be 12.41 SGD, concluding that adopting a hybrid grid could potentially lead to cost savings by 14% with day-ahead energy scheduling even under uncertainty. The comparative results of the base case and stochastic optimized model (hybrid grid) along with–potential cost savings are shown in **Figure 2**. The data-driven optimization presented in this study essentially captures significant uncertainty in the data, as it is trained on big data (hourly interval data for 3 years) – which also make the optimization more reliable.

One observed aspect in the creation of the uncertainty set via the k-means clustering algorithm was\s the variation of the uncertainty set on repeated implementation. This is due to the randomized component in k-means clustering. Although the results obtained from the k-means clustering varied in every iteration, the expected value of the uncertainty sets over every iteration remained the same. Hence, the final energy schedule resulting from the optimization problem did get affected by the random initialization in k-means clustering and gave consistently similar distribution mix and cost savings.

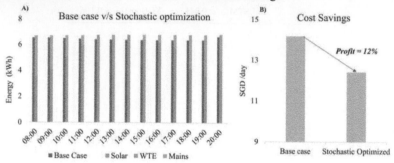

Figure 2. A) Comparative evaluation of the base case energy consumed with respect to hybrid grid B) 14% savings in terms of cost is observed with data-driven optimization under uncertainty

3. Conclusion

In this study, we first used various supervised algorithms to disaggregate the central load of the MF to four individual machines in the shop floor, where the LightGBM was found to be the most accurate in terms of the predictive as well generalization ability – as it was trained on a big data. In another case study, we showcased the application of unsupervised learning for creating scenarios essential to solve stochastic optimization. The approach is effective as was observed from the fact that this data-driven optimization resulted in cost savings of 14% for day-ahead energy scheduling for facility management. Both the approaches are purely data-driven and cross deployable in any manufacturing setting provided is availability of historical data in the plant.

References

ASHRAE, 2020. 2020 ASHRAE Handbook -- HVAC Systems and Equipment. ASHRAE.

Mehar, A.M., Matawie, K., Maeder, A., 2013. Determining an optimal value of K in K-means clustering, in: 2013 IEEE International Conference on Bioinformatics and Biomedicine.

Ning, C., You, F., 2019. Optimization under uncertainty in the era of big data and deep learning: When machine learning meets mathematical programming. Computers & Chemical Engineering 125, 434–448.

Suvarna, M., Büth, L., Hejny, J., Mennenga, M., Li, J., Ng, Y.T., Herrmann, C., Wang, X., 2020. Smart Manufacturing for Smart Cities—Overview, Insights, and Future Directions. Advanced Intelligent Systems 2, 2000043.

Suvarna, M., Yap, K.S., Yang, W., Li, J., Ng, Y.T., Wang, X., 2021. Cyber–Physical Production Systems for Data-Driven, Decentralized, and Secure Manufacturing—A Perspective. Engineering.

Tan, D., Suvarna, M., Shee Tan, Y., Li, J., Wang, X., 2021. A three-step machine learning framework for energy profiling, activity state prediction and production estimation in smart process manufacturing. Applied Energy 291, 116808.

Tao, F., Qi, Q., Liu, A., Kusiak, A., 2018. Data-driven smart manufacturing. Journal of Manufacturing Systems, Special Issue on Smart Manufacturing 48, 157–169.

Proceedings of the 14th International Symposium on Process Systems Engineering – PSE 2021+
June 19-23, 2022, Kyoto, Japan © 2022 Elsevier B.V. All rights reserved.
http://dx.doi.org/10.1016/B978-0-323-85159-6.50284-0

Adaptive least-squares surrogate modeling for reaction systems

Robert E. Franzoi[a], Brenno C. Menezes[a*], Jeffrey D. Kelly[b], Christopher L. E. Swartz[c]

[a]*Division of Engineering Management and Decision Sciences, College of Science and Engineering, Hamad Bin Khalifa University, Qatar Foundation, Doha, Qatar*
[b]*Industrial Algorithms Ltd., 15 St. Andrews Road, Toronto M1P 4C3, Canada*
[c]*Department of Chemical Engineering, McMaster University, 1280 Main St W, Hamilton ON L8S 4L7, Canada*
bmenezes@hbku.edu.qa

Abstract

Surrogate modeling has been increasingly used to predict the behavior of a given system as an alternative to complex formulations that often lead to time consuming solutions and convergence issues. Surrogates are addressed herein to replace complex formulations for reactor systems within optimization problems. An adaptive sampling algorithm explores the solution space by iteratively building surrogates. Latin Hypercube Sampling is used for the experimental design (input data), and a first principles reaction formulation calculates the output data. Then, discrete least-squares regression minimizes the deviation between the surrogate response and the function being approximated. An optimization problem based on a reaction system is formulated, in which complex first principles equations are successfully replaced by the surrogates. The results indicate highly accurate predictions and near optimal solutions. Therefore, the surrogates can replace the rigorous model without significant loss in the solution quality and objective function. This methodology can potentially provide several benefits and improvements for real-time applications and for integrated optimization environments, in which the use of complex or rigorous models is not suitable.

Keywords: Surrogate modeling, adaptive sampling, optimization, data-driven, machine learning.

1. Introduction

Commercial tools for rigorous simulation have been widely used to provide highly accurate solutions for industrial problems. However, they are typically not suitable for large-scale optimization applications due to the expensive computational burden and convergence issues arising from their rigorous high-fidelity formulation. An alternative to overcome these mathematical and computational challenges derived from a detailed and complex modeling is the use of surrogate models. Several benefits have been reported in the use of surrogates in multiple applications, and the interest in developing reduced-size formulations for industrial applications has recently increased. Surrogates are built using data generated from the original model or any reliable source, and several aspects are important when designing a surrogate building strategy. First, there is a trade-off between model accuracy and computational tractability. In general, surrogates should be as accurate as possible given the availability of time and effort for their application (Mencarelli et al., 2020). Second, their functional form should be selected considering the problem characteristics, availability of data, and dimensionality (Hüllen et al., 2019). Third, the design of experiments is chosen to generate the samples needed to train the surrogates (Simpson, 2001). Fourth, a performance method is selected to measure the fit from the surrogate to the data set (Alizadeh et al., 2020).

Previous literature on the topic has shown significant benefits in the use of surrogates in optimization problems that involve complex models and the use of rigorous simulation tools (Yang et al., 2016; Franzoi et al., 2020; Franzoi et al., 2021). In this work, we are particularly interested in building simple surrogate models for replacing complex first principles equations from reaction systems. The surrogate functions are required to have high accuracy to achieve high-quality predictions and to lead to near optimum solutions, while providing simplified formulations that are faster to solve and easier to converge.

The contribution of this work relies on the implementation of an adaptive sampling framework that iteratively builds accurate and small in size surrogates for reaction systems. Each iteration consists of bounds tightening, sampling selection, surrogate building, system optimization with surrogates, and system simulation with rigorous model. Latin Hypercube Sampling (LHS) (McKay, 1979) is used for experimental design with data generated from a rigorous formulation. The surrogates are multivariable second-order polynomial functions and include constant, linear, and bilinear terms. Discrete least squares regression minimizes the deviations between the surrogate response and the function being approximated. The proposed methodology can introduce useful applications in further embedding the simplified (yet accurate) reaction formulation into further optimization decision-making environments (e.g., scheduling, control), which can be extended to other processes and applications as well.

This paper is structured as follows. In Section 2 we present the problem statement and the mathematical formulation. In Section 3 we introduce the surrogate model building framework. The case study and the respective results are presented and discussed in Section 4. Finally, we highlight the main findings of this work in Section 5.

2. Problem statement

The case study chosen to illustrate the surrogate model methodology is proposed by (Williams and Otto, 1960). The Williams-Otto plant is considered at steady state operation and it is illustrated in Figure 1. The problem is formulated using the unit-operation-port-state-superstructure (UOPSS) formulation (Kelly, 2005), in which the mass balance consistency is ensured throughout the process.

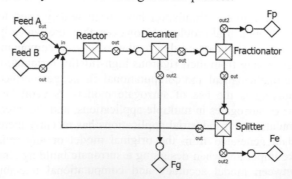

Figure 1: Williams-Otto plant flowsheet.

Although small in size, this problem is highly nonlinear due to the complex reaction system and the reflux rate that typically imposes convergence issues in the optimization. Hence, the system is very sensitive, and inaccurate formulations often lead to infeasibilities, which provides a good case study for testing and tuning surrogate model building strategies.

Two feed streams and a recycle stream enter the continuous stirred tank reactor, in which there are three exothermic reactions, with an Arrhenius temperature dependence. Eqs.(1) to (6) calculate the outlet flows of each individual component.

$$F_A^{reactor,out} = F_A^{reactor,in} - W_r f_1 e^{-\left(\frac{E_1}{T_R}\right)} Xr_A^{reactor} Xr_B^{reactor} \tag{1}$$

$$F_B^{reactor,out} = F_B^{reactor,in} - W_r f_1 e^{-\left(\frac{E_1}{T_R}\right)} Xr_A^{reactor} Xr_B^{reactor} + W_r f_2 e^{-\left(\frac{E_2}{T_R}\right)} Xr_B^{reactor} Xr_C^{reactor} \tag{2}$$

$$F_C^{reactor,out} = F_C^{reactor,in} + 2W_r f_1 e^{-\left(\frac{E_1}{T_R}\right)} Xr_A^{reactor} Xr_B^{reactor} - 2W_r f_2 e^{-\left(\frac{E_2}{T_R}\right)} Xr_B^{reactor} Xr_C^{reactor}$$
$$+ W_r f_3 e^{-\left(\frac{E_3}{T_R}\right)} Xr_C^{reactor} Xr_P^{reactor} \tag{3}$$

$$F_E^{reactor,out} = F_E^{reactor,in} + 2W_r f_2 e^{-\left(\frac{E_2}{T_R}\right)} Xr_B^{reactor} Xr_C^{reactor} \tag{4}$$

$$F_G^{reactor,out} = F_G^{reactor,in} + 1.5W_r f_3 e^{-\left(\frac{E_3}{T_R}\right)} Xr_C^{reactor} Xr_P^{reactor} \tag{5}$$

$$F_P^{reactor,out} = F_P^{reactor,in} + W_r f_2 e^{-\left(\frac{E_2}{T_R}\right)} Xr_P^{reactor} Xr_E^{reactor} - 0.5W_r f_3 e^{-\left(\frac{E_3}{T_R}\right)} Xr_C^{reactor} Xr_P^{reactor} \tag{6}$$

In the above, W_r is the reactor total mass content, f_1, f_2, f_3 are the frequency factors, E_1, E_2, E_3 are the activation energies, and $Xr^{reactor}$ are the mass fractions of each component inside the reactor. The reaction optimization problem maximizes the profit from revenue with products P and E by subtracting feedstock costs and treatment cost of by-product G in Eq.(7). Additional details of the problem and its mathematical formulation can be found in previous works (Chaudhary et al., 2009).

$$\text{Max } Z = 8400 \left(0.6614 F_p + 0.0150 F_e - 0.0441 F_a - 0.0661 F_b - 0.0220 F_g\right) \tag{7}$$

A mixed-integer quadratic programming (MIQP) technique determines optimizable surrogates to correlate variations of independent X variables to dependent Y variables. Seven input variables (reactor temperature, flow, and five inlet compositions) and six output variables (reactor outlet compositions) are used. We choose the form of each equation to account for linear, bilinear, and quadratic coefficients, as shown in Eq.(8), in which DV and IV are the sets for the dependent and independent variables, respectively.

$$Y_i = a_i + \sum_{j \in IV} b_{ij} X_j + \sum_{j \in IV} \sum_{k \in IV} c_{ijk} X_j X_k \qquad \begin{array}{l} \forall\ i \in DV, \\ \\ \forall\ j \leq k \end{array} \tag{8}$$

The coefficients a_i, b_{ij} and c_{ijk} are the parameters to be estimated within the surrogate model building strategy that minimizes the least squares error in Eq.(9), where y_{ip} and Y_{ip} are the real and the estimated value for the dependent variable i at point p.

$$\text{Minimize } \sum_{p=1}^{n} (y_{ip} - Y_{ip})^2 \tag{9}$$

The model is subject to Eqs.(10) to (13), that limit the values of the coefficients in the surrogates, where M is a large enough number, z_0, z_j, and z_{jk} are auxiliary binary variables, and K is the maximum number of coefficients.

$$-Mz_0 \leq a_i \leq Mz_0 \qquad \forall\ i \in DV \tag{10}$$

$$-Mz_j \leq b_{ij} \leq Mz_j \qquad \forall\ j \in IV, i \in DV \tag{11}$$

$$-Mz_{jk} \leq c_{ijk} \leq Mz_{jk} \qquad \forall\ \{(j,k) \in IV, j \leq k\}, i \in DV \tag{12}$$

$$z_0 + \sum_{j \in IV} z_j + \sum_{j \in IV, k \in IV, j \leq k} z_{jk} \leq K \qquad z_0, z_j, z_{jk} \in \{0,1\} \tag{13}$$

3. Surrogate model building: an adaptive sampling algorithm

Three decisions should be made upon designing the surrogate model building strategy, which concern the surrogate functional form, data generation, and quality of fit. To keep the model simple, we choose to build the surrogate as second-order polynomial functions, which have been often used due to their robust performance and computational efficiency (Yang et al., 2016). The LHS technique randomly samples points for the input variables (building the independent X data set); then, a first principles model calculates the output variables for each point to build the dependent (Y) data set. Lastly, least squares regression is employed within an MIQP formulation using Eqs.(9) to (13) to find the optimal coefficients that form the surrogate.

Building global surrogates for the entire solution space may introduce difficulties due to the wide search space in the optimization, which potentially leads to convergence issues. Therefore, an adaptive sampling framework is used to build locally accurate surrogates within their respective trust regions in an iterative fashion. Our algorithm selects the sampling bounds for each independent variable, and one surrogate is built for each dependent variable at each iteration, until a convergence criterion is met.

The algorithm is implemented in Python 3 using Microsoft Visual Studio 2019 in the industrial modeling and programming language (IMPL) platform. The optimizations are carried out by CPLEX 12.10.0, and the non-linearities are handled by a sequential linear programming strategy. The machine used was an Intel Core i7 with 2.90 GHz and 16 GB RAM. The following steps explain how the algorithm works.

1) Generate data set: Data are generated by sampling points for the input variables using LHS and evaluating each point using the rigorous model to calculate the output variables.

2) Update sampling survey: At each iteration, the sampling survey is updated around the incumbent optimal values of each variable. For that, a parameter δ is introduced, so that the sampling survey of an independent variable Xr is defined as $(Xr^{opt} - \delta) \leq Xr \leq (Xr^{opt} + \delta)$. We initially set $\delta = 0.10$, which is updated upon some shrinkage criteria.

3) Build surrogate model: The surrogate model comprises a group of six surrogate functions, one for each output variable. These functions are obtained through MIQP optimizations to identify their optimal coefficients.

4) Solve optimization problem using the Surrogate Model (SM): The surrogates are embedded in the reaction optimization problem, which is solved to local optimality.

5) Simulate system using the First Principles (FP) model: After optimizing the problem using the incumbent surrogate model, the optimal values of the decision variables (F_A, F_B, T_R) are fixed, and the system is simulated using the rigorous reaction model.

6) Convergence criteria: The algorithm terminates if there is no improvement in three sequential objective functions (with 0.01% tolerance) or at a maximum of 20 iterations.

7) Final surrogate model: The best surrogate is selected based on the objective function.

4. Example

The proposed case study is the Williams-Otto plant shown in Figure 1. Our framework iteratively builds surrogates for the reactor unit and solves the reaction optimization problem using the surrogates. The results are presented in Figure 2 (left plot). The line with circular markers represents the SM objective function, the line with square markers represents the FP objective function, and the straight line is the best objective function found by optimizing the model using the rigorous blending formulation. As the framework moves across iterations, there are improvements in the optimization search space chosen to build the surrogates, and a smooth convergence is achieved, reaching a high-quality solution within 11 iterations (beyond which there is no significant improvement in the surrogates). The best objective function found for this instance is \$1683.81, while the best surrogate model leads to an objective function of \$1683.71.

Although achieving highly accurate surrogates, reducing the size of the model would be recommended for larger problems to keep its simplicity. Thus, additional tests were performed considering different numbers of terms. For that, we run the framework multiple times by setting the maximum number of terms to be $K = \{8, 12, 18, 24, 30, 36\}$. Figure 2 (right plot) shows the objective function of the surrogates built within the framework for each scenario.

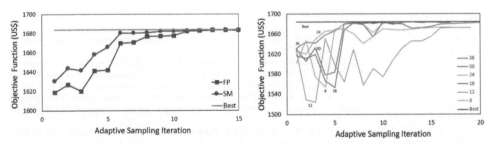

Figure 2: Adaptive sampling algorithm results.

All scenarios performed well, achieved high-quality predictions, and provided good solutions. The best performance is achieved by using 36 coefficients (0.01% lower than the best solution known), while the least accurate performance was achieved by using $K = 8$ coefficients, which resulted in a gap of 0.75%. Using less than 8 coefficients does not provide sufficiently accurate results. These results demonstrate that although a large number of terms can provide higher accuracy, sufficiently accurate surrogates of smaller size can also be built, which provides benefits in terms of reduced size, lower computational effort, and better convergence within simulation/optimization problems. In most instances, the non-improvement termination criterion is met prior to reaching the maximum number of surrogates. This provided an average reduction of over 20 % in the computational time with no loss in the objective function.

The methodology employed herein has shown to be efficient in terms of accuracy when building surrogates for highly nonlinear and complex systems, such as the Williams-Otto plant. The nonlinear Arrhenius-based equations were efficiently approximated by second-order polynomial functions, and no convergence issues are detected when embedding the surrogates in the original reaction system. The adaptive sampling algorithm successfully explores the optimization search space to find more promising regions to build the surrogates within a few iterations of the method. High-quality solutions and smooth convergence are achieved in the case study tested.

5. Conclusions

Surrogate modeling has been increasingly used to predict the behavior of a system as an alternative to complex formulations that often lead to time consuming solutions and to both convergence and calibration difficulties. In this work, an adaptive sampling algorithm iteratively explores the solution space, whereby the incumbent surrogate is embedded into an optimization problem to assure feasibility and to collect feedback for the following iteration. The methodology is applied to a reaction system network and the surrogates are built to predict the reactor outputs within optimization environments. The results indicate that the surrogates are properly built, have high accuracy, and can effectively replace the first principles model in the optimization without significant loss in the objective function. The effectiveness of the method is also demonstrated in building smaller surrogates by limiting the maximum number of coefficients, which also provides high-quality predictions. We believe this methodology is appropriate for other reaction systems and can be useful for handling data-driven black-box nonlinear formulations. Moreover, several benefits and improvements can be achieved for time-limited applications and for integrated optimization environments, in which the use of complex or rigorous models might not be suitable.

References

R. Alizadeh, J. K. Allen, and F. Mistree, 2020, Managing computational complexity using surrogate models: a critical review, Research in Engineering Design, 31(3), 275-298.

M. N. R. Chaudhary, 2009, Real time optimization of chemical processes, Doctoral dissertation, Curtin University.

R. E. Franzoi, T. Ali, A. Al-Hammadi, B. C. Menezes, 2021, Surrogate Modeling Approach for Nonlinear Blending Processes, 1st International Conference on Emerging Smart Technologies and Applications (eSmarTA), 1-8.

R. E. Franzoi , B. C. Menezes, J. D. Kelly, J. A. W. Gut, I. E. Grossmann, 2020. Cutpoint temperature surrogate modeling for distillation yields and properties, Industrial and Engineering Chemistry Research, 59 (41), 18616–18628 .

G. Hüllen, J. Zhai, S. H. Kim, A. Sinha, M. J. Realff, and F. Boukouvala, 2019, Managing Uncertainty in Data-Driven Simulation-Based Optimization, Computers and Chemical Engineering, 136, 106519.

J. D. Kelly, 2005, The Unit-Operation-Stock Superstructure (UOSS) and the Quantity-Logic-Quality Paradigm (QLQP) for Production Scheduling in the Process Industries, Multidisciplinary International Scheduling Conference Proceedings: New York, United States, 327.

J. D. Kelly, B. C. Menezes, 2019, Industrial Modeling and Programming Language (IMPL) for Off- and On-Line Optimization and Estimation Applications. In: Fathi M., Khakifirooz M., Pardalos P. (eds) Optimization in Large Scale Problems, Springer Optimization and Its Applications, 152, 75-96.

M. D. McKay, R. J. Beckman, and W. J. Conover, 1979, A comparison of three methods for selecting values of input variables in the analysis of output from a computer code, Technometrics, 21, 239-245.

L. Mencarelli, A. Pagot, and P. Duchêne, 2020, Surrogate-based modeling techniques with application to catalytic reforming and isomerization processes, Computers and Chemical Engineering, 135, 106772.

T. W. Simpson, D. K. Lin, and W. Chen, 2001, Sampling strategies for computer experiments: design and analysis, International Journal of Reliability and Applications, 2(3), 209-240.

T. J. Williams, and R. E. Otto, 1960, A generalized chemical processing model for the investigation of computer control, Transactions of the American Institute of Electrical Engineers, Part I: Communication and Electronics, 79(5), 458-473.

L. Yang, S. Liu, S. Tsoka, and L. G. Papageorgiou, 2016, Mathematical programming for piecewise linear regression analysis, Expert systems with applications, 44, 156-167.

Proceedings of the 14th International Symposium on Process Systems Engineering – PSE 2021+
June 19–23, 2022, Kyoto, Japan
http://dx.doi.org/10.1016/B978-0-323-85159-6.50285-2

Machine Learning and Inverse Optimization Approach for Model Identification of Scheduling Problems in Chemical Batch Plants

Hidetoshi Togo[a], Kohei Asanuma[b], Tatsushi Nishi[a*]

[a]*Graduate School of Natural Science and Technology, Okayama University, 3-1-1 Tsushima-naka, Kita-ku, Okayama City, Okayama 700-8530, Japan*
[b]*Graduate School of Engineering Science, Osaka University, 1-3 Machikaneyama-cho, Toyonaka City, Osaka 560-8531, Japan*

nishi.tatsushi@okayama-u.ac.jp

Abstract

Scheduling problems are widely used in recent production systems. In order to create an appropriate modeling of a production scheduling problem more effectively, it is necessary to build a mathematical modeling technique that automatically generates an appropriate schedule instead of an actual human operator. This paper addresses two types of model estimation methods for weighting factors in the multi-objective scheduling problems from input-output data. The one is a machine learning-based method, and the other one is the parameter estimation method based on an inverse optimization. These methods are applied to multi-objectives parallel machine scheduling problems. The accuracy of the proposed machine learning and inverse optimization methods is evaluated. A surrogate model that learns input-output data is proposed to reduce the computational efforts. Computational results show the effectiveness of the proposed method for weighting factors in the objective function from the optimal solutions.

Keywords: Inverse Optimization, Machine Learning, Multi-Objective Optimization, Production Scheduling, Weighting Factors, Model Identification

1. Introduction

Scheduling problems are widely used in chemical batch plants in current production systems. In recent years, the real-world scheduling problem is so large and it becomes so complicated. Therefore, it is required to aid the decision-makers to model the problem that enables the efficiency and flexibility of production systems. Data-driven optimization methods have expected to build appropriate optimization model from historical data. For multi-objective scheduling problem, a mathematical model that reflects the human operator's decision making is required based on the selection of multi-criteria optimization. However, it is not easy to set the appropriate weighting factors that indicate the importance of each objective function. If the weighting factors do not reflect the worker's intention, the desired solution cannot be obtained. Therefore, human operators must manually fine-tune the schedule.

Some related works have been addressed for estimating weighting factors in the multi-objective scheduling problems (Watanabe et al. (2002), Kobayashi et al. (2018)). Matsuoka et al. (2019) developed a machine learning approach for the identification of the objective function for parallel machine scheduling problems. Asanuma and Nishi (2020) addressed machine learning and inverse optimization approach for estimating weighting factors from historical data (Asanuma and Nishi (2020)). The exact solution algorithm is adopted to solve the inverse optimization problem. Togo et al. (2021) reported an approximate solution approach for the inverse optimization problem. However, the applicability of the practical scheduling data has not been studied in conventional works. Even if the desired schedule is obtained, the situation surrounding production environment changes. Therefore, it is necessary to repeatedly correct the weighting factors for multi-objective scheduling problem. This paper presents an inverse-optimization approach for model identification of production scheduling problem using historical data. The proposed approach is applied to real data of chemical batch plants. Various approximate solutions have been proposed for solving scheduling problems and they have advantages and disadvantages (Kise et al. (1995)). If the approximate solution method is used, the solutions are sometimes different from those obtained by the exact solution method. In this paper, we propose machine learning and inverse optimization methods to estimate weighting factors in the objective function. In order to apply to large scale problems, a simulated annealing method is proposed to derive near-optimal solutions for the multi-objective parallel machine scheduling problems. In the machine learning method, we try to extract features to improve the estimation accuracy. From the result of numerical experiments, it is confirmed that the estimation accuracy is improved by adding the feature of the errors in due date setting for each machine in the machine learning for the parallel machine scheduling problem. This paper consists of the following sections. Section 2. explains the problem definition. Section 3. introduces our proposed approach for estimating weighting factors in the objective function. Section 4. provides computational experiments. Section 6. states our conclusion and future works.

2. Problem description

2.1. *Estimation problem for weighting factors in the objective function*

We consider a parallel machine scheduling problem in a chemical batch plant. The plant consists of several batch units and several tanks. The plant configuration, the demand and recipe information are given. The plant has a number of daily schedules that are created by human operators. The problem treated in this study is to estimate of the objective functions and weighting factors for the scheduling problem by using the input and output data. Given a set of problem instance data and the solutions of the scheduling problem under the condition that the weighting factors of the scheduling problem are unknown, the problem is to estimate appropriate weighting factors of the objective function.

2.2. *Production scheduling problem*

We consider a parallel machine scheduling problem. This scheduling problem is the determination of the allocation of jobs to multiple machines and the processing order of jobs under the condition that each job is processed by a single machine. The following constraints are considered.

Machine Learning and Inverse Optimization Approach for Model
Identification of Scheduling Problems in Chemical Batch Plants

1713

Constraints
1. Each machine has no idle time.
2. One machine can only handle one job at a time.
3. Each job cannot be interrupted or divided once the processing has started.
Two types of objective functions (the sum of delivery delays f_1 and the sum of setup costs f_2) are considered in this study.

3. Proposed approach

The outline of the proposed approach is shown in Fig. 1. In this approach, historical data is used to estimate the weighting factors of the objective function. The details are explained in the following sections.

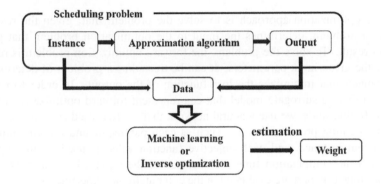

Figure 1: Outline of the proposed method

3.1. *Machine Learning Algorithm*

We use a supervised learning method of machine learning. First, we prepare a large number of problem examples of scheduling problems and a large set of outputs are obtained by solving them exactly via an approximate solution algorithm such as simulated annealing or genetic algorithm. In the case study, the weight, delivery date, processing time, label, and setup cost of each job are given as the parameters. In addition, the output includes the starting time of each operation and the type of machine that performs the processing. The weighting factor of each objective function actually used is taken as the correct answer. Preprocessing is applied to the prepared problem examples and output, and the feature is extracted. The square error between the estimated value of the obtained weighting coefficient and the weighting factor of the correct answer is obtained. The machine learning model trained from the process is evaluated. A recurrent random forest is used in the machine learning method. Random forest is learning using a large number of decision trees that improve generalization performance by using the representative values of these output results as the overall output (Breiman et al. (2001)).

In order to develop an accurate prediction model in a random forest, it is necessary to extract effective features. It is difficult to improve the prediction accuracy by using only the input / output data. On the contrary, if unnecessary features or meaningless features are included, noise may occur during learning, which may significantly increase the learning time or lower the prediction accuracy. From here, we describe the data generation method for use as features.

We consider the following features in the random forest.
Feature 1 The value of the objective function
Feature 2 Rank correlation coefficient
Feature 3 The error of delivery date setting for each machine
Feature 4 The sum of the processing completion times of each machine

3.2. Inverse Optimization Approach

The inverse optimization approach is to solve the problem that, given the result of an optimization problem, determines the input to the optimization problem that generates the correct result. To solve the problem, the forward optimization problem is repeatedly solved for the given input parameters. It can be treated as a problem of determining the input parameter that minimizes the loss function of the output. In order to reduce the computation time, a surrogate model that can represent forward optimization model has been used. In this study, we use a neural network that is often used as a surrogate model. During the learning process, the data such as each parameter and the weighting factor of the problem instance, and each objective function value is used as an output. The accuracy and the computation time performance of the replaced neural network is the forward optimization part are evaluated in the computational experiments.

4. Computational experiments

4.1. Randomly generated instances

We consider a parallel machine scheduling problem with 5 machines and 50 jobs. 100 problem examples are generated   by setting parameters at random. Schedules are generated by changing the weighting factors of the objective function by using the approximate solution method. Random forest is used for machine learning. Four types of features (Feature 1 only, Features 1 and 2, Features 1 and 3, Features 1 and 4) are utilized. Table 1 shows the mean square error of the proposed machine learning method and the inverse optimization method. The computational results show the effectiveness of using Features 1 and 3. The results of the inverse optimization are more effective than those of the machine learning method.

Table 1: MSE of the machine learning method and inverse optimization method

Feature 1 only	Features 1, 2	Features 1, 3	Features 1, 4	Inverse optimization
4.78e-3	4.75e-3	4.20e-3	4.33e-3	6.66e-4

Machine Learning and Inverse Optimization Approach for Model
Identification of Scheduling Problems in Chemical Batch Plants

1715

5.   Application to petrochemical batch plant

The proposed approach is applied to a lubricant oil production plant. The lubricant process consists of proportion mixing and additive oil mixing process, filling process into several batch tanks from feed oil. The feed oil is obtained by hydrogenation and solvent desulfurization for low, medium and high viscosity oils which are extracted from distilation columns from feed crude oils. The batch plant treated in this study is the filling process. There are six parallel filling units in the batch plant. The scheduling problem for filling process of the lubricant oil production plant is regarded as a parallel machine scheduling problem with several practical constraints.

5.1.   Case study

For the input data, the amount of oil to be filled and the filling speed for each tank are given. A week of job data is utilized. For the input data, production schedules are generated by changing the weighting factor by 0.1 for each objective function. For the output data, the starting time and ending time of each filling operation, allocation to the tank, and each objective function value are obtained. The weighting factors of each objective function are estimated from these input / output data. Two types of objective functions (the sum of delivery delays f_1 and the sum of setup costs f_2) are considered in the problem for filling process of the lubricant oil production plant. In machine learning, the value of the objective function (Feature 1) and the value of the objective function and the sum of the processing completion times of each machine (Features 1, 4) are used in the random forest.

In the inverse optimization, the weighting factors are updated so that the gradient between the result obtained with the given weighting factor and the correct result becomes smaller. The algorithm finishes when it is repeated a certain number of times. Table 2 shows the estimation results when the weighting factor of f_1 and f_2 are (0.2, 0.8), (0.4, 0.6), (0.9, 0.1). The estimation result of (0.9, 0.1) of inverse optimization was far from the correct answer. In the actual schedule data, the sum of setup costs increased significantly when the weight of f_2 changed from 0.2 to 0.1 and the accuracy of the proposed method is not good. For randomly generated instances, the changes in the objective function value were little. Due to the sudden increase in the objective function value, the weights updates were increased. Then the MSE of the proposed method is not better in those situations.

Table 2: Weighting factors estimation results for machine learning (ML) and inverse optimization

	ML using Feature 1	ML using Features 1, 4	Inverse optimization
(0.200, 0.800)	(0.252, 0.748)	(0.222, 0.778)	(0.266, 0.734)
(0.400, 0.600)	(0.402, 0.598)	(0.408, 0.592)	(0.370, 0.630)
(0.900, 0.100)	(0.842, 0.159)	(0.847, 0.153)	(0.991, 0.009)

Table 3 shows the mean square error (MSE) of the proposed machine learning method and inverse optimization method. The results show the effectiveness of using feature the sum of the processing completion times of each machine. Comparing with the three methods, the machine learning provides better results. The inverse optimization is more susceptible to the range of change in the objective function value than machine learning. It is considered that the cause of the actual schedule is that the range of change of the objective function value is different from that of the randomly generated schedule.

Table 3: MSE of the machine learning (ML) and inverse optimization method

	ML using Feature 1	ML using Features 1,4	Inverse optimization
MSE	5.96e-3	4.90e-3	2.61e-2

6. Conclusion and future works

We have studied the application of machine learning and inverse optimization method for estimating weighting factors from thepractical petrochemical scheduling problem. The effectiveness of our proposed approach to real chemical batch plants has been confirmed from the computational results.

References

K. R. Baker, J.C.Smith, A Multiple-Criterion Model for Machine Scheduling. *Journal of Scheduling* **6**(1): 7–16.

S. Watanabe, T. Hiroyasu and M. Miki (2002). Neighborhood Cultivation Genetic Algorithm for Multi-objective Optimization Problems. *Journal of Information Processing* **43**(10): 183–198.

H. Kobayashi, R. Tachi and H. Tamaki (2018). Introduction of Weighting Factor Setting Technique for Multi-Objective Optimization. *Journal of the Institute of Systems, Control and Information Engineers* **31**(8): 281–294.

K. Asanuma and T. Nishi (2020). Estimation of weights of multi-objective production scheduling problems - an inverse optimization approach -, *Proceedings of 2020 International Conference on Industrial Engineering and Engineering Management.*

H. Togo, K. Asanuma, T. Nishi, Estimating weighting factors using approximate solutions of multi-objective scheduling problems, *Proceedings of International Symposium on Scheduling 2021*: 30–33.

H. Kise (1995). Simulated Annealing Method for Scheduling Problems. *Operations Research* **5**(8): 268–273.

Y. Matsuoka, T. Nishi, K. Tierney, Machine learning approach for identification of objective function in production scheduling problems, Proceedings of 2019 IEEE International Conference on Automation Science and Engineering, 679-684 (2019)

L. Breiman (2001). Random Forest. *Kluwer Academic Publishers* (45): 5–32.

Proceedings of the 14th International Symposium on Process Systems Engineering – PSE 2021+
June 19–23, 2022, Kyoto, Japan ©2022 Elsevier B. V. All rights reserved.
http://dx.doi.org/10.1016/B978-0-323-85159-6.50286-4

Decision-Focused Surrogate Modeling with Feasibility Guarantee

Rishabh Gupta[a], Qi Zhang[a]*

[a]*Department of Chemical Engineering and Materials Science, University of Minnesota, Minneapolis, MN 55455, USA*

qizh@umn.edu

Abstract

Surrogate models are commonly used to reduce the computational complexity of solving difficult optimization problems. In this work, we consider decision-focused surrogate modeling, which focuses on minimizing decision error, which we define as the difference between the optimal solutions to the original model and those obtained from solving the surrogate optimization model. We extend our previously developed inverse optimization framework to include a mechanism that ensures feasibility (or minimizes potential infeasibility) over a given input space. The proposed method gives rise to a robust optimization problem that we solve using a tailored cutting-plane algorithm. In our computational case study, we demonstrate that the proposed approach can correctly identify sources of infeasibility and efficiently update the surrogate model to eliminate the found infeasibility.

Keywords: surrogate modeling, learning for optimization, inverse optimization, feasibility guarantee.

1. Introduction

A common strategy for solving difficult optimization problems, especially in real-time applications, is to develop surrogate models of reduced computational complexity. In particular, data-driven surrogate modeling methods have become very popular with the opportunity to leverage recent advances in machine learning. Here, one uses the original model to generate data, which are used to fit the surrogate model that can then be embedded in the optimization problem. A key challenge in surrogate modeling is the balance between model accuracy and computational efficiency. As a result, much of the research effort in this area has focused on developing surrogate models that have simple functional forms or specific structures such that the optimization problems are easier to solve using the surrogate models [Cozad et al., 2014, Zhang et al., 2016].

The vast majority of existing surrogate modeling methods construct models that are given as systems of equations, which represent all or part of the equality constraints of the original optimization model [Bhosekar and Ierapetritou, 2018]. The goal of these surrogate modeling algorithms is to minimize the prediction error with respect to the original systems of equations. However, as we found in our recent (not yet published) work, a low prediction error in this kind of surrogate models does not necessarily lead to a low *decision error*, which is defined as the difference between the optimal solutions of the original and the surrogate optimization models. Yet arguably, decision accuracy is what the user primarily cares about once the optimization model is deployed as a decision-making tool. We developed a data-driven inverse optimization approach to construct surrogate models that take the form of simpler optimization models and directly minimize the decision error; hence, we refer to it as *decision-focused surrogate modeling*.

Decision-focused surrogate modeling focuses on the set of optimal solutions rather than the larger set of feasible solutions. As such, it is prone to generating surrogate models that violate constraints in the original model. In this work, we address this issue by extending our inverse optimization framework to construct surrogate models with feasibility guarantees. We propose a robust optimization approach where we treat the set of possible inputs as an uncertainty set, and we develop a tailored cutting-plane algorithm to solve the resulting extended inverse optimization problem. Results from our computational case study show that using the proposed approach, we can construct surrogate optimization models with feasibility guarantees without substantial sacrifice of decision accuracy.

2. Mathematical Formulation

We consider an original optimization problem of the following general form:

$$
\begin{aligned}
\underset{x \in R^n}{\text{minimize}} \quad & f(x, u) \\
\text{subject to} \quad & g(x, u) \leq 0,
\end{aligned}
\tag{1}
$$

which is a, possibly nonconvex, nonlinear program (NLP). Here, x and u denote the decision variables and model input parameters, respectively. We assume that solving problem (1) requires more time than what is allowed in our desired online application; however, we can solve it offline to generate data in the form of (u_i, x_i)-pairs, where x_i is the optimal solution to problem (1) given the input u_i.

Given a set of data points $\mathcal{I}$, the goal is to generate a surrogate optimization model that is easier to solve but achieves the same or almost the same optimal solutions as the original model. We postulate a surrogate optimization model of the following form:

$$
\begin{aligned}
\underset{x \in R^n}{\text{minimize}} \quad & \hat{f}(x, u; \theta) \\
\text{subject to} \quad & \hat{g}(x, u; \omega) \leq 0,
\end{aligned}
\tag{2}
$$

where $\hat{f}$ and $\hat{g}$ are parameterized by θ and ω, respectively, and are constructed to be convex in x, which renders problem (2) a convex NLP.

The decision-focused surrogate modeling problems attempts to directly learn an optimization model from data that are assumed to be optimal solutions to this model. As such, it gives rise to a data-driven inverse optimization problem (IOP) [Gupta and Zhang, 2021], which can be formulated as follows:

$$
\underset{\theta \in \Theta, \, \omega \in \Omega, \, \hat{x}}{\text{minimize}} \quad \sum_{i \in \mathcal{I}} \|x_i - \hat{x}_i\|
\tag{3a}
$$

$$
\text{subject to} \quad \hat{x}_i \in \underset{\tilde{x} \in R^n}{\arg\min} \left\{ \hat{f}(\tilde{x}, u_i; \theta) : \hat{g}(\tilde{x}, u_i; \omega) \leq 0 \right\} \quad \forall i \in \mathcal{I},
\tag{3b}
$$

where $\hat{x}_i$ denotes the solution predicted by the surrogate model. The objective is to determine the surrogate model parameters θ and ω that minimize the decision error defined in (3a) as the difference between the optimal solution to the original problem x_i and $\hat{x}_i$ across the given data set. Constraints (3b) state that for each $i \in \mathcal{I}$, $\hat{x}_i$ is the optimal solution to the surrogate optimization model with input u_i.

One potential issue with the IOP formulation (3) is that a predicted solution $\hat{x}_i$ is not guaranteed to be feasible in the original model (1). In addition, assuming that the input u can be chosen from a set $\mathcal{U}$, the optimal solution to the surrogate model is not guaranteed to be feasible in (1) for all $u \in \mathcal{U}$ even if $\hat{x}_i$ is feasible in (1) for all $i \in \mathcal{I}$. Hence, to ensure feasibility, we add the following constraints to problem (3):

$$\left. \begin{array}{c} \bar{x} \in \underset{\tilde{x} \in R^n}{\arg \min} \left\{ \hat{f}(\tilde{x}, u; \theta) : \hat{g}(\tilde{x}, u; \omega) \leq 0 \right\} \\ g(\bar{x}, u) \leq 0 \end{array} \right\} \quad \forall u \in \mathcal{U}, \tag{4}$$

which state that given a surrogate model defined by θ and ω, the optimal solution to the surrogate model for any $u \in \mathcal{U}$, $\bar{x}$, also has to satisfy the original constraints $g(\bar{x}, u) \leq 0$.

3.   Solution Strategy

The extended IOP is a bilevel semi-infinite program. To solve this problem, we propose a cutting-plane algorithm that iterates between a master problem and a cut-generating separation problem. The master problem is formulated as follows:

$$\begin{aligned} \underset{\theta \in \Theta, \, \omega \in \Omega, \, \hat{x}, \, \bar{x}}{\text{minimize}} \quad & \sum_{i \in \mathcal{I}} \|x_i - \hat{x}_i\| \\ \text{subject to} \quad & \hat{x}_i \in \underset{\tilde{x} \in R^n}{\arg \min} \left\{ \hat{f}(\tilde{x}, u_i; \theta) : \hat{g}(\tilde{x}, u_i; \omega) \leq 0 \right\} \quad \forall i \in \mathcal{I} \\ & \bar{x}_j \in \underset{\tilde{x} \in R^n}{\arg \min} \left\{ \hat{f}(\tilde{x}, u_j; \theta) : \hat{g}(\tilde{x}, u_j; \omega) \leq 0 \right\} \quad \forall j \in \mathcal{J} \\ & g(\bar{x}_j, u_j) \leq 0 \quad \forall j \in \mathcal{J}, \end{aligned} \tag{5}$$

which is a relaxation of the extended IOP since the semi-infinite constraints (4) have been replaced by a finite number of constraints defined over a discrete set $\mathcal{J}$. For each $j \in \mathcal{J}$, we have a specific input u_j and the corresponding predicted solution $\bar{x}_j$. If the optimal solution to (5) satisfies constraints (4), then it is also optimal for the extended IOP. Otherwise, we solve the following separation problem for each constraint function g_k to identify inputs for which the solutions of the surrogate model violate the original constraints:

$$\begin{aligned} \underset{u \in \mathcal{U}, \, \bar{x}}{\text{maximize}} \quad & g_k(\bar{x}, u) \\ \text{subject to} \quad & \bar{x} \in \underset{\tilde{x} \in R^n}{\arg \min} \left\{ \hat{f}(\tilde{x}, u; \theta) : \hat{g}(\tilde{x}, u; \omega) \leq 0 \right\}. \end{aligned} \tag{6}$$

If the optimal value of (6) is greater than zero (or some defined feasibility threshold ϵ), we add the corresponding input u to the set $\mathcal{J}$ and re-solve problem (5). By doing so, we iterate between the master and the separation problems until no more constraint violations can be found, which indicates that we have solved the extended IOP.

Both problems (5) and (6) are bilevel optimization problems. To solve them, we first reformulate them into single-level problems by replacing the lower-level problems with their KKT conditions, which is possible since the surrogate optimization model is designed to be convex. The resulting formulations generally do not satisfy common regularity conditions, which makes their direct solution using standard NLP solvers difficult. Instead, we solve an exact penalty reformulation, which we do not describe here in detail due to space limitations. Note that while a local solution to problem (5) is usually enough to provide good results, problem (6) has to be solved to global optimality to guarantee feasibility.

4.   Computational Case Study

In our case study, we consider the heat exchanger network shown in Figure 1, which is adopted from Biegler et al. [1997]. Here, the inlet temperature of stream H2, T_5, has a nominal value of 583 K but is subject to random disturbances. Whenever there is a change in T_5, we optimize the operation of the heat exchanger network by solving the following NLP in which we can adjust the cooling duty Q_c and the heat capacity flowrate F_{H2}:

$$\underset{Q_c, F_{H2}}{\text{minimize}} \quad 10^{-2} Q_c + 4 (F_{H2} - 1.7)^2 \tag{7a}$$

$$\text{subject to} \quad 0.5 Q_c + 165 \geq 0 \tag{7b}$$

$$-10 - Q_c + (T_5 - 558 + 0.5 Q_c) F_{H2} \geq 0 \tag{7c}$$

$$-10 - Q_c + (T_5 - 393) F_{H2} \geq 0 \tag{7d}$$

$$-250 - Q_c + (T_5 - 313) F_{H2} \geq 0 \tag{7e}$$

$$-250 - Q_c + (T_5 - 323) F_{H2} \leq 0 \tag{7f}$$

$$Q_c \geq 0, \ F_{H2} \geq 0, \tag{7g}$$

which is nonconvex due to the bilinear term in constraint (7c).

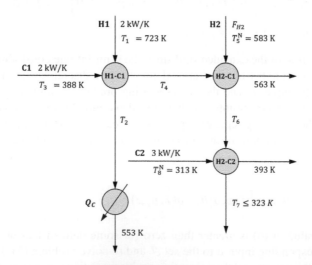

Figure 1: Given heat exchanger network.

We employ the proposed decision-focused surrogate modeling approach to replace the bilinear term $Q_c F_{H2}$ in constraint (7c) with the following approximation:

$$Q_c F_{H2} \rightarrow a(T_5) Q_c + b(T_5) F_{H2}, \tag{8}$$

where a and b are some functions of the input parameter T_5. This change, together with estimating the objective function $\hat{f}$ as a convex quadratic function and keeping all linear constraints, results in a surrogate convex QP for problem (7) that is much easier to solve.

We obtain the initial surrogate model by randomly sampling ten values of T_5 in the range [573 K, 593 K] and solving problem (3) with the corresponding global optimal solutions of (7). Here, we assume a and b in (8) to be cubic polynomials in T_5. The result is depicted in Figure 2a, which shows, for each chosen T_5, the true optimal Q_c and the Q_c obtained from solving the surrogate optimization model. In addition, it shows the sets of feasible Q_c for the original (red area) and surrogate (blue area) models. One can observe that while the feasible regions are quite different, there is very good agreement in the true and predicted optimal solutions, which can be attributed to the decision-focused nature of our approach.

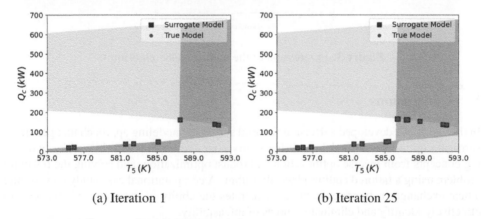

(a) Iteration 1   (b) Iteration 25

Figure 2: Comparison between the original model and the surrogate optimization model.

Next, we solve the extended IOP to minimize the violation of constraint (7c) at the optimal solutions of the surrogate model. We perform 25 iterations of the proposed cutting-plane algorithm. Figure 3 shows the maximum constraint violation, which is the optimal value of problem (6) solved for constraint (7c), and the corresponding violated input temperature T_5 that is then added to set $\mathcal{J}$ in problem (5) at each iteration. One can see that as the algorithm progresses, violations across the entire input range are detected until from iteration 13 onward, the algorithm only detects constraint violation in the region around $T_5 = 586.3$ K. This can be explained by Figure 2b, which shows all training data points accumulated over the 25 iterations and the feasible regions of the true and surrogate models. We see that for $T_5 \geq 586.3$ K, part of the feasible region of the surrogate model is infeasible in the true model. While the surrogate model achieves a very good fit for almost all optimal solutions in this region, there seems to be always some point at $T_5 \approx 586.3$ K that is infeasible, which is where we see a "transition" in the feasible region of the surrogate model. This indicates that the proposed cubic approximation of constraint (7c) is not sufficient to achieve feasibility across the entire input range, resulting in the algorithm focusing on minimizing infeasibility by repeatedly sampling the area around 586.3 K. However, our algorithm correctly identifies the main source of infeasibility. In this particular case, the result instructs a simple remedy of the problem, which is to create two surrogate models, one for $T_5 < 586.3$ K and one for $T_5 \geq 586.3$ K. Then, with the same training data points, solving the corresponding IOPs directly returns two surrogate optimization models whose optimal solutions are feasible for the entire input space.

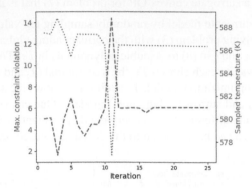

Figure 3: Progression of the cutting-plane algorithm.

5.   Conclusions

In this work, we developed a decision-focused surrogate modeling approach that generates surrogate optimization models with feasibility guarantees. This is achieved by combining concepts from inverse optimization and robust optimization, and solving the resulting problem using a tailored cutting-plane algorithm. A computational case study considering a heat exchanger network example demonstrates the ability of the proposed approach to effectively identify and eliminate sources of infeasibility.

References

A. Bhosekar and M. Ierapetritou, 2018. Advances in surrogate based modeling, feasibility analysis, and optimization: A review. *Computers and Chemical Engineering*, 108:250–267.

L. T. Biegler, I. E. Grossmann, and A. W. Westerberg, 1997. Systematic methods for chemical process design.

A. Cozad, N. V. Sahinidis, and D. C. Miller, 2014. Learning Surrogate Models for Simulation-Based Optimization. *AIChE Journal*, 60(6):2211–2227.

R. Gupta and Q. Zhang, 2021. Decomposition and adaptive sampling for data-driven inverse linear optimization. *INFORMS Journal on Computing*, forthcoming.

Q. Zhang, I. E. Grossmann, A. Sundaramoorthy, and J. M. Pinto, 2016. Data-driven construction of Convex Region Surrogate models. *Optimization and Engineering*, 17(2): 289–332.

Proceedings of the 14th International Symposium on Process Systems Engineering – PSE 2021+
June 19-23, 2022, Kyoto, Japan © 2022 Elsevier B.V. All rights reserved.
http://dx.doi.org/10.1016/B978-0-323-85159-6.50287-6

Grade transition optimization by using gated recurrent unit neural network for styrene-acrylonitrile copolymer process

Shi-Chang Chang[a], Chun-Yung Chang[a], Hao-Yeh Lee[a]*, I-Lung Chien[b]

[a] Dept Chemical Engineering, National Taiwan University of Science and Technology, No.43, Sec. 4, Keelung RD., Da'an Dist.,10607 Taipei City, Taiwan;
[b] Dept Chemical Engineering, National Taiwan University, No.1, Sec. 4, Roosevelt Road, Da'an Dist.,10617 Taipei City, Taiwan;
haoyehlee@mail.ntust.edu.tw

Abstract

The melt index (MI) of polymer products is an important quality reference for the product properties. However, MI cannot be measured in real-time, and the current value of MI can only be obtained by laboratory analysis after several hours, which leads to unsatisfactory quality control results. To solve the problem, this paper adopts the styrene-acrylonitrile (SAN) copolymers process as a target process and uses the Gated Recurrent Unit (GRU) to establish a MI dynamic prediction model for different grades of SAN copolymer to estimate the current and future MI values, which ultimately improve the MI quality control performance. In addition, to solve the quality fluctuation caused by the difficulty of fine tune the chain modifier feed flow during the grade transition. Therefore, this paper also combines the GRU dynamic model and a virtual controller to provide recommended operating values for the chain modifier to reduce the transient time during grade transition. The simulation results in this paper show that the predicted value of MI is in agreement with the actual measured value. In addition, the recommended value of the chain modifier feed flow rate in comparison to actual manual control can significantly reduce about 28.6 hours of the grade transition time.

Keywords: GRU, soft sensor, melt index, control, grade transition.

1. Introduction

SAN resin is composed of 70~80 wt% styrene (SM) and 20~30 wt% acrylonitrile (AN). The property of polymer products is usually adjusted according to market demand. To meet the different final product of physical or chemical properties requirements, the MI of polymer needs to be adjusted, and each MI value corresponds to each grade. The transfer of polymer products from one grade to another is called grade transition. Because no sensor can measure the MI value, it is through low-frequency manual sampling and laboratory analysis. Therefore, it will take about 4 to 8 hours or even 1 day to have one MI measurement data. This large delay measurement not only makes MI more difficult to control but also requires additional manpower for quality analysis. Therefore, a model needs to be established to simulate the dynamic behavior of the process to estimate the accurate real-time MI value, also known as a soft sensor. However, due to the complexity of the copolymer polymerization reaction and the lack of a complete reaction kinetic formula, it is difficult to establish a first-principle model of the system, so this paper uses a data-driven GRU (Cho et al., 2014) model is used to estimate the dynamic behavior of

MI of different grades of SAN copolymers. In addition, various fine-tune policies are often based on the operator experiences to adjust the feed flow rate of the chain modifier during the grade transition. It will cause quality fluctuations. Therefore, this paper designs a virtual velocity form of PI controller through the GRU dynamic model and calculates the recommended value of chain modifier feed flow rate in real-time, which can ultimately reduce transient time, the waste of raw materials, the output of secondary products.

Neural networks have found widespread usage in modeling the complex and dynamic behavior of various polymer processes. Lee et al. (2009) used the first-principle EVA copolymerization reactor model to generate data, then established an artificial neural network (ANN) model to estimate the MI of the EVA process, and used PI controller for controlling melting index. Noor et al. (2010) reviewed various cases of ANNs used in polymer process simulation and emphasized the advantages of ANNs for fitting highly nonlinear systems. Jumari and Yusof (2017) used ANN and the first-principle model to simulate the MI of the polypropylene process, and compared the prediction performance of the two models. Compared with the past, this paper uses more novel deep learning modeling technology and GRU model to try to obtain a more accurate MI prediction value and combined with the virtual controller to provide the recommended value of the chain modifier feed flow rate. Finally, the model and virtual controller are applied to the SAN copolymer process.

2. Process description

The production process of SAN is shown in **Figure 1**. Fresh SM and AN monomers after being mixed with the recycled monomers in a mixing tank flow into the reactor for polymerization. An additional amount of chain modifier is also added into the feed stream before entering the reactor. The material inside the reactor is mixed at a constant mixing speed, while the reactor temperature is controlled through an external cooling utility. The reactor effluent is preheated and enters the devolatilizer from the top to separate unreacted monomers, solvent, and chain modifier. The molten resin is then passed through an extruder to obtain the final product, SAN. The gas phase material from the devolatilizer is first condensed, followed by the extraction of acrylonitrile by fresh styrene. The extracted acrylonitrile is then routed to the recycle tank. The sampling result of the recycle tank concentration is used to determine the amount of fresh monomer addition and the recycle flow to the mixing tank. After the initial adjustment to the flow of fresh monomers during grade transition, the chain modifier is mainly used for fine-tuning the MI to achieve new grade specifications. Low product MI means high molecular weight and low fluidity which necessitate increasing the chain modifier, and vice versa.

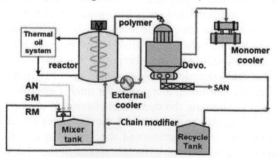

Figure 1. SAN process flowsheet

3. Data-driven GRU dynamic model

3.1. Data collection and pre-processing

All the data in this paper is obtained from the plant distributed control system (DCS) and the quality measurement results. The DCS data sampling time is 10 minutes, and the quality data (MI) is 4 to 8 hours. There is a total of 6 product grades in the data, corresponding to grade A~F, with each grade having its unique MI. Since each process variable has different units, the obtained data is first normalized between 0 and 1 to prevent the convergence problem of the neural network.

The data is divided into 8 data sets after excluding the abnormal data from a total of 48,541 DCS data sets and 2,105 MI measurement points. The training set accounts for 51%, the validation set accounts for 27%, and the testing set accounts for 22% of all data sets. The training set and the validation data set contain all grades due to the model training requirements. Due to the insufficiency of the data, only 4 types of grades that are frequently produced are covered in the testing data set.

3.2. Variable selection

This paper attempts to quantitatively analyze the importance of all the variables affecting MI. Due to the nonlinear characteristics of the system, this paper uses eXtreme Gradient Boosting (XGBoost) (Chen and Guestrin, 2016) for analysis, and combines chemical engineering background knowledge to select the input variables of the model. There are 15 DCS measurement variables in the SAN process, for which the importance analysis result is shown in **Figure 2**. The red bar represents the selected variable while the values on the y-axis signify the importance score. It is important to note that variables $f10$ and $f13$ which correspond to the total feed flow and reactor temperature, respectively, were ranked low despite their importance in estimating the MI. The reason is that the feed flow is dictated by the market demand, and in the case of fixed demand the feed flow usually remains constant. The temperature, on the other hand, is controlled by the temperature controller. Therefore, it is difficult for the XGBoost model to analyze the dynamic relationship between these two variables and MI.

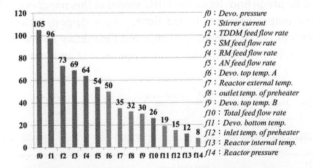

Figure 2. XGBoost variable importance analysis result

3.3. MI simulation method and results

This paper adopts Bayesian optimization (BO), a more efficient optimizer, as the hyperparameter tuning method. However, to prevent the model from overfitting, the hidden layer of the model is fixed at 3 layers, and the early stopping method is added. Furthermore, the model training in this paper uses RAdam (Liu et al., 2019) combined with Lookahead (Zhang et al., 2019) optimizer, also known as Ranger optimizer. After selecting the hyperparameters, the iterative method based on Mini-Batch is used to optimize the parameters of the GRU model with the Ranger optimizer and the mean square error (MSE) as the objective function. An epoch represents one iterative loop over the entire training dataset. After completing an epoch training, the model is verified and it is decided whether to stop the training process (Early Stop) according to the validation result. The simulation results of each grade of the training set are found to be in good agreement to the actual MI having Mean Absolute Percent Error (MAPE) of each data set within 5%; As illustrated in **Figure 3**, the MAPE of the test set is 3.629%.

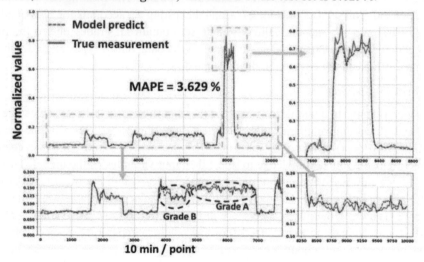

Figure 3. MI simulation result of the test data set

3.4. MI real-time prediction

To avoid data storage and calculations from increasing with time, this paper uses a rolling algorithm for online prediction. Since the GRU model in this paper considers time delay, it can predict the output response in the future. As a demonstration in **Figure 4**, it combines the concept of GRU input time step and time delay so that it can predict not only the current output y(t) but also the future output y(t+1) and y(t+2).

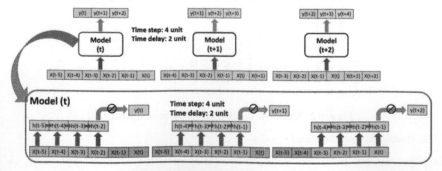

Figure 4. Model online rolling prediction

4. Improvement of grade transition

4.1. Controller design

In this paper, the Relay auto-tuning method (Åström and Hägglund, 1984) is used to determine the PI controller parameters, and the controller parameters are tuned according to a different grade. When the Relay auto-tuning method is used for tuning, the controller is temporarily replaced with a relay, and the chain modifier feed flow is passed through the relay to convert its feedback value into an up-and-down oscillation. Other input variables remain at the steady-state values corresponding to the current tuning grade. The dynamic parameters between chain modifier flow and MI obtained as a result of the relay feedback test are used to calculate the ultimate gain and period. Finally, the parameters of the controller are calculated through the Ziegler-Nichols tuning relation (Ziegler and Nichols, 1942).

The controller parameter of each grade does not apply for the situation of grade transition. Therefore, during the grade transition, the MI setpoint is set to the target MI of the next grade, while the virtual controller parameters are taken to be the average of current and next grades. If the current MI prediction value is already within the control range of the next grade, the controller parameters are set to the values of the next grade.

4.2. Grade transition result

Figure 5 is the result of the grade A to grade B transition when the virtual controller and the MI prediction model form a closed loop. The transition from grade A to grade B is a case of shifting from high MI to low MI. Therefore, the controller reduces the chain modifier flow to a minimum at the beginning of the transition, causing the MI to drop rapidly. The first yellow arrow in **Figure 5** represents the time point when the MI simulation result reaches the product set point, while the second yellow arrow represents the time point when the measured MI reaches the set point. It is observed that the transition time reduces by 28.6 hours for the MI simulation when compared to actual data. Therefore, after adopting the recommended value of chain modifier flow, the MI simulation result during grade transition is found to be smoother, reaching the set value faster. In addition, transitions between the other products have similar results.

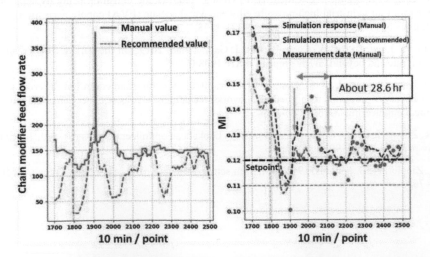

Figure 5. Chain modifier feed flow rate and MI simulation result (grade A to B)

5. Conclusions

In this paper, a GRU dynamic prediction model for the SAN production process is established to estimate the current and future MI values and combine with virtual PI controller provide the recommended value of chain modifier feed flow rate and consequently reduce the transient time during the grade transition.

The GRU model is used to build the MI prediction model having MAPE of each training and validation dataset within 5%. The MAPE of the test dataset is 3.629%, which represents good simulation results. The model also shows an accuracy of about 95% when tested online. The actual MI values are also found to be in good agreement with the predicted values from this model.

The model is then used to build a virtual controller that can suggest the values for chain modifier flow rate. The offline simulation results show that the MI simulation results are closer to the set value, and the product transitional time can be shortened by up to 28.6 hours. The results of this research can reduce substandard products in the process of product transition in polymer plants to reduce costs. Moreover, this research has already been implemented in an actual plant, and indeed shorten the time for grade transition.

References

K. J. Åström, T. Hägglund, 1984, Automatic tuning of simple regulators with specifications on phase and amplitude margins, Automatica, *20(5)*, 645-651.

T. Chen, C. Guestrin, 2016, Xgboost: A scalable tree boosting system, In proceedings of the 22nd acm sigkdd international conference on knowledge discovery and data mining, 785-794.

K. Cho, B. Van Merriënboer, C. Gulcehre, D. Bahdanau, F. Bougares, H. Schwenk, Y. Bengio, 2014, Learning phrase representations using RNN encoder-decoder for statistical machine translation, arXiv preprint arXiv:1406.1078.

N. F. Jumari, K. M. Yusof, 2017, Comparison of melt flow index of propylene polymerisation in loop reactors using first principles and artificial neural network models, Chemical Engineering Transactions, *56*, 163-168.

H. Y. Lee, T. H. Yang, I. L. Chien, H. P. Huang, 2009, Grade transition using dynamic neural networks for an industrial high-pressure ethylene–vinyl acetate (EVA) copolymerization process, Computers & Chemical Engineering, *33*, 1371-1378.

L. Liu, H. Jiang, P. He, W. Chen, X. Liu, J. Gao, J. Han, 2019, On the variance of the adaptive learning rate and beyond, arXiv preprint arXiv:1908.03265.

R. M. Noor, Z. Ahmad, M. M. Don, M. H. Uzir, 2010, Modelling and control of different types of polymerization processes using neural networks technique: A review, The Canadian journal of chemical engineering, *88(6)*, 1065-1084.

M. R. Zhang, J. Lucas, G. Hinton, J. Ba, 2019, Lookahead optimizer: k steps forward, 1 step back, arXiv preprint arXiv:1907.08610.

J. G. Ziegler, N. B. Nichols, 1942, Optimum settings for automatic controllers, Trans. ASME, 64, 759–768.

Proceedings of the 14th International Symposium on Process Systems Engineering – PSE 2021+
June 19-23, 2022, Kyoto, Japan © 2022 Elsevier B.V. All rights reserved.
http://dx.doi.org/10.1016/B978-0-323-85159-6.50288-8

Development of Estimating Algorithm for Biodegradation of Chemicals Using Clustering and Learning Algorithm

Kazuhiro Takeda[a*], and Kazuhide Kimbara[a]

[a] Department of Engineering, Shizuoka University, Shizuoka 432-8561, Japan
*takeda.kazuhiro@shizuoka.ac.jp

Abstract

Chemical substances should be assessed for biodegradation in environment. The biodegradation tests usually require 28 days of continuous testing and expensive costs. To reduce the cost, some software has been proposed to estimate the biodegradation of chemicals. Although, the software has not enough performance. Therefore, we develop a new estimation algorithm for the biodegradation of chemicals using clustering and machine learning algorithms. The combination of the Birch clustering algorithm and the XGboost learning algorithm is proposed to estimate the biodegradation of chemicals. Using 4200 real tested chemicals, the proposed algorithm was examined. The Birch algorithm might collect chemicals that are clear relationship between the explanatory variables and biodegradation.

Keywords: Biodegradation test; Chemical descriptor; Machine learning; Clustering.

1. Introduction

Chemicals should be assessed for biodegradation in environment. The collection of ready biodegradability tests is defined by the OECD (Organization for Economic Co-operation and Development) (OECD, 1992). The OECD 301C (MITI (Ministry of international Trade and Industry, Japan) I) test, one of the biodegradation tests defined by the OECD, usually requires 28 days of continuous testing and expensive costs. To reduce the cost, some software has been proposed to estimate the biodegradation of chemicals (Tunkel, 2000; Jaworska, 2002). Many tools for estimating the biodegradation of chemicals have been reviewed (Singh, 2021). The CATALOGIC (Laboratory of Mathematical Chemistry, 2021) is one of the most well-known software for estimating the biodegradation of chemicals, with 94% correct answer rate for not readily biodegradable products and 77% correct answer rate for easily decomposable products (NITE, 2020). Therefore, we develop a new estimation algorithm for the biodegradation of chemicals using clustering and machine learning algorithms.

2. Methods

Since biodegradation tests are affected by various factors such as bacteria and enzymes, it is a very difficult task to estimate the biodegradation of chemical substances. This paper proposes a new estimation algorithm for the biodegradation of chemicals using clustering and machine learning algorithms. This section briefly describes the different types of clustering and machine learning algorithms.

2.1. Clustering algorithm

There are several algorithms for clustering which automatically generate clusters from unlabelled data. The clustering algorithms such as K-Means (Sculley, 2010), Meanshift (Comaniciu, 2002), Ward (Ward, 1963), and OPTICS (Ester, 1996) use distances between points to make clusters. The K-Means algorithm clusters the data by trying to separate the samples into n groups of homoscedasticity, minimizing a criterion known as inertia or sum of squares within the cluster. The mini-batch K-Means algorithm is a variant of the K-Means algorithm. The Meanshift algorithm aims to find blobs with a smooth density of samples. The Ward algorithm is one of the hierarchical clustering algorithms. The Ward algorithm minimizes the sum of the squares of the differences in all clusters. The OPTICS algorithm creates a reachability graph to determine cluster membership. The DBSCAN algorithm (Schubert, 2017) is similar with the OPTICS algorithm but uses distances between nearest points. The Affinity propagation (Kettani, 2014) and the Spectral clustering algorithm (David, 2017) use graph distance. The Affinity propagation algorithm creates a cluster by sending a message between a pair of samples until it converges. The Spectral algorithm performs low-dimensional embedding of affinity matrices between samples. The Agglomerative clustering algorithm (Kettani, 2014) and Birch algorithm (Zhang, 1996) use any pairwise distance. The Agglomerative clustering algorithm objects use a bottom-up approach to perform hierarchical clustering. The Birch algorithm builds a tree called the clustering feature tree for the given data. The Gaussian mixture algorithm (Duda, 1973) is a probabilistic algorithm that assumes all the data points are generated from a mixture of a finite number of gaussian distributions with unknown parameters.

2.2. Machine learning algorithm

Many kinds of machine learning algorithms for regression have been proposed. Two well-known algorithms are examined. One is linear and the other is non-linear. The Lasso algorithm (Kim, 2007) is a linear model that estimates sparse coefficients and effectively reducing the number of features. On the other hand, the XGboost algorithm (Tianqi, 2016) works as Newton Raphson in function space unlike gradient boosting that works as gradient descent in function space, a second order Taylor's approximation is used in the loss function to make the connection to Newton Raphson method.

2.3. Objective data

The predictive performances by several clustering and machine learning algorithm were evaluated by the biodegradation data for 4200 chemicals, which were tested under MITI I 301C test condition. The explanatory variables for estimating the biodegradation were chemical descriptors that contain 2D and 3D structural and charge information generated by the Gaussian software (Gaussian, 2019) using the PM6 method. The chemical descriptors were generated from a SMILES specification (SMILES, 2016) of each chemical by using of the AlvaDesc software (Alvadesc, 2019). After cleaning the generated data, 4293 chemical descriptors were obtained.

3. Results and Discussions

The result of clustering the explanatory variables by the explanatory variable was demonstrated in Fig. 1. The performance of various clustering algorithms was indicated in scores. The score is coefficient of determination for validation data by the Lasso for the largest cluster by each clustering algorithm. As shown in Fig.1, the mini-batch K-Means and the Birch algorithms were higher performance than other algorithms.

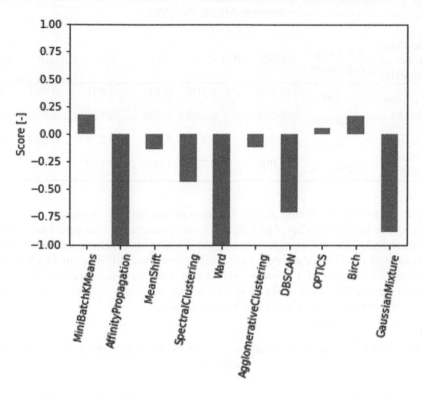

Figure 1 Classification scores of various clustering algorithms.

Then, clusters generated by the mini-batch K-Means and the Birch algorithms were examined. The other clustering algorithm cannot generate large clusters to train. Table 1 shows the evaluation results for various datasets of chemicals in or out of cluster by the mini-batch K-Means algorithm, where RMSE is rooted mean square error for training data, RMSE_V is rooted mean squared error for validation data, R^2 is coefficient of determination for training data, Q^2 is coefficient of determination for validation data, classR^2 is correct answer rate for readily biodegradable product in training data, and classQ^2 is correct answer rate for not readily biodegradable product in validation data. The results were averages of three trials. The biggest cluster generated by the algorithm contained 3718 chemicals. The Lasso or the XGboost algorithm applied to the biggest cluster or the other chemicals. The XGboost algorithm took lower RMSE and RMSE_V, and higher R^2, Q^2, classR^2, and classQ^2 than those of the Lasso algorithm. However, due to overfitting of the training data, the differences between the training data results and the validation data results of the XGboost algorithm were larger than those of the Lasso algorithm.

Table 1 Evaluation results for various datasets of chemicals in/out cluster by the mini-batch K-Means algorithm.

Used Data (number of chemicals)	Learning Algorithm	RMSE	RMSE_V	R^2	Q^2	classR^2	classQ^2
1st cluster (3718)	Lasso	0.550	0.706	0.700	0.491	0.907	0.856
	XGboost	0.006	0.660	1.000	0.567	0.999	0.862
Except for 1st cluster (482)	Lasso	0.314	0.817	0.888	0.325	0.980	0.888
	XGboost	0.000	0.727	1.000	0.510	1.000	0.911

Table 2 shows the evaluation results for various datasets of chemicals in or out of cluster by the Birch algorithm. The biggest cluster generated by the algorithm contained 1878 chemicals. The differences between the training data results and the validation data results of the Birch algorithm were smaller than those of the mini-batch K-Means algorithm. The results of 1st cluster were better than the other clusters results. It is indicated that the 1st cluster might be able to collect chemicals that are clear relationship between the explanatory variables and biodegradation.

Table 2 Evaluation results for various datasets in/out cluster by the Birch clustering algorithm.

Used Data (number of chemicals)	Learning Algorithm	RMSE	RMSE_V	R^2	Q^2	classR^2	classQ^2
1st cluster (1878)	Lasso	0.494	0.733	0.761	0.429	0.960	0.895
	XGboost	0.127	0.671	0.984	0.547	0.992	0.909
Except for 1st cluster (2332)	Lasso	0.755	0.787	0.431	0.378	0.781	0.780
	XGboost	0.181	0.684	0.967	0.534	0.986	0.832
2nd cluster (1239)	Lasso	0.679	0.755	0.528	0.456	0.810	0.801
	XGboost	0.477	0.696	0.773	0.513	0.914	0.823
3rd cluster (955)	Lasso	0.727	0.812	0.469	0.341	0.806	0.764
	Xgboost	0.047	0.768	0.998	0.428	0.997	0.800
4th cluster (39)	Lasso	0.900	1.037	0.179	-0.080	0.716	0.722
	XGboost	0.000	1.093	1.000	-0.232	1.000	0.667

Table 3 shows comparing among various cluster results. The cluster results of the mini-batch K-Means algorithm were a little better than those of whole sample chemicals. On the other hand, the cluster results, especially classQ^2, of the Birch algorithm were better than those of whole sample chemicals. The classR^2 and classQ^2 of proposed combination algorithm were higher than those of the CATALOGIC. As control experiments, 1800 chemicals were randomly sampled. The results of the control experiment led to the Birch algorithm results not being due to sample size.

Table 3 Evaluation results for various clustering and learning algorithm.

Clustering Algorithm (number of chemicals)	Learning Algorithm	RMSE	RMSE_V	R^2	Q^2	classR^2	classQ^2
Whole sample (4200)	Lasso	0.639	0.711	0.594	0.488	0.895	0.866
	XGboost	0.134	0.649	0.982	0.579	0.992	0.881
1st cluster by mini-batch K-Means (3718)	Lasso	0.550	0.706	0.700	0.491	0.907	0.856
	XGboost	0.006	0.660	1.000	0.567	0.999	0.862
1st cluster by Birch (1878)	Lasso	0.494	0.733	0.761	0.429	0.960	0.895
	XGboost	0.127	0.671	0.984	0.547	0.992	0.909
Random sampling (1800)	Lasso	0.730	0.767	0.471	0.366	0.850	0.836
	XGboost	0.223	0.670	0.944	0.525	0.983	0.872

4. Conclusions

The combination of the Birch clustering algorithm and the XGboost learning algorithm is proposed to estimate the biodegradation of chemicals. The mini-batch K-Means and the Birch algorithms were higher performance than other algorithms. Then, clusters generated by the mini-batch K-Means and the Birch algorithms were examined. The Lasso or the XGboost algorithms applied to the biggest cluster or the other chemicals. The correct answer rate for not readily biodegradable products and correct answer rate for easily decomposable products of proposed combination algorithm were higher than those of the CATALOGIC. The Birch algorithm might be able to collect chemicals that are clear relationship between the explanatory variables and biodegradation. The cluster results of the Birch algorithm were better than the other results. In future works, the characteristics of chemicals contained in the Birch algorithm will be investigated.

Acknowledgements

This study was conducted under the METI contract research "Research on the Introduction of Weight of Evidence Approach to the Evaluations of Biodegradation and Bioaccumulation of Chemicals (2020)". Deliverables from the cooperative research between Shizuoka University and NITE" Development of Biodegradation Predictive QSAR Method Using AI (2019-2022)" were used in this study. The authors wish to thank Dr. M.Takatsuki (Waseda Univ.) for helpful discussions.

References

AlvaDesc, 2019, https://www.alvascience.com/alvadesc/

J. F. Brendan et.al., 2007, Clustering by Passing Messages Between Data Points, Science Feb.

T. Chen, et al., 2016, XGBoost: A Scalable Tree Boosting System, https://dl.acm.org/doi/epdf/10.1145/2939672.2939785

D. Comaniciu, et.al., 2002, Mean shift: A robust approach toward feature space analysis, IEEE Transactions on Pattern Analysis and Machine Intelligence

Z. David, et. al., 2017, Preconditioned Spectral Clustering for Stochastic Block Partition Streaming Graph Challenge, 2017 IEEE High Performance Extreme Computing Conference

R.O. Duda, el. al., 1973, Pattern Classification and Scene Analysis, Wiley, New York

M. Ester, et. al., 1996, A density-based algorithm for discovering clusters in large spatial databases with noise, Proceedings of the Second International Conference on Knowledge Discovery and Data Mining, 226–231

Gaussian, 2019, https://gaussian.com/

J. Jaworska, et al., 2002, Probabilistic assessment of biodegradability based on metabolic pathways: CATABOL System, SAR QSAR Environ. Res. 13, 2, 307-323

O. Kettani, et. al., 2014, Agglomerative clustering via maximum incremental path integral, Pattern Recognition, 46, 11, 3056–65

S. J. Kim, et. al., 2007, An Interior-Point Method for Large-Scale L1-Regularized Least Squares, in IEEE Journal of Selected Topics in Signal Processing

Laboratory of Mathematical Chemistry, 2021. http://oasis-lmc.org/products/software/catalogic.aspx.

NITE, 2020, https://www.nite.go.jp/data/000111311.pdf

OECD, 1992, http://www.oecd.org/chemicalsafety/risk-assessment/1948209.pdf

E. Schubert, et. al., 2017, DBSCAN Revisited, Revisited: Why and How You Should (Still) Use DBSCAN, ACM Trans. Database Syst., 42, 3, 19:1–19:21

D. Sculley, 2010, Web Scale K-Means clustering, Proceedings of the 19th international conference on World wide web

A. K. Singh, et al., 2021, Trends in predictive biodegradation for sustainable mitigation of environmental pollutants: Recent progress and future outlook, Science of The Total Environment, 770, 144561

SMILES, 2016, http://opensmiles.org/opensmiles.html

J. Tunkel, et al., 2000, Predicting ready biodegradability in the Japanese ministry of international trade and industry test, Environ. Toxicol. Chem. 19, 10, 2478 -2485

J. H.Ward, 1963, Hierarchical Grouping to Optimize an Objective Function, Journal of the American Statistical Association, 58, 301, 236–244

T. Zhang, et al., 1996, BIRCH: An efficient data clustering method for large databases. https://www.cs.sfu.ca/CourseCentral/459/han/papers/zhang96.pdf

Proceedings of the 14th International Symposium on Process Systems Engineering – PSE 2021+
June 19-23, 2022, Kyoto, Japan © 2022 Elsevier B.V. All rights reserved.
http://dx.doi.org/10.1016/B978-0-323-85159-6.50289-X

Surrogate Classification based on Accuracy and Complexity

Maaz Ahmad[a], Iftekhar A. Karimi[a,*]

[a]*Department of Chemical & Biomolecular Engineering, 4 Engineering Drive 4, National University of Singapore, Singapore-117585, Singapore*

cheiak@nus.edu.sg

Abstract

The prevalence of various machine-learning modeling techniques and numerous possible model configurations generates a long list of unique surrogate forms. Exhaustive enumeration and search for the best surrogate form from a large pool of candidate forms is a non-trivial task. In this work, we aim to assess similarities in modeling capabilities among many different surrogate forms. We examine modeling capabilities for noisy and non-noisy data based on two different surrogate performance metrics. We use a similarity metric to identify similar pairs of surrogate forms, and then group mutually similar forms into distinct families. The similarities among various forms vary depending on the quality of data set and choice of performance metric. This work enables us to exploit families of similar forms to create a reduced search set of contrasting surrogate forms, and facilitate surrogate form selection.

Keywords: machine-learning, surrogate, surrogate selection.

1. Introduction

With growing digitalization in industrial process operations, the use of digital twins for decision making has been on the rise. While high-fidelity models provide detailed understanding and analysis, they are limited by computational expense. Computationally cheaper, data-driven surrogate models offer a viable alternative. They simply learn the correlations between the input-output data generated by the system under study. The development of a surrogate model starts with the identification of an appropriate modeling technique or a learning algorithm. Artificial Neural Network (ANN), Radial Basis Function (RBF), Support Vector Regression (SVR) are examples of modeling techniques. The next step is to create a surrogate form by choosing its configurations and analytical functions. For instance, an ANN technique using a sigmoid activation function and having one hidden layer with ten nodes defines an ANN form. Similarly, an RBF technique with a linear basis function creates an RBF surrogate form. The last step involves constructing the final surrogate model with known parameters by training the form on a data set. Various possible modeling techniques and associated model configurations can generate innumerable distinct surrogate forms. Determining the best form is demanding not only due to the expansive search required over many candidate forms, but also due to the importance of searching across various diverse forms. While one would expect unique forms to behave differently from the other, recent works (Ahmad and Karimi, 2021; Bhosekar and Ierapetritou, 2018; Garud et al., 2018) have indicated similarities in modeling capabilities among different surrogate forms. Clearly, it would be more rational to search across a set of contrasting surrogate forms to expedite the selection process. To this end, in this work, we aim to assess and analyse the similarities in performance across many surrogate forms. We group forms with similar performance into distinct families such that all forms within a family are mutually similar. In the subsequent sections, we discuss our scope of work, detail our numerical methodology to construct families of similar surrogate forms, and highlight some key observations. Finally, we conclude in section 5 with the scope of future work.

2. Data sets, Surrogate Forms, and Performance Metrics

For the sake of consistency, let us denote the kth sample point of any data set with N-dimensional inputs as $x_n^{(k)}$ $(n = 1, 2, ..., N)$. In this work, we only consider single response at any sample point. Let $y^{(k)}$ denote the response at the kth sample point. We train any surrogate form on a given data set to obtain the final surrogate model $S(x)$.

2.1. Data sets

We gather diverse data sets from various sources. The presence of noise in response affects the predictive performance of different surrogate models. Hence, we aim to study similarities in surrogate performances for non-noisy and noisy data sets separately. We gather various data sets from 93 analytical test functions (Surjavonic and Bingham, 2013), 20 simulation runs (Coimbatore Meenakshi Sundaram and Karimi, 2021), and plant observations. In total, we have 1508 non-noisy data sets and 1591 noisy data sets, identical to those used in our previous work as well (Ahmad and Karimi, 2021). The maximum input dimension over all non-noisy and noisy data sets was 20.

2.2. Surrogate Forms

We considered 49 surrogate forms from eight modeling techniques: Polynomial Response Surface Model (PRSM), Kriging (KRG), RBF, SVR, Multivariate Adaptive Regression Spline (MARS), ANN, Gaussian Kernel Regression (GKR), and Power Law (PL). Each technique offers various options for functional forms and model configurations to generate many possible distinct surrogate forms. The 49 surrogate forms and their notations are listed in Table 1 in the next page.

2.3. Performance Metric

We use one accuracy-based performance metric and one hybrid accuracy-complexity-based performance metric (PM) to evaluate each surrogate's performance. We consider R^2 or coefficient of determination as the accuracy-based PM, and SQS or Surrogate Quality Score (Ahmad and Karimi, 2021) as the hybrid PM. SQS balances accuracy and complexity by favouring an accurate model based on R^2 while penalizing a complex model based on extent of freedom (eof) or number of independent model parameters.

$$R^2 = 1 - \sum_{k=1}^{K'} \left(y^{(k)} - S(x^{(k)}) \right)^2 \bigg/ \sum_{k=1}^{K'} (y^{(k)} - \bar{y})^2 \tag{1}$$

$$SQS = (1 - R^2) \times \left\{ \frac{\ln(1 + eof)}{\ln(1 + K)} \right\}^{0.5} \tag{2}$$

$\bar{y}$ in Eq. 1 is the mean response, while K' is the number of sample points over which PM is computed. SQS can be used to compare across qualitatively different modeling techniques efficiently.

Table 1: Modeling Techniques, Surrogate Forms, and their Notations used in this work.

Modeling Techniques	Surrogate Forms (shorthand notation in parenthesis)	Notations for Forms
PRSM	1st order and 2nd order PRSM	PRSM1, PRSM2
KRG	Regression Functions: constant (0), PRSM1 (1), PRSM2 (2)	K0e, K1e, K2e, K0g, K1g, K2g, K0l, K1l, K2l, K0s, K1s, K2s, K0c, K1c, K2c
	Correlation Functions: Exponential (e), Gaussian (g), Linear (l), Spherical (s), Cubic (c)	
RBF	Basis Functions: Bi-harmonic (BH), Multi-quadratic (MQ), Inverse Multi-quadratic (IMQ), Thin Plate Spline (TPS), Gaussian (G)	RBH0, RBH1, RBH2, RMQ0, RMQ1, RMQ2, RIMQ0, RIMQ1, RIMQ2, RTPS0, RTPS1, RTPS2, RG0, RG1, RG2
	Tail Functions: constant (0), PRSM1 (1), PRSM2 (2)	
SVR	Kernel functions: Linear (lin), 3rd order Polynomial (poly), Gaussian (gauss)	SVRlin, SVRpoly, SVRgauss
MARS	{Max interactions, Max basis functions}: {2,5N} (1) and {3,10N} (2)	MARS1, MARS2
ANN	Activation functions: tansig (T), logsig (L), radial basis (R)	ANN1TN, ANN1T2N, ANN2TN, ANN1LN, ANN1L2N, ANN2LN, ANN1RN, ANN1R2N, ANN2RN
	{Number of hidden layers, Number of nodes in each layer}: {1,N}, {1,2N}, {2,N}	
GKR	Projected dimension $R = 2N$	GKR
PL	Sum and Product of $x_n, n = 1, 2, ..., N$ terms	APL, MPL

3. Similarity Assessment and Families Identification

Since the performance of any surrogate model would depend on the quality of data (noisy vs non-noisy) and performance metric (R^2 vs SQS), we aim to assess similarities among various forms for the four cases separately. We denote them as NNR2, NNSQS, NR2, and NSQS, where NN and N denote data quality (non-noisy and noisy respectively) while R2 and SQS denote PM (R^2 and SQS respectively). Furthermore, for noisy data, KRG and RBF are not the ideal techniques since they always fit noise. Hence, we consider 49 forms for non-noisy data and 19 forms (excluding 15 KRG and 15 RBF forms) for noisy data.

The first step to initiate our assessment and analysis involves constructing and evaluating all surrogate models for all data sets. A surrogate is trained on a few input-output points which constitute the train set, while it is evaluated over a few additional new sample points which constitute the test set. To this end, for synthetic and simulated data, we generate additional K sample points for each data set to make up a test set with $2K$ sample points (train set + K new points). For real-world data, we randomly select a few points ($ceil(K/2)$) as the train set, while evaluate at all K points. This gives us a performance vector (PV) for each surrogate. A surrogate's PV consists of its PM values for all data sets. The next step involves identification of similar pairs of surrogates. For this, we used the concept of concordance correlation coefficient (ρ_C) (Lin, 1989) that measures similarity between two vectors of data based on the expected value of squared deviation between them. ρ_C considers the direction of Pearson's correlation and is a signed metric. However, in our case, we simply need to check for resemblance between the performance vectors of any two surrogates. Hence, we develop a metric Similarity Index (SI) by modifying the normalization factor used by ρ_C such that $SI \in [0,1]$. $SI = 1$ for identical performance vectors.

$$SI\big(S_1(\boldsymbol{x}), S_2(\boldsymbol{x})\big) = 1 - \frac{E\big(PV_{S_1(x)} - PV_{S_2(x)}\big)^2}{E\left[\big(PV_{S_1(x)} - PV_{S_2(x)}\big)^2 \big| \rho = -1\right]} \tag{3a}$$

$$SI\big(S_1(\boldsymbol{x}), S_2(\boldsymbol{x})\big) = \frac{2(1 + \rho)\sigma_{S_1(x)}\sigma_{S_2(x)}}{\big(\mu_{S_1(x)} - \mu_{S_2(x)}\big)^2 + \big(\sigma_{S_1(x)} - \sigma_{S_2(x)}\big)^2} \tag{3b}$$

ρ, μ, and σ denote Pearson's correlation coefficient, mean, and standard deviation for the two PVs for $S_1(\boldsymbol{x})$ and $S_2(\boldsymbol{x})$. We compute SI between all pairs of surrogates for each of NNR2, NNSQS, NR2, and NSQS. The final step involves grouping mutually similar forms into distinct families. We use $SI \geq 0.90$ to consider two forms to be similar. Then, we create families of similar forms by solving an MILP formulation. Considering I forms, let us consider a similarity matrix with $a_{ij} = 1$ if $SI(i,j) \geq 0.9$, otherwise $a_{ij} = 0$, for $i = 1, 2, \ldots, I; j > i$. We use a binary variable f_i to indicate whether form i belongs to a family ($f_i = 1$) or not ($f_i = 0$). First, we find the family with maximum size, or most number of mutually similar forms. This entails solving $\max \sum_{i=1}^{I} f_i$ subject to the following similarity constraint:

$$f_i + f_j \leq 1 + a_{ij}, \quad j > i \tag{4}$$

Eq. 4 ensures that if forms i and j belong to a family, then they must be similar. Solving this problem gives us the largest family with size S. Then, starting with $s = S$, we identify all distinct families of size $s = S, S - 1, S - 2, \ldots, 1$ one at a time. This essentially reduces to a feasibility problem to identify a family of a given size, if it exists. This can be achieved as follows.

$$\min obj = 1 \tag{5a}$$

$$\sum_{i=1}^{I} f_i = s \tag{5b}$$

Solving Eq. 5a subject to constraints Eq. 4 and Eq. 5b gives us one family of size s. Since, we get one family on each run and multiple families of size s may exist, we re-solve the MILP after adding the previously obtained solution (z_i^*) as a new cut-constraint (Eq. 5c).

$$\sum_{i=1}^{I} z_i^* \leq s - 1 \tag{5c}$$

This would eliminate the problem to yield previously found families or their subsets in subsequent runs. In case the problem is infeasible, we search for the next smaller sized family with size $s - 1$. This procedure is repeated for each $s = S, S - 1, S - 2, \ldots, 1$ to get a complete set of families.

4. Families for NNR2, NNSQS, NR2, and NSQS

Based on the detailed methodology discussed in the previous section, we obtain different sets of families for NNR2, NNSQS, NR2, and NSQS (Tables 2 – 5).

Table 2: Families of surrogate forms for NNR2

	Surrogate Forms		Surrogate Forms
F1	K0l, K1l, K0s, K1s, K0c, K1c	F6	RIMQ1, RIMQ2
F2	K0e, K1e, K0l, K1l, K0s, K1s	F7	RTPS1, RTPS2, RG0
F3	K0g, K1g	F8	MARS1, MARS2, ANN1T2N
F4	RG1, RG2	F9	ANN1TN, ANN1T2N, ANN2TN, ANN1LN, ANN2LN, ANN2RN
F5	K2e, K2l, K2s, K2c, RBH0, RBH1, RBH2, RMQ0, RMQ1, RMQ2, RIMQ0, RTPS0	F10	PRSM2, SVRpoly, SVRgauss, ANN1TN, ANN1T2N, ANN2TN, ANN1LN, ANN1L2N, ANN2LN, ANN1RN

Table 3: Families of surrogate forms for NNSQS

	Surrogate Forms		Surrogate Forms
F1	K0l, K1l, K0s, K1s, K0c, K1c	F8	MARS1, MARS2
F2	K0e, K1e, K0l, K1l, K0s, K1s	F9	ANN2TN, ANN2LN, ANN2RN
F3	K0g, K1g	F10	ANN1TN, ANN1T2N, ANN2TN, ANN1LN, ANN1L2N, ANN2LN, ANN1RN
F4	RG1, RG2	F11	PRSM2, SVRgauss, ANN1TN, ANN1T2N, ANN1LN, ANN1L2N
F5	K2e, K2l, K2s, K2c, RBH0, RBH1, RBH2, RMQ0, RMQ1, RMQ2, RIMQ0, RTPS0	F12	PRSM2, ANN1TN, ANN1T2N, ANN1LN, ANN1L2N, ANN2LN, ANN1RN
F6	RIMQ1, RIMQ2	F13	PRSM2, K2l, K2s, K2c, RBH0, RBH1, RBH2
F7	RTPS1, RTPS2, RG0		

Table 4: Families of surrogate forms for NR2

	Surrogate Forms		Surrogate Forms
F1	SVRgauss, SVRpoly, ANN1TN, ANN1L2N	F6	SVRgauss, ANN1T2N, ANN1LN, ANN2LN
F2	SVRgauss, MARS1, ANN1TN, ANN1L2N	F7	SVRgauss, ANN1T2N, ANN1L2N, ANN2LN
F3	SVRgauss, MARS1, ANN1T2N, ANN1L2N	F8	ANN2TN, ANN2LN
F4	SVRgauss, ANN1TN, ANN1LN, ANN2LN	F9	PRSM2, SVRpoly
F5	SVRgauss, ANN1TN, ANN1L2N, ANN2LN	F10	SVRgauss, MARS1, MARS2

Table 5: Families of surrogate forms for NSQS

	Surrogate Forms		Surrogate Forms
F1	MARS1, MARS2	F3	ANN1TN, ANN1T2N, ANN1LN, ANN1L2N, ANN2LN
F2	ANN2TN, ANN2LN		

The derived sets of families reveal some important observations. For NNR2 and NNSQS, KRG forms with a constant or linear regression function show similarities irrespective of the correlation function, with the exception of Gaussian correlation. Similarly, K2e, K2l, K2s, and K2c show similarities, but K2g is not similar to any of them. This indicates a significantly contrasting behaviour of KRG using Gaussian correlation functions than others. Many RBFs show similarities with KRG using PRSM2 regression function for NNR2 and NNSQS. Both KRG and RBF techniques precisely learn the responses, and hence can potentially make similar predictions. Since KRG and RBF have identical and maximum complexity ($eof = K$), SQS penalizes both models identically. Hence, we observe the same sets of families F1-F7 for NNR2 and NNSQS. For all cases, we see that MARS1 and MARS2 are similar as they only differ in their hyperparameter settings. Also, ANNs using *tansig* and *logsig* activations readily show similarities irrespective of the network depth and width. This may be attributed to similar analytical forms of the two activation functions, and simple, shallow networks considered in our work. However, ANNs using radial basis functions differ based on their network configurations. We also observe that SVRgauss shows similarities with many ANN forms for NNR2, NNSQS, and NR2. Both SVRs and ANNs offer sufficient fidelity to model complex data sets effectively. One final observation from all families is that certain forms such as PRSM1, SVRlin, GKR, APL, and MPL do not belong to any family. In other words, such forms do not show similarities with any other form for either non-noisy or noisy data based on either PM.

The key application of families of similar forms is to facilitate and expedite surrogate form selection. In our future work, we aim to extract one representative surrogate form from each family to obtain a reduced search space of contrasting forms. We then search among the surrogate forms in this reduced search space. Based on the performance of the true best surrogate form and the form identified from the reduced space, we can validate the efficacy of our proposed surrogate form selection procedure.

5. Conclusions

This work aims to compare performances of various surrogate forms for modeling non-noisy and noisy data sets based on two performance metrics. Based on our numerical evaluation of similarity, we group mutually similar surrogate forms into families. Certain forms having different modeling techniques show similar performance, while certain others do not resemble with any other form. This work may act as an important prelude to our future work on expediting the surrogate form selection process by identifying and searching across contrasting surrogate forms.

References

Ahmad, M., Karimi, I.A., 2021. Revised learning based evolutionary assistive paradigm for surrogate selection (LEAPS2v2). Computers & Chemical Engineering 152, 107385.

Bhosekar, A., Ierapetritou, M., 2018. Advances in surrogate based modeling, feasibility analysis, and optimization: A review. Computers & Chemical Engineering 108, 250–267.

Coimbatore Meenakshi Sundaram, A., Karimi, I.A., 2021. State transients in storage systems for energy fluids. Computers & Chemical Engineering 144, 107128.

Garud, S.S., Karimi, I.A., Kraft, M., 2018. LEAPS2: Learning based Evolutionary Assistive Paradigm for Surrogate Selection. Computers & Chemical Engineering 119, 352–370.

Lin, L.I.-K., 1989. A Concordance Correlation Coefficient to Evaluate Reproducibility. Biometrics 45, 255.

Surjavonic, S., Bingham, D., 2013. Virtual Library of Simulation Experiments: Test Functions and Datasets. [WWW Document].

Proceedings of the 14th International Symposium on Process Systems Engineering – PSE 2021+
June 19-23, 2022, Kyoto, Japan © 2022 Elsevier B.V. All rights reserved.
http://dx.doi.org/10.1016/B978-0-323-85159-6.50290-6

Training Stiff Dynamic Process Models via Neural Differential Equations

William Bradley[a], Gabriel S. Gusmão[a], Andrew J. Medford[a], Fani Boukouvala[a]*

[a]*Georgia Institute of Technology, School of Chemical & Biomolecular Engineering, 311 Ferst Dr., Atlanta, GA, 30332, USA*
fani.boukouvala@chbe.gatech.edu

Abstract

A common step in developing generalizable, dynamic mechanistic models is to fit unmeasured parameters to measured data. Fitting differential equation-based models can be computationally expensive due to the presence of nonlinearity and stiffness. This work proposes a two-stage indirect approach where Neural ODEs approximate state derivatives, which are used to estimate the parameters of a differential model. In addition to its computational efficiency, the proposed method demonstrates the ability to work in concert with direct methods to accurately estimate parameters, even in the case of stiff systems. The method is shown here for the training of a microkinetic model.

Keywords: Neural Networks, Parameter Estimation, Stiff ODEs, Neural ODEs.

1. Introduction

The task of finding parameter values of a differential equation (DE) model to explain available experimental data is ubiquitous throughout engineering. The physical meaning of these DE models (also referred to here as a mechanistic model) permit the modeler to predict a system's behavior in unexplored experimental spaces, assuming the parameters have been estimated correctly. However, due to the complexity of DE systems, methods that automate their parameter estimation must often balance efficiency and accuracy. Gradient-based 'direct' methods either rely on repeated integration of the ODEs being regressed, or formulating a constrained nonlinear program discretizing the system of ODEs to solve for the parameter values (Li et al. 2005, Hamilton 2011). Both 'direct' methods face computational tractability issues, which become more severe when the initial parameter estimates are far from the true values, or the ODEs are nonlinear with respect to their parameters. Another problem, common to reaction systems, is the presence of rate terms which vary over large orders of magnitude, resulting in a system with fast and slow dynamics (i.e., at different timescales). Ultimately, to make these regression problems tractable for direct methods, a modeler may need to apply model reduction strategies, ranging from setting tight bounds on parameters to fixing insensitive parameters. Such strategies require domain expertise, which may not be available, as well as user-intervention, preventing automation of the parameter estimation process.

As an alternative to the direct approach, an indirect parameter estimation approach has been proposed, which avoids discretizing the mechanistic model (Swartz and Bremermann 1975, Brunel 2008). In this 2-stage approach, the experimental data is interpolated by a data-driven model, which is differentiated to obtain system derivative estimates. Those derivative estimates combined with state estimates of the interpolating model can be used to estimate the parameters of the mechanistic DEs via nonlinear programming (NLP). The indirect 2-stage approach is so named since it breaks up a single regression problem into two regression problems whose combined computational cost is generally less than that of the direct approaches. Yet despite having the advantage of being computationally cheap, this method is often limited in accuracy due to the difficulty in accurately estimating a system's derivative information.

Recently, we proposed using Neural ODEs (NODEs) as the data-driven surrogate to interpolate measurement data (Figure 1) and for estimating system derivatives (Bradley and Boukouvala 2021). In that work, NODEs compared favorably with other methods for automated parameter estimation of a nonlinear mechanistic DE. However, one class of DEs not covered in that work were those with 'stiff' dynamics. This class of problems can be particularly challenging for parameter estimation methods. One reason for this is the need for numerical methods that balance the number of functional evaluations (i.e., computation) and stability (i.e.. accuracy). Recent work has evaluated numerical techniques for fitting Neural ODEs to stiff

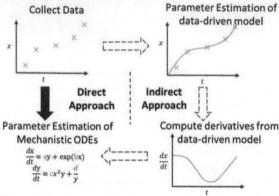

Figure 1. Depiction of the direct vs indirect approach to parameter estimation

system data, and for parameter estimation of stiff systems (Kim et al. 2021), however further work is needed to develop methods that are both general and accurate.

2. Methods

In this work, several approaches, and potential combinations thereof, are compared for the parameter estimation of stiff DEs. To start, direct approaches find the parameters p to a mechanistic model $f(x,p)$ by minimizing the following discrepancy function:

$$\min \sum (x_{k,j,meas} - x_{k,j,pred})^2 \tag{1}$$

$$s.t. \ \frac{dx_{k,MM}}{dt} = f(x_k, p) \tag{2}$$

Here, K state variables x_k, where $k = 1, ..., K$, are measured and predicted at time points j, where $j = 1, ..., J$, by integrating the mechanistic model (MM) with respect to independent variable t. Though statistically robust, this method can be computationally intensive. For such cases, a 2-stage indirect approach can be attractive.

As illustrated in Figure 1, the 2-stage indirect approach fits the parameters of the mechanistic model by solving 2 separate regression problems. In the first stage, the parameters of the data-driven model are fitted using the original measurement data. In the second stage, the parameters of the mechanistic ODE are found using the state and derivative estimates of the data-driven model. The data-driven model used in our work is a NODE model. This is done by first solving Eq.(1) subject to Eq.(3):

$$s.t. \ \frac{dx_{k,NODE}}{dt} = NN(x_k, w) \tag{3}$$

Neural Network parameters w are fitted to minimize an objective function equal to the sum of squared errors between the model prediction and measured state data. Once the

NODE is trained, derivative estimates are obtained by evaluating the trained NODE at times where measured data was collected using the same process conditions of the measured data. Following the procedure of (Bradley and Boukouvala 2021), we exclude derivative estimates at time t=0, which tend to be less reliable, to improve parameter estimates of the mechanistic DE. For stage two, an NLP is formulated as in Eq.(4) and (5) to find the parameters of the mechanistic DE without integrating the mechanistic DE.

$$\min \sum \left(\frac{dx_{j,k,NODE}}{dt} - \frac{dx_{j,k,MM}}{dt}\right)^2 \tag{4}$$

$$s.t. \ \frac{dx_{j,k,MM}}{dt} = f(x_{j,k,NODE}, p) \tag{5}$$

Depending on the required accuracy, the indirect 2-stage approach may be sufficient for the needs of the model-building problem at hand. However, if increased accuracy is required, we hypothesized a more robust fit would require including the mechanistic model constraints when fitting the measured state data. A tempting option would be a simultaneous approach, which combines the objective functions of the 2-stage approach into a single hybrid objective function:

$$\min \sum (x_{k,j,meas} - x_{k,j,pred})^2 + \lambda \sum \left(\frac{dx_{j,k,NODE}}{dt} - \frac{dx_{j,k,MM}}{dt}\right)^2 \tag{6}$$

Like the indirect approach, Eq.(6) uses the data-driven Neural ODE of Eq.(2) to fit the state data and provide derivative estimates. However, in the hybrid objective both the Neural ODE fit and mismatch between NODE and mechanistic DE are minimized simultaneously, their relative weights controlled by the hyperparameter lambda, λ.

A final alternative to increasing model fidelity is to fit the mechanistic DE directly (i.e., minimize Eq.(1) subject to Eq.(2)). However, as mentioned earlier this incurs an increased compute overhead. In the case of stiff systems, the increased compute cost comes from the finer discretization required to stably integrate the mechanistic model. To reduce compute costs, the direct approach can use parameter estimates informed by the indirect approach. Specifically, the parameters estimated from the 2-stage fitting are used as an initial guess for the DE of Eq.(2). A single application of the indirect followed by the direct approach is herein referred to as the incremental approach.

Throughout this work, we use PyTorch's LBFGS solver and IPOPT within PYOMO as the nonlinear optimizers of the stage 1 and stage 2 regression problems, respectively. For the sake of consistency, the structure of the NODE is fixed to a single hidden layer with tanh activation function and 15 hidden nodes. Further, we assume minimal knowledge of the true parameters prior to model-fitting, and thus all parameters are initialized to the same order of magnitude, specifically a value of 2, for the direct and indirect approaches.

3. Results

To demonstrate the effectiveness of the 2-stage approach, we chose as an example a microkinetic model (MKM) for heterogeneous catalysis (Gusmão et al. 2020). MKMs represent a large class of coupled differential equations which exhibit stiffness due to the presence of both slow and fast rate terms caused by parameter values varying over large orders of magnitude. The MKM system of ODEs governed by a material balance and rate equations are outlined in Figure 2. Table 1 lists the true parameter values.

$$\frac{d[A]}{dt} = -r_1 \qquad \frac{d[C]}{dt} = -r_3 \qquad r_1 = k_3[A][*] - k_4[A*]$$

$$\frac{d[A*]}{dt} = r_1 - r_5 \qquad \frac{d[C*]}{dV} = r_3 + r_5 \qquad r_2 = k_5[B][*] - k_6[B*]$$

$$\frac{d[B]}{dt} = -r_2 \qquad \qquad \qquad r_3 = k_7[C][*] - k_8[C*]$$

$$\frac{d[D*]}{dt} = 2r_4 - r_5 \qquad r_4 = k_{11}[B*][*] - k_{12}[D*]^2$$

$$\frac{d[B*]}{dt} = r_2 - r_4 \qquad \frac{d[*]}{dt} = -r_1 - r_2 - r_3 - \qquad r_5 = k_{13}[A*][D*] - k_{14}[C*][*]$$
$$r_4 + r_5$$

Figure 2. Full MKM system of ODE equations

In this process, gaseous reactants A and B adsorb to a solid surface to form intermediate species before the final product C desorbs into the gas phase. Reactants bound to a catalyst surface site [*] are indicated by an asterisk '*'. All reactions are reversible.

Two datasets were used to represent possible fitting scenarios for the 2-stage approach, each comprising data simulated from two sets of initial conditions. In one dataset, state variables are sampled 15 times logarithmically for each run in the range t = [10e-3, 0.5], amounting to a sample size of 30 datapoints. The second dataset includes the same number of points sampled linearly from time t=0 to t=0.5. At first, the 2-stage approach was applied on the linear dataset. Specifically, the data was used to fit a NODE whose derivatives were then used to solve for the parameters of the mechanistic DE. The parameters found through this approach are compiled in Table 1, column labelled 'Linear Indirect'. Results show that some of the parameters found differ significantly from the true parameters. This is not surprising since data is available only sparsely at earlier times where state values change rapidly due to the stiffness of the system.

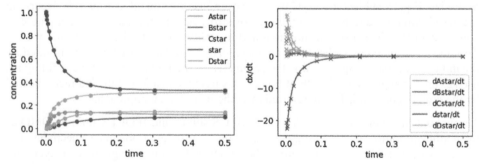

Figure 3. NODE fit (solid lines) to log-sampled data (dots). True derivative shown with x's.

The fitting procedure was repeated with data sampled logarithmically with respect to time. The fit of the Neural ODE to the log-sampled data is presented in Figure 3 for the adsorbed species. Noticeably, despite the NODE fitting the state outputs perfectly (effectively to machine precision), the data-driven model does not capture the exact profile of the derivatives. This result is believed to be due to the inherent flexibility of NODEs, which are not as constrained in outcomes as the simulating mechanistic model. The results of the 2-stage regression including the mean absolute error (MAE) of the fitted model on the log-sampled data are compiled in Table 1 ('Log Indirect' column).

Aiming to improve the accuracy of the fitted mechanistic model, the simultaneous approach was applied using various values for lambda. However, minimizing the hybrid objective function did not result in significantly improved parameter estimates vs the indirect approach, notwithstanding its higher compute cost. This finding was again attributed to the flexibility of NODEs, their being able to interpolate state data despite estimating derivatives that may not exactly match the 'true' derivatives. Due to their low accuracy, results of the simultaneous approach were not included in Table 1.

Instead, the remaining columns in Table 1 include the computational cost and model accuracy from integrating the mechanistic DE during training, either using an uninformed initial guess (i.e., the direct approach) or initializing the mechanistic parameters with the parameters found by the 2-stage methods (i.e., the incremental approach). Figure 4 displays the mechanistic model fit to the linearly-sampled data via the indirect and direct approach. The direct and incremental approaches gave similar simulated trajectories so only the results of the direct method are plotted.

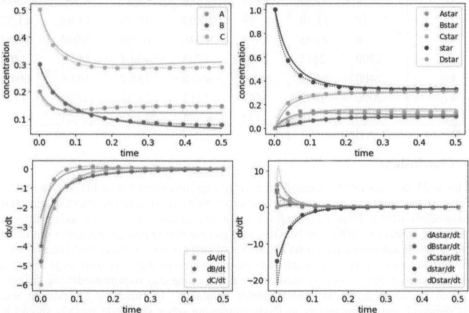

Figure 4. State and derivative estimates of the mechanistic model after parameter estimation via indirect (solid line) and direct approach (dotted line) on linearly-sampled data (solid dots).

A couple trends are worth noting. Firstly, the MAE of the final simulation is lower after using the direct approach, regardless of sampling strategy, indicating increased accuracy can be gained via the direct approach. What's more, applying the incremental approach offers compute savings over the direct approach with an uninformed initial guess, at least for the log-sampled case. However, when fitting the linearly-sampled data, the compute savings from incremental approach are negligible. At least two factors are believed to cause this discrepancy. First, the parameters found through the indirect approach on the linear data were further from the true parameters than for the log-sampled case, offering a poorer initial guess. Second, an increased number of Euler steps were required between datapoints for integrating mechanistic model on the log-sampled data (n=256 vs 56 in the linear case) to avoid divergence issues near the equilibrium region, exacerbating the computational load in the log-sampled case when uninformed initial estimates are used. Ultimately, this indicates that, given a sufficiently sampled experimental space, the incremental approach can merge the direct and indirect approaches in ways that balance both accuracy and efficiency.

Table 1. Table of compute times, parameters estimated, and model errors for different approaches

	True params	Log Indirect	Log Incr.	Log Direct	Linear Indirect	Linear Incr.	Linear Direct
Fit (s) Time	N/A	15.21	137.7	362.31	12.62	74.22	80.94
MAE	N/A	2.46E-3	6.05E-4	4.52E-4	1.45E-2	4.70E-4	2.69E-4
k_3	20	20.46	19.98	19.97	12.82	19.67	19.80
k_4	8	9.061	7.983	7.994	4.556	7.854	7.917
k_5	16	16.50	15.69	15.81	13.49	16.07	15.84
k_6	4	3.490	3.825	3.886	2.637	4.051	3.938
k_7	12	11.38	11.99	12.09	10.19	11.89	11.92
k_8	8	7.695	7.998	8.048	6.719	7.940	7.957
k_{11}	1200	2615	2607	1793	400.4	446.5	1809
k_{12}	400	849.3	871.6	600.8	138.7	147.6	604.0
k_{13}	2000	1672	1662	1117	38.28	2999	1745
k_{14}	1600	1320	1332	890.3	24.05	2401	1395

*Abbreviations: Incr. (Incremental Approach)

4. Conclusions

This work demonstrated a method for accelerating the regression of mechanistic ODEs for stiff systems and evaluated the ability of NODEs to estimate mechanistic ODE parameters with a large magnitude of variability in their true values using different sampling strategies. While the NODE-based incremental approach presents a promising step towards automated parameter estimation of stiff systems, several challenges remain. Neural Networks have limited ability to make predictions that vary over large orders of magnitude, common to stiff systems, which to overcome may require modifying the data-driven model structure to enable greater accuracy. In future, a comparison with incremental and simultaneous methods employing other surrogate models should be performed to assess the Neural ODE's suitability as a general-purpose DE estimator.

References

Bradley, W. and F. Boukouvala (2021). "Two-Stage Approach to Parameter Estimation of Differential Equations Using Neural ODEs." Industrial & Engineering Chemistry Research.

Brunel, N. J. B. (2008). "Parameter estimation of ODE's via nonparametric estimators." Electron. J. Statist. **2**: 1242-1267.

Gusmão, G. S., A. P. Retnanto, S. C. da Cunha and A. J. Medford (2020). "Kinetics-Informed Neural Networks." arXiv preprint arXiv:2011.14473.

Hamilton, F. (2011). "Parameter Estimation in Differential Equations: A Numerical Study of Shooting Methods." SIAM Undergraduate Research Online **Volume 4**.

Kim, S., W. Ji, S. Deng and C. Rackauckas (2021). "Stiff Neural Ordinary Differential Equations." Chaos **31 9**: 093122.

Li, Z., M. R. Osborne and T. Prvan (2005). "Parameter estimation of ordinary differential equations." IMA Journal of Numerical Analysis **25**(2): 264-285.

Swartz, J. and H. Bremermann (1975). "Discussion of parameter estimation in biological modelling: Algorithms for estimation and evaluation of the estimates." Journal of Mathematical Biology **1**(3): 241-257.

Proceedings of the 14th International Symposium on Process Systems Engineering – PSE 2021+
June 19-23, 2022, Kyoto, Japan © 2022 Elsevier B.V. All rights reserved.
http://dx.doi.org/10.1016/B978-0-323-85159-6.50291-8

Wiz 4.0: A Novel Data Visualisation and Analytics Dashboard for a Graphical Approach to Industry 4.0

Louis Allen,[a†] Jack Atkinson,[a†] Joan Cordiner,[a] Mohammad Zandi,[a] and Peyman Z. Moghadam[a,*]

aDepartment of Chemical and Biological Engineering, The University of Sheffield, Sheffield S1 3DJ, UK
p.moghadam@sheffield.ac.uk

[†] These authors contributed equally.

Abstract

Proliferation of data owing to the onset of Industry 4.0 (I4.0) has led to many traditional data analysis approaches becoming redundant. Novel and innovative solutions are required to facilitate the new era of data-driven manufacturing characteristic of I4.0. This work demonstrates one such approach in the formation of a bespoke web-based visualisation and machine learning analytics platform, designed to bridge the gap between the old ways and new. Our unique I4.0 data analytics platform, called Wiz 4.0, enables advanced big data analytics in conjunction with user-friendly features and multivariate data visualisations. This allows for both a holistic overview of manufacturing processes as well as detailed analysis of data. Wiz 4.0 lays the foundations of an industry defining software to grant deep insight into the inner relationships between process variables to the everyday user. The software provides the ability to analyse data using a variety of machine learning algorithms and plot the data in high dimensional space through the innovative no-code platform hosted on the Siemens MindSphere. This approach is set to revolutionise the value creation of data in the new IoT and smart factory paradigms emerging from the transition towards I4.0.

Keywords: visualization, analytics, dashboard, IoT, Industry 4.0

1.   Background

Widespread development of technologies such as cloud computing, Internet of Things (IoT), Automated Intelligence (AI), additive manufacturing (AM) and digital twins are driving the adoption of new working paradigms in manufacturing. These have the potential to revolutionise the sector when implemented effectively to create large autonomous systems and smart factories. At the root of this is an increased reliance on data. The transition is seen as a new age in manufacturing and wider industry; this fourth industrial revolution (Industry 4.0, I4.0) is associated with increased productivity, efficiency, and profit.

A joint 2019 study by Deloitte and the Manufacturers Alliance for Productivity and Innovation (MAPI) (Wellener et al. 2019) presented key findings about the benefits of smart factory initiatives. While a direct connection was found between these smart factory initiatives and business value via an increase in metrics such as manufacturing output, labour utilisation and labour productivity, only 3% of the sample indicated full scale smart factory adoption. These conflicting findings indicate a barrier preventing the manufacturing industries shift to I4.0. An aging workforce (Oppert & O'Keeffe, 2019) and lack of I4.0 expertise (Raj et al., 2020) is likely to decrease the agility of small and medium enterprise (SME) manufacturers to changes needed and will hinder the adoption of technology associated with I4.0 and smart factory initiatives. In comparison, larger firms who because of their size have an increased budget for retraining and upskilling of work, as well as a higher turnover and intake of younger staff, are more adaptable to change. Organisations struggling to adopt I4.0 require a stepping-stone approach toward digitalization, with a shallow learning curve and low barrier-for-entry.

Data visualization techniques offer an attractive way of jump-starting this shift. Effective visualizations offer universal understanding of complex data and relationships. If this can be combined with manufacturing specific analytics, utilising machine learning (ML) for applications such as predictive maintenance or autonomous control, far-reaching business value can be realized at a fraction of the cost of full-scale digitalisation. Furthermore, visualization of data will aid the operator in the role of abstract critical thinking and diagnoses, something that will certainly become more prominent as increasing levels of automation are used for ever more complex tasks.

Here, we have demonstrated the shift to digital manufacturing at the University of Sheffield's multi-million-pound Diamond Pilot Plant (DiPP), initially installed in 2017. The continuous powder processing plant equipment, shown in Figure 1, can produce 20

Figure 1. The DiPP's ConSigma tabletting line equipment. Includes: (a) twin screw granulator, (b) 6-chamber fluidised bed dryer, (c) mill and (d) continuous tableting press.

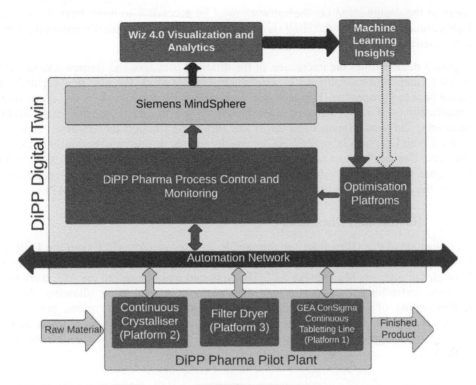

Figure 2. A schematic showing its digitization architecture. Wiz 4.0 can utilise Siemens MindSphere for data collection to create intuitive visual and analytical insight. The software has been made modular to allow the integration of future analytics such as process optimisation, fault detection, and predictive maintenance. This insight can affect the overall process decision making and is indicated by the dashed arrow. Figure adapted from Zandi, 2019.

kg/h of tablets and includes an AWL filter dryer, Nitech continuous crystalliser and GEA ConSigma tabletting line. These were retrospectively digitized through the combined efforts of The University of Sheffield, Siemens, and Perceptive Engineering, resulting in an IoT enabled world-leading industrial demonstrator. Figure 2 shows how this digitalisation has allowed a data-driven approach to advanced process control and automation due to the plant's ability to collect, align and centrally store data in Siemens MindSphere: Siemens' IoT platform. In this industrial microcosm, these developments presented an exciting and unique opportunity to engineer Wiz 4.0.

2.   Methods

When developing the software, careful consideration was given to the user interface to directly address the barrier to entry and expertise challenges identified in I4.0. In addition, an emphasis was placed on an effective integration of high-dimensional visualisations for the dashboard interface to maintain power in real-world manufacturing scenarios where analysis and exploration of results is routinely multidimensional. With this in mind, the objectives of the software we developed were i) to map and understand complex relationships between process variables in an easy and intuitive way and ii) to consider users in the design stage i.e. the software must be accessible to those from a range of backgrounds across manufacturing shop floors, including those typically unfamiliar with data science tools.

Figure 3 demonstrates the interaction of the different packages and frameworks utilised in the formation of the software. In general terms, we can bind the analytical capabilities to the backend processing. This includes cloud data storage using AWS, and regression, classification and outlier detection algorithms from SciKitLearn (Pedregosa et al., 2011). Similarly, the visualisation capabilities are expressed through the frontend. Plotly.js utilises Web Graphics Library (WebGL) enabling the high fidelity visualisation of large datasets. Use of React enables dynamic dashboard functionality, enabling the user to configure their own individual space to analyse and observe their data. Figure 3 also details the link of the software to Siemens MindSphere via the MindSphere API.

3.   Results and Discussion

Visualisation: Data taken from programmable logic controllers (PLCs) in the DiPP is uploaded to the Siemens MindSphere cloud foundry, where the application can access it through the Siemens MindSphere API. The front end makes a call to the API every two seconds, refreshing user-constructed visualisations in real time allowing for an understanding of conditions within the process equipment at any given point. The extended range of visualisations the software is capable of grants versatility, allowing for both a holistic approach as well as enabling detailed insight. Figure 4 demonstrates this with both a

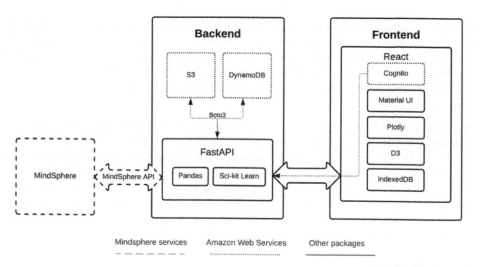

Figure 3. Schematic detailing interaction of fundamental packages and frameworks used in the construction of the software including front end construction in React JavaScript utilising Plotly.js charting library, and back-end construction utilising various AWS packages and analytics modules, constructed in Python, using SciKitLearn ML packages.

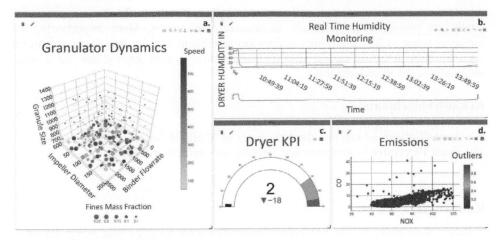

Figure 4: Wiz dashboard interface developed in Siemens MindSphere: (a) 5-dimensional plot showing conditions within the twin screw granulator, (b) real-time monitoring of the humidity in the fluidised bed dryer, (c) gauge plot showing dryer temperature as a key performance indicator (KPI) showing how the data is trending, and (d) scatter plot visualisation of local outlier factor (LOF) unsupervised machine learning algorithm showing outlier data detection.

detailed 5-dimensional plot of twin-screw granulator dynamics (Figure 4a) and a gauge showing purely KPIs for the spray dryer (Figure 4c).

Analytics: The software also acts as a no-code platform for low-level machine learning. This means users can construct, train, and store their own machine learning algorithms with a view to applying these to live data. So far, unsupervised (Principal Component Analysis (PCA) and outlier detection) and supervised (classification) algorithms have been used successfully to analyse uploaded datasets. As well as storing the models, the insights from the models created can be displayed intuitively across the dashboard environment allowing for fast and effective decision making from machine learning - all without ever having to write a line of code. This distillation of complex data analysis methods into a familiar graphical interface is where this software holds its value in industry.

4.   Conclusion

Here, we present Wiz 4.0, a data analytics/visualisation platform with a rich variety of features including real-time high-dimensional scatter plots, conducive to displaying relationships in multivariate data common in manufacturing. In addition, analysis tools including configurable ML algorithms have been added to give users access to additional insight and importantly these have been unified with the dashboard and plotting features.

Wiz 4.0 forms part of a more general trend in industry towards data-dashboards as a way to effectively communicate the larger amounts of data that are being used at all decision-making levels in real Industry 4.0 settings. Due to the software's modular dashboard interface, information pertinent to all levels of an organization can be collected, for example for the creation of a management level dashboard. This could provide less detailed information about the equipment but could still leverage important data collected by Siemens MindSphere to inform managers of important statistics such as operational run time, energy consumption, efficiency, and product output. In this way, Wiz 4.0 retains its power for a multitude of scenarios and users. More importantly, the implementation of ML algorithms enables key intelligent decision-making technologies such as fault detection and predictive maintenance: crucial capabilities for a truly successful smart manufacturing.

Acknowledgements:

We thank Wellcome Leap and the University of Sheffield for funding the project. We also thank Perceptive Engineering and Siemens for useful discussions.

References

C. Balzer, R. Oktavian, M. Zandi, D. Fairen-Jiminez, P. Moghadam, 2020, Wiz: A Web-Based Tool for Interactive Visualization of Big Data, Patterns, 1, 8, 100-107

M. L. Oppert, V. O'Keeffe, 2019, The future of the ageing workforce in engineering: relics or resources?, Australian Journal of Multidisciplinary Engineering, 15, 1, 100-111

F. Pedregosa, G. Varoquax, A. Gramfort, V. Michel, B. Thirion, M. Blondel, 2011, SciKit-learn Machine learning in Python, the Journal of machine Learning research, 12, 2825-2830

A.Raj, G. Dwivedi, A. Sharma, A. B. Lopes de Sousa Jabbour, S. Rajak, 2020, Barriers to the adoption of industry 4.0 technologies in the manufacturing sector: An inter-country comparative perspective, International Journal of Production Economics, 224

P. Wellener, S. Shepley, B. Dollar, S. Lapper, H. Manolian, 2019, 2019 Deloitte and MAPI Smart Factory Study, Research Centre for Energy & Industrials Group, Deloitte.

M. Zandi, 2019, Preparing students for 21st Century Chemical Industry, Perspectives from academia and industry for dealing with a digital world, University College London, UK

Proceedings of the 14th International Symposium on Process Systems Engineering – PSE 2021+
June 19-23, 2022, Kyoto, Japan © 2022 Elsevier B.V. All rights reserved.
http://dx.doi.org/10.1016/B978-0-323-85159-6.50292-X

About data reduction techniques and the role of outliers for complex energy systems

Luise Middelhauve[a], François Maréchal[a]

[a] *École polytechnique fédérale de Lausanne, Rue de l'Industrie 17, 1951 Sion, Switzerland, luise.middelhauve@epfl.ch*

Abstract

Optimal design and scheduling of energy systems with a high share of renewables is a complex and computationally demanding task. The mismatch of supply and demand of energy requires the consideration of timeseries with a granularity of a few minutes, which is in contrast to the lifetime of the system of multiple decades. This paper proposes an algorithm for systematically reducing the input data and computational effort in mixed integer linear programming (MILP) of energy systems. Unlike the state- of- the-art, the influence of different numbers of typical periods is not examined on the on the quality of the clustering algorithm but on the objective function and the integer decisions. The issue is addressed by exploiting the two-stage nature of the optimal design and planning of the system by sequentially performing k-medoids clustering. The demonstration of the proposed algorithm shows that very few typical periods are sufficient to achieve near optimal decisions. The proposed approach is outperforming algorithms for time series aggregation (TSA) in this field by reducing CPU time by more than 40 %. The inclusion of the integer decision in the algorithm allows the application to multi objective optimization (MOO). The case study demonstrates that the runtime of the MOO can be reduced by approximately 90 %, while diverting less than 2 % on Pareto optimal solutions. Outliers have no impact on the techno-economic analysis but may lead to significant electricity peaks in energy systems with a high share of renewables.

Keywords: Energy system design, Renewable energies, Mixed integer linear programming, Data reduction, k medoids clustering, outliers

1. Introduction

Daily and seasonal cycles lead to reoccurring patterns in the supply and demand of energy. Hence, it is popular to aggregate yearly time series to typical periods, in order to reduce computational effort for optimization problems of energy systems. Hoffman et al. (2020) have reviewed Time Series Aggregation (TSA) methods for modelling energy systems applied in 130 different publications. The authors conclude that for the same computational time, the more intuitive aggregation of seasons or months result in insignificantly larger errors than machine learning techniques. TSA methods of latter category performed similarly well, although k-medoids were most reliable for approximating costs.

In current TSA methods, the identification of the optimal number of typical periods is the first challenge. State-of-the-art approaches in modelling complex energy systems almost exclusively base their decision on Key Performance Indicators (KPI) of the clustering algorithm itself. The performance of the intended application to the energy system optimization is not considered in the selection process. Schütz et al. (2018) have demonstrated that the KPI *sum of square error*, typically applied when evaluating the clustering is not suitable for the application in energy system optimization problems. Therefore, Bahl (2018) developed a systematic method, which evaluates the error in the objective function of the energy system optimization for different TSA aggregation lengths. The impact of the unit decisions taken in the optimization problem is not considered in their approach.

The previous paragraph allows the conclusion that machine learning algorithms for data reduction techniques are not very well integrated in the optimization technique of complex energy systems. Thus, this work presents an iterative method for the TSA for mixed integer linear programming (MILP) of energy systems.

2. Method

An overview of the usual steps involved in TSA is provided in Figure 1. The steps, which are modified in comparison to the current state-of-the-art are highlighted.

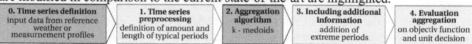

| 0. Time series definition input data from reference weather or measurement profiles | 1. Time series preprocessing definition of amount and length of typical periods | 2. Aggregation algorithm k - medoids | 3. Including additional information addition of extreme periods | 4. Evaluation aggregation on objectiv function and unit decision |

Figure 1 Procedure of time series aggregation (TSA) with relevant pictures in light grey.

Schütz et al. (2018) as well as Hoffman et al. (2020) have identified that the k- medoids algorithm is most reliable for techno- economic evaluation of energy systems. Hence, k-medoids algorithm is chosen as aggregation method in step 2.

Another challenge, which is generally overlooked in TSA for complex energy systems is the role of extreme periods. These extreme periods serve as protection or guarantee that the energy system can still provide required services, even in these extreme situations (Kotzur et al. (2018)). Extreme periods are often given by national regulations, are added in a postprocessing step (Stadler et al. (2018)) and are not further analysed. This work additionally investigates the role of outliers, which serve as extreme situations.

In the following, the aggregated problem (AP) is the energy system optimization based on typical periods, whereas the operating problem (OP) is the optimization of only the operation, with fixed unit decisions, on a full timeseries.

2.1. Time Series Aggregation

The first part of the TSA (Figure 2) is an iterative process, testing two convergence criteria on the objective function. After the first convergence criterion $|\varepsilon_a| \geq \frac{AP_n - AP_{n-1}}{AP_n}$ is valid, the OP is solved. The second convergence criterion $|\varepsilon_b| \geq \frac{OP_n - AP_n}{AP_n}$ is evaluated. In case both criteria are true, the sizing variables of the unit design are added to the solution space.

The second part is initiated with the first element, characterized by n clusters in the solution space. Additionally, the AP is executed with n+m steps. This way, different energy system designs with similar value of the objective function are identified and added to the solution space. If the unit decisions remain unchanged, the TSA is terminated with n typical periods.

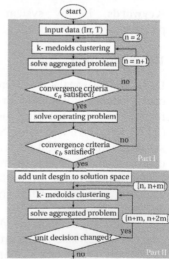

Figure 2 Overview of the iterative timeseries aggregation algorithm.

2.2. Outlier detection

Similar to the method developed by Liu et al. (2021), outliers are detected during the process of clustering. The procedure is as follows:

1. Clustering data set
2. Calculation of Euclidian distance from all points to their centroid
3. Removal of o outlier periods with the largest distance from data set
4. Repetition of steps 1-3 until centroids do not change anymore.

The detected o outliers are added as individually occurring typical period to the optimization problem. In contrast, demand peaks remain as extreme period. As current standard in optimization of energy systems, one cold weather and one hot weather period is added, both consisting of one single timestep (Stadler et al., (2018)).

3. Case study

The data reduction technique is demonstrated on optimizing the energy system of a typical residential building located in the climatic zone of Geneva, Switzerland. The building is a single- family home with 2 floors and in total 250 m^2 heated surface from around 1980. The considered energy demands are: electricity demand, thermal demand for space heating and hot water. The optimization is formulated as a MILP problem with the aim to find the optimal sizing and operation among nine energy conversion and storage technologies. For further insights on the modelling approach of the building energy system, the reader is kindly referred to Middelhauve et. al (2021). On a full timeseries of one year, this case study leads to over 840 thousand constrains and 790 thousand variables, among which are almost 10'000 binaries. The tuning parameter of the algorithm are set to be $\varepsilon_a = 5\%$, $\varepsilon_b = 5\%$ and $m = 3$.

The k-medoids clustering with the R package *wcKMedoids* is performed for aggregating one typical year (DOE, (2020)). The problem is formulated in AMPL Version 20191001 and solved with CPLEX 12.9.0.0 on a local machine with following processor details: Intel(R) Core (TM) i7-8559U CPU @ 2.70GHz. The relative tolerance between relaxed linear problem and best integer solution is set to *mipgap*=5e-7. The remaining CPLEX settings are equal to the default settings reported at (CPLEX, (2020)).

4. Results and Discussion

In a pre-processing step, global irradiation and the ambient temperature are clustered to increasing number of k-medoids, each typical period is chosen to be the length of one full day. The state-of-the-art procedure, which bases the selection of optimal number of periods at KPI slope thresholds, would lead to around 10 typical periods. Thus, the result of the proposed TSA is compared to 10 typical periods in the following.

4.1. Time Series Aggregation

First, the AP with an increasing number of clusters is solved. The objective function is total expenses, which is equally weighting two conflicting objectives, the capital expenses (CAPEX) and operational expenses (OPEX). Part one of the proposed TSA algorithm is demonstrated in Figure 3. After already three clusters, the first convergence is approximately 5%. The unit decisions are fixed and the OP is solved. The second convergence criterion is also below 5 % and therefore three typical periods are chosen as result of Part 1. Both problems the AP and the OP are solved for up to 12 typical days for demonstration purposes. The difference of the objective functions, which is the second convergence criterion, remains below 1% after 7 typical periods.

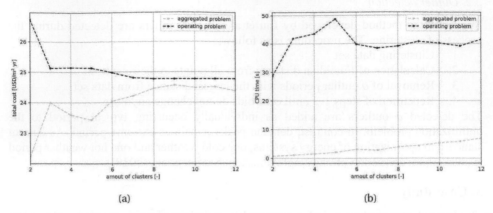

(a)   (b)

Figure 3 Demonstration the Part 1 of the proposed TSA algorithm. Impact of different numbers of clusters on a) the objective function, b) the runtime.

The aim of Part 2 of the algorithm is to confirm that the unit decisions are sufficiently taken with the detected number of typical periods (compare Figure 2). For two clusters, an air-water heat pump in combination with electrical heater and thermal storage tanks is detected as most economical decision. For three clusters, photovoltaic modules are chosen additionally. This configuration remains unchanged for not only n+m = 6 clusters, but for all 12 clusters examined. Therefore, the three typical periods can be confirmed and further used during Multi objective optimization (MOO).

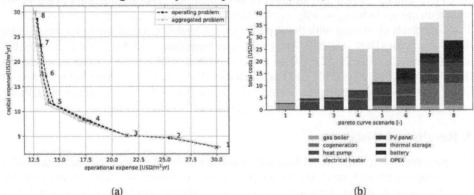

(a)   (b)

Figure 4 Results of multi objective operation a) Pareto frontier with 10 (x) and 3 (•) typical periods. b) unit decisions along Pareto frontier.

One requirement of appropriate TSA in energy system optimization is that the aggregation is valid also in MOO problems. Therefore, Figure 4 compares the Pareto curves of the AP to the associated OP for the previously resulting 3 typical periods and compares them to the state-of the-art choice of 10 typical periods. The unit decision of Scenario 1 is based on a natural gas boiler and small thermal storage tanks. In this case, both APs as well as their linked OPs result in almost identical points on the Pareto frontier. The energy system is more diverse in higher investment scenarios. It includes renewable energy sources such as PV panels, an air-water heat pump and thermal and electrical storage in Scenario 8. The performance of this energy system is more depending on the weather data or, in terms of the storage, the optimal scheduling within one typical period.

Nevertheless, comparing the OPEX of the AP to the related problem on the fulltime series, the OP shows a deviation by only 3%. The difference among the AP with 3 typical periods and with 10 typical periods is less than 2%, maximum occurring in higher investment scenarios.

4.2. Outlier detection

K-medoids clustering is a method, which is itself robust to outliers. Therefore, the centroid is not changing during the detection of outliers. Nevertheless, seven outliers have been identified and added to the typical periods. Figure 5 demonstrates the impact of outliers on the MOO of the energy system. Along the pareto curve, identical unit decisions are taken and the pareto curve with and without 7 outliers are identical.

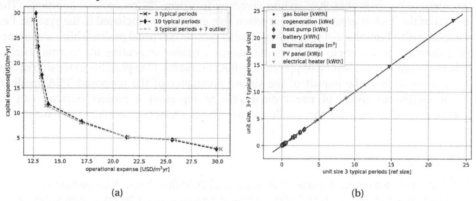

(a)   (b)

Figure 5 Comparison of multi objective optimization with and without detected outliers. a) Pareto curve b) Parity plot of all unit decisions along the pareto curve.

In contrast to the detected scenarios along the pareto curve, Figure 6 shows an important impact of outliers.

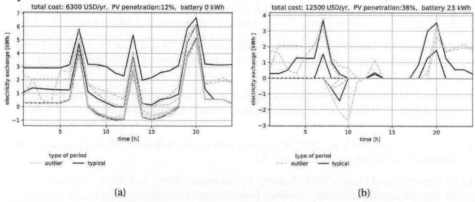

(a)   (b)

Figure 6 Electricity exchange for energy system designs with a) a lower investment and b) higher investment and greater share of renewable energy integration

Outliers may not be relevant for the thermal energy side or the unit decisions. But in future energy systems with a high share renewables, outliers reveal electricity peaks of the grid exchange, which are higher than during typical periods.

5. Conclusion

This work proposes a novel method for TSA of complex energy systems. Compared to state-of-the-art approaches in this field the CPU time was reduced by 40% in the presented case study. This is achieved by using two convergence criteria, which avoids the computationally intensive computation of the OP at each iteration step. Additional innovation is to take unit decisions into account, which allows the application of the TSA method to MOO problems.

In contrast to comparable work in MOO of energy systems, the selection of the appropriate number of typical days is not based on KPIs evaluating the underlying machine learning algorithm. Hence, presented TSA method allows to significantly reduce the runtime by more than 90%, while diverting less than 2% on optimal solutions. The impact of the TSA is greater, the more renewable energies are included in the system. For systems with a high share of renewables, outliers reveal electrical peaks, which are greater than during typical periods. Outliers are however neither impacting the thermal energy side nor economic evaluation of presented MOO problems. One possible extension of this work is to analyse the impact of the tuning parameter of the proposed TSA method. Additionally, the usage of one typical year to represent a project horizon can be questioned. This includes the challenge to predict changing weather data subject to climate change.

References

T. Schütz, M.H. Schraven, M. Fuchs, P. Remmen, and D. Müller, 2018, "Comparison of clustering algo-rithms for the selection of typical demand days for energy system synthesis", Renewable Energy vol.129, pp. 570–582.

P. Stadler, L. Girardin, A. Ashouri, and F. Maréchal, 2018, "Contribution of Model Predictive Controlin the Integration of Renewable Energy Sources within the Built Environment", Frontiers in Energy Research, vol. 6

M. Hoffmann, L. Kotzur, D. Stolten, and M Robinius, 2020, "A Review on Time Series Aggregation Methods for Energy System Models", :Energies,vol. 13.3, p. 641

L. Middelhauve, L. Girardin, F. Baldi, and F. Maréchal. 2021 "Potential of Photovoltaic Panels on Building Envelopes for Decentralized District Energy Systems", Frontiers in Energy Research p. 689-781

L. Kotzur, P Markewitz, M. Robinius, and D. Stolten. 2018, "Impact of different time series aggregation methods on optimal energy system design", Renewable Energy, vol. 117, pp. 474–487

H. Liu, J. Li, Y. Wu, and Y.Fu, 2021 "Clustering with Outlier Removal", IEEE Transactions on Knowledge and Data Engineering, vol. 33

DOE. Department of Energy, BTO. Building Technologies Office, and NREL. National Renewable Energy Laboratory, Weather Data, EnergyPlus, https://energyplus.net/weather accessed Mar. 2020

CPLEX Options for AMPL. https://ampl.com/products/solvers/solvers-we-sell/cplex/options/, accessed Oct. 2021.

B. Bahl, 2018, "Optimization-Based Synthesis of Large-Scale Energy Systems by Time-Series Aggregation" ; RWTH Aachen University. PhD thesis, 148 pages

Proceedings of the 14th International Symposium on Process Systems Engineering – PSE 2021+
June 19-23, 2022, Kyoto, Japan © 2022 Elsevier B.V. All rights reserved.
http://dx.doi.org/10.1016/B978-0-323-85159-6.50293-1

DeepGSA: Plant Data-Driven Global Sensitivity Analysis using Deep Learning

Adem R.N. Aouichaoui[a], Resul Al[b], Gürkan Sin[a*]

[a]*Process and Systems Engineering Center (PROSYS), Department of Chemical and Biochemical Engineering, Technical University of Denmark, Kgs.Lyngby, DK-2800, Denmark*
[b]*Novo Nordisk A/S, Bagsvaerd, DK-2880, Denmark*
gsi@kt.dtu.dk

Abstract

Data-driven modeling provides a viable alternative for process modeling especially in applications where mechanistic modeling falls short of explaining the underlying phenomena. The increasing amount of plant data collected through various sensors and lab tests lays the foundation for various data-driven modeling approaches such as Deep Neural Networks (DNN). In this work, we present a new software tool, named deepGSA, incorporating well-established variance-decomposition and derivative-based global sensitivity analysis (GSA) methods, such as Sobol sensitivity indices, with the plant data-driven deep learning modeling techniques. The deepGSA aims at enabling non-specialist practitioners to leverage DL-based models for GSA application purposes. The tool is successfully applied on a benchmark case study as well as the case of modeling liquid nitrous oxide concentration in a wastewater treatment plant to highlight its capabilities. The deepGSA toolbox, documentation, installation guide, and several examples are freely available on GitHub through the link: https://github.com/gsi-lab/deepGSA.

Keywords: Deep-Learning, Big Data, Sensitivity Analysis, Wastewater Treatment

1. Introduction

Digitalization has become an increasingly hot topic in various fields across academia and the chemical industry issuing a new industrial revolution under the guise of "industry 4.0". At the core of this digital transformation, we find Digital Twins, which are virtual mock models capable of simulating/describing real-life physical processes (de Beer and Depew, 2021). These models often rely on first principle mechanistic models particularly in cases where the phenomena occurring are well understood e.g. distillation and extraction processes. However, this process understanding is not as mature in many life science applications especially if it combines a wide range of different processes e.g. chemical, biological, and mechanical processes. A prime example of this is the modeling wastewater treatment plants (WWTP) (Sin and Al, 2021). Mathematical models have been applied to model the microbial activity in WWTP in the form of Activated Sludge Models (ASMs)(Sin and Al, 2021). The original ASM models have been modified and expanded in the past decades to include additional microbial conversion processes such as single-pathway and two-pathway models for N_2O production (Sin and Al, 2021). However, these efforts resulted in very complex mathematical models that are difficult to apply in practice due to the high number of model parameters that need to be calibrated (Chen *et al.*, 2019). Data-driven modeling of industrial processes provides an alternative approach to construct digital twins by leveraging the vast amount of plant data collected through sensors to some extent considered "Big data". Data-driven models such as Deep Neural Networks (DNNs) although providing good predictive performance across many fields and applications (Hwangbo *et al.*, 2021), lack the aspect of transparency and interpretability due to their black-box nature. Global Sensitivity Analysis (GSA) could potentially help elucidate the workings of these models. In this work, we present a Matlab-based toolbox, deepGSA, with the aim of streamlining data-driven modelling to non-specialist practitioners and enabling them to leverage their plant data to gain process understanding through various GSA methods implemented in the toolbox.

2. Methods

Global sensitivity analysis aims at quantifying the effect of an independent variable while all other variables are also varied. In doing so, the GSA takes into account interactions among the variables (in contrast to One-At-a-Time methods) and does not depend on the choice of a nominal point. Two GSA techniques are integrated into the tool: the Sobol method and the derivative-based global sensitivity method (DGSM).

2.1. Sobol Method

Sobol's method is based on variance decomposition of Monte-Carlo simulations. The method provides both first order (S_i) and total order indices (S_{Ti}) for the input parameter x_i. S_i measures the individual contribution of input x_i to the total output variance, while S_{Ti} measures the total contribution to the total output variance including those resulting from interactions with other inputs (Saltelli *et al.*, 2007). The expressions for S_i and S_{Ti} are shown in Eq.(1), where $V[E(y|x_i)]$, $V[E(y|x_{-i})]$ and $V(y)$ are the conditional variance, the conditional variance derived from all variables but x_i varied, and the unconditional total output variance respectively, all of which are numerically estimated through Monte-Carlo simulations. The difference between the two indices (S_{Ti}-S_i) is a direct indicator of the strength of interactions variable x_i is involved in.

$$S_i = \frac{V[E(y|x_i)]}{V(y)} \qquad S_{T_i} = 1 - \frac{V[E(y|x_{-i})]}{V(y)} \tag{1}$$

2.2. Derivative-Based Global Sensitivity Method (DGSM)

DGSM uses the second moment of the model derivatives to measure the importance of the input parameter x_i as formulated in Eq.(2) (Kucherenko *et al.*, 2009). The aim of this is to avoid canceling off negative and positive impacts of the local sensitivity (E_i) over the entire input domain H_n.

$$v_i = E\left[\left(\frac{\partial f(X)}{\partial x_i}\right)^2\right] = \int_{H_n} \left(\frac{\partial f}{\partial x_i}\right)^2 dx \text{ with } E_{i,n} \frac{\partial f}{\partial x_i} \tag{2}$$

A mean measure for the global sensitivity can then be obtained by Eq.(3).

$$\mu_i = \sqrt{v_i} \tag{3}$$

The mean measure is often used to screen unimportant factors in the input domain for which μ_i is very small or negligible. Additionally, the method gives insights into the directional pull of the input factor (positive or negative) by capturing the local sensitivity.

Although empirical, there is a link between the Sobol S_{Ti} and v_i as expressed in Eq.(4), where D is the total variance. The relation can be interpreted as the upper bound of S_{Ti}.

$$S_{T_i} \leq \frac{v_i}{\pi^2 D} = S_{Ti}^{DGSM} \tag{4}$$

3. DeepGSA: Toolbox for plant data-driven and DL-assisted GSA

Data-driven modeling

Given a large dataset X representing inputs and the corresponding response variable vector y, deepGSA is able to develop a DNN model to predict y from the supplied information X. The DNN model hyperparameters are optimized using a grid-search tuning policy for which a user can supply a pool of architectures, a set of training functions, and a set of activation functions. Each possible combination is trained on 80% of the supplied data, while the remaining data are equally split between validation and test purposes to avoid data leakage and overfitting. The toolbox also leverages Matlab's parallel computation using high-performance GPUs and computer clusters (if present) to speed up model development and GSA calculations. The optimal model can be selected based on various criteria evaluated on the validation set. The user can choose between: the coefficient of determination (R^2), the mean squared error (MSE), the root mean squared error (RMSE), the Bayesian/Akaike information criterion (BIC/AIC).

Inference of input distributions

Sampling from the input data is an important step to perform data-driven GSA. In some cases, the distributions of the input factors are known a priori, thus can be readily supplied to the toolbox. However, in many engineering applications, the underlying distribution of X is unknown. By supplying X, the toolbox can also infer the distributions of variables by fitting a kernel distribution to the data, which then allows sampling from a distribution similar to the input factor distributions to preserve the characteristics of the system.

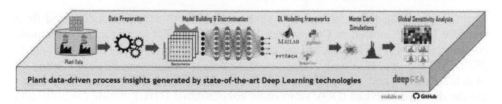

Figure 1: deepGSA Framework

4. Case Study

4.1. Benchmark Case Study: Sobol g-function

The g-function of Sobol, as given in Eq.(2), is often used as a benchmark function for numerical experiments in sensitivity analysis literature. The reason being the theoretical first-order Sobol indices are analytically available (Marrel *et al.*, 2009).

$$f(x) = \prod_{i=1}^{d} \frac{|4x_i - 2| + a_i}{1 + a_i} \text{, with } a_i = \frac{i-2}{2} \text{ and } x_i \sim U(0,1) \text{ for all } i = 1,..,d \quad (2)$$

Through Monte-Carlo simulations, a sample of size N = 5,000 is generated and used to evaluate the corresponding f(x) values following Eq.(2) and perform GSA. Figure 2 shows the sensitivities indices obtained: analytically, by evaluating the g-function, and by constructing DNN. All methods produce the same sensitivity indices which validate not only the implementation of the GSA techniques but also that DNN models retain the dynamics of the process it models. The obtained DNN was trained with the Bayesian regularization backpropagation algorithm in MATLAB and had the layers [32,16,8,4] with hyperbolic tangent sigmoid transfer function.

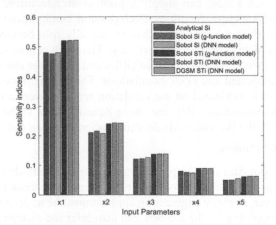

Figure 2: Sensitivity indices for the Sobol g-function estimated through various methods

4.2. Plant Data Case Study: Modelling liquid N_2O concentration in WWTP

DeepGSA is applied on a second case study using various sensor data collected from Avedøre WWTP located in Copenhagen Denmark to predict the liquid Nitrous oxide concentration (N_2O). The data and the plant are described in detail by (Hwangbo *et al.*, 2020, 2021). In this case study, a sample of the data spanning over 90 days starting on March 26[th,] 2018, and ending on June 24[th,] 2019 was selected. The input factors chosen are the dissolved oxygen (DO), ammonium concentration (NH_4), nitrate concentration (NO_3), temperature (T), the airflow for aeration (Q_{air}), and the influent flow (Q_{inf}). A total of 25,500 data points were used and all inputs were smoothed using the moving median method with a moving window of 12 timesteps and scaled between 0 and 1. The grid search included the following architectures: [128,64,32,16,8,4], [64,32,16,8,4], [32,16,8,4] with two possible training functions: '*trainlm*' and '*trainbr*' and two possible transfer functions: the log-sigmoid and positive linear functions. The selection criteria chosen is the R^2, the distributions of the input factors are inferred by the tool and a sample of size N=100,000 was used to conduct GSA. The sampling number was set very high since the computational solution of DNN is very affordable. The output layer activation function adapts to the characteristics of the target output e.i. constrained between [0, inf] through the positive linear (poslin) activation function. The selection criteria in this case study were chosen to be the R^2 metric. The best DNN achieved R^2 of 0.81 with a hidden configuration of [12,8,64,32,16,8,4] trained with 'trainbr' with poslin activation function.

Table 1: Inference statistics and sensitivity measures related to the DGSM and Sobol methods

	DGSM			Sobol	
Variable	μ (E)	σ (E)	S_{Ti}^{DGSM}	S_i	S_{Ti}
Q_{inf}	00.2	63.2	2.121	0.0000	0.3502
DO	00.1	46.9	1.169	0.0083	0.3238
NH_4	15.3	96.1	5.018	0.0317	0.4553
NO_3	11.4	103.2	5.718	0.1087	0.5848
T	01.1	109.2	6.332	0.0329	0.6579
Q_{air}	04.4	79.9	3.398	0.0529	0.5821

Table 1 provides various inferred statistics related to the DGSM and Sobol. The Sobol method identifies T, the NO_3, NH_4 concentrations as well as the influent and air flowrates as the most important factors. This is in line with the results obtained by (Hwangbo *et al.*, 2021). There is a large difference between S_i and S_{Ti}, with the Si values being very close to zero and the sum of S_{Ti} exceeding unity. These are all evidence of the presence of strong interaction effects of the input factors on the response variable (N_2O liquid concentration). These observations are consistent with the process insight of the N_2O dynamics in a WWTP and are explained in detail by (Hwangbo *et al.*, 2021). The S_{Ti}^{DGSM} is qualitatively valid for all input factors. Furthermore, the ranking is almost identical to the one obtained by Sobol indices except for two factors for which the ranking is swapped (Q_{air} and NH_4). This is in line with observations by (Sobol and Kucherenko, 2009) that for highly non-linear systems the ranking might not hold, and instead, the Sobol indices are more reliable. The statistics of the local sensitivity reported in Table 1 and the visualization of the empirical cumulative distribution function for selected input factors in Figure 3 suggest that the factors influence the output both positively and negatively. This is in fact due to data coming from a closed-loop (controlled) plant and it shows the effects of the inputs are nonlinear and therefore the effects depend on relative values of other inputs (input space has both regions where a factor produces a negative impact and positive impacts). Figure 3 further shows that the distribution exhibits some heavy-tailed distribution known as "fat tails", making reporting the mean and standard deviation misleading.

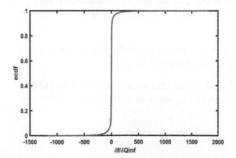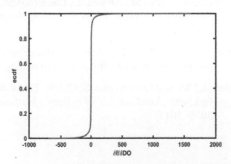

Figure 3: Empirical Cumulative distribution function of the local sensitivities of T and DO revealing a heavy-tailed distribution. In the tails, one observes a small change input has a disproportionately high effect on the output which is N_2O emissions.

5. Conclusions & Future work

DeepGSA provides an easy-to-use, plug and play tool to perform data-driven modeling using deep neural networks and perform global sensitivity analysis using both variance decomposition-based and derivative-based sensitivity analysis methods. The tool constructs a search space to tune the hyperparameters of the neural network and provide various metrics to select the optimal network. The GSA performed on the model is then used to provide process insights that conform with domain knowledge for two cases studies presented in this work. Future improvements to the toolbox include implementing more GSA techniques (Morris screening, standard regression coefficient (SRC), and GSA for dependent inputs) and the addition of new features such as an internal plotting tool, data cleaning, and scaling options. Finally, a wrapper function will be developed to allow users to plug in DNN models developed in open source DL frameworks such as TensorFlow and Pytorch.

References

de Beer, J. and Depew, C. (2021) 'The role of process engineering in the digital transformation', *Computers and Chemical Engineering*, 154, p. 107423.

Chen, X., Mielczarek, A.T., Habicht, K., Andersen, M.H., Thornberg, D. and Sin, G. (2019) 'Assessment of Full-Scale N 2 O Emission Characteristics and Testing of Control Concepts in an Activated Sludge Wastewater Treatment Plant with Alternating Aerobic and Anoxic Phases', *Environmental Science & Technology*, 53(21), pp. 12485–12494.

Hwangbo, S., Al, R., Chen, X. and Sin, G. (2021) 'Integrated Model for Understanding N2O Emissions from Wastewater Treatment Plants: A Deep Learning Approach', *Environmental Science and Technology*, 55(3), pp. 2143–2151.

Hwangbo, S., Al, R. and Sin, G. (2020) 'An integrated framework for plant data-driven process modeling using deep-learning with Monte-Carlo simulations', *Computers & Chemical Engineering*, 143, p. 107071.

Kucherenko, S., Rodriguez-Fernandez, M., Pantelides, C. and Shah, N. (2009) 'Monte Carlo evaluation of derivative-based global sensitivity measures', *Reliability Engineering and System Safety*, 94(7), pp. 1135–1148.

Marrel, A., Iooss, B., Laurent, B. and Roustant, O. (2009) 'Calculations of Sobol indices for the Gaussian process metamodel', *Reliability Engineering and System Safety*, 94(3), pp. 742–751.

Saltelli, A., Ratto, M., Andres, T., Campolongo, F., Cariboni, J., Gatelli, D., Saisana, M. and Tarantola, S. (2007) *Global Sensitivity Analysis. The Primer*. Chichester, UK: John Wiley & Sons, Ltd.

Sin, G. and Al, R. (2021) 'Activated sludge models at the crossroad of artificial intelligence—A perspective on advancing process modeling', *npj Clean Water*, 4(1), p. 16.

Sobol, I.M. and Kucherenko, S. (2009) 'Derivative based global sensitivity measures and their link with global sensitivity indices', *Mathematics and Computers in Simulation*, 79(10), pp. 3009–3017.

Proceedings of the 14th International Symposium on Process Systems Engineering – PSE 2021+
June 19-23, 2022, Kyoto, Japan © 2022 Elsevier B.V. All rights reserved.
http://dx.doi.org/10.1016/B978-0-323-85159-6.50294-3

Analysing Different Dynamically Modelled Data Structures and Machine Learning Algorithms to Predict PM2.5 Concentration in China

Danny Hartanto Djarum[a], Nur Hidanah Anuar[a], Zainal Ahmad[a*], and Jie Zhang[b]

[a] School of Chemical Engineering, Universiti Sains Malaysia, Engineering Campus, 14300, Nibong Tebal, Penang, Malaysia.
[b] School of Engineering, Merz Court, Newcastle University, Newcastle upon Tyne NE1 7RU, United Kingdom.
chzahmad@usm.my

Abstract

Harmful air pollutant such as $PM_{2.5}$ is still a major concern in many countries. At high concentrations, it could lead to adverse health effect on human, escalating the risk of cardiovascular and respiratory diseases. In order to mitigate this issue, continuous air quality monitoring systems have been deployed to alert the general public of high $PM_{2.5}$ level. However, such monitoring system requires substantial budget and resources to construct, thus may not be accessible in some regions especially developing countries. Therefore, it is important to develop a high performance $PM_{2.5}$ prediction model that only employs easily attainable input parameters as a more cost-effective alternative. In this study, common meteorological data from five different cities in China were utilized for the $PM_{2.5}$ prediction model. Dynamic model such as the nonlinear autoregressive network with exogenous inputs (NARX) with different input/output time lag were applied to transform training dataset into different data structures. Additionally, machine learning algorithms were analysed and evaluated to predict $PM_{2.5}$, namely: multi-linear regression (MLR), and feed-forward artificial neural network (FANN). The results shows that FANN model with 10 hidden neurons using NARX-2 data structure is the best model combination with an R^2 values of up to 0.973.

Keywords: $PM_{2.5}$, NARX, MLR, FANN.

1. Introduction

The world population has grown significantly in the past decades, as a result, this causes the rise in rapid industrialization and urbanization to cope with the high demand of food and energy resources (Mobaseri et al., 2021). The excessive rate of industrialization consequently translates to generation of pollutants leading to extreme deterioration in global air quality. $PM_{2.5}$ which stands for particulate matter 2.5, is considered as one of the major and most dangerous atmospheric pollutants in the air. Any airborne particle that has an aerodynamic diameter of less than 2.5 μm is categorized as $PM_{2.5}$. Various researchers have shown in their study that $PM_{2.5}$ is associated with serious health diseases such as strokes, chronic heart disease, lung cancer, respiratory infections, and premature death (Gakidou et al., 2017). In a 2019 study, Chen et al. (2019) discovered that an increase of 10 μg/m³ in atmospheric $PM_{2.5}$ levels would lead to a reduction in adult life expectancy by approximately 0.8 years.

With the growing concern over the dangerous health hazard poses by PM$_{2.5}$, continuous PM$_{2.5}$ monitoring stations have been proposed by many to alert the general public. However due to the high initial investment required, such system is only available in specified locations. In fact, the world health organization (WHO) reported that approximately 9 out of 10 people in the world still resides in area with PM$_{2.5}$ level exceeding the WHO air quality guidelines (WHO, 2021). In search for a more cost-effective alternative, numerous studies have been carried out to develop prediction models that can accurately reflect PM$_{2.5}$ concentrations. A feed forward artificial neural network (FANN) model was developed by (Perez et al., 2020) to predict PM$_{2.5}$ concentrations in Coyhaique, Chile. Their results shows that the model was able to capture the instance when PM$_{2.5}$ falls under critical region 85 % of the time. (Biancofiore et al., 2017) compared three different models to predict the levels of PM$_{2.5}$ for multiple days ahead (recursive neural network, feed forward neural network, and multiple linear regression). They concluded that recursive neural network outperforms the other models with correlations coefficients up to 0.89.

In recent years, artificial neural network model seems to be of the up most interest for PM$_{2.5}$ forecasting. However, the number of research that analysed the effect of time series data structure on the ANN model seems pale in comparison. Therefore, in this study, a dynamic time series model with different input/output time lag were utilized to transform the data structure, namely nonlinear autoregressive network with exogenous inputs (NARX). Additionally, a feed forward artificial neural network (FANN) model utilizing different form of input data structure were analysed to predict PM$_{2.5}$ concentration. Furthermore, the resulting performance were compared against widely popular multi linear regression (MLR) model that utilized untransformed input data.

2. Methodology

2.1. Dataset

The data used for this study is an hourly air quality and meteorological measurements recorded for six years between 2010 and 2015 in five Chinese cities, which are Beijing, Chengdu, Guangzhou, Shanghai, and Shenyang. These data were obtained from UCL Machine Learning Repository published by (Liang et al., 2016). There are 52,584 rows of data and with a total of 10 parameters this translates to about 525,840 individual data points for each city. The list of the input and output parameters for the PM$_{2.5}$ prediction model could be found in Table 1.

Table 1: List of input and output features for PM$_{2.5}$ prediction model.

List of input parameters	Seasons, temperature, pressure, dew point, humidity, combined wind direction, cumulated wind speed, hourly precipitation, cumulated precipitation
List of output parameters	PM$_{2.5}$ concentration

2.2. Data Pre-processing

Multiple pre-pre-processing techniques were performed on the original dataset before transformation on the data structure and development of the prediction model. In the missing data analysis, it was found that more than half of the dataset comprised of missing values. Due to the high degree of missing data, median value substitution could not be carried out as it will generate unwanted bias in the training data. Therefore, the missing data were instead discarded from the dataset, with approximately 20,000 rows of data remaining for each city. Normalization was then performed on the data, since the dataset has parameters with highly varying magnitude which similarly could lead to bias when training the model. Lastly, label encoding was performed to convert the parameter with categorical value such as wind direction into numeric value as most machine learning algorithms prefer to work with numerical attributes.

2.3. Data Structures

A dynamic time series model called nonlinear autoregressive network with exogenous inputs (NARX) was used to transform the training data set into different data structure. In this study, three type of data structures were analyzed, namely: The base structure without any transformation, NARX-1 (NARX with time lag of 1), and NARX-2 (NARX with time lag of 2). The complete equations for the different transformation on the data structure could be observed from Eq.(1) for base structure, Eq.(2) for NARX-1 and Eq.(3) for NARX-2.

$$y(t) = Fn(x_t) \tag{1}$$

$$y(t) = Fn(y_{t-1}, x_{t-1}) \tag{2}$$

$$y(t) = Fn(y_{t-1}, y_{t-2}, x_{t-1}, x_{t-2}) \tag{3}$$

Where $y(t)$ represent the output and $x(t)$ is the input of the model at time t. 1 and 2 is the maximum input and output time lag. Fn is the nonlinear function.

NARX is one of the more robust time series models that has exogeneous inputs, it has the ability to generate forecast of future output value based on the past input and output values with varying degree of time lag. In theory, by transforming the training data structure with NARX model, the learning rate of the resulting trained neural network model should be more effective and generate better prediction performance. Additionally, the neural networks model should converge to the desired solution much faster, leading to an overall better training efficiency and generalization capability (Lin et al., 1996). A two-step combined algorithm based on NARX neural network were compared by (Buevich et al., 2021) against other models to predict various greenhouse gases (CH_4 and H_2O) concentrations. They found that the algorithms coupled with NARX model consistently outperform other models in predicting the greenhouse gaseous concentrations with correlation coefficient up to 0.87 when validated against the test dataset. In another study, Mohebbi et al. (2019) analyzed NARX modelled artificial neural network model against static neural network model to predict carbon monoxide concentration in Shiraz city. Their results, shows that the NARX-ANN model significantly improve the model performance with an R^2 values of up to 0.72 compared to that of static neural network model that has an R^2 value of only 0.31. In our study the feed forward artificial neural network (FANN) model will utilized the NARX transformed data as the input to the model with maximum input and output lag time of 1 and 2.

2.4. Prediction Model

Multiple linear regression (MLR) and Feed forward artificial neural network (FANN) model were analyzed and evaluated to predict the $PM_{2.5}$ concentration. MLR is one of the most popular machine learning regression technique due to its simplicity, fundamentally, it explains a linear relationship between multiple input features to predict a single output feature. Eq.(4) below describe the universal formula for multiple linear regression model:

$$y = \beta_0 + \beta_1 X_1 + \cdots + \beta_n X_n + \varepsilon \tag{4}$$

Where y is the independent variable, β_i is the coefficient, X_i is the independent variable, and ε is the residual error between the real value and forecasted value. When performing MLR modelling it is assumed that the data is normally distributed, achieved linearity, without extreme values, and input parameters must be independent of each other.

FANN were then analyzed which theoretically has better capability to capture the dynamic nonlinear relationship between the input parameters. Being one of the earliest classes of artificial neural network, FANN has been widely popular amongst the scientific community especially since it is generally superior compared to traditional statistical technique such as multi regression and ensemble decision tree technique (Grivas and Chaloulakou, 2006). Generally, the FANN algorithm has multiple neurons grouped within three types of layers: Input layers, hidden layers, and output layers. The input layer consists of various neurons that receive information from the dataset, each neuron is connected to every single neuron in next layer (hidden layers) with its own associated weight. The weight determines the strength between the two connected neurons. Each neuron sums all of the information received and pass it on to the output layer where the output value will be computed based on predefined activation function or transfer function (Kim et al., 2010).

Levenberg-Marquardt algorithms was selected to train the network, as it converges faster to the desired solution compared to the more popular error backpropagation algorithm (EBP). Early stopping mechanism was also utilized, where training will be stopped immediately once the changes in validation errors no longer improve, thus avoiding overfitting and poor generalization capability. From early analysis, it was discovered that the best training parameter for the FANN model is 1 hidden layer and 10 neurons. Therefore, there are three type of FANN model that will analyzed, the base FANN model, FANN-NARX(1), and FANN-NARX(2)

The training dataset is randomly split into 70% training set, 15% test set, and 15% validation set. The performance of the model is compared and evaluated based on the coefficient of determination (R^2) values and root mean squared error (RMSE).

Analysing Different Dynamically Modelled Data Structures and Machine
Learning Algorithms to Predict PM2.5 Concentration in China

1769

3. Results and Discussion

Meteorological data from five different cities in china were utilized to predict $PM_{2.5}$ concentration, where the performance of four different machine learning models were analysed, namely: MLR, FANN, FANN-NARX(1), and FANN-NARX(2). Initially, MLR and FANN model were trained using the original untransformed data structure. The performance of these base models in predicting the $PM_{2.5}$ concentration for the test data set could be observed from Table 2. Due to the slight difference in scale between dataset from different cities, the RMSE values may slightly varies.

Table 2: $PM_{2.5}$ prediction performance of base MLR and FANN models for test dataset.

Model	Beijing		Chengdu		Guangzhou		Shenyang		Shanghai	
	R^2	RMSE	R^2	RMSE	R^2	RMSE	R^2	RMSE	R^2	RMSE
MLR	0.29	0.873	0.27	0.840	0.20	0.899	0.26	0.881	0.22	0.858
FANN	0.69	0.528	0.64	0.586	0.44	0.737	0.54	0.695	0.53	0.688

The results from Table 2 shows that the FANN model consistently outperform the MLR model with much higher R^2 values and lower RMSE values throughout all five cities. It could be observed that even the best performing MLR model (Beijing dataset) could only score an R^2 values of 0.29 which is about 58% lower than that of the best performing FANN model that scored an R^2 values of 0.69. The RMSE is also lower at only 0.528 compared to MLR model at 0.873. This shows that despite of only using the base untransformed data structure, FANN model is far superior to MLR model in predicting the $PM_{2.5}$ concentration. However, as described in earlier section, the performance of the FANN model could be further improved by training it with NARX transformed data. Table 3 below depicts the performance of FANN model coupled with NARX with time lag of 1 and 2 on test data.

Table 3: $PM_{2.5}$ prediction performance of FANN-NARX(1) and FANN-NARX(2)

Model	Beijing		Chengdu		Guangzhou		Shenyang		Shanghai	
	R^2	RMSE	R^2	RMSE	R^2	RMSE	R^2	RMSE	R^2	RMSE
FANN-NARX(1)	0.96	0.084	0.97	0.047	0.93	0.148	0.96	0.069	0.97	0.406
FANN-NARX(2)	0.96	0.080	0.97	0.044	0.94	0.136	0.97	0.063	0.97	0.360

By utilizing the NARX transformed data structure on the FANN model, significant improvement in prediction performance could be observed across all five datasets. In term of time lag, the result shows that FANN-NARX(2) has an overall slight edges in performance compared to FANN-NARX(1) model, this is especially evident through the lower RMSE scores. The largest increase in performance could be observe from the Guangzhou dataset, where the FANN-NARX(2) model scored an R^2 values of 0.94 which is 53% higher than the base FANN model and a staggering 79% higher than the base MLR model.

4. Conclusion

In this study, nonlinear autoregressive network with exogenous inputs (NARX) were analysed in transforming the training data structure for $PM_{2.5}$ predictions. Two variations were examined NARX(1) with time lag of 1 and NARX(2) with time lag of 2. The models utilised for the prediction models are multi linear regression (MLR) and feed forward artificial neural network (FANN). When only trained with the base structure, the results shows than the FANN model is vastly better with an R^2 values of up to 0.69 (Beijing dataset) compared to that of MLR at 0.29. However, when coupled with NARX transformed data structure, the performance of the FANN model significantly improved. The best performing FANN-NARX(2) model achieved an R^2 values of up to 0.97 (Chengdu dataset) which is about 34% higher than base FANN model and 72% higher than MLR model. In future study, the generalization capability of the model could be further improved by training the model with a combined dataset instead of individual dataset and testing the model with real world external data.

Acknowledgement

This work was supported by Universiti Sains Malaysia (USM), special gratitude to Department of Environmental (DOE) Malaysia for providing the air quality data for this study and Kementerian Pendidikan Malaysia (KPM) through Fundamental Research Grant Scheme (FRGS) grant number FRGS/1/2018/TK02/USM/02/10.

References

Mobaseri, M., Mousavi, S.N., Mousavi Haghighi, M.H., 2021. Causal effects of population growth on energy utilization and environmental pollution: A system dynamics approach. Caspian Journal of Environmental Sciences 19, 601–618.

Gakidou, E., Afshin, A., Abajobir, A. A., Abate, K. H., Abbafati, C., et al., 2017. Global, regional, and national comparative risk assessment of 84 behavioural, environmental and occupational, and metabolic risks or clusters of risks, 1990–2016: a systematic analysis for the Global Burden of Disease Study 2016. The Lancet 390, 1345–1422.

Chen, C.-C., Chen, P.-S., Yang, C.-Y., 2019. Relationship between fine particulate air pollution exposure and human adult life expectancy in Taiwan. Journal of Toxicology and Environmental Health, Part A 82, 826–832.

World Health Organization (WHO). 2021, New WHO global air quality guidelines aim to save millions of lives from air pollution. Available at: https://www.who.int/news/item/22-09-2021-new-who-global-air-quality-guidelines-aim-to-save-millions-of-lives-from-air-pollution (Accesed: 30/10/2021)

Perez, P., Menares, C., Ramírez, C., 2020. PM2.5 forecasting in Coyhaique, the most polluted city in the Americas. Urban Climate 32, 100608.

Biancofiore, F., Busilacchio, M., Verdecchia, M., Tomassetti, B., Aruffo, E., Bianco, S., di Tommaso, S., Colangeli, C., Rosatelli, G., di Carlo, P., 2017. Recursive neural network model for analysis and forecast of PM10 and PM2.5. Atmospheric Pollution Research 8, 652–659.

Liang, X., Li, S., Zhang, S., Huang, H. & Xi Chen, S. 2016. PM2.5 data reliability, consistency and air quality assessment in five chinese cities. Journal of Geophysical Research: Atmospheres, 121, 10,220-10,236. https://ieeexplore.ieee.org/document/548162

Buevich, A., Sergeev, A., Shichkin, A., Baglaeva, E., 2021. A two-step combined algorithm based on NARX neural network and the subsequent prediction of the residues improves prediction accuracy of the greenhouse gases concentrations. Neural Computing and Applications 33, 1547–1557.

Mohebbi, M.R., Karimi Jashni, A., Dehghani, M., Hadad, K., 2019. Short-Term Prediction of Carbon Monoxide Concentration Using Artificial Neural Network (NARX) Without Traffic Data: Case Study: Shiraz City. Iranian Journal of Science and Technology, Transactions of Civil Engineering 43, 533–540.

Grivas, G. & Chaloulakou, A. 2006. Artificial neural network models for prediction of PM10 hourly concentrations, in the greater area of Athens, Greece. Atmospheric Environment, 40, 1216-1229.

Kim, M.H., Kim, Y.S., Lim, J., Kim, J.T., Sung, S.W., Yoo, C., 2010. Data-driven prediction model of indoor air quality in an underground space. Korean Journal of Chemical Engineering 27, 1675–1680.

Proceedings of the 14th International Symposium on Process Systems Engineering – PSE 2021+
June 19-23, 2022, Kyoto, Japan © 2022 Elsevier B.V. All rights reserved.
http://dx.doi.org/10.1016/B978-0-323-85159-6.50295-5

A multi-output machine learning approach for generation of surrogate models in process engineering

Jimena Ferreira[a,b], Martín Pedemonte[b], Ana Inés Torres[a*]

[a] *Instituto de Ingeniería Química, IIQ, Facultad de Ingeniería, Universidad de la República, Montevideo, 11600, Uruguay*
[b] *Instituto de Computación, INCO, Facultad de Ingeniería, Universidad de la República, Montevideo, 11600, Uruguay*
aitorres@fing.edu.uy

Abstract

This work discusses a multi-output strategy to generate surrogate models in the chemical engineering context. The rationale behind our approach is that in a chemical process many outputs, such as concentrations of different species, are highly related by the underlying physicochemical phenomena. Hence, the expressions that model them should have terms that are common to all. In here, we extend our previous work on surrogate modelling to develop an algorithm that, in the same execution, learns models for several related output variables. We evaluate the algorithm by taking different combinations of CSTR arrangements as case studies, that represent structures with different levels of sharing. We conclude that the multi-output strategy is successful in building models that share common functionalities with adequate errors.

Keywords: Machine Learning; Surrogate model; Multi-output regression; Kaizen Programming; Evolutionary Algorithms.

1. Introduction

The use of surrogate models is related to the need of simple but accurate representations of the variation of output variables with several different input variables. Common techniques to build surrogate models including Gaussian Process (Kriging, Krige, 1951), Support Vector Machine (Smola and Scholkop, 2004) or ALAMO (Cozad et al., 2014), make use of a set of predefined functional bases to generate single-output models.

However, chemical processes are rarely single-output; many outputs such as concentrations of different species are highly related by the underlying physicochemical phenomena. For this reason, it is desirable to have techniques that can construct multi-output surrogate models. We hypothesize that in these multi-output models, outputs that we know have the same underlying physical-chemical phenomena, need to have models with terms that are common.

In this work, we extend the algorithm in Ferreira et al. (2019) to allow for the generation of several surrogate models concurrently for the different outputs of a system in one execution of the algorithm.

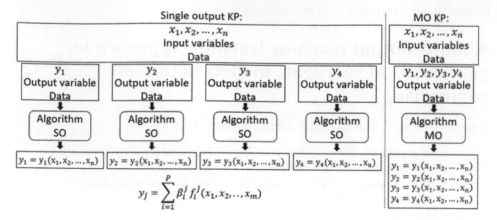

Figure 1 – Generation of surrogate models for several outputs with the same input data. Left: the single-output strategy repeatedly executes the algorithm to learn one model at a time. Right: the multi-output strategy learns all the models in the same run.

As before, the models are learned based on our implementation of Kaizen programming (KP, Ferreira et. al, 2021a). KP is an iterative algorithm for solving symbolic regression problems as a linear combination of nonlinear bases, with no a priori assumption of the functional bases. Thus, in the proposed multi-output extension of KP, models that share functional bases are favoured. A first approach to multi-output KP applied to the benchmark functions in Veloso et al. (2018) was included in Ferreira et al. (2021b).

The paper is organized as follows: section 2 presents the multi-output strategy, section 3 discusses the case studies, section 4 presents the numerical results.

2. Multi-output machine learning strategy

Figure 1 schematizes the general problem we are solving and the difference with the current approach. The general problem is finding surrogate models for several related output variables. Currently, most machine learning methods solve this problem by repeatedly executing the same algorithm using the same inlet data (represented by several vectors x) and the corresponding outlet data (one of the vectors y). We refer to this approach as the single-output (SO) strategy. By doing this, we obtain a set of outlet surrogate models for y, where each y_j is computed as a combination of the nonlinear basis f_i^j In KP these f_i^j are learned from the data, thus not necessarily the same set of f_i^j is obtained if a different execution is run for each output y_j. We will refer to the single-output strategy using Kaizen Programming as SO-KP.

If, the output variables y_j are related to each other, as is the case with many chemical engineering applications, it is desirable to learn the surrogate models for all the output variables in the same execution, as this allows for using the same functional bases for all the models that are being learned. We will refer to this multi-output strategy using Kaizen Programming as MO-KP.

Recalling that that there are two main steps in SO-KP: one where the functional bases (f_i^j) are created and modified using genetic programming, and another one where the

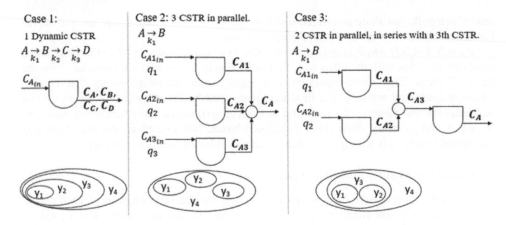

Figure 2 – Case studies and the scheme of shared functional bases. Output variables are in bold.

coefficients (β_i^j) are calculated by ordinary least squares (OLS), the modifications required for the MO-KP algorithm were at this second step and are as follows:

(i) We substituted the SO-OLS step, by a multi-output linear regression based on OLS (MO-OLS). At the MO-OLS step we first specified that the same set of functional bases f_i^j was considered for all outputs y_j; then, we added a step in which the functional bases were ranked according to the *p-value* (this parameter is used to measure the relevance of a particular functional basis in each final expression for y_j).

(ii) We used a different fitness function as a criterion for selection of the model: in SO-KP an adjusted R^2 of the output variable model was used, whereas in MO-KP, the minimum of all the adjusted R^2 of the output variables was used. In both cases this adjusted R^2 refers to training data.

It is important to note that these changes do not force the final expressions of y_j to all have the same functional bases but enhances the probability of that happening. More details of the algorithms can be found in Ferreira et al. (2019) and Ferreira et al. (2021).

3. Case studies

The proposed MO-KP algorithm was evaluated by considering three levels of overlap in the terms of the output variable models y_j, that may arise in chemical engineering settings (see Figure 2):

Case study 1-Large overlap (CS1): considers the dynamic model of a CSTR with three first order reactions in series (A→B→C→ D). By deriving the analytical solution for this system (for a constant volume and inlet flowrate of reactant A) it can be shown that the functional bases of y_1 (in this case C_A (t, C_{Ain})) are included in the functional bases of y_2, (in this case C_B (t, C_{Ain})), those of y_2 in y_3 (in this case C_C (t, C_{Ain})), and so on.
Case study 2-Least overlap (CS2): considers three steady state CSTRs in parallel, with a mixing point to combine the outlets. A first order reaction (A→B) occurs in each reactor. This could model, for example, three municipal wastewater treatment plants that receive different loads and are combined before final disposal. In the analytical solution of this system, y_1 (in this case $C_{A1}(q_1)$), y_2 (in this case $C_{A2}(q_2)$) and y_3 (in this

case $C_{A3}(q_3)$) do not share any functional bases, and y_4 (in this case $C_A(q_1, q_2, q_3)$) shares functional bases with the other three output variables.

Case study 3 (CS3)-Medium overlap: considers two steady state CSTRs in parallel and a third one in series with the other two; first order reactions (A→B) are considered for the three of them. This could represent also waste water treatment plants that receive different loads and are sent to a final treatment plant before disposal. In the analytical solutions for this case, y_1 ($C_{A1}(q_1)$) and y_3 ($C_{A3}(q_1, q_2)$) share some functional bases, y_2 ($C_{A2}(q_2)$) and y_3 share some other functional bases, but y_1 and y_2, do not share any functional bases; y_4 ($C_A(q_1, q_2, q_3)$) shares functional bases with the other three.

4. Numerical results and discussion

Input-output datasets were obtained by considering the analytic solutions of each output variable for each case study in Fig. 2. The following values were used: V=1 L, k_1=3.7 min^{-1}, k_2=4 min^{-1}, k_3=3 min^{-1}, q=2 Lmin^{-1}, $C_A(0)$=$C_B(0)$=$C_C(0)$=$C_D(0)$=0, C_{A1in}=1 molL^{-1}, C_{A2in} =2 molL^{-1}, and C_{Ain} =3 molL^{-1}. 100 points were randomly generated using a uniform distribution with the following ranges: $C_{Ain} \in$ [0.5-3] molL^{-1}, $t \in$ [0-3] min, $q \in$ [0.5-4] Lmin^{-1}. Each case study was executed 100 times, with a different training set, the parameters of the algorithm were as in Ferreira et al. (2021), except the number of iterations, which is set in 3000. The expressions with the best adjusted R^2 of MO algorithm over the learning set are as follows:

Case study 1:

$$y_1 = 41.38f_1 + 0.026f_2 + 0.075f_3 + 0.11f_4 + 0.31f_5 + 15.28f_6 + 4.01f_7 + 40.91f_8$$
$$y_2 = 28.70f_1 + 0.012f_2 + 0.69f_3 + 0.40f_4 - 0.66f_5 + 8.86f_6 + 4.15f_7 - 28.58f_8$$
$$y_3 = -33.67f_1 + 0.074f_2 - 1.98f_3 - 0.53f_4 + 1.51f_5 - 11.50f_6 - 4.36f_7 + 34.34f_8$$
$$y_4 = -64.59f_1 - 0.32f_2 + 4.87f_3 - 0.10f_4 - 0.26f_5 - 24.70f_6 - 5.33f_7 + 63.07f_8$$

With $f_1 = \dfrac{C_{Ain}t}{(t + 0.46)}$, $f_2 = \dfrac{C_{Ain}t}{(0.087t^2 + 1.03)}$, $f_3 = \dfrac{C_{Ain}t}{(1.73\,t + 0.087t^2 + 0.66)}$, $f_4 = \dfrac{C_{Ain}}{(0.28/t + 0.50\,t)}$

$f_5 = \dfrac{C_{Ain}}{(0.63/t + t)}$, $f_6 = \dfrac{C_{Ain}}{(t + 0.50)}$, $f_7 = \dfrac{C_{Ain}}{(t + 0.087t^2 + 0.38)}$, $f_8 = C_{Ain}$

Case study 2:

$$y_1 = 1.49f_1 + 0.034\,f_2 + 1.59\,f_3 - 0.0097f_4 + 0.034f_5 + 0.86f_6 + 0.14f_7 - 0.92$$
$$+ 0.031q_1 + 0.031q_2$$
$$y_2 = 0.32f_1 + 0.049\,f_2 + 3.04\,f_3 - 0.049f_5 + 0.18f_6 + 0.13f_7 - 0.59 + 0.073q_1 +$$
$$0.073q_2 - 0.0039\,q_3$$
$$y_3 = -0.0058\,f_1 - 0.0058\,f_2 + 1.32\,f_4 + 0.010\,f_7 - 0.34 + 0.20\,q_3$$
$$y_4 = -2.57\,f_1 + 0.66\,f_2 - 2.37f_3 - 0.36f_4 + 0.66f_5 - 1.48\,f_6 - 0.76\,f_7 + 2.09$$
$$+ 0.068\,q_1 + 0.068\,q_2 + 0.081\,q_3$$

With $f_1 = \dfrac{q_1}{(q_1 + q_2 + 1.59)}$, $f_2 = \dfrac{q_1}{(q_1 + 2\,q_2 + 1.38)}$, $f_3 = \dfrac{q_2}{(q_1 + 2\,q_2 + 1.40)}$, $f_4 = \dfrac{q_3}{(q_3 + 0.65)}$,

$f_5 = \dfrac{q_3}{(q_1 + 2q_2 + 1.38)}$, $f_6 = \dfrac{1}{(q_1 + q_2 + 1.59)}$, $f_7 = \dfrac{1}{(q_1 q_2 + 0.58\,q_1 + q_2q_3)}$

Case Study 3:

$$y_1 = -0.0015f_1 - 0.00078f_2 - 0.25f_3 - 0.34f_4 + 0.0070\,f_5 - 1.51f_6 + 0.63$$
$$+ 0.18q_1 - 0.071q_2$$
$$y_2 = -0.0028\,f_1 - 0.0014\,f_2 - 0.39\,f_3 + 1.04\,f_4 - 0.54 + 0.28\,q_1$$
$$y_3 = 0.0062\,f_1 + 0.0031\,f_2 + 1.017\,f_3 - 1.18\,f_4 - 0.023\,f_5 - 3.83\,f_6 + 1.73$$
$$- 0.66\,q_1 + 0.26\,q_2$$

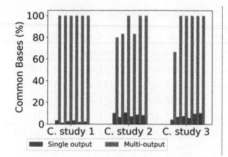

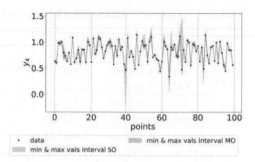

Figure 3 - Pairwise comparison of shared terms for each outlet expression, ordered as: y_1-y_2, y_1-y_3, y_1-y_4, y_2-y_3, y_2-y_4, y_3-y_4.

Figure 4 – Validation data points vs predictions for MO-KP and SO-KP for 100 executions. Results correspond to CS2.

$$y_4 = 0.00042\, f_1 + 0.00021\, f_2 + 0.43 f_3 - 0.70\, f_4 - 0.0069\, f_5 - 1.40\, f_6 + 0.60 - 0.24\, q_1 + 0.17\, q_2$$

With $f_1 = q_1 q_2$, $f_2 = q_1\, q_2^2$, $f_3 = \dfrac{q_1^2}{(1.56\, q_1 + q_2 + 1.26)}$, $f_4 = \dfrac{q_2}{(0.67 q_1 q_2 + 1.02\, q_2 + 2.17)}$,

$f_5 = q_2^2$, $f_6 = \dfrac{1}{(0.85\, q_1 + q_2 + 2.17)}$

A graphical comparison of the number of terms shared by the different models in each case study, together with a comparison with the results from the SO-KP algorithm, is depicted in Figure 3. From here we conclude that MO-KP successfully finds all the overlaps that should be found (e.g., y_4 with all the other y_i in every case study, y_3 with y_1 and y_2 in CS3). However, it is also noticeable that expressions with overlapping terms for cases where there should not be any, are also returned. Examples of the latter are y_1-y_2-y_3 in CS2, and y_1-y_2 in CS3. This is a side effect of tuning the algorithm to favor common bases in all cases, regardless of the expected degree of sharing. On the other hand, SO-KP rarely finds a common term between expressions that do have them.

A comparison of the average and median RMSE (over validation sets) for the 100 executions is shown in Table 1. As seen both MO-KP and SO-KP provide accurate models. As expected, as every function is learned independently, SO-KP generally performs better in terms of the median of the error. However, it is interesting to note that in CS1, where the overlap of the functions is the largest, the error of MO-KP is similar or even better than that of SO-KP. Figure 4 shows a typical comparison between the output data points and the distribution of responses for both algorithms. As seen, despite lower overall errors, the SO-KP produces a few but significant outliers, which suggests that the model overfitted the learning data, thus did not generalize well to validation data. This was not seen in MO-KP, hence suggesting that MO-KP models should be preferred for applications as there is a lower risk of a highly erroneous prediction. However, preventing overfitting was not the main objective of this work, thus further studies on this direction may be needed.

5. Conclusions and future directions

In this work we analyse a multi-output strategy for learning surrogate models for several related output variables. In the analysis we consider three arrangements that are commonly found in Chemical Engineering settings and have different structures in terms of expected share of functional bases. We conclude that in terms of error in validation data the multi-output strategy builds models that are competitive with those

Table 1 - Average and median of RMSE distribution for 100 executions over validation sets

Output	Alg.	Case study 1		Case study 2		Case study 3	
		Average	Median	Average	Median	Average	Median
y_1	SO	3.16E+00	1.59E-03	9.02E-05	1.93E-05	2.83E-05	4.31E-06
	MO	8.01E-02	4.41E-04	8.34E-03	8.27E-03	1.73E-03	1.89E-03
y_2	SO	4,64E-01	1,74E-03	6,41E-02	2,26E-05	4,02E-05	6,50E-06
	MO	4,04E-01	1,18E-03	1,61E-02	1,55E-02	3,59E-03	3,72E-03
y_3	SO	1,22E+03	1,90E-03	2,46E-04	5,24E-05	1,83E-03	1,39E-03
	MO	2,89E-01	1,13E-03	2,48E-02	2,50E-02	2,73E-03	2,69E-03
y_4	SO	6,02E-01	9,58E-04	1,52E-02	1,16E-02	6,32E-04	5,39E-04
	MO	4,73E-01	1,66E-03	1,95E-02	1,95E-02	1,52E-03	1,48E-03

obtained with single-output strategy, and even better in the case study whose structure has a large overlap of terms. In addition, the multi-output strategy performs better than the single-output in terms of the presence of outliers. However, the strategy tends to find shared terms even in those cases where the structure prevents so. Thus, future versions of the algorithm should include steps that allow to find a balance in term-sharing based on the structure of the problem. Another future direction includes learning expressions that resemble the underlying physicochemical phenomena. In this sense, the expressions learned for CS2 and CS3 already include terms that have similarities with the analytical ones. This feature should be further exploited so that the algorithm returns models that are closer to the theoretically expected expression.

Acknowledgements

J. Ferreira thanks the Agencia Nacional de Investigación e Innovación of Uruguay for financial support for her graduate studies (Award No. POS NAC 2018 1 152185).

References

A. Cozad, N. V. Sahinidis and D.C. Miller, 2014, Learning surrogate models for simulation-based optimization. AIChE Journal. 60, 2211–2227.

V. V. de Melo and W. Banzhaf, 2018, Automatic feature engineering for regression models with machine learning: An evolutionary computation and statistics hybrid, Information Sciences, 430, 287-313.

J. Ferreira, M. Pedemonte and A. I. Torres, 2021a, Development of a Machine Learning-based Soft Sensor for an Oil Refinery's Distillation Column. Submmited.

J. Ferreira, A. I. Torres and M. Pedemonte, 2021b, Towards a Multi-Output Kaizen Programming Algorithm, 2021 IEEE Latin American Conference on Computational Intelligence (LA-CCI), Temuco, Chile. Accepted.

J. Ferreira, A. I. Torres and M. Pedemonte, 2019, A Comparative Study on the Numerical Performance of Kaizen Programming and Genetic Programming for Symbolic Regression Problems, 2019 IEEE Latin American Conference on Computational Intelligence (LA-CCI), Guayaquil, Ecuador, 1-6.

D.G. Krige, 1951, A statistical approach to some mine valuations and allied problems at the Witwatersrand. Master's thesis of the University of Witwatersrand.

A. Smola and B. Scholkop, 2004, A tutorial on support vector regression, Statistics and Computing, 14, 199–222.

Proceedings of the 14th International Symposium on Process Systems Engineering – PSE 2021+
June 19-23, 2022, Kyoto, Japan © 2022 Elsevier B.V. All rights reserved.
http://dx.doi.org/10.1016/B978-0-323-85159-6.50296-7

Practical Human Interface System for Transition Guidance in Chemical Plants using Reinforcement Learning

Shumpei Kubosawa[a,b*], Takashi Onishi[a,b], Yoshimasa Tsuruoka[a,c],

Yasuo Fujisawa[d], Masanori Endo[d], Atsushi Uchimura[d], Masahiko Tatsumi[d],

Norio Esaki[d], Gentaro Fukano[e], Tsutomu Kimura[e], Akihiko Imagawa[e],

Takayasu Ikeda[e]

[a]NEC-AIST AI Cooperative Research Laboratory, National Institute of Advanced Industrial Science and Technology (AIST), Tokyo, 135-0064 Japan
[b]Data Science Research Laboratories, NEC Corporation, Kanagawa, 216-8666, Japan
[c]Department of Information and Communication Engineering, The University of Tokyo, Tokyo, 113-0033, Japan
[d]Production and Technology Center, Mitsui Chemicals, Inc., Tokyo, 105-7122, Japan
[e]Omega Simulation Co., Ltd., Tokyo, 169-0051, Japan
kubosawa@nec.com

Abstract

In chemical plants, transition operations, such as changing the production load from 100% to 80%, are commonly performed to satisfy production needs. As plant models used in conventional automatic control methods (e.g. step response models) cannot predict non-steady states, these transition operations warrant manual control. Previously, we proposed an automatic optimal control method using dynamic simulators and reinforcement learning, a machine learning method in artificial intelligence (AI), for transition operations. We implemented this existing AI system in an actual industrial plant and determined that further improvements were required in the interaction between the system and human operators for reliable and acceptable guidance. In this paper, we propose a human interface system for realising optimal transition operations by enabling AI to cooperate with human operators. To validate and authorise the AI-proposed manipulations performed by human operators, the interface system presented the entire procedure and sensors influencing the AI decision for online disturbance rejection prior to actual manipulations. The interface, coupled with the control method, was evaluated experimentally in an actual chemical plant. The proposed system demonstrated optimised transition operations for producing purity changes under abrupt heavy rain disturbance in terms of guidance.

Keywords: chemical plant, reinforcement learning, optimisation, explainable AI.

1. Introduction

Modern chemical plants are commonly equipped with advanced controllers, such as model predictive control (MPC), to maintain steady states and stable production. In addition to steady operations, transition operations, such as changing production loads, are frequently performed in chemical plants to satisfy production needs. Although several automation methods, including sequence control, have been introduced to aid these transition operations, limited methods exist for their optimal control. Previously, we proposed an automatic optimal control system for transition operations that leverages dynamic simulation and reinforcement learning (RL) (Kubosawa et al., 2021a and 2021b). However, in addition to control methods, the human interface is crucial for realising acceptable and reliable operations. This particularly applies to chemical plants, wherein any unforeseen incident can significantly impact society. In this paper, we propose a human interface system to guide transition operations in chemical plants. The interface system coupled with the control method was evaluated experimentally in an actual plant, which demonstrated that the optimised operations were efficiently performed with guidance from the system.

2. Related work

2.1. Simulation-based optimal control

Kubosawa et al. (2021a) proposed a method to optimise and control the transition operation of an actual plant by leveraging dynamic simulation, tracking simulation, RL and domain randomisation. The proposed method utilised dynamic simulators for training human plant operators as the plant model. They further improved the state identification performance using RL and proposed an RL-based disturbance rejection method (Kubosawa et al., 2021b). Both improvements were aimed at reducing the gaps between simulated and real states. These gaps can be triggered by multiple factors, including modelling errors, incorrect identification of states and external disturbances, such as changes in weather. If the difference in behaviours can be reduced by adjusting the parameter values of the model with time, then RL can potentially improve the simulation behaviour and enhance the state identification accuracy of the tracking simulation. As the simulation-to-reality gaps can also be caused by disturbances, human operators are required to handle disturbances by leveraging domain knowledge and plant dynamics based on their experiences. This defines the quantitative relationship between manipulation and response. Previously, we focused on this aspect and designed the RL task to minimise the gaps between simulated states and real states online. These methods were evaluated in an actual chemical plant, and the results verified the optimisation of transition operations under abrupt heavy rain disturbances.

Notably, the objectives of the proposed framework and nonlinear MPCs can be identical, that is, optimal control; however, their approaches are significantly different. MPCs commonly adopt online (on-site) optimisation methods, whereas RL is an offline optimisation method (i.e. prior optimisation to actual control). In addition, RL is a variant of machine learning methods, which consist of a field of artificial intelligence (AI). In this light, we distinguish the two approaches and refer the proposed method as an AI.

2.2. Explainable AI

As machine learning and AI applications include mission-critical systems, performance improvement alone is insufficient to determine whether AI can be adopted in actual operations. Therefore, the Defense Advanced Research Projects Agency initiated the Explainable AI (XAI) program in 2017. This program aimed to develop AI systems capable of explaining their rationale, strengths, weaknesses and future behaviour. As the explainability of a policy is essential in RL, one concept of the program explored the development of explainable policy functions, whereas the other focused on developing an analysis method for existing black-box policies (Gunning and Aha, 2019).

2.3. Sensitivity analysis

Sensitivity analysis involves investigating the behaviours of mathematical models considering the changes in variables, including parameters and inputs (Pannell, 1997).

$$g(\boldsymbol{x}, i) = \left| \frac{\partial f(x_1, x_2)}{\partial x_i} \right| \tag{1}$$

Eq. (1) is an example index $g(\boldsymbol{x}, i)$ for the analysis of a model $f: \mathbb{R}^2 \to \mathbb{R}$ using a partial derivative of the i-th input variable x_i. This index describes the effect of a minor change in x_i on the output. In other words, it indicates the importance of each input in the model prediction (i.e. weight of the nonlinear model at the input). Sakahara and Kubosawa (2021) proposed the directions of applying these analysis methods to RL models.

3. Proposed method

3.1. Operation flow

The transition operations of chemical plants involve non-steady states and unfamiliar situations. Therefore, human operators conduct these procedures cautiously to handle unforeseen situations during operation. To support human operators anticipating various situations, AI-based procedures should be validated by human operators before acceptance or rejection. To develop validation methods, we leveraged the two concepts of XAI. One concept presents the future behaviour of the AI, whereas the other analyses the AI's perspective by determining which sensor affects AI decisions.

Figure 1 depicts the workflow of the human operator, overall architecture of the AI controller and proposed interface system. The thick and thin arrows indicate the process and data flows, respectively. The AI system comprises three RL agents: First, the

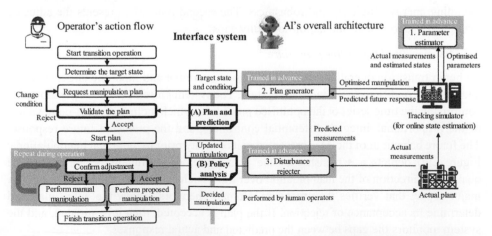

Figure 1. Operation flow and AI architecture with the proposed interface system.

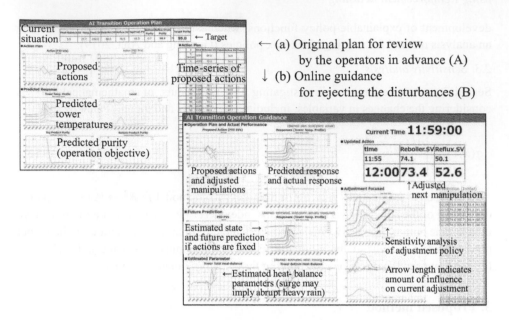

Figure 2. Interface system screenshots.

parameter estimator provides optimised parameter values to the tracking simulator, which receives real-time manipulations from the actual plant and estimates the plant state online. Second, the plan generator presents an optimised procedure that begins with the current plant state and finally achieves the required target state. Third, the disturbance rejecter adjusts the planned manipulation values and presents optimal values if the gap between the actual and predicted states is unexpectedly widened (Kubosawa et al., 2021b).

As indicated in Figure 1, two major interaction points exist between human operators and AI in this interface. The first point (A) presents the time-series of the optimised procedure and predicted future response of the total operation to a specified target state and condition, such as target production load of 80%, or the priority of the procedure, including energy efficiency and robustness. The second point (B) presents the adjusted manipulations to reduce simulation-to-reality gaps.

3.2. Assessment of the offline procedure

To begin transition operations, human operators determine and input the target situation and desired conditions to the interface system. The system presents (A) the operation plan comprising the time series of the optimised procedure, which includes the set-point values of the proportional–integral–differential controllers, and the predicted future response. The future behaviour of the AI is presented in its entirety prior to the actual manipulations. Figure 2(a) depicts a screenshot of the plan presented during the experiment. The qualitative direction of the manipulation over time, such as increasing or decreasing, is a major aspect that verifies the procedure. Human operators review the entire plan and determine its acceptance or rejection. If the plan is accepted, then it is initiated, and the system monitors the gaps between the predicted and actual responses.

3.3. Assessment of the online action

During the procedure, gaps may emerge between the predicted and actual responses owing to a mismatch in the dynamics, such as temporal changes in model parameters, and external disturbances. To handle such abrupt changes in actual situations, the disturbance rejection agent works online. The agent monitors the gaps and proposes the adjusted and updated manipulation values to cancel the gap periodically. We used a control interval of 5 min in the experiment. To explain the adjustment and review the manipulations before performing them, the interface system analyses the behaviour and presents the result (B) to the operators 3 min prior to the subsequent manipulation, during which the operators review the manipulation and determine its acceptance or rejection. Figure 2(b) illustrates a screenshot of the guidance for rejection during the experiment. The system presents the importance of the sensors in the adjustment using the sensitivity analysis of the adjusting policy regarding the current state, which is a partial derivative of the policy function of states or each sensor. If the AI-focused sensors, which are typically mismatched, are considered inappropriate in the current situation, then the operators can reject the proposal.

4. Experiment

We evaluated our system at an actual methanol distillation plant for training human operators, as described in previous studies (Kubosawa et al., 2021a and 2021b). The plant separates methanol and water from its liquid mixture. The experimental task involves the transition operation of producing purity changes (downgrade and upgrade) under abrupt disturbances caused by heavy rains. A detailed description of the plant and its model used in this experiment is presented in Kubosawa et al. (2021b).

In the experiment presented in Figure 3, all proposals were accepted by the human operators. Figure 3 illustrates the responses obtained during the downgrade and upgrade operations. The dashed lines indicate the originally predicted actions or responses. In Figure 3(a), the gaps between the solid (adjusted) and dashed (originally planned) lines

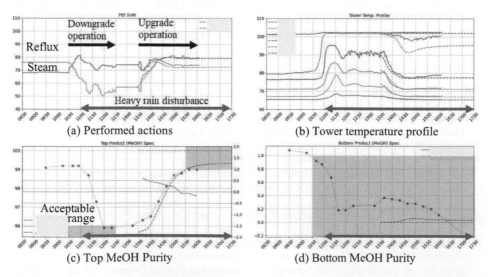

Figure 3. Experimental results of the purity transition operation using the proposed

indicate the adjustment of actions. Figure 3(b) depicts the oscillation phenomena observed in the response, which were suppressed owing to the adjustments. Figure 3(c) illustrates the time series of the top purity, and the points indicate the actual sampled purity. The shaded region in the figure indicates the acceptable purity. Both operations achieved acceptable states when counteracting the heavy rain disturbance, which began shortly after the first downgrade operation.

In the other experiment of a downgrade operation, the transmission of the sensor readings to the system was abruptly suspended (i.e. the sensor readings were fixed since then), and the excessive action values has been proposed. This is because the temperatures of the actual plant were changing, whereas, as for the system, the response of the proposed corrected action seemed to be stayed; thus, the system significantly increased the action values. In this case, the human operators rejected the proposal during 3 min for the judgement and selected the originally planned values instead.

5. Conclusions

As transition operations in chemical plants involve non-steady states, unforeseen situations may be triggered. Therefore, explainable interfaces for AI guidance systems are required. In this study, we proposed a practical explanatory interface system using an existing automatic control system. The interface system proposes optimal procedures that can be performed by human operators with reliability and acceptability. We evaluated the entire system considering the transition operation for producing purity changes in an actual chemical plant and demonstrated the optimisation of the operations. In the future, we intend to investigate the application and installation of the proposed system to industrial chemical plants to optimise the actual production.

Acknowledgements

The authors thank Kazumi Yamamoto, Toshikazu Tanaka, Kiyoshi Sugawara and Katsushige Sumida (Mitsui Chemicals Inc.) for evaluating the system as expert operators and providing general useful comments on chemical plant operations.

References

Gunning, D., and Aha, D. (2019). DARPA's explainable artificial intelligence (XAI) program. AI Magazine, Vol. 40, No. 2, pp. 44–58.

Kubosawa, S., Onishi, T., and Tsuruoka, Y. (2021a). Computing operation procedures for chemical plants using whole-plant simulation models. Control Engineering Practice, Vol. 114, p. 104878.

Kubosawa, S., Onishi, T., and Tsuruoka, Y. (2021b). Non-steady state control under disturbances: navigating plant operation via simulation-based reinforcement learning. In Proceedings of the 60th Annual Conference of the Society of Instrument and Control Engineers of Japan (SICE), pp. 796–803.

Pannell, D.J. (1997). Sensitivity analysis: strategies, methods, concepts, examples. Agricultural Economics, No. 16, pp. 139–152.

Sakahara, M., and Kubosawa, S. (2021). Future directions in sensitivity analysis on deep reinforcement learning for process control via Fisher information matrix. In Proceedings of the 60th Annual Conference of the Society of Instrument and Control Engineers of Japan (SICE), pp. 804–807.

Proceedings of the 14th International Symposium on Process Systems Engineering – PSE 2021+
June 19-23, 2022, Kyoto, Japan © 2022 Elsevier B.V. All rights reserved.
http://dx.doi.org/10.1016/B978-0-323-85159-6.50297-9

Surrogate modeling for nonlinear gasoline blending operations

Tasabeh H. M. Ali[a], Robert E. Franzoi[a], Brenno C. Menezes[a*]

[a]*Division of Engineering Management and Decision Sciences, College of Science and Engineering, Hamad Bin Khalifa University, Qatar Foundation, Doha, Qatar*
bmenezes@hbku.edu.qa

Abstract

The application of surrogate modeling in engineering is surging recently for predicting the functional behavior of a system using analytical formulations as an alternative to complex models that often lead to non-convergence issues and not sufficiently accurate solutions in decision-making problems. The surrogate model building procedure addressed in this paper consists of four major steps to be applied in nonlinear blending of gasoline streams. The first is the input (x) dataset generation, performed using the Latin Hypercube Sampling (LHS) technique, which is coupled with a rescaling strategy, and used for evaluating the output (y) dataset. Secondly, the generated data are improved with a normalization procedure to mitigate numerical issues and to avoid biased surrogates. Thirdly, mixed integer quadratic programming (MIQP) formulation based on the least-squares regression is employed to build an optimizable surrogate function for each variable of interest. Fourthly, smaller and simpler surrogates are established and selected to be employed for gasoline blending operations by substituting the complex nonlinear and nonconvex rigorous formulation in an optimization case.

Keywords: surrogate modeling, blending operations, optimization, machine learning.

1. Introduction

Quality specifications, operational complexity, and environmental regulations are the most challenging obstacles in chemical processes, especially in crude-oil refineries, which affect the supply-chain profitability (Lotero et al., 2016). Most blend properties of final products are nonlinear and non-convex. Hence, estimating data-driven final product properties from computer-controlled in-line databases can be a quite complex process (Ounahasaree et al., 2016). The optimization addressed herein uses predicted formulas based on amounts of inlet streams to replace nonlinear complexities for optimizing the final products' amounts by matching blended properties with regulated specifications. Rigorous models of first principles, mechanistic, physics and engineering-based techniques can be reduced to black-box surrogate models to predict causation and correlation in blended properties of amounts of intermediate streams in scheduling, planning, multi-unit coordinating, and real-time optimization problems. There are previous works on surrogate-based optimization addressing MIQP (mixed-integer quadratic programming). Straus et al. (2018) applied MIQP on ammonia reaction processes to predict the optimal selection matrix that reduces the sampling-space in a self-optimizing variable surrogate modeling. Franzoi et al. (2020) used MIQP-based surrogate modeling to predict blending calculations of distillation unit outputs considering multiple feedstocks and operational variables. Franzoi et al. (2021) proposed an adaptive sampling MIQP-based surrogate modeling to predict reaction system conversions. Moreover, surrogates can also be used as factor-flow balances in blenders for blend scheduling optimization (Kelly et al., 2018). In general, surrogate modeling is employed for representing algebraic formulas $Y = f(x)$ that formulates the causation relationships between the input of independent variable (x) and the output of dependent variable (Y) to model complex, unmodeled, or unknown systems.

2. Problem statement

Blending operations are continuous processes of feedstocks entering a blender to produce a final product with determined specification on qualities. In this paper, the gasoline blending operation is addressed. There are nine feedstocks (x) in the process presenting the most common refinery blends, which are shown in Table 1, with their respective nomenclatures for j $\in$ {1..9}. Similarly, yields of i $\in$ {1..7} for properties (y) are considered, namely, Reid vapor pressure (RVP), aromatic content (ARO), olefin content (OLE), specific gravity (SG), sulfur content (SUL), research octane number (RON), and motor octane number (MON). These are the most important properties in operational and economical levels for enhancing the gasoline octane rating, which is especially important to increase fuel performance, suitability, and efficiency. These properties are also relevant for the fuel volatility, combustibility, and level of pollution.

Table 1: Gasoline blending feedstocks and their properties (Menezes et al., 2014).

Properties (y_i) / Feedstocks (x_j)	RVP (kPa)	RON	MON	ARO (vol%)	OLE (vol%)	SG (g/cm³)	SUL (ppmw)
Hydrotreated Light Cracked Naphtha	55	93	82	25	10	0.729	0.005
Hydrocracked Naphtha	52	94.05	81.78	45	29	0.758	0.002
Hydrotreated Cocker Light Naphtha	56	83	76	2	1	0.718	0.005
Light Naphtha	90.24	69.1	67.1	0.001	0.00001	0.699	0.0059
Reformate	85	98	90	54	20	0.79	0.005
Ethanol	17	109	90	0.1	0.02	0.79	0
Isomerate	69	106	100	0.1	0.02	0.85	0.005
Alkalyte	85	96	92	54	20	0.85	0.005
Butane	51.52	93.8	90	0	0	0.601	0

3. Surrogate modeling methodology

1) Data Generation: the independent input (x) variables of compositions are generated using the Latin Hypercube Sampling (LHS) technique, a statistical method of near-random distribution sampling. Each point in the data set is constrained by Eq.(1) to sum to 1; each sampled x' of feedstock j in IV (independent variables) is constrained to produce a final x of j. This ensures the mass balance across the blending operations.

$$x_j = \frac{x'_j}{\sum_{j \in IV} x_j} \quad \forall \; j \in IV \tag{1}$$

The dependent output (y) variables of yields are generated using the formulas from Eq. (2) to Eq.(6) for each sampled point of the dataset (Ounahasaree et al., 2016), which are regarded as the nonlinear properties. The subscripts j refer to the feedstock of the stream IN resembled in a data point entering the blender, and G is the gasoline product. The terms Q_j, sul_j, and sg_j respectively resemble the volumetric flowrate, sulfur content, and specific gravity of feedstock j. The octane properties of gasoline (RON and MON) are calculated using Eq.(7) to Eq.(11). Index v is the volume-based property, ARO_{VQ} is the volume-based property for ARO^2, and coefficients a to g are experimentally estimated values from real data retrieved from Menezes et al. (2014).

$$RVP_G = \frac{\left(\sum_{j \in IN}(Q_j RVP_j^{1.25})\right)^{0.8}}{\sum_{j \in IN} Q_j} \tag{2}$$

$$ARO_G = \frac{\sum_{j \in IN} Q_j ARO_j}{\sum_{j \in IN} Q_j} \tag{3}$$

$$OLE_s = \frac{\sum_{j \in IN} Q_j OLE_j}{\sum_{\in IN} Q_j} \tag{4}$$

$$sg_s = \frac{\sum_{j \in IN} Q_j sg_j}{\sum_{j \in IN} Q_j} \tag{5}$$

$$sul_G = \frac{\sum_{j \in IN} Q_j sg_j sul_j}{\sum_{j \in IN} Q_j sg_j} \tag{6}$$

$$RON_{jv} = RON_j + a[(RON_j - RON_v)(J_j - J_v)] + b(ARO_j - ARO_v)^2 + c(OLE_j - OLE_v)^2 + d[(ARO_j - ARO_v)(OLE_j - OLE_v)] \quad \forall \; j \in IN \tag{7}$$

$$MON_{jv} = MON_j + e[(MON_j - MON_v)(J_j - J_v)] + f(ARO_j - ARO_v)^2 + g[2(OLE_j - OLE_v)^2(ARO_{VQ} - ARO_v^2) - (ARO_{VQ} - ARO_v^2)^2] \quad \forall \; j \in IN \tag{8}$$

$$J_{j,v} = RON_{j,v} - MON_{j,v} \quad \forall \; j \in IN \tag{9}$$

$$RON_G = \frac{\sum_{j \in IN} Q_j RON_{jv}}{\sum_{j \in IN} Q_j} \tag{10}$$

$$MON_G = \frac{\sum_{j \in IN} Q_j MON_{jv}}{\sum_{j \in IN} Q_j} \tag{11}$$

The data set points are partitioned into 50% training and 50% testing, the former for building the surrogate functions, and the latter to evaluate their performance, reliability, and robustness.

2) Data improvement: the normalization of the generated x and y variables (Var) using Eq.(12) is performed to prevent the data from being biased or running across numerical issues when building the surrogates through MIQP optimizations. The terms of $min(VAR_{i,j}^{tr})$ and $max(VAR_{i,j}^{tr})$ refer to the minimum and maximum values of the training set array. Also, DV and IV refer to dependent and independent variables, respectively.

$$VAR_{i,j} = \frac{VAR_{i,j} - min(VAR_{i,j}^{tr})}{max(VAR_{i,j}^{tr}) - min(VAR_{i,j}^{tr})} \quad \forall\, j \in IV, i \in DV \tag{12}$$

3) *Surrogate Model Building:* surrogates are built following a bilinear functional form estimation as in Eq.(13) that correlates the normalized x and y variable sets, with coefficients of b_{ij} and c_{ij} to be predicted using this correlation. The intercepts are eliminated to avoid any multi-collinearity issues. The estimation uses MIQP optimizations, utilizing the objective function of Eq.(14) that minimizes the least-squares error (LSE) for linearly regressing the y into Y, subject to constraints Eq.(15) to Eq.(17), using $n_{training}$ points only, with z as a binary decision variable and M as a sufficiently large number (Franzoi et al., 2020).

$$Y_i = \sum_{j \in IV} b_{ij} X_j + \sum_{j \in IV} \sum_{k \in IV} c_{ijk} X_j X_k \quad \forall\, \{(j,k) \in IV, j \leq k\}, i \in DV \tag{13}$$

$$Minimize \sum_{p=1}^{n_{training}} (y_{ip} - Y_{ip})^2 \tag{14}$$

$$-Mz_j \leq b_{ij} \leq Mz_j \qquad\qquad \forall\, j \in IV, i \in DV \tag{15}$$

$$-Mz_{jk} \leq c_{ijk} \leq Mz_{jk} \qquad\qquad \forall\, \{(j,k) \in IV, j \leq k\}, i \in DV \tag{16}$$

$$\sum_{j \in IV} z_j + \sum_{j \in IV, k \in IV, j \leq k} z_{jk} \leq K \qquad \forall\, z_j, z_{jk}\, \{0,1\} \tag{17}$$

4) *Performance Check:* surrogates are used to-recalculate the generic dependent variables of Y, which are regarded as bilinear qualities of surrogates. The model performance is investigated by calculating the prediction errors and carrying out statistical analysis to ensure good predictability and reliability of results.

4. Results and discussion

Two distinct case studies are established for constructing the surrogate models and investigating their performance. For simplicity, only blended properties of the Y2 and Y3 plots are shown in Figure 1, Figure 2, and Figure 3, as they represent the most complex and difficult to predict variables considered in the formulation.

Case I: The model is built using data sizes $N \in \{100, 1000\}$; to testify the effect of the data set size in building it by comparing the MAE, referred to as prediction errors. Figure 1 illustrates that increasing the set size enhances the model overall performance, with lower testing and training MAE errors. In contrary, Table 2 reveals comparable model performance for both N trials, that shows prediction errors not exceeding $MAE \approx 10^{-4}$ for both trials. In contrast, the percentual errors did not exceed $MAE\% \approx 10^{-3}$ and $MAE\% \approx 10^{-4}$, respective to the trials, proving that prediction errors are enhanced with the data size increment. The overall results account for satisfying model performance. However, this is not sufficient for evaluating the model robustness and reliability. Hence, statistical analysis of MAE average, variance, standard deviations, and confidence intervals are also carried out to complement the analysis. Thus, 5%

uncertainty is accounted for testing the poor prediction regions (reliability) and the extent of applicability (robustness). With using 95% confidence interval probability, Figure 2 illustrates the estimation of the statistical analysis arranged in ascending order. Consequently, it resulted in base points plot, lying between upper and lower bounds curves of MAE added to and subtracted from the confidence intervals respectively, showing narrow convergence between all the curves. This illustrates that the variances and the standard deviations are too small, and all regions' MAE are approximate to the punctual values, accounting for satisfying model reliability and robustness.

Table 2: Mean absolute errors for the surrogate predictions.

	Y1 (RVP)	Y2 (RON)	Y3 (MON)	Y4 (ARO)	Y5 (OLE)	Y6 (SG)	Y7 (SUL)
MAE (N=100)	$6.10 \; 10^{-4}$	$7.28 \; 10^{-8}$	$8.98 \; 10^{-5}$	$8.94 \; 10^{-8}$	$1.09 \; 10^{-7}$	$3.43 \; 10^{-10}$	$1.16 \; 10^{-8}$
MAE% (N=100)	$1.07 \; 10^{-3}$	$7.81 \; 10^{-8}$	$1.09 \; 10^{-4}$	$3.97 \; 10^{-7}$	$1.09 \; 10^{-6}$	$4.62 \; 10^{-8}$	$2.70 \; 10^{-4}$
MAE (N=1000)	$3.19 \; 10^{-4}$	$6.12 \; 10^{-8}$	$8.08 \; 10^{-5}$	$1.25 \; 10^{-7}$	$5.44 \; 10^{-8}$	$3.20 \; 10^{-10}$	$9.36 \; 10^{-9}$
MAE% (N=1000)	$5.68 \; 10^{-4}$	$6.57 \; 10^{-8}$	$9.75 \; 10^{-5}$	$5.70 \; 10^{-7}$	$5.48 \; 10^{-7}$	$4.30 \; 10^{-8}$	$2.21 \; 10^{-4}$

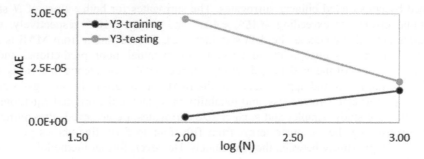

Figure 1: MAE of bilinear surrogates versus the number of sample points.

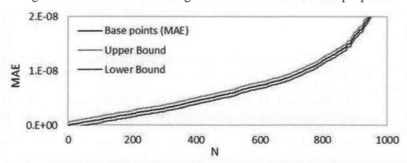

Figure 2: MAE with confidence interval range for the sample points of Y2.

Case II: Bilinear surrogates with simpler formulation and smaller size for $N \in 1000$ are identified by limiting the number of terms (K) in the surrogate (originally 54). Thus, the surrogates are tested by plotting the MAE against K as shown in Figure 3 to identify whether a lower number of coefficients with provides sufficient accuracy. As a result, the plot confirmed that using up to 45 simplified terms, the surrogates are predicted to generate the same original model accuracy. Using fewer coefficients compromises the quality of the surrogate. In general, this procedure does not give the exact optimal value of K, especially as it depends on different aspects, such as the problem type, dimensionality, functional form, data size, etc.

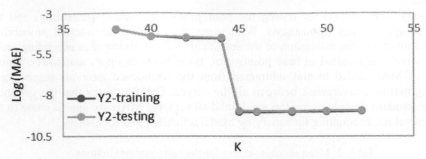

Figure 3: MAE versus different number of coefficients for Y2 surrogate.

5. Conclusion

Surrogate modeling is utilized to predict analytical formulations for complex systems. In this work, nonlinear equations used to predict gasoline blending properties are replaced by constructed bilinear surrogates. The surrogates for both values of N show percentual errors not exceeding $MAE\% \approx 10^{-3}$ and $MAE\% \approx 10^{-4}$ respectively, with noticeable error decrement as the data size increases. The deviation from MAE is also calculated using 5% uncertainty to account for eventual poor predictions, and to measure the extent of the model applicability. Hence, a 95% confidence interval is used to calculate the lower and upper bounds of the MAE. As a result, the surrogates show sufficient performance, robustness, and reliability in replacing the original equations. In the second case study, simpler and smaller size surrogates are achieved. Presenting Y2 results, decreasing the surrogate terms from fifty-four to forty-five shows prediction errors of high proximity between the two models. However, further methodologies need to be developed in future to properly adjust the surrogate model size within an automatic and systematic fashion.

References

R. E. Franzoi, J. D. Kelly, B. C. Menezes, C. L. E. Swartz, 2021, An adaptive sampling surrogate model building framework for the optimization of reaction system, Computers and Chemical Engineering, 152, 107371.

R. E. Franzoi, B. C. Menezes, J. D. Kelly, J. A. W. Gut, I. E. Grossmann, 2020, Cutpoint temperature surrogate modeling for distillation yields and properties, Industrial and Engineering Chemistry Research, 59 (41), 18616–18628.

J. D. Kelly, B. C. Menezes, I. E. Grossmann, 2018, Successive LP approximation for nonconvex blending in MILP scheduling optimization using factors for qualities in the process industry, Industrial and Engineering Chemistry Research, 57(32), 11076-11093.

I. Lotero, F. Trespalacios, I. E. Grossmann, D. J. Papageorgiou, M. S. Cheon, 2016, An MILP-MINLP decomposition method for the global optimization of a source based model of the multiperiod blending problem, Computers and Chemical Engineering, 87, 13–35.

B. C. Menezes, L. F. L. Moro, W. O. Lin, R. A. Medronho, F. L. P. Pessoa, 2014, Nonlinear production planning of oil-refinery units for the future fuel market in Brazil: Process design scenario-based model, Industrial and Engineering Chemistry Research, 53(11), 4352–4365.

Y. Ounahasaree, U. Suriyapraphadilok, M. Bagajewicz, 2016, global optimization of gasoline blending model using bound contraction technique, Computer Aided Chemical Engineering, 38, 1293–1298.

J. Straus, S. Skogestad, 2018, Surrogate model generation using self-optimizing variables, Computers and Chemical Engineering, 119, 143–148.

Proceedings of the 14th International Symposium on Process Systems Engineering – PSE 2021+
June 19-23, 2022, Kyoto, Japan © 2022 Elsevier B.V. All rights reserved.
http://dx.doi.org/10.1016/B978-0-323-85159-6.50298-0

Continuous Manufacturing Process Sequential Prediction using Temporal Convolutional Network

Haoran Li[a,b], Tong Qiu[a,b*]

[a] *Department of Chemical Engineering, Tsinghua University, Beijing 100084, China*
[b] *Beijing Key Laboratory of Industrial Big Data System and Application, Tsinghua University, Beijing 100084, China*
qiutong@tsinghua.edu.cn

Abstract

In the era of intelligent manufacturing, the continuous manufacturing industry will benefit from digitalization technologies such as digital twins. This paper proposes a temporal convolutional network sequence-to-sequence (TCN-StS) model as a data-driven simulation tool for the construction of digital twins. The proposed model captures time delay information through temporal convolution operation and thus better predicts the process state variations than recurrent neural networks on an actual industrial sintering dataset and shows good robustness over time. This study sheds new light on process sequence-to-sequence modelling through convolutional networks.

Keywords: sequence-to-sequence; temporal convolution; digital twin

1. Introduction

The continuous process is ubiquitous in steel, chemical, pharmaceutical, and other manufacturing industries. Prediction and operation decision-making in traditional continuous manufacturing processes rely on the knowledge reserve and cognitive level of operators, which severely restricts the safe and efficient operation of the production process. Over the decades, the soaring development of big data and artificial intelligence has brought transformational opportunities for the digitization of the process industry. The concept of digital twin (Glaessgen and Stargel, 2012; Gockel et al., 2012), initially proposed by the National Aeronautics and Space Administration (NASA), has recently been transplanted and deemed as the future solution to the manufacturing industry (Rosen et al., 2015).

Sequence-to-sequence modelling is the closest approach as a digital twin, as it uses the historical operation sequences to capture the dynamics of the process and predict the future evolution. Chou et al. first designed a sequence-to-sequence soft sensor model and performed excellently for product impurity predictions of an industrial distillation column (Chou et al., 2020). Kang et al also built a sequence-to-sequence model and achieved rolling predictions in the process of vapor-recompression C3 (Kang et al., 2021). Although canonical recurrent neural networks such as LSTMs and GRUs are considered by most deep learning practitioners synonymous with sequence modelling, Bai et al indicated that temporal convolutional networks (Lea et al., 2016) outperformed recurrent neural networks across a diverse range of tasks and datasets, while demonstrating longer effective memory (Bai et al., 2018).

These recent researches illustrate the importance of sequence-to-sequence modelling for the manufacturing industry and the potential of convolutional neural networks as a positive option for sequential modelling. This paper proposes a temporal convolutional network sequence-to-sequence (TCN-StS) model to achieve continuous manufacturing process sequential prediction. The model is applied to the sintering process in the iron-making industry and tested in an actual industrial dataset.

2. Methodology

2.1. Process description

The sintering process is an important thermochemical process in the blast furnace ironmaking system. It involves the heating of fine iron ore with flux and coke fines or coal to produce a semi-molten mass that solidifies into porous pieces of sinter with the size and strength characteristics necessary for feeding into the blast furnace.

This process is a typical continuous manufacturing process (Figure 1). Firstly, iron ore, coke, limestone, and returning sinter are mixed and then fed in a moving trolley to form a uniform sintering bed. Next, the igniter ignites the surface of the bed, and the blower under the moving trolley generates negative pressure in the bellow below the bed through the exhaust. As the trolley gradually moves to the end of the sintering machine, the combustion continues to develop downward (Zhou et al., 2019). In the end, the raw ore powder will gradually form sintered ore with a certain particle size, which will enter the subsequent blast furnace ironmaking production as iron material.

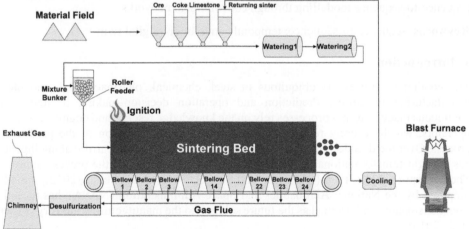

Figure 1. Sintering process schematic.

The sintering process owns the following two characteristics:

Time delay. There is a time interval between the change of process variables in the sintering system and its downstream variables, which is called mechanism time lag. In addition, for the same batch of raw materials, because different variables are measured at different times, there will be a technical time lag. The coexistence of the two types of time delays makes the sintering process exhibit strong time-delay characteristics.

Non-linearity. In the sintering process, a large number of chemical reactions such as coke combustion and limestone decomposition and a two-dimensional three-phase complex heat and mass transfer relationship exist at the same time, so the sintering system variables present obvious nonlinear relationship characteristics.

An industrial sintering process dataset of 27,000 samples was obtained and used in this study. The dataset was collected from 2019/12/05 to 2019/12/24 with a sampling frequency of 1 min. There are 15 key variables including 9 operating variables (OVs) and 6 state variables (SVs) in the process (Table 1).

Table 1. Key variables of the process.

Variable	Notation
Sintering bed thickness	OV1
Ignition intensity of row A	OV2
Ignition intensity of row B	OV3
Ignition temperature	OV4
Trolley speed	OV5
Round roller speed	OV6
Seven roller speed	OV7
Frequency of No.1 exhaust fan	OV8
Frequency of No.2 exhaust fan	OV9
Bellow 14 negative pressure	SV1
Bellow 22 negative pressure	SV2
Bellow 14 gas temperature	SV3
Bellow 22 gas temperature	SV4
End point of sintering	SV5
End point temperature of sintering	SV6

2.2. Temporal convolutional network sequence-to-sequence (TCN-StS) Model

The basic temporal convolutional network is a one-dimensional fully convolutional network with zero padding applied to make sure that the output sequence has the same length as the input sequence. To keep the convolution operation causal, which means for every i in $\{0, \ldots, \text{input\_length} - 1\}$, the i-th element of the output sequence only depends on the elements of the input sequence with indices $\{0, \ldots, i\}$, zero-padding is applied only on the left side of the input tensor.

Nevertheless, it is very challenging to apply basic causal convolution directly to long-term sequence problems due to its ability to look back only in a linear order in the depth of the network. Dilated convolution, which enables an exponentially large receptive field, eliminates this problem. Formally, for a one-dimensional sequence input $x \in R^n$ and a filter $f: \{0, \ldots, k-1\} \to R$, the dilated convolution operation F on elements s of the sequence is defined as Eq. (1).

$$F(s) = (\mathbf{x} *_d f)(s) = \sum_{i=0}^{k-1} f(i) \cdot \mathbf{x}_{s-d \cdot i} \tag{1}$$

where d denotes the dilation factor, k is the filter size, and $s - d \cdot i$ accounts for the direction of the past.

Besides, to avoid the gradient exploding/vanishing problems in deep neural networks, residual blocks with skip connections initially designed in ResNet (He et al., 2015) are used in TCN. The skip connection is a branch leading out to a series of transformations *F*, whose outputs are added to the input **x** of the block as Eq. (2).

$$o = \text{Activation}(\mathbf{x} + (\boldsymbol{F}(\mathbf{x}))) \tag{2}$$

Within a residual block, two layers of dilated causal convolution, rectified linear unit (ReLU) activation, weight normalization, and spatial dropout are stacked. Figure 2(a) illustrates the TCN architecture.

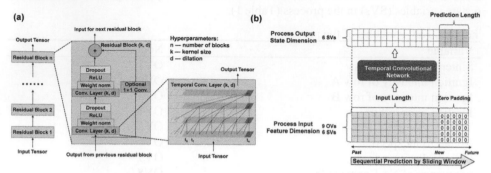

Figure 2. Proposed TCN-StS model. (a) Temporal convolutional network (TCN) architecture. (b) Sequence-to-sequence prediction demonstration.

A sequence-to-sequence prediction manner is proposed in Figure 2(b). At time T, a sequence of shape (input length, feature dimensions) are combined with zero values of shape (output length, feature dimensions). This sequence represents the history from time $(T - \text{input length})$ to T. The model output is a sequence of shape (output length, output dimension) predicting the process state from time $T + 1$ to $(T + \text{output length})$. Formally, the TCN-StS model produces the mapping as Eq. (3).

$$\hat{y}_{T+1}, \hat{y}_{T+2}, \cdots, \hat{y}_{T+\text{out\_len}} = \text{TCN\_StS}\left(x_{T-\text{in\_len}}, \cdots, x_{T-1}, x_T\right) \tag{3}$$

3. Result and discussion

The dataset is split into 70 % training, 10 % validating, and 20 % testing. Each feature of the original dataset is standardized separately. The input feature dimension is 15 containing 9 OVs and 6 SVs while the output dimension is 6. An input length of 40 minutes is set according to the time delay of the sintering process and predictions are made by the TCN-StS model for the time length of 5, 10, 15, and 20 minutes.

Figure 3(a) presents a snapshot of sequential prediction for SV6. The prediction sequence shows good coincidence with the true sequence, especially at shorter prediction lengths such as 5 minutes and 10 minutes. The predictions shift away from the true values at longer prediction lengths. Two canonical recurrent neural networks, RNN and LSTM are adopted for comparison. The mean squared errors (MSEs) and mean absolute errors (MAEs) of the three models are given in Table 2. The TCN-StS model has lower MSEs and MAEs at all prediction lengths. The results indicate TCN-StS model outperforms recurrent neural networks at the sintering process sequential prediction.

Table 2. Sequential prediction accuracy of RNN, LSTM, and TCN-StS model.

Model	5 min		10 min		15 min		20 min	
	MSE	MAE	MSE	MAE	MSE	MAE	MSE	MAE
RNN	0.32	0.38	0.46	0.48	0.55	0.54	0.62	0.57
LSTM	0.30	0.37	0.38	0.44	0.50	0.52	0.54	0.55
TCN-StS	**0.26**	**0.33**	**0.36**	**0.40**	**0.41**	**0.44**	**0.46**	**0.47**

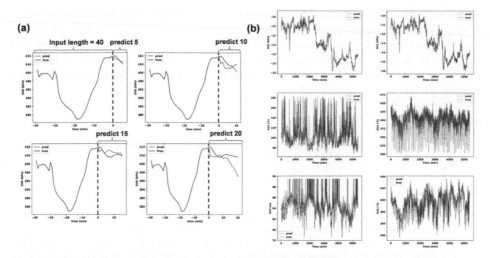

Figure 3. Prediction results. (a) A sequential prediction snapshot for SV6 at 5, 10, 15, and 20 minutes prediction length. (b) Prediction results at 10 minutes time points in the 10 minutes prediction sequence.

To better present the prediction performance over time, prediction results for each time point are extracted separately from the sequence. Figure 3(b) shows prediction results at 10 minutes time points in the 10 minutes prediction sequence. Prediction accuracies for each point in 10 minutes are shown in Figure 4. It can be found that as the prediction length increases, the prediction accuracy of all three models will become worse. For example, the TCN-StS model, with the average MSE of 0.36 and MAE of 0.40, shows the MSE and MAE of 0.24 and 0.33 at 1 min point, and 0.43 and 0.45 at 10 min point, respectively. The same growing trends are also found in RNN and LSTM models. Nevertheless, the TCN-StS model still has lower prediction errors than RNN and LSTM models at nearly all given time points, showing good robustness over time.

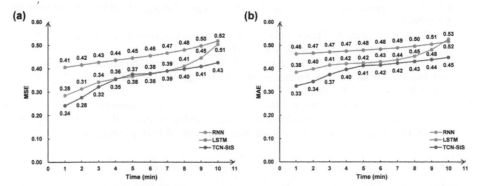

Figure 4. Prediction accuracy for each time point in the sequence. (a) Mean squared errors (MSEs). (b) Mean absolute errors (MAEs).

4. Conclusions

This paper designed a new convolutional-based sequence-to-sequence model architecture for continuous manufacturing process sequential prediction. Comparted to recurrent neural networks, the proposed TCN-StS model demonstrates better prediction accuracy at all given time lengths on an actual industrial dataset as well as a robust prediction capability over time. This study addresses the effectiveness of convolutional networks for sequence modelling and gives insights into utilizing sequence-to-sequence modelling as an effective simulation tool for constructing digital twins in the continuous manufacturing process.

Acknowledgment

The authors gratefully acknowledge Technological Innovation 2030—"New Generation Artificial Intelligence" Major Project (2018AAA0101605) for its financial support.

References

Bai, S., Kolter, J.Z., Koltun, V., 2018. An Empirical Evaluation of Generic Convolutional and Recurrent Networks for Sequence Modeling. arXiv:1803.01271 [cs].

Chou, C.-H., Wu, H., Kang, J.-L., Wong, D.S.-H., Yao, Y., Chuang, Y.-C., Jang, S.-S., Ou, J.D.-Y., 2020. Physically Consistent Soft-Sensor Development Using Sequence-to-Sequence Neural Networks. IEEE Transactions on Industrial Informatics 16, 2829–2838. https://doi.org/10.1109/TII.2019.2952429

Glaessgen, E., Stargel, D., 2012. The Digital Twin Paradigm for Future NASA and U.S. Air Force Vehicles, in: 53rd AIAA/ASME/ASCE/AHS/ASC Structures, Structural Dynamics and Materials Conference. American Institute of Aeronautics and Astronautics. https://doi.org/10.2514/6.2012-1818

Gockel, B., Tudor, A., Brandyberry, M., Penmetsa, R., Tuegel, E., 2012. Challenges with Structural Life Forecasting Using Realistic Mission Profiles, in: 53rd AIAA/ASME/ASCE/AHS/ASC Structures, Structural Dynamics and Materials Conference. American Institute of Aeronautics and Astronautics. https://doi.org/10.2514/6.2012-1813

He, K., Zhang, X., Ren, S., Sun, J., 2015. Deep Residual Learning for Image Recognition. arXiv:1512.03385 [cs].

Kang, J.-L., Mirzaei, S., Lee, Y.-C., Chuang, Y.-C., Frias, M., Chou, C.-H., Wang, S.-J., Wong, D.S.H., Jang, S.-S., 2021. Digital Twin Model Development for Chemical Plants Using Multiple Time-Steps Prediction Data-Driven Model and Rolling Training, in: Computer Aided Chemical Engineering. Elsevier, pp. 567–572. https://doi.org/10.1016/B978-0-323-88506-5.50090-5

Lea, C., Vidal, R., Reiter, A., Hager, G.D., 2016. Temporal Convolutional Networks: A Unified Approach to Action Segmentation. arXiv:1608.08242 [cs].

Rosen, R., von Wichert, G., Lo, G., Bettenhausen, K.D., 2015. About The Importance of Autonomy and Digital Twins for the Future of Manufacturing. IFAC-PapersOnLine 48, 567–572. https://doi.org/10.1016/j.ifacol.2015.06.141

Zhou, K., Chen, X., Wu, M., Cao, W., Hu, J., 2019. A new hybrid modelling and optimization algorithm for improving carbon efficiency based on different time scales in sintering process. Control Engineering Practice 91, 104104. https://doi.org/10.1016/j.conengprac.2019.104104

Proceedings of the 14th International Symposium on Process Systems Engineering – PSE 2021+
June 19-23, 2022, Kyoto, Japan © 2022 Elsevier B.V. All rights reserved.
http://dx.doi.org/10.1016/B978-0-323-85159-6.50299-2

Surrogate modeling for mixed refrigerant streams in the refrigeration cycle of an LNG plant

Aisha A. Al-Hammadi[a,b], Robert E. Franzoi[a], Omar E. Ibrahim[a], Brenno C. Menezes[a*]

[a]Division of Engineering Management and Decision Sciences, College of Science and Engineering, Hamad Bin Khalifa University, Qatar Foundation, Doha, QATAR
[b]Operations Support Department, Qatargas, Doha, QATAR
bmenezes@hbku.edu.qa

Abstract

Given the importance of liquefaction processes in the LNG value chain, it is necessary to model the complexity of such process. A key stage is the mixed refrigerant (MR) cycle used to liquify the natural gas in the liquefaction plant. The MR refrigeration cycle consists of compressors and heat exchangers in different compression stages that affect the MR properties in terms of temperature and pressure. In this work, the use of surrogate models is addressed for the compressor's power consumption and efficiency formulations along with the heat exchanger's performance in terms of heat duty after each compression stage. A training data set containing 500 points is used for building the surrogates, while a testing data set of 500 points verifies their accuracy. The surrogates built herein are shown to be sufficiently accurate to be further employed in decision-making industrial applications such as simulation, optimization, and control.

Keywords: Surrogate modeling, refrigeration cycle, machine learning, liquefaction.

1. Introduction

The natural gas liquefaction plant is the most essential and critical process in the liquefied natural gas (LNG) value chain. In this process, the natural gas is cooled and liquified to a cryogenic temperature of around -162 ∘C. To achieve good quality of liquified natural gas, a comprehensive understanding of the process is required for improved design and operations of liquefaction plants. A key component in such processes is the mixed refrigerant (MR) stream, which plays a significant role in precooling, subcooling, and liquifying the natural gas to produce LNG. The MR streams undergo certain processing in the refrigeration cycle that consists mainly of refrigerant compressors and cooler type heat exchangers. In this work, a MR network is considered, in which propane (C3) and MR (C1, C2, C3, and N2) are used to cool and liquify natural gas. The mixed refrigeration process controls the LNG production outputs, while the propane refrigeration system aims to provide cooling to the MR (Mokhatab et al., 2014). Since the MR is precooled and liquified using propane during the process, the criticality of this operation arises from maintaining optimal MR temperature in the refrigeration cycle. Very high temperature of MR results in lower LNG production. A main reason of operating with warm MR is due to the high-power consumption of the compressor and its lower efficiency in addition to the low heat duty of heat exchangers. Therefore, it is important to include details of rigorous liquefaction process to better optimize the power consumption, the compressor efficiency, and the heat duty to produce cooler MR, and hence, more efficient LNG production. However, given its thermodynamics complexity, it is intractable to introduce such nonlinearities in optimizable and controllable environments. Therefore, additional methods are required for achieving simplified correlations that simultaneously provide good accuracy and can be further embedded in decision-making problems.

Recently, there has been an extensive use of surrogate modeling approaches for predicting the behavior of complex systems to provide improved computational tractability and avoid convergence issues in decision-making problems (Franzoi et al., 2021a). This work aims to build surrogate models from supervised simulated data of an MR refrigeration cycle unit that calculate the power consumption, efficiency of compressors, and heat duty of heat exchangers based on variable inputs. This method relies on three steps: 1) define base functions to be selected by a coefficient-setup approach that minimizes the variable outputs as the difference between $y_{predicted}$ and $y_{experimental}$; 2) the design of experiments including the sampling method and the data generation approach; and 3) the regression or identification method to determine the selected base functions and their respective coefficients (Tran and Georgakis, 2018; Hullen et al., 2020). To build the surrogate model, the data shall be accurate and can be determined in rigorous simulation software, collected from experiments in the plant, etc. Moreover, a proper balance between the quality of the model accuracy and the computational effort must be considered depending on the processing time and resources for a certain application (Mencarelli et al., 2020).

In this paper, a surrogate modeling approach is used to handle the compressor's power consumption and efficiency formulations as well as the heat exchanger performance after each compression stage in the liquefaction of natural gas, in which complex nonlinear equations are transformed into simpler bilinear and trilinear equations. The surrogate models correlate variations of independent X variables to dependent Y variables, which can be used to model different types of complex and unknown systems.

2. Problem Statement

The type of problem addressed herein concerns the cooling of mixed refrigerant, used to cool and liquefy the LNG in the main cryogenic heat exchanger. MR is widely employed within the natural gas liquefaction. The complexity of this process arises from the nonlinear terms that calculate the compressors efficiency and power and heat exchangers heat duty. The case study proposed in this work addresses the MR refrigeration cycle that undergoes three compression stages with intermediate cooling processes. Figure 1 illustrates the MR refrigeration network, in which the MR flow is recycled. It exits the main cryogenic heat exchanger (MCHE) top side in gaseous form and low pressure; throughout the network, the MR pressure is increased, liquefied in the C3 refrigeration cycle, and enters the MCHE again for cooling and liquefying the natural gas.

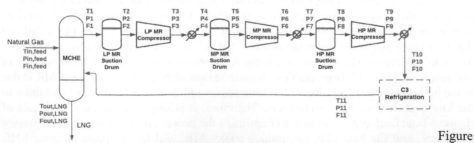

Figure 1: MR refrigeration cycle network.

This type of process is nonlinear and nonconvex because of the fluctuating temperatures and pressures after each stage and the requirement of calculating related variables such as the density and heat capacity for each temperature and pressure points. For each compression stage, Eqs.(1) and (2) are used to calculate the efficiency and power consumption for the compressors and the heat duty for the heat exchangers. The variables T_{i-1}, T_i, and T_i' are the inlet, outlet, and ideal discharge temperatures from the compressors, respectively, and η_{cj} is the compressor's efficiency. The indices i and j respectively belong to the stream flows (before and after each compressor (SFC)) and to the compression stage (CS). The compressor power consumption is calculated using Eq.(3), in which W_j is the power consumption for each compression stage and C_{pj} is the heat capacity for each compression stage.

$$T_i' = T_{i-1} \times (r_p^{(\frac{y-1}{y})}) \qquad \forall\ i \in \text{SFC} \qquad (1)$$

$$\eta_{cj} = \frac{T_i' - T_{i-1}}{T_i - T_{i-1}} \qquad \forall\ j \in \text{CS}, i \in \text{SFC} \qquad (2)$$

$$W_j = C_{pj}(T_i - T_{i-1}) \qquad \forall\ j \in \text{CS}, i \in \text{SFC} \qquad (3)$$

The heat duty for each heat exchanger is calculated using Eqs.(4) and (5).

$$V_{actual} = \left[\frac{P_{normal} \times V_i}{T_{normal}}\right] \times \frac{T_{i-1}}{P_{i-1}} \qquad \forall\ i \in \text{SFH} \qquad (4)$$

$$Q_j = [\rho_j \times V_{actual}] \times C_{pj} \times [T_i - T_{i-1}] \qquad \forall\ j \in \text{CS}, i \in \text{SFH} \qquad (5)$$

In which Q_j is the heat duty for each compression stage j, ρ_j is the density for each compression stage, V_i is the flow at each stream flow i before and after each heat exchanger (SFH), while $P_{normal} = 101,325\ PA$ and $T_{normal} = 293\ K$.

3. Methodology

The main purpose of this paper is to build surrogate models that can effectively replace the complex thermodynamic equations. The surrogates are built to represent three dependent variables of interest Y, which are the properties of the compressors and heat exchangers shown in Figure 1 and represented by Eqs.(2), (3), and (5). Independent-dependent variations correlate as $Y_i = f(X_j)$. The input variables considered to form the surrogates represent the stream properties (i.e., temperature, pressure, and flow). When building surrogate models, it is important to consider some important aspects concerning the required data for training and testing, functional form of the surrogates and proper evaluation to verify their performance. A simplified framework methodology proposed by Franzoi et al.(2021b) is used in this work, as shown in Figure 2.

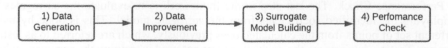

Figure 2: Framework for the surrogate model building strategy (adapted from Franzoi et al. 2021b).

1) *Data Generation*: Latin Hypercube Sampling (LHS) method is used to generate the required data for training the surrogates. This generates random sample points for the input (independent variables) dataset that includes temperatures, pressures, and flows. There are boundaries for each independent variable in each compression stage based on experiments. The independent dataset is used to calculate the output values (dependent variables) from compressors and heat exchangers related formulas in Eqs.(2), (3), and (5) for each sample point. Then, a complete X-Y input-output data set is constructed to estimate the behavior of the surrogates. The dataset is equally split into training and testing data. The former focuses on building the surrogates by identifying their bases and coefficients, while the latter is used to fairly evaluate the accuracy and performance of the surrogates.

2) *Data Improvement*: This step is performed to avoid any biased surrogates or numerical issues due to the different units and magnitudes used in the model, mostly because variables that are too small or too large can affect obtaining the accurate and reliable coefficients. Hence, data normalization is used to normalize all training input and output data sets for each variable using Eq.(6), which results in minimum and maximum values to be within 0 to 1 range.

$$x_{jp} = \frac{x_{jp} - min(x_j^{tr})}{max(x_j^{tr}) - min(x_j^{tr})} \qquad\qquad \forall\ p \in N_{training} \qquad (6)$$

In which the minimum and maximum values of the training data x_j correspond to $min\ (x_j^{tr})$ and $max(x_j^{tr})$.

3) *Surrogate Model Building*: To estimate the behavior of the MR refrigeration cycle processes for the compressors and heat exchangers, trilinear surrogate models are used, as shown in Equation (7) to calculate the dependent variable Y_i.

$$Y_i = I_i + \sum_{j \in DV} b_{ij} X_j + \sum_{j \in IV} \sum_{k \in IV} c_{ijk} X_j X_k \qquad\qquad \forall\ i \in DV,$$
$$+ \sum_{j \in IV} \sum_{k \in IV} \sum_{n \in IV} t_{ijkn} X_j X_k X_n \qquad\qquad \begin{matrix} \forall\ j \le k, \\ \\ \forall\ k \le n \end{matrix} \qquad (7)$$

In which I_i is the intercept of each point i within dependent variables (DV), and b_{ij}, c_{ijk}, and t_{ijkn} are the coefficients to be determined or estimated during the building process of the surrogate model by evaluating the accuracy of the input-output data to achieve the target of minimizing the prediction error for each independent variable point (IV). This optimization target is presented in Eq.(8) for minimizing the least-squares error (LSE).

$$Minimize\ \sum_{p=1}^{n_{training}} (y_{ip} - Y_{ip})^2 \qquad\qquad (8)$$

In which y_{ip} is the actual calculated value using thermodynamics equations for the variable i at each point p within the training data set $n_{training}$, whereas Y_{ip} represents the calculated values using the surrogate model estimated coefficients for each independent variable i at each point p.

4) Performance Check: The last step in the framework is to evaluate the surrogates by employing the identified coefficients from the surrogate model. This is done by using different set of points from the calculated ones previously, which are referred to as testing data set. Testing data sets are the points that were not used in training the surrogate, which can identify the accuracy of the predictions. The evaluation is done via calculating the mean absolute error using the testing data that indicate the surrogate accuracy.

4. Results

The surrogate models built are tested for three compression stages in the MR refrigeration cycle including compressors and heat exchangers, as shown in Figure 1. The size of the dataset considered in the surrogates is set to be $N \in \{1000\}$, which is equally split between training and testing data sets. Bilinear and trilinear surrogates are built for each compressor and heat exchanger. For the compressors, surrogates are built testing the accuracy of efficiency and power consumption equations, whereas heat duty is considered for the heat exchangers. The mean absolute error is calculated for each system to illustrate the difference between the calculated and predicted values, as shown in Table 1. Trilinear surrogates provide higher accuracy, mostly because the higher amount of coefficients and higher predictability power given their complexity. The methodology employed herein has proved its efficiency in terms of accuracy for nonlinear and complex systems, as those surrogates can successfully be employed in replacement of the thermodynamic equations and can be potentially used in further optimization and decision-making processes.

Table 1: Mean absolute error (MAE) for bilinear and trilinear surrogates.

		Bilinear					
		LP- MAE		MP- MAE		HP- MAE	
System	SM	Training	Testing	Training	Testing	Training	Testing
Compressors	Y1-Efficiency	$2.83 \ 10^{-3}$	$7.48 \ 10^{-3}$	$1.73 \ 10^{-2}$	$3.35 \ 10^{-2}$	$2.11 \ 10^{-2}$	$5.28 \ 10^{-2}$
	Y2-Power	$1.22 \ 10^{-5}$	$1.77 \ 10^{-5}$	$2.00 \ 10^{-3}$	$2.18 \ 10^{-6}$	$2.63 \ 10^{-6}$	$2.68 \ 10^{-5}$
Heat Exchangers	Y3-Heat Duty	$9.45 \ 10^{-4}$	$9.12 \ 10^{-4}$	$8.14 \ 10^{-4}$	$7.04 \ 10^{-4}$	$8.14 \ 10^{-4}$	$7.04 \ 10^{-4}$
		Trilinear					
		LP- MAE		MP- MAE		HP- MAE	
System	SM	Training	Testing	Training	Testing	Training	Testing
Compressors	Y1-Efficiency	$2.03 \ 10^{-4}$	$5.64 \ 10^{-4}$	$3.26 \ 10^{-3}$	$7.68 \ 10^{-3}$	$3.91 \ 10^{-3}$	$1.18 \ 10^{-2}$
	Y2-Power	$2.92 \ 10^{-6}$	$2.94 \ 10^{-6}$	$5.71 \ 10^{-7}$	$1.52 \ 10^{-6}$	$3.72 \ 10^{-7}$	$2.81 \ 10^{-6}$
Heat Exchangers	Y3-Heat Duty	$1.73 \ 10^{-4}$	$1.64 \ 10^{-4}$	$1.55 \ 10^{-4}$	$1.33 \ 10^{-4}$	$3.12 \ 10^{-4}$	$2.77 \ 10^{-4}$

5. Conclusion

The cooling and liquefaction of natural gas involves complex networks that require the use of refrigerants within a rigorously controlled environment. Those refrigerants shall meet certain temperature and pressure points via undergoing cooling cycle, called refrigeration cycle. This work considered the MR refrigeration cycle with three compression stages, LP, MP, and HP. This cycle consists mainly of compressors and heat exchangers, whereby such a system is studied considering certain variables of interest associated with the compressors, such as the efficiency and power consumption, in addition to the heat duty for heat exchangers. Complex and nonlinear thermodynamics equations are typically used to calculate those variables. However, this work addresses the use of the surrogate models for the purpose of simplifying the formulas into bilinear and trilinear terms, which can open opportunities for integrating such complex modeling systems with further optimization decision-making approaches. The methodology employed uses a surrogate model building framework comprised of four steps. First, the input dataset is generated experimentally from the Latin Hypercube Sampling method, which is utilized to calculate the required output variables using the first principles thermodynamics equations. Then, datasets are enhanced by normalization, in which the data points are normalized to lie between 0 and 1 aiming to avoid numeric issues. The third step builds the surrogates using input-output datasets and generates bilinear and trilinear coefficients used to calculate the required variables. Finally, performance check is carried out to determine the accuracy of the surrogates using testing datasets.

Depending on the complexity of the equations to be predicted, bilinear surrogates can be sufficiently accurate. However, trilinear surrogates typically provide significant higher accuracy than their bilinear counterparts. Overall, both provide good efficiency and accuracy for replacing the thermodynamics equations and can be further utilized for simulation, control, and optimization cases. The proposed surrogate modeling approach builds a data-driven model that can be used in different optimization applications, such as in: a) optimizing the power consumption of compressors considering certain design boundaries for each parameter (e.g., allowable temperature and pressure for the compressors); b) optimizing the efficiency of the compressors, which lead to colder MR after the C3 refrigeration cycle resulting in a higher production of LNG with lower temperature; and c) optimizing the heat exchanger duty after each compression stage aiming to achieve better usage of MR in terms of liquifying the natural gas without requiring makeup MR to provide a better cooling and liquefaction to the natural gas. Such approach would allow the utilization of optimization-based tools instead of their simulation counterparts that rely on complex thermodynamics equations.

References

R. E. Franzoi, T. Ali, A. Al-Hammadi, B. C. Menezes, 2021b, Surrogate Modeling Approach for Nonlinear Blending Processes, 1st International Conference on Emerging Smart Technologies and Applications (eSmarTA), 1-8.

R. E. Franzoi , J. D. Kelly, B. C. Menezes, C. L. E. Swartz, 2021a, An adaptive sampling surrogate model building framework for the optimization of reaction systems. Computers and Chemical Engineering, 152, 107371.

G. Hullen, J. Zhai, S. H. Kim, A. Sinha, M. J. Realff, F. Boukouvala, 2020, Managing uncertainty in data-driven simulation-based optimization, Computers and Chemical Engineering, 136, 106519.

L. Mencarelli, A. Pagot, P. Duchêne, 2020, Surrogate-based modeling techniques with application to catalytic reforming and isomerization processes, Computers and Chemical Engineering, 135, 106772.

S. Mokhatab, J. Y. Mak, J. V. Valappil, D. A. Wood, 2014, In Handbook of liquefied natural gas, pp. 297-300, Gulf Professional Publishing.

D. P. Penumuru, S. Muthuswamy, P. Karumbu, 2020, Identification and classification of materials using machine vision and machine learning in the context of industry 4.0. Journal of Intelligent Manufacturing, 31(5), 1229–1241.

A. P. Tran, C. Georgakis, 2018, On the estimation of high-dimensional surrogate models of steady-state of plant-wide processes characteristics, Computers and Chemical Engineering, 116, 56–68.

Proceedings of the 14th International Symposium on Process Systems Engineering – PSE 2021+
June 19-23, 2022, Kyoto, Japan © 2022 Elsevier B.V. All rights reserved.
http://dx.doi.org/10.1016/B978-0-323-85159-6.50300-6

Prediction for heat deflection temperature of polypropylene composite with Catboost

Chonghyo J.[a,b] , Hyundo P.[a,c] , Seokyoung H.[b] , Jongkoo L.[d] , Insu H.[d] ,
Hyungtae C.[a] , Junghwan K.[a,*]

[a] *Green Materials and Processes R&D Group, Korea Institute of Industrial Technology, Ulsan 44413, Republic of Korea*
[b] *Department of Chemical Engineering, Konkuk University, Seoul 05029, Republic of Korea*
[c] *Department of Chemical and Biomolecular Engineering, Yonsei University, Seoul 03722, Republic of Korea*
[d] *Research & Development Center, GS Caltex Corporation, Daejeon 34122, Republic of Korea*

kjh31@kitech.re.kr

Abstract

Recently, polypropylene composites (PPCs) are in the spotlight because of their versatilities in composite industries. Properties of PPCs are determined by numerous physical property values (PPV), among which heat deflection temperature (HDT), polymer's resistance to distortion, is a key indicator. However, enormous trial and error is required to produce PPCs with desired PPV because there is no theoretical equations between material composition and PPV. Hence, to reduce the cost and time of finding material composition to meet the desired PPV, we proposed a machine learning-based PPV prediction model. However, some categorical data which can have an influence on the prediction model performance are included in the dataset, because some of data were from repeated experiments. Therefore, algorithm case study (Multiple linear regression (MLR), XGBoost, and CatBoost) was conducted to develop the optimal HDT prediction model which could process the normal data as well as the categorical data. The performances of each prediction model were evaluated with R^2 and RMSE. As a result, the CatBoost-based HDT prediction model was proposed as the optimal model to solve the trial and error problem.

Keywords: PP composites; Categorical data; Machine learning; Catboost

1. Introduction

In recent years, polypropylene composites (PPCs) have been highlighted owing to their versatility in composite industries. PPCs exhibit excellent physical properties, such as high strength, light weight, and high impact resistance. These excellent physical property values are afforded by the addition of additives that can improve the physical properties of polypropylene (PP). For example, fillers such as talc are added to improve the rigidity and toughness of PP, whereas rubber is added to improve its flexibility and ductility. Hence, because the physical properties depend on the type and composition of PP and the additives, i.e., the "recipe," before synthesizing PPCs, the appropriate recipe should be selected based on the target physical property values.

Among the various physical properties of PPC products, specific heat deflection temperature (HDT) is one of the key indicators in the design of PPC products. HDT provides an indication of the temperature at which materials begin to soften when exposed to a fixed load at elevated temperatures. Hence, HDT is an important physical property in PPC applications because it allows engineers to determine the temperature limit above which the material is not appropriate for a structure. Despite the importance of the HDT, achieving the required HDT remains challenging. The required HDT for applications is achieved via numerous trials and errors because the HDT cannot be calculated using the recipe before synthesizing the PPC specimen and testing the HDT. The numerous experiments revealed time-and cost-consuming problems that should be solved to ensure the efficiency of the PPC development process.

As an alternative to the trial-and-error approach, machine learning (ML) has been proposed as it can reduce the number of trials and errors based on the use of a data-driven model that can predict specific values. The data-driven model extracts the relationship between the input and output data, analyzes the relationship, and predicts the output data using the input data. By predicting the output values, the number of trials and errors can be reduced as additional experiments are not required. However, one of the most significant problems in ML is that the prediction performance of a data-driven model depends on the data quality. For instance, if some data have the same input values with different output values in the dataset, then the predictive performance of the model will be low because of overfitting; this is because most ML algorithms replace different output values with the mean value in the categorical data. Among many ML-based algorithms, the CatBoost algorithm was developed to manage categorical data in data-driven modeling. Unlike other regression algorithms, CatBoost can manage categorical data without overfitting by considering various values instead of replacing the output values with the mean value.

Herein, we propose a CatBoost-based model for HDT prediction to reduce the number of trials and errors in the PPC development process and to solve overfitting by categorical data. First, we discovered that some categorical data existed in the dataset because the HDT values depended on the experimental environment, such as temperature, person, machine, and humidity. The categorical data were extracted by comparing them with a dimensionless number "*A*," which we defined. Second, three data-driven models were developed using multiple linear regression (MLR), XGBoost, and CatBoost, separately, to compare their predictive performances. Finally, their R^2 and RMSE were compared to identify the best data-driven model for HDT prediction.

2. Method

2.1. Categorical data treatment

Because some categorical data existed in the recipe dataset, a method was suggested to analyze them. The method comprised two steps: First, "same recipes" which imply the categorical data in a recipe dataset, are defined; second, a dimensionless number, denoted as "*A*," is calculated to obtain the distribution of the same recipes. The same recipes were detected in two steps, i.e., encoding and detecting, as shown in Figure 1. Because the recipes contain information regarding the materials and their weight percentages, it was challenging to count the number of same recipes. Therefore, encoding to assign codes to the recipes was performed to count the number of same recipes. In this step, all of the recipes were assigned codes that contained information regarding the weight percentage of the materials. By comparing the codes, the same recipes were detected, as shown in Figure 1.

	P001	P002	...	F001	F002	...	R001	R002	...	OTH1	OTH2	...	OTH9
Recipe A	0	70	...	15	0	...	15	0	...	0	0	...	0
Recipe B	0	30	...	35	0	...	35	0	...	0	0	...	0
Recipe C	5	15	...	15	15	...	20	10	...	5	0	...	15
Recipe D	5	15	...	15	15	...	20	10	...	5	0	...	15
Recipe C	5	5	...	30	10	...	15	15	...	10	0	...	10

Encoding for counting same recipes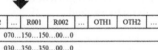

	P001	P002	...	F001	F002	...	R001	R002	...	OTH1	OTH2	...	OTH9
Recipe A						070...150...150...00...0							
Recipe B						030...350...350...00...0							
Recipe C						515...1515...2010...50...15							
Recipe D						515...1515...2010...50...15							
Recipe C						55...3010...1515...100...10							

	P001	P002	...	F001	F002	...	R001	R002	...	OTH1	OTH2	...	OTH9
Recipe A						070...150...150...00...0							
Recipe B						030...350...350...00...0							
Recipe C						515...1515...2010...50...15							
Recipe D						515...1515...2010...50...15							
Recipe C						55...3010...1515...100...10							

} Same recipe

Figure 1 Procedures to detect and count same recipes in dataset

After detecting the same recipes, the HDT distribution of the same recipes was obtained by calculating the dimensionless number "*A*" using Eq. (1). Subsequently, "*A*" was used to compare the differences in the HDT for the same recipe dataset. Using the minimum HDT in the same recipe as the denominator and the HDT of a recipe as the numerator, the differences in the same recipes can be obtained, as listed in Table 1.

$$A = \frac{HDT \ of \ the \ recipe}{The \ minimum \ HDT \ of \ same \ recipes} \tag{1}$$

Table 1 Example of calculating A from the same recipe

	Materials (wt%)				HDT (℃)	A
	P006	R013	OTH1	OTH7		
Recipe 369	87.977	9.775	1.955	0.293	79.5	1.174 (79.5/67.7)
Recipe 378	87.977	9.775	1.955	0.293	80.4	1.188 (80.4/67.7)
Recipe 617	87.977	9.775	1.955	0.293	67.7	1.000 (67.7/67.7)
Recipe 679	87.977	9.775	1.955	0.293	79.2	1.170 (79.2/67.7)

2.2. Catboost

Unlike other regression algorithms, because CatBoost uses ordered encoding instead of mean encoding, it can solve overfitting when a categorical dataset are used for data-driven modeling. The two encoding methods for categorical data are presented in Figure 2. The regression algorithms using mean encoding replace all categorical data with the mean value of the data. Overfitting is incurred in this process because the mean value is proposed as the criterion when the loss of the regression is calculated in every iteration. By contrast, CatBoost, which uses ordered encoding, proposes various mean values that are calculated using random samples in a categorical dataset. Using different loss criteria in each iteration, regression can be generalized.

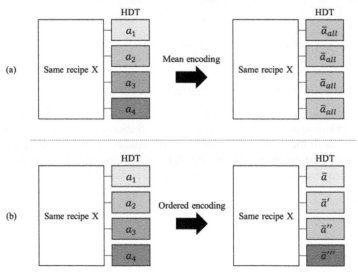

Figure 2 Procedure to process categorical data: (a) mean encoding and (b) ordered target encoding

3. Results and discussions

3.1. Dimensionless number "A" for categorical data treatment

By encoding to count the number of same recipes, 199 recipes among 993 recipes were detected. Because each recipe belonged to a different categorical data group and the groups had different HDT ranges, a criterion to normalize the categorical values was required to visualize the differences in the HDT. Therefore, a dimensionless number "A" was calculated in this study. The calculation results are shown in Figure 3. As shown, the minimum and maximum A values were from 1 to 2 which implies that some recipes have twice as large HDT values in the same recipe groups at the maximum. In the same recipe groups, 59 recipes were randomly selected and then used as the test dataset to compare the predictive performances of the three ML algorithms in Section 3.2. The remaining recipes were segregated for data-driven modeling, and the ratio of the segregated data is shown in Table 2.

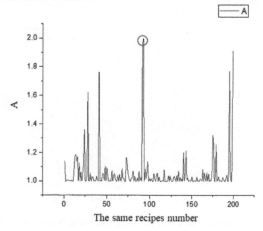

Figure 3 Calculation result of A

Table 2 Ratio of data segregated for data-driven modelling

	Number of recipes	Number of the same recipes
Train dataset	638	119
Validation dataset	152	21
Test dataset	203	59
Total	993	199

3.2. Comparison of the three algorithms

In this study, three different algorithms (MLR, XGBoost, and CatBoost) were used to perform data-driven modeling to predict HDT values. MLR is a simple linear regression algorithm that uses mean encoding. XGBoost and CatBoost are classification and regression tree (CART)-based nonlinear regression algorithms based on the boosting method. However, CatBoost uses ordered encoding, whereas XGBoost uses mean encoding.

The predictive performances of the three models are shown in the evaluation results in Figure 5. The R^2 and RMSE were calculated to compare the model performance quantitatively. We discovered that the MLR-based model yielded the lowest R^2 (0.8162) and the highest RMSE (9.7934) when all the recipes in the test dataset were tested. However, the XGBoost-based model indicated a higher R^2 (0.8578) and a lower RMSE (8.7007) than the MLR-based model. Meanwhile, the CatBoost-based model yielded the highest R^2 (0.8965) and lowest RMSE (7.3477). Moreover, the R^2 and RMSE were calculated for only 59 same recipes. The MLR-based model indicated the lowest R^2

(0.8256) and the highest RMSE (7.8899) for 59 same recipes. By contrast, the XGBoost-based model indicated a higher R^2 (0.9690) and a lower RMSE (2.6105) than the MLR-based model. Meanwhile, the CatBoost-based model indicated the highest R^2 (0.9801) and lowest RMSE (2.6105). This shows that the CART-based nonlinear regression algorithms are more appropriate for the data-driven modeling of HDT prediction than the linear regression algorithm, and that the CatBoost-based model can predict the HDT with better performance than the XGBoost-based model even when the dataset includes some categorical data with a high value of A.

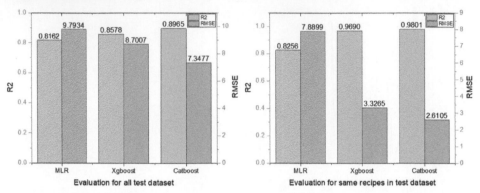

Figure 4 Evaluation result for all recipes in test dataset (left), and for only the same recipes in test dataset (right)

4. Conclusion and future work

In this study, we proposed a CatBoost-based HDT predictive model to reduce the number of trials and errors in the PPC development process. In addition, to detect categorical data, a new approach was proposed, where a recipe is encoded to a code, and the difference in the HDT is calculated for the same recipe group using a dimensionless number "A." The results indicated that although the HDT in the same recipes was different, the CatBoost-based predictive model performed better than the MLR and XGBoost-based models. Therefore, if the proposed model is applied to the PPC development process, then the number of trials and errors can be reduced.

In future studies, we will use Shapley additive explanations to further explain the model as well as extend this study to other properties of the PPC.

Acknowledgements

This study has been conducted with the support of the Korea Institute of Industrial Technology as "Development of AI Platform Technology for Smart Chemical Process (kitech JH-21-0005)" and "Development of Global Optimization System for Energy Process (kitech EM-21-0022)"

References

K. Harutun, Impact Behavior of Polypropylene, Its Blends and Composites. Handb. Polypropyl. Polypropyl. Compos. Revis. Expand., CRC Press; 2003, p. 150–213

L. Prokhorenkova, G. Gusev, A. Vorobev, A. Dorogush, A. Gulin, Catboost: Unbiased boosting with categorical features, Adv Neural Inf Process Syst 2018, 6638–43

Proceedings of the 14th International Symposium on Process Systems Engineering – PSE 2021+
June 19-23, 2022, Kyoto, Japan © 2022 Elsevier B.V. All rights reserved.
http://dx.doi.org/10.1016/B978-0-323-85159-6.50301-8

A New Machine Learning Framework for Efficient MOF Discovery: Application to Hydrogen Storage

Teng Zhou[a,b,*], Zihao Wang[b], Kai Sundmacher[a,b]

[a] *Process Systems Engineering, Otto-von-Guericke University Magdeburg, Universitätsplatz 2, D-39106 Magdeburg, Germany*

[b] *Process Systems Engineering, Max Planck Institute for Dynamics of Complex Technical Systems, Sandtorstr. 1, D-39106 Magdeburg, Germany*

zhout@mpi-magdeburg.mpg.de

Abstract

Metal-organic frameworks (MOFs) are recognized as promising materials for gas storage and separation due to their structural diversity, high porosity, and tailorable functionality. Considering the large number of possible MOFs, an integrated machine learning framework is proposed to discover promising candidates with desirable adsorption properties. The framework consists of structure decomposition, feature integration, and predictive modelling. Unlike most of the previous studies employing solely structural or geometric descriptors, our method integrates both structural and chemical features of MOFs for adsorption property prediction using the graph convolutional network (GCN) and feed-forward neural network (FNN) approaches. The machine learning framework is first introduced and then applied to hydrogen storage. Promising MOF candidates exhibiting respectable hydrogen storage capacities are successfully identified, which potentially outperform the existing porous materials for hydrogen storage.

Keywords: machine learning, metal-organic framework discovery, graph convolutional network, hydrogen storage

1. Introduction

Metal-organic frameworks (MOFs) are an important type of porous materials with large structural diversity, high porosity, and tailorable functionality. In the past two decades, MOFs have been attracting wide attentions in many applications, especially gas storage and separation (Gándara et al., 2014; Cui et al., 2016; Zhou et al., 2020; Zhang et al., 2021; Chen et al., 2018). Through the combinations of numerous metal nodes and organic linkers under specific topologies, we can in principle synthesize an infinite number of different MOFs. This makes MOF discovery via experimental trail-and-error extremely challenging. High-throughput screening techniques in tandem with molecular simulations or ab-initio calculations are being used to calculate properties of MOFs. Although the grand canonical Monte Carlo (GCMC) simulation has shown remarkable accuracy (Moghadam et al., 2018; Chung et al. 2016) for MOF adsorption property prediction, it is computationally inefficient for finding the best MOFs from a large number of candidates.

Over the past few years, many researchers have applied various supervised machine learning (ML) methods to predict MOF properties (Chong et al., 2020; Altintas et al., 2021; Zhou et al., 2019) from available data. With the established ML models, one can perform in-silico predictions on a large number of new MOFs and quickly find the best candidates. So far, most of the previous ML works have employed only geometric descriptors, e.g., void fraction, surface area and pore diameters, to correlate MOF adsorption properties (Fernandez et al., 2013; Shi et al., 2020; Yuan et al., 2021; Thornton et al., 2017). Although geometric descriptors largely affect the adsorption performance, chemical diversity of building blocks can also play a crucial role (Moosavi et al., 2020). Unfortunately, given the massive chemical descriptors (e.g., number of atom, atomic charge and dipole moment), applying them for ML requires a lot of domain knowledge and labour for descriptor (or feature) generation and optimal selection. In contrast, representation learning allows the machine to automatically learn important features directly from material structures without any human input.

In this contribution, we propose an end-to-end ML framework where the MOF structure is imported as input, chemical and structural features are learned automatically, and the corresponding adsorption property is finally predicted in one step. The framework is trained with a relatively small amount of GCMC-derived data and the resulting ML model can then be used for a fast MOF discovery from a much larger set of candidates. The proposed method is applied to hydrogen storage aiming to find MOFs with superior volumetric storage capacity.

2. Methods

Prior to GCMC simulations and ML modelling, MOF database pre-treatment is conducted. We select the hypothetical MOF (hMOF) database (Wilmer et al., 2012) as our basis. First, MOFid and MOFkey identifiers are obtained for each hMOF in the database based on their CIF files (Bucior et al., 2019). Second, the metal nodes, organic linkers, and underlying topological networks are extracted from the identifiers. Finally, data cleaning is performed to remove those MOFs sharing duplicate and incomplete identifiers, with invalid organic linkers, and consisting of more than three types of linkers. This finally results in 9156 unique MOFs.

Figure 1 summarizes the integrated ML framework. As indicated, features of the organic molecule are automatically generated by the graph convolutional network (GCN) approach. Initially, each atom in the molecule is assigned with a fixed-length vector of randomized features. Afterwards, atoms learn their representations by aggregating features from their neighbors (i.e., connected atoms). With two GCN layers, two unconnected atoms can learn features from each other if both of them are attached to the same atom. In this way, with multiple GCN layers, all the atomic features are iteratively updated several times to capture the local and global information about the whole molecular structure. Finally, a global pooling is performed on the graph to generate the overall molecular features by aggregating all the atomic features with a certain mathematical rule. After obtaining the organic linker's features, a feedforward neural network (FNN)-based ML model can be finally constructed to predict MOF adsorption uptakes using all the chemical and structural features as inputs. The chemical features include directly embedded metal node and GCN-based organic linker's features, while the structural features incorporate both embedded topology and five key MOF geometric descriptors. Notably, the initial atomic features, GCN layers, and the FNN model are optimized or trained simultaneously to minimize the overall prediction error of the model. After the ML model is built and successfully validated, it can replace the traditional molecular simulation for providing in-silico predictions on thousands or millions of new MOFs and quickly find the best candidates possessing the most desirable adsorption properties.

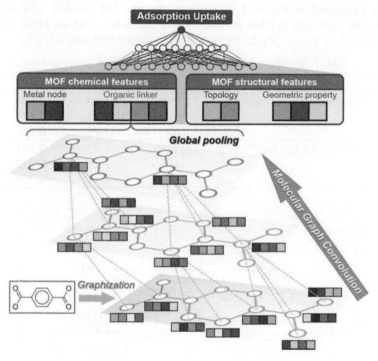

Figure 1. Schematic diagram of the proposed ML framework

3. Case Study

Hydrogen is an appealing energy carrier due to its high gravimetric energy density and low environmental impact. It is industrially stored at around 700 bar, requiring substantial amount of energy and special safety considerations (Gómez-Gualdrón et al., 2016). Recently, MOFs have emerged as promising materials for adsorption-based hydrogen storage at a much moderate pressure around 100 bar. The objective of this work is to use the ML method to find potential MOF candidates that possess high volumetric hydrogen storage capacity. To do so, GCMC simulations are first carried out to compute the hydrogen uptakes at 100 bar/77 K and 2 bar/77 K for the 9156 MOFs. Based on the obtained data, two different ML models are established according to the framework in Figure 1 to predict the two uptakes separately. Using the ML models, the hydrogen storage capacity (maximal amount of hydrogen that can be stored/released for one charging/discharging process), is directly calculated as the difference between the two uptakes.

In order to obtain reliable ML models, the entire dataset is divided into training, validation and test sets accounting for 80%, 10% and 10% of the 9156 data points. These three sets are used for model training, hyper-parameter optimization and early stopping, and model assessment, respectively. Considering all possible hyper-parameter combinations, the optimal ML configuration is first determined by the grid search method using the validation set. After determining the best hyper-parameters, the ML models are trained with the training and assessed with the test data. Model performance is evaluated with mean absolute error (MAE) and coefficient of determination (R^2), as summarized in Table 1. The parity plot and error distribution of the obtained ML models is visualized in Figure 2. In general, the two ML models achieve accurate predictions for hydrogen uptakes at both 100 bar and 2 bar, with an MAE of 1.04 g/L (1.25 g/L) and 1.03 g/L (1.29 g/L) for the training set (test set), respectively.

Table 1. Model performance in the prediction of hydrogen uptakes

Prediction target	Dataset	MAE (g/L)	R^2
H_2 uptake at 100 bar/77 K	Training	1.04	0.984
	Validation	1.24	0.975
	Test	1.25	0.976
H_2 uptake at 2 bar/77 K	Training	1.03	0.961
	Validation	1.29	0.927
	Test	1.29	0.928

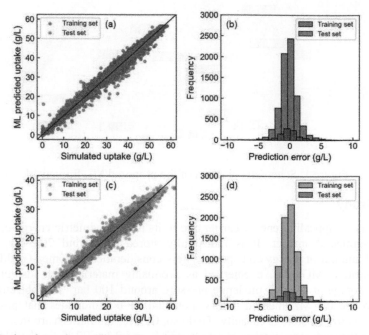

Figure 2. Parity plot and error distribution of the simulated and ML predicted hydrogen uptakes at 100 bar/77 K (a, b) and 2 bar/77 K (c, d)

The ultimate goal is to employ the ML models to discover potential MOFs for efficient hydrogen storage. For this purpose, we collect another much larger MOF database, consisting of 21,384 new MOFs. The two established ML models are then used to predict hydrogen uptakes for these new MOFs, based on which the best candidates possessing the highest storage capacities are screened out. Top 100 candidates are identified, whose storage capacities are between 45.44 g/L and 47.20 g/L. Verification on these top 100 MOFs by GCMC simulations leads to similar capacities. The best two MOFs showing the highest GCMC-derived capacities are illustrated in Figure 3. As indicated, their computed hydrogen storage capacities are higher than the best experimentally verified MOF that shows a capacity of ~ 42 g/L (Ahmed et al., 2017). This proves the great potential of the identified two MOFs for practical applications. As indicated in Figure 3, these two MOFs have similar structures. For instance, the same topology *pcu* is shared and three types of similar organic linkers are found as well. This provides some useful insights for optimal MOF synthesis.

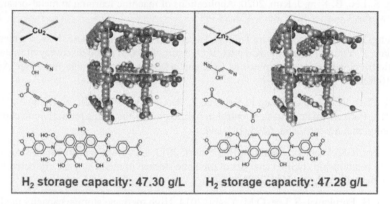

Figure 3. Top two MOF candidates identified for hydrogen storage

4. Conclusions

An integrated ML framework is proposed for the prediction of gas adsorption capacities using both chemical information and structural characteristics of MOFs. The method has been successfully applied to a hydrogen storage case study. High-potential MOFs are successfully identified by large-scale database screening using the obtained ML models. Their superior performances have been validated by GCMC simulations and should be further verified by experiment. It is anticipated that the approach can be used for other applications, e.g., CO_2 capture and methane storage, with a variety of porous materials including zeolites, porous polymers, and covalent-organic frameworks.

Unlike other works using engineered features to train ML models, our approach automatically learns useful features most relevant to adsorption properties. This is efficient, reliable, and can help to discover promising materials. However, it is difficult to draw useful insights on how MOF chemical and structural characteristics influence their performance. This knowledge can be acquired by using the so-called interpretable ML technique, which certainly deserves future studies. Besides, our MOF discovery is achieved by large-scale screening on existing databases. An alternative and probably more efficient way for direct targeting of optimal MOFs is to formulate and solve an optimization-based reverse design problem based on the ML model.

References

A. Ahmed, Y. Liu, J. Purewal, L.D. Tran, A.G. Wong-Foy, M. Veenstra, A.J. Matzger, D.J. Siegel, 2017, Balancing gravimetric and volumetric hydrogen density in MOFs, Energy Environ. Sci., 10, 11, 2459-2471.

C. Altintas, O.F. Altundal, S. Keskin, R. Yildirim, 2021, Machine learning meets with metal organic frameworks for gas storage and separation, J. Chem. Inf. Model., 61, 5, 2131-2146.

B.J. Bucior, A.S. Rosen, M. Haranczyk, Z. Yao, M.E. Ziebel, O.K. Farha, J.T. Hupp, J.I. Siepmann, A. Aspuru-Guzik, R.Q. Snurr, 2019, Identification schemes for metal-organic frameworks to enable rapid search and cheminformatics analysis, Cryst. Growth Des., 19, 11, 6682-6697.

Y. Chen, Z. Qiao, H. Wu, D. Lv, R. Shi, Q. Xia, J. Zhou, Z. Li, 2018, An ethane-trapping MOF PCN-250 for highly selective adsorption of ethane over ethylene, Chem. Eng. Sci., 175, 110-117.

S. Chong, S. Lee, B. Kim, J. Kim, 2020, Applications of machine learning in metal-organic frameworks, Coord. Chem. Rev., 423, 213487.

Y.G. Chung, D.A. Gómez-Gualdrón, P. Li, K.T. Leperi, P. Deria, H. Zhang, N.A. Vermeulen, J.F. Stoddart, F. You, J.T. Hupp, O.K. Farha, R.Q. Snurr, 2016, In silico discovery of metal-organic frameworks for precombustion CO2 capture using a genetic algorithm, Sci. Adv., 2, 10, e1600909.

X. Cui, K. Chen, H. Xing, Q. Yang, R. Krishna, Z. Bao, H. Wu, W. Zhou, X. Dong, Y. Han, B. Li, 2016, Pore chemistry and size control in hybrid porous materials for acetylene capture from ethylene, Science, 353, 6295, 141-144.

M. Fernandez, T.K. Woo, C.E. Wilmer, R.Q. Snurr, 2013, Large-scale quantitative structure-property relationship (QSPR) analysis of methane storage in metal-organic frameworks, J. Phys. Chem. C, 117, 15, 7681-7689.

F. Gándara, H. Furukawa, S. Lee, O.M. Yaghi, 2014, High methane storage capacity in aluminum metal-organic frameworks, J. Am. Chem. Soc., 136, 14, 5271-5274.

D.A. Gómez-Gualdrón, Y.J. Colón, X. Zhang, T.C. Wang, Y.S. Chen, J.T. Hupp, T. Yildirim, O.K. Farha, J. Zhang, R.Q. Snurr, 2016, Evaluating topologically diverse metal-organic frameworks for cryo-adsorbed hydrogen storage, Energy Environ. Sci., 9, 10, 3279-3289.

P.Z. Moghadam, T. Islamoglu, S. Goswami, J. Exley, M. Fantham, C.F. Kaminski, R.Q. Snurr, O.K. Farha, D. Fairen-Jimenez, 2018, Computer-aided discovery of a metal-organic framework with superior oxygen uptake, Nat. Commun., 9, 1, 1-8.

S.M. Moosavi, A. Nandy, K.M. Jablonka, D. Ongari, J.P. Janet, P.G. Boyd, Y. Lee, B. Smit, H.J. Kulik, 2020, Understanding the diversity of the metal-organic framework ecosystem, Nat. Commun., 11, 1, 1-10.

Z. Shi, H. Liang, W. Yang, J. Liu, Z. Liu, Z. Qiao, 2020, Machine learning and in silico discovery of metal-organic frameworks: Methanol as a working fluid in adsorption-driven heat pumps and chillers, Chem. Eng. Sci., 214, 115430.

A.W. Thornton, C.M. Simon, J. Kim, O. Kwon, K.S. Deeg, K. Konstas, S.J. Pas., M.R. Hill, D.A. Winkler, M. Haranczyk, B. Smit, 2017, Materials genome in action: identifying the performance limits of physical hydrogen storage, Chem. Mater., 29, 7, 2844-2854.

C.E. Wilmer, M. Leaf, C.Y. Lee, O.K. Farha, B.G. Hauser, J.T. Hupp, R.Q. Snurr, 2012, Large-scale screening of hypothetical metal-organic frameworks, Nat. Chem., 4, 2, 83-89.

X. Yuan, X. Deng, C. Cai, Z. Shi, H. Liang, S. Li, Z. Qiao, 2021, Machine learning and high-throughput computational screening of hydrophobic metal-organic frameworks for capture of formaldehyde from air, Green Energy Environ., 6, 5, 759-770.

T. Zhou, Z. Song, K. Sundmacher, 2019, Big data creates new opportunities for materials research: A review on methods and applications of machine learning for materials design, Engineering, 5, 6, 1017-1026.

Y. Zhou, T. Zhou, K. Sundmacher, 2020, In silico screening of metal-organic frameworks for acetylene/ethylene separation, In: S. Pierucci, F. Manenti, G.L. Bozzano, D. Manca, eds., Comput.-Aided Chem. Eng., Elsevier, 48, 895-900.

X. Zhang, T. Zhou, K. Sundmacher, 2022, Integrated metal-organic framework and pressure/vacuum swing adsorption process design: Descriptor optimization, AIChE J., e17524.

Proceedings of the 14th International Symposium on Process Systems Engineering – PSE 2021+
June 19-23, 2022, Kyoto, Japan © 2022 Elsevier B.V. All rights reserved.
http://dx.doi.org/10.1016/B978-0-323-85159-6.50302-X

Data-driven Modeling for Magma Density in the Continuous Crystallization Process

Nahyeon An[a, b], Hyukwon Kwon[a] ,Hyungtae Cho[a] and Junghwan Kim[a*]

[a]*Green Materials and Processes R&D Group, Korea Institute of Industrial Technology, Ulsan 44413, Republic of KOREA*
[b]*Department of Chemical and Biomolecular Engineering, Yonsei University, Seoul 03722, Republic of KOREA*
kjh31@kitech.re.kr

Abstract

Crystallization processes have been widely used for separation in many fields, such as food, pharmaceuticals, and chemicals. The crystallization process is a highly nonlinear system, owing to complex crystallization dynamics; therefore, it is difficult to model the process to control the crystal product quality. In this study, a data-driven neural network was implemented to predict the magma density of the continuous crystallization process that produces maleic acid crystals from the mother liquor. Three neural network algorithms, namely deep neural network, long short-term memory, and gated recurrent unit (GRU), were applied for magma density prediction. Process variables, such as the feed flow rate, pressure, and steam flow rate were defined as input, while magma density, the most important control variable in continuous crystallization, was defined as an output variable. The grid search method was used to select suitable hyperparameters for each method, and the predictive accuracy of the models was compared with the root mean square error (RMSE). The GRU-based model afforded the best prediction accuracy among the applied models, with an RMSE of 2.04. Consequently, the developed predictive model can be used as a proper control strategy.

Keywords: Crystallization predictive model; product density prediction; artificial neural network; machine learning and big data

1. Introduction

The crystallization process is used in many fields to produce high-purity products (Velásco-Mejía et al., 2016). Indeed, the quality of the produced crystals has a significant influence on the efficient operation of the downstream process. Therefore, it is crucial to maintain a high and stable quality of the crystallization products. However, the crystallization process consists of several complex mechanisms, such as nucleation, crystal growth, and agglomeration. Moreover, it is difficult to solve the model equation because the mechanisms consist of nonlinear algebraic and partial differential equations. Thus, reliable modeling of the crystallization process remains challenging (Griffin et al., 2016).

To overcome the limitations of equation-based process modeling, the application of artificial neural networks (ANNs) to crystallization process modeling has been studied. Thus, ANNs have been used to model the nonlinear relationship between the input and output variables with high performance. Meng et al. (2021) attempted to monitor the process using a hybrid soft sensor model capable of predicting the mother liquor purity, supersaturation, particle size distribution, and crystal content, which are difficult to

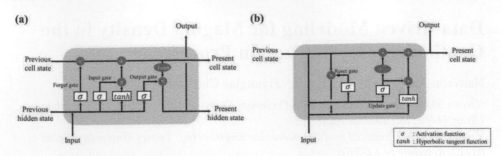

Figure 1 Cell structure of (a) LSTM and (b) GRU

measure in continuous cane sugar crystallization. On the other hand, Manee et al. (2019) developed a model to measure the particle size distribution (PSD) in batch crystallizers through deep learning. Furthermore, in many studies related to the crystallization process, models have been developed to predict the mother liquor purity, particle size, and distribution.

Although the size of each particle is important in the batch crystallization process, it is more crucial to produce products with stable magma density in a continuous process for high productivity. Therefore, in continuous crystallization, the magma density, rather than the PSD, is used as the primary process control variable. However, it is difficult to control the magma density because of the nonlinearity and instability of the process. To solve this problem, this study developed a dynamic prediction model based on an artificial neural network (ANN) to predict the magma density in the continuous crystallization process. Magma density predictive models were developed using three algorithms: deep neural network (DNN), long short-term memory (LSTM), and gated recurrent unit (GRU), wherein LSTM and GRU are recurrent neural network algorithms that reflect the time series of the process. The most suitable model for the continuous crystallization process was selected by comparing the accuracies of the three algorithms for magma density prediction. In addition, the developed data-driven model was applied to control the steam flow rate, and its applicability was verified. Thus, the off-spec of the process was significantly reduced, and the crystals were produced more stably.

2. Preliminaries

2.1. Deep neural network

A DNN is an ANN with multiple layers consists of input, output, and hidden layers. DNNs have been applied to forecast many problems with relatively high performance. Each layer is given the output from the previous layer and transfers it to the next layer. The hidden layers are trained by a backpropagation stochastic gradient descent. The model accuracy highly depends on the algorithms, hyperparameters, the property of the data, and the learning scheme. The outputs (**h**) of the first, hidden, and output layers are expressed as

$$\mathbf{h}_i = \sigma(\mathbf{W}_i^T \mathbf{x} + \mathbf{b}_i) \tag{1}$$

$$\mathbf{h}_n = \sigma(\mathbf{W}_n^T \mathbf{h}_{n-1} + \mathbf{b}_n) \tag{2}$$

$$\hat{\mathbf{y}} = \mathbf{W}_o^T \mathbf{h}_N + \mathbf{b}_o \tag{3}$$

where $\mathbf{W}$ and $\mathbf{b}$ represent the weight matrix and bias vector of the nth hidden layer, respectively. For the input layer, the input variable vector ($\mathbf{x}$) is used instead of $\mathbf{h}_{n-1}$, while for the output layer, the predicted values of the output layer ($\hat{\mathbf{y}}$) are used instead of $\mathbf{h}_n$.

2.2. Recurrent neural network : LSTM and GRU

An RNN stores the past data and forwards the information to calculate the output of the next step. Unlike the DNN, temporal dynamics can be considered, which is commonly used for time series prediction. Two RNNs, LSTM and GRU, were developed to solve the gradient vanishing problem of standard RNNs. Figure 1 shows the cell structure of the developed LSTM and GRU, comprising three and two gates, respectively, for long-term memory to be efficiently stored. Both algorithms are described in detail in the literature (Hochreiter and Urgen Schmidhuber, 1997).

3. Development of the magma density prediction model

3.1. Process and data description

The target process of this study (Figure 2) is a continuous crystallization process in which an approximately 60% maleic acid–water mixture is concentrated to 78% maleic acid, and 30% of the feed flow rate is crystallized (Ulsan, Republic of Korea). A forced circulation crystallizer was used for crystallization, in which a feed enters the equipment after being heated by 45 °C in a heat exchanger. The crystallizer was vacuumed to 50 mbar using a vacuum ejector. The maleic acid crystal was discharged with the mother liquor as magma, where the density of magma was used as a control variable in the process. For stable product production, the magma density was maintained constant at 1,330 kg/m^3 as the set point.

In this study, 10 process variables collected every hour from January to June 2020 were used to develop the data-driven model. Because missing values and outliers in the data

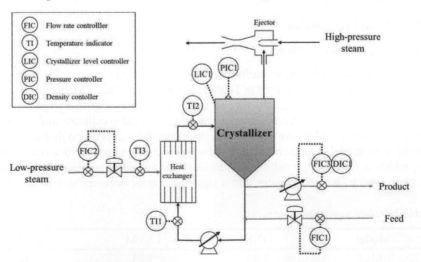

Figure 2 Target process description

Table 1 Parameters of each network

	DNN	LSTM	GRU
Hidden layers	2	1	3
Hidden neurons	20	100	90
Batch size		64	
Early stopping patience		30	
Loss function		Mean square error	

due to instrumental failure adversely affect predictive model training, they were removed and used. The model was developed and evaluated using 80% of 3,300 data as learning data and 20% as test data.

3.2. Model structure

Three algorithms, DNN, LSTM, and GRU, were used to develop a magma density prediction model for the continuous crystallization process. The input variables used in each model were the feed flow rate, low-pressure steam flow rate, pressure, and temperature instruments measured for process monitoring to control the density. The predicted performance was compared using RMSE and calculated using Equation (4). The RMSE, defined as the average absolute ratio error, indicates a higher performance as it approaches 0. The structures of the three predictive models were optimized using grid research, and each model structure is listed in Table 1. In addition, the model with the best performance among the three was derived by analyzing the prediction error distribution.

$$RMSE = \sqrt{\frac{1}{N}\sum_k (y_k - \hat{y}_k)^2} \tag{4}$$

where y_k and $\hat{y}_k$ are the actual and predicted data, respectively, and N is the total number of data samples.

4. Results and discussion

4.1. Density prediction accuracy

The magma density prediction performance in the continuous crystallization process using the three models is presented in Table 2. Comparison of the RMSEs reveals that the difference in accuracy between the models is not significant. Figure 3 shows the results predicted by each model, wherein the red line represents the conditions under which the predicted and actual values are equal. Thus, the closer the points are to this line, the more accurate is the model. A point marked in red refers to a data sample with a relatively large error, which is far from the red line. In the three models, the positions marked with red dots were similar, which indicates that a significant error occurred due to an abnormal

Table 2 Prediction accuracy

Model	DNN	LSTM	GRU
RMSE	2.02	2.23	2.04

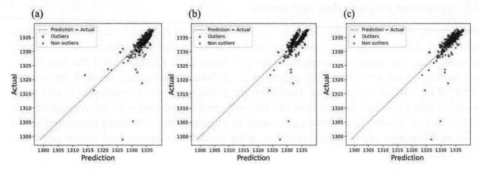

Figure 3 Predicted vs. actual value plots of (a) DNN, (b) LSTM, and (c) GRU models

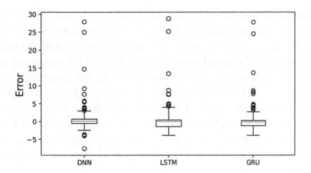

Figure 4 Error distribution boxplot

state of the process. As shown in Figure 3, the DNN and GRU models afforded better predictions than did the LSTM model. Additionally, the box plots in Figure 4 show the distribution of absolute errors in each model, wherein the model accuracy increases as the box at the 0 point becomes narrower. The plot reveals that the absolute error of the GRU is distributed in a narrower area than that of the DNN. Therefore, based on the above analysis, we concluded that the most accurate model for magma density prediction was the GRU model.

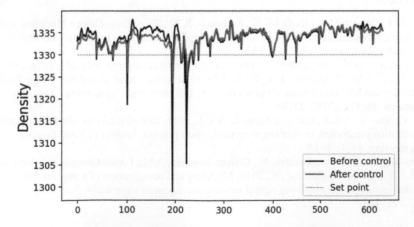

Figure 5 Magma density of before and after steam flow rate control

4.2. Application for steam flow rate control

The density prediction model developed in this study was applied to the steam flow rate control of the target process and compared with the actual steam data. In the current process state, the steam flow rate was changed from -30% to +30%, and the steam flow rate satisfying the set point of the process was derived. In the actual process, an average of 897.2 kg/h of steam was used. However, the prediction model developed in this study revealed that a product that satisfies the setpoint can be produced with an average of only 771.9 kg/h of steam. In addition, as shown in Figure 5, the off-specs generated in the process were also significantly reduced, even though the amount of steam used was decreased by 14%.

5. Conclusion

In this study, we developed an artificial neural network-based dynamic prediction model that can predict the magma density of a continuous crystallization process, which we subsequently applied to control steam usage. We determined that the most suitable model in this process was an RNN-based model using GRU, which presented an RMSE of 2.04. Furthermore, by controlling the steam flow rate using the developed model, 14% of the existing steam usage could be reduced. The dynamic prediction model developed in this study exhibits a high performance, but still has room for improvement. Data used for model development cannot reflect real-time information owing to long sampling rate. Therefore, in future research, if a model is developed by reducing the data sampling interval through data interpolation, it can be applied to an actual process with a more reliable prediction.

Acknowledgement

This study has been conducted with the support of the Korea Institute of Industrial Technology as "Development of digital-based energy optimization platform for manufacturing innovation (KITECH IZ-21-0063)" and "Development of Global Optimization System for Energy Process (KITECH IZ-21-0052, IR-21-0029, EM-21-0022)"

References

Griffin, D.J., Grover, M.A., Kawajiri, Y., Rousseau, R.W., 2016, Data-Driven Modeling and Dynamic Programming Applied to Batch Cooling Crystallization. Industrial and Engineering Chemistry Research. 55(5), 1361–1372.

Hochreiter, S., Urgen Schmidhuber, J., 1997, Long Shortterm Memory. Neural Computation.

Manee, V., Zhu, W., Romagnoli, J.A., 2019, A Deep Learning Image-Based Sensor for Real-Time Crystal Size Distribution Characterization. Industrial and Engineering Chemistry Research. 58(51), 23175–23186.

Meng, Y., Yao, T., Yu, S., Qin, J., Zhang, J., Wu, J., 2021, Data-driven modeling for crystal size distribution parameters in cane sugar crystallization process. Journal of Food Process Engineering. 44(4), 1–15.

Velásco-Mejía, A., Vallejo-Becerra, V., Chávez-Ramírez, A.U., Torres-González, J., Reyes-Vidal, Y., Castañeda-Zaldivar, F., 2016, Modeling and optimization of a pharmaceutical crystallization process by using neural networks and genetic algorithms. Powder Technology. 292, 122–128.

Proceedings of the 14th International Symposium on Process Systems Engineering – PSE 2021+
June 19-23, 2022, Kyoto, Japan © 2022 Elsevier B.V. All rights reserved.
http://dx.doi.org/10.1016/B978-0-323-85159-6.50303-1

Gaussian Process Regression Machine Learning Models for Photonic Sintering

Ke Wang[a], Mortaza Saeidi-Javash[b], Minxiang Zeng[b], Zeyu Liu[b], Yanliang Zhang[b], Tengfei Luo[b], Alexander W. Dowling[a*]

[a]*Department of Chemical and Biomolecular Engineering, University of Notre Dame, Notre Dame, IN 46556, USA*
[b]*Department of Aerospace and Mechanical Engineering, University of Notre Dame, Notre Dame, IN 46556, USA*
[*]adowling@nd.edu

Abstract

Novel solid-state thermoelectric (TE) materials have the potential to improve energy efficiency by converting waste heat into electricity. However, the performance of many state-of-the-art TE materials remains inadequate for adoption beyond niche applications. Current efforts to optimize photonic sintering, an important step in additive manufacturing of TE devices, rely on expert-driven trial-and-error search which is often extremely time-consuming and without the guarantee of improvement. Emerging Bayesian optimization frameworks offer a principled approach to intelligentially recommend optimized experimental conditions by balancing exploitation and exploration. In this paper, we develop a Gaussian Process Regression (GPR) machine learning model to predict the thermoelectric power factor of aerosol jet printed n-type $Bi_2Te_{2.7}Se_{0.3}$ TE films. We compare hyperparameter tuning methods and perform retrospective analysis to quantify the predictivity of GPR. Finally, we discuss the challenges and opportunities of adopting Bayesian optimization for photonic sintering and fabrication of high-performance TE devices.

Keywords: Additive Manufacturing; Data Science; Bayesian Optimization; Machine Learning; Gaussian Process Regression

1. Introduction

1.1. Background

Discover functional materials with desired properties is a central goal of material science and engineering; yet materials discovery and optimization is often slow and expensive. For example, out of the 10^{23} possible drug-like molecules, only 10^8 have been synthesized. (Elton et al., 2019) Computer-aided molecular design (CAMD) is frequently used to design new functional material, however, its success is usually limited by the accuracy and efficiency of the physical models. (Austin et al., 2016) Supervised machine learning has demonstrated great promise for predicting the physical properties of material and revolutionizing the design process. (Lookman et al., 2019) For example, Gaussian Process Regression (GPR) and Bayesian optimization have been shown over the past decade to accelerate the design and manufacturing of new functional material. (Wang et al., 2022)

In this paper, we apply GPR to model the thermoelectric power factor of aerosol jet printed n-type $Bi_2Te_{2.7}Se_{0.3}$ TE films under different photonic sintering variables. With the urgent demand for waste energy recovery and wearable electronic devices, developing high-performance TE generators has attracted much attention. For example, TE devices can harvest heat from the automobile exhaust to improve its overall efficiency. (Han et al., 2018) They can also convert heat generated by the human body into energy for the electronic device, e.g., wearable medical monitors. (Jiang et al., 2020) The thermoelectric performance of TE materials is typically evaluated by a dimensionless figure of merit, $zT = \frac{S^2\sigma T}{\kappa}$, which depends on the electrical conductivity (σ), Seebeck coefficient (S), thermal conductivity (κ), and absolute temperature (T). Among these, the power factor, ($S^2\sigma$), is usually of the utmost importance in developing new TE materials platforms.

2. Method

2.1. Decision Variables and Data

Photonic sintering is a well-known technique for fabricating high-performance thin-film TE material. (Yu et al., 2017) For photonic sintering, there are four key process variables to optimize: voltage (x_{i1}), pulse duration (x_{i2}), number of pulses (x_{i3}), and pulse delay (x_{i4}). For each experiment i, photonic sintering is performed at conditions x_i and the power factor (y_i) is measured. Saeidi-Javash et al. (2019) described the experimental procedure used to collect the thermoelectric properties under various photonic sintering conditions in this paper. Figure 1 summarizes the photonic sintering dataset. In Group 0, seven experiments were performed with one pulse, and voltage and pulse duration systematically varied based on expert intuition. The remaining groups, $1 - 4$, are one dimension sensitivity analysis in which voltage (x_{i1}), pulse duration (x_{i2}), and pulse delay (x_{i4}) are fixed in each group and the number of pulses (x_{i3}) is systematically varied. The GPR machine learning models are then trained using these data in this work.

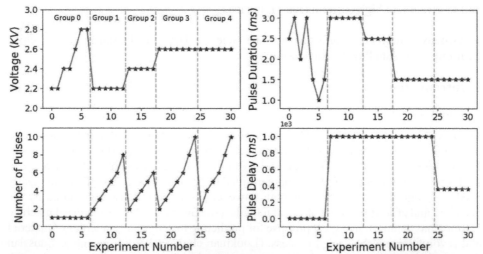

Figure 1: The four photonic sintering optimization variables were systematically varied, based on expert intuition, across 31 experiments which are divided into five groups.

Let $D = \{(x_i, y_i), |x_i \in R^4, y_i \in R, i \in 1,...,31\}$ be a collection of 31 photonic sintering experiments (Figure 1). For convenience, we denote the data $D = (X, y)$ using matrix $X = (x_1,...,x_n)^T$ and vector $y = (y_1,...,y_n)^T$. However, each dimension of x_i, as well as y_i, has different units. To address this, each dimension of D is standardized using the mean (expected value) and standard deviation where x_j is the jth column of X:

$$y \leftarrow \frac{(y - E(y))}{\sqrt{Var(y)}}, \qquad x_j \leftarrow \frac{\left(x_j - E(x_j)\right)}{\sqrt{Var(x_j)}} \qquad (1)$$

2.2. Gaussian Process Regression

Gaussian Processes (GPs) are non-parametric probabilistic models that are well-known to emulate expensive continuous functions, $f(.)$, by interpolating between training data.

$$f \sim GP\left(m(x), k(x, x')\right) \qquad x, x' \in R^p \qquad (2)$$

The output of a GP model is a normally distributed random variable fully specified by the mean function, $m(x) = E[f(x)]$, and the kernel function $k(x, x') = E\left[\left(f(x) - m(x)\right)\left(f(x') - m(x')\right)\right]$. (Rasmussen, 2003) The kernel function determines how the GP model interpolates between the encoded data D. In doing so, the kernel function also specified the uncertainty in GP predictions. $k(.,.)$ contains hyperparameters which often include the length-scales, denoted by $l \in R^p$, for each dimension of the input data X. The smaller the l_j, the more important that corresponding feature (x_j). Training the hyperparameters is often performed by using log marginal likelihood or cross-validation methods. The optimal length-scales l help identify which features are most important. For simplicity, we set $m(x)$ to zero and use Radial Basis Function (k_{RBF}) defined in Eq. (3), where $l = (l_1, l_2, l_3, l_4)^T$.

$$k_{RBF}(x, x') = e^{-\frac{1}{2}\sum_{j=1}^{p}\left(\frac{x_j - x'_j}{l_j}\right)^2} \qquad \theta = l \qquad (3)$$

We define new inputs values X_* with corresponding prediction f_*. Given training data (X, y) and values of the hyperparameters θ, we can write the outputs y and f_* as a multivariate normal (Gaussian) distribution, Eq. (4), where $K(.,.)$ kernel function $k(.,.)$ is evaluated elementwise. Moreover, we assume each measurement is corrupted by normally distributed observation error ε with zero mean and variance σ^2, $\varepsilon \sim N(0, \sigma^2)$.

$$\begin{bmatrix} y \\ f_* \end{bmatrix} \sim N\left(\begin{bmatrix} m(X) \\ m(X_*) \end{bmatrix}, \begin{bmatrix} K(X,X) + \sigma^2 I & K(X, X_*) \\ K(X_*, X) & K(X_*, X_*) \end{bmatrix}\right) \qquad (4)$$

The conjugacy properties of multivariate Gaussian distribution give (Bishop, 2006):

$$E(f_*) = m(X_*) + K(X_*, X)[K(X, X) + \sigma^2 I]^{-1}(y - m(X)) \qquad (5a)$$

$$Var(f_*) = K(X_*, X_*) - K(X_*, X)[K(X, X) + \sigma^2 I]^{-1}K(X, X_*) \qquad (5b)$$

2.3. Hyperparameter Tuning

A key step in GP modeling is training the hyperparameters. We start by comparing the performance of log marginal likelihood (LML) Eq.(6) and cross-validation (CV) Eq.(7) for training the length scales of each dimension (l_1, l_2, l_3, l_4) and the optional observation error σ as hyperparameters. LML (a.k.a. maximum likelihood estimation, MLE) uses all the training data D to find the hyperparameter values which maximize the log-likelihood function:

$$\log p(y|X,\theta) = -\frac{1}{2}y^T[K(X,X|\theta) + \sigma^2 I]^{-1} - \frac{1}{2}\log|K(X,X|\theta) + \sigma^2 I| - \frac{n}{2}\log 2\pi \quad (6)$$

In contrast, CV uses only a subset of the data to reduce the variance of the prediction evaluation. The LML is computed with data $D_{-i} = (X_{-i}, y_{-i})$ where $-i$ denotes all data except sample i:

$$\log P(y_i|X_{-i}, y_{-i}, \theta) = -\frac{1}{2}\log \sigma_i^2 - \frac{(y_i - \mu_i)^2}{2\sigma_i^2} - \frac{1}{2}\log 2\pi \quad (7a)$$

The conjugacy property of GPR greatly reduces the computation cost of evaluating Eq. (7a). The overall leave-one-out CV (Loo-CV) likelihood function is computed by averaging all the leave-one-out samples:

$$L_{Loo-CV}(X, y, \theta) = \frac{1}{n}\sum_{i=1}^{n} \log P(y_i|X_{-i}, y_{-i}, \theta) \quad (7b)$$

The domain knowledge of experimentalists believed that the four proposed variables are all influential for determining power factor (y_i). To incorporate this prior knowledge, we bounded hyperparameter values, including l and σ, between 0 and 1 for this preliminary analysis, since a large value for l_j would imply that dimension j is not important.

3. Results

3.1. Log Marginal Likelihood (LML) and Leave-one-out Cross-Validation (Loo-CV) identify similar hyperparameter values

We start by comparing LML and Loo-CV hyperparameter training approaches for the photonic sintering data. Table 1 shows LML and Loo-CV identify identical optimal hyperparameters using grid search. The first two rows correspond to optimizing l with fixed $\sigma = 0.1$ which is informed by the experimental observation error. The optimal l obtaining with LML and Loo-CV methods are the same which suggests the simpler method, LML, is adequate for this photonic sintering dataset. Conversely, the third and fourth rows consider both l and σ as optimized hyperparameters. With σ considered as a tuneable hyperparameter, σ increases from 0.1 to 0.2, and l_2 increases from 0.635 to 0.687. These changes reflect the trade-off between bias and variance (Bishop, 2006) and correspond to the conclusion that relatively more complicated model (e.g., $l_2 = 0.635$) usually obtaining low observation error (e.g., $\sigma = 0.1$), while a simpler model (e.g., $l_2 = 0.687$) has higher observation error (e.g., $\sigma = 0.2$).

Table 1: Comparison of hyperparameter values from LML and Loo-CV training

	l_1	l_2	l_3	l_4	σ
LML with σ fixed	1	0.635	0.322	1	0.1
Loo-CV with σ fixed	1	0.635	0.322	1	0.1
LML with σ tuned	1	0.687	0.322	1	0.2
Loo-CV with σ tuned	1	0.687	0.322	1	0.2

Next, we consider the predictive uncertainty of the GP model. Figure 2 is a parity plot for LML optimal hyperparameters ($l_1 = 1, l_2 = 0.687, l_3 = 0.322, l_4 = 1, \sigma = 0.2$). This plot shows the leave-one-out predictions with the GP model. The x-axis and y-axis are experimental and predicted power factor, respectively. The five symbols demark groups of experiments. The error bars show one prediction standard deviation from the GP model. We observed that 27 out of 31 predictions are within one prediction standard deviation of the parity line, suggesting the GPR model successfully emulates the relation between experimental conditions x_i and target y_i.

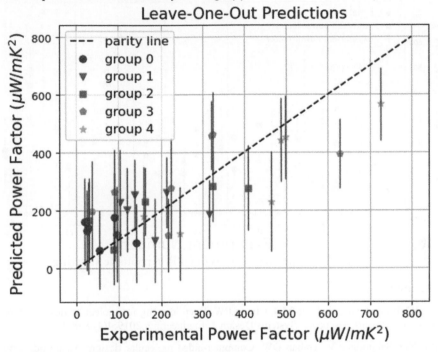

Figure 2: Parity plot of GPR prediction and experimental power factor R in photonic sintering.

3.2. Retrospective Analysis

Figure 3 illustrates the retrospective analysis of GPR prediction in the photonic sintering dataset. The predicted power factors (squares) are generated iteratively (with hyperparameters $l_1 = 1, l_2 = 0.687, l_3 = 0.322, l_4 = 1, \sigma = 0.2$ fixed) using all previous data (to the left of each square). For example, the GPR prediction for experiment 5 uses 5 prior observations for training. The red diamonds show the experimentally measured power factor, and the dashed lines demark each experimental group. In this analysis, we observed 25 out of 30 samples fallen into the predicted (within one standard deviation) bounds. This result demonstrates the GPR model predictions improve as additional data are incorporated into the model.

4. Conclusion

In this work, we successfully develop a GP model to predict the power factor of sintered n-type $Bi_2Te_{2.7}Se_{0.3}$ TE films as a function of four photonic sintering variables. This analysis shows that LML and Loo-CV hyperparameter tuning methods identify the same optimal hyperparameters. Through both the parity plot (Figure 2) and retrospective analysis (Figure 3), we show the accuracy of the GPR predictions (both mean and uncertainty estimates). These results suggest that the GP models can be integrated into a Bayesian optimization framework to identify photonic sintering experimental conditions to maximize the thermoelectric power factor. We thank the U.S. Department of Energy's Office of Energy Efficiency and Renewable Energy (EERE) and Advanced Manufacturing Office Award DE-EE0009103 for funding.

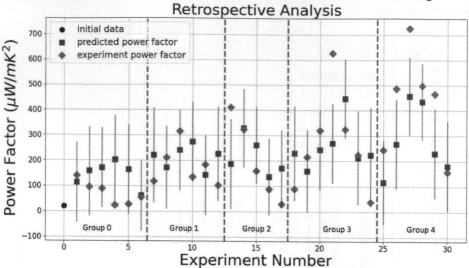

Figure 3: Retrospective analysis of GPR in photonic sintering

References

Elton DC, Boukouvalas Z, Fuge MD, Chung PW. Deep learning for molecular design—a review of the state of the art. Molecular Systems Design & Engineering. 2019;4(4):828-49.

Austin ND, Sahinidis NV, Trahan DW. Computer-aided molecular design: An introduction and review of tools, applications, and solution techniques. Chemical Engineering Research and Design. 2016 Dec 1;116:2-6.

Jiang C, Ding Y, Cai K, Tong L, Lu Y, Zhao W, Wei P. Ultrahigh performance of n-type Ag2Se films for flexible thermoelectric power generators. ACS applied materials & interfaces. 2020 Feb 3;12(8):9646-55.

Saeidi-Javash M, Kuang W, Dun C, Zhang Y. 3D Conformal Printing and Photonic Sintering of High-Performance Flexible Thermoelectric Films Using 2D Nanoplates. Advanced Functional Materials. 2019 Aug;29(35):1901930.

Bishop CM. Pattern recognition. Machine learning. 2006 Feb;128(9).

Rasmussen CE. Gaussian processes in machine learning. InSummer school on machine learning 2003 Feb 2 (pp. 63-71). Springer, Berlin, Heidelberg.

Yu M, Grasso S, Mckinnon R, Saunders T, Reece MJ. Review of flash sintering: materials, mechanisms and modelling. Advances in Applied Ceramics. 2017 Jan 2;116(1):24-60.

Lookman T, Balachandran PV, Xue D, Yuan R. Active learning in materials science with emphasis on adaptive sampling using uncertainties for targeted design. npj Computational Materials. 2019 Feb 18;5(1):1-7.

Ke Wang, Alexander W Dowling, Bayesian optimization for chemical products and functional materials, Current Opinion in Chemical Engineering, Volume 36, 2022, 100728

Han C, Tan G, Varghese T, Kanatzidis MG, Zhang Y. High-performance PbTe thermoelectric films by scalable and low-cost printing. ACS Energy Letters. 2018 Feb 23;3(4):818-22.

Proceedings of the 14th International Symposium on Process Systems Engineering – PSE 2021+
June 19-23, 2022, Kyoto, Japan © 2022 Elsevier B.V. All rights reserved.
http://dx.doi.org/10.1016/B978-0-323-85159-6.50304-3

Development of Dye Exhaustion Behavior Prediction Model using Deep Neural Network

Jonghun Lim[a,b§], Soohwan Jeong[a,c§], Sungsu Lim[c], Hyungtae Cho[a], Jae Yun Shim[d], Seok Il Hong[d], Soon Chul Kwon[d], Heedong Lee[d], Il Moon[C] , Junghwan Kim[a*]

[a]*Green Materials and Process R&D Group, Korea Institute of Industrial Technology, 55, Jonga-ro, Jung-gu, Ulsan 44413, Republic of Korea*
[b]*Department of Chemical and Biomolecular Engineering, Yonsei University, 50, Yonsei-ro, Seodaemun-gu, Seoul 03722, Republic of Korea*
[c]*Department of Computer Science Engineering, Chungnam National University, 99, Daehak-ro, Yuseong-gu, Daejeon 34134, Republic of Korea*
[d]*ICT Textile and Apparel R&BD Group, Korea Institute of Industrial Technology, Ansan, 15588, Republic of Korea*

[§]*Lim J. and Jeong S. contributed equally to this work as first authors.*
kjh31@kitech.re.kr

Abstract

The textile dyeing process consumes a significant quantity of energy as it is necessary to maintain the water temperature between 60–120 °C during the dyeing of reactive dyes. Therefore, to reduce the overall cost of the process through reducing the quantity of waste energy, it is crucial to increase the right first time (RFT) rate, which corresponds to the rate at which the target color is imparted through a single dyeing process (Park et al., 2009). To improve the RFT rate, the proper operation with following the optimal dye exhaustion behavior in consideration of the color difference and dyeing uniformity is a critical factor. The color difference is determined according to maximal absorption and the dyeing uniformity is decided by dye exhaustion behavior [Bouatay et al., 2016]. In this study, we developed a model for predicting dyeing exhaustion behavior, and utilized the model to predict optimal dye exhaustion behavior under various dyeing conditions. A deep neural network-based on the dye exhaustion behavior prediction model was developed through regression analysis, the model was further developed and evaluated by dividing the entire dataset into learning and evaluation data. The model's performance was evaluated using the root mean square error (RMSE) parameter alongside the coefficient of determination (R^2) which acted as performance evaluation metrics. Using these performance metrics, it was found that the proposed DNN regression exhibited the highest performance and the smallest error in comparison with established models, with root mean square error RMSE and R^2 values of 0.016 and 0.994, respectively. The results reported in this study demonstrate that the proposed model exhibits superior performance in predicting the dye exhaustion behavior.

Keywords: Textile industry, re-dyeing, right-first-time, deep neural network-based prediction

1. Introduction

Within the textile industry, dyeing forms a cost-intensive process requiring considerable volumes of hot water and the chemicals required for reactive dyeing. In addition, this process generates significant volumes of wastewater, which forms a severe environmental pollutant. The exhaust method is frequently used within the textile industry to dye cellulose fibers with reactive dyes as it achieves the highest productivity over a short period of time. However, the exhaust method exhibits a disadvantage in that the quality of the final product exhibits significant variations as a result of slight fluctuations in dyeing conditions, such as the dye ratio, temperature, and the Na_2SO_4 and Na_2CO_3 concentrations. Therefore, to obtain the product's target quality, a re-dyeing procedure is often required, which increases the overall dyeing cost by 98–169 % in comparison to a one-time dyeing process, while environmental pollution is also increased owing to the increased wastewater volume. Thus, to address this problem, it is crucial to increase the right first time (RFT) rate, which is the rate at which the target color is reproduced with only one dyeing cycle in the dyeing machine. To increase the RFT rate during the textile dyeing process, active research has recently been conducted to increase the compatibility of the dye and also to determine the optimal dyeing conditions to increase the final exhaustion rate, which impacts the target color change. To increase the dye compatibility, Kim et al. determined the optimal pH conditions for mononicotinic acid triazine-type dyes. It was found that the highest final exhaustion rate was achieved at pH values greater than 9. Kim et al. proposed the optimal dyeing conditions for the application of both reactive and acidic dyes to Angora fibers to improve the final exhaustion rate. The optimal dyeing conditions were reported to be a dye concentration of 8 % o.w.f. for both the reactive and acidic dyes, a pH of 3–4 at 110 °C for reactive dyes, and a pH of 3–4 at 70 °C for acidic dyes.

Despite the considerable number of reports aiming to increase the RFT rate, a significant limitation still remains. During the textile dyeing process, the dyeing quality was determined based on the color difference relative to the target color and dyeing uniformity. If the dye is adsorbed into the fabric at a greater rate than the optimal exhaustion behavior, the dye alkalizes rapidly during the reaction stage. This results in uneven fixation and dyeing, which degrades the dyeing uniformity. Thus, in order to increase the RFT rate, it is crucial to consider dyeing uniformity, which simultaneously determines both the exhaustion behavior and color difference. At present, no relevant studies have been conducted in this respect.

To address these limitations, we propose a dye exhaustion behavior prediction model, utilizing a deep neural network (DNN), to determine the optimal dyeing conditions. The aim of this particular study was to increase the RFT rate by deriving the optimal dyeing conditions, which are obtained through a determination of the optimal exhaustion behavior which considers both the color difference and dyeing uniformity to overcome the significant limitations imparted by excessive costs and environmental contamination.

2. Process description

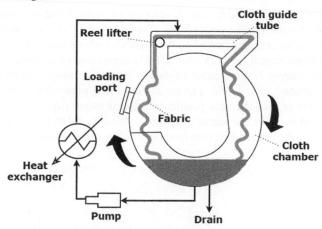

Figure 1. Schematic diagram of a round-configuration jet dyeing machine

Figure 1 shows a schematic diagram of a round-configuration jet-dyeing machine. In this particular apparatus, dyeing proceeds under high temperature and pressure to facilitate the reactive dyeing procedure. In general, the reactive dyeing process proceeds as follows. Initially, the dye is ejected through a nozzle attached to the cloth guide tube within the closed tube system of the cloth chamber. During the continuous dye ejection through the nozzle, the dye is absorbed into the fabric, which is rotated over the reel lifter through the application of the injection pressure. The resultant turbulence facilitates dye penetration into the fabric, while simultaneously reducing the mechanical impact on the fabric. Finally, the dyeing solution containing the reactive dye is heated using a heat exchanger to obtain a suitable temperature for reactive dyeing.

3. Methodology

3.1. Data generation and preprocessing

In order to develop the dye exhaustion behavior prediction model with respect to the dyeing procedure, 615 datasets were extracted detailing time, temperature, and Na_2SO_4 and Na_2CO_3 input quantities. The time values ranged between 0 and 120 min at 3 min intervals, and the temperature fluctuated according to certain set values over this time period. Three Na_2SO_4 input quantities were evaluated, specifically, 10, 30, and 50 g; while five Na_2CO_3 input quantities were evaluated, specifically, 0, 5, 10, 15, and 20 g. Following this, the extracted datasets were preprocessed in two steps. First, any datasets that were not required for the dye exhaustion behavior prediction model were removed. As a result, since no trend was observed at an Na_2CO_3 input quantity of 0 g, the corresponding dataset was eliminated. Secondly, each dataset was normalized as the units ascribed to each data point vary, which, in turn, hinders the learning process. In this study, the z-score normalization method was applied, while the mean and standard deviation for each value of the same parameter were used to scale the data to exhibit a normal distribution with a mean of 0 and a standard deviation of 1. In addition, any time warping or alignment of the data sets were ignored.

3.2. Development of DNN-based prediction model

A DNN-based prediction model was developed to predict the exhaustion behavior exhibited during textile dyeing procedures. Typically, a DNN is a machine learning algorithm based on an artificial neural network (ANN) which mimics the principles and structure of a human neural network. An ANN is composed of an input layer, a hidden layer, and an outer layer. If the number of hidden layers is greater than or equal to three, the system is denoted as a DNN. Figure 2 shows a typical DNN structure. During regression analysis or classification problems, a linear estimation function of the type y = $w^T x$ + b is used to solve the linear problem. Typically, a DNN applies an activation function to a linear estimation function to solve nonlinear problems.

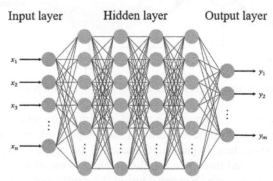

Figure 2. Typical DNN structure

In order to develop the DNN-based dye exhaustion behavior prediction model, the datasets obtained after preprocessing were used. These were the exhaustion rates according to time, temperature, and the Na_2SO_4 and Na_2CO_3 input quantities. The pre-processed datasets were divided into training (75 %) and test (25 %) sets. With respect to the DNN hyper-parameter, the model consisted of three hidden layers (h_1, h_2, and h_3), with the unit corresponding to each hidden layer set to 100. ReLU was used as the activation function for h_1 and h_2, while a sigmoid was used for h_3.

$$ReLU(x) = \max(0, x) \tag{1}$$

$$sigmoid(x) = (1 + e^{-x})^{-1} \tag{2}$$

Finally, Adam was applied as the optimizer function, while the mean squared error (MSE) was determined to form an appropriate loss function for this regression analysis.

$$MSE = \frac{\sum_{i=1}^{n}(y_i - \hat{y}_i)^2}{n} \tag{3}$$

4. Results and discussion

4.1. Performance of the DNN-based prediction model

This section discusses the performance of the exhaustion behavior prediction model developed using a DNN-based regression analysis, in which the MSE and R^2 were used as performance indicators. To evaluate the performance of the DNN model, other regression analysis methods, such as lasso, ridge, and support vector regression, were applied to identical datasets. Each model was subjected to 100 regression experiments, and the training and test data were varied randomly each time. Table 1 lists the performance of each model.

Table 1. Performance of each regression model

Regression model	R^2	$RMSE$
Lasso regression	0.726 ± 0.040	0.110 ± 0.015
Ridge regression	0.723 ± 0.037	0.273 ± 0.020
Support vector regression	0.902 ± 0.035	0.066 ± 0.017
DNN regression	0.994 ± 0.004	0.016 ± 0.006

From the data shown in table 1, it can be seen that the DNN regression exhibited the highest performance and the smallest error, with root mean square error (RMSE) and R^2 values of 0.016 and 0.994, respectively.

4.2. Prediction results of the dye exhaustion behavior

Based on the proposed DNN-based prediction model, the dye exhaustion behavior as a function of time, temperature, and the Na_2SO_4 and Na_2CO_3 input quantities was predicted. Figures 3 correspond to the cases for which the Na_2CO_3 input quantities were 5 and 10 g, and 15 and 20 g, respectively.

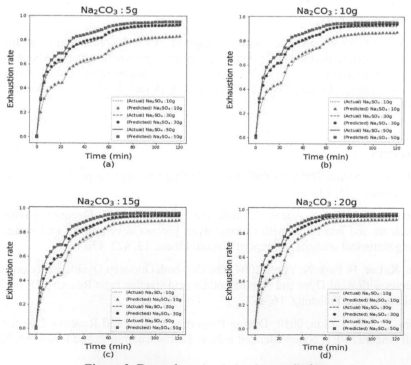

Figure 3. Dye exhaustion behavior prediction

It is clear that the data obtained from the DNN-based prediction model strongly agrees the exhaustion behavior exhibited over time. Hence, the prediction model which is defined using the true dataset can be considered reliable. Through the application of the proposed DNN-based prediction model, the optimal dye exhaustion behavior can be predicted with high accuracy under various dyeing conditions. Therefore, it is possible to derive the optimal dyeing conditions that are derived from the targeted optimal dye exhaustion behavior without the need to perform the dyeing process several times.

5. Conclusions

In this study, we developed a dye exhaustion behavior prediction model which utilized a DNN to determine the optimal dyeing conditions. As the proposed DNN-based prediction model was used to predict the dye exhaustion rate under various dyeing conditions, it was then possible to derive the optimal dyeing conditions through an evaluation of the optimal dye exhaustion behavior. Thus, it will be possible to increase the RFT rate by considering both the color difference and dyeing uniformity to overcome the significant hinderances of excessive cost and environmental contamination through the generation and release of wastewater. As a result, the application of the proposed DNN-based prediction model reduces the re-dyeing rate through increasing the RFT rate. Therefore, the proposed model facilitates a significant improvement in the environmental and economic impact imparted during the dyeing process, while also providing valuable insight into the textile dyeing process.

Acknowledgements

This study has been conducted with the support of the Korea Institute of Industrial Technology with respect to "Development of complex parameter smart analysis modules for color customering (kitech EH-21-0008)" and "Development of Global Optimization System for Energy Process (kitech EM-21-0022), (kitech IZ-21-0052), (kitech IR-21-0029)".

References

J. Park, and J. Shore, 2009, Evolution of right-first-time dyeing production, Coloration technology, 125, 133-140

F. Bouatay, N. Meksi, S. Adeel, F. Salah, and F. Mhenni, 2016, Dyeing behavior of the cellulosic and jute fibers with cationic dyes: process development and optimization using statistical analysis, Journal of Natural Fibers, 13, 423-436

J. Kim, K. Lee, H. Park, N. Yoon, 2004, The One-bath One-step Dyeing of Nylon / Cotton Blends with Acid Dyes and Mononicotinic acid-triazine type Reactive Dyes, Textile Coloration and Finishing, 16, 1-7

Y. Kim, S. Lee, Y. Son, 2010, Dyeing Properties of Acid and Reactive Dye for Super Soft Angora / PET, Nylon Blended Fabric. Textile Coloration and Finishing, 22, 332-40

Proceedings of the 14th International Symposium on Process Systems Engineering – PSE 2021+
June 19-23, 2022, Kyoto, Japan © 2022 Elsevier B.V. All rights reserved.
http://dx.doi.org/10.1016/B978-0-323-85159-6.50305-5

Guaranteed Error-bounded Surrogate Modeling and Application to Thermodynamics

Ashfaq Iftakher, Chinmay M. Aras, Mohammed Sadaf Monjur, M. M. Faruque Hasan*

Artie McFerrin Department of Chemical Engineering, Texas A&M University, College Station, TX 77843-3122, USA.
hasan@tamu.edu

Abstract

We present a data-driven surrogate modeling technique and demonstrate its applicability to replace complicated thermodynamic models for efficient process simulation and synthesis. We employ data-driven edge-concave underestimators and edge-convex overestimators to provide guaranteed error-bounded approximation over the entire domain. A surrogate model is then achieved by performing a parameter estimation that ensures the approximation to be bounded between the vertex polyhedral under- and over-estimators of the original model. We also present GEMS (Guaranteed Error-bounded Modeling of Surrogates) framework, which is a package with automated dataflow for sample evaluation, Hessian bound estimation and parameter estimation to obtain the surrogate models. We apply the technique to predict the solubility of hydrofluorocarbon (HFC) refrigerants in ionic liquids.

Keywords: Surrogate Modeling, Simulation-based Optimization, Data-driven Modeling

1. Introduction

Thermodynamic models are key to design realistic chemical processes to address climate change, decarbonization and other grand challenges. For example, ionic liquid (IL)-assisted innovative separation processes using extractive distillation require rigorous solubility modeling. Also, due to the nature of ILs as 'designer solvents', selection of optimal IL from many candidate solvents for the same separation task requires the understanding of phase behavior. In general, computer aided process intensification guides the discovery of innovative process units that may result in dramatic performance improvement. To attain confidence in the proposed design, the underlying mathematical model must be able to sufficiently capture the physical phenomena. To achieve this, rigorous thermodynamic models are incorporated while solving process synthesis problems. However, attaining globally optimal solutions has been a challenge due to the nonlinear and nonconvex nature or large size of thermodynamic models.

In simulation based optimization approach, a thermodynamic model is treated as black-box. The sampling data is generated over the operating domain and the output data is utilized by an optimizer which performs derivative-free optimization (Bajaj et al., 2021). To obtain guaranteed convergence to optimal solution, one requires dense sampling. If the black-box model is very large, the samplings can become computationally expensive. To increase efficiency, a set of sampling data can be used to generate surrogate (reduced order) models (Cozad et al., 2014; Boukouvala et al., 2017). Such simpler models can achieve computational efficiency by replacing the original computationally expensive models. However, the prediction accuracy remains a challenge. For example, single/piecewise linear approximations, polynomial response surfaces, Artificial neural networks (ANN) suffer from not being able to reliably predict the type of approximation (under vs overestimation) (Jones, 2001). Also, Kriging (Jones et al., 1998), radial basis functions (RBF) and other interpolating surrogates (Bhosekar and Ierapetritou, 2018) exactly predict the training points (Wang et al., 2014) while providing no guarantee on the quality of prediction over the entire domain of interest.

It has been recently shown that theoretically guaranteed lower bounds can be tractably obtained just by data-driven black-box sampling (Bajaj and Hasan, 2019). The only information required is the global upper bound on the diagonal Hessian elements which can be obtained by either physical intuition, solving an NLP procedure or automatic differentiation over the whole discretized domain. Specifically for the application of thermodynamics, since it is possible to derive the bounds on the diagonal Hessian elements for known models, the edge concavity-based relaxation provides an attractive way towards developing surrogate thermodynamic models with theoretically guaranteed bounds on the prediction errors. In this work, we extend the underestimator formulation and propose a new data-driven surrogate modeling technique that provide theoretically guaranteed tight error bounding (under and overestimation) of blackbox models over the entire domain. We also present a framework that performs sampling of the blackbox models, calculates Hessian via automatic differentiation, and performs globally bounded parameter estimation through GAMS thereby facilitating the data transfer and allowing a single flexible and user friendly package.

2. THEORETICAL BOUNDED APPROXIMATION OF BLACKBOX MODELS

2.1. Edge-concave underestimation and edge-convex overestimation based bounding

We adopt the edge-concave underestimator (Hasan, 2018; Bajaj and Hasan, 2019) and utilize vertex polyhedral property (Tardella, 2004) to construct linear facets of the convex envelope solely based on evaluation of a given greybox/blackbox function $f(\mathbf{x})$ at the domain bounds and interior (sampled) points. Assuming twice-differentiability of $f(\mathbf{x})$, its edge-concave underestimator, $L(\mathbf{x})$ is given by:

$$L(\mathbf{x}) = f(\mathbf{x}) - \sum_{i=1}^{n} \theta_i^L (x_i - x_i^{Int})^2 \tag{1}$$

where, x_i^{Int} is the value of sampled variable x_i, and the parameter θ_i^L is defined as:

$$\theta_i^L = \max \left\{ 0, \frac{1}{2} \left[\frac{\partial^2 f}{\partial x_i^2} \right]^U \right\} \tag{2}$$

Similarly, edge-convex overestimator, $U(\mathbf{x})$ is expressed as follows:

$$U(\mathbf{x}) = f(\mathbf{x}) + \sum_{i=1}^{n} \theta_i^U (x_i - x_i^{Int})^2 \tag{3}$$

where, the parameter θ_i^U is defined as:

$$\theta_i^U = \max\left\{0, \frac{1}{2}\left[\frac{-\partial^2 f}{\partial x_i^2}\right]^U\right\} \tag{4}$$

For an n-dimensional problem, J simulations result in J underestimators and J overestimators. The linear facets of which result in $2J$ number of $n+1$ dimensional simplices (polytopes) each having 2^n vertices at the domain bounds and 1 interior vertex (simulation point) pertaining to that particular simplex j.

2.2. Bounded Surrogate Model Parameter Estimation Formulation

We limit our focus on generating regression based non-interpolating polynomial surrogates. To attain guarantee in the type of approximation, we bound the surrogate prediction using the linear facets of the estimators, i.e., enforce necessary constraints so that the surrogate prediction lies within a prescribed error bound. To that end, we generate the under and over-estimators from f_i^L and f_i^U which correspond to the shifted points (below and above respectively) of the sampled point f_i. The mathematical formulation is as follows:

$$\min \sum_{n=1}^{N}\left[\left(\frac{\hat{f}_i - s_i}{\hat{f}_i}\right)^2 + \left(\frac{f_i^U - \hat{f}_i}{\hat{f}_i}\right)^2 + \left(\frac{\hat{f}_i - f_i^L}{\hat{f}_i}\right)^2\right] \tag{5}$$

s.t.

$$s_i = \alpha + \sum_{n=1}^{N}\beta_n\hat{x}_{i,n} + \sum_{n=1}^{N}\sum_{m=n}^{N}\gamma_{n,m}\hat{x}_{i,n}\hat{x}_{i,m} \tag{6}$$

$$f_i^L - \sum_{n=1}^{N}\theta_n^{f,L}\cdot(x_{i,n,v} - x_i^{Int})^2 = s_i - \sum_{n=1}^{N}\theta_n^{s,L}\cdot(x_{i,n,v} - x_{i,n}^{Int})^2 \qquad i\in I, v\in V \tag{7}$$

$$f_i^U + \sum_{n=1}^{N}\theta_n^{f,U}\cdot(x_{i,n,v} - x_i^{Int})^2 = s_i + \sum_{n=1}^{N}\theta_n^{s,U}\cdot(x_{i,n,v} - x_{i,n}^{Int})^2 \qquad i\in I, v\in V \tag{8}$$

$$\theta_n^{s,L} \geq \gamma_{n,n} \qquad n\in N \tag{9}$$

$$\theta_n^{s,U} \geq -\gamma_{n,n} \qquad n\in N \tag{10}$$

The first term in the objective function (Eq. 5) minimizes the error of the surrogate fit from the sampled data while the second and third terms reduce the shifts of f_i^L and f_i^U from the original sampled point. Eq. 6 denotes the general form of an n-dimensional quadratic function where α, β_n and $\gamma_{n,m}$ are the estimated parameters. Eqs. 7 and 8 ensure that the linear facets of the under- and over-estimator of the surrogate fit are bounded by the linear facets of the under- and over-estimator of the original function shifted by some value at each sampled point. Here, $\theta_n^{f,L}$ and $\theta_n^{f,U}$ are the parameters required to construct the under- and over-estimators of the original function respectively. Similarly, $\theta_n^{s,L}$ and $\theta_n^{s,U}$ are the variables required to generate the estimators of the surrogate function. Eqs. 9 and 10 guarantee the edge-concavity and edge-convexity of the under and over-estimators of the surrogate function respectively. For estimating n-dimensional functions using a higher order (> 2) polynomial, the model formulation essentially remains the same. However, in this case, the upper bounds of the second derivatives of the surrogate function depend upon the sign of the estimated parameters. This justifies incorporating mixed-integer logic in Eqs. 9 and 10 for evaluating $\theta_n^{s,L}$ and $\theta_n^{s,U}$.

3. GEMS FRAMEWORK

For a specified model defined by closed functional forms or a system of equations, and the bounds on the independent variables, the GEMS framework (shown in Figure 1) allows data sampling (e.g, via Latin Hypercube sampling). To generate the upper bound of the Hessian (θ), the framework utilizes two different approaches: 1) If the model is of closed form and the symbolic differentiation is tractable, the NLP with an objective function of maximizing the second derivatives subject to the domain bounds, can be solved through GAMS. 2) If the model is not of closed form, or the number of independent variables is high, then the GEMS framework allows efficient computation of the Hessian by Automatic differentiation (AD). AD employs techniques similar to backpropagation and provides numerical values of derivatives (Griewank, 2003).

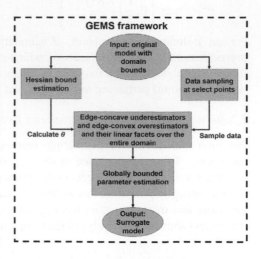

Figure 1: GEMS framework.

The calculated θ values, together with the sampled data are utilized to solve the globally bounded parameter estimation problem (see Section 2.2). The output is a reduced order surrogate polynomial that is guaranteed to not overshoot above and below a certain threshold defined by the relaxed piecewise linear bounding of the edge-concave under and edge-convex overestimators of the original blackbox model. The main advantage of the framework is that it allows the linking between Python (PyTorch), C++ (ADOL-C) for calculating Hessians via automatic differentiation, sampling of the blackbox model via either uniform sampling or Latin hypercube sampling and parameter estimation in GAMS thereby handling the required data transfer and facilitating a user friendly package.

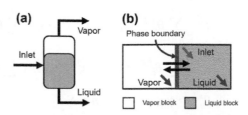

Figure 2: Flash separator: (a) conventional representation and (b) building block-based representation.

4. APPLICATION TO SURROGATE THERMODYNAMIC MODELING

To combat global warming, ILs have garnered significant attention as a potential solvent for the separation of high global warming potential refrigerants, such as hydrofluorocarbons (HFCs). To design innovative and intensified processes for such separation task, one must accurately represent the HFC/IL binary system through thermodynamic models. Rigorous models such as Gamma-Phi or equation of state (EOS) have higher prediction accuracy. However, these models are highly non-linear and nonconvex, which significantly increase the computational burden for any process synthesis and optimization frameworks. Therefore, a surrogate model with a simpler functional form can evade the model complexity. To that end, we focus on generating surrogate for the well-known Gamma-Phi thermodynamic model, which for the prediction of the amount of HFC absorbed in IL, can be simplified as follows:

$$f(P,T,\widetilde{x}_1) = \ln P + \frac{(P - P_1^s)(B_1 - \widetilde{V}_1)}{RT} - \ln P_1^s - \ln \widetilde{x}_1 - \ln \gamma_1 \tag{11}$$

Here, P_1^s is the saturated vapor pressure, B_1 is the second virial coefficient, $\widetilde{V}_1$ is the molar volume of the ionic liquid, P is the total pressure of the system, $\widetilde{x}_1$ is the equilibrium liquid phase composition of HFC. Activity coefficient model for component i, (γ_i) can be represented via well-known Margules or NRTL model. More details on the thermodynamic modeling can be found elsewhere (Shiflett and Yokozeki, 2006).

Here we consider R-32/[bmim][PF$_6$] binary system. We derive the analytical expression of the Hessian from the thermodynamic model (see Eq. 11.) Given the variable bounds, i.e., $[P^L, P^U], [T^L, T^U], [\widetilde{x}_1^L, \widetilde{x}_1^U]$, we determine the required $\theta_P^L, \theta_T^L, \theta_{\widetilde{x}_1}^L, \theta_P^U, \theta_T^U, \theta_{\widetilde{x}_1}^U$ (see Eq. 2 and 4) which are used to construct the piecewise linear

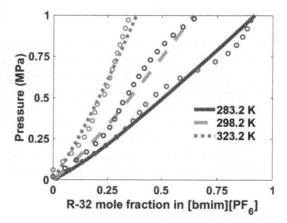

Figure 3: Solubility isotherms of R-32 with Margules activity coefficient model. The lines represents the solubility isotherms by the Gamma-phi based method, the symbol (o) represent the solubility predicted by the surrogate model.

bounding of the Gamma-Phi model. Since the goal is to predict mole fraction, $\widetilde{x}_1$, for a given P, T; we divide the entire P, T space into four subregions as follows: $R1 = \{P, T : P \in \{0.01, 0.5\}, T \in \{280, 330\}\}, R2 = \{P, T : P \in \{0.5, 1\}, T \in \{280, 330\}\}, R3 = \{P, T : P \in \{0.01, 0.5\}, T \in \{330, 375\}\}, R4 = \{P, T : P \in \{0.5, 1\}, T \in \{330, 375\}\}$. For each of the subregions, we apply the model formulation as described in Section 2.2, and obtain a cubic surrogate polynomial.

After that, we incorporate the surrogate to a flash-separator (see Figure 2) through SPICE (Monjur et al., 2021a,b) framework that leverages building-block based representation followed by superstructure optimization. To represent the vapor-liquid phases using building blocks, we require two blocks. The phase contact is represented by a semi-restricted boundary depicted by the thick vertical line in Figure 2b. Mass transfer between the phases takes place through this boundary which is represented by the VLE model (see Eq. 11).

When the surrogate prediction from the flash separator is compared to the Gamma-Phi prediction (see Figure 3), we observe that the averaged prediction error lies within $\approx$ 8.45%. We obtain four different best fitted surrogates for each of the subregions which explains the discontinuity at P = 0.5 MPa for each of the isotherms in Figure 3. It can be seen that the surrogates intersect the Gamma-Phi based isotherms more than once. This may suggest that the surrogate model may be of higher order than the original function. Since we assumed a simple polynomial form, i.e. cubic, the possibility of exponential or logarithmic terms in the actual function may also lead to this behavior. The main takeaway point is that through the GEMS framework, we are able to replace the nonlinear thermodynamics by a simpler surrogate with guaranteed error bounds, allowing us to use the same problem formulation for process synthesis applications.

5. CONCLUSIONS

The data-driven approach presented through the GEMS framework can be efficiently applied for the approximation of VLE models. It also shows a promising pathway for solving general data-driven global optimization problems. Incorporation of automatic differentiation in the framework allows the calculation of the upper bound on the Hessian even if a system of equations rather than an explicit functional form of the model is available. The approach could be an efficient way for accelerating computationally demanding process simulations employing complex models by proposing simple and more computationally favorable accurate enough surrogates, thereby providing a means to attain the global convergence of data-driven process optimization problems.

6. Acknowledgment

The authors gratefully acknowledge support from the NSF EFRI DCheM (2029354) and NSF CAREER (CBET-1943479) grants.

References

I. Bajaj, A. Arora, M. F. Hasan, 2021. Springer, pp. 35–65.

I. Bajaj, M. F. Hasan, 2019, 1–16.

A. Bhosekar, M. Ierapetritou, 2018. Computers Chemical Engineering 108, 250–267.

F. Boukouvala, M. F. Hasan, C. A. Floudas, 2017. Journal of Global Optimization 67 (1-2), 3–42.

A. Cozad, N. V. Sahinidis, D. C. Miller, 2014. AIChE Journal 60 (6), 2211–2227.

A. Griewank, 2003. Acta Numerica 12, 321–398.

M. M. F. Hasan, 2018. Journal of Global Optimization 71 (4), 735–752.

D. R. Jones, Dec 2001. Journal of Global Optimization 21 (4), 345–383.

D. R. Jones, M. Schonlau, W. J. Welch, Dec 1998. Journal of Global Optimization 13 (4), 455–492.

M. S. Monjur, S. E. Demirel, J. Li, M. M. F. Hasan, 2021a. Comput. Aided Chem. Eng 50, 287–293.

M. S. Monjur, S. E. Demirel, J. Li, M. M. F. Hasan, 2021b. Ind. Eng. Chem. Res.

M. B. Shiflett, A. Yokozeki, 2006. AIChE Journal 52 (3), 1205–1219.

F. Tardella, 2004. Frontiers in Global Optimization, 563–573.

C. Wang, Q. Duan, W. Gong, A. Ye, Z. Di, C. Miao, 2014. Environmental Modelling Software 60, 167–179.

Proceedings of the 14th International Symposium on Process Systems Engineering – PSE 2021+
June 19-23, 2022, Kyoto, Japan© 2022 Elsevier B.V. All rights reserved.
http://dx.doi.org/10.1016/B978-0-323-85159-6.50306-7

Development of an ANN-based soft-sensor to estimate pH variations in Intelligent Packaging Systems with visual indicators

Isadora F. Brazolin[a]*, Felipe Matheus Mota Sousa [b]; Flavio Vasconcelos Silva [b]; Viktor O. C. Concha[a]; Cristiana M. P. Yoshida[a]

[a]*UNIFESP - Federal University of São Paulo, Institute of Ambiental, Chemical, and Pharmaceutical Science, São Paulo, BRAZIL*
[b]*UNICAMP – University of Campinas, School of Chemical Engineering, São Paulo, BRAZIL*
ibrazolin@unifesp.com.br

Abstract

Based on the experimental data of colorimetric indication, an artificial neural network was first established to classify the pH ranges of the intelligent food packaging device. An intelligent packaging system monitors the package product's condition to provide information about the quality and/or safety during transport, distribution, and storage. The intelligent packaging senses and informs the conditions of the product in an easy and accessible manner, without opening the package. Food pH is strongly related to the quality of food packaged products, indicating deterioration, microbial growth, and adulteration. In the case study, the development and training of an artificial neural network (ANN) aimed to easy quality control of food products that can present alterations/adulterations from pH variation reactions, based on a functional colorimetric indicators' response from a sustainable, intelligent packaging device (biopolymeric chitosan films) of easy and renewable source manufacturing. Chitosan intelligent films were formulated with different chitosan and natural colorimetric indicator (anthocyanin) concentrations, forming the intelligent device. The intelligent devices were immersed in a wide pH range (1.0 to 13.0) solutions, and color parameters (L*, a*, b*) variations were measured. An empirical multivariable model was developed based on artificial intelligence (ANN) to classify pH ranges through the indicator's color variation and the chitosan and anthocyanin concentrations. The ANN of chitosan intelligent films device could ensure acceptable food quality and safety levels to provide adequate protection for consumers and facilitate trade.

Keywords: Intelligent Packaging; Machine Learning; IA; colorimetric indicator

1. Introduction

Packaging technologies are being developed to improve products preservation, quality, and safety. Among recent technologies, intelligent packaging is products' condition monitoring systems, providing quality information during transportation, distribution, and storage. The intelligent packaging device senses the environment inside or outside the package and informs the manufacturer, retailer, and consumer regarding the product's condition (Kuswandi et al., 2011).

Food products' shelf-life tests demand time and cost, while colorimetric indicators can assist the consumer or retailer when buying the products, ensuring quality and food safety. Supply chain management based on pH measurements can significantly decrease food waste, a critical environmental and social concern. An efficient supply chain management could save food disposal, water, energy, increase return-on-investment, improve consumer satisfaction and support regulatory requirements (Mercier & Uysal, 2018). Yoshida et al. (2014) developed an easy-manufacturing and sustainable colorimetric indicator using anthocyanin as a pH colorimetric indicator incorporated into a natural polymer matrix (chitosan). The intelligent films presented final properties to be applied as intelligent devices material, with a color variation due to pH range, which was observed pink color to acid pH, green-blue color to neutral pH, and yellow color to basic pH.

The food industry, like many others, benefits a lot from modernization and the use of technology. Industry 4.0 brings connections and interactions between machines and operators from different sectors using artificial intelligence, including artificial neural networks (ANN). ANN is a computational model built from several simple processing units (neurons) capable of assimilating data and information presented and, from the acquired knowledge, estimating solutions that were not known until now. In this way, an ANN simulates the nervous system behavior of a living being. The definitions involved make an analogy to the components and processes related to the functioning of the human brain (Haykin, 2009).

This work aimed to apply artificial intelligence, specifically artificial neural networks, to determine the pH value range displayed on the sustainable colorimetric indicating device. The empirical pH model was obtained by varying the formulation (concentrations of chitosan and anthocyanin) and measuring the respective color parameters measured in a colorimeter.

2. Materials and methods

2.1. Chitosan Intelligent films

The chitosan intelligent films were obtained accordingly to (Yoshida et al., 2014), using different anthocyanin (Cath, 0.5, 1.0 e 2.0 % m/m) and chitosan (Cch, 0.5, 1.0 e 2.0 % m/m) concentrations.

2.2 Color indication intelligent device

The different pH values were measured using standard solutions in a wide range of pH from 1.20 (HCl 1 mol/L) to 13.29 (NaOH 1 mol/L) using MilliQ water, generating the data required to train the artificial neural network. Buffer solutions were prepared to obtain intermediate pH solutions: McIlvaine, Kolthoff, boric acid-potassium chloride-sodium hydroxide. Chitosan intelligent films were immersed in pH solutions for ninety seconds. Instrumental color parameters (L*, a*, b*) were measured using a portable colorimeter (Konica Minolta, CR-400, Osaka, Japan) in three random positions.

2.3 Development of empirical mathematical models using Artificial Neural Networks

Experimental tests were carried out to create a database used to determine the empirical model of the pH of the chitosan film. Chitosan and anthocyanin concentrations and measured color parameters (L *, a *, and b *) were used as ANN input variables. The pH value of the chitosan film was used to create three distinct classes for the network output. The database was divided into three sets for use in the training, validation, and testing phases of the ANNs. The database consists of 430 data, each one containing 5 inputs (Cch, Cath, L *, a *, and b *) and 1 output (pH class).

The ANNs were developed using the dedicated libraries Tensorflow and Keras. Adam optimization was used as an optimizer, and the categorical cross-entropy was chosen as the objective function. The experimental database was divided randomly into training and test datasets with 80 and 20 % of the original set. The input variables were normalized in the range of -1 and 1 to ensure unbiased models. Three pH ranges were selected as labels for the outputs corresponding to pH under 4, between 4 and 8, and over 8. Early-stopping was introduced during the training stage to avoid model overfitting, using 10 % of the training dataset after performing cross-validation with the selected hyper-parameters summarized in Table 1. It was used 10-fold cross-validation to improve the ANN model's choosing, eliminating usually found consequences of random choosing.

Different architectures of multilayer feedback artificial neural networks (hidden layers and an output layer) were proposed. Python, an interpreted language guaranteeing a free tool with easy installation and platform independence, was used to implement computational models. The hyper-parameters were optimized using a grid-search algorithm. The output layer is a three-neuron-densely-connected layer with a SoftMax activation function to determine the likelihood of each category. For each input vector, the output with the highest value was selected as the predicted output for direct comparison with the real data.

The methodology applied to this case study is illustrated in Figure 1.

Table 1. Hyper-parameters for this classification problem.

Hyper-parameter	Values
Hidden layers	1 and 2
Neurons	10, 20, 30, 40, 50, 60, 70, 80, 90, 100
Weight initialization	Identity, random normal, and random uniform
Activation function	Hyperbolic tangent, sigmoid and ReLU

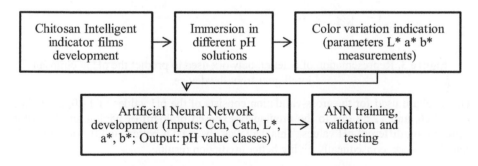

Figure 1. Scheme of the methodology adopted.

3. Results

Chitosan intelligent films containing different concentrations of anthocyanin presented a violet color. Films with lower chitosan concentration were more transparent and flexible, facilitating visual color variation. The chitosan intelligent films were immersed in different solutions in a wide pH range (1.20 to 12.58), from acidic to alkaline (Figure 2). The chitosan intelligent film device presented a reddish color for the pH range from 1 to 5 (acid condition), a blue-greenish color for the range 6 to 10, changing to a yellowish color in a more alkaline range (pH > 12). For the pH values of 7.77 and 8.87 and 3.79 and

5.87, the films presented a very similar color even though representing different pH values.

(a) 1.20   (b) 3.79   (c) 5.87   (d)7.77   (e) 8.87   (f) 9.96   (g)11.40   (h)12.58

Figure 2. Visual color variation of chitosan intelligent film device in contact with different pH conditions (a-h).

In this study, an ANN was designed to classify the pH range of a chitosan film from the concentrations of chitosan and anthocyanin used in the film formulation and from the measured color parameters (Cath, Cch, L *, a *, b *), as seen in Figure 3.

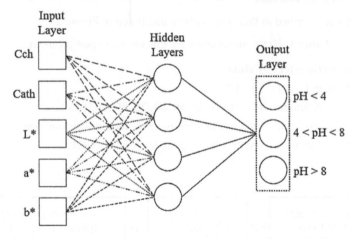

Figure 3. Representation of a neural network used to predict the pH at chitosan intelligent from the device.

The database used for the cross-validation consists of the pH values of 1.20, 3.79, 5.87, 7.77, 9.96, and 12.58 due to the previously exposed problem of having similar colors representing different pH values. This way was observed a better-quality training of the neural network, consequently giving better prediction results.

Several models were generated involving changes in the number of hidden layers (1 or 2), number of neurons (10 to 100), weight initialization (identity, random normal and random uniform), and activation function (hyperbolic tangent "tanh," rectified linear unit "ReLU" and sigmoid). Table 2 summarizes the best model regarding test accuracy.

Table 2. Best generated model

Hidden Layers	Neurons 1st layer	Neurons 2nd layer	Weight initialization	Activation function	Training accuracy	Precision Mean
2	80	20	Random normal	tanh	0.79	0.62

The following results represent the best model illustrated in Table 2. The confusion matrix is shown in Figure 4.

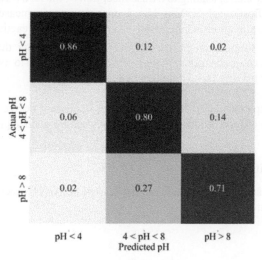

Figure 4. Confusion matrix: a summary of the prediction results of this study's best model.

The accuracy for the first and second classes was over 80 %, while the "pH > 8" class was misclassified as the "4 < pH < 8" class at a maximum rate of 27 %. This misclassification can result from similar color parameters (inputs L*, a*, b*) of the intelligent films immersed in pH values solutions closest to the edge of the two last classes (pH values 7 to 9, as seen in Figure 2.). Previous types show more accurate results due to more significant color changes in the films, and consequently, distinct color parameters, allowing the network to distinguish the pH ranges more efficiently. Again, the accuracy presented can result from similar film colors on the edge of both pH ranges.

Due to the relatively low test accuracy of the model (79 %) and the favorable misclassification rate of the third class, classification algorithms such as decision trees and support vector machines could be tested to obtain better pH-classification models. Mainly, the ANN applied to intelligent packaging is unexplored. The devices are based on biopolymer matrix formation that could not form a standard linkage between the chains. It is necessary to get more experimental data for successful of AI techniques. However, this is a very promising area in the future.

4. Conclusions

The artificial neural network was developed with data obtained from the color parameters for all formulations of chitosan intelligent films devices in contact with different pH solutions (from 1.20 to 12.58) and chitosan and anthocyanin concentration values. The chitosan intelligent films formulation with the most efficient colorimetric results were Cch = 0.5 % and Cath = 0.5 % (w/w).

A classification model to identify the devices out of the non-spoilage range (pH values from 4 to 8) ensures a better standard of quality and safety of the food product during the supply chain, also allowing the use of the model as a software sensor, assisting in the decision-making of changes. The best ANN showed a decent generalization accuracy, about 79 %, but still below the desirable rate. This could have resulted from similar sensor colors on the classes' edges, leading to a misconception of the ANN. However, the results showed that classification algorithms based on colorimetric measurements could be explored to indicate alterations/adulterations from pH variation reactions.

The sensor device formulation needs to be improved to show the most significant distinction between color changes; that way, it would be possible to use a more extensive database with more pH values, leading to a better trained ANN.

The commercial implementation of the sustainable, intelligent device is still a challenge, but a global market of food products with strict laws can be good support.

Acknowledgments

Capes - Coordenação de Aperfeiçoamento de Pessoal de Nível Superior and FAPESP – Fundação de Amparo a Pesquisa

References

Haykin, S. (2009). *Neural networks and learning Third Edition. Institute of Physics Conference Series* (Vol. 127).

Kuswandi, B., Wicaksono, Y., Jayus, Abdullah, A., Heng, L. Y., & Ahmad, M. (2011). Smart packaging: Sensors for monitoring of food quality and safety. *Sensing and Instrumentation for Food Quality and Safety*, 5(3–4), 137–146. https://doi.org/10.1007/s11694-011-9120-x

Mercier, S., & Uysal, I. (2018). Neural network models for predicting perishable food temperatures along the supply chain. *Biosystems Engineering, 171,* 91–100. https://doi.org/10.1016/j.biosystemseng.2018.04.016

Yoshida, C. M. P., Maciel, V. B. V., Mendonça, M. E. D., & Franco, T. T. (2014). Chitosan biobased and intelligent films: Monitoring pH variations. *LWT - Food Science and Technology*, 55(1), 83–89. https://doi.org/10.1016/j.lwt.2013.09.015

Proceedings of the 14th International Symposium on Process Systems Engineering – PSE 2021+
June 19–23, 2022, Kyoto, Japan ©2022 Elsevier B. V. All rights reserved.
http://dx.doi.org/10.1016/B978-0-323-85159-6.50307-9

Process Systems Engineering Guided Machine Learning for Speech Disorder Screening in Children

Farnaz Yousefi Zowj[a], Kerul Suthar[a], Marisha Speights Atkins[b], Q. Peter He[a]*

[a]*Department of Chemical Engineering, Auburn University, Auburn, AL 36849, USA*
[b]*Communication Sciences & Disorders, Northwestern University, Evanston, IL 60201, USA*

qhe@auburn.edu

Abstract

Auditory perceptual analysis (APA) is the primary method for clinical assessment of speech-language deficits, one of the most prevalent childhood disabilities. Due to multiple limitations of APA including being susceptible to intra- and inter-rater variabilities, automated methods such as Landmark (LM) analysis that quantify speech patterns for diagnosing speech disorders in children are developed. This work investigates the utilization of LMs for automatic speech disorder detection in children. Leveraging the similarities between disease detection in medical/clinical research and fault detection in process systems engineering (PSE), we propose to improve the detection of speech disorder in children via PSE principles. Specifically, the parsimony principle is followed for reducing feature and parameter spaces. Domain knowledge is utilized for generating a set of novel knowledge-based features to address the challenge of large within-class variations in LM measurements. A systematic study and comparison of different linear and nonlinear machine learning classification techniques are conducted to assess the effectiveness of the novel features in classifying speech disorder patients from normal speakers.

Keywords: systems engineering, machine learning, feature engineering, child speech disorder, landmark

1. Introduction

Speech-language deficits are one of the most prevalent childhood disabilities affecting about 1 in 12 children between the ages of three and five years old. Approximately 40 % of children with speech and language disorders do not receive intervention because their impairment goes undetected (Nelson et al., 2006). Auditory perceptual analysis (APA) is the main method for clinical assessment of disordered speech; however, results from APA are susceptible to intra- and inter-rater variabilities. Another factor to consider is that some children may be reluctant to participate in long testing sessions, and even if they do, transcription of large data sets of audio recordings is time-consuming and requires a high level of expertise from therapists. These limitations of manual or hand transcription based diagnostic assessment methods have led to an increasing need for automated methods to quickly and consistently quantify child speech patterns. Landmark (LM) analysis is such an approach that characterizes speech with acoustic markers that are developed based on the LM theory of speech perception. LM analysis has been suggested as the basis for automatic speech analysis (Ishikawa et al., 2017). Therefore, in this work we focus on the utilization of LMs for automatic speech disorder detection in children, with LMs extracted using SpeechMark® toolbox (Boyce et al., 2012). The description of each landmark detected by this tool and used in this study are presented in Table 1.

Different LM-based features have been proposed in the literature. The most common one is the counting of individual LMs, a.k.a. unigrams, which does not consider the specific order or sequence of the LMs. n-gram, which is a generalization of unigrams and is defined as a sequence of n consecutive LMs, takes the specific LM order into consideration when $n \geq 2$. It was found that n-gram counts (n=1, 2, 3, 4) were good features for depression detection. Besides n-gram count, time based LM features have also been proposed in the literature. These time based LM features include durations of the bigrams (i.e., 2-grams) and LM pairs (i.e., onset and offset of a LM as defined in Table 1) (Huang et al., 2019), and speech rate, which is defined as the number of phonetic units, such as syllables or words, uttered per unit time (Huici et al., 2016). Syllabic cluster (SC) analysis, which clusters LMs into syllabic units, has also been found to be an important feature for speech disorder detection (Boyce et al., 2012; Atkins et al., 2019).

Systems engineering principles have been instrumental in analyzing various biological data. For instance, we have recently reviewed a large body of research that utilizes PSE principles and techniques to address some of the technical challenges in Big Data analytics for biological, biomedical and healthcare applications, including the principle of parsimony in addressing overfitting, the dynamic analysis of biological data and the role of domain knowledge in biological data analytics (He and Wang, 2020). This work aims to improve speech disorder detection following PSE principles. The basic idea is that, despite the extremely different physical implementations, a human body can be viewed as a complicated biochemical plant and they share many common features at the system level. For example, a human disease or disorder can be viewed as an anomaly (or "fault" in a PSE term) in a human body. As a result, the principles and techniques developed for fault detection and diagnosis in the PSE community can be adopted to address some of the challenges in disease detection. The first step when developing a model is to select input features or variables from the PSE perspective. In particular, many PSE applications have demonstrated that raw features are often not the best features for capturing process characteristics, while engineered features with statistical and/or physical meanings are often more informative and robust in characterizing process behavior, and therefore are more effective in fault/disease detection (He and Wang, 2011, 2020; Lee et al., 2020; Shah et al., 2020; Suthar and He, 2021; Wang and He, 2010). Specifically, we propose novel knowledge-based features that are the ratios of the count of n-grams ($n \geq 2$) to that of unigrams. Ratios are usually better features than the absolute individual values in addressing the individual variations of samples within the same class. In addition, the parsimony principle of PSE leads us to develop robust models by reducing feature space (through feature selection) and parameter space (e.g., through the utilization of simple linear models). The final contribution of this work is a systematic study and comparison of different linear and nonlinear machine learning classification techniques and their effectiveness in classifying speech disorder patients from normal speakers.

The remainder of this work is organized as follows: Section 2 describes materials used in this study, the features proposed in this work, and introduces the analytical methods. Section 3 presents results and discussions of this work, and Section 4 draws conclusions.

2. Materials and Methods

2.1. Data description

The speech of 51 children ages 33 - 94 months was retrieved from the Speech Evaluation and Exemplars Database (SEED) (Speights Atkins et al., 2020). 39 were typically developing, and 12 with speech sound disorder without language impairment. Speech samples retrieved for this study were recordings of children uttering the word "flower" which is one of the 11 triage words of the set from Anderson and Cohen (2012).

2.2. Methodology

The raw features extracted from audio recordings using the SpeechMark MATLAB Toolbox include time stamp and strength of each LM listed in Table 1, plus SC count. We have adopted all LM and SC based features proposed in the literature, including n-gram counts, and duration and rate features based on LMs and n-grams. In addition, we explore LM strength based features and propose n-gram ratio based features to better address within-class variations. We have 189 engineered features after removing illegitimate or trivial features (e.g., n-gram counts that are all zeros, or ratios with a denominator of zero) and redundant features (i.e., the features that are highly correlated with an existing feature- Pearson correlation coefficient of 0.99). Recursive feature elimination with cross-validation (RFECV) from scikit-learn is utilized for feature selection with the default 5-fold cross-validation with a linear discriminant analysis (LDA) model as the classifier. Result indicates that only 10 features are needed to obtain the optimal cross-validation score. Nine out of the ten features (seven ratio-based and two strength-based features) are new features proposed in this work that have not been utilized before.

There is approximately a 3:1 class imbalance between the normal speaker samples and the disordered ones. The synthetic minority over-sampling technique (SMOTE) is utilized in this work, in which new samples are synthesized from the existing samples. Once the training set is balanced using SMOTE, we train four different classification algorithms, namely linear discriminant analysis(LDA), support vector machine (SVM), extreme gradient boosting (XGBoost), and random forest (RF), tune their hyperparameters with a

Table 1: Description of landmarks used in this study

Landmark	Description
g (glottis)	Onset (+) and offset (-) of sustained motion of vocal fold
b (burst)	Onset (+) and offset (-) of frication or bursts in an unvoiced segment
s (syllabicity)	Release (+) and closure (-) of sonorant consonant in a voiced segment
f (unvoiced frication)	Onset (+) and offset (-) of frication in an unvoiced segment
v (voiced frication)	Onset (+) and offset (-) of frication of in a voiced segment

10-fold stratified cross-validation (CV), apply the models to the left-out test samples, and report the sensitivity and specificity which are two most commonly used critical metrics when dealing with binary classification problems in healthcare. This whole procedure is referred to as one Monte-Carlo validation and testing (MCVT). We report the mean and standard deviation of sensitivity and specificity of 50 such MCVT runs, which is a robust way of comparing different modeling techniques and assessing their performances. Sensitivity is the true positive rate, i.e., the classifier's ability to detect diseased patients correctly, and specificity is the true negative rate, i.e., the classifier's ability to detect normal controls (i.e., the ones without diseases) correctly. We also use accuracy as a single measure when we need to evaluate the overall performance of a classifier. Throughout the modeling procedure, grid search and random search are used for hyperparameter tuning.

3.   Results and discussion

In this work, we conduct investigation from two perspectives: (1) comparing classification performance when different feature sets are used, and (2) comparing classification performance when different classification techniques are used. When comparing different features, the following two feature sets are studied: (a) the original 21 features directly obtained from the SpeechMark Toolbox, which include the counts and strengths of the ten LMs (listed in Table 1, considering both onset and offset) for each sample, plus one syllabic count per sample; and (b) The ten selected features selected via RFECV.

As shown in Table 2, when the 21 raw features are used, SVM with RBF kernel provides the best overall classification performance with 75.0 % accuracy (i.e., 75.0 % of the samples are classified correctly) and has somewhat a balanced specificity and sensitivity of 80.0 % and 70.0 %, respectively. LDA provides the second-best result with 71.0 % accuracy. The overall performances of all methods, linear or nonlinear, are relatively poor, indicating that the raw features are not very informative in classifying the two classes.

Next, we apply different classification methods to the selected ten features obtained through rational feature engineering and selection. The results are listed in Table 3 and shown in Figure 1. By comparing Table 2 and 3, we can see that the performances of all methods have significantly improved classification accuracy. The significantly improved performance with these features across all classification methods demonstrates that the proposed features are more informative than the raw features. Since there are seven features that are

Table 2: Comparison of classification performance based on raw features

Method	Sensitivity	Specificity	Accuracy
LDA	64	54	59
SVM (Linear)	68	74	71
SVM (Poly)	78	32	55
SVM (RBF)	70	80	75
SVM (Sigmoid)	80	54	67
XGBoost	50	86	68
RF	28	76	52

Table 3: Comparison of classification performance based on rationally engineered and selected features

Method	Sensitivity	Specificity	Accuracy
LDA	94.0	92.0	93.0
SVM (Linear)	86.0	92.0	89.0
SVM (Poly)	84.0	94.0	89.0
SVM (RBF)	72.0	94.0	83.0
SVM (Sigmoid)	88.0	88.0	88.0
XGBoost	60.0	88.0	74.0
RF	76.0	94.0	85.0

ratio based, the improved performance is most likely due to our hypothesis that ratio-based features are better at addressing individual variations of samples from the same class. In particular, LDA classifier achieves 94.0 %, 92.0 % and 93.0 % in sensitivity, specificity, and overall accuracy respectively. Several other methods also achieve nearly 90.0 % in sensitivity, specificity and overall accuracy, including SVM with linear, polynomial and sigmoid kernels. In comparison to the raw features, the sensitivity and specificity based on the selected engineered features are much more balanced.

4. Conclusions

In this work, we propose an automated computer-assisted screening method for detecting speech disorder in children following PSE principles. Specifically, the proposed knowledge-based features have been found particularly informative in characterizing audio recordings for speech disorder detection, and the parsimonious models derived from these features are found to be not only accurate but also robust. It is demonstrated that, with raw features, all classification methods fail to achieve high classification performance. In comparison, with the ten selected features, which contain nine features proposed in this work, the performances of all classification methods are significantly improved indicating that the proposed features are more effective for characterizing speech disorder using speech LMs. With knowledge-based features, LDA achieves a classification sensitivity of 94.0 %, speci-

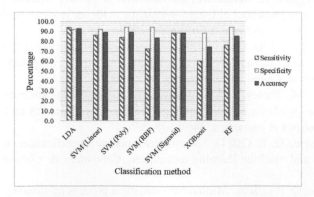

Figure 1: Comparison of classification performance when selected features are used

ficity of 92.0 %, and overall accuracy of 94.0 % compared to the SVM with RBF kernel using raw features reaching 70.0 % sensitivity, 80.0 % specificity, and 75.0 % overall accuracy. This work demonstrates that integration of domain knowledge into ML techniques can significantly improve the performance of purely data-driven or data-centric methods.

References

Anderson, C. and Cohen, W. (2012). Measuring word complexity in speech screening: single-word sampling to identify phonological delay/disorder in preschool children. *International journal of language & communication disorders*, 47(5):534–541.

Atkins, M. S., Boyce, S. E., MacAuslan, J., and Silbert, N. (2019). Computer-assisted syllable complexity analysis of continuous speech as a measure of child speech disorders. In *Proceedings of the 19th International Congress of Phonetic Sciences,(ICPhS 2019), Melbourne, Australia*, pages 4–10.

Boyce, S., Fell, H., and MacAuslan, J. (2012). Speechmark: Landmark detection tool for speech analysis. In *Thirteenth Annual Conference of the International Speech Communication Association*.

He, Q. P. and Wang, J. (2011). Statistics pattern analysis: A new process monitoring framework and its application to semiconductor batch processes. *AIChE journal*, 57(1):107–121.

He, Q. P. and Wang, J. (2020). Application of systems engineering principles and techniques in biological big data analytics: A review. *Processes*, 8(8):951.

Huang, Z., Epps, J., and Joachim, D. (2019). Investigation of speech landmark patterns for depression detection. *IEEE Transactions on Affective Computing*.

Huici, H.-D., Kairuz, H. A., Martens, H., Van Nuffelen, G., and De Bodt, M. (2016). Speech rate estimation in disordered speech based on spectral landmark detection. *Biomedical signal processing and control*, 27:1–6.

Ishikawa, K., MacAuslan, J., and Boyce, S. (2017). Toward clinical application of landmark-based speech analysis: Landmark expression in normal adult speech. *The Journal of the Acoustical Society of America*, 142(5):EL441–EL447.

Lee, J., Kumar, A., Flores-Cerrillo, J., Wang, J., and He, Q. P. (2020). Feature based fault detection for pressure swing adsorption processes. *IFAC-PapersOnLine*, 53(2):11301–11306.

Nelson, H. D., Nygren, P., Walker, M., and Panoscha, R. (2006). Screening for speech and language delay in preschool children: systematic evidence review for the us preventive services task force. *Pediatrics*, 117(2):e298–e319.

Shah, D., Wang, J., and He, Q. P. (2020). Feature engineering in big data analytics for iot-enabled smart manufacturing–comparison between deep learning and statistical learning. *Computers & Chemical Engineering*, 141:106970.

Speights Atkins, M., Bailey, D. J., and Boyce, S. E. (2020). Speech exemplar and evaluation database (seed) for clinical training in articulatory phonetics and speech science. *Clinical linguistics & phonetics*, 34(9):878–886.

Suthar, K. and He, Q. P. (2021). Multiclass moisture classification in woodchips using iiot wi-fi and machine learning techniques. *Computers & Chemical Engineering*, 154:107445.

Wang, J. and He, Q. P. (2010). Multivariate statistical process monitoring based on statistics pattern analysis. *Industrial & Engineering Chemistry Research*, 49(17):7858–7869.

Proceedings of the 14th International Symposium on Process Systems Engineering – PSE 2021+
June 19-23, 2022, Kyoto, Japan © 2022 Elsevier B.V. All rights reserved.
http://dx.doi.org/10.1016/B978-0-323-85159-6.50308-0

Emission and mitigation of CO_2 and CH_4 produced by cattle: a case study in the Brazilian Pantanal

Victor G. Moretti[a], Celma O. Ribeiro[a*], Claudio A. Oller do Nascimento[b], Julia Tomei[c], Alison J. Fairbrass[c]

[a] *Department of Production Engineering, EPUSP - Polytechnic School of the University of São Paulo, SP, Brazil*
[b] *Department of Chemical Engineering, EPUSP - Polytechnic School of the University of São Paulo, SP, Brazil*
[c] *UCL Institute for Sustainable Resources, University College London, London, UK*
v.gmoretti94@gmail.com

Abstract

This study aims to analyse the effects on emissions and land use resulting from the increase of productivity in cattle farming in Brazil, focusing on the Brazilian Pantanal region. Considering public data from the municipalities comprising the Pantanal, the study considers alternative scenarios for the reduction of ranching areas through sustainable intensification and technological improvements, identifying the effects of alternative policies on emissions and natural vegetation preservation. By employing a System Dynamics model, the study analyses the effects of these policies and identifies the relationship between cattle intensive practices, land use distribution and CH_4 emissions.

Keywords: Energy, Food and Environmental Systems, Emissions, Brazilian Pantanal, System dynamics

1. Introduction

The increase in the demand for agricultural commodities and meat products in the world has a significant impact on the carbon footprint and in the land use. According to FAO, the global livestock is responsible for 7.1 Gigatons of CO2-equiv per year, with cattle being responsible for about 65 % of the livestock sector's emissions (Gerber, 2013). Global methane emissions were addressed at the COP-26 event, held in November/2021 in Glasgow, Scotland, where the Methane Agreement was signed, establishing the global commitment to cut the gas emissions by 30% by 2030, bringing great challenges to the Brazilian cattle production chain.

Brazil is the second largest producer and the largest beef exporter in the world, with large part of the livestock's herd raised on pasture. The use of pastures reduces financial costs for the producers but results in a significant reduction of natural vegetation and productivity. The evolution of Brazilian cattle ranching practices in recent decades has increased productivity mainly through technological improvements, but the numbers are still far from ideal. The increase of pasture area from 135 million hectares in 1990 to 167 million hectares in 2019 (MapBiomas, 2020) indicates that there is room for significant improvement in the country.

The Pantanal, surrounded by the Amazonia, Cerrado, and Chaco biomes, is a megadiverse tropical wetland. It supports numerous valuable ecosystem services including the provision of wild foods, environmental regulation, and maintenance services such as carbon storage and sediment retention, a diverse tourism industry, water supply and a growing cattle ranching industry. The biome has recently received global attention due to an alarming increase in the frequency and extent of wildfires, the causes of which are diverse and complex. A challenge for decision-makers is how to report on the environmental and socio-economic impacts of these competing human uses of the biome, to inform decisions that maximize benefits to local, regional, and global communities while maintaining ecosystem integrity.

Based on system dynamics modelling, this study indicates how technological changes in cattle ranching can improve productivity and can lead to a decrease in emissions, and simultaneously to the preservation on the concept of natural capital.

2. Livestock in the Brazilian Pantanal region

The Pantanal encompasses an area of 150,355 km², occupying 1.76% of the Brazilian territory and comprising about 3% of the entire world's wetlands (IBGE, 2004). Brazilian beef cattle development is historically based on the expansion of the agricultural frontier, through the deforestation of regions without infrastructure and lands depleted by agriculture. Brazil has the second largest cattle herd in the world and is the largest meat exporter, exporting around 20% of its production. Even so, it still has productivity rates below other great producers' countries. In Pantanal, the cattle herd had an impressive increase in the last three decades, growing from around 3.5 million to 4.2 million heads (21.8 %) from 1990 to 2018 in the municipalities covered by the bioma (IBGE, 2004), resulting in a conversion of around 2.1 Mha of land to new pastures, that went from 2.6 Mha to 4.7 Mha. The decrease in the savanna and forest formations during the period, from 9,0 Mha to 7.0 Mha (MapBiomas, 2020), indicates the use of part of these lands to ranching activities through deforestation. Although cattle raising methods have improved in some municipalities with intensification of cattle practices, use of new technologies and high-quality production, the efforts have not been enough to assure the Pantanal natural resources conservation.

This study focuses on livestock activities as one of the major factors driving land change and emissions in the Brazilian Pantanal. Negative impacts on ecosystem services resulting from the decrease in native vegetation, methane and CO_2 emissions and socio-economic effects are assumed to be the main results from the significant increase of new pastures of the region. Enteric methane emissions from cattle are responsible for an expressive amount of world's GHG emissions and improving animal productivity is recognized as an important pathway to achieve global sustainable goals.

Alternative public policies can modify the global greenhouse emissions resulting from livestock as presented by Avery Cohn *et al.* (2014), who examined policies to encourage cattle ranching intensification in Brazil as a strategy to reduce GHG emissions. Semi intensive practices may not be the best alternative for Pantanal region, but several alternatives related to productivity increase and emission reductions can be adopted. The strength of the policy depends on wellbeing and the economic value of ecosystem services.

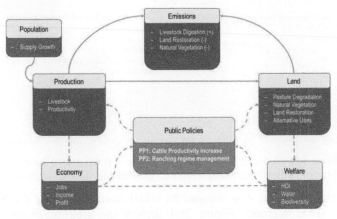

Figure 1 – Main subsystems. Source: Own Elaboration

3. The system dynamics (SD) model

System Dynamics model applies for dynamics complexity related problems in which links of different natures are made through various types of variables, flows, auxiliary variables, and parameters, with simulation modelling based on feedback systems theory. In this paper, the interaction of seven subsystems (land, economy, policies, emissions, welfare, livestock, and population) were considered, based in Fiddaman (2012).

The proposed system dynamics (SD) model illustrates the stock and flow of cattle pasture and pastureland, focusing on the emissions subsystems and measuring the relationships among indicators identified in each subsystem, as showed is Figure 1. For that, it is assumed that the CH$_4$ resulting from enteric fermentation of the digestive process of the cattle is the main source of emissions in Pantanal. Additionally, the study considers the pastures as the major cause of natural vegetation reduction. Water and biodiversity changes are considered exogenous, possibly resulting from climatic changes. The main driving factors for a policy shift from the Business as Usual (BAU) model to a climate policy focusing on emission reduction through the adoption of good ranching practices and restoration of natural vegetation are assumed to result from social and environmental pressures.

4. Material and methods

For this research, the Insight Maker®, a free simulation tool that runs in the web browser, was used to support the system dynamics application, enabling diagramming, and modelling features and creating representations of the system. The SD model is presented with emphasis on sustainable cattle ranching practices to cover issues on Pantanal over the period of 1990-2019. Through the SD simulation, we can analyse the impact of productivity policies on CH$_4$ emission and land use and the relationship between cattle intensive practices, natural vegetation restoration and land use distribution.

Historical time series from municipalities comprising the Brazilian Pantanal were gathered from public databases, from 1990 to 2019 in a yearly basis, considering information related to land use, cattle production, and regional socioeconomic indicators. Parameters related to jobs, income, cattle productivity and CH$_4$ emissions, among others, were obtained from literature.

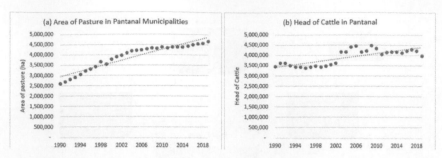

Figure 2 - Main data used in the BAU scenario. Figure 2(a) represents the area of pasture in Pantanal Municipalities. Figure 2(b) represents the Pantanal head of cattle supply growth

Figure 2 represents actual data adopted as initial assumptions for the SD model, to illustrate the BAU scenario. Figure 2(a) depicts the area of pasture in Pantanal municipalities, while Figure 2(b) gives rise to the supply growth of head of cattle in Pantanal, which is assumed to be linear for this study, from 1990 to 2019.

A reduction proxy of 38.5 % in the use of area (ha/head of cattle) was adopted to illustrate a pasture intensification scenario, based on an experience with cattle in the Mato Grosso state. The productivity gain when applying intensive cattle in comparison to the traditional one gives rise to a great number of recovered pastures (Fabiano Alvim Barbosa, 2015).

A report by EMBRAPA (2014) showed that the intensification of cattle, despite increasing methane emissions, increases carbon sequestration in the recovered pastures, resulting in a positive net balance in the total emissions. According to their study, the average annual net emission of enteric methane (CH_4) per animal is 57 CH_4 kg per year per animal and that, through better cattle management practices, such as ranching and feeding conditions, this number can reach 37.7 kg, representing a 33.9 % emission reduction of this gas, which is one of the main global greenhouse gases.

To simulate the impact of cattle intensification policies under emission and land use behaviours, two hypothetical public policies (PP1 and PP2) were proposed. With supply growth and pressure for more sustainable cattle, these policies aim to increase productivity, reconciling rural development, supply growth and environmental conservation, represented here by CH_4 emissions and land use from cattle. PP1 assumes that all supply growth shall be attended by new ranching practices with higher productivity (heads/ha). PP2 considers that, after its implementation, a percentage of the existing cattle will have their ranching regime changed, moving from extensive to semi-intensive cattle practices within a fixed horizon (5 years).

The SD model was tested to evaluate its accuracy in interpreting the actual scenario (BAU – Business as Usual model). The policies were assumed to be adopted in 2000 (year 10) and with a linear ramp-up of 5 years for full implementation of policy PP2 in year 2005 (year 15).

5. Results

By employing the System Dynamics modelling, the study analyses the results of the proposed policies and identifies the relationship between cattle intensive practices, land use distribution and CH4 emissions. For that, three scenarios were considered. Scenario 1 considers the Business as usual (BAU), which replicates what happened with the number of cattle heads and land use between 1990 and 2019 in Pantanal; Scenario 2 contemplates the application of PP1 and PP2 policies disregarding any supply growth. This is a theoretical scenario aimed at controlling and isolating the effect of the policies and illustrating how supply growth plays an important role in land use and global emissions, when compared to the Scenario 3, which applies, in addition to the policies, an annual supply growth.

The application of policies PP1 and PP2 reflects how ranching time optimization and better management practices impacts on cattle productivity and CH$_4$ emissions. In comparison to the BAU scenario, it is possible to identify an improvement in the land use area, as showed in Figures 3(a) and 3(b). By comparing Scenario 3 with Scenario 1 (BAU), the natural vegetation, which in 2019 totalled 15.1 Mha, would reach 16.2 Mha (an increase of 6.8 %). Supply growth is responsible for more than 10 % of natural vegetation consumption with pasture, by comparing Scenario 2 and 3. The pastureland would experience a significant reduction of almost 28 % in Scenario 3.

Regarding the number of heads, considering the policies application in Scenario 3, in year 30, intensive cattle practices would represent more than 35 % of the region's cattle supply. The total emissions per year, in tons of CH$_4$, by year 30, went from 244,100 tons of CH$_4$ in the BAU scenario to 203,000, by applying policies PP1 and PP2 (Scenario 3), representing an important reduction of 20.4 %. When analyzing the number of tons of CH$_4$ emitted per ton of meat, considering cattle with average weight of 400 kg, there is a significant reduction of 26.4% (from 2.28 ton of CH$_4$ per ton of meat to 1.67 ton of CH$_4$ per ton of meat). Even with supply growth (Scenario 3), this number remains descending in the analyzed horizon, as shown in figure 3(d).

From the perspective of socioeconomic impact, there was an improvement in HDI index. However, it is not possible to correlate this improvement with livestock practices in the region, since it proportionally follows the growth of the country's HDI, resulting from federal socio-economic measures implemented in the last 30 years. In addition, the income, with the new practices would increase by 50%, according to specialists, since intensive techniques would require greater professional qualification of ranching employees, contributing to technical development and quality of life improvement for the region.

6. Conclusion

This study aimed to analyze the impacts of productivity increase in cattle farming on the Brazilian Pantanal region on CH4 emissions and land use. It was considered the application of public policies (PP1 and PP2) that only admitted intensification practices for new cattle and proposed a change in a percentual of existing cattle, from extensive to intensive regime, by year 10. The proposed SD model compared the actual scenario, based on actual data (Scenario 1), with two theoretical scenarios (Scenario 2 and 3), that contemplated these new policies, from 1990 to 2019. The employment of a System dynamics model connects and gives rise to the system's behavior, enabling better policy decisions and design.

It was found that the application of public policies that are not so restrictive can effectively reduce the number of emissions and offer more efficient livestock practices in the Pantanal. Even considering supply growth, the total emissions per year went from 244,100 ton of CH4 in the actual scenario to 203.000 ton of CH4 in Scenario 3, in year 30 (a reduction of more than 20 %).

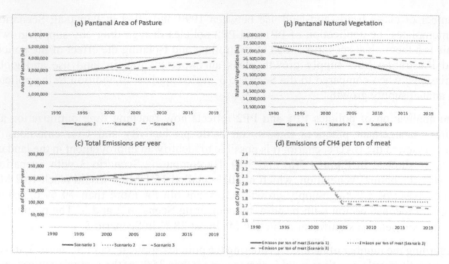

Figure 3 - Results from scenarios analysis for Pantanal cattle case. Figure 3(a): land used for pasture; Figure 3(b): natural vegetation available in Pantanal; Figure 3(c): total CH4 emissions per year, in tons of CH4; Figure 3(d): total emissions of CH4 per ton of meat of cattle

A reduction of 26.4 % in tons of CH4 per ton of meat could also be observed, (from 2.28 ton of CH4 per ton of meat to 1.67 ton of CH4 per ton of meat). The natural vegetation could experience an important achievement, by applying these policies, reaching an increase of almost 7 % (from 15.1 Mha to 16.2 Mha) in comparison to the actual scenario.

From a socioeconomic point of view, these policy changes would lead to an increase in the region's income and a need for more professional qualification. An in-depth study is recommended to study how improvements in livestock production practices could lead to an improvement in socioeconomic conditions in the region.

These important results ensure that improvements in land use and in CH_4 emissions can be made with easy-to-implement measures, helping Brazilian institution to meet the objective of COP-26.

7. References

Avery Cohn, A. M. (2014). Cattle ranching intensification in Brazil can reduce global greenhouse gas emissions by sparing land from deforestation. Proceedings of the National Academy of Sciences, 25.

EMBRAPA. (2014). Revista XXI - Ciência para a vida. 27-29, 14.

Fabiano Alvim Barbosa, B. S. (2015). Cenários para a Pecuária de Corte Amazônica. Belo Horizonte.

Fiddaman, T. S. (2012). Exploring policy options with a behavioral climate–economy model. System Dynamics Review, 18(2), 243-267.

IBGE. (2020). PPM - Pesquisa da Pecuária Municipal.

MapBiomas. (2020). MapBiomas Project - Collection 5.0 of Brazilian Land Cover & Use Map Series. Acesso em 26 de 06 de 2021.

P. J. Gerber, H. S. (2013). Tackling climate change through livestock: A global assessment of emissions and mitigation opportunities. FAO.

Proceedings of the 14th International Symposium on Process Systems Engineering – PSE 2021+
June 19–23, 2022, Kyoto, Japan ©2022 Elsevier B. V. All rights reserved.
http://dx.doi.org/10.1016/B978-0-323-85159-6.50309-2

Promoting phosphorus recovery at livestock facilities in the Great Lakes region: Analysis of incentive policies

Edgar Martín Hernández[a,b]*, Yicheng Hu[c], Victor M. Zavala[c], Mariano Martín[a], Gerardo J. Ruiz-Mercado[d,e]

[a]*Department of Chemical Engineering, University of Salamanca, Plza. Caídos 1-5, 37008 Salamanca, Spain*
[b] *Oak Ridge Institute for Science and Education, hosted by Office of Research & Development, US Environmental Protection Agency, 26 West Martin Luther King Drive, Cincinnati, Ohio 45268, Unied States*
[c] *Department of Chemical and Biological Engineering, University of Wisconsin-Madison, Madison, Wisconsin 53706, Unied States*
[d] *Center for Environmental Solutions and Emergency Response (CESER), U.S. Environmental Protection Agency, 26 West Martin Luther King Drive, Cincinnati, Ohio 45268, Unied States*
[e] *Chemical Engineering Graduate Program, University of Atlántico, Puerto Colombia 080007, Colombia*

emartinher@usal.es

Abstract

Intensive farming activities release large amounts of phosphorus into the environment in the form livestock manure, contributing to the eutrophication of waterbodies, and can lead to algal bloom episodes. This work conducts a study on the design and analysis of incentive policies to promote the implementation of phosphorus recovery systems at intensive livestock facilities minimizing their negative impact on the economy of livestock operations. The Great Lakes area is used as case study, analyzing the economic impact of the implementation of phosphorus recovery systems, either considering the deployment of standalone phosphorus recovery processes, or integrated systems combining nutrient recovery with anaerobic digestion for the production of electricity. Moreover, the fair allocation of monetary resources when the available budget is limited has been studied using the Nash allocation scheme.

Keywords: Environmental Policy, Circular Economy, Resource Recovery, Nutrient Pollution, Organic Waste

1. Introduction

Since the 19th century, the agricultural sector has experienced an accelerated industrialization pursuing its intensification to satisfy the demand of food and agricultural products, i.e., increasing the agricultural production per unit of input resources (land, labor, time, fertilizer, seeds, and investment) (FAO, 2004). However, multiple environmental challenges must be faced as a consequence of the industrialization of the agricultural and farming activities. One of the main sources of concern are the nutrient releases from intensive livestock facilities in the form of manure, which contribute to the nutrient pollution of waterbodies, and contribute to eutrophication and harmful algal blooms (HABs) episodes. Therefore, nutrient recovery and recycling is not only a desirable but also a necessary approach to develop a more sustainable agricultural paradigm.

In this work, the effect of different incentive policies to promote the implementation of P recovery systems at concentrated animal feeding operations (CAFOs) is evaluated. Since P recovery systems can be implemented either as standalone systems, or integrated with biogas production and upgrading processes, the combination of incentives for the recovery of both phosphorus and electricity has also been considered. In addition, we study the allocation of limited monetary resources using a Nash scheme; this determines the break-even point for the allocation of monetary resources based on the availability of incentives.

2. Framework for the assessment of incentive policies

A two-stage framework is proposed for the evaluation of incentive policies, as shown in Figure 1. In the first stage, the size and geographical location of the studied CAFOs are analyzed, selecting the most suitable P recovery process for each CAFO assessed from a pool of six P recovery technologies. We note that these technologies can be implemented either standalone, or integrated with anaerobic digestion (AD) for biogas production. The P recovery selection stage is composed of different models that are fed with data regarding the type and number of animals in the studied CAFO, as well as its geographical location (box *a*). The assessment of the regional environmental vulnerability to nutrient pollution is performed through a geographical information system (GIS) model (box *b*). Additionally, the techno-economic assessment of the different phosphorus recovery technologies, and biogas production in those cases where this process is considered, is performed based on the characteristics of each CAFO under evaluation in parallel (box *c*). The information returned by these models is normalized and aggregated in a multi-criteria decision analysis (MCDA) model to select the most suitable nutrient management technology for the evaluated livestock facility (box *d*). A detailed description of decission-support framework used for the assessment and selection of P recovery systems considering the environmental vulnerability to nutrient pollution can be found in Martín-Hernández et al. (2021).

In a second stage, the effect of incentives on the economic performance of the P recovery systems selected in the first stage is evaluated. This study is performed through an economic model that estimates the profit of the P recovery systems implemented and the total cost of phosphorus recovery. Additionally, a cost-benefit analysis comparing the recovery cost and the economic looses due to nutrient pollution is performed (box *e*).

The incentives considered in this study to promote the implementation of nutrient recovery systems are phosphorous credits (P credits) and renewable electricity certificates (REC). P credits can be articulated as a system for the acquisition of phosphorus emission allowances, or conversely, as an income obtained by recovering phosphorus, which is the P credits definition considered in this work. In addition, in those scenarios where biogas production is integrated, REC are also considered. REC incentives provide a fixed remuneration for the electricity produced, which can result in a higher transaction price of electricity to cover the extra production costs and guarantee long-term price stability.

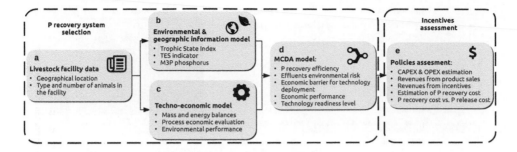

Figure 1: Flowchart of the models for selection, sizing, and evaluation of nutrient recovery systems at livestock facilities.

The states of the Great Lakes area, i.e., Minnesota, Indiana, Ohio, Pennsylvania, Wisconsin, and Michigan, are the study region considered to analyzed the impact of incentive policies on P recovery. The CAFOs considered for the deployment of livestock waste treatment processes are those livestock facilities with more than 300 animal units reported in the National Pollutant Discharge Elimination System (NPDES) by the U.S. Environmental Protection Agency (US EPA) in the states under evaluation. An animal unit is defined as an animal equivalent of 1,000 pounds live weight. 2,217 CAFOs are considered in total.

3. Results and discussion

The results of the implementation and allocation of incentives for phosphorus recovery are shown in this section. The results regarding the techno-economic assessment of the different processes, technology selection, and phosphorus recovered can be found in Martín-Hernández et al. (2021).

3.1. Combined effect of incentives for phosphorus and renewable electricity recovery

The net processing costs obtained for different scenarios combining multiple values of P credits and REC incentives are shown in Figure 2. They show a base cost for the recovery of phosphorus between 5.81 and 12.47 USD per ton of processed manure if no incentives or anaerobic digestion stages are considered. The installation of biogas processes is not profitable by itself, increasing the processing costs by 1.2-1.9 times over the base case, and it is only beneficial for large size CAFOs unser specific scenarios combining moderate P credits (>3 USD/kgP recovered) and electricity incentives (>60 USD/MWh). The scenarios combining states with large CAFOs and high value for P credits, and the optional production of renewable energy from biogas result in negative processing costs, i.e., they are profitable. Since the analysis of the different scenarios is carried out at the state level, this means that the profitable P recovery processes are able to balance out the non-profitable ones in the state.

3.2. Environmental cost-benefit analysis

The cost-effectiveness of the total cost involved in phosphorus recovery, including the amortization of the investment, operating costs, and total cost of incentives for each sce-

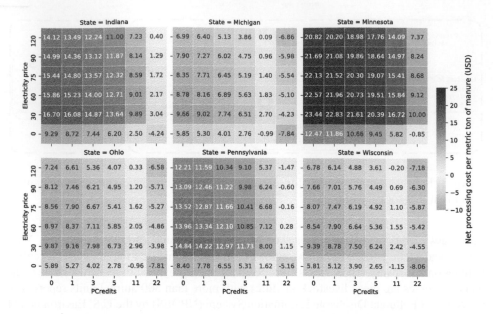

Figure 2: Net processing cost per metric ton of manure (USD).

nario under evaluation has been studied to determine the long-term economic benefits of phosphorus recovery.

Figure 3 shows the total cost of phosphorus recovery under different policies, including all the items previosuly described. The economic losses due to phosphorus releases have been estimated in 74.5 USD per kg of phosphorus released by Sampat et al. (2021). It can be observed that P recovery is economically beneficial in all scenarios considered, even those resulting in the largest P recovery costs as a result of high incentive values. The role of the size of CAFOs in the cost of phosphorus recovery can be also observed in this study. Those states with larger average size of CAFOs, such as Wisconsin, Ohio, and Michigan, have recovery costs significantly lower than the states where medium and small size CAFOs are predominant.

3.3. Fair distribution of incentives

In those scenarios where the available budget is not sufficient to cover the operating expenses of the unprofitable P recovery processes, the fair distribution of incentives is a relevant problem. For the CAFOs in the study region considered in this work, the necessary budget to cover the economic losses of the unprofitable P recovery systems is 222.6 MM USD. We note that Due to the marginal benefits obtained by installing AD processes, as described in Section 3.1., the implementation of only P recovery systems is assumed in both studies, and therefore only incentives for P recovery are considered.

The fair allocation of limited incentives has been addressed by using the Nash allocation scheme. This approach has been selected because this scheme is able to capture the scales of the different stakeholders (CAFO facilities) in order to achieve a fair distribution of a certain resource (incentives), as it was demonstrated by Sampat and Zavala (2019).

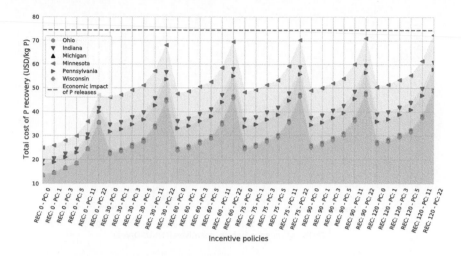

Figure 3: Comparison of the total cost of phosphorus recovery for each scenario assessed and the environmental remediation cost due to phosphorus releases. *REC* denotes the electricity incentive values considered in USD/MWh, and *PC* denotes the value of phosphorus credits in USD/$kg_{P\ recovered}$. The red dotted line represents the economic losses due to phosphorus releases to the environment.

Figure 4 illustrates the distribution of incentives as a function of the net revenues of the P recovery system installed in each CAFO c before any incentive is applied. The cases where the available budget are the 10%, 30%, 50%, 70%, and 100% of the incentives needed to cover the economic losses of unprofitable CAFOs are analyzed (22.3, 66.8, 111.3, 155.8 and 222.6 MM USD respectively). Since the available incentives are limited, a break-even point determining the profitability level of the P recovery systems below which the incentives should be allocated is set for each scenario. As a result, the fewer incentives available, the more restrictive the break-even point is. Additionally, it can be observed that the displacement of the break-even points is progressively reduced as the available incentives increase, resulting in a marginal improvement between the scenarios considering the 50% and 70% of the economic resources needed to guarantee the economic neutrality of the nutrient management systems.

4. Conclusions

This work aims at analyzing incentive policies for the implementation of phosphorus recovery systems for the abatement of nutrient releases from CAFOs. The deployment of phosphorus recovery processes is self-profitable through struvite sales only for the largest P recovery processes, which represent less than the 5% over the total CAFOs in all the studied states. However, the application of P credits increases the fraction of profitable processes around to 100% in the states with large-size CAFOs (Michigan, Ohio and Wisconsin), and up to 80% for the states with medium-size CAFOs (Indiana and Pennsylvania). The incentives necessary for covering the economic losses of unprofitable CAFOs estimated in 222.6 million USD. The integration of phosphorus recovery technologies with

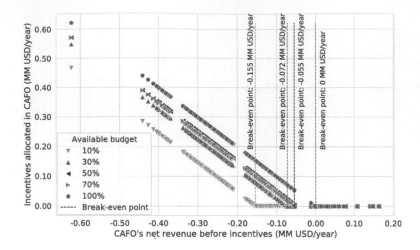

Figure 4: Distribution of incentives considering the Nash allocation scheme. Scenarios assuming available incentives equal to the 10%, 30%, 50%, 70%, and 100% of the incentives needed to cover the economic losses of unprofitable P recovery systems in the Great Lakes area are illustrated.

anaerobic digestion and biogas upgrading processes does not result in any practical improvement in terms of economic performance. The total cost of phosphorus recovery, including the investment amortization, operating costs, and total cost of incentives is lower than the long-term economic losses due to phosphorus pollution for all the evaluated states and policies, proving that sustainable nutrient management systems are economically and environmentally beneficial. Additionally, the fair distribution of limited incentives has been studied, determining the break-even point for the allocation of monetary resources based on the availability of incentives.

Disclaimer: The views expressed in this article are those of the authors and do not necessarily reflect the views or policies of the U.S. Environmental Protection Agency. Mention of trade names, products, or services does not convey, and should not be interpreted as conveying, official U.S. EPA approval, endorsement, or recommendation.

References

FAO, 2004, The ethics of sustainable agricultural intensification, FAO.

E. Martín-Hernández, M. Martín, G.J. Ruiz-Mercado, 2021, A geospatial environmental and techno-economic framework for sustainable phosphorus management at livestock facilities, Resources, Conservation and Recycling, 175, 105843.

A.M. Sampat, A. Hicks, G.J. Ruiz-Mercado, V.M. Zavala, 2021, Valuing economic impact reductions of nutrient pollution from livestock waste, Resources, Conservation and Recycling, 164, 105199.

A.M. Sampat, V.M. Zavala, 2019, Fairness measures for decision-making and conflict resolution, Optimization and Engineering 20, 1249–1272.

Proceedings of the 14th International Symposium on Process Systems Engineering – PSE 2021+
June 19-23, 2022, Kyoto, Japan © 2022 Elsevier B.V. All rights reserved.
http://dx.doi.org/10.1016/B978-0-323-85159-6.50310-9

Production of ethanol, xylitol and antioxidants in a biorefinery from olive tree wastes: process economics, carbon footprint and water consumption

Luis David Servian-Rivas[a], Ismael Díaz[a], Elia Ruiz Pachón[b], Manuel Rodríguez[a], María González-Miquel[a], Emilio J. González[a]

[a]*Departamento de Ingeniería Química Industrial y del Medioambiente, Escuela Técnica Superior de Ingenieros Industriales, Universidad Politécnica de Madrid, C/ José Gutierrez Abascal 2, 28006 Madrid, Spain*

[b]*Bioenergy and Energy Planning Research Group, EPFL, Switzerland*
luisdavid.servian.rivas@upm.es

Abstract

This work focuses on the evaluation of the economic performance and carbon and water-related environmental impacts of a biorefinery scheme using olive tree pruning wastes as feedstock. The considered process is based on a multiproduct biorefinery producing 0.66 m^3/h of ethanol, 114 kg/h of xylitol and 144.4 kg/h of antioxidants valorising both the cellulose/hemicellulose and the extractives fraction. The plant is not energetically self-sufficient (even considering the combined heat and power production from the combustion of the waste solids fraction) requiring a supply of natural gas. Nonetheless, the plant shows a positive investment balance with a net present value of 11.56 M€ in a 20-year period being the most important product the antioxidant which represent 66.2 % of total revenues. In addition, the biorefinery also shows a better environmental profile in comparison to the business-as-usual production of ethanol, xylitol and antioxidant.

Keywords: Lignocellulosic biorefinery, Technoeconomic Analysis, Life Cycle Assessment.

1. Introduction

Europe established in the European Green Deal an impulse on circular bioeconomy, replacing fossil-based materials and energy by biobased solutions (European Commission, 2019), in which biowaste is the principal feedstock. This strategy helps to mitigate the problem of the generation of high-volume residues, and at the same time products of great interest are obtained. Furthermore, several biorefinery projects in Europe using biowastes as feedstock have already shown a substantial reduction of energy and greenhouse gas emissions (GHGs) with respect to the reference fossil-based processes (IEA, 2021).

Mediterranean countries in Europe could take advantage of incorporating this circular bioeconomy, because one of the most promising feedstocks for the obtention of a wide range of chemical products is olive tree waste (Lo Giudice et al., 2021). Those countries produce 70 % of the total world olive oil production (21 million t), being Spain the main world producer (1.8 million t) (FAOSTAT, 2021). Furthermore, the olive oil industry (olive cultivation and olive oil production) produces 10 million tonnes of residues per year, with a great potential to be valorised by means of different biorefinery processes.

One of the residues is olive tree pruning (OTP), which is an abundant lignocellulosic residue (1-2 t/ha each year) from olive cultivation. This residue must be removed from the field to prevent the propagation of vegetal pests (Contreras et al., 2020), but it is usually burned in the field, being this burning the main source of fine aerosol in winter in Mediterranean countries (Kostenidou et al., 2013).

OTP has been used to obtain marketable products such as ethanol and xylitol and it is also a good source of natural antioxidants such as hydroxytyrosol, tyrosol, and oleuropein (Conde et al., 2009a). There are previous works that have studied the economic feasibility of plants that manufacture those products (Susmozas et al., 2019); however, an assessment of the associated environmental impacts (carbon and water) has not been carried out so far.

This work focuses on the economic performance and the carbon and water-related impact assessment of a biorefinery plant using OTP as feedstock. The biorefinery was modelled using commercial process simulation software (Aspen Plus). The subsequent economic and environmental analysis carried out were rooted on data from the process simulations. Economic performance is studied using the net present value as economic metric. The carbon footprint and water consumption were accounted using a life cycle assessment (LCA) methodology to compare the impacts of the biorefineries with respect to the business-as-usual solution.

2. Materials and Methods.

2.1. Feedstock

The OTP composition %wt (dry basis)is the same described by Ballesteros et al.(2011): 28.0 cellulose (as cellulose); 20.6 hemicellulose (as xylan); 25.2 lignin; 2.7 acetic groups; 5.9 Ash (as CaO); 7.9 glucose, 0.1 arabinose, 0.1 mannose, 1 galactose, 0.7 xylose, 4 mannitol, and 3.8 antioxidant. (Ballesteros et al.(2011). The OTP is assumed to be already crushed (this will affect the price).

2.2. Process modelling.

Materials and energy balances of the considered scheme were computed using Aspen Plus v.11. Two thermodynamic packages (NRTL and UNIF-LL) were used in this simulation. The semi-empirical NRTL model was used as the default method in the simulation, with the only exception the liquid-liquid extraction of antioxidants. For this purpose, the UNIF-LL package was used to estimate the activity coefficients required in calculations of fluid phase equilibria of the mixtures involving antioxidants, ethyl acetate, sugars, and water. Missing pure component parameters were estimated using the built-in property estimation models (group contribution models from molecular structure). The antioxidant is simulated as hydroxytyrosol because it represent the 90 % of total compounds in this fraction (Conde et al., 2009b). The process is design to work with 96 t/day of OTP, which represent the 0,7 % of total available OTP residue at year. Process parameters such as stoichiometric conversion and operating conditions for pre-treatment, saccharification, and fermentation were retrieved from literature.

2.3. Process description.

In figure 1 can be observed that OTP is firstly subjected to a hot water extraction process with water at 393.15 K and 5 bar, to extract 90 % of the extractives. A mixture of solids and liquids is obtained in this unit operation: a liquid fraction with the extractives and a solid fraction (SF) with the insoluble solids. The liquid fraction is quenched to 308.15 K and pumped to a liquid-liquid extraction process using ethyl acetate solvent in a 3:1 (v/v) liquid-solvent ratio. Ethyl acetate selectively extracts 99 % antioxidants from the feed to the organic liquid, while sugars remain in the aqueous fraction.

On the other hand, a significant amount of ethyl acetate dissolves in the aqueous stream so it must be separated and sent back to the process. This is required to improve the economy of the process (recovering as much solvent as possible) and to reduce the organic fraction sent to the wastewater treatment plant.

The organic liquid with the antioxidant dissolved is sent to a flash vessel at 373.15 K and 0.1 bar, where the solvent is recovered by evaporation (light component) and the extracted antioxidants are collected from the bottom with a purity of 97.33 wt %. A total recovery of 90 % of the incoming antioxidants from OTP is obtained.

The SF is sent to a steam explosion process with steam at 468.99 K and 14 bar and phosphoric acid (1 %) as catalyst. This operation breaks lignin and eases the subsequent saccharification and fermentation of cellulose. Two fractions are obtained after filtration: water-insoluble solids (WIS) and water-soluble solids (WSS).

The WSS fraction is composed of xylose as the main organic constituent and is sent to the xylitol production route. The WSS is cold down at 323 K and mixed with lime to remove furans and phenolic compounds, followed by neutralization with H_2SO_4. The resulting solids are filtered and disposed. The filtered liquid is fermented at 303 K, obtaining a yield of 75 % of xylose to xylitol, then subjected a filtration to remove yeast, and evaporation process at 313.15 k and 0.5 bar to concentrate xylitol up to 50 wt%. Finally, xylitol is mixed with ethanol to decrease xylitol solubility at a 0.13:1 (w/w) ethanol-dissolution ratio. Finally, a crystallization process takes place at 268.15 K and 1 bar to obtain xylitol crystals with 99 % purity.

The WIS mainly contains insoluble cellulose and lignin. This fraction is converted to ethanol by a conventional sequence of pre-saccharification, saccharification, and fermentation steps (PSSF). Pre-saccharification is carried out at 323 K. Then the resulting stream is further cooled to 308 K and fermented to ethanol (70% of the theoretical yield is obtained). The resulting ethanol stream is firstly purified in a beer column which increases ethanol concentration up to 50 %wt and further concentrated in a purification column which increases ethanol concentration to 93 %. Finally, 99% ethanol is obtained using molecular sieves.

The aqueous wastes from the process are sent to the wastewater treatment system (anaerobic and aerobic processes) where methane, sludge, and biogas are produced and then burned in a combined heat and power section to partially compensate the plant energy requirements.

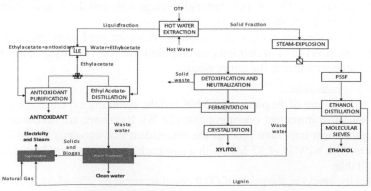

Figure 1. Biorefinery process flow diagram

2.4. Economic evaluation

Net present value (NPV) has been used as the financial metric to measure the profitability of the plant with a lifetime of 20 years, an interest rate of 15 % and 25 % of taxes. The Inside battery limit (ISBL) plant cost is obtained by means of Aspen Economic Analyzer (Aspen Technologies Inc., USA). The cost of all equipment has been increased by 30% to cover possible uncertainties.

A sensitivity analysis is also studied, varying the prices of the OTP and the three products until the NPV is equal to zero. The base prices of the products are: 0.58 €/L ethanol, 3170 €/t xylitol, 11000 €/t antioxidants and the OTP 44.77 €/t (Susmozas et al., 2019)

2.5. Carbon footprint and water consumption

The carbon and water footprints analysis were calculated in line with the Life Cycle Assessment (LCA) principles, which is an internationally standardized methodology that helps to quantify the environmental impact of processes and products.

First, the main goal is to quantify the environmental impacts of producing ethanol, xylitol, and antioxidants from OTP, compared to the conventional counterparts of biorefinery products (reference system). The reference system is made up of ethanol from ethylene, xylitol from corncob, and propyl gallate from spruce bark was considered as the counterpart reference for antioxidant production. The overall functional unit of the whole system is 1 kg of OTP.

As the studied biorefinery is a multiproduct system, an economic allocation approach has been adopted following the value-based methodology proposed by Gnansounou et al.(2015).

Data to build the OTP biorefinery inventory are obtained from the process simulations carried out in this work, this information is quantified and transformed into environmental impacts categories ReCiPe 2016 impact assessment methodology. SimaPro v11 software was used to assist in the calculation of the four environmental impacts selected: global warming (kg CO_2) and water consumption (m^3), The database Ecoinvent 3.5 is used.

3. Results and Discussion

3.1. Process design and simulation.

The modelled biorefinery produces 0.66 m^3/h of ethanol, 114 kg/h of xylitol and 144 kg/h of antioxidants with a raw material consumption of 4040 kg/h. The thermal need of the plant is 61.2 GJ/h, but the plant just produces 23.9 GJ/h from the combustion of the solid organic residues and biogas of the plant, so the plant needs an external supply of energy by means of natural gas. This plant is thermally integrated, and the consumption of heat is already optimized using the Aspen Energy Analyzer tool.

The main areas of thermal energy consumption are steam explosion pretreatment and antioxidant extraction (63 % of thermal consumption). Besides, these processes consume 73 % of total process water in the entire plant.

3.2. Economic evaluation.

The capital expenditure of the plant is 39.5 M€ and the operating expenses are 7.14 M€/y, being OTP the most important operating cost (23 %). Revenues are accounted for 20.3 M€/y, being the most important product the antioxidant that accounts for 66.2 % of the total revenues. The NPV is 11.53 M€, with a pay-back time of 4 years. Meaning that the plant is profitable.

In the sensitivity analysis, the price of OTP could increase by 224.45 % with respect its original price, even being the most important operating cost, the plant could resist price variations keeping positive profits. Furthermore, the price of ethanol could be reduced by a maximum of 98 %, and xylitol could be sold for free, and the plant would reach a NPV of 1.5 M€. On the other hand, the antioxidant price just could be reduced by 27 %.

This means that the plant profits are supported by the antioxidant market, making this part of the production scheme the most important in terms of process economics. So future optimization of the plant should focus on the improvement of the antioxidant production or even on the consideration of a standalone antioxidant production plant.

3.3. Carbon footprint and water consumption

The indirect and direct emissions in terms of kg CO_2 equivalent (carbon footprint) and the total water consumptions were calculated for both the considered biorefinery and the reference system (Table 2). It can be seen a clear reduction in the carbon footprint of the biorefinery products with respect to the individual reference counterparts. The global warming impact is reduced from 8.17 to 0.42 kg CO_2 equivalent. However, in terms of water consumption the advantages of the biorefinery are not so clear. On the one hand, there is a global reduction of water use which is a consequence of the higher water consumption of the reference antioxidant production process (propyl gallate). On the other hand, both ethanol and xylitol produced in the biorefinery have a higher water footprint than the individual reference systems.

Table 1. LCIA results of the OTP biorefinery and the reference system for 1 kg of OTP.

Biorefinery		Reference System	
Global warming (kg CO_{2eq})			
Total	0.42	8.17	Total
Antioxidant	0.16	7.62	Propyl Gallate
Ethanol	0.14	0.16	Ethanol (fossil)
Xylitol	0.12	0.38	Xylitol (corncob)
Water consumption (m³)			
Total	0.12	0.16	Total
Antioxidant	0.071	0.150	Propyl Gallate
Ethanol	0.026	0.006	Ethanol (fossil)
Xylitol	0.019	0.001	Xylitol (corncob)

As stated above, antioxidants are the most economically appealing product in the considered multiproduct biorefinery scheme. However, in terms of environmental aspects they are also responsible for the highest impacts of the plant in comparison to ethanol and xylitol. This is mainly due to the high water and energy consumptions required in the antioxidant hot water extraction process (Section 3.1). At the same time, the antioxidants produced from OTP are the only product of the plant which reduces both the carbon and water footprints with respect to its reference (propyl gallate).

4. Conclusions

From the results shown, it can be clearly seen that the use of OTP as feedstock for the production of ethanol, xylitol and antioxidants is both economically profitable and less environmentally harmful in terms of carbon footprint and water consumption than the business-as-usual solution. At the same time, antioxidants are the most economically interesting product of the plant (66.2 % of revenues). From an environmental point of view, they also reduce drastically all the impacts with respect to the reference propyl gallate production.

References

Ballesteros, I., Ballesteros, M., Cara, C., Sáez, F., Castro, E., Manzanares, P., Negro, M. J., & Oliva, J. M. (2011). Effect of water extraction on sugars recovery from steam exploded olive tree pruning. *Bioresource Technology*, *102*(11), 6611–6616. https://doi.org/10.1016/j.biortech.2011.03.077

Comission, E. (2019). *A European Green Deal | European Commission*. European Commission.

Conde, E., Cara, C., Moure, A., Ruiz, E., Castro, E., & Domínguez, H. (2009a). Antioxidant activity of the phenolic compounds released by hydrothermal treatments of olive tree pruning. *Food Chemistry*, *114*(3), 806–812. https://doi.org/10.1016/j.foodchem.2008.10.017

Conde, E., Cara, C., Moure, A., Ruiz, E., Castro, E., & Domínguez, H. (2009b). Antioxidant activity of the phenolic compounds released by hydrothermal treatments of olive tree pruning. *Food Chemistry*, *114*(3), 806–812. https://doi.org/10.1016/j.foodchem.2008.10.017

Contreras, M. del M., Romero, I., Moya, M., & Castro, E. (2020). Olive-derived biomass as a renewable source of value-added products. *Process Biochemistry*, *97*(March), 43–56. https://doi.org/10.1016/j.procbio.2020.06.013

FAOSTAT. (2021). *Crops*. http://www.fao.org/faostat

Gnansounou, E., Vaskan, P., & Pachón, E. R. (2015). Comparative techno-economic assessment and LCA of selected integrated sugarcane-based biorefineries. *Bioresource Technology*, *196*, 364–375. https://doi.org/10.1016/j.biortech.2015.07.072

IEA. (2021). *Net Zero by 2050 A Roadmap for the Global Energy Sector*. 222.

Kostenidou, E., Kaltsonoudis, C., Tsiflikiotou, M., Louvaris, E., Russell, L. M., & Pandis, S. N. (2013). Burning of olive tree branches: A major organic aerosol source in the Mediterranean. *Atmospheric Chemistry and Physics*, *13*(17), 8797–8811. https://doi.org/10.5194/acp-13-8797-2013

Lo Giudice, V., Faraone, I., Bruno, M. R., Ponticelli, M., Labanca, F., Bisaccia, D., Massarelli, C., Milella, L., & Todaro, L. (2021). Olive trees by-products as sources of bioactive and other industrially useful compounds: A systematic review. *Molecules*, *26*(16). https://doi.org/10.3390/molecules26165081

Susmozas, A., Moreno, A. D., Romero-García, J. M., Manzanares, P., & Ballesteros, M. (2019). Designing an olive tree pruning biorefinery for the production of bioethanol, xylitol and antioxidants: A techno-economic assessment. *Holzforschung*, *73*(1), 15–23. https://doi.org/10.1515/hf-2018-0099

Proceedings of the 14th International Symposium on Process Systems Engineering – PSE 2021+
June 19-23, 2022, Kyoto, Japan © 2022 Elsevier B.V. All rights reserved.
http://dx.doi.org/10.1016/B978-0-323-85159-6.50311-0

Application of CAPE Tools into Prospective Life Cycle Assessment: A Case Study in Acetylated Cellulose Nanofiber-Reinforced Plastics

Yasunori Kikuchi[a,b,c*] and Yuichiro Kanematsu[b]

[a] *Institute for Future Initiatives, The University of Tokyo, Tokyo 113-8654, Japan*
[b] *Presidential Endowed Chair for "Platinum Society", The University of Tokyo, Tokyo 113-8656, Japan*
[c] *Department of Chemical System Engineering, The University of Tokyo, Tokyo 113-8656, Japan*
ykikuchi@ifi.u-tokyo.ac.jp

Abstract

In this study, we are tackling systems design with assessments for emerging technologies. Computer-aided process engineering (CAPE) tools such as process design heuristics, process simulation, optimization, parametric analysis for characterizing sensitivity and alternative generation, and decision making with uncertainties have huge potential to compensate the data limitation of emerging technology and jump up to the deep technology assessments with quantified results. A case study on the application of CAPE tools into prospective life cycle assessment was conducted for the acetylated cellulose nanofiber-reinforced plastics, which has been developed to replace the conventional structural materials, e.g., steel or fossil-based plastics, in applications automobile or home appliances. We performed simulation-based life cycle inventory analysis to reveal the environmental and economic performance of CNF-reinforced plastics considering the future scale-up of production processes. Through this case study, it was demonstrated that the application of CAPE tools into prospective LCA enables the strategic technology assessments for systems design. Especially in the proof of concept on technology implementation can be verified and validated with the ranged values of uncertainties in emerging technology under development.

Keywords: LCA, CNF, greenhouse gas emission, production cost

1. Introduction

In order to achieve the decarbonization target by 2050 with defossilization, we must focus on the early introduction and diffusion of state-of-the-art elemental technologies. However, many promising elemental technologies are still under development, and even if they are expected to be commercialized, there is uncertainty about their decarbonizing effects in implementation. Therefore, for these promising elemental technologies, an early technology assessment on the economic and environmental aspects of the technology should be carried out before large-scale implementation, and a roadmap for the diffusion of the technology should be formulated, taking into account technological characteristics such as the maturity of technological development, technological change, and economies of scale, while limiting uncertainties. In recent years, there has been an increase in the number of case studies using prospective life cycle assessment (LCA), which take into account the future potential of the technology and aim to predict the environmental impacts on the technology under development. (Arvidsson et al., 2018; Moni et al., 2020; Thonemann et al., 2020)

In this study, we are tackling systems design with assessments for emerging technologies. Because of the data limitation on the systems and processes adopting emerging technologies, their design and assessments have uncertainties and difficulties to implement them into society smoothly. Computer-aided process engineering (CAPE) tools such as process design heuristics, process simulation, optimization, parametric analysis for characterizing sensitivity and alternative generation, and decision making with uncertainties have huge potential to compensate such data limitation and jump up to the deep technology assessments with quantified results. In this paper, we examine the applicability of CAPE tools for systems design and assessment adopting emerging technologies with a case study in acetylated cellulose nanofiber-reinforced plastics (AcCNF-RP). AcCNF-RP has been developed to replace the conventional structural materials, e.g., steel or fossil-based plastics, in applications automobile or home appliances. Cellulose nanofibers (CNF) can be produced from plant-derived renewable resources and have advantage of mechanical properties in lightness and strength when it was applied as the filler of the composites. Examining the proof-of-concept, mitigating fossil use and greenhouse gas (GHG) emissions, is strongly needed before such emerging technologies spread to the market and society. In this study, we performed simulation-based life cycle inventory analysis to reveal the environmental and economic performance of AcCNF-RP considering the future scale-up of production processes.

2. Materials and methods

2.1. Application of CAPE tools into prospective LCA

Figure 1 shows the description of systems assessments applying CAPE tools for prospective LCA. In management activity and resource provider, data estimation and interpretations are assigned to CAPE tools considering the conditions in prospective LCA.

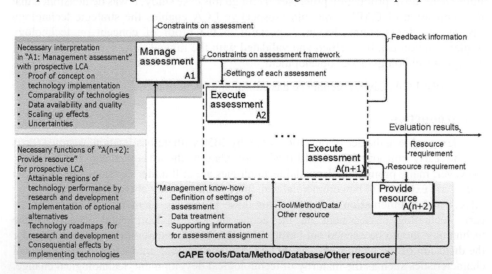

Figure 1 Description of multiple assessment activities with the necessary conditions for prospective LCA. n is the number of assessment methods. (Modified from previous studies (Kikuchi, 2014; Kikuchi et al., 2010; Kikuchi and Hirao, 2009))

Conventional LCA does not take into account changes in technology level, because it refers to information on the current technology level and specifically estimates the environmental impacts of each process related to the provision of products and services. On the other hand, efforts to tackle climate change have become more active in recent years, and new products and technologies are changing concepts and models more rapidly, making the transition to a low-emission society more urgent.

The significance of conducting a strategic LCA of emerging technologies for the 30-year time horizon up to the target year of 2050 arose regarding the issues on the climate change. Emerging technologies, as defined by Rotolo et al. (2015), are; "innovative and rapidly growing technologies that have the potential to have a significant social and economic impact in the domains in which they are structured, with some degree of persistent coherence, actors, institutions, ways of interacting with them and related knowledge production processes. It is characterised by its potential to have significant social and economic impacts. However, its most prominent impact lies in the future and is therefore somewhat uncertain and ambiguous at the stage at which the technology emerges." These technologies are characterized as "innovative", "rapid growth", "consistent", "significant impact" and "uncertain", which makes technology assessment difficult due to lack of existing data and knowledge.

Four main issues were identified as needing to be addressed in conducting prospective LCAs of emerging technologies (Thonemann et al., 2020; Moni et al., 2020). (1) comparability of technologies; (2) availability and quality of data; (3) scale-up challenges; and (4) uncertainty of assessment results. Process modeling and simulation are effective in estimating the missing process inventories in industrial scale production, because these technologies are under development in lab or pilot scale.

2.2. Case study: Acetylation of pulp for AcCNF-RP

Figure 2 shows the boundaries examined in this study. AcCNF-RP have been developed as substitutes for conventional structural materials (Eichhorn et al., 2010). Although kneading with polymers is required for pulp disintegration into nanofibers, it was excluded in this paper to focus on the chemical processes of acetylation applying CAPE tools (see also the previous paper (Kanematsu et al., 2021) on kneading process).

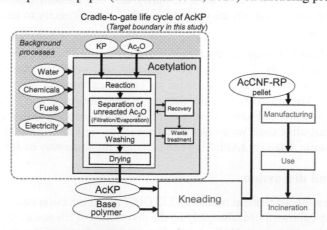

Figure 2 Related boundaries for AcCNF-RP life cycles in this study. KP: kraft pulp, AcKP: acetylated KP, AcCNF-RP: acetylated CNF-reinforced plastic, Ac₂O: acetic anhydride. (Modified from the previous literature (Kanematsu et al., 2021))

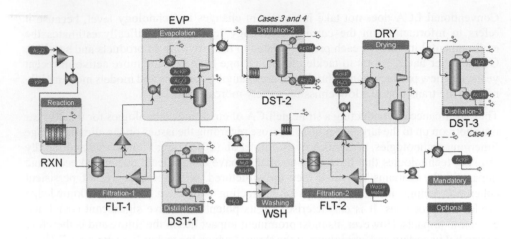

Figure 3 Process flow including all designed alternatives of the AcKP production process at an industrial scale. DST-1, -2, and -3 are optional processes, and the effects of their addition were compared through the process simulation. All reflux ratios in distillation column were set as the 1.3 times the theoretical minimum reflux ratio. This process consists of the process sections of reaction (RXN), filtration (FLT), evaporation (EVP), distillation (DST), washing (WSH), and drying (DRY). (Modified from the previous literature (Kanematsu et al., 2021))

As shown in Figure 2, Acetylation was defined as foreground processes, and kraft pulp (KP) obtained from a paper mill was used as pulp feedstock for CNF. The chemical modification process was examined in a scaled-up industrial process system implemented in the process simulator AspenPlus™ as well as in a lab-scale production (Kanematsu et al., 2021). The model constructed on process simulator is shown in Figure 3. The process simulation enables the evaluation of the process system in the actual production, which is not considered in the lab-scale production. For example, acetic anhydride (Ac$_2$O) can be recovered from the mixture of unreacted Ac$_2$O and acetic acid (AcOH) after the reactor and can be reused (DST-1 and -2). This is not normally done in the laboratory, but is a recycling process that is always considered in chemical plants. It is also possible to purify the byproduct, i.e., AcOH, to a level of purity that can be sold externally (DST-3). However, these unit operations are only optional and it is necessary to analyse the effect of introducing them.

LCA is carried out by combining the foreground data obtained by process simulation with the background data. In this study, the functional unit is the production of 1 kg of acetylated KP (AcKP), and the greenhouse gas emissions from the life cycle (LC-GHG) are calculated. The production scale was set at 100 tonnes/day of hydrous pulp as feedstock. Capital expenditure for equipment and other costs including staff costs and general administrative costs were calculated for each process alternative using the Aspen Process Economic Analyzer (APEA™) and costed in the same way as for LC-GHG.

3. Results and discussion

Figure 4 shows the assessment results on LC-GHG and production cost for unit amount of AcKP. Case 1 shows the results applying the inventories of lab-scale experiment. Cases 2 to 4 shows the results with CAPE tools. The dominant factor of the difference of case 1 against cases 2 to 4 was the recovery of Ac$_2$O, due to the relatively high cradle-to-gate LC-GHG of Ac$_2$O. In experiment, excess Ac$_2$O were consumed rather than raw materials.

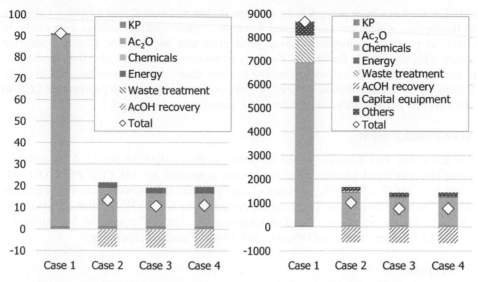

(a) Cradle-to-gate LC-GHG [kg-CO₂eq/kg-AcKP]   (b) Production cost [JPY/kg-AcKP]

Figure 4 Assessment results based on process simulation by CAPE tools. (110 JPY/USD)

Even for the results in Cases 2 to 4, the contribution of Ac_2O is significant rather than the other factors. As it is important to reduce the consumption of Ac_2O, it is effective to mitigate the reaction of Ac_2O with the water content in the KP, e.g., 27.8 wt% of water in this paper. However, it is essential to consider the effect of moisture in KP on the kneading mechanism of friction and disintegration. In addition, although there is a benefit from the recovered AcOH, Case 4 is not the best case because the increase in LC-GHG and production costs due to energy consumption is greater than the increase in AcOH recovery by building more distillation columns.

The results of Case 1 are based on the assumption that the production is carried out as it was in the laboratory. Such calculations have been adopted if environmental impacts and costs are needed for technologies with low technology readiness level (TRL), especially in the evaluation of emerging technologies. The CAPE tool particularly enables analyses of the additional energy input required to recover unreacted material, which is usually not done in the laboratory. The change from Case 1 to Case 2 can therefore be easily analysed using the CAPE tool, which also simulates the relationship between the recovery of by-products, i.e., AcOH, and the additional energy required as shown in the results of Cases 2 to 4. In Figure 1, the CAPE tool can be used for the activity: "Provide resource", which allows a certain analysis of the low TRL emerging technology.

4. Conclusions

In this study, we performed simulation-based life cycle inventory analysis to reveal the environmental and economic performance of AcCNF-RP considering the future scale-up of production processes. Through this case study, it was demonstrated that the application of CAPE tools into prospective LCA enables the strategic technology assessments for systems design. Especially in the proof of concept on technology implementation can be verified and validated with the ranged values of uncertainties in emerging technology under development.

CAPE tools have huge potentials for systems design and assessment adopting emerging technologies, which are necessitated towards carbon neutral society. Especially in chemical production, biomass-derived production can become one of the production routes with sustainable feedstocks. Not only conversion routes, but also the acquisitions of feedstocks from agriculture or forestry are now under development and construction. Before their huge installation, CAPE tools should be combined with prospective LCA to visualize the performances of such low TRL emerging technologies.

Acknowledgement

We thank Mr. Dai Kikuchi for the supports on data collection for LCA of cellulose nanofiber-reinforced plastics. This work was supported by MEXT/JSPS KAKENHI Grant Number JP21H03660. Activities of the Presidential Endowed Chair for "Platinum Society" at the University of Tokyo are supported by the KAITEKI Institute Incorporated, Mitsui Fudosan Corporation, Shin-Etsu Chemical Co., ORIX Corporation, Sekisui House, Ltd., the East Japan Railway Company, and Toyota Tsusho Corporation.

References

R. Arvidsson, A.M. Tillman, B.A. Sandén, M. Janssen, A. Nordelöf, D. Kushnir, S. Molander. 2018, Environmental assessment of emerging technologies: recommendations for prospective LCA. J Ind Ecol, 22(6), 1286-1294.

S.J. Eichhorn, A. Dufresne, M. Aranguren, N.E. Marcovich, J.R. Capadona, S.J. Rowan, C. Weder, W. Thielemans, M. Roman, S. Renneckar, W. Gindl, S. Veigel, J. Keckes, H. Yano, K. Abe, M. Nogi, A.N.. Nakagaito, A. Mangalam, J. Simonsen, A.S. Benight, A. Bismarck, L.A.. Berglund, T. Peijs. 2010. Review: Current International Research into Cellulose Nanofibres and Nanocomposites. J. Mater. Sci., 45, 1–33.

Y. Kanematsu, Y. Kikuchi, H. Yano. 2021. Life Cycle Greenhouse Gas Emissions of Acetylated Cellulose Nanofiber-reinforced Polylactic Acid Based on Scale-up from Lab-scale Experiments, ACS Sustainable Chem Eng, 9(31), 10444-10452

Y. Kikuchi, M. Hirao. 2009, Hierarchical Activity Model for Risk-Based Decision Making Integrating Life Cycle and Plant-Specific Risk Assessments, J Ind Ecol, 13(6) 945-964.

Y. Kikuchi, K. Mayumi, M. Hirao. 2010. Integration of CAPE and LCA Tools in Environmentally-Conscious Process Design: A Case Study on Biomass-Derived Resin, Computer-Aided Chemical Engineering, 27, 1051-1056.

Y. Kikuchi. 2014. Activity and Data Models for Process Assessment Considering Sustainability, Kagaku Kogaku Ronbunshu, 40(3) 211-223.

Kyoto University. 2019. Entrusted Work for Performance Evaluation Project of Cellulose Nanofiber (Proving, Evaluation, and Verification for Introduction of CNF Materials for Social Implementation: Automotive Field), 2020. http://www.env.go.jp/earth/mat49_kyoto-univH31.pdf (accessed November 2021).

S.M. Moni, R. Mahmud, K. High, M. Carbajales-Dale. 2020. Life cycle assessment of emerging technologies: A review. J Ind Ecol, 24, 52-63.

D. Rotolo, D. Hicks, B.R. Martin. 2015. What is an emerging technology? Res. Policy, 44(10), 1827-1843.

N. Thonemann, A. Schulte, D. Maga. 2020. How to Conduct Prospective Life Cycle Assessment for Emerging Technologies? A Systematic Review and Methodological Guidance. Sustainability, 12(3), 1192.

Proceedings of the 14th International Symposium on Process Systems Engineering – PSE 2021+
June 19-23, 2022, Kyoto, Japan © 2022 Elsevier B.V. All rights reserved.
http://dx.doi.org/10.1016/B978-0-323-85159-6.50312-2

Climate Control in Controlled Environment Agriculture Using Nonlinear MPC

Wei-Han Chen[a*], Fengqi You[a]

[a]Cornell University, Ithaca, New York, 14853, USA
wc593@cornell.edu

Abstract

The climate in controlled environment agriculture (CEA) is a highly nonlinear complex system that contains nonlinearity. In addition, there are dependencies between each system state. In order to simultaneously control multiple system states in CEA climate, this paper develops a nonlinear model predictive control (NMPC) framework for the CEA climate control to minimize the total control cost and the constraint violation probability. The nonlinear dynamic model of the CEA climate, including temperature, humidity, CO_2 level, and lighting, will be first constructed. After constructing all dynamic models, historical weather data is gathered to identify the system parameters for the nonlinear CEA climate model. A nonlinear optimization problem can then be developed to obtain the optimal control inputs. A case is used to demonstrate the performance of the proposed NMPC framework.

Keywords: Controlled environment agriculture, temperature, humidity, CO_2, lighting.

1. Introduction

Because the controlled environment agriculture (CEA) climate is a multi-input multi-output system, model predictive control (MPC) has advantages over other classical control methods (e.g., On-Off control and proportional–integral–derivative (PID) control). In some past studies, linear MPC is adopted for CEA climate control (Piñón et al., 2005). However, CEA climate contains nonlinearity due to the complex system itself and the relationship between each system state (Chen and You, 2021), which makes nonlinear MPC (NMPC) a suitable approach to deal with the nonlinearity within CEA climate control (Ding et al., 2018). Several system states should be considered in CEA climate control. Among the studies that adopted NMPC, many only control one or two system states instead of simultaneously controlling all four system states mentioned (Blasco et al., 2007; Gruber et al., 2011; Liang et al., 2018; Lin et al., 2021). So far, there is still a lack of comprehensive studies that integrate temperature, humidity, CO_2 concentration, and lighting using the NMPC framework. Therefore, in this work, we propose a novel NMPC framework for CEA climate control to minimize the total control cost and the constraint violation probability.

2. Dynamic model formulation

Within an MPC framework for CEA climate control, a dynamic model is required in order to predict CEA climate (e.g., temperature, humidity, CO_2 concentration, and lighting) as a function of control inputs and certain disturbances, to minimize the control cost and avoid the CEA climate from becoming harmful to crop growth.

In this work, the states we consider are indoor air temperature, relative humidity, CO_2 concentration, and photosynthetically active radiation (PAR). Control actuators are fans, pad cooling, CO_2 injection, supplemental lightings, and blinds. The disturbances considered are ambient temperature, ambient relative humidity, ambient CO_2 concentration, and solar radiation. The structure of the proposed dynamic model for CEA climate control is shown in Figure 1.

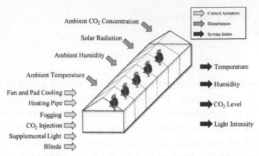

Figure 1: CEA structure that shows control actuators, disturbances, and system states.

The differential equation can be determined by analyzing the energy and mass balance between objects. The continuous-time CEA temperature model can be generally described as the following form:

$$\rho V C_a \frac{dT_i}{dt} = q_{sol} + q_{pipe} + q_{light} - q_{cover} - q_{vent} - q_{pad} \tag{1}$$

where ρ is the air density, V is the volume of the CEA, C_a is the air specific heat, T_i is the indoor temperature, q_{sol} is the net solar radiation, q_{pipe} is the heat flux from heating pipes, q_{light} is the heat flux from lighting, q_{cover} is the heat flux through the cover, q_{vent} is the heat flux from ventilation, and q_{pad} is the heat flux from the pad.

The humidity inside the indoor can also be modeled by differential equations. The absolute humidity is first modeled by using the mass balance equation. The relative humidity is then calculated from absolute humidity and indoor temperature (Chen and You, 2022). The mass balance equation of water, including the net flow from ventilation, evapotranspiration, and the fogging system, is shown as:

$$\rho V \frac{dh_i}{dt} = m_{vent} + m_{trans} + m_{fog} \tag{2}$$

where h_i is the absolute humidity, m_{vent} is the water net flow from ventilation, m_{trans} is the water net flow from transpiration of the plants, and m_{fog} is the water net flow from the fogging system.

In the CO_2 mass balance equation, the photosynthesis process consumes CO_2. The mass balance of CO_2 level is shown as:

$$\rho V \frac{dX_i}{dt} = X_v - X_{pho} + X_{inj} \tag{3}$$

where X_i is the indoor CO_2 concentration level, X_v is the CO_2 net flow from ventilation, X_{pho} is the net consumption from photosynthesis, and X_{inj} is the control input of CO_2 injection. The total light intensity in the CEA is the sum of the PAR from the sun and PAR provided by the supplemental lightings. The light intensity model can be shown as:

$$I = I_a K_a \left(1 - \tau_s u_{blind}\right) + \tau_l u_{light,max} u_{light} \tag{4}$$

where I_a is the outdoor global radiation, K_a is the coefficient of the solar equation, τ_s is the shading percentage, u_{blind} is the control input of blind ranging from 0 to 1, τ_l is the energy to light conversion percentage, $u_{light,max}$ is the maximum energy input of the supplemental light system, and u_{light} is the control input of the supplemental light ranging from 0 to 1.

3. Nonlinear model predictive control

NMPC is used for controlling the CEA climate in this work. The system dynamic models are discretized using the Euler method. Under a given length of prediction horizon H, a compact form of the dynamic CEA climate model can be expressed as:

$$\mathbf{x} = \mathbf{f}\left(x_0, \mathbf{u}, \mathbf{v}\right) \tag{5}$$

where $\mathbf{x}$, $\mathbf{u}$, and $\mathbf{v}$ are the system state, control input, and disturbance sequences vectors, respectively, and x_0 is the initial system state.

After constructing the dynamic model, we could then develop the nonlinear optimization problem to be solved at each time step. For the optimization problem, there are constraints on system states and control inputs. The constraints are defined for control inputs and system states throughout the entire prediction horizon H. The control inputs should be between the minimum and maximum values. A CEA environment should also be maintained within a specified range to facilitate the growth of plants and fruits and prevent them from being damaged by harsh climate conditions. The compact form of constraints for control inputs and system states can be represented as:

$$\mathbf{G}_x \mathbf{x} \le \mathbf{g}_x, \quad \mathbf{G}_u \mathbf{u} \le \mathbf{g}_u \tag{6}$$

where $\mathbf{G}_x$ and $\mathbf{g}_x$ are vectors that define the system state constraints in compact form. $\mathbf{G}_u$ and $\mathbf{g}_u$ are vectors that represent the control input constraints in compact form.

Once the dynamic model and the constraints are prepared, the optimization problem can then be written out. The objective function is to minimize the total control cost. **cc** represents the cost coefficient for different control actuators, and the coefficients are higher for actuators using energy with higher energy costs. A vector of slack variables ε is added to the objective function because there are limitations on control inputs which could cause the optimization problem to become infeasible. In order to penalize the constraint violation, the penalty weight $\mathbf{S}$ is added to the objective function. Since the slack variables are always positive, the state constraints are therefore softened (Lu et al., 2020). In this control framework, the optimization problem is solved to get the optimal control inputs for each time step. The system states at the next time step can be updated by using the dynamic models.

$$\min_{u} J = \mathbf{cc}^T \mathbf{u} + \boldsymbol{\varepsilon}^T \mathbf{S} \boldsymbol{\varepsilon}$$

$$\text{s.t.} \quad \mathbf{x} = \mathbf{f}\left(x_0, \mathbf{u}, \mathbf{v}\right)$$

$$\mathbf{G}_x \mathbf{x} \le \mathbf{g}_x + \boldsymbol{\varepsilon} \tag{7}$$

$$\mathbf{G}_u \mathbf{u} \le \mathbf{g}_u$$

$$\boldsymbol{\varepsilon} \ge 0$$

4. Case studies on simulated CEA

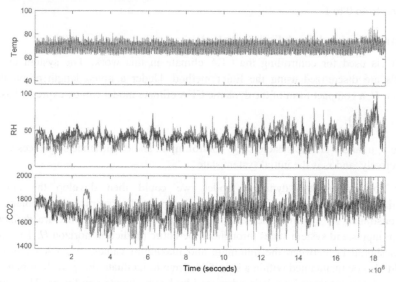

Figure 2: System identification results for greenhouse climate. The darker lines are the predicted trajectories by system identification; the lighter ones are measurement data.

In this work, a CEA located in Ithaca, New York, USA, for tomato production is simulated for closed-loop temperature, humidity, CO_2 concentration level, and lighting control under the NMPC control framework. The system states controlled in this work are indoor temperature, relative humidity, CO_2 level, and PAR. System identification is first conducted to obtain the undetermined parameters in the CEA climate model. The weather data and CEA indoor climate data from November 1, 2019, to May 31, 2020, are gathered to conduct the system identification. The simulation is performed for one week in winter during December 17-23, 2019. The weather forecast data from December 17-23 are collected for the optimization problem. The actual measurement data at the same period are also collected to reveal the system states at the next time step. The sampling interval is 15 min, and the control horizon is 5 hours. The average CPU time for solving the optimization problem is 2.53 seconds on a computer with an Intel Core i7-6700 processor at 3.40 GHz and 32 GB of RAM.

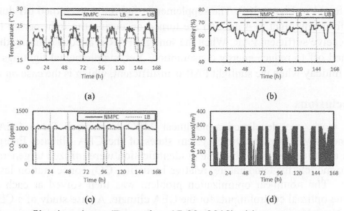

Figure 3: The profiles in winter (December 17-23, 2019), (a) temperature profile, (b) humidity profile, (c) CO_2 level profile, (d) lamp PAR profile.

Figure 2 shows the system identification results of indoor temperature, relative humidity, and CO_2 concentration. The darker lines are the trajectories predicted by system identification, and the lighter lines are the trajectories of CEA measurement data. The figure shows that the temperature model and relative humidity model predict better than the CO_2 model with around 78% and 65% accuracy compared to the 43% accuracy of the CO_2 model. The CO2 model is not as accurate as of the other two models due to the lack of CO_2 injection data. The way to overcome the lack of CO_2 input data is by estimating data using back-calculation. The CO_2 model can still be used to minimize the total control cost in practice because the system states would be updated in each time step. The effect of model error will then be reduced.

Figure 3(a) shows the temperature profile in winter during December 17-23, 2019. The lower and upper bound are set differently throughout the day for the light period and dark period of the photosynthesis. The light period starts from 4 am to midnight and the dark period is between midnight and 4 am. When the CEA changes from light period to dark period or vice versa, the lower and upper bound are set to be gradually increased or decreased to avoid the abrupt changes of the indoor temperature. There is a clear diurnal pattern, and the profile could be maintained within the region between lower and upper bounds. However, the constraint violation still occurs sometimes due to forecast errors. The relative humidity profile in winter during December 17-23, 2019, is presented in Figure 3(b). The lower bound and upper bound are set as 50% and 70%. The humidity profile is better maintained within the region between lower and upper bounds compared to the temperature profile in winter. The cold outdoor air is drier than indoor air so that the ventilation system could help draw in the outdoor air when the humidity level is about to surpass the upper bound. Figure 3(c) depicts the CO_2 level profile in winter during December 17-23, 2019. The CO_2 injection occurs from 4 am to midnight for the light period. During the light period, the CO_2 level is maintained above 1000 ppm to stimulate tomato growth. CO_2 is not necessary during the dark period, so the CO_2 injection is set to zero. The CO_2 level would gradually drop to the ambient level because of the ventilation. The lamp PAR result in winter during December 17-23, 2019, can be found in Figure 3(d). The supplemental lightings are turned off during the dark period from midnight to 4 am and are turned on during the light period to ensure the plants receive sufficient PAR. When the sunlight PAR is adequate during the day, lamps are turned off to reduce energy consumption. Yet, supplemental lightings are required even at midday when the sunlight PAR is insufficient, which is the case on the last day.

5. Conclusions

In this work, we developed a nonlinear MPC framework for a CEA that could simultaneously control multiple system states of the CEA climate. Energy and mass balance equations followed by system identification were utilized to generate nonlinear dynamic models for temperature, relative humidity, CO_2 concentration level, and light intensity. The nonlinear optimization problem was then solved at each time step to obtain the optimal control inputs for the CEA climate. A case study of a CEA located in Ithaca, New York was conducted. The results showed the NMPC framework could efficiently minimize total control cost and constraint violation. Future extensions of this work could include irrigation control (Shang et al., 2020), sensor integration (Chen et al., 2021), and accounting for uncertainty in weather forecast (Shang et al., 2019).

References

X. Blasco, M. Martínez, J. M. Herrero, C. Ramos, and J. Sanchis, 2007, Model-based predictive control of greenhouse climate for reducing energy and water consumption, Computers and Electronics in Agriculture, 55(1), 49-70.

W.-H. Chen, F. You, 2021, Smart greenhouse control under harsh climate conditions based on data-driven robust model predictive control with principal component analysis and kernel density estimation. Journal of Process Control, 107, 103-113.

W.-H. Chen, C. Shang, S. Zhu, et al., 2021, Data-driven robust model predictive control framework for stem water potential regulation and irrigation in water management. Control Engineering Practice, 113, 104841.

W.-H. Chen, F. You, 2022, Semiclosed Greenhouse Climate Control Under Uncertainty via Machine Learning and Data-Driven Robust Model Predictive Control. IEEE Transactions on Control Systems Technology, DOI:10.1109/tcst.2021.3094999.

Y. Ding, L. Wang, Y. Li, and D. Li, 2018, Model predictive control and its application in agriculture: A review, Computers and Electronics in Agriculture, 151, 104-117.

J. K. Gruber, J. L. Guzman, F. Rodriguez, C. Bordons, M. Berenguel, and J. A. Sanchez, 2011, Nonlinear MPC based on a Volterra series model for greenhouse temperature control using natural ventilation, Control Engineering Practice, 19(4), 354-366.

M.-H. Liang, L.-J. Chen, Y.-F. He, and S.-F. Du, 2018, Greenhouse temperature predictive control for energy saving using switch actuators, IFAC-PapersOnLine, 51(17), 747-751.

D. Lin, L. Zhang, and X. Xia, 2021, Model predictive control of a Venlo-type greenhouse system considering electrical energy, water and carbon dioxide consumption, Applied Energy, 298, 117163.

S. Lu, J.H. Lee, F. You, 2020, Soft-constrained model predictive control based on data-driven distributionally robust optimization. AIChE Journal, 66, e16546.

S. Piñón, E. F. Camacho, B. Kuchen, and M. Peña, 2005, Constrained predictive control of a greenhouse, Computers and Electronics in Agriculture, 49(3), 317-329.

C. Shang, W.-H. Chen, A.D. Stroock, F. You, 2020, Robust Model Predictive Control of Irrigation Systems With Active Uncertainty Learning and Data Analytics. IEEE Transactions on Control Systems Technology, 28, 1493-1504.

C. Shang, F. You, 2019, A data-driven robust optimization approach to scenario-based stochastic model predictive control. Journal of Process Control, 75, 24-39.

Proceedings of the 14th International Symposium on Process Systems Engineering – PSE 2021+
June 19-23, 2022, Kyoto, Japan © 2022 Elsevier B.V. All rights reserved.
http://dx.doi.org/10.1016/B978-0-323-85159-6.50313-4

Thermodynamic Analysis of an Integrated Renewable Energy Driven EWF Nexus: Trade-off Analysis of Combined Systems

Jamileh Fouladi, Ahmed AlNouss, Yusuf Bicer, Tareq Al-Ansari*

College of Science and Engineering, Hamad Bin Khalifa University, Qatar Foundation, Doha, Qatar

*talansari@hbku.edu.qa

Abstract

The utilization of energy, water and food resources across multiple technologically driven sub-systems has attracted much attention in the literature. In this work, a solar energy-based system integrated with a utility unit is proposed to generate fresh water from seawater by desalination, power, ammonia/urea, and syngas from biomass utilisation. The main objective is to design a combined integrated system (solar/biomass system) based on solar energy, and to study the dependence of the Energy-Water-Food nexus performance on the solar capacity. The core components of the developed system include solar thermal collectors, Rankine cycle, reverse osmosis (RO) desalination unit, food/agriculture sector, biomass gasification process, ammonia and urea production units. The beneficial uses of reverse osmosis brine streams in agriculture sector is considered. The syngas produced from gasification process is used for Ammonia/Urea production as well as in the Rankine cycle to generate electricity. A comphrensive thermodynamic model and energy-exergy balances are used to assess the performance of the proposed system using the Engineering Equation Solver (EES). Different scenarios are solved to capture the trade offs amongst different technologies and explore the optimum EWF interlinkages. Furthermore, the effects of different load changes such as solar radiation and ambient temperature on some of the outputs of the system are investigated. In addition, the energy and exergy efficiencies of the system are calculated and compared.

Keywords: Multi-generation system, EWF Nexus, Solar, Biomass, Exergy.

1. Introduction

As the global population increases towards 9 billion in 2050, the need for energy, water and food (EWF) resources will increase accordingly. It is expected that the demand for food and water resources will rise by 50 % (Karan et al., 2018). Incidentally, there are inherent inter-linkages between EWF resources. As such, the EWF nexus concept was developed at the Bonn Nexus Conference in 2011 as a consequence to the realization of these inter-dependencies. The underlying analysis within the EWF Nexus concept enables the identification of the inter-linkages amongst different resources, and as such supports the identification of synergies and trade-offs (Al-Ansari et al., 2017; 2015). Multi-generation systems, which are integrated resource systems can support the development of EWF nexus systems. For instance, Luqman et al. (2020) presented a multigeneration system that explores the thermodynamics of oxy-hydrogen combustor based on wind and solar energy. The useful products of the developed system are freshwater, power, hydrogen, cooling, and domestic water heating. Energy and exergy balances are used to evaluate the performance, where the overall respective efficiencies of the system were 50 % and 34 %, respectively. Nazari and Porkhial (2020) developed a multi-generation system that integrates solar energy and a biomass utilization unit to generate heating, cooling, freshwater, and electricity. The thermodynamic and economic model of the system was conducted, which demonstrates that through biomass integration, an exergy efficiency of 21.48 % can be achieved. Moreover, a multi-objective optimization problem considering second law efficiency maximization and total product cost minimization were solved. The cost of the optimum solution is decreased by 10 %, while the exergy efficiency was 0.2% higher than the base case results. Ghasemi et al.(2018) analyzed a multi-generation system operating through solar-biomass energy using a thermodynamic and thermo-economic approach. The energy system includes both the desalination process and liquefaction of natural gas (LNG). Results indicate that the proposed system has energy and exergy efficiencies of approximately 46 % and 11 % respectively.

Evidently, there are numerous multi-generation systems, which study the utilization of waste streams, demonstrating that integrating renewable resources can enhance sustainability, decentralization and resilience of integrated systems. Incidentally, in most previous studies, biomass utilization was studied without considering the wider EWF nexus elements. The hybrid EWF nexus and multigeneration concepts can be further expanded to include biomass utilization, associated with food production units and the energy-water nexus in an integrated manner (Fouladi et al., 2021). The novelty of this work is the trade-off analysis of an integrated solar-biomass energy system driven by the EWF nexus combined with a fertilizer production unit. The main objectives of this study are to design an integrated renewable EWF nexus system, and to capture the trade-offs between multiple resources to optimize the EWF interlinkages.

2. System Description

Figure 1 illustrates a representation of the integrated EWF multigeneration system. In this multigeneration system, the main sources of energy are from the syngas generated from biomass gasification process and solar energy. A Rankine cycle is used within the system to produce power and to be utilized within the existing energy sinks. Parabolic Trough Collectors (PTCs) are integrated to utilize solar energy to satisfy the required demand. The main components of the system include an agriculture unit, a reverse osmosis desalination plant, a biomass utilization process, and an ammonia/urea plant for fertilizer production. Moreover, the brine stream from the desalination plant is integrated to enhance the system and reduce. Finally, to improve the environmental emissions of the overall system, the potential for CO_2 reduction via capture technology from the Rankine cycle for reuse within urea production is considered.

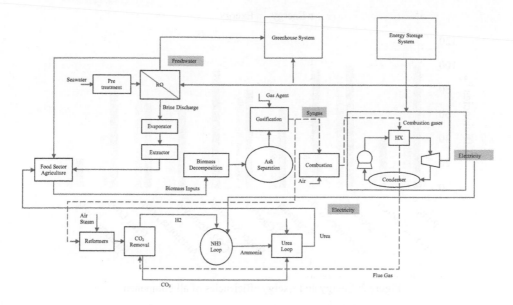

Figure 1: Proposed Integrated System.

Multiple assumptions and input data have been used to simulate the integrated system using EES. Table 1 indicates the main parameters used in this study.

Table 1: Input data for the proposed system.

Parameters	Value
Reference Temperature, T_0 and Pressure, P_0	25 °C and 101 kPa
Biomass gasifier operating conditions	888 °C and 101 kPa
Reverse osmosis recovery ratio	0.4
Seawater salinity	35000 ppm
Fresh water salinity	450 ppm
Isentropic efficiencies of pump and turbine	85 %
Rankine cycle pressure ratio	100
Surface temperature of the Sun	5500 °C

3. Thermodynamic Analysis & Results

The thermodynamic analysis of the integrated system is performed using mass, energy, entropy, and exergy balances (using the first and second laws of thermodynamics). Therefore, to evaluate the performance of every component and the overall system, the energy and exergy efficiencies of all units in the proposed system are calculated. Figure 2 illustrates the obtained efficiency values. Overall, the energy and exergy efficiencies of the system is found to be approximately 46 % and 19 %, respectively. Furthermore, the maximum exergy destruction rates correspond to the gasifier and ammonia process.

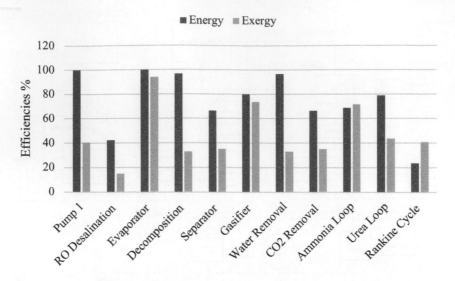

Figure 2: Energy and Exergy efficiencies of all components.

The trade-offs analysis of the integrated system demonstrates that the syngas produced from the biomass utilization process offsets the natural gas requirement as input to other processes, which results in reduced energy consumption. Figure 3 indicates that by optimizing the energy-food nexus segment and increasing the biomass flow rate, the overall energy and exergy efficiencies of the system reduce. The large exergy destruction rates of these processes are the main reason of this behaviour. The most common sources of irreversibility in the gasifier and ammonia unit are the chemical reactions occurring within the process, which lead to destruction of chemical exergies. Moreover, physical exergy destruction are also associated with the expansion and compression unit within the plants.

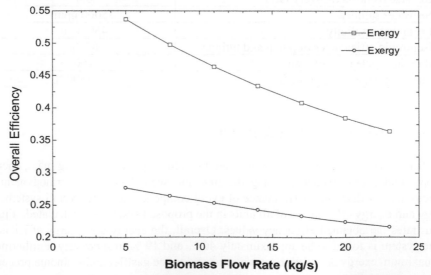

Figure 3: Effect of input biomass rate variation on overall system efficiencies.

Furthermore, the impact of changing the solar radiation on the performance of the turbine and seawater supply flow are illustrated in Figure 4 and Figure 5. The solar radiation linearly affects the outputs of the combined system. By increasing the solar radiation value, the net power generated by the Rankine cycle turbine increases linearly. This is due to the higher outlet temperature of the PTCs, which causes a higher temperature at the inlet of the Rankine cycle as well.

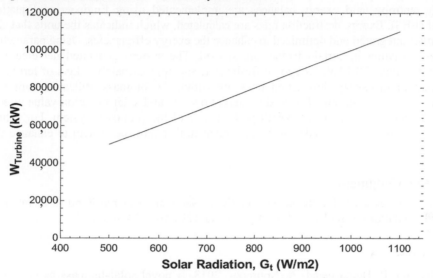

Figure 4: Effect of solar radiation variation on work generated by turbine.

From Figure 5, it is observed that the mass flow rate of seawater in the desalination unit increases as the solar radiation rate increases. The parametric studies demonstrate that the variation of some parameters has direct impact on the overall efficiencies of the system, which further integrates the system by optimizing the operational conditions.

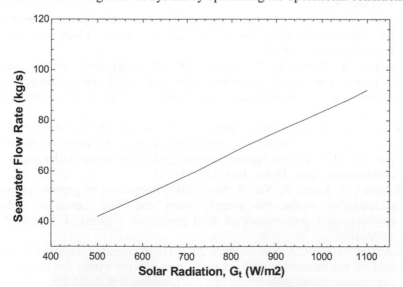

Figure 5: Effect of solar radiation variation on seawater flowrate

4. Conclusion

This study developed an integrated renewable energy driven EWF nexus system by combining various units, such as those within food systems, Rankine cycle, biomass utilization, ammonia/urea process and desalination. Electricity, freshwater, and fertilizer are the main outputs of the proposed system. Thermodynamic analysis including mass, energy, entropy, and exergy balances are implemented using Engineering Equation Solver (EES). Exergy destruction rates are calculated, which indicates the units that can be further integrated and optimized to enhance the exergy efficiencies. Using parametric studies, variation in certain factors are studied. The system generated net electrical power of almost 78 MW, 28 kg/s of freshwater, and approximately 7 kg/s of fertilizer. The maximum exergy destruction rates are within the biomass utilization unit and Ammonia production unit. The feed biomass flow rate and solar radiation value affects the different outputs and the overall performance of the proposed system. For future studies, cooling/heating loads can be integrated further into the system to enhance the resilience.

5. Acknowledgment

The authors acknowledge the support of Qatar National Research Fund (a member of Qatar Foundation) provided by GSRA grants no. GSRA6-1-0416-19014.

References

A. Ghasemi, P. Heidarnejad, A. Noorpoor, 2018, A novel solar-biomass based multi-generation energy system including water desalination and liquefaction of natural gas system: Thermodynamic and thermoeconomic optimization, J. Clean. Prod, 196, 424–437.

E. Karan, S. Asadi, R. Mohtar, M. Baawain, 2018, Towards the optimization of sustainable food-energy-water systems: A stochastic approach, J. Clean. Prod, 171, 662–674.

J. Fouladi, A. AlNouss, T. Al-Ansari, 2021, Sustainable energy-water-food nexus integration and optimisation in eco-industrial parks. Computers & Chemical Engineering, 146, 107229.

M. Luqman, Y. Bicer, T. Al-Ansari, 2020, Thermodynamic analysis of an oxy-hydrogen combustor supported solar and wind energy-based sustainable polygeneration system for remote locations, Int. J. Hydrogen Energy, 45, 3470–3483.

N. Nazari, S. Porkhial, 2020, Multi-objective optimization and exergo-economic assessment of a solar-biomass multi-generation system based on externally-fired gas turbine, steam and organic Rankine cycle, absorption chiller and multi-effect desalination, Appl. Therm. Eng, 179, 115521.

T. Al-Ansari, A. Korre, Z. Nie, N. Shah, 2017, Integration of greenhouse gas control technologies within the energy, water and food nexus to enhance the environmental performance of food production systems, J. Clean. Prod. 162, 1592–1606.

T. Al-Ansari, A. Korre, Z. Nie, N. Shah, 2015, Development of a life cycle assess- ment tool for the assessment of food production systems within the energy, water and food nexus, Sustainable Production and Consumption, 2, 52–66.

Proceedings of the 14th International Symposium on Process Systems Engineering – PSE 2021+
June 19-23, 2022, Kyoto, Japan © 2022 Elsevier B.V. All rights reserved.
http://dx.doi.org/10.1016/B978-0-323-85159-6.50314-6

Low-Carbon Hydrogen Production in Industrial Clusters

Yasir Ibrahim, Mohammad Lameh, Patrick Linke, Dhabia M. Al-Mohannadi*

Departement of Chemical Engineering, Texas A&M University at Qatar, Education City, PO Box 23874, Doha, Qatar
dhabia.al-mohannadi@qatar.tamu.edu

Abstract

Hydrogen has gained a huge hype as future fuel that might aid in the transformation to a zero-carbon energy system. In this work, a dynamic model of green hydrogen production is proposed to assess the trade-offs between the economic and environmental impacts of incorporated green hydrogen in energy-intensive chemical processes. The model considers the intermittency of renewable energy sources and their effect on the sizing of the renewable energy and hydrogen production units, as well as on the size required for H_2 storage. The validity of the model was tested in the decarbonization of two energy-intensive processes namely, ammonia and methanol. The results outputs clearly show the trade-offs between the economic and environmental performances at various decarbonization targets based on hourly solar availability from a time horizon of one year.

Keywords: Renewable Hydrogen, Decarbonized Economy, Solar Electrolysis

1. Introduction

In light of climate change, the Intergovernmental Panel on Climate Change (IPCC) establishes a protocol to cut down the Greenhouse Gas (GHG) emissions in order to achieve so-called carbon neutrality by 2050 (IPCC 2018). Unfortunately, the majority of the global energy is generated by carbon-based resources, which leads to the continuous release of CO_2 emissions. Most of these emissions are driven from hard-to-abate sectors, such as petrochemical production, steel, oil refining, etc.In this context, Hydrogen (H_2) would play a substantial role to decarbonize the global energy system since it is a clean-burning molecule. Many factors are reinforcing why hydrogen is the key block in energy transition, which are, H_2 can solve the renewable energy intermittency issue by utilizing H_2 as a cleaner, affordable, and available storage. In addition, H_2 can deliver a deep reduction in CO_2 emissions specifically in the hard-to-abate sectors. In these sectors, renewable energy may not contribute significantly to decarbonize of these sectors as much as it can contribute to utility and power sectors. Although there many factors favoring a sustainable uprising in the investment of H_2 , which are substantially stronger than any period, significant challenges are yet to be overcome. The major challenge to be addressed is the enormous emissions associated with current H_2 production. The majority of H_2 production accounting for 95% of the world capacity is produced from fossil fuels via Steam Methane Reforming (SMR), resulting in so-called grey H_2 (Renewable and Agency 2019). One promising solution is to produce the H_2 from a wide variety of renewable resources, which is labelled as green H_2. The main technology to produce green H_2 is water electrolysis in where water decomposed into H_2 and Oxygen (O_2) molecules in presence of electric current. For green H_2 to become a major energy carrier, production scale will need to be increased and in so partly address the production cost.

Notwithstanding, there has been little quantitative analysis of green H_2 from variable renewable energy in off-grid connection. (Schnuelle et al. 2020) developed a simulation model to determine the economic performance based on determined operating characteristics of onshore and offshore wind, as well as Photovoltaics (PV) plants. (Decker et al. 2019) evaluated the cost breakdown of a baseline case for PEM in on-grid settings. (Glenk and Reichelstein 2019) proposed a techno-economic model to investigate the H_2 produced from variable renewable energy considering the scalability of the electrolyzer in their analysis. (Mallapragada et al. 2020) developed a framework to decide the plant size and operating condition through the optimization of the size of the components taking into account hourly solar availability, and production requirement. However, it did not consider the utilization of the H_2 produced. (Koleva et al. 2021) established a mathematical model to evaluate PV-powered water electrolysis from an economic point of view. The model was tested under different weather conditions. The majority of the previous work has been focused on the economics of green H_2; nevertheless, there is no detailed investigation of dynamic modelling of PV-powered electrolysis. Therefore, a comprehensive dynamic model is developed to assess the trade-offs between the economic and environmental impacts of the green H_2. Furthermore, the dynamic model was formulated based on the intermittency of renewable energy that can affect the size of renewable energy units, the H_2 production and utilization facilities, as well as the H_2 storage. The mass and energy balance is calculated based on the hourly solar availability. The economic and environmental impacts are characterized by two metrics, which are total cost and CO_2 emissions saved respectively. A detailed discussion on the mechanism of the proposed dynamic model is present in the following section.

2. Methodology

The aim of this study is to analyze the economic and environmental impacts of utilizing green hydrogen on the performance of chemical processes that act as H_2 sinks. The study considers the intermittency of renewable energy sources and their effect on the sizing of the renewable energy and hydrogen production units, as well as on the size required for H_2 storage. The economic impact of introducing green H_2 is characterized by the total cost of establishing and operating the different units. Figure 1 shows the flow diagram with the different components considered in the evaluation. H_2 can be supplied to the chemical process either through green hydrogen production, or through grey hydrogen production. The production of grey hydrogen is accompanied with a relatively high level of CO_2 emissions that can be avoided by using green hydrogen with a low environmental footprint. The dynamics of the varying renewable energy source are considered by discretizing the annual operation into hourly time steps; in each time step, mass and energy balance calculations determine the renewable power generated, the rate of production of green hydrogen, and the flowrate of H_2 deliver to and from storage. These outcomes depend on the size of the different units, and on different technical parameters that are used as inputs. The hourly capacity factor of the renewable energy production unit ($CF_{RE}(t)$) can be determined based on the radiation data of the geographic location chosen, and the expected losses in energy transformation. The green hydrogen production unit is characterized by its efficiency-ε (the amount of electricity required to produce H_2 – kWh/kgH_2) and cost-C_{green} ($/kW). The hourly production of green H_2 is determined as shown in Eq.(1)

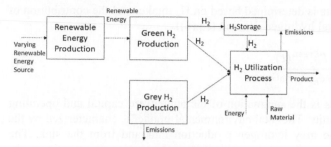

Figure 1 Flow diagram of the considered process

$$F_{H_2}^{Green}(t) = \frac{Size_{RE} \times CF_{RE}(t) \times \Delta t}{\varepsilon} \tag{1}$$

The size of the renewable energy production unit is selected so that the total annual production of green hydrogen meets the annual demand for green hydrogen, which is set based on the H_2 intake of the utilization unit (kgH_2/kgProduct), and the contribution of the green hydrogen to the total demand (%Green). This is shown in Eq.(2). Two different cases for addressing the temporal variation in green hydrogen production are investigated in this work: introducing H_2 storage and oversizing the H_2 utilization unit.

$$\sum_{t=1}^{t=t_f} F_{H_2}^{Green}(t) = intake_{H_2} \times Annual\ Product\ Rate \times \%Green \tag{2}$$

Note that the rate of H_2 production from the grey H_2 unit is equal to the difference between the demand for H_2 and the supply of green H_2. The introduction of H_2 storage allows the utilization of the excess H_2 when the rate of green H_2 production is higher than the demand for H_2 by the sink, while maintaining a consistent production rate by the sink. In this case, the production rate of the sink in each timestep is equal to the annual production rate divided by the number of timesteps. The storage capacity is tracked to determine the initial H_2 storage requirement and the total size of the storage. Eq (3), (4), and (5) show the equations that describe the storage.

$$H_2 toStr\ (t) = (F_{H_2}^{Green}(t) - H_2 demand(t)) \times x(t) \tag{3}$$

$$H_2 fromStr\ (t) = (H_2 demand(t) - F_{H_2}^{Green}(t)) \times (1 - x(t)) \tag{4}$$

$$H_2 inStor\ (t) = H_2 inStr\ (t-1) - H_2 fromStr\ (t) + H_2 toStr\ (t) \tag{5}$$

Note that x(t) is a binary variable that is equal to one when there is excess H_2, and it is zero otherwise. The initial mass of H_2 in the storage ($H_2 inStor\ (0)$) is selected such that the minimum content of the storage is zero. The size of the storage is set equal to the maximum $H_2 inStor\ (t)$ achieved throughout the year. The other option of addressing the dynamic variations in the hydrogen production is to allow the dynamic variation in the production of the H_2 utilization unit. This will result in varying product flowrate, and the size of the sink to allow the utilization of all the hydrogen produced. Hence, for each time step, the produced hydrogen ($F_{H_2}^{Green}(t)$) is determined (equation (1)), and the corresponding production rate is determined based on H_2 intake, and the contribution of the green hydrogen to the total hydrogen demand (eq.(6)).

$$Product\ Rate\ (t) = \frac{F_{H_2}^{Green}(t)}{intake_{H_2} \times \%Green} \qquad (6)$$

The total cost of the process is the summation of the annualized capital and operating costs of all the considered units. The total environmental impact is characterized by the emissions flowrate from the grey hydrogen production unit and from the sink. The assessment is conducted for different H_2 utilization technologies to investigate the variation in the impact of introducing green hydrogen to different processes on their environmental and economic performance. This can be reflected in the marginal abatement cost (MAC) of introducing green hydrogen, which is defined as below

$$MAC = \frac{Total\ cost\ with\ green\ H_2 - Cost\ of\ Base\ Case}{Emissions\ in\ the\ Base\ Case - Emissions\ with\ green\ H_2} \qquad (7)$$

3. Results & Discussion

In this section, the versatility of the dynamic model was tested to assess the trade-offs between the economic and environmental impact of decarbonizing specific energy-intensive chemical processes. Two chemical processes were analyzed namely, ammonia and methanol. The decarbonization is achieved through incorporating direct green H_2. The proposed dynamic model will provide insights into these trade-offs at various decarbonization targets. These targets are varied from 50% to 100% in 10% intervals. In each interval, the dynamic model will determine the size of the production and utilization of the green H_2, and the flowrate of H_2 deliver to and from storage, adhering to the hourly solar availability from a time horizon of one year. For the environmental impacts, the green H_2 contribution is compared with blue H_2, which is grey H_2 accompanied with Carbon Capture and Storage (CCS) based on their MAC. Integrating CCS with grey hydrogen can cut down emissions up 90% (IEA 2019a). The MAC of blue H_2 is not dependent on H_2 utilization process, and it depends only on grey H_2 and H_2 storage. It should be noted that this work uses Proton Exchange Membrane PEM electrolysis for green hydrogen production for its flexibility with the variation in renewable energy and SMR for grey and blue hydrogen. The techno-economic data assumptions are summarized in Table 1. The results of the analysis described in the methods section are shown in figuresFigure 2 andFigure 3.

Table 1 Techno-economic data assumptions

Parameter	Value	Reference
Ammonia Cost ($/ton)	901	(IEA 2019b)
Methanol Cost ($/ton)	392	(Al-Mohannadi et al. 2017)
PV Installation cost ($/kW)	714	(Agency 2020)
PEM Electrolysis Cost ($/kW)	1,100	(IEA 2019b)
Grey H_2 Cost ($/kg H_2)	348	(IEA 2019b)
H_2 Storage Cost ($/kg H_2)	615	(Nordin and Rahman 2019)
OPEX of % CAPEX	2.5	(IEA 2019b)
CCS Cost ($/ton H_2)	80.2	(Ahmed et al. 2020)

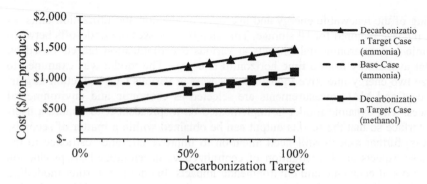

Figure 2 The effect of the decarbonization target on the total cost of utilization plant

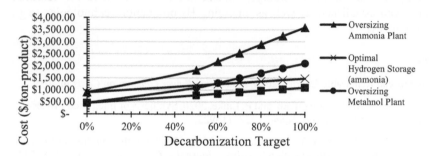

Figure 3 Options comparison between oversizing the utilization plant and the optimal H_2 storage

It can be seen clearly from Figure 2 that the decarbonization target significantly affects the total cost since the cost of ammonia and methanol produced. At 100% green H_2 is 1.6 and 2.4 times that of ammonia and methanol produced at 0% green H_2 respectively. It should be emphasized that the option of oversizing the utilization plant to address the variation of the production of the green H_2 is not attractive from an economic point of view. As shown in Figure 3, there is less variation in the cost for the two options (optimal H_2 storage and oversizing the utilization plant) at a lower decarbonization target. However, at a higher decarbonization target, the variation in the cost is significant since the total cost of oversizing the utilization plant is increased by 143% and 91% for ammonia and methanol respectively. For the environmental impacts, as mentioned in the methodology section MAC metric is selected to assess the environmental performance of incorporating green H_2. The MAC of ammonia and methanol is $304, and $802/ton of CO_2 saved respectively. It can be noted that MAC of methanol is higher than MAC of ammonia due to the high emissions associated with methanol production. As expected, the MAC of blue H_2 is lower than the MAC of green H_2 as it was estimated at $110/ton of CO_2 saved, considering 90% can be captured from both production process and energy emission streams.

4. Conclusions

A dynamic model to evaluate the economic and environmental impacts of incorporating green H_2 for decarbonizing energy-intensive chemical processes. The proposed dynamic model takes into account the intermittency of renewable energy sources and their effect on the sizing of the renewable energy and hydrogen production and utilization units, as well as on the size required for H_2 storage. This analysis assesses the trade-offs between the economic and environmental impacts at various decarbonization targets based on hourly solar availability from a time horizon of one year. The model was examined to decarbonize two energy-intensive industrial processes at various CO_2 reduction targets. The total cost and MAC measurements are selected as economic and environmental performances. The dynamic model is implemented on a spreadsheet to enable a user-friendly interface so that the results output can be obtained within a matter of seconds. Nevertheless, further aspects still need attention to improve and add resilience to the model. These aspects are investigation of centralized vs decentralized H_2 production facility in terms of economic and environmental impacts. In addition, further modelling work is needed to determine the optimum decisions related to capacity of H_2 production and utilization plants, size of the H_2 storage, and size of PV. Moreover, further research might explore well-established as well as emerging H_2 production technologies (e.g., Photoelectrocatalysis).

References

Agency, International Renewable Energy. 2020. *Renewable Power Generation Costs in 2019*.

R. Ahmed, S. Shehab, D. M. Al-Mohannadi, and P. Linke. 2020, Synthesis of Integrated Processing Clusters, *Chemical Engineering Science* 227, 115922.

D. M. Dhabia M., K. Abdulaziz, S. Y. Alnouri, and P. Linke. 2017. On the Synthesis of Carbon Constrained Natural Gas Monetization Networks. *Journal of Cleaner Production,* 168, 735–45.

M. Decker, F. Schorn, R. Can Samsun, R. Peters, and D. Stolten. 2019, Off-Grid Power-to-Fuel Systems for a Market Launch Scenario – A Techno-Economic Assessment." *Applied Energy* 250, 1099–1109

G. Glenk, and S. Reichelstein. 2019, Economics of Converting Renewable Power to Hydrogen, *Nature Energy* 4(3), 216–22.

IEA. 2019. *World Energy Outlook 2019*.

IPCC. 2018. *Global Warming of 1.5°C*.

M. Koleva, O. J. Guerra, J, Eichman, B,Hodge, and J. r Kurtz. 2021,Optimal Design of Solar-Driven Electrolytic Hydrogen Production Systems within Electricity Markets, *Journal of Power Sources* 483:229183.

D. Mallapragada, S. Emre Gençer, P. Insinger, D. William Keith, and F. Martin O'Sullivan, 2020, ,Can Industrial-Scale Solar Hydrogen Supplied from Commodity Technologies Be Cost Competitive by 2030?, *Cell Reports Physical Science* 1(9)

Renewable, International, and Energy Agency, 2019, *HYDROGEN : A RENEWABLE*.

Schnuelle, Christian, T. Wassermann, D. Fuhrlaender, and E. Zondervan. 2020,Dynamic Hydrogen Production from PV & Wind Direct Electricity Supply – Modeling and Techno-Economic Assessment, *International Journal of Hydrogen Energy* 45(55),29938–52.

Proceedings of the 14th International Symposium on Process Systems Engineering – PSE 2021+
June 19-23, 2022, Kyoto, Japan © 2022 Elsevier B.V. All rights reserved.
http://dx.doi.org/10.1016/B978-0-323-85159-6.50315-8

Thermoacoustic Flow-Through Cooler for Cryogenic Hydrogen

Konstantin I. Matveev[*], Jacob W. Leachman

School of Mechanical and Materials Engineering, Washington State University, Pullman, WA, 99164, USA
matveev@wsu.edu

Abstract

As hydrogen is becoming an increasingly important energy carrier for renewable energy systems, a need for efficiency improvements of hydrogen cooling and liquefaction rises as well. An additional challenge associated with these processes in cryogenic conditions is the exothermic conversion between ortho and para isomers of hydrogen, which requires removing extra heat and accelerating this reaction with catalysis. In this paper, a new system with flowing-through hydrogen and elegantly combined thermoacoustic heat pump and catalytic regenerator is analyzed using thermoacoustic theory. Calculations with variable channel sizes in a regenerator and acoustic impedances are conducted for standing-wave and travelling-wave variants of this system. The corresponding optimal second-law efficiencies of these setups are estimated to be about 0.3 and 0.6. The throughput of cooled hydrogen is assessed for systems with catalyzed and non-catalyzed regenerators.

Keywords: Hydrogen; Thermoacoustics; Cryogenics; Cooling.

1. Introduction

Hydrogen is one of the most promising fuels for the future "green" economy, as it does not produce harmful emissions when reacting with oxygen. As a liquid, hydrogen is one of the most energy-dense carriers. However, the process of cooling hydrogen at cryogenic temperatures either for subsequent liquefaction or to reduce boil-off during storage and transportation is a challenging problem, especially in small- and medium-scale systems (Rivard et al., 2019), which are needed to broaden applications of hydrogen fuel.

In this study, a novel system for cooling cryogenic hydrogen is analyzed. This system involves hydrogen flowing through a porous medium where both thermoacoustic heat pumping and conversion between hydrogen isomers are taking place (Figure 1). Acoustic oscillations can produce heat flux in fluids near solid surfaces that can be utilized for refrigeration (Swift, 2002). To achieve significant thermoacoustic cooling, porous materials with large surface-to-volume ratio are required.

A peculiar feature of cooling cryogenic hydrogen is a spin conversion of hydrogen molecules from predominantly ortho- to para-state that happens primarily below 100 K prior to condensation (Pedrow et al., 2021). When unassisted, this conversion is a very slow process. Moreover, this reaction is exothermic, thus requiring removal of additional heat. If this conversion does not happen during cooling processes, then the liquefied hydrogen evaporates more intensely during storage or transport, and thus, it will be lost. To accelerate the ortho-to-para conversion, surface catalysis can be employed, which requires porous materials to be effective. The main idea explored in this study is to utilize a single porous matrix for both thermoacoustic heat pumping and as a catalytic bed. Such a system can be more efficient, compact and reliable than competing technologies.

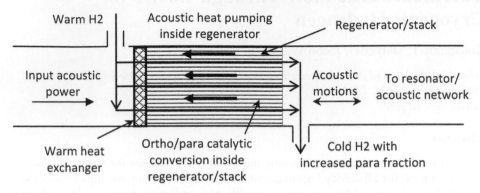

Figure 1 Schematic of thermoacoustic-catalytic system for cooling cryogenic hydrogen.

A device that accommodates both thermoacoustic and catalytic processes is shown in Figure 1. Acoustic power comes from either a linear motor or thermoacoustic engine. At appropriate phasing between acoustic pressure and velocity, heat is pumped from one side of the porous material (also called regenerator or stack) to the other side. Hydrogen entering at the warmer end cools down when flowing toward the cold end and exits the system at lower temperature. If a regenerator matrix is covered with a catalyst boosting ortho-to-para conversion, some hydrogen molecules will not only cool down but will also change their spin orientation. Hence, both a catalytic converter and a cooling device are efficiently combined in a single setup. Additional advantages include elimination of the cold heat exchanger, minimization of moving parts, and intrinsically efficient continuous heat transfer at smaller temperature differences along the stack. A few exploratory studies with flow-through thermoacoustic systems were done in the past, but not in cryogenic conditions or spin-transforming fluids (Hiller and Swift, 2000; Reid and Swift, 2000).

This paper outlines a simplified thermoacoustic model that can be used for initial assessment of the system operation in standing- and travelling-wave configurations. Amounts of hydrogen that can be cooled down are estimated in setups with catalyzed and non-catalyzed stacks. The presented model and results can benefit practitioners working on hydrogen systems and implementation of sustainable energy concepts.

2. Mathematical Model

A simplified approach to evaluate thermoacoustic heat transport in a stack or regenerator follows a theory developed for thermoacoustic devices (Swift, 2002). It is assumed that a stack contains a number of narrow channels between hot and cold ends. Other stack geometries, including random porous materials, can also be considered (Matveev, 2010), and basic thermoacoustic effects will remain similar.

Fluid inside the stack performs oscillations with the primary motions along the channels, while acoustic power comes from an external source, such as a motor or thermoacoustic engine. Due to thermal and acoustic interactions, heat can be transported along the channels. In case of large temperature gradients imposed in the stack, acoustic power can be generated. In case of relatively low temperature gradients, heat can be pumped from colder to warmer space, while acoustic power will be consumed in the process.

At the starting point of computing the heat flow and acoustic dissipation in stacks of thermoacoustic systems, one need to define acoustic pressure fluctuation p' and volumetric velocity fluctuation U'. The common notation involves complex numbers,

$$p'(x,t) = Re[p_1(x)e^{i\omega t}] \tag{1}$$

$$U'(x,t) = Re[U_1(x)e^{i\omega t}] \tag{2}$$

where p_1 and U_1 are the complex amplitudes of acoustic pressure and volumetric velocity fluctuations, i is the imaginary unity, ω is the angular frequency of oscillations, and t is the time. The x-axis is directed along the primary orientation of gas particle motions in the acoustic wave (along the stack). A relation between acoustic pressure and volumetric velocity at the stack depends on the entire system (not just stack), so that the system can be designed to achieve desirable impedances at the stack location (Matveev et al., 2006).

The heat transport rate produced by this thermoacoustic mechanism with a correction for ordinary heat conduction can be calculated as follows (Rott, 1975),

$$Q = \frac{1}{2} Re\left[p_1 \tilde{U}_1 \frac{\tilde{f}_v - f_k}{(1+\sigma)(1-\tilde{f}_v)}\right] + \left[\frac{\rho_m c_p |U_1|^2}{2\omega A(1-\sigma^2)|1-f_v|^2} Im(f_k + \sigma \tilde{f}_v) - (Ak + A_{sol}k_{sol})\right]\frac{\Delta T_m}{\Delta x} \tag{3}$$

where f_k and f_v are the thermoacoustic functions that depend on viscous and thermal penetration depths, $\delta_v = \sqrt{2\mu/(\omega\rho_m)}$ and $\delta_k = \sqrt{2k/(\omega\rho_m c_p)}$, and the channel thickness h; A and A_{sol} are the cross-sectional areas of the stack occupied by gas and solid, respectively; tilde indicates complex conjugate; σ is the Prandtl number; μ, ρ_m and c_p are the gas viscosity, mean density and specific heat, respectively; k and k_{sol} are the heat conductivities of gas and solid, respectively, and ΔT_m is the variation of temperature in the x-direction over distance Δx (stack length).

The pumped heat already accounts for heat conduction along the stack in Eq. (3). However, besides providing cooling power Q_C, this pumped heat must also remove acoustic power W_a dissipated in the stack, which can be estimated as follows,

$$W_d = \frac{1}{2} Re[\tilde{U}_1 \Delta p_1 + \tilde{p}_1 \Delta U_1] \tag{4}$$

where Δp_1 and ΔU_1 are relatively small changes of acoustic pressure and velocity amplitudes over the stack. As additional dissipation will occur in the other parts of the system (outside stack), the total acoustic power that need to be supplied is estimated as $W_a = bW_d$, where b is a given resonator loss correction. Then, the coefficient of performance and the second-law efficiency can be calculated as follows,

$$COP = \frac{Q_C}{W_a} = \frac{|Q| - W_d}{bW_d} \tag{5}$$

$$\eta_{II} = \frac{Q_C}{W_a} \frac{T_H - T_C}{T_C} \tag{6}$$

where T_H and T_C are the temperatures of the hot and cold ends of the stack, and ΔT_m in Eq. (3) is the difference between these temperatures, $T_C - T_H$.

3. Sample Results

Two examples of thermoacoustic coolers are presented below. The first setup involves a standing-wave phasing, whereas the second system is of the traveling-wave type. Their selected parameters, corresponding to typical thermoacoustic and hydrogen systems (aiming at H_2 flow rates of the order of 1 g/s), are listed in Table 1. The desired temperature drop is initially specified as 15 K. The inlet temperature of 77 K is chosen as the boiling temperature of liquid nitrogen, which permits economical cooling of hydrogen down to this temperature. The variable parameters used for optimization included a spacing distance h between plates in the stack and a magnitude of normalized acoustic impedance $|z|$ at the stack location. The normalized impedance is defined as follows,

$$|z| = \frac{p_1 A}{U_1 \rho a} \tag{7}$$

where a is the speed of sound. Additional requirements imposed in this optimization study are the cooling capacity of at least 200 W (needed for intended applications) and acoustic power input within 300 W, whereas the optimized parameter is the second-law efficiency.

Table 1. Selected system parameters.

Mean pressure	10^6 Pa	Stack length	2 cm
Warm temperature	77 K	Porosity	0.8
Cold temperature	62 K	Acoustic frequency	350 Hz
Stack plate material	steel	Acoustic pressure amplitude	$2 \cdot 10^5$ Pa
Resonator diameter	5 cm	Resonator loss correction	1.5

The calculated performance metrics for the standing-wave system, with velocity lagging pressure by about 90°, are shown in Figure 3. These metrics include the cooling power Q_C, coefficient of performance COP, and the 2nd-law efficiency η_{II}. This efficiency peaks at a certain value of the channel thickness normalized by the thermal penetration depth, h/δ_k. Among the considered variations of the plate spacing and acoustic impedance, the configuration with $h/\delta_k \approx 2.9$ and $|z| = 8$ produces the highest $\eta_{II} \approx 0.340$, while providing 207 W of cooling power and requiring 147 W of power input, thus satisfying criteria for the minimum cooling capacity and maximum acoustic power.

For the found optimal values of the stack-plate spacing and acoustic impedance, calculations have been also conducted to determine the amount of hydrogen flowing through the stack that can be cooled down to 65 K, 62 K, and 59 K, which correspond to temperature drops in the stack of 12 K, 15 K, and 18 K. Two situations were considered. First, no ortho-para conversion of hydrogen was assumed, implying a non-catalyzed stack. In the second scenario, a catalytic stack was used, and the complete conversion down to the equilibrium ortho-para ratio at the exit temperature was assumed. The results are shown in Figure 4. The dependence of the input acoustic power on flow rate of hydrogen is linear, as higher-order phenomena were neglected in this study. The larger the required temperature differential, the less hydrogen can be cooled.

The equilibrium ortho-para fraction ratio at the entrance temperature of 77 K is about 0.49, whereas this ratio decreases down to 0.39, 0.36 and 0.33 at the three exit temperatures considered here. When the maximum possible ortho-para conversion is achieved, the amounts of cooled hydrogen will be smaller (thin lines in Figure 4), since some of the cooling power has to compensate for the heat released during this conversion.

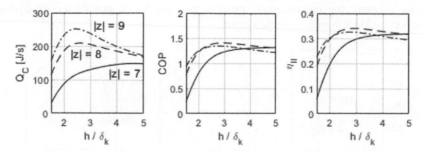

Figure 3 Characteristics of standing-wave system: cooling power Q_C, COP, and 2nd-law efficiency η_{II}. Normalized impedance $|z|$: 7, solid line; 8, dashed, 9 dash-dotted line.

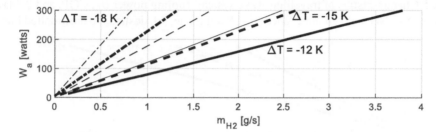

Figure 4 Acoustic power required to cool down flowing hydrogen in standing-wave system. Bold lines correspond to no para-to-ortho conversion, and thin lines to complete conversion. Temperature drops: 12 K, solid lines; 15 K, dashed; 18 K, dash-dotted lines.

Results of the optimization study conducted for the travelling-wave system, where pressure acoustic velocity and pressure fluctuations are in phase, are shown in Figure 5. Narrower channels and higher impedances are needed in such systems in comparison with standing-wave setups (Swift, 2002). The second-law efficiency for the travelling-wave reaches 0.574 for the relative plate spacing $h/\delta_k \approx 0.4$ and acoustic impedance $|z| = 40$, although the difference between peaks of the efficiency curves is small (Figure 5). The cooling power in this state is 538 W, while acoustic power of 227 W is required. One can note significantly higher efficiency is attained in the travelling-wave setup.

Results of calculations for the amount of flowing-through hydrogen in the optimized travelling-wave system are given in Figure 6 for the same temperature drops. Due to higher efficiency of this configuration, roughly twice larger flow rates of the cooled hydrogen are possible. Again, in case of catalyzed regenerators enabling ortho-para conversion, the amounts of cooled hydrogen are lower (thin lines in Figure 6).

4. Conclusions

A novel approach to cool cryogenic hydrogen, involving a flow-through thermoacoustic system, has been analyzed. Using thermoacoustic theory, high-performing geometrical and acoustic parameters of this device were determined under given operational conditions. The second-law efficiencies around 0.3 and 0.6 for standing-wave and travelling-wave setups were estimated. Mass flow rates of hydrogen flowing through the system and undergoing ortho-para conversion were evaluated. The possible theoretical extensions of this study can include modeling of the entire apparatus, accounting for finite convection heat transfer rate and ortho-para conversion rate, and modification of the model to account for condensation process to consider a possibility of hydrogen liquefaction inside a thermoacoustic system.

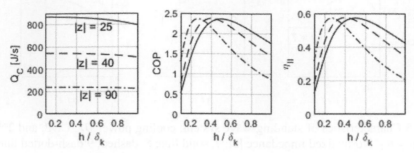

Figure 5 Characteristics of travelling-wave system: cooling power Q_C, COP, and 2nd-law efficiency η_{II}. Normalized impedance $|z|$: 25, solid line; 40, dashed; 90 dash-dotted lines.

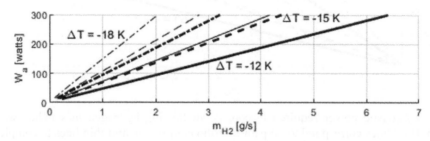

Figure 6 Acoustic power required to cool down flowing hydrogen in traveling-wave system. Bold lines correspond to no para-to-ortho conversion, and thin lines to complete conversion. Temperature drops: 12 K, solid lines; 15 K, dashed; 18 K, dash-dotted lines.

References

R.A., Hiller, G.W. Swift, 2000, Condensation in a steady-flow thermoacosutic refrigerator, Journal of the Acoustical Society of America, 108, 1521-1527.

K.I. Matveev, 2010, Thermoacoustic energy analysis of transverse-pin and tortuous stacks at large acoustic displacements, International Journal of Thermal Sciences, 49, 1019-1025.

K.I. Matveev, G.W. Swift, S. Backhaus, 2006, Temperatures near the interface between an ideal heat exchanger and a thermal buffer tube or pulse tube, International Journal of Heat and Mass Transfer, 40, 868-878.

B. P. Pedrow, S.K. Muniyal Krishna, E.D. Shoemake, J.W. Leachman, K.I. Matveev, 2021, Parahydrogen-orthohydrogen conversion on catalyst-loaded scrim for vapor-cooled shielding of cryogenic storage vessels, Journal of Thermophysics and Heat Transfer, 35, 142-151.

R.S. Reid, G.W. Swift, 2000, Experiments with a flow-through thermoacosutic refrigerator, Journal of the Acoustical Society of America, 108, 2835-2842.

E. Rivard, M. Trudeau, K. Zaghib, 2019, Hydrogen storage for mobility: a review, Materials, 12, 1973.

N. Rott, 1975, Thermally driven acoustic oscillations, part III: second-order heat flux, Z. Angew. Math. Phys., 26, 43-49.

G.W. Swift, 2002, Thermoacoustics: a Unifying Perspective for Some Engines and Refrigerators, Acoustical Society of America, Melville, NY, USA.

Proceedings of the 14th International Symposium on Process Systems Engineering – PSE 2021+
June 19-23, 2022, Kyoto, Japan © 2022 Elsevier B.V. All rights reserved.
http://dx.doi.org/10.1016/B978-0-323-85159-6.50316-X

Life Cycle Assessment of Green Hydrogen Transportation and Distribution Pathways

Malik Sajawal Akhtar, J. Jay Liu*

Department of Chemical Engineering, Pukyong National University, 48513 Busan, South Korea
jayliu@pknu.ac.kr

Abstract

The production of green hydrogen from renewable energy produced from wind and solar resources is deemed a more promising solution due to high energy quality, comparatively easy storage compared to electricity, and the prospect of using it at the time of use. Hydrogen has increasingly emerged as a potential energy carrier, making a global hydrogen mobility infrastructure essential to accelerating the transition to a hydrogen economy. Therefore, this work presents a cradle-to-gate life cycle assessment (LCA) for four hydrogen delivery pathways: compressed gas via tube trailers (CGH_2-TT), liquid hydrogen (LH_2), liquid organic hydrogen carrier (LOHC), liquid ammonia (LNH_3). The LCA results depict that for short distance of 100 km CGH_2-TT is the most eco-friendly option with the lowest global warming potential (GWP) of 1.81 $kgCO_2$-eq/kgH_2. Whereas, the LOHC pathway has shown the worst results with the highest GWP of 3.58 $kgCO_2$-eq/kgH_2. Likely, delivery via LNH_3 also showed significant emissions of 3.14 $kgCO_2$-eq/kgH_2 and remained the second worst candidate or hydrogen delivery.

Keywords: Green hydrogen, Life cycle assessment, Hydrogen transportation and distribution, Liquid organic hydrogen carriers, Green ammonia.

1. Introduction

The intermittency of solar and wind power results in the overproduction of electricity at times or in less production than needed at other times. Therefore, it is deemed important to store the overproduced electricity and use it when needed. Large scale energy mobility infrastructure is other challenge that needs to be addressed by finding the most sustainable energy carrier. Hydrogen owing to its high gravimetric energy density (120 MJ/kg) is being considered as a game changer. However, its low volumetric energy density hinders its mobility on a large scale and long distances (Akhtar and Liu, 2021a). Hydrogen can be stored as gaseous state as a compressed gas in high-pressure vessels or medium pressure pipelines, as a liquid state in cryogenic tanks, and as a liquid state in material based hydrogen storage (e.g., chemical hydrides and metal hydrides) (Abdin et al., 2020; Niermann et al., 2019; Wan et al., 2021).

So far, high pressure compressed hydrogen gas (CGH_2) and liquid hydrogen (LH_2) are the most hydrogen storage forms. However, both methods are not efficient on economic perspective depending on the current technological conditions, as CGH_2 storage uses around 15% of the stored energy of hydrogen to achieve 700 bar compression and LH_2 uses as much as 30% for the process of liquefaction (based on lower heating value of 120 MJ/kg) (Felderhoff et al., 2007). Liquid organic hydrogen carriers (LOHCs), a material-based hydrogen storage, are gaining much importance owing to easy handling, transportation, and no CO_2 emissions during hydrogenation or dehydrogenation.(Aakko-Saksa et al., 2018; Niermann et al., 2019; Preuster et al., 2017). LOHCs have hydrogen storage capacities of 6%–8% and can generally store hydrogen at atmospheric pressure and temperature.(Aakko-Saksa et al., 2018). Hydrogen storage and transportation using LOHCs can therefore be more cost-effective and environmentally friendly than conventional methods, which require high-pressure vessels or cryogenic tanks. A second advantage of the LOHCs is that they are chemically similar to gasoline and diesel, therefore can be transported at a larger scale while using the existing infrastructure for petroleum processing and transport. Recently, hydrogen transportation via liquid carriers was studied and LOHC was declared as a favorable solution from an economic perspective (Wulf and Zapp, 2018). Moreover, in another study, LOHCs were compared with CGH_2 and LH_2 with regard to hydrogen transportation and were declared as feasible solution for hydrogen transportation compared with CGH_2 and LH_2 (Reuß et al., 2017).

Ammonia (NH_3) is also considered a potential candidate for hydrogen storage owing to its high hydrogen content of 17.8% and existing transportation infrastructure. (Akhtar and Liu, 2021b) In a recent study, Aziz et al. studied the transportation of hydrogen from Australia to Japan in the form of LOHC, LH_2, and NH_3 and concluded that NH_3 is the most cost effective solution for hydrogen transportation over long distances (Aziz et al., 2019). In another study, Akhtar and Liu presented a comparative feasibility study on NH_3 as a hydrogen carrier via a techno-economic analysis of transporting NH_3 from Australia to Korea and concluded that using imported green NH_3 is an economically viable alternative compared with the domestic production and transportation of hydrogen (Akhtar and Liu, 2021b).

It is important to note that not all economically optimal solutions are environmentally sustainable. Hydrogen supply chain has been thoroughly studied on economic perspective but a comprehensive analysis on environment sustainability perspective is still needed. Recent studies have exclusively focused on the gaseous transportation of hydrogen, while others have looked at the liquid transport, concluding the best and worst methods. However, afterward, the goal of what to do next was left as a question for research. Therefore, in this work, a comprehensive life cycle assessment (LCA) on a cradle-to-gate approach has been performed for four pathways:

1. Hydrogen delivery as highly compressed (500 bar) gas via tube trailers (CGH_2-TT).

2. Hydrogen delivery as liquid in cryogenic tanks via liquid trucks (LH_2).

3. Hydrogen delivery in dibenzyl toluene (DBT) with the natural gas-assisted dehydrogenation process (LOHC).

4. Hydrogen delivery by liquid NH_3 (LNH_3).

2. Methodology

Life cycle assessment (LCA) is a universal tool to access or quantify the environmental impacts associated with a product throughout its life cycle based on ISO 14040 and 14044 (Akhtar and Liu, 2021a). In LCA, all energy and material flow that occur during upstream, midstream, and downstream stages including recycling or disposal of the analyzed products are quantified and evaluated. Simapro 9.1.1.1 is used to evaluate the environmental impacts of the entire lifecycle of the above-mentioned hydrogen delivery pathways.1kg of hydrogen gas is used as a functional unit and CML-IA baseline V3.06

method is used for the evaluation of the environmental impacts. A case study for the city of Perth, which is located in Western Australia and has an urban market and a population of 2.1 million is presented. The four hydrogen delivery pathways as mentioned above are shown in **Fig. 1**. The hydrogen delivery network consists of (1) hydrogen production, (2) pre-treatment and storage of the hydrogen gas, (3) hydrogen gas/carrier transportation from production facility to hydrogen refueling station (HRS), (4) post-treatment of hydrogen gas, and (5) dispensing of the hydrogen gas to fuel cell vehicle. In all pathways, hydrogen is produced via alkaline water electrolysis using the electricity from an on-site wind power plant.

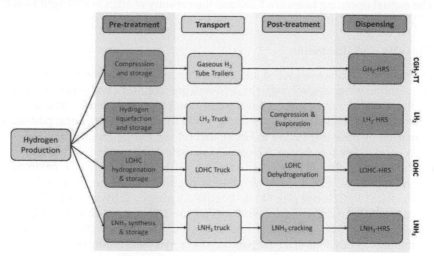

Fig. 1. Hydrogen delivery pathways.

Table 1. Main process conditions and life cycle inventory.

Process	Parameter	Units	CGH₂	LH₂	LOHC	LNH₃
Pre-treatment	Pressure	bar	500	1	1	150
	Temperature	°C	-	-253	150	450
	Electricity	kWh/kgH₂	2.2	10	0.7	4.2
Transport	Capacity per truck per trip	kgH₂	1100	4300	1800	7200
Post-treatment	Pressure	bar	-	-	350	1
	Temperature	°C	-	-	-	400
	Electricity	kWh/kgH₂	-	0.5	0.4	0.2
	Heat	kWh/kgH₂	-	-	10.5	14.3
Dispensing	Pressure	bar	700	700	700	700
	Electricity compression	kWh/kgH₂	1.6	0.5	3.96	3.45
	Electricity pre-cooling	kWh/kgH₂	4	-	4	4

At the pre-treatment stage, different processes such as compression, liquefaction, catalytic hydrogenation of LOHC, or NH_3 synthesis occur to increase the volumetric density of the hydrogen gas. Whereas, compression, pumping and evaporation, dehydrogenation of LOHC, and NH_3 cracking are the post-treatment processes occurring at HRS. For the analysis of hydrogen delivery by LOHC, dibenzyl toluene (DBT) is used, since it is the most promising hydrogen carrier.(Teichmann et al., 2012) The distance for transportation is taken as 100 km for the base case. However, in order to see the impact of distance variation, the distance is varied from 100 km to 500 km. The LCI and details of pre-treatment and post-treatment processes for all pathways is presented in **Table 1**.The annual operating hours are 8300 and the capacity of HRS is 850 kg/d (Reuß et al., 2017).

3. Results and discussion

The entire supply chain is divided into three stages: (a) *production* (hydrogen production using alkaline water electrolysis), (b) *delivery* (pre-treatment: compression, liquefaction or hydrogenation, and storage in underground salt caverns or liquid tanks; transport: via liquid carriers in trucks, pressurized transports in trucks, or by pipelines; post-treatment: evaporation or dehydrogenation at HRS), and (c) *dispensing* (operational activities like compression to bring hydrogen/NH_3 at FCV pressure, followed by filling into vehicles). **Fig. 2** illustrates that hydrogen delivery via CGH2-TT is the most environmentally friendly option, as it resulted in the lowest contribution to global warming potential (GWP) of 1.81 kgCO2-eq/kgH2. On the contrary, hydrogen delivery via LOHC is the least environmentally friendly option with a GWP of 3.57 kgCO2-eq/kgH2. **Fig. 2** further shows that hydrogen production and hydrogen delivery are the two main stages contributing to emissions in all the pathways. For hydrogen production, the key driver of environmental emissions is the windmill-derived electricity used for electrolysis.

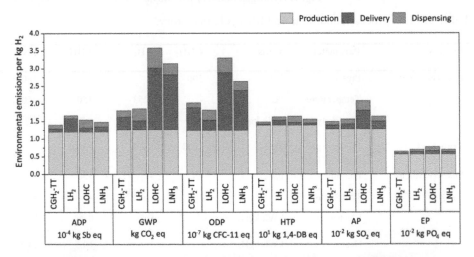

Fig. 2. Breakdown of LCA results for a transport distance of 100 km.

The delivery through LOHC presented the highest number of emissions because the post-treatment process, i.e., dehydrogenation at the HRS, is very energy-intensive, as shown in **Fig. 3(c)**. Following LOHC, the delivery via LNH3 results in the highest CO_2 emissions. A temperature of over 500°C is required for releasing hydrogen from NH_3. More than 80% of the CO_2 delivery emissions through LNH3 are caused by the post-treatment process of NH_3 cracking, and 99% of the 86% are produced by the energy required to achieve the required reaction temperature, whereas NH_3 synthesis only contributes 11% to the CO_2 emissions, as shown in **Fig. 3(d)**.

With the increase in the distance of transportation from 100km to 400km, the highest increase of 75% is observed for transporting hydrogen via CGH2-TT with a distance increase from 100 to 400 km as shown in **Fig. 4**. With the distance increase from 100 to 400 km, hydrogen delivery via LNH_3 showed the lowest increase for transporting hydrogen by trucks compared with hydrogen delivery via LOHC since the amount of hydrogen transported per trip is much higher (7200 kg for LNH_3 trucks compared with LOHC (1800 kg)) owing to the high volumetric densities of 682 kg/m^3 for NH_3 and 57 kg/m^3 for DBT.

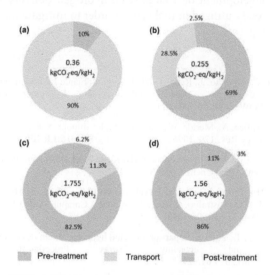

Fig. 3. Breakdown of LCA results for the impact category GWP for the delivery of hydrogen via (a) CGH$_2$-TT, (b) LH$_2$, (c) LOHC, and (d) LNH$_3$.

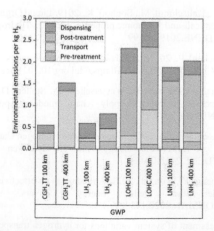

Fig. 4. Comparison of LCA results for GWP of all pathways.

4. Conclusions

In order to use hydrogen as an energy carrier for intermittent power sources, such as wind and solar, a sustainable, safe, and efficient method for storing and delivering hydrogen is needed. This paper presented a LCA for four hydrogen delivery pathways for short and long transport distances for the city of Perth, located in Western Australia. The results show that for short distances CGH_2-TT is the most responsible candidate on environmental perspective. However, with the increase in distance of transportation from 100km to 400km the highest increase in GWP is observed for CGH_2-TT. On the contrary the lowest increased is observed for the case of LHN_3 when the distance is increased to 400km. Therefore, for long distances of transportation at a larger scale, the delivery via CGH_2-TT would not be an ecologically responsible option. On the contrary, LNH_3 can play a vital role in development of a large-scale hydrogen delivery infrastructure if in future NH_3 can be directly utilized in fuel cells in order to mitigate the significant impact related to NH_3 cracking.

References

Aakko-Saksa, P.T., Cook, C., Kiviaho, J., Repo, T., 2018. Liquid organic hydrogen carriers for transportation and storing of renewable energy – Review and discussion. J. Power Sources 396, 803–823. https://doi.org/10.1016/J.JPOWSOUR.2018.04.011

Abdin, Z., Zafaranloo, A., Rafiee, A., Mérida, W., Lipiński, W., Khalilpour, K.R., 2020. Hydrogen as an energy vector. Renew. Sustain. Energy Rev. 120, 109620. https://doi.org/10.1016/J.RSER.2019.109620

Akhtar, M.S., Liu, J., 2021a. Life Cycle Assessment of Hydrogen Production from Imported Green Ammonia: A Korea Case Study. Comput. Aided Chem. Eng. 50, 147–152.

Akhtar, M.S., Liu, J., 2021b. Process Design and Techno-economic analysis of Hydrogen Production using Green Ammonia Imported from Australia- A Korea Case Study. Comput. Aided Chem. Eng. 50, 141–146. https://doi.org/10.1016/B978-0-323-88506-5.50023-1

Aziz, M., Oda, T., Kashiwagi, T., 2019. Comparison of liquid hydrogen, methylcyclohexane and ammonia on energy efficiency and economy. Energy Procedia 158, 4086–4091. https://doi.org/10.1016/j.egypro.2019.01.827

Felderhoff, M., Weidenthaler, C., Von Helmolt, R., Eberle, U., 2007. Hydrogen storage: The remaining scientific and technological challenges. Phys. Chem. Chem. Phys. 9, 2643–2653. https://doi.org/10.1039/b701563c

Niermann, M., Beckendorff, A., Kaltschmitt, M., Bonhoff, K., 2019. Liquid Organic Hydrogen Carrier (LOHC) – Assessment based on chemical and economic properties. Int. J. Hydrogen Energy 44, 6631–6654. https://doi.org/10.1016/J.IJHYDENE.2019.01.199

Preuster, P., Papp, C., Wasserscheid, P., 2017. Liquid organic hydrogen carriers (LOHCs): Toward a hydrogen-free hydrogen economy. Acc. Chem. Res. 50, 74–85. https://doi.org/10.1021/acs.accounts.6b00474

Reuß, M., Grube, T., Robinius, M., Preuster, P., Wasserscheid, P., Stolten, D., 2017. Seasonal storage and alternative carriers: A flexible hydrogen supply chain model. Appl. Energy 200, 290–302. https://doi.org/10.1016/J.APENERGY.2017.05.050

Teichmann, D., Arlt, W., Wasserscheid, P., 2012. Liquid Organic Hydrogen Carriers as an efficient vector for the transport and storage of renewable energy. Int. J. Hydrogen Energy 37, 18118–18132. https://doi.org/10.1016/J.IJHYDENE.2012.08.066

Wan, Z., Tao, Y., Shao, J., Zhang, Y., You, H., 2021. Ammonia as an effective hydrogen carrier and a clean fuel for solid oxide fuel cells. Energy Convers. Manag. 228, 113729. https://doi.org/10.1016/J.ENCONMAN.2020.113729

Wulf, C., Zapp, P., 2018. Assessment of system variations for hydrogen transport by liquid organic hydrogen carriers. Int. J. Hydrogen Energy 43, 11884–11895. https://doi.org/10.1016/j.ijhydene.2018.01.198

Proceedings of the 14th International Symposium on Process Systems Engineering – PSE 2021+
June 19-23, 2022, Kyoto, Japan © 2022 Elsevier B.V. All rights reserved.
http://dx.doi.org/10.1016/B978-0-323-85159-6.50317-1

Sector coupling of green ammonia production to Australia's electricity grid

Nicholas Salmon[a], René Bañares-Alcántara[a*]

[a]Department of Engineering, University of Oxford, Parks Road, Oxford, OX1 3PJ, UK
[*]rene.banares@eng.ox.ac.uk

Abstract

'Sector coupling' of large industries to national power networks has been identified as a technique to stabilise energy systems as they electrify, and hence to reduce the cost of decarbonisation. This work explores sector coupling of green ammonia production, for which the grid can provide a stable backup power supply to reduce the costs of energy and hydrogen storage; in turn, green ammonia plants can provide renewable electricity to displace fossil fuels in the grid. We present a model which minimises the levelised cost of ammonia at 701 locations across Australia; it finds that in almost 50% of cases, paying for a grid connection reduces the cost of ammonia production, with savings of more than 10% realisable in some locations. We further show that a grid connection creates a relationship between LCOA and production scale, and improves operational stability.

Keywords: Green Fuels, Levelised Cost of Ammonia, Electricity Arbitrage, MILP

1. Introduction

Green ammonia is a derivative of green hydrogen, produced from renewable electricity, water and air. It is carbon-free, and has applications as an energy transport and storage vector, as a shipping fuel and as a fertiliser (Nayak-Luke and Bañares-Alcántara, 2020). Because of Australia's reliable renewable energy resource, large land availability, and its proximity to large future markets in East Asia, green ammonia production represents a significant economic opportunity (Srinivasan et al., 2019).

In this work, we explore how a further strategic advantage – namely, Australia's high reliability electricity network – can further improve the economic potential of green ammonia production. This coupling of a large export industry to the national electricity system has been identified as a promising opportunity to decarbonise affordably without threatening the stability of energy systems (Bloomberg New Energy Finance, 2020).

We extend on the work of previous authors who have optimised green ammonia plant designs (Fasihi et al., 2021, Nayak-Luke and Bañares-Alcántara, 2020) by developing a MILP model which includes the opportunity for a grid connection (if that connection reduces overall costs), which we solve using the Gurobi optimisation solver. Some non-linear optimisation techniques, such as a brute force calculation and genetic algorithms, are considered in the literature; however, because they take much longer to converge, and do not significantly improve solution accuracy, these techniques are not appropriate for this application.

This work further considers the relationship between a grid connection and the optimum plant scale, as well as exploring the modifications which might be required for plant design in order to achieve stable operation with and without grid connectivity.

2. Methodology

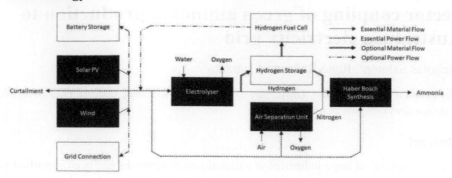

Figure 1 - Green ammonia production flowchart.

Figure 1 shows the green ammonia production process from renewable electricity. As described in detail in other works, green ammonia production ceases to benefit from economies of scale at a production rate of around 1 MMTPA (Salmon et al., 2021). This eliminates nonlinearities, meaning MILP is suitable to solve this problem.

In <u>design mode</u>, the solver uses location-specific weather and grid data to minimise the levelized cost of ammonia (LCOA) production by deciding (i) whether the plant should invest in a grid connection; (ii) the size in MW (or MWh for storage equipment) of the non-grid units shown on Figure 1; and (iii) the plant's operating behaviour. The design is subject to physical (i.e. mass and energy balances over each unit) and technical (i.e. minimum operating rates and maximum ramp rates of the HB plant) constraints.

The capital cost of grid connection is estimated based on the distance between the site and the nearest electricity transmission line. AC connections are used for distances of <40 km; HVDC connections are used for larger distances. The plant can import electricity using the live power price for the state in which it connects to the grid, and export excess electricity back to the grid. Imported electricity carries a transmission usage cost estimated to be 10 AUD/MWh. Costs of Australian equipment were estimated using the IRENA database (for renewable energy/electrolysers); CSIRO estimates (for grid connections/battery storage) and data from Nayak-Luke and Bañares-Alcántara (2020) (for the ammonia plant/air separation unit). Operating costs are calculated from the power withdrawn from the electricity grid, water consumption (estimated at 2 USD/kL), and an operations and maintenance (O&M) fraction of 2%. The LCOA is estimated from the CAPEX and OPEX using a discount rate of 7%.

In <u>operation mode</u>, the model takes as inputs both the plant design and a different year of weather/grid electricity data, and selects operating conditions which maximise cash flow. As an approximation, the costs of water and O&M are neglected, and the cash flow is calculated simply as ammonia sales minus net electricity cost. The sale price of ammonia is estimated as 500 USD/t, which is at the higher end of spot prices from the last decade.

The design problems have 157,684 constraints and 122,650 variables; all but one of these variables are continuous, the binary variable being used to decide if a grid connection is suitable. The root relaxation and node relaxation are solved using the concurrent and barrier algorithms respectively, since these were found to converge the fastest.

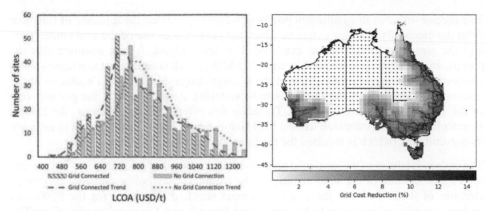

Figure 2 - Impacts of grid connection compared to a no grid case. (a) - Left: Distribution of LCOAs with grid connection compared to the same sites without grid connection. (b) - Cost reduction achieved from grid connection (Sites with a black dot and no shading did not connect to the grid).

3. Grid Connectivity Results

Over 701 cases, the average time taken for the model to solve on an i7 processor with 8 cores and 16 GB of RAM was 70 s/problem; it solves almost twice as quickly if the binary variable is fixed to a value of 0 (i.e. the grid connection is disallowed). In all cases, the model solved to within a tolerance of 10^{-4}.

The benefits of connecting to the grid are meaningful in many locations. Figure 2 (a) shows the cost distribution shifts to the left when a grid connection is included; on average, this reduction was 6%; however, as Figure 2 (b) demonstrates, there is a correlation between the distance of the site from the grid and the cost benefit. Less than 50 km from the grid, the cost reduction averages more than 8%. The best site, in Tasmania, is cheapest in both grid and non-grid cases; it sees a cost reduction from grid connection of almost 60 USD/t. Some sites become active, bi-directional participants in the market: at 63 sites, the revenue from power sales exceeds the cost of power purchased. The main cause of cost reduction is the replacement of energy storage equipment. No site with a grid connection requires battery storage or hydrogen fuel cells, and the hydrogen storage equipment reduces in size by ~30% compared to a no-grid case. Figure 2 (b) shows that sites which are grid connected are more concentrated near population centres on the coastline. This geographical spread could enable both (a) power generation at non-coastal sites that is transmitted through the grid; and (b) access to infrastructure, including a skilled workforce, spare equipment, water for desalination, and ports for export.

4. Impacts of Scale

Without grid connection, the production LCOA at industrial scale has limited dependence on production rate; while downstream infrastructure may benefit from economies of scale, most equipment required for ammonia production is modular. However, including a grid connection introduces scale dependency. At small scales (<0.1 MMTPA), the fixed cost of grid connection is not worthwhile; at large scales (i.e. ~10 MMTPA), a grid connection will not be sufficient to supply a meaningful amount of electricity to the ammonia plant without impacting the local network (meaning the grid connection is small relative to overall plant size). Between these extremes lies a minimum production cost, at which the plant makes maximum use of its investment in the electricity grid.

The precise value of this minimum production cost depends on the distance of the plant from the electricity grid, the extent to which the plant makes use of the grid connection, and the size of grid connection available. For this analysis, it was assumed that the maximum size of the grid connection was ~175 MW, which is equivalent to a large LVAC connection, and equivalent to ~10% of the average demand in Australia's smaller states (meaning larger sizes are very unlikely to be suitable). At small scales, the per unit cost of the Haber-Bosch plant increases. To factor this effect into the analysis, the Haber-Bosch plant size was estimated using a load factor of 80% before optimisation to estimate its per-unit cost (which is required for the linear optimisation problem).

The carbon intensity of the local grid must also be considered; if grid electricity represents too large a fraction of the input electricity, the ammonia may not be considered 'green'. Because of its limited use, there is no common standard at present for the maximum carbon intensity of ammonia to be considered 'green'; one European model, CertifHy, specifies a maximum of 36 g CO_2-e/MJ of fuel for hydrogen; that limit is marked on Figure 3 (b). For carbon accounting, we here assume that green electricity sold onto the electricity grid displaces fossil fuels, and therefore counts as a carbon credit.

Figure 3 shows the impact of scale on production in two locations. At the first location, which uses more grid electricity than it exports, the minimum production cost corresponds to a high carbon intensity; to reduce these emissions, a larger plant scale is required at which the grid provides less of the power, slightly increasing costs. On the other hand, at the second location in Tasmania (the cheapest identified for 2019 data), which sells more electricity than it purchases, the ammonia is carbon negative at all scales. At this location, it is profitable to sell electricity from the renewable energy production; that sale effectively subsidises the cost of green ammonia, and therefore benefits from very small scales – the larger the production rate, the smaller the subsidy from electricity export per ton.

The different behaviour at the two locations is caused both by different weather patterns, and different electricity grids; Western Australia has a stable, carbon intensive grid powered dominantly by gas; Tasmania's grid uses mostly hydropower and wind, enabling more opportunity for cost arbitrage and low-carbon grid connection.

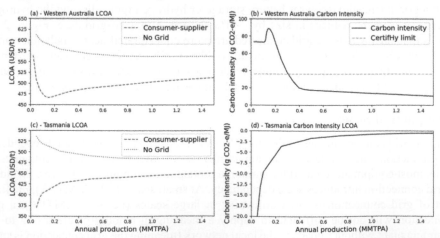

Figure 3 - Impact of scale on LCOA and carbon intensity at two low-cost sites, one in Western Australia (top), and the other in Tasmania (bottom).

5. Operating Considerations

Because the plant is optimised to minimise the LCOA, the plant (grid-connected or not) may not operate as efficiently in different weather conditions than in those under which it was designed. To some extent, grid-connected sites can use back-up power to maintain stable operation, but doing so increases costs, so electricity inputs should be minimised. Non-grid connected (islanded) sites rely on energy and hydrogen storage when they cannot generate power; if storage is too small, there is a risk of system failure (shutdown).

The main operating challenge is the requirement of the Haber-Bosch (HB) plant to operate above a minimum rate. Ambitious estimates put this minimum rate at around 20% of rated capacity. If the designed plant cannot maintain this rate for a given weather profile, regardless of how it is operated, the operating model will fail to converge, which occurs for many grid-connected and islanded cases. To reduce the likelihood of failure, the plant was overdesigned by imposing increasingly tight restrictions on the green ammonia minimum rate (i.e. > 20%) (in design mode only), but allowed to operate at the most flexible minimum rate (i.e. = 20%); this reduces failure frequency, but increases costs.

Two sets of islanded sites are compared to grid-connected sites: The Islanded (I) set refers to sites in the same location as grid-connected sites at which the model was re-run without the grid; the Islanded (II) set refers to different sites where grid connection is not optimal, whether or not it is allowed. At islanded sites, costs are mostly capitalised; at grid-connected sites, electricity costs are operational, and therefore impact cash flow. For fair comparison between sets, we report the "Cash flow delta", which is given by the cash flow in the operating year minus the cash flow that was anticipated in the design year.

Both grid-connected and islanded sites require overdesign to reduce the plant failure risk, but grid-connected sites still outperform islanded sites on two fronts. Firstly, the cost to overdesign is higher at both sets of islanded sites than at grid-connected sites – see Figure 4 (a). Secondly, while imposing stricter overdesign requirements reduces the failure rate at all sites, there are fewer failures at grid connected sites than at either set of islanded sites – see Figure 4 (b). Figure 4 (c) also indicates that sites are more likely to connect to the grid during the design process if the constraints imposed on the HB plant are tighter.

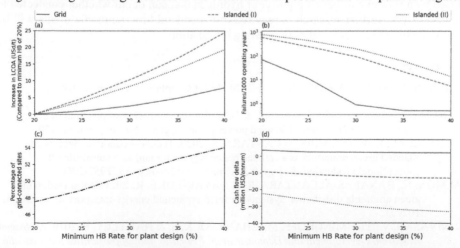

Figure 4 - Plots showing operating performance of the three sets of sites at different minimum HB rates in plant design. No operating failures were recorded for grid sites with minimum rates of 35% or greater, which is recorded as a rate of 0.5/thousand years so it can be read on a log scale.

With no overdesign, the average cash flow delta overall is below 0, which indicates that performance is generally worse during operation than was anticipated during system design, because the plant is not optimised for the new weather or grid electricity profile. However, grid-connected sites on average had a cash flow delta slightly greater than 0, even though the operating timeframe included both years in which the grid was cheaper and years in which it was more expensive than the design year. The cash-flow delta worsens for islanded sites as the plant is more overdesigned, while for grid-connected sites it stays relatively constant: this is an artefact of cases which were previously non-converging being factored into the averages.

6. Conclusions

This research explores the benefits of sector coupling for green ammonia production. It demonstrates that significant reductions in the LCOA are achievable using a grid connection. The most substantial cost reductions, which are in the order of 10%, occur when the plant is located near the electricity grid. In Australia, these sites are mostly coastal, which will locate them close to other supporting industry and to export ports. Further cost reductions are achievable by optimising the plant scale relative to the maximum allowable size of the grid connection. In some locations, ammonia production can be significantly subsidised by profitable participation in the grid.

When a site connects to the grid, it is less likely to fail during operation, and will generate more cash flow than if a grid connection is not used. Regardless of whether a site is grid-connected, it requires some overdesign, which can be achieved by designing with tighter limitations on the minimum rate of the HB plant than are achievable during operation; the cost of overdesign at grid-connected sites is less than at islanded sites.

The integration of optimised green ammonia production and grid electricity is a first step in understanding how sector-coupling in the energy system can reduce the costs of decarbonisation. Further research should consider other industries which may have synergies with electricity grids, and how electricity grids themselves will transform over time. Additionally, the operating model demonstrated that there is a risk of plant failure caused by a shortage of back-up power or hydrogen that can occur whether connected to the grid or not; further research is required to understand how ammonia plants will be operated with imperfect weather forecasting information.

References

BLOOMBERG NEW ENERGY FINANCE 2020. Sector Coupling in Europe: Powering Decarbonization.

FASIHI, M., WEISS, R., SAVOLAINEN, J. & BREYER, C. 2021. Global potential of green ammonia based on hybrid PV-wind power plants. *Applied Energy,* 294, 116170.

NAYAK-LUKE, R. & BAÑARES-ALCÁNTARA, R. 2020. Techno-economic viability of islanded green ammonia as a carbon-free energy vector and as a substitute for conventional production. *Energy & Environmental Science,* 13, 2957-2966.

SALMON, N., BAÑARES-ALCÁNTARA, R. & NAYAK-LUKE, R. 2021. Optimization of green ammonia distribution systems for intercontinental energy transport. *iScience,* 24, 102903.

SRINIVASAN, V., TEMMINGHOFF, M., CHANRNOCK, S. & HARTLEY, P. 2019. *Hydrogen Research, Development and Demonstration: Priorities and Opportunities for Australia, CSIRO* [Online]. Available: https://www.csiro.au/en/Do-business/Futures/Reports/Energy-and-Resources/Hydrogen-Research [Accessed December 2020].

Proceedings of the 14th International Symposium on Process Systems Engineering – PSE 2021+
June 19-23, 2022, Kyoto, Japan © 2022 Elsevier B.V. All rights reserved.
http://dx.doi.org/10.1016/B978-0-323-85159-6.50318-3

Development of Multi-Purpose Dynamic Physical Model of Fuel Cell System

Shigeki Hasegawa[a*,b], Shun Matsumoto[b], Yoshihiro Ikogi[b],

Tsuyoshi Takahashi[b], Sanghong Kim[c], Miho Kageyama[a], and Motoaki Kawase[a]

[a] *Department of Chemical Engineering, Kyoto University, Kyoto 615-8510, Japan*
[b] *Commercial ZEV Product Development Div. Toyota Motor Corporation, Aichi 471-8571, Japan*
[c] *Department of Applied Physics and Chemical Engineering, Tokyo University of Agriculture and Technology, Tokyo 184-8588, Japan*
s.hasegawa@cheme.kyoto-u.ac.jp

Abstract

1-dimentional (1D) physical modeling methods of the fuel cell (FC) system including the FC stack, air system, H_2 system, and cooling system were investigated. To ensure the simulation of life-long system operation (> 100,000 km) of a vehicle with the allowable calculation time, the proper model resolution was selected and the in-house high-speed numerical solvers were developed. The acceptable accuracy was confirmed by the comparison between the model outputs and the actual FC-system data collected with 2nd-generation MIRAI under a variety of operating condition.

Keywords: Fuel cell system; Model-based development; Physical modeling;

1. Introduction

Hydrogen energy is regarded as one of the most promising alternative energies to fossil fuels from the view of CO_2 emission and energy efficiency. The FC-system manufacturers are required to develop the products for the wide range of applications such as passenger vehicles, commercial vehicles of buses and tracks, railways, marine vessels, aviation, and stationary power generator purposes.

On the other hand, due to the complexity in the hardware and software configuration of the FC systems, neither a system model nor a systems approach for the FC-system simulation, analysis, optimization, and manufacturing has not been proposed, in spite of intensive investigation on the simulation of fuel cell itself (Weber et al., 2014). Thus a significant effort and cost of trail-and-error for the development of each commercial application is required to the FC-system manufacturers, which is one of the largest barriers of entry to the fuel cell industry. The purpose of this research is the development of a state-of-the-art FC-system model, with which the FC-system manufacturers can do the model-based design, evaluate the system, and reduce the difficulty described above.

In this study, the FC system implemented to 2nd-generation MIRAI (Takahashi and Kakeno, 2021) shown in Fig. 1, is taken as an example of the application of the developed model to describe the modeling strategies and detailed implementation methods.

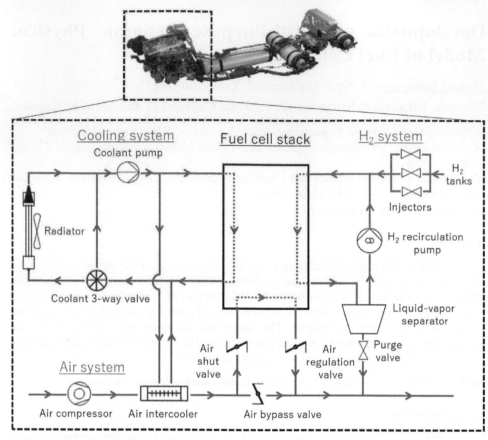

Fig. 1. Flow diagram of the FC system implemented in 2nd generation MIRAI

The FC system in Fig. 1 consists of the FC stack as main engine, air system, H_2 system and cooling system, and the components in each system and their functions are listed in Table 1.

Table 1. The components and their functions in the FC system shown in Fig. 1

System	Component	Function
Air	Air compressor	Air supply
	Air intercooler	Air cooling for the FC-stack materials protection
	Air shut-valve	Seal-up during system shut-down condition
	Air regulation valve	Pressure control, seal-up during system shut-down condition
	Air bypass valve	Control of air flowrate to the FC stack and the exhaust line
H_2	Injectors	Hydrogen supply from high-pressure H_2 tank
	Liquid-vapor separator	Separation of exhausted liquid water
	Purge valve	Exhaust of gaseous impurities (N_2, O_2, and H_2O), and liquid water
	H_2 recirculation pump	Recirculation of unreacted H_2 to inlet-side
Cooling	Coolant pump	Coolant supply
	Radiator	Heat exhaust from the coolant
	Radiator fan	Enhancement of heat exchange rate at the radiator
	Coolant 3-way valve	Control of coolant flowrate to the radiator and the bypass lines

2. Physical modeling methods of FC system

The modeling strategies and implementation methods of the FC stack and H_2 system, whose configuration is shown in Fig. 1, are explained. The system configurations are described as the function-block diagram shown in Fig. 2(b), where the overall system configuration is broken-down to the component level. In each function block, state variables (pressure, flowrate, temperature, and gas composition) and individual component models are encapsulated. In more detail, physical models of mass-transfer and electrochemistry in the FC stack are developed (Hasegawa *et al.*, 2021) and implemented in '(5) aFC' block, and physical models of fluid and thermal dynamics of various system components (Bird et al, 2006) are implemented in the other blocks.

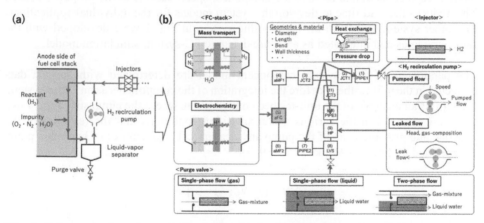

Fig. 2. Schematic drawings of the configurations of the FC stack and the H_2 system in a (a) flow-diagram and (b) function-block diagram

The dynamic relationships between state variables in each function-block are described as the algebraic equations of pressure balance, material balance, and energy balance. An example of pressure balance is shown in Eq. (1),

$$
\begin{array}{l}
\text{(1) PIPE1} \\
\text{(2) JCT1} \\
\text{(3) JCT2} \\
\text{(4) aMF1} \\
\text{(5) aFC} \\
\text{(6) aMF2} \\
\text{(7) PIPE2} \\
\text{(8) LVS} \\
\text{(9) HP} \\
\text{(10) PIPE3} \\
\text{(11) JCT3}
\end{array}
\begin{pmatrix}
a_{1,1} & a_{1,2} & 0 & 0 & 0 & 0 & 0 & 0 & 0 & 0 & 0 \\
a_{2,1} & a_{2,2} & a_{2,2} & 0 & 0 & 0 & 0 & 0 & 0 & 0 & 0 \\
0 & a_{3,2} & a_{3,3} & a_{3,4} & 0 & 0 & 0 & 0 & 0 & 0 & a_{3,11} \\
0 & 0 & a_{4,3} & a_{4,4} & a_{4,5} & 0 & 0 & 0 & 0 & 0 & 0 \\
0 & 0 & 0 & a_{5,4} & a_{5,5} & a_{5,6} & 0 & 0 & 0 & 0 & 0 \\
0 & 0 & 0 & 0 & a_{6,5} & a_{6,6} & a_{6,7} & 0 & 0 & 0 & 0 \\
0 & 0 & 0 & 0 & 0 & a_{7,6} & a_{7,7} & a_{7,8} & 0 & 0 & 0 \\
0 & 0 & 0 & 0 & 0 & 0 & a_{8,7} & a_{8,8} & a_{8,9} & 0 & 0 \\
0 & 0 & 0 & 0 & 0 & 0 & 0 & 0 & a_{9,9} & a_{9,10} & 0 \\
0 & 0 & 0 & 0 & 0 & 0 & 0 & 0 & a_{10,9} & a_{10,10} & a_{10,11} \\
0 & 0 & a_{11,3} & 0 & 0 & 0 & 0 & 0 & 0 & a_{11,10} & a_{11,11}
\end{pmatrix}
\begin{pmatrix}
P_{tot,1} \\
P_{tot,2} \\
P_{tot,3} \\
P_{tot,4} \\
P_{tot,5} \\
P_{tot,6} \\
P_{tot,7} \\
P_{tot,8} \\
P_{tot,9} \\
P_{tot,10} \\
P_{tot,11}
\end{pmatrix}
=
\begin{pmatrix}
b_1 \\
b_2 \\
b_3 \\
b_4 \\
b_5 \\
b_6 \\
b_7 \\
b_8 \\
b_9 \\
b_{10} \\
b_{11}
\end{pmatrix}
\quad (1)
$$

where $P_{\text{tot},i}$ is the total pressure in function-block i [Pa], $a_{i,j}$ is the (i, j) element of the coefficient matrix, and b_i is constant term of the function-block i. The coefficient matrix is sparse since only the elements of connected function blocks have non-zero values. The relationship between the function-block diagrams and the equations is considerably simple and clear. Because these equations are derived by implicit methods (Patankar, 1980), the solutions of these equations directly result in the pressure, concentration, and temperature distribution, throughout the entire FC system. To reduce the error caused by the linearization of the physical models, which is necessary to implement the physical models in the Eq. (1) expression, the numerical solver for error-convergence was implemented. Owing to the function-block-diagram modeling method and numerical implementation methods of in the linear algebraic equations, it is remarkably easy to modify, replace, add, and remove the component specifications in the proposed model. This reduces the lead-time to develop the system models for the individual applications. The other system models of air and cooling system in Fig. 1 were developed with the same strategies and integrated as an entire dynamic FC-system simulation model.

The parameters of each system component model are determined with the test data collected in the unit-testbeds before the integration of the components as an entire system. Fig. 3(a) is an example of the unit-testbed configuration, where the pump speed and valve angle are changed in the various gas composition conditions and the response of the flowrate and pressure head of the pump are measured. The dynamics of hydrogen pump is expressed by Eq. (2) (Akaike et al., 1983),

$$\dot{v} = C_1 N - \left(C_2 N C_3 \frac{\Delta P^{C_4}}{\mu^{C_5} \rho^{C_6}} \right) \tag{2}$$

where $\dot{v}$ is volumetric flowrate [m³/s], N is rotational speed [rad/s], ΔP is pump head [Pa], μ and ρ are fluid viscosity [Pa·s] and density [kg/m³], and $C_1 - C_6$ are the tuning parameters defined by pump geometries. The parameters $C_1 - C_6$ were determined by non-linear least-square method to fit the experimental data in Fig. 3(b). Such parameter determination procedures require only unit-testbed data. In other words, no system-testbed data, and the effort of data collection can be considerably reduced. The parameters for the other system component models were determined by such simple methodologies with the unit test data.

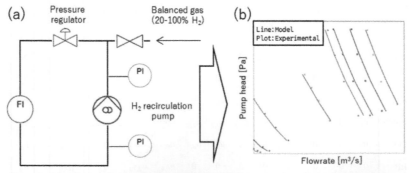

Fig. 3. (a) The unit-testbed configuration and (b) collected test data for parameter determination for H₂ recirculation pump

3. Model validation and verification

For the model validation and verification (V&V), the special test vehicles were manufactured by attaching many sensors in addition to the original ones throughout the entire FC system as shown in Fig. 4. Within the FC system, the H_2 system has a unique hardware configuration and controller specification compared with other powertrains as internal combustion engine, and it was necessary to develop special measurement instruments to collect V&V data efficiently. Moreover, ultra-compact and highly-integrated sensors were essential to collect the V&V data in the realistic dynamic vehicle operating conditions due to the limitation in the packaging space of the test vehicle. These sensors also have to be durable even in high humidity condition where large amount of liquid water exists around the sensors. To meet such requirements, H_2 concentration sensor and liquid-water level sensor were newly developed and utilized in V&V data collection by the test vehicles (Hasegawa *et al*, 2021). As described in the previous section, these data were not used for the parameter determination but only for V&V of the integrated model. When the experimental data and model output do not agree with each other, the possibilities of the missing physics or the deficiencies of parameter determination procedures were investigated repeatedly.

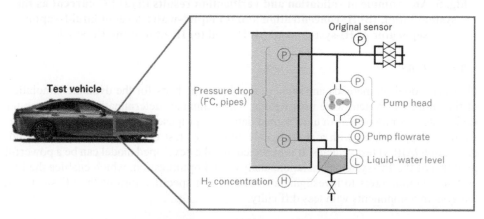

Fig. 4. Schematic drawing of the experimental configuration of data collection for the validation and verification of the developed models

Fig.5 is an example of the results of V&V of the models. A considerable amount of test data was collected by the test vehicle under a wide range of operating condition of low to high loads, operating temperatures, and atmospheric pressures. The same inputs were given to the model and the model output data were compared with the experimental data. It was confirmed that the simulation results and the measured fuel cell voltage were in good agreements within 10 % error as shown in Fig. 5 (e) and the other model outputs in Fig. 5 agreed with the experimental data, although the deviations between measured and calculated liquid-water levels in the liquid-vapor separator were observed during 300-400 s and 500-550 s in Fig. 5 (c). The deviations were not mainly caused by modeling error, but by erroneous measurements from the liquid-water level sensor in Fig. 4, whose accuracy is not guaranteed when FC-current is low.

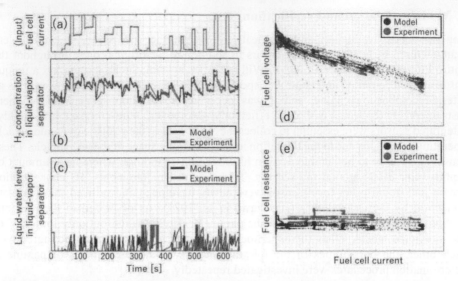

Fig. 5. An example of validation and verification results in (a) FC current as the system input, (b) H₂ concentration and (c) liquid-water level in liquid-vapor separator of H₂-system, and (d) IV and (e) IR plot of the FC stack

Conclusions

Physical modeling methods and numerical solving methods for the dynamic simulation of the entire FC system were investigated. The parameter determination procedures for each system component with unit-testbed data were proposed. The integrated model was validated and verified with a considerable amount of test data collected with the 2nd-generation MIRAI test vehicles. It is expected that the developed model can be a powerful platform of the FC-system simulation, analysis and optimization, which enables the FC-system manufacturers to investigate a wide range of specifications of the FC stack and the system components with less difficulty.

Acknowledgement

This work was supported by the FC-Platform Program: Development of design-for-purpose numerical simulators for attaining long life and high performance project (FY 2020–2022) conducted by the New Energy and Industrial Technology Development Organization (NEDO), Japan.

References

A. Z. Weber et al., 2014, *Journal of The Electrochemical Society*, **161** (12) F1254-F1299

S. Akaike and M. Nemoto, 1983, *Mechanical Engineering Journal of Japan*, Vol.49, No.447.

S. Hasegawa, M. Kimata, Y. Ikogi, M. Kageyama, S. Kim, and M. Kawase, 2021, *ECS Transactions*, 104 (8) 3-26

S. Hasegawa *et al.*, 2021, *EVTeC 2021 Proceedings*, No.B1.2.

S. V. Patankar, 1980, Numerical Heat Transfer and Fluid Flow, *McGraw–Hill, New York*

T. Takahashi and Y. Kakeno, 2021, *EVTeC 2021 Proceedings*, No.B1.1.

Proceedings of the 14th International Symposium on Process Systems Engineering – PSE 2021+
June 19-23, 2022, Kyoto, Japan © 2022 Elsevier B.V. All rights reserved.
http://dx.doi.org/10.1016/B978-0-323-85159-6.50319-5

Embracing the era of renewable energy: model-based analysis of the role of operational flexibility in chemical production

Thomas Knight[a], Chao Chen[a], Aidong Yang[a]

[a] *Department of Engineering Science, University of Oxford, Oxford OX1 3PJ, UK*
aidong.yang@eng.ox.ac.uk

Abstract

The chemical industry, like other industrial sectors, is expected to embrace renewable resources as replacement for fossil fuels in the coming decades. Focusing on the special challenges arising from the use of intermittent and variable renewable energy in large-scale chemical production, this paper presents two case studies, one of methanol and another of aluminium, to explore the role of operational flexibility in mitigating the burden of energy storage requirement. Whole-system optimisation and unit-level dynamic simulation were applied in the two cases, respectively, with both results showing potential benefits of shifting process operation from a constant-load mode to one that accommodates a certain degree of flexibility.

Keywords: Renewable energy, chemical production, aluminium, methanol, operational flexibility

1. Introduction

Grand environmental challenges such as climate change, ecosystem deterioration and resource depletion have greatly re-shaped the landscape of energy supply in recent decades, with the production and use of renewable energy emerging clearly as the preferred direction. The transition to renewable energy systems will contribute greatly to the decarbonisation of the chemical industry which conventionally depends on energy and feedstock derived from fossil sources.

To date, much work has been done on incorporating renewable feedstock, in particular biomass, based on the concept of biorefinery (Kokossis and Yang, 2010). On the energy front, considerations have been given to the utilisation of renewable energy such as wind and solar power in the framework of "power-to-X" (Sternberg and Bardow, 2015), where "X" typically represents hydrogen and chemicals and materials that can be derived from hydrogen in combination of other molecules such as CO_2. Currently, most large-scale chemical production processes have been designed to operate with a relatively stable load, supported with stable energy input from either the grid or on-site generation by consuming fossil fuels. In contrast, renewable energy generation is often intermittent and variable, which requires costly energy storage to supply to chemical processes with a rigid demand profile. To overcome this barrier, it is desirable to explore the possibility of operating a chemical process with a certain degree of flexibility in its load, to reduce the energy storage requirement via a closer match between the energy supply and demand profiles. While the perspective of flexible operation has been discussed recently in the context of power to ammonia (Cheema and

Krewer, 2018), methane (Matthischke et al., 2018) and methanol (Hank et al., 2018), detailed model-based assessment of the potential of this strategy is still rather limited.

In this paper, we present two distinctive case studies to demonstrate the potential of using operational flexibility as a tool to reduce the burden of energy storage when powering large-scale chemical production with variable renewables. The first study extends our earlier work (Chen and Yang, 2021) on methanol production based on CO_2 hydrogenation, which uses an optimisation model of the combined system of energy supply, storage and chemical production to reveal the potential of a holistically optimised system enabled by different levels of operational flexibility. This study considers H_2 for both energy and feedstock storage. It focuses on the impact (i.e. "what-if"), not the realisation (i.e. "how"), of process flexibility in a complex system comprising a number of subsystems. In contrast, the second study offers a detailed, process unit-level analysis of aluminium production, considering the replacement of grid electricity with power from a wind energy facility to run the smelter, where the energy storage to be tackled is in the form of batteries. It is based on dynamic simulation of the smelter to predict (1) the extent to which flexible load can be achieved within operational constraints and (2) the corresponding impacts.

2. Case study 1: methanol production

2.1. System overview and modelling approach

As shown in Fig. 1, the system consists of electrolytic H_2 production, CO_2 capture, methanol synthesis and purification. Compressed H_2 storage and fuel cell-based H_2-to-power conversion are included to reconcile the mismatch between variable renewable energy generation and the demands for H_2 and power by methanol production, while supplementary power supply from a dispatchable source (e.g. grid) is also available. In terms of operational flexibility, the electrolyser was assumed to be fully flexible (i.e. with no restrictions with respect to the minimum load). In the methanol production subsystem, the methanol synthesis reactor was assumed to be flexible to a certain degree (specified by a minimum load), while the other components, namely CO_2 capture (by an amine-based absorption-desorption cycle) and methanol purification (by distillation) were assumed to operate at a constant load, thus giving rise to the potential need for CO_2 storage and raw methanol storage. Further details of the system can be found in Chen and Yang (2021).

For a specified minimum load of the methanol synthesis reactor and a targeted level of annual methanol production rate, an optimisation model developed in our previous work (Chen and Yang, 2021) can be used to identify the design and operational decisions leading to the minimisation of the levelized methanol production cost (L_{MeOH}), which comprises the cost for power supply and all the components for conversion and storage of energy and chemicals. A key factor to be explored is the trade-off between the oversizing of the methanol reactor and the reduction of H_2 storage requirements, varying with the assumed degree of flexibility of the synthesis reactor. Our previous study investigated two specific geographical locations, namely Norderny in Germany and Kramer Junction in the US, which are known to have superior wind and solar energy sources, respectively. The optimisation model has now been further applied to more locations to assess the impact of process flexibility under a wider range of conditions, including particularly an area near Tokyo (latitude 35.75; longitude 140.75) for which the results are presented below for a system producing 400,000 t/y of methanol. The global horizontal radiation and wind speed data were obtained from NASA's worldwide energy resource (POWER) database; monthly averages were taken over Jul 1983 - Jun 2005 and Jan 1984 - Dec 2013 for solar and wind, respectively. Either wind, solar or a combination of the two may be utilised to power the production.

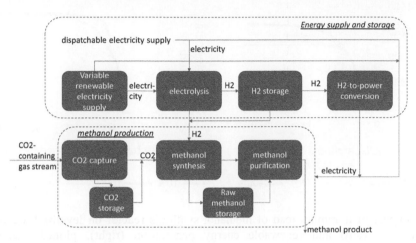

Figure 1. Methanol production based on CO_2 hydrogenation with renewable energy.

2.2. Results

As shown in Fig. 2 (left), a system with the greatest operational flexibility in the methanol reactor (with minimum load = 10%, referred to as "flexible design") could achieve a L_{MeOH} up to 12% lower than that with no flexibility in the load of the reactor (with minimum load = 100%, referred to as "nonflexible design"), which occurs at the higher end of the dispatchable energy price, a circumstance where the use of (variable) renewable energy is highly preferred. The detailed optimisation results (not shown) reveal that under such circumstance, the flexible design for a system fully powered by renewables would need to increase the annualised capital cost of the methanol reactor from ~$28m (of the nonflexible reactor) to ~$40m due to a larger reactor size needed to compensate for operating not always at its full load, but the corresponding annualised cost of hydrogen storage would reduce dramatically from ~$46m to ~$6m, which was predicted to be economically advantageous even after adding the costs of ~$11m for intermediate storage of CO_2, raw methanol and heat (the latter is omitted in Fig. 1 for clarity) needed to accommodate flexibility of the reactor. However, compared to the cost reduction (in terms of L_{MeOH}) previously predicted of Norderny (~20%) and of Kramer Junction (~30%), the benefit of process flexibility predicted of this near-Tokyo site is lower. The discrepancy is due to the difference in the renewable energy profiles between these sites in terms of the combination of diurnal and seasonal variations which affect the relative importance of energy storage in the system for meeting a specified annual methanol output, which in turn affects the impact of process flexibility.

In addition to the reduction in L_{MeOH}, Fig. 2 (right) shows that in a cost-optimized system incorporating process flexibility, the proportion (indicating the penetration level) of renewable energy in total energy supply also increases particularly when the dispatchable energy price is in its mid-range, which is particularly ideal from the perspective of carbon emission reduction if the dispatchable energy is generated with a high carbon intensity. Note that in this case study, the range of flexibility tested by simulation was an assumption; its operational feasibility has not been evaluated. In principle, such evaluation could be carried out using detailed dynamic simulation, as demonstrated for a different process in the second case study.

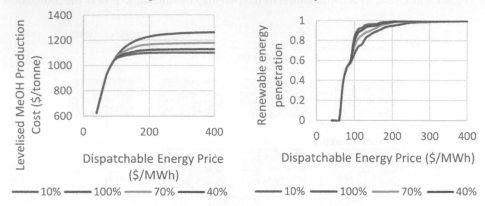

Figure 2: Impact of minimum load of methanol synthesis reactor on levelized MeOH production cost (left) and renewable energy penetration (right), plotted against dispatchable energy price. The four series indicate levels of flexibility, which is measured in terms of the minimum load of the reactor.

3. Case study 2: aluminium production

3.1. System overview and modelling approach

This second study considers aluminium production by the typically adopted Hall Héroult process, in which alumina is dissolved in molten cryolite (as the electrolyte) and electrolysed with a direct current (Fig. 3, right). The overall chemical reaction, taken into account the loss of current to side reactions, is

$$(\eta + 1)\, Al_2O_3 + 3\, C \rightarrow 2(\eta + 1)\, Al + 3\eta\, CO_2 + 3(1 - \eta)\, CO \qquad \text{Eq. (1)}$$

where η is the current efficiency. The (simplified) energy balance equation for the smelter bath is

$$m \cdot c_p \frac{dT_{bath}}{dt} = I \cdot V_{bath} - \Delta H_R - \Delta H_{feed} - Q \qquad \text{Eq. (2)}$$

where m, c_p and T_{bath} are the mass, heat capacity and temperature of the bath, respectively, I is the current, V_{bath} is the bath voltage, ΔH_R, ΔH_{feed}, and Q are the net enthalpy of all the reactions in the bath, the thermal energy required to heat and dissolve the feed alumina and the heat loss from the bath, respectively. The bath voltage has multiple components including the reversible voltage of the electrolysis reaction, resistive voltage of the bath and the bubble layer, and voltage drops at the anode and the cathode. The detailed model of these components and that of the other terms in Eq. (2), together with the detailed mass balance model and the model for estimating current efficiency, can be found in Knight (2021).

Aluminium reduction cells are operated within a very tight region which provides the best conditions for efficiency, alumina dissolution and cell life. This means the reaction has to take place at approximately 970°C with less than 40°C tolerance. Due to these constraints, most aluminium smelters operate under a constant energy input to help control the bath conditions. To allow a smelter to run with a more flexible load, the use of a shell heat exchanger (SHE) as developed by Energia Portior (Depree *et al.*, 2016) was considered. A SHE can be installed on the outside of a cell by acting as either an insulator or conductor (hence regulating Q in Eq. (2)) to give smelter operators a new tool to aid in control of cell conditions. As part of the temperature regulation, the use of a SHE also offers extra control to facilitate the maintenance of another operational constraint, the minimum thickness of the cell's ledge (consisting of cryolite) which melts and freezes during operation.

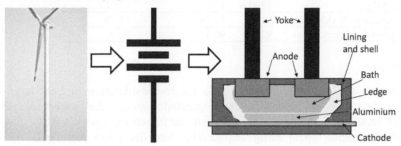

Figure 3. Structure of a system comprising an aluminium smelter (right) powered by variable renewable energy (left), mediated by a battery-based electricity storage (middle).

Dynamic simulation was performed to a system with a battery-based electricity storage to connect a wind farm and a smelter (Fig. 3). The degree of operational flexibility of the smelter was represented by a simple approach that manipulates the deviation between the current profile provided by the wind farm (I_{wind}) and that taken up by the smelter ($I_{smelter}$):

$$I_{smelter}(t) = I_{constant} - b(I_{wind}(t) - I_{constant}) \qquad \text{Eq. (3)}$$

where b is an adjustable parameter ranging from 0 to 1; these two extremes represent an operation with no flexibility (where the smelter takes a constant current, $I_{constant}$, hence requiring greatest energy storage) and one completely following the current output from the wind farm (hence requiring no energy storage), respectively. The feasibility of a chosen value for b is predicted by the dynamic simulation of the smelter, against the permitted range of bath temperature and minimum ledge thickness.

3.2. Results

Table 1 summarises representative simulation results for a 24-hour operation feeding on wind power with a profile scaled from German national data (https://www.amprion.net/Netzkennzahlen/ Windenergieeinspeisung/) recorded on 5 April 2020. The bath temperature was restricted to the range of 955°C to 985°C and the minimum ledge thickness set to 5 cm (i.e. 1/3 of the initial thickness) which equates to ~2500 kg of frozen cryolite for the system considered. One can see that a "conventional" smelter (without using a SHE) can already accommodate a certain degree of load flexibility. However, the use of the SHE, thanks to its extra regulation of heat loss, allows the system to implement greater load fluctuation (reflected by a larger b value) and hence achieve greater reduction in electricity storage size. Although a more robust comparison needs to balance the storage cost with the cost of the SHE and the slight penalty in aluminium production rate and current efficiency (shown in Table 1), the more than 25% of battery size reduction due to the use of a SHE suggests that technologies enhancing the operational flexibility of aluminium smelters may hold considerable potential.

Table 1. Results of applying different load modulation strategies to the smelter.

Current modulation	b = 0	b = 0.14 (without SHE)	b = 0.34 (with SHE)
Maximum temperature (°C)	965.9	982.3	984.9
Minimum ledge mass (kg)	3780	2685	2559
Battery size (kWh)	2599	2001	1489
Al produced (kg)	1399.6	1382.7	1359.7
Energy to produce Al (kWh/kg)	11.48	11.76	11.78
Average current efficiency (%)	92.0	90.1	90.3

4. Conclusions

Transitioning to the era of renewables calls for re-consideration of how large-scale chemical processes should be designed and operated. Through the two complementary case studies presented in this work, it is evident that increasing operational flexibility to allow load fluctuation could bring considerable benefits when feeding on variable renewable energy, in terms of reducing the electricity or chemical storage burden arising from the mismatch in the "rhythm" of energy supply and process operation. In this context, PSE tools such as detailed dynamic simulation and whole-system optimisation can play an important role, respectively, in assessing flexibility-accommodating operational feasibility of existing and adapted process units and revealing best system designs corresponding to location-specific energy supply patterns.

References

I. Cheema, U. Krewer, 2018. Operating envelope of Haber–Bosch process design for power-to-ammonia. RSC Adv 8:34926–36.

C. Chen, A. Yang, 2021. Power-to-methanol: The role of process flexibility in the integration of variable renewable energy into chemical production. Energy Conversion and Management 228: 113673

N. Depree, R., Düssel, P. Patel, T. Reek, 2016. The 'Virtual Battery' — Operating an Aluminium Smelter with Flexible Energy Input. Light Metals, 2016: 571-576.

C. Hank, S. Gelpke, A. Schnabl et al., 2018. Economics & carbon dioxide avoidance cost of methanol production based on renewable hydrogen and recycled carbon dioxide – power-to-methanol. Sustain Energy Fuels 2:1244–61.

T. Knight, 2021. The role of operational flexibility in aluminium production driven by variable renewable power. Research project report, Department of Engineering Science, University of Oxford.

A.C. Kokossis, A. Yang, 2010. On the use of systems technologies and a systematic approach for the synthesis and the design of future biorefineries, Computers & Chemical Engineering, 34 (9), 1397-1405.

S. Matthischke, S. Roensch, R. Güttel, 2018. Start-up time and load range for the methanation of carbon dioxide in a fixed-bed recycle reactor. Ind Eng Chem Res 57(18):6391–400.

A. Sternberg, A. Bardow, 2015. Power-to-What? — Environmental assessment of energy storage systems. Energy and Environmental Science, 8 (2): 389–400.

Proceedings of the 14th International Symposium on Process Systems Engineering – PSE 2021+
June 19-23, 2022, Kyoto, Japan © 2022 Elsevier B.V. All rights reserved.
http://dx.doi.org/10.1016/B978-0-323-85159-6.50320-1

Hollow Fiber-based Rapid Temperature Swing Adsorption (RTSA) Process for Carbon Capture from Coal-fired Power Plants

Kuan-Chen Chen[a], Jui-Yuan Lee[b], Cheng-Liang Chen[a*]

[a]*Department of Chemical Engineering, National Taiwan University, Taipei 10617 Taiwan*
[b]*Department of Chemical Engineering and Biotechnology, National Taipei University of Technology, Taipei 10608, Taiwan*
*CCL@ntu.edu.tw

Abstract

Post-combustion carbon capture is one of the feasible methods to reduce emission of carbon dioxide (CO_2) from coal-fired power plants. This work proposes a hollow fiber based rapid temperature swing adsorption (RTSA) method for capturing CO_2 from a typical 550 MW coal-fired power plant. The proposed RTSA approach can shorten the operating time and using low-grade energy for regeneration of adsorption elements.

This work studies the impact of using low-grade steam extracted from a low-pressure turbine as the heating source of the dual column vacuum RTSA (DC-vRTSA). The DC-vRTSA at 120 °C-55 °C will reduce the efficiency of the coal-fired power plant by 8.2 % to 1.9 %. The lowest CO_2 capture cost, 19.20 US dollars per tons of captured CO_2, is located at 60 °C desorption temperature.

Keywords: CO_2 capture; Coal-fired power plant; Hollow fiber; Rapid temperature swing adsorption (RTSA).

1. Introduction

1.1. Global Warming and Carbon Capture

Carbon dioxide emissions are one of the major contributors to greenhouse effects and global warming. IEA (2021) indicates that there are 31.5 Gt of CO_2 emissions worldwide in 2020 where fossil-fired power generation plays the dominant role in manmade CO_2 emissions. The carbon capture technique becomes a key to reduce the greenhouse gas effect in the near future.

In the carbon capture adsorption process, there are several available methods that can be used to desorb CO_2, such as Pressure Swing Adsorption (PSA), Temperature Swing Adsorption (TSA), and Vacuum Swing Adsorption (VSA). Haghpanah et al. (2013) presented a systematic analyses of several VSA cycles with Zeochem zeolite 13X as the adsorbent to capture CO_2 from dry flue gas. Joss et al. (2017) studied the design of TSA cycles and analyzed how individual steps within TSA cycles would affect the purity and recovery of the CO_2. Liu et al. (2019) investigate the relationship between CO_2 recovery, productivity rate, purity, specific energy consumption, and second-law efficiency based on experimental data. The effect of CO_2 concentration, desorption duration, adsorption temperature and desorption temperature has been considered in a lab-scale 4-step TSA system. However, the pressurizing cost of PSA or long heating time for regeneration of TSA is a limitation for large-scale carbon capture (Rezaei and Webley, 2010). For the carbon capture technology mentioned above, there are many literature that compared the cost for different capture technologies. Wang et al. (2017) have reviewed the economic performance of the post-combustion CO_2 capture technologies from a coal-fired power plant, including chemical absorption and membrane-based separation.

1.2. Hollow Fiber-Based Rapid Temperature Swing Adsorption

Lively et al. (2009) proposed a novel porous hollow fiber to overcome the temperature swing limitation. The hollow fiber has a large contact area to speed up the heat transfer rate much faster than the conventional packed bed tower. Therefore hollow fiber based adsorption column is expected to realize the Rapid Temperature Swing Adsorption process. This process utilizes hollow fiber morphology to pass cooling water through the pores to maximize the adsorption capacity, and steam to pass through the pores to effectively desorb CO_2. Rezaei et al.(2014) developed a two-dimensional mathematical model of a rapid temperature swing adsorption (RTSA) process for the first time to predict polymer supported amine hollow fiber sorbent performance during post-combustion CO_2 capture from flue gas. This work focuses on manufacturing a single fiber model to simulate a four step RTSA process. Also, the sensitivity analysis to parametrric values such as gas and water velocities and initial temperatures are evaluated. Hosseini et al. (2017) developed a two-dimensional mathematical model to analyze the effects of operating variables on RTSA performance (recovery, purity, productivity, the amount of separated pure carbon dioxide in 24 h, and specific energy consumption).

However, the existing work is limited to small-scale single column simulation for the RTSA process. The cost analysis of the RTSA process has not been thoroughly discussed. This paper proposes a large-scale hollow fiber for the dual-column RTSA process to capture CO_2 from a 550 MW coal-fired power plant. In order to enhance the performance of the RTSA process, the vacuum system is used in the desorption step.

2. Process Description

The process configuration and mathematical model details are described in previous work (Chen et al., 2020). In this paper, the research focus will be economic analysis. The DC-vRTSA (Dual Column vacuum Rapid Temperature Swing Adsorption) is evaluated to a typical coal-fired power plant. The stream data of the illustrated coal-fired power plant come from Urueli (2010). Figure 1(a) shows approximate flowsheet diagram of steam from the coal-fired power plant. The high pressure water first sends to steam generator to generate high pressure steam and then electricity by pass thought the high-pressure (HP) turbine. It is then sent to steam generator again for reheating, and goes thought intermedium-pressure (IP) turbine and low-pressure (LP) turbine for generation of more electricity. The outlet of low pressure turbine is sent to condenser using cooling water to condense. Finally, it passes through condensate pump, de-aerator and feed water pump to make high pressure water than recycle to steam generator. Figure 1(b) also shows the process detail stream data of the typical power plant with the DC-vRTSA, which includes steam temperature, pressure, enthalpy, flow rate, turbine energy output, condense heat, cooling water flow rate and the power plant efficiency.

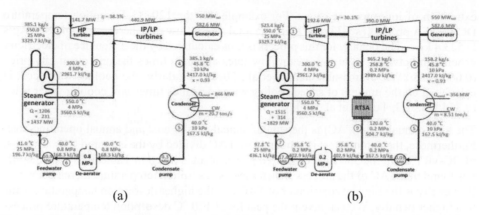

(a)   (b)

Figure 1. The illustrative coal-fired power plant (a) without carbon capture and (b) with the DC-vRTSA at 120 °C desorption temperature.

The steam extracted from the low-pressure turbine is used to provide the heat required by the DC-vRTSA process. In the first step of this simulation, stream enthalpy, temperature, pressure and DC-vRTSA desorption temperature data are given. Then saturated temperature and pressure are searched from steam table. Using saturated pressure to find the superheated temperature that should be used to provide heat for the DC-vRTSA process extracted from the low pressure turbine. Then one can find the superheated steam enthalpy. The appropriate amount of superheated steam will be extracted to provide sensible heat and latent heat for desorbing CO_2 from the DC-vRTSA process. For example, in the 120 °C desorption temperature, the extracted superheated steam temperature is 258.8 °C. This will provide the sensible heat of the superheated steam from 258.8 °C to 120 °C, then the latent heat from saturated 120 °C steam to 120 °C liquid water. Therefore the inlet is 258.8 °C superheated steam, and the outlet is 120 °C saturated liquid water. The latter will be mixed with the condensate water and returned to the boiler. Next, one makes a guess of the total steam flow rate to estimate the heat duty and the CO_2 emission (88 kg of CO_2 per GJ) from the steam generator. The DC-vRTSA is set to capture 90 % of CO_2 emission from the steam generator. Following one can calculate steam flow rate to the DC-vRTSA process and all stream data for the power plant. One calculates a new total steam flow rate which can provide 582.6 MW (including 550 MW output and total auxiliaries 32.6 MW, Zoelle et al., 2015) power and see if the new total steam flow rate equals to the guessed value afterward. If not, one uses the new one as an updated guess value and calculates again until the new value is equal to the previous one, which means the simulation is completed. Then the simulation results will be used in economic analysis.

3. Economic Analysis

For economic analysis, the detailed model of cost estimation can be found elsewhere. The desorption temperature from 120 °C to 55 °C DC-vRTSA has been used to analyze the total cost (50 °C cannot achieve 90 % purity and 90 % capture ratio).

The total capital cost (TCC) includes the RTSA material and frame, vacuum pump (for low desorption pressure). The annual capital cost (ACC) is the total capital cost divided by the designed power plant lifetime (25 years). The annual operating cost (AOC) includes the penalty, vacuum pump electricity, cooling water and RTSA material replacement. The penalty is estimated by the electricity lost in the power plant due to the extraction of low pressure steam. For example, for the 120 °C desorption temperature DC-vRTSA process (see Figure 1(b)), the total steam flow rate increases from 385.1 kg/s to 523.4 kg/s. The penalty for this process is thus calculated by the difference of electricity generation between these two steam flow rates and then times the unit electricity price (0.065 US$/kWh, Ramasubramanian et al., 2012). Similarly, the cooling water cost is calculated by the amount of saved cooling water demands times unit cooling water price (0.354 US$/GJ, Turton et al., 2008).

The total annual cost (TAC) is the sum of annual capital cost and annual operating cost. Furthermore, the capture cost is estimated by TAC divided by the captured CO_2. The cost of DC-vRTSA in the 550 MW coal-fired power plant is listed in Table 1. From the table, it is found that TAC of the process is increased as desorption temperature increases, where the penalty is the biggest contributor of TAC and the higher desorption temperature leads to a higher penalty. We can invent the penalty of 120 °C desorption temperature process (90.67 US$/year) is about four times of 55 °C desorption temperature process. But the lower desorption temperature process has a lower gas volume flow rate (see Table 1). It needs more DC-vRTSA unit to capture 90 % of carbon emission. The DC-vRTSA unit will affect the DC-vRTSA price and the RTSA replacement price.

Table 1. The cost summary of DC-vRTSA in a 550 MW coal-fired power plant

Desorption temperature(°C)	120	110	100	90	80	70	60	55
Capital cost							(million US$)	
DC-vRTSA unit								
Material	9.13	9.63	9.61	11.14	9.76	12.25	16.21	21.89
Frame	97.61	101.09	100.93	111.09	101.98	118.19	141.75	172.34
Total	106.74	110.72	110.54	122.23	111.74	130.44	157.96	194.23
Vacuum pump	65.96	64.22	78.48	76.87	112.83	110.09	108.38	108.33
TCC	172.70	174.94	189.02	199.10	224.57	240.53	266.34	302.56
ACC (/25 years)	6.91	7.00	7.56	7.96	8.98	9.62	10.65	12.10
Operating cost							(million US$/year)	
Penalty	90.67	77.53	62.37	52.79	37.45	27.58	21.28	20.88
Electricity (vacuum pump)	16.08	15.65	19.13	18.74	27.50	26.83	26.42	26.40
Cooling water	3.22	2.73	2.18	1.82	1.23	0.83	0.58	0.57
Rplacement	1.46	1.54	1.54	1.78	1.56	1.96	2.59	3.50
AOC	111.43	97.45	85.22	75.13	67.74	57.20	50.87	51.35
TAC	118.34	104.45	92.78	83.09	76.72	66.82	61.52	63.45
Captured CO_2 (Mton/year)	3.74	3.64	3.53	3.45	3.34	3.25	3.21	3.20
CO_2 capture cost (US$/ton)	31.68	28.71	26.32	24.07	23.00	20.53	19.20	19.81
(US$/kWh)	0.0289	0.0255	0.0226	0.0203	0.0187	0.0158	0.0150	0.0154

Table 1 shows the lowest desorption temperature process (55 °C) has the highest DC-vRTSA capital cost and replacement cost. The vacuum pump price mainly depends on the desorption pressure, lower desorption pressure leads to higher vacuum pump price and electricity demand. Also, the desorption temperature will affect the cooling water cost, too. The best total annual cost and CO_2 capture costs of these processes is 61.52 million US dollars and 19.20 US dollars per tons of captured CO_2. The capture cost that needs to be shared for each kWh also be calculated. For 120 °C and 55 °C desorption temperature process, each kWh of electricity needs to bear 0.0289 and 0.0154 US$, respectively.

3.1. Sensitivity of Price Annualize Factor

For the economic analysis above, the price annualize factors are fixed. In this section, we will investigate how the material price of DC-vRTSA, power plant lifetime and electricity price affect the capture cost.

Figure 2(a) shows the effect of material price on the capture cost. For those process with high material area, applying high material price brings the addition in the capture cost; on the other hand, it reduces the difference of capture cost between high and low desorption temperature processes. Figure 2(b) shows the effect of power plant lifetime on capture cost. The price of electricity will mainly affect the penalty and the vacuum pump electricity in the annual operating cost. In the above economic analysis, the electricity price is 0.07 US$/kWh. Figure 2(c) illustrates how the price of electricity affects the capture cost.

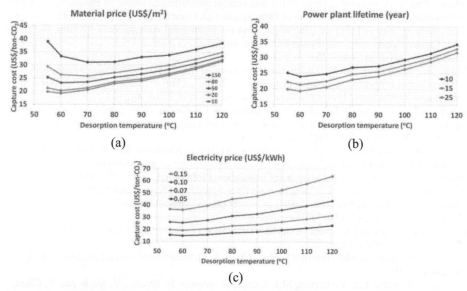

Figure 2. Sensitivity analysis of (a) material price, (b) power plant life time and (c) electricity price.

4. Conclusions

In this study, the Rapid Temperature Swing Adsorption (RTSA) process for capturing CO_2 from the coal-fired power plants has been simulated. With considering the possibility of using steam extracted from the low-pressure turbine as the heat source for the DC-vRTSA process, the impact on the efficiency and stream data of a typical coal-fired power plant were studied. The DC-vRTSA at 120 °C-55 °C reduces the efficiency of the coal-fired power plants by 8.2 % to 1.9 %. The economic analysis of the DC-vRTSA (120 °C-55 °C desorption temperature) used in the coal-fired power plant was performed. The best total annual cost and CO_2 capture costs of these processes located at 60 °C desorption temperature, which are 61.52 million US dollars and 19.20 US dollars per tons of captured CO_2. The sensitivity analysis shows that material price and power plant lifetime have significant effect on lower desorption temperature area.

References

IEA (2021), Global Energy Review: CO2 Emissions in 2020, IEA, Paris
 https://www.iea.org/articles/global-energy-review-co2-emissions-in-2020

R. Haghpanah, R. Nilam, A. Rajendran, S. Farooq, and I. Karimi, Cycle Synthesis and
 Optimization of a VSA Process for Postcombustion CO_2 Capture. Vol. 59. 2013.

L. Joss, M. Gazzani, and M. Mazzotti, Rational design of temperature swing adsorption cycles for
 post-combustion CO2 capture. Chemical Engineering Science, 2017. 158, 381-394.

B. Liu, Y. Lian, S. Li, S. Deng, L. Zhao, B. Chen, and D. Wang, Experimental investigation on
 separation and energy-efficiency performance of temperature swing adsorption system for
 CO2 capture. Separation and Purification Technology, 2019. 227, 115670.

F. Rezaei and P. Webley, Structured adsorbents in gas separation processes. Vol. 70. 2010. 243-
 256.

Y. Wang, L. Zhao, A. Otto, M. Robinius, and D. Stolten, A Review of Post-combustion CO_2
 Capture Technologies from Coal-fired Power Plants. Energy Procedia, 2017. 114, 650-665.

M.R.M. Abu-Zahra, J.P.M. Niederer, P.H.M. Feron, and G.F. Versteeg, CO_2 capture from power
 plants: Part II. A parametric study of the economical performance based on mono-
 ethanolamine. International Journal of Greenhouse Gas Control, 2007. 1(2), 135-142.

A. Rao and E. Rubin, A Technical, Economic, and Environmental Assessment of Amine-Based
 CO_2 Capture Technology for Power Plant Greenhouse Gas Control. Environmental science &
 technology, 2002. 36, 4467-75.

P. Maas, N. Nauels, L. Zhao, P. Markewitz, V. Scherer, M. Michael, D. Stolten, and J.F. Hake,
 Energetic and economic evaluation of membrane-based carbon capture routes for power plant
 processes. International Journal of Greenhouse Gas Control, 2016. 44, 124-139.

R.P. Lively, R.R. Chance, B.T. Kelley, H.W. Deckman, J.H. Drese, C.W. Jones, and W.J. Koros,
 Hollow Fiber Adsorbents for CO2 Removal from Flue Gas. Industrial & Engineering
 Chemistry Research, 2009. 48(15), 7314-7324.

S. Fatemeh Hosseini, M. Reza Talaie, S. Aghamiri, M. Hasan Khademi, M. Gholami, and M.
 Nasr Esfahany, Mathematical modeling of rapid temperature swing adsorption; the role of
 influencing parameters. Separation and Purification Technology, 2017. 183, 181-193.

K.-C. Chen, J.-Y. Lee, and C.-L. Chen, Hollow fiber-based rapid temperature swing adsorption
 process for carbon capture from coal-fired power plants. Separation and Purification
 Technology, 2020. 247, 116958.

I. Urieli, Case Study - The General James M. Gavin Steam Power Plant, Chapter 8, Applied
 Engineering Thermodynamics,.
 http://www.ent.ohiou.edu/~thermo/Applied/Chapt.7_11/SteamPlant/GavinCaseStudy.html,
 accessed., 2010.

A. Zoelle, D. Keairns, L.L. Pinkerton, M.J. Turner, M. Woods, N. Kuehn, V. Shah, and V. Chou,
 Cost and Performance Baseline for Fossil Energy Plants Volume 1a: Bituminous Coal (PC)
 and Natural Gas to Electricity Revision 3. United States: N. p., 2015.

K. Ramasubramanian, H. Verweij, and W.S. Winston Ho, Membrane processes for carbon
 capture from coal-fired power plant flue gas: A modeling and cost study. Journal of Membrane
 Science, 2012. 421-422, 299-310.

R. Turton, R.C. Bailie, W.B. Whiting, and J.A. Shaeiwitz, Analysis, Synthesis and Design of
 Chemical Processes. 3rd Ed. Pearson Education, 2008.

W.L. Luyben, Comparison of extractive distillation and pressure-swing distillation for
 acetone/chloroform separation. Computers & Chemical Engineering, 2013. 50, 1-7.

S. Swernath, K. Searcy, F. Rezaei, Y. Labreche, R.P. Lively, M.J. Realff, and Y. Kawajiri,
 Chapter 10 - Optimization and Technoeconomic Analysis of Rapid Temperature Swing
 Adsorption Process for Carbon Capture from Coal-Fired Power Plant, in Computer Aided
 Chemical Engineering, Y. Fengqi, Editor. 2015, Elsevier. p. 253-278.

T.C. Merkel, H.Q. Lin, X.T. Wei, and R. Baker, Power plant post-combustion carbon dioxide
 capture: An opportunity for membranes. Journal of Membrane Science, 2010. 359(1-2), 126-
 139.

L. Zhao, E. Riensche, L. Blum, and D. Stolten, Multi-stage gas separation membrane processes
 used in post-combustion capture: Energetic and economic analyses. Journal of Membrane
 Science, 2010. 359(1-2), 160-172.

Proceedings of the 14th International Symposium on Process Systems Engineering – PSE 2021+
June 19-23, 2022, Kyoto, Japan © 2022 Elsevier B.V. All rights reserved.
http://dx.doi.org/10.1016/B978-0-323-85159-6.50321-3

Determining Accurate Biofuel System Outcomes: Spatially Explicit Methods for Combined Landscape-Feedstock and Supply Chain Design

Eric G. O'Neill[a,b*], Christos T. Maravelias[a,b,c]

aDepartment of Chemical and Biological Engineering, Princeton University, Princeton, 08540, USA
bDOE Great Lakes Bioenergy Research Center, USA
cAndlinger Center for Energy and the Environment, Princeton University, Princeton, NJ 08540, USA
eoneill@princeton.edu

Abstract

The adoption of sustainably produced second generation biofuels will rely heavily on an optimized and integrated biofuel supply chain (SC) system from field to product, and a large amount of land will need to be converted to dedicated bioenergy crops to support sufficient economies of scale. Efficient models are needed to determine the optimal upstream 'landscape design' decisions that balance trade-offs with more commonly studied SC decisions. Landscape design optimization, deciding where in the landscape to plant bioenergy crops and how to manage that land (e.g. fertilization), has been shown to improve the environmental impact of farm-scale biomass production (including soil carbon (C) sequestration), but has been studied largely separately from biofuel SC network design (SCND). In this paper we present a model for landscape design optimization and a model/data integration strategy that enables the use of both high spatial resolution crop simulations and simultaneous optimization of the downstream biofuel SC. Using crop simulations that include realistic yield and environmental data, we present an illustrative case study in Michigan, USA and highlight the benefits of the model formulation and insights from simultaneously optimizing the SC and the landscape.

Keywords: Sustainable Supply Chains, MILP, Biofuels

1. Background

In previous research related to biofuel supply chain (SC) optimization, researchers assume fixed locations and availability for biomass feedstock and do not explicitly consider design decisions related to the crop and landscape system as decision variables (Ghaderi et al., 2016; O'Neill and Maravelias, 2021). Contrary to this assumption and because dedicated bioenergy crops have yet to be planted in large quantities, decisions related to landscape design and crop management can have downstream effects on the optimal configuration and operation of the biofuel SC. There is an opportunity to extend mathematical optimization approaches for biofuel supply chain network design (SCND) to include the simultaneous optimization of the upstream landscape decisions to find sustainable and economically favorable solutions for both SC and landscape design and operation.

There are two major challenges to integrating a landscape optimization model with the biofuel SCND. First, the yield and soil carbon (C) sequestration potential has been shown to be highly field specific which necessitates landscape optimization models with a high spatial resolution (Field et al., 2018). Second, accurately predicting the field-specific yield and soil C potential for crops like switchgrass is difficult and relies on biogeochemical crop model simulations for high-resolution data (Basso and Ritchie, 2015). A strategy for efficient model/data integration between crop simulations and optimization models is needed to preserve the computational tractability of the optimization model without sacrificing landscape design accuracy.

Accordingly, in this paper we propose a computationally efficient model formulation designed with the data in mind, and a model/data integration strategy. We then incorporate the landscape design model into a biofuel SCND model and present a case study to demonstrate the benefit from making simultaneous landscape and SC decisions at a high spatial resolution.

2. Landscape Design Model

The landscape design model makes decisions regarding the establishment, fertilization, and harvesting of biomass. Harvest decisions correspond to a multi-period ($t \in \mathbf{T}$) model to remain general to crops having multiple harvests. The primary outputs are the yield at each field and the amount of soil C that is sequestered. An overview of the integrated model and data approach is shown in Figure 1.

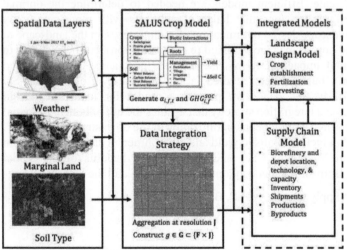

Figure 1. Overview of the integrated model and data approach. Parameter definitions are described below, and the data integration strategy is discussed in Section 3.

First, we introduce the set of fields $f \in \mathbf{F}$. The fraction of field f established with crop $i \in \mathbf{I}^F$ is controlled by decision variable $E_{i,f}$ in Eq.(1).

$$0 \leq E_{i,f} \leq 1 \tag{1}$$

To capture the effect of nitrogen (N) fertilization we also introduce continuous variable $R_{i,f} \in [0,1]$, which represents the fraction of fertilization to apply between 0 and the maximum, ω (kgN/ha). The variable $R_{i,f}$ is not used in the actual formulation, instead we construct auxiliary variable $D_{i,f} = E_{i,f}R_{i,f}$ which is used to calculate the additional yield from fertilization and its effect on soil C in Eq.(3) and Eq.(4). Because $E_{i,f}$ is bounded by one, $D_{i,f}$ is bounded as in Eq.(2).

$$0 \leq D_{i,f} \leq E_{i,f} \tag{2}$$

The biomass yield $Y_{i,f,t}$ (Mg) of each field f is given by Eq.(4) where $\alpha_{i,f,t}$ is the potential yield (Mg) at time period $t \in \mathbf{T}$ with zero fertilization if the field had been fully planted with crop i. The second term represents the change from fertilization where $\varsigma^1_{i,f,t}$ is the additional yield (Mg) achieved from fertilization (calculated as the potential yield at full fertilization minus the potential yield at zero fertilization). The auxiliary variable is used because a fraction of the additional fertilization yield can be equivalently attained by planting or fertilizing a fractional amount. The soil C sequestration, $GHG^{SOC}_{i,f}$, is calculated similarly (Eq.(4)) with $\Gamma^{SOC}_{i,f}$ and $\varsigma^2_{i,f}$ being the annualized soil C sequestered at 0 kgN/ha and the change from full fertilization respectively.

$$Y_{i,f,t} = E_{i,f}\alpha_{i,f,t} + D_{i,f}\varsigma^1_{i,f,t} \tag{3}$$

$$GHG^{SOC}_{i,f} = E_{i,f}\Gamma^{SOC}_{i,f} + D_{i,f}\varsigma^2_{i,f} \tag{4}$$

The actual area of field f that is planted $A_{i,f}$ (ha) and the actual amount of fertilizer applied $F_{i,f}$ (kgN) are given by Eq.(5) and Eq.(6) respectively. Where σ_f is the full area.

$$A_{i,f} = E_{i,f}\sigma_f \tag{5}$$

$$F_{i,f} = D_{i,f}\omega\sigma_f \tag{6}$$

In a realistically sized SC, there are a very large number of potential fields. To connect the output of the landscape optimization model to a SC model while maintaining computational tractability, we consider transportation at a coarser spatial resolution represented by a set of 'harvesting sites' $j \in \mathbf{J}$. Multidimensional set $g \in \mathbf{G} \subset (\mathbf{F} \times \mathbf{J})$ describes the membership of fields to harvesting sites. The amount of harvested biomass $H_{i,j,t}$ transported to downstream SC nodes is constrained in Eq.(7).

$$H_{i,j,t} \leq \sum_{f \in G_j} Y_{i,f,t} \tag{7}$$

By modelling transportation at the harvesting site level, we preserve landscape design detail but avoid adding flow variables for every field.

The costs associated with landscape design are given in Eq.(8) where λ_i are the per-Mg costs from harvesting, ρ is the cost of fertilization per kgN, and ϕ_i is the combined annualized cost of establishment and per hectare cost of land management.

$$C^{LAND} = \sum_{i,j,t} \lambda_i H_{i,j,t} + \sum_{i,f} \rho F_{i,f} + \sum_{i,f} A_{i,f}\phi_i \tag{8}$$

Similarly, the emissions from landscape activities are given by Eq.(9) where Γ_i^{MG} are the per-Mg emissions, Γ^N are the emissions from fertilization, and Γ_i^{HA} are the annualized per-hectare establishment and management emissions.

$$GHG_s^{LAND} = \sum_{i,j,t} \Gamma_i^{MG} H_{i,j,t,s} + \sum_{i,f} \Gamma^N F_{i,f} + \sum_{i,f} A_{i,f} \Gamma_i^{HA} + \sum_{i,f} GHG_{i,f}^{SOC} \tag{9}$$

For brevity, the full mixed-integer linear SC model is not presented here, but a similar model without extensions to measure SC specific GHG emissions or landscape design is presented by (Ng et al., 2018). The modified SC model is a multi-period MILP formulation that minimizes the total annualized cost. Key decisions include the location, technology, and capacity of pre-processing depots and biorefineries, inventory levels, and shipment and production schedules. GHG emissions are considered in the case study via the ϵ-constraint method and include emissions from shipment, production, landscape, and soil C sources, but do not include any credit for replacing fossil fuels.

3. Model and Data Integration

Because biomass yield and soil C sequestration potential are field specific and depend on local weather and soil quality, modelling landscape design at high resolution is critical. The model presented in section 2 uses large amounts of spatial data for parameters $\alpha_{i,f,t}$, $\Gamma_{i,f}^{SOC}$, $\varsigma_{i,f,t}^1$, and $\varsigma_{i,f}^2$ and requires a convenient way to integrate the model and the data.

Biogeochemical crop models such as SALUS (Basso and Ritchie, 2015) are tools that simulate crop growth and capture the spatial variability of crop yields and soil C sequestration. The underlying data that SALUS uses to capture spatial variability is a weather layer (gridMET (Abatzoglou, 2013)), a soil layer (SSURGO 30m resolution), and a geographic raster defining the available fields for planting biomass. Identifying potential fields is outside the scope of this paper, interested readers can refer to (Lark et al., 2020). Holding all other inputs constant, simulations with the same soil type and weather will result in the same yield and soil C sequestration. With this observation, we define the weather grid's 4x4 km pixels as harvesting sites and the soil types as fields. For a given weather grid, if there are fields with identical soil types, the fields are aggregated into a single element of **F** without losing landscape design accuracy.

The model data integration strategy allows for the aggregation of 'identical' fields which reduces the model size without aggregating unlike fields which could reduce the impact of the landscape design decisions. The model data integration strategy can also be extended to a user defined harvesting site resolution. A coarser resolution may be of interest because it can reduce the model size and enable the modelling of larger study areas. A coarser harvesting site grid can be produced by overlaying an arbitrary grid, labelling the fields enclosed in each cell with a 'grid label', then aggregating (summing yield, soil C, and field area) fields with identical grid labels and soil types.

4. Results

4.1. Case Study Description

We consider 24 potential biorefineries and 500 potential pre-processing depots in the lower peninsula of the state of Michigan, USA. SALUS simulations were performed for switchgrass grown on recently abandoned, expanded, and intermittent cropland identified by (Lark et al., 2020). There are originally 275,502 fields in the study area.

Table 1. Model size and error in yield from applying the model/data integration procedure for various harvesting site resolutions in Michigan, USA.

Harvesting Site Resolution	$|F|$	$|J|$	Mean error (std.) in yield (Mg/ha)
4x4 km	69,835	5,720	reference
8x8 km	41,120	1,656	0.57 (0.59)
12x12 km	29,874	794	0.58 (0.59)
16x16 km	22,801	453	0.60 (0.62)

Results from applying the data integration strategy for several harvesting site resolutions are shown in Table 1. The error in yield for each soil type is calculated as the absolute value of the yield at the 4x4 km resolution minus the aggregated yield at the coarser resolution for that same soil type. The mean and standard deviation are taken for all soils in the study area.

4.2. Landscape Design Benefit

Figure 2 shows the benefit of designing the landscape simultaneously with the supply chain. In panel (a) we use the ϵ-constraint method to find the minimum supply chain cost (Eq.(10)) while constraining the GHG emissions. Note that the GHG emissions are negative due to a sequestration of soil C which offsets the other sources of CO_2 (CO_2 equivalents from transportation, production, and land management) and does not include any credits from bioethanol replacing fossil fuels. Panels (b) and (c) correspond to the configuration of the minimum cost SC for a high and low amount of GHG emissions respectively.

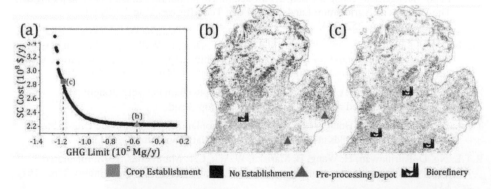

Figure 2. (a) The pareto frontier for the ϵ-constraint method constraining GHG emissions. (b) The SC configuration at a high emission level. (c) The SC configuration at a low emission level.

As the upper bound on GHG emissions decreases (becomes more restrictive), the landscape design becomes more important. Panel (c) shows that the optimal configuration avoids planting biomass at fields which are 'poor' at sequestering soil C and instead transports biomass further for the added benefit of sequestering carbon at more distant fields, opposed to panel (b) where biomass is planted close to the refinery to lower transportation costs. Furthermore, at lower GHG solutions, additional biorefineries are constructed to reduce the emissions from transportation, and depots are avoided which have a GHG penalty from using local grid energy to pelletize biomass. Interestingly, by designing the SC and landscape simultaneously, significant reductions in GHG emissions are possible for only a marginal increase in costs as shown by the flat region in the bottom right of Figure 2(a).

5. Conclusions

The high-resolution landscape design model described in section 2 allows the incorporation of the data integration strategy and results in a computationally efficient way to consider landscape design and management simultaneously with SC optimization. The large amount of spatial data needed for modelling dedicated bioenergy crops is aggregated in a way that maintains the separation of un-like fields (allowing the model greater control over the outcomes from landscape design) while ensuring a tractable model. We showed that by introducing the upstream landscape design model, integrated solutions could be found that leverage crop establishment locations to find attractive GHG solutions for only a relatively small increase in SC costs. The flexible model and data integration strategy can be used to model biofuel SCs with landscape design considerations on a much larger scale. Decision makers could use the integrated model to analyze the environmental and economic trade-offs between land quality, land distribution, SC configuration, and SC operation and the influence of key parameters on the integrated system.

References

J.T. Abatzoglou, 2013. Development of gridded surface meteorological data for ecological applications and modelling. Int. J. Climatol. 33, 121–131.

B. Basso, J.T. Ritchie, 2015. Simulating Crop Growth and Biogeochemical Fluxes in Response to Land Management Using the SALUS Model, in: Hamilton, S., JE, D., Robertson, P. (Eds.), The Ecology of Agricultural Landscapes: Long Term Research on the Path to Sustainabiliy. pp. 252–274.

J.L. Field, S.G. Evans, E. Marx, M. Easter, P.R. Adler, T. Dinh, B. Willson, K. Paustian, 2018. High-resolution techno–ecological modelling of a bioenergy landscape to identify climate mitigation opportunities in cellulosic ethanol production. Nat. Energy 3, 211–219.

H. Ghaderi, M.S. Pishvaee, A. Moini, 2016. Biomass supply chain network design: An optimization-oriented review and analysis. Ind. Crops Prod.

T.J. Lark, S.A. Spawn, M. Bougie, H.K. Gibbs, 2020. Cropland expansion in the United States produces marginal yields at high costs to wildlife. Nat. Commun. 11, 1–11.

R.T.L. Ng, D. Kurniawan, H. Wang, B. Mariska, W. Wu, C.T. Maravelias, 2018. Integrated framework for designing spatially explicit biofuel supply chains. Appl. Energy 216, 116–131.

E.G. O'Neill, C.T. Maravelias, 2021. Towards integrated landscape design and biofuel supply chain optimization. Curr. Opin. Chem. Eng.

Proceedings of the 14th International Symposium on Process Systems Engineering – PSE 2021+
June 19-23, 2022, Kyoto, Japan © 2022 Elsevier B.V. All rights reserved.
http://dx.doi.org/10.1016/B978-0-323-85159-6.50322-5

Assessing the Environmental Potential of Hydrogen from Waste Polyethylene

Cecilia Salah[a], Selene Cobo[a], Gonzalo Guillén-Gosálbez[a*]

[a] Institute for Chemical and Bioengineering, Department of Chemistry and Applied Biosciences, ETH Zürich, 8093 Zürich, Switzerland
gonzalo.guillen.gosalbez@chem.ethz.ch

Abstract

In 2019, nearly 370 million tonnes of waste plastic were generated, an amount that has been steadily increasing over the years. Here we assess hydrogen production from waste polyethylene in the context of a circular economy of plastics. Based on the gasification of polyethylene waste (wPG), we performed a Life Cycle Assessment (LCA) study following the ReCiPe method. Our results show that the wPG process coupled with carbon capture and storage (CCS) performs very well environmentally relative to other H_2 production routes, outperforming steam methane reforming (SMR) with and without CCS and biomass gasification (BG) in the three endpoint impact categories.

Keywords: Hydrogen; Waste polyethylene; Circular economy; Life cycle assessment

1. Introduction

Every year, the global demand for single-use polymers increases and, with it, the generation of plastic waste. According to Geyer et al. (2017), as of 2015, 79 % of all plastic ever made had been disposed of in landfills or the environment. Although the percentage of polymer waste destined for recycling has been increasing over the years, millions of tonnes of residues are annually mismanaged globally. This continuous accumulation underlines the need for a circular economy that valorizes polymer residues.

The circular economy of plastics is based on the recycling, repurposing, refurbishing, and revalorization of the generated waste. Notably, valuable feedstocks to the chemical industry could be produced from waste polymers through chemical recycling, as pointed out by Pacheco-Lopez et al. (2021). For instance, waste polyethylene and polypropylene can be processed to recover their respective monomers through pyrolysis or to produce synthesis gas (syngas), a key feedstock for chemical production, through gasification (Saebeaa et al. 2020). While the first route would be ideally preferred due to the higher value of monomers, the second is significantly easier to implement due to the more mature gasification technologies.

Here we explore the benefits of recycling the hydrogen chemically stored in polymers via gasification of waste polyethylene (wPE). Hydrogen has attracted increasing attention as an energy carrier and low-carbon feedstock for various fields. Steam methane reforming (SMR), currently the standard and cheapest route for hydrogen production, relies on natural gas (Parkinson et al. 2019), which leads to significant carbon emissions. Alternatives with lower emissions include SMR with carbon capture and storage (CCS) and electrolysis powered by renewable energies. Based on process modeling and LCA, here we investigate whether H_2 from recycled plastics is environmentally appealing, which at present remains unclear.

2. Methods description

2.1. Process description

We consider a hydrogen production process from waste polyethylene, alone or coupled with CCS, based on data collected from Saebeaa et al. (2020), Luyben (2018), and Susmozas (2015). The resulting block flow diagram is shown in Figure 1. The waste-polymer gasification (wPG) process consists of two main parts: syngas generation through steam gasification of wPE, followed by H_2 production and purification through water-gas shift (WGS) and pressure swing adsorption (PSA). CCS (wPG+CCS) is done on a CO_2-rich stream (95.6 mol%), which is compressed to 150 bar prior to injection.

The gasification takes place at 800 °C and 1.013 bar with steam as the gasifying agent to generate syngas with 67.25 mole% H_2, 25.24 mole% CO, 7.33 mole% CO_2, and 0.18 mole% CH_4, as in Saebeaa et al. (2020). The stream is then compressed to 32.5 bar with intercooling before entering the WGS section to meet the conditions in Luyben (2018). The syngas undergoes a high-temperature water-gas shift (HT-WGS) at 400 °C with 88 % conversion and a low-temperature water-gas shift (LT-WGS) at 250 °C with 95 % conversion, following the reaction in Eq.(1).

$$CO + H_2O \leftrightarrow CO_2 + H_2 \tag{1}$$

The H_2-enriched stream is cooled down to 40 °C and flashed before being sent to the pressure swing adsorption (PSA) unit, obtaining H_2 at 99.9 mole% purity based on Susmozas (2015). The tail gas is decompressed and undergoes combustion at 1300 °C and 1.5 bar, covering the energy needs of the gasification. Combustion is performed with air, and the post-combustion stream is vented after cooling and flashing at 40 °C.

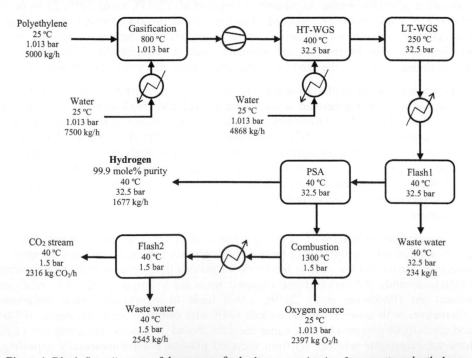

Figure 1. Block flow diagram of the process for hydrogen production from waste polyethylene.

For the process with CCS (wPG+CCS), oxy-combustion with a near stoichiometric O_2 ratio is considered at the same conditions as in wPG, obtaining a stream 95 mole% in CO_2 after cooling and flashing, which is compressed to 150 bar in four steps with inter-cooling and flashing, requiring 0.09 kW/kg CO_2 (Pipitone and Bolland of 2009). We considered that the flashed streams from this multi-stage compression are decompressed to 1.013 bar and undergo a final flash, recycling the gaseous stream to the combustion reactor. Therefore, no direct emissions are produced in the wPG+CCS process. The waste water streams from all flash units are sent to water treatment. Only cooling utilities are required.

2.2. Life cycle assessment (LCA) and scenarios definition

The environmental assessment was performed according to the ISO 14040 (2006). For each technology, the endpoint impacts on human health (HH), ecosystems quality (EQ), and resource depletion (RD) were calculated for 1 kg of hydrogen with the ReCiPe2016 method (Huijbregts et al. 2017) using data of global activities from Ecoinvent 3.7 (Wernet et al. 2016) in Simapro 9.2. The assessment follows a cradle-to-gate approach that considers the impacts from raw materials, electricity, process utilities, products, and direct emissions at point of substitution (APOS), disregarding the end-of-life of the plant infrastructure and use phase of the produced hydrogen. The ReCiPe2016 methodology quantifies the impacts on HH in disability-adjusted life years (DALY), which are the number of years during which individuals are not in total health; the effects on EQ are measured as the fraction of species that may be lost over time due to changes in environmental systems (species.y); RD in USD 2013 represents the extra cost required for the exploitation of resources in the future.

We quantified the life cycle inventories (LCI) for wPG and wPG+CCS from the mass and energy balances of the process in section 2.1. Moreover, we expanded the system boundaries to account for the treatment of waste polyethylene. Hence, we assume that our process avoids the landfilling and incineration of wPE, considering the proportion destined to each alternative worldwide in 2015, showcased by Geyer et al. (2017): 55 % of wPE to landfills and 25.5 % to incineration, using processes from Ecoinvent v3.7. The impacts of the remaining 19.5 % of waste, which is recycled, were omitted.

The wPG and wPG+CCS processes were compared to other hydrogen production routes, of which the inventories were taken from literature: SMR with and without CCS, following Dufour et al. (2012); biomass steam gasification (BG) with and without CCS, based on Susmozas et al. (2016); proton exchange membrane (PEM) electrolysis powered by various energy sources, following Lee et al. (2010), i.e., bioenergy with CCS (BECCS), hydropower, nuclear power, solar power from photovoltaic cells, wind power and the electricity mix from the 2018 power grid.

3. Results and Discussion

Figure 2 displays the endpoint environmental impacts of 1 kg of H_2 for the 12 scenarios studied here. The total values per impact category are also available in Table 1.

wPG and wPG+CCS perform differently for various reasons. wPG+CCS requires 14 % more electricity and 78 % more cooling water per kg of H_2 than wPG due to the CO_2 compression unit. Additionally, wPG+CCS includes the impacts embodied in the oxygen feedstock for oxy-combustion, avoided in the standard wPG route that uses excess air as an oxygen source and vents the post-combustion stream. Overall, wPG+CCS is worse in RD and better in HH and EQ.

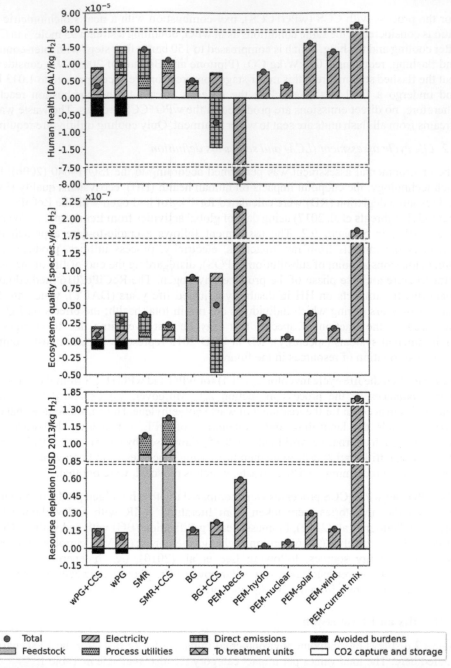

Figure 2. Endpoint environmental impacts of the hydrogen production routes assessed in this study, broken down by process components. The electricity term corresponds to the impacts of the energy sources used in the process (different for each PEM scenario). The following acronyms are employed: wPG: waste polyethylene gasification; CCS: carbon capture and storage; SMR: steam methane reforming; BG: biomass gasification; PEM: proton exchange membrane electrolysis; beccs: bioenergy with CCS; hydro: hydropower; nuclear: nuclear power plant; solar: photovoltaic energy; wind: wind power; current mix: electricity from the power grid of 2018.

Table 1. Total endpoint environmental impacts of the different technologies per impact category.

	Human health [DALY/kg H_2]	Ecosystems quality [species.y/kg H_2]	Resource depletion [USD 2013/kg H_2]
wPG+CCS	3.62×10^{-6}	9.16×10^{-9}	0.13
wPG	9.84×10^{-6}	2.92×10^{-8}	0.10
SMR	1.46×10^{-5}	3.93×10^{-8}	1.09
SMR+CCS	1.13×10^{-5}	2.44×10^{-8}	1.24
BG	5.19×10^{-6}	9.43×10^{-8}	0.17
BG+CCS	-7.43×10^{-6}	5.33×10^{-8}	0.23
PEM-beccs	-7.88×10^{-5}	2.13×10^{-7}	0.62
PEM-hydro	7.95×10^{-6}	3.51×10^{-8}	0.02
PEM-nuclear	3.95×10^{-6}	6.63×10^{-9}	0.06
PEM-solar	1.66×10^{-5}	4.12×10^{-8}	0.32
PEM-wind	1.40×10^{-5}	1.91×10^{-8}	0.18
PEM-current mix	8.62×10^{-5}	1.82×10^{-7}	1.79

In terms of impacts on HH, technologies involving biomass coupled to CCS (PEM-beccs and BG+CCS) are the most favorable scenarios, with negative impacts. wPG+CCS, PEM-nuclear, BG, PEM-hydro, wPG, SMR+CCS, PEM-wind, SMR, PEM-solar, and PEM-current mix follow. In terms of EQ, PEM-nuclear has the least impact, followed by wPG+CCS, PEM-wind, SMR+CCS, wPG, PEM-hydro, SMR, PEM-solar, BG+CCS, BG, PEM-current mix, and PEM-beccs. As for RD, PEM-hydro and PEM-nuclear are the least impactful technologies. wPG, wPG+CCS, BG, PEM-wind, BG+CCS, PEM-solar, PEM-beccs, SMR, SMR+CCS, and PEM-current mix follow.

wPG H_2 with and without CCS outperforms SMR H_2 in all three endpoint categories. Moreover, except for technologies involving biomass coupled with CCS (PEM-beccs and BG+CCS), wPG+CCS has the lowest impact on HH among the studied routes.

In terms of EQ, wPG+CCS is the second least harmful route, only behind PEM-nuclear. The technologies involving biomass and CCS, which performed very well in HH, are the most damaging to the ecosystems. This is due to the significant land requirements of biomass plantations (i.e., poplar).

As for RD, the PEM-current mix presents the highest impact because the 2018 energy mix is heavily reliant on fossil resources, as reported by the IEA 2019 World Energy Outlook. SMR with and without CCS follow as they require natural gas as feedstock, representing 83 % of the total RD impact for SMR and 73 % from SMR+CCS.

Interestingly, PEM-solar and PEM-wind are not the preferred option in any category. This is due to their life-cycle impacts, linked to the manufacture of photovoltaic panels and wind turbines, which are higher than those associated with other energy sources. For photovoltaic panels, these impacts mostly come from material and the energy provision for manufacturing, while for wind turbines, it is the material provision and construction. These results are consistent with the ones presented by Turconi et al. (2013).

4. Conclusions

Our work assessed the potential environmental benefits of producing hydrogen from waste polyethylene (wPE). Our results show that wPG+CCS outperforms SMR (business as usual) and SMR+CCS in the studied impact categories. Moreover, wPG+CCS performs better in human health than biomass (BG) and electrolytic H_2 (PEM) from renewables, excluding BECCS. It also shows lower impacts on ecosystems quality than most processes (except for PEM-nuclear). The processes using wPE as a feedstock also display lower impacts on resource availability than SMR and BG with and without CCS.

Overall, our results suggest that hydrogen production based on plastic waste via wPG+CCS is environmentally appealing. This technology would help realize the circular economy concept in chemicals production by recycling polymer residues that would otherwise end up in landfills or incineration facilities.

Acknowledgments

This publication was created as part of NCCR Catalysis, a National Centre of Competence in Research funded by the Swiss National Science Foundation.

References

A. Pacheco-López, A. Somoza-Tornos, A. Espuña, M. Graells, 2021, Systematic generation and targeting of chemical recycling pathways: A mixed plastic waste upcycling case study, Computer Aided Chemical Engineering, 50, 1125-1130.

A. Susmozas, 2015, Analysis of energy systems based on biomass gasification (Doctoral thesis, Universidad Rey Juan Carlos), retrieved from BURJC-Digital database.

A. Susmozas, D. Iribarren, P. Zapp, J. Linßen, Javier Dufour, 2016, Life-cycle performance of hydrogen production via indirect biomass gasification with CO_2 capture, International Journal of Hydrogen Energy, 41, 42, 19484-19491.

B. Parkinson, P. Balcombe, J.F. Speirs, A.D. Hawkes, K. Hellgardt, 2019, Levelized cost of CO_2 mitigation from hydrogen production routes, Energy and Environmental Science, 12, 1, 19-40

D. Saebeaa, P. Ruengritb, A. Arpornwichanopc, Y. Patcharavorachot, 2020, Gasification of plastic waste for synthesis gas production, Energy Reports, 6, 202-207

G. Pipitone, O. Bollav, 2009, Power generation with CO_2 capture: Technology for CO_2 purification, International Journal of Greenhouse Gas Control, 3, 5, 528-534.

G. Wernet, C. Bauer, B. Steubing, J. Reinhard, E. Moreno-Ruiz, B. Weidema, 2016, The ecoinvent database version 3 (part I): overview and methodology, International Journal of Life Cycle Assessment, 21, 1218–1230.

IEA, World Energy Outlook 2019, 2019, International Energy Agency.

ISO, 2006, ISO 14040:2006(E): Environmental Management – Life Cycle Assessment – Principles and Framework, International Standards Organization.

J. Dufour, D.P. Serrano, J.L. Gález, A. González, E. Soria, J.L. Fierro, 2012, Life cycle assessment of alternatives for hydrogen production from renewable and fossil sources, International Journal of Hydrogen Energy, 37, 2, 1173-1183.

J. Lee, S. An, K. Cha, T. Hur, 2010, Life cycle environmental and economic analyses of a hydrogen station with wind energy, International Journal of Hydrogen Energy, 35, 6, 2213-2225.

M. Huijbregts, Z. Steinmann, P. Elshout, G. Stam, F. Verones, M. Vieira, A. Hollander, M. Zijp, R. Van Zelm, 2017, Recipe2016: a harmonized life cycle impact assessment method at midpoint and endpoint level, International Journal of Life Cycle Assessment, 22, 138-147.

Plastics Europe, 2021, Plastics – the facts 2020, Plastics Europe.

R. Geyer, J. R. Jambeck, K. L. Law, 2017, Production, use, and fate of all plastics ever made, Science Advances, 3, 7.

R. Turconi, A. Boldrin, T. Astrup, 2013, Life cycle assessment (LCA) of electricity generation technologies: Overview, comparability and limitations, Renewable and Sustainable Energy Reviews, 28, 555-565.

W. L. Luyben, 2018, Plantwide control of a coupled/reformer ammonia process, Chemical Engineering Research and Design, 134, 518-527.

Proceedings of the 14th International Symposium on Process Systems Engineering – PSE 2021+
June 19-23, 2022, Kyoto, Japan © 2022 Elsevier B.V. All rights reserved.
http://dx.doi.org/10.1016/B978-0-323-85159-6.50323-7

A Systematic Comparison of Renewable Liquid Fuels for Power Generation: Towards a 100% Renewable Energy System

Antonio Sánchez[a]*, Elena C. Blanco[a], Mariano Martín[a], Pastora Vega[b]

[a]*Department of Chemical Engineering. University of Salamanca. Plz. Caidos 1-5. 37008. Salamanca (Spain)*
[b]*Department of Computer Science and Automatic. Univeristy of Salamanca. Plz. Caidos 1-5. 37008. Salamanca (Spain)*
antoniosg@usal.es

Abstract

Energy store will be essential in the future power system due to the inherent fluctuations of the renewable resources. The use of liquid fuels, such as methanol or ammonia, allows for seasonal storage and, as energy carriers, can be used for different energy applications. In this work, a systematic evaluation of different alternatives to produce power from methanol and ammonia has been performed. Particularly, for each of them, the thermochemical (combustion) and electrochemical (fuel cells) routes are evaluated. The operating conditions of each of the alternatives are optimized yielding energy efficiencies between about 15-40 %. From an economic point of view, the cost of electricity for the proposed power facilities is around 0.8 €/kWh for the fuel cell based production and 0.25 €/kWh for the combustion one, with lower prices for ammonia based alternatives. The fuel cells are more suitable for small scale applications, hence, with lower economies of scale. Therefore, this techno-economic assessment demonstrates the feasibility of liquid fuels to provide a robust power grid based on renewable resources meeting sustainable development.

Keywords: Ammonia, Energy Storage, Methanol, Power-to-X, Renewable fuels

1. Introduction

Renewable energy is a vital resource for the energy transition and sustainable development, in order to achieve the UN target of ensuring access to affordable, reliable, sustainable, and modern energy for all. Therefore, an increase in the share of renewable energy sources (RES) is expected in the coming years. IRENA (2018) forecasts that, by 2050, renewable generation will account for about 85 % of the global power generation, and around 60 % of the total final energy consumption. However, this increase in renewable penetration threatens the stability of the power system due to the inherently stochastic nature of the two main RES, wind and solar. At this point, energy storage will be crucial to ensure demand satisfaction at every time regardless of weather conditions in an optimal way. In this area, two alternatives emerge as the most promising in the future scenario, lithium-ion batteries and hydrogen and its derivatives (Schmidt et al., 2019). This latter option is particularly suitable for long-term energy storage (more than 700h). Apart from the power system, the share of renewables in other activities must be increased to achieve a high sustainable final energy consumption rate. In this field, energy carriers arise as one of the main tools for different applications where direct electrification is difficult.

Regarding energy storage/carriers, liquid fuels, such as methanol or ammonia, emerge as one of the most attractive options. Liquid fuels have a high volumetric energy density, a scalable and flexible behavior for different time scales and power capacities, and easy conditions of storage and transportation. Several works in the literature have analysed Power-to-liquids (PtL) technology taking into account the variability of the solar/wind availability and its influence on the chemical production. Once the fuel has been synthesised, the second stage will be the generation of electricity when it is needed, for instance, when renewable generation is low. At this point, two different alternatives are available for this purpose. The first one is based on the combustion of the liquid fuels to produce power using, mainly, a gas turbine. On the other side, liquid fuels can be converted into power using a fuel cell. Direct methanol/ammonia fuel cells are being developed, but another alternative is a first step based on the reforming of the fuels and, subsequently, using a hydrogen fuel cell which is a more mature technology. Different experimental works have evaluated the transformation of liquid fuels to power at laboratory scale from different perspectives. However, a process-scale analysis is required, including all the stages involved in the transformation, determining the energy efficiency of the process, and computing the cost of electricity for the different alternatives. This step is essential for planning an orderly introduction of these liquid fuels in real applications such as grid storage, mobility, etc.

In this work, a holistic comparison of the two main hydrogen liquid carriers, methanol and ammonia, is performed. For each of them, two different transformation routes have been studied: thermochemical (combustion) and electrochemical (fuel cell). The preparation of the raw material, the power conversion, and the different gas treatment operations have been included in this assessment. With this work, a fair comparison of both liquid fuels is conducted from a technical and economic perspective.

2. Process Description

Two different alternatives are possible to transform methanol or ammonia into power: thermochemical or electrochemical (Figure 1). The combustion of methanol alone is difficult due to the cold start (related to the high latent heat of vaporization of methanol) and the associated emissions. Therefore, the methanol/hydrogen blends have been proposed as one of the most promising fuels in order to overcome the issues associated with methanol combustion. Hence, the first step of the process is the preparation of the mixture, in this case, a proportion of 85 % methanol and 15 % of hydrogen is used (Zhen & Wang, 2015). The hydrogen required is generated using methanol steam reforming in a catalytic reactor using $Cu/ZnO/Al_2O_3$ as catalyst. In the combustion of the blend, undesirable NO_x is generated. Its concentration is computed using an experimental-based correlation. The maximum temperature in the combustion chamber is limited to 1873 K due to material limitations. To produce power using the gases from the combustion chamber, a combined cycle is employed with a first step based on a gas turbine and a second part including a Rankine cycle.

Ammonia combustion follows a similar pattern than methanol. Due to issues related to ammonia combustion, a mixture of 70 % of ammonia and 30 % of hydrogen is used as fuel (Sánchez et al., 2021). Hydrogen is produced through ammonia decomposition in a membrane catalytic reactor. As in the methanol alternative, a combined cycle is set up to generate power. To overcome the maximum gas turbine operating temperature, an inert component (argon) is introduced into it. Due to the combustion characteristics, different gas clean-up operations are required. Firstly, NO_x removal can be introduced if necessary to meet the environmental restrictions. Secondly, hydrogen is leaving the combustion chamber and must be recovered using a selective membrane. After the gas clean-up section, a mixture of N_2 and Ar is obtained and, to reuse these components, a distillation column is used to separate them.

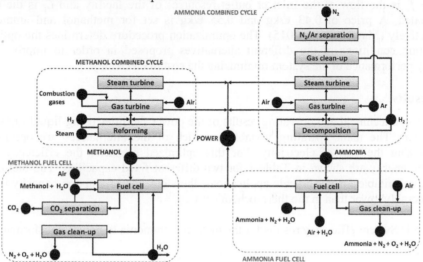

Figure 1: Process superstructure diagram of power production using methanol or ammonia

The electrochemical alternative is also evaluated. Methanol can be transformed into power in a fuel cell. An aqueous solution of methanol (1 mol/L) is fed to the unit along with air as oxidizer (Goor et al., 2019). From the anode of the cell, carbon dioxide, water and traces of methanol are obtained and, from the cathode, water, carbon dioxide, oxygen and nitrogen leave the cell. The operating conditions of the fuel cell (mainly voltage and intensity) determine the amount of power that can be produced from methanol. The anode stream is separated to recycle methanol and water into the fuel cell. Finally, the CO_2 from the cathode stream is also recovered by a zeolite system for synthesis purposes.

The last analysed system in this work is the electrochemical conversion of ammonia into power (Siddiqui & Dincer, 2019). In this alternative, ammonia is fed to the fuel cell together with wet (50 %) air. The main products of the reaction are nitrogen and water. From the anode, a mixture of nitrogen, water and traces of ammonia is produced. In the cathode, nitrogen, water, oxygen from the air but also NO_x is obtained that must be removed using a selective catalytic removal (SCR).

All the units and operations of each section in Figure 1 have been modelled using an equation based approach. Different modelling techniques have been used including mass and energy balances or experimental correlations. The entire superstructure is decomposed as a set of non-linear programming problems, one for each of the alternatives and fuels (up to 2000 equations per alternative). The formulation is implemented in GAMS and CONOPT 3.0 has been used as preferred solver in a multistart optimization approach. As objective function, a simplified operating cost of the facility is used such that (Eq.(1)):

$$Z = \sum_{i \in IN} f_i C_i - \sum_{j \in OUT} f_j C_j \qquad (1)$$

where f_i is the flow of the inlet or outlet resources of the facility and C_i is the cost associated. A price of 0.43 €/kg and 0.55 €/kg is set for methanol and ammonia respectively (Matzen et al., 2015). The optimization procedure determines the optimal operating conditions of the different alternatives proposed in order to improve the energy performance of the system minimizing the cost of electricity.

3. Results

First, some technical and operating results of the power facilities using liquid fuels are presented. The main operating variables of each of the alternatives are optimized minimizing the cost of electricity. For this optimized scenario, the efficiency and specific energy are shown in Table 1 for two different fixed capacities. The fuel cells are more suitable for small scale applications; therefore, the production capacity is set considerably lower than in the thermochemical way.

Table 1: System efficiency results for the methanol/ammonia to power transformation

Capacity (MW)	Technology alternative	Efficiency (%)	Specific energy (kWh/kg)
100.0	Methanol combined cycle	38.08	2.122
	Ammonia combined cycle	33.95	1.768
1.0	Methanol fuel cell	22.99	1.032
	Ammonia fuel cell	15.00	0.710

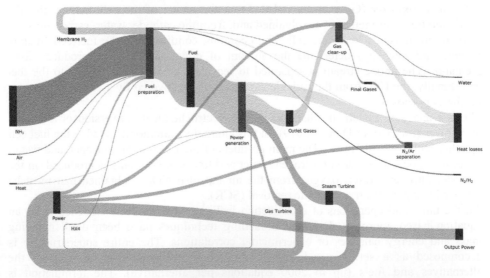

Figure 2: Sankey diagram for the energy flows in thermochemical ammonia transformation

In general, the thermochemical alternatives present better efficiencies than the electrochemical options with differences of almost double. The better efficiency is achieved in the methanol combined cycle reaching about 38 %. The thermochemical processes are based on the combination of a gas turbine and a Rankine cycle, achieving, a good performance in efficiency terms. Nevertheless, the main limitation to improve the efficiency values of these alternatives is the maximum temperature allowed in the gas turbine due to material restrictions. If this constraint is removed, a significant increase in the efficiency values could be expected. Therefore, a detailed research in the gas turbine construction should be made to widen the operating conditions of these units. Methanol combined cycle shows better efficiency results than ammonia based alternatives. The ammonia transformation is a more complex process that includes gas clean-up operations and N_2/Ar separation. These operations, particularly the last one, are energy demanding reducing the energy performance of the whole system.

The electrochemical systems show a drastically reduction in the efficiency values, about 15-20 %. These technologies are still under development at laboratory scale and, therefore, further research is required to improve the energy efficiency of these devices. These values are significantly lower than those obtained in the hydrogen alternative because the liquid fuels studied have a lower electrochemical reactivity. The lower values of efficiency in the methanol/ammonia fuel cells are due to the low operating voltage of these devices. At this point, the temperature is limiting the cell voltage to avoid the damage of the electrochemical catalyst. A better design of the cell membrane could improve the performance of these systems reducing the fuel crossover or improving the catalyst features. Methanol fuel cell shows better efficiency due to, according to experimental results, can operate with higher voltage and intensity and, therefore, with higher power density.

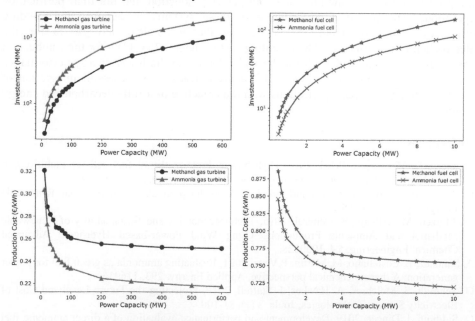

Figure 3: Capital and production costs of the different methanol/ammonia power generation system

The performance of the entire system, including all the operations involved, is analysed to determine the flows of energy in each section. For instance, in Figure 2, a Sankey diagram is included to analyse the energy operation of the thermochemical conversion of ammonia. A significant amount of energy is required in the different operations in the facility reducing the total performance of the system. For example, around 50 % of the produced power is devoted to internal operations. Similar results are obtained for the rest of the alternatives, however, for the sake of brevity, only one is included in this chapter. This holistic analysis is necessary to a fair evaluation of the system because including only the fuel cell or the combustion system is not enough to determine the real energy operation.

An economic evaluation of this power generation system based on liquid fuels is also performed in this work (as shown in Figure 3). The thermochemical alternatives exhibit a production cost about 0.2-0.3 €/kWh with an associated investment of about 3-5 MM€/MW depending on the capacity. For the electrochemical alternatives, the production costs significantly increase to values around 0.75-0.85 €/kWh requiring an investment of around 10 MM€/MW due to the lower production capacities than the previous alternatives. The lower energy efficiency of these systems and the lower power capacities (with reduced economies of scale) determine the worst economic performance of this electrochemical route.

4. Conclusions

In this work, a systematic evaluation of the transformation of methanol/ammonia into power is performed. Two different main routes have been assessed: thermochemical (combustion) and electrochemical (fuel cell). The entire process is analysed including the preparation of the raw materials, the power production and the final purification operations. The technical performance of these systems is demonstrated yielding efficiency values of about 35 % for the thermochemical and 18 % for the electrochemical alternatives. The operating cost is about 0.25 €/kWh for the combustion based processes and 0.8 €/kWh for the fuel cells due to the lower energy performance and the reduced power capacities. This analysis leads to a successful integration of these technologies in real applications with the objective of a fully decarbonized energy system.

References

M. Goor, S. Menkin, E. Peled, 2019, High power direct methanol fuel cell for mobility and portable applications, International Journal of Hydrogen Energy, 22, 3138-3143.

IRENA, 2018, Global Energy Tranformation: A roadmap to 2050, International Renewable Energy Agency, Abu Dhabi.

M. Matzen, M., Alhajji, Y. Demirel, 2015, Technoeconomics and Sustainability of Renewable Methanol and Ammonia Productions Using Wind Power-based Hydrogen. Advanced Chemical Engineering, 5,4, 1000128.

A. Sánchez, E. Castellano, M. Martín, P.Vega, 2021, Evaluating ammonia as green fuel for power generation: A thermo-chemical perspective, Applied Energy, 293, 116956.

O. Schmidt, S. Melchior, A. Hawkes, I. Staffell, 2019, Projecting the Future Levelized Cost of Electricity Storage Technologies, Joule, 3 (1), 81-100.

O. Siddiqui, I. Dincer, 2019, Development and performance evaluation of a direct ammonia fuel cell stack, Chemical Engineering Science, 200, 285-293.

X. Zhen, Y. Wang, 2015, An overview of methanol as internal combustion engine fuel, Renewable and Sustainable Energy Reviews, 52, 477-493.

Proceedings of the 14th International Symposium on Process Systems Engineering – PSE 2021+
June 19–23, 2022, Kyoto, Japan ©2022 Elsevier B. V. All rights reserved.
http://dx.doi.org/10.1016/B978-0-323-85159-6.50324-9

Guiding innovations and Value-chain improvements using Life-cycle design for Sustainable Circular Economy

Vyom Thakker, Bhavik R. Bakshi*

William G. Lowrie Department of Chemical and Biomolecular Engineering, The Ohio State University, Columbus, Ohio 43210, USA

bakshi.2@osu.edu

Abstract

While the current linear state of the economy has led to large scale natural-resource exploitation and pollution, circular economy can also lead to unexpected harm to environmental sustainability. Thus, there is a need to design product value-chains to achieve a Sustainable and Circular Economy (SCE). Previous work has focused on developing a systems engineering framework using life-cycle assessment with 'superstructure' network optimization to find optimal value-chain pathways while considering product life-cycles. However, the role of innovations in the form of novel technologies, societal action and new policy action has become increasingly crucial to establish SCE. In this work, we propose a sensitivity optimization framework to find the most attractive innovation directions within the value-chain using parameter perturbations as additional decision variables over pathway choice. The objectives include maximizing circularity and minimizing carbon dioxide emissions. We quantify the trade-off between these objectives and determine win-win innovative solutions using pareto and perturbation fronts. The method is demonstrated for an illustrative example, and its applicability to real value-chain networks has been probed.

Keywords: Life-cycle design, Sustainable and circular economy, Multi-objective optimization, Innovation modeling

1. Introduction

The current 'linear' state of the economy is contributing to many man-made disasters like climate change, plastic oceanic gyres, resource scarcities, harmful algal blooms in lakes, etc. While 'circular economy' is expected to bring about reduction in waste and pollution, it may not always be aligned with sustainability requirements such as curtailing climate change and respecting nature's carrying capacity. Progress toward a Sustainable Circular Economy (SCE) is crucial to mitigate large-scale exploitation of natural resources and pile-up of man-made materials like plastics in the environment. For achieving a SCE there is a need to holistically design entire value-chains of products and services while considering the environmental, economic, and social implications of potential alternatives. The field of Process Systems Engineering (PSE) has the potential to contribute towards establishing SCE for material life-cycles provided it expands its system boundary to account for the life cycle, economy, and ecosystems (Bakshi, 2019). This work is aimed at expanding PSE models and methods toward Sustainable Engineering to find optimal value-chain reforms and discover most attractive innovation directions.

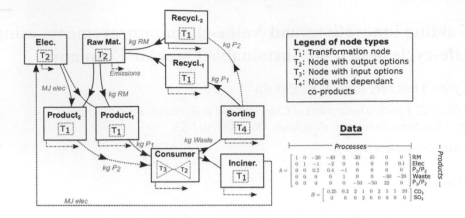

Figure 1: Illustrative example of a typical circular value-chain network

We construct the connection between SCE and PSE using the framework of life-cycle assessment (LCA), which focuses on calculating net environmental impact of 'singular' product value-chains, while considering entire life-cycles of the product - right from natural resource extraction to disposal. The computational structure of LCA involves solving a linear programming (LP) problem on life-cycle inventory data of value-chains (Heijungs and Suh, 2002). In our previous work (Thakker and Bakshi, 2021a), we have developed the SCE design framework which uses optimization to evaluate multiple alternative value-chains by creating a 'superstructure' network of alternatives. In this work, we expand the scope of design to identify the optimal perturbations in technology efficiencies, supply chains, policies and behavior that can be brought about by innovations. These perturbations do not consider systemic disruptions to value-chains that can be brought about by innovations, which need to be modeled as separate processes in the superstructure network. Sections 2 and 3 of this paper are devoted to finding SCE optimal value-chain pathways from the illustrative example network in figure 1 using previous work. Section 4 then describes the novel sensitvity optimization framework developed for innovation guidance, which is followed by insights on potential applications of the framework in conclusions.

2. Life Cycle Assessment (LCA) framework

LCA is used to find the impact of catering to a particular demand of a product, from the processes in its entire life cycle. The input, output and emissions data for each process is found from national averages, and are included within columns of the technology and intervention matrices (A and B). These matrices are available from national agencies and commercial organizations. LCA method consists of the equations, $As = f$, $g = Bs$ which ensures flow conservation of all products (rows), while meeting the final demand 'f', specified for the LCA study. 's' is the scaling vector, which represents the scale of operation of each process in A to meet the demand. 'g' is a vector of total resource use and emissions for meeting the demand, and is found by scaling the interventions B with the same 's'. One short-coming of LCA is that it requires separate analysis for each alternative value-chain pathway, which is rectified using the SCE design method described below.

3.   SCE Design Framework

In our previous work (Thakker and Bakshi, 2021a), we have developed a multi-objective superstructure optimization method to find optimal value-chain pathways for SCE objectives, and develop pareto fronts to quantify trade-offs between these objectives.

3.1.   *Node-alphabet representation of superstructure networks*

The framework is generally applicable to any circular system owing to the node-alphabet representation. Nodes in a typical life-cycle network are classified into 4 types (T_{1-4}) according to the substitutability of inputs and outputs, and whether the streams undergo transformations. Any SCE network may be represented as a combination of these node types, as shown for the illustrative example in figure 1.

3.2.   *Illustrative example*

This paper describes the foundational work and methodological developments using an illustrative example, shown in figure 1. This example involves finding the optimal-value chain to meet a consumer demand from one of the two-products (P_1 & P_2), which are sent either to segregation (and recycling) or to incineration. The goal is to find optimal pathways to meet SCE requirements. This illustration is chosen since the network is representative of typical product life-cycles, e.g. plastic containers, semi-conductors, laptops, etc., thereby highlighting the wide applicability of this work to relevant SCE problems.

3.3.   *Constraints*

The decision variables of the SCE design problem are the scaling factors 's' denoting the pathway selection, i.e. s_j is 0 if value-chain process j is inactive. Since the technology matrix A represents a 'superstructure' network of alternatives, A is a rectangular matrix (not full rank) and pathways design for an arbitrary objective $Z(s)$ is possible. However, optimizing value-chains requires flow conservation with the life-cycle which is established by specifying the LCA equations as constraints on the decision variables.

$$\min_{s} z := Z(s)$$

$$\text{Subject to:} \quad As = f \tag{1}$$
$$g = Bs$$

In addition, the network needs to be scaled to meet consumer demand, which is added as a constraint on the life-cycle flows (Hs). Furthermore, governing equations such as material, energy and component balances are specified as balance constraints ($\mathcal{F}_B$). Node efficiencies are also specified as constraints ($\mathcal{F}_n$) for each node based on its type (T_{1-4}).

$$Hs \geq u \tag{2}$$
$$\mathcal{F}_B(s) \leq 0 \tag{3}$$
$$\mathcal{F}_n(s) \leq 0 \tag{4}$$

Finally, non-negative scaling ($s \geq 0$) and the net-zero final demand of intermediate flows ($f_i = 0$) are ensured using variable bounds. All these constraints yield a feasible design space of pathway choices and technology options in a non-linear problem (NLP), which is optimized for various SCE objectives and to characterize the trade-offs between them.

3.4. Objectives and Pareto-front generation

SCE objectives must comprise of Environmental, Economic and Circularity aspects. The emission and effluent flows are captured within the g vector and can be used as environmental objectives. In the illustrative example, there are two emission flows, i.e. carbon dioxide (CO_2) and sulfur dioxide (SO_2), which need to be minimized. However, real life-cycle inventory data contains hundreds of emission flows, which can be aggregated into midpoint indicators such as global warming, acidification and eutrophication potential.

The economic objective can be formulated as life cycle cost (LCC), which would consider the cost of directly and indirectly used natural resources. For simplicity, this objective has been excluded from the illustrative example. Within the circularity domain, we formulate a novel metric θ using life-cycle flows of the network to quantify the circularity of the network. It is calculated as the ratio of the value of circular flows within the system to the value of manufactured products ($M.s$). Circular flows comprise of recycling, refurbishment, down-cycling and up-cycling in the technological system ($C \in A$) and valuable effluents such as compost g_c to the environment. The general expression for θ is as follows,

$$\theta = \frac{\Sigma \gamma_i C.s + \Sigma \gamma_k g_c}{\Sigma \gamma_i M.s} \quad i \in \text{products}, k \in \text{emissions}, C \in A_{\text{circular}}, M \in A_{\text{manufacturing}} \quad (5)$$

γ denotes the value function, typically in monetary, exergetic or physical units, and it determines the nature of the circularity metric, θ. In this example, we consider monetary circularity with both recycled raw material and electricity generating monetary value. Since we consider multiple objectives of SCE, there are bound to be trade-offs and win-win solutions. These are quantified using pareto optimal solutions, found using the ϵ-constraint method. The pareto-optimal solutions form a front which represent the best possible solutions without bias to any one of the three objective domains. The points lying above the front are sub-optimal and the ones lying below are infeasible. Thus, pareto front generation provides quantification and visualization of trade-offs.

3.5. SCE designs for the illustrative example

The SCE design solutions for the illustrative example are shown on the extremities of the black solid line pareto front in figure 2. The P_1-C-S-R_1 pathway corresponds to minimum CO_2 emissions, whereas the P_2-C-S-R_2 pathway has maximum circularity. ϵ-constraint is used to find points on the pareto front (solid line) which correspond to impartial compromise solutions between the objectives. The objective space below the front is sub-optimal, whereas the space above corresponds to win-win solutions which can only be achieved by innovations and pathways outside the superstructure network.

4. Sensitivity optimization for innovation discovery

It is crucial to identify the most attractive directions for innovations in the value-chain to improve SCE objectives. This is achieved by modifying the SCE design framework to

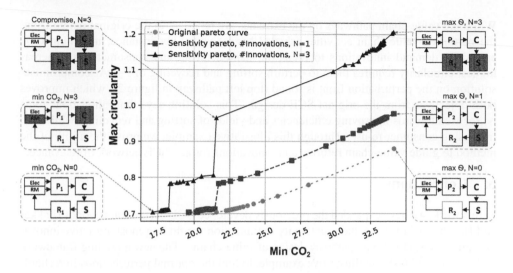

Figure 2: Pareto and Perturbation fronts for SCE design of network in figure 1.

include additional decision variables corresponding to the sensitivity of the parameters in the technology matrix(A). Each element in A (a_{ij}) represents a particular property of a life-cycle activity (j). We introduce binary variables y_{1ij} and y_{2ij} to assume the value '1' if a_{ij} can be increased or decreased (respectively), by a factor of $\Gamma \in [0,1]$. These perturbations to a_{ij} are assumed to be brought about by innovations, better technologies, or improved policy, and the number of such permitted perturbations (N) are set by the user of the framework. Flow conservation and material-energy balances need to hold despite these perturbations, which is done by changing the original design formulation to the following.

$$\min z := \text{kgCO}_2, \ \text{kgSO}_2$$

$$\max_{s, y_1, y_2} z := \text{Circularity} \ (\theta)$$

$$\text{Subject to:} \quad A \odot (1 + \Gamma y_1 - \Gamma y_2)s_j = f$$

$$y = y_1 + y_2 \tag{6}$$

$$\sum_i \sum_j y_{ij} \leq N, \ \text{\# of permitted perturbations}$$

$$G(s) \geq 0, \ \text{other constraints on } s$$

$$y_1, \ y_2, y \in \mathbb{Z}^{i \times j}; \ \Gamma \in [0,1]; \ s \geq 0$$

Here, '$\odot$' represents element-wise multiplication, and (y_1, y_2) are binary variables to identify the optimal innovative perturbations in the positive and negative directions. The resulting optimization is a MINLP that finds best value-chain pathways and most attractive perturbations within a user-defined range (Γ=0.1). The pareto fronts for SCE objectives using this formulation are referred to as 'perturbation fronts'. These fronts provide win-win solutions over the original pareto front, and greater win-win is obtained when more innovation perturbations (N) are allowed. For the illustrative example, two perturbation fronts are developed; with $N = 1$ (dashed line), $N = 3$ (dotted line), as shown in figure 2.

While the value-chain pathways on the extremes are identical to the original pareto front, the perturbed value-chain activities (shaded boxes) vary. For instance, minimum CO_2 emissions demand innovations to focus on manufacturing and consumer use, whereas highest circularity requires them to perturb sorting and recycling of P_2. A 'compromise' solution on the perturbation front is found (top left pathway in figure 2), which improves both objectives from the original SCE design. This solution says that innovations must be focused towards improving efficiencies and yields of sorting and recycling, while also increasing consumer re-use. Through this illustrative example, we prove the utility of the method to guide value-chain reforms and innovations for any SCE network of relevance.

5. Conclusions

In this work, we have expanded the previously developed SCE design framework (Thakker and Bakshi, 2021a) to include sensitivity optimization for finding most attractive innovation directions, along with pathway design of value-chains. The new modeling framework is demonstrated upon an illustrative example, to find the optimal perturbations in technology and societal parameters that can lead to win-win solutions from circularity and CO_2 emissions viewpoint. Pareto front generation allows quantification of trade-offs and selection of pareto-optimal solutions, which can inform new research directions based on a reasonable 'compromise' between SCE objectives. Future work will pertain to application of this methodology to a real-life value-chains of products, such as plastic-containers, windmills, etc. While the general applicability of the framework is established in section 3.1., the tractability of sensitivity optimization for large value-chain network is currently being explored using a case study on plastic grocery bags. In addition, it may be needed to introduce a physico-chemical transformation network using a multi-scale approach (Thakker and Bakshi, 2021b) to provide a realistic constraints on allowable perturbations.

6. Acknowledgments

The authors acknowledge the support from "The Global KAITEKI Center" at Arizona State University (ASU), and the National Science Foundation (NSF) grant EFMA-2029397.

References

Korhonen, J., Honkasalo, A. and Seppälä, J., 2018. Circular economy: the concept and its limitations. Ecological economics, 143, pp.37-46.

Bakshi, B.R., 2019. Toward sustainable chemical engineering: the role of process systems engineering. Annual review of chemical and biomolecular engineering, 10, pp.265-288.

Thakker, V. and Bakshi, B.R., 2021. Toward sustainable circular economies: A computational framework for assessment and design. *J of Cleaner Production*, 295, p.126353.

Heijungs, R. and Suh, S., 2002. The computational structure of life cycle assessment (Vol. 11). *Springer Science & Business Media*.

Thakker, V. and Bakshi, B.R., 2021. Multi-scale Sustainable Engineering: Integrated Design of Reaction Networks, Life Cycles, and Economic Sectors. *Computers & Chemical Engineering*, p.107578.

Proceedings of the 14th International Symposium on Process Systems Engineering – PSE 2021+
June 19-23, 2022, Kyoto, Japan © 2022 Elsevier B.V. All rights reserved.
http://dx.doi.org/10.1016/B978-0-323-85159-6.50325-0

Simultaneous Optimal Operation and Design of a Thermal Energy Storage Tank for District Heating Systems with Varying Energy Source

Caroline S. M. Nakama[a], Agnes C. Tysland[a], Brage R. Knudsen[b], Johannes Jäschke[a*]

[a]*Department of Chemical Engineering, Norwegian University of Science and Technology (NTNU), Sem Sælandsvei 4, Trondheim 7491, Norway*
[b]*SINTEF Energy Research, Kolbjørn Hejes vei 1B, Trondheim, 7491, Norway*

johannes.jaschke@ntnu.no

Abstract

District heating systems based on industrial waste heat play an important role in using energy efficiently. Combined with a thermal energy storage technology, such as pressured-water tanks, they have the potential of significantly reducing greenhouse gas emissions as well. However, installing thermal energy storage requires capital and, therefore, it is important to find an optimal design that balances the benefits of energy storage with the costs of installing such system. In this work we formulate a dynamic optimization model for designing a thermal energy storage tank based on operational conditions and apply it to a case study using historical data from a district heating system that recovers heat from an industrial plant in Norway. We found that a relatively large tank (greater than 5000 m3) would be necessary to store all excess energy provided by the plant that cannot be immediately used for the period and input data considered. However, the results can be used to investigate uncertainties and their effects on the optimal tank volume and return of investment.

Keywords: Energy systems; thermal energy storage; optimal operation; optimal design.

1. Introduction

Environmental, energetic and climate issues of today require a shift from society's fossil fuel dependency to renewable energy sources. The pace of this change must accelerate, and significant measures are taken to increase the development and use of renewable-energy-based technologies (Mirandola and Lorenzini, 2016), and environmental policies implemented by governments. For such shifted scenario, decarbonized energy system, district heating (DH) systems and thermal energy storage (TES) can play a critical role and contribute significantly to Europe's 2050 emission goals (Connolly el al., 2014). An important DH system type is those utilizing industrial waste heat; however, due to the commonly high variation of the waste heat availability, its combination with TES is of interest to further reduce the use of peak-heating sources. Pressured-water tanks are the most suitable TES technology for DH systems, yet they can be very costly and space availability may be limited (Knudsen et al., 2021).

In this work we focus on the optimal *operation and design* of a TES tank for utilization in a DH system based on waste–heat recovery. Integrating operation into the sizing problem is important, as operational conditions have a significant impact on how efficiently the waste heat is utilized, which in turn can influence the size of the TES tank. We present an approach that formulates a single nonlinear dynamic optimization problem that accounts for optimal operation and sizing simultaneously, as opposed to combined optimization/simulation-based methods previously proposed, e.g., Knudsen et al., 2021; Li et al., 2021. We demonstrate this method on a historical data set from a DH plant in Norway that recovers heat from a ferrosilicon plant.

2. Case Study

We consider a case study for designing a TES tank for the heating plant of the DH system of Mo i Rana in Norway. The DH plant is located inside Mo Industry Park and receives waste heat from a ferrosilicon plant. The objective of the TES is to increase the waste-heat utilization and thereby reduce necessary peak-heating.

The DH system has 6 boilers heating up the water that is sent back to the city. Two of them use waste-heat from the industrial park and four of them are peak-heating boilers. They run primarily on electricity or CO-gas as energy source, the latter being a by-product from a manganese plant in the industry park and thus with varying availability. Since today waste-heat availability does not exactly match demand, excess heat is dumped, and deficit heat is supplied by the peak-heat boilers. Figure 1 shows a simplified diagram of the process with a TES tank; the waste-heat boilers (WHB) and peak-heat boilers (PHB) are lumped together and represented as one unit. Nodes A and B represent split or merging of the main water flow, depending on whether the TES tank is charging or discharging, since there is no variation of volume in the TES tank. A description of the variables is presented in the Modelling section.

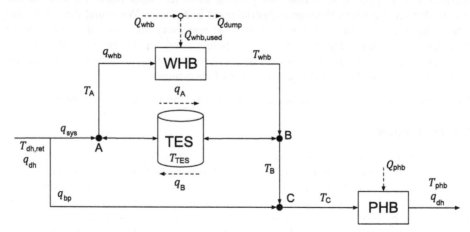

Figure 1. Flow diagram of the DH system of Mo i Rana.

2.1. Historical Data

For this case study, we selected March of 2019 as a representative month in which waste-heat availability oscillates from shortage to excess when compared against the heat demand from the city, as seen in Figure 2. This behaviour, usually seen during the transition months between summer and winter, has a potential for short-term savings, as opposed to long periods of shortage (winter) or excess (summer) of heat availability that would require long-term storage. From the DH system, we also have given the return and supply temperatures and mass flow rate of water for every hour available as input data; the temperatures are shown in the top Figure 3.

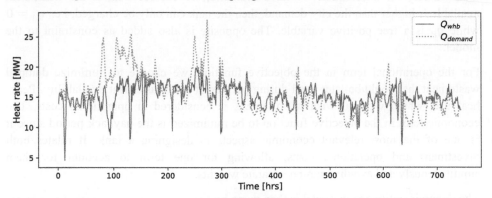

Figure 2. Historical waste-heat and heat demand data from Mo i Rana district heating system for March 2019.

3. Methodology

We formulate an optimization model to obtain the optimal volume of a TES tank for the Mo i Rana DH system taking operational conditions into account. For that, we need mass and energy balances of the process, as well as operational and cost functions that can be minimized to express our main goal. The mass and energy balances act as constraints in the model and are as follows

$$q_{\text{dh}}(t) - q_{\text{sys}}(t) - q_{\text{bp}}(t) = 0 \tag{1a}$$

$$q_{\text{sys}}(t) - q_{\text{whb}}(t) - q_A(t) + q_B(t) = 0 \tag{1b}$$

$$q_{\text{sys}}(t)C_pT_{\text{dh,ret}}(t) + q_B(t)C_pT_{\text{TES}}(t) - q_{\text{whb}}(t)C_pT_A(t) - q_A(t)C_pT_A(t) = 0 \tag{1c}$$

$$q_{\text{whb}}(t)C_pT_{\text{whb}}(t) + q_A(t)C_pT_{\text{TES}}(t) - q_{\text{sys}}(t)C_pT_B(t) - q_B(t)C_pT_B(t) = 0 \tag{1d}$$

$$q_{\text{bp}}(t)C_pT_{\text{dh,ret}}(t) + q_{\text{sys}}(t)C_pT_B(t) - q_{\text{dh}}(t)C_pT_C(t) = 0 \tag{1e}$$

$$Q_{\text{phb}}(t) - q_{\text{dh}}(t)C_p\left(T_{\text{phb}}(t) - T_C(t)\right) = 0 \tag{1f}$$

$$Q_{\text{whb,used}}(t) - q_{\text{whb}}(t)C_p\left(T_{\text{whb}}(t) - T_A(t)\right) = 0 \tag{1g}$$

$$Q_{\text{whb}}(t) - Q_{\text{whb,used}}(t) - Q_{\text{dump}}(t) = 0 \tag{1h}$$

$$\frac{d}{dt}\left(\rho V_{\text{TES}}C_pT_{\text{TES}}(t)\right) = q_A(t)C_p\left(T_{\text{TES}}(t) - T_A(t)\right) - q_B(t)C_p\left(T_{\text{TES}}(t) - T_B(t)\right) \tag{1i}$$

where q. are flow rates in kg/s, T. corresponds to the temperature at the outlet of the subscript reference in °C, C_p is the specific heat capacity of the water in kJ/(kgK), Q. are heat rates in W, ρ is the density of the water in kg/m³, and V_{TES} is the volume of the TES tank in m³. It is important to point out that q_A and q_B correspond to the same flow but in opposite direction. For example, when the TES tank is charging, $q_B > 0$ and q_A must be zero, and vice versa. If we enforced this condition in the optimization model, we would get a mathematical program with complementarity constraints, which is a class of nonconvex optimization models that can be particularly challenging to solve. To avoid that, we rely on information we have available; we enforce that, if the waste-heat available is higher than the city demand, then the tank can only be charged, i.e., $q_A = 0$ while q_B is a free positive variable. The opposite is also added as constraint to the model.

For the operational term in the objective function, we choose to minimize dumped waste-heat that could be later used during periods of low waste-heat availability. Peak-heat use, which we also wish to minimize, is considered in operational costs. The economic term in the objective function to be minimized is the payback period since it is one of the most relevant economic aspects in designing a tank. It relates both investment and operational costs, allowing for one term to account for them simultaneously and avoiding tuning separate weights.

The dynamic optimization model is then given by

$$\min_{q, V_{\text{TES}}} \quad N + C \int_0^T Q_{\text{dump}}(t)dt + 10^{-7} \int_0^T q_{\text{whb}}dt + 10^{-5} \int_0^T q_{\text{bp}}dt \tag{2a}$$

$$\text{s.t.} \quad N = \frac{ln(S/(S - I\{V\}r))}{ln(1+r)} \tag{2b}$$

$$S = n\, C \int_0^T \big(Q_{\text{phb,noTES}}(t) - Q_{\text{phb}}(t)\big)dt \tag{2c}$$

$$I(V) = 4.7V^{0.6218} \tag{2d}$$

$$q_A(t) = 0 \quad \text{if } Q_{\text{whb}}(t) < Q_{\text{demand}}(t) \tag{2e}$$

$$q_B(t) = 0 \quad \text{if } Q_{\text{whb}}(t) \geq Q_{\text{demand}}(t) \tag{2f}$$

$$x_{\text{lb}} \leq x \leq x_{\text{ub}} \tag{2g}$$

The model in Eq. (1)

where N is the payback period in years, T is the total length of the considered period in hours, $I(V)$ is an expression describing initial investment cost in 10^3 euros as a function of the volume of the tank in m^3 (Li et al., 2021), r is the annual interest rate, S is financial savings in 10^3 euros/year, n is the number of representative periods in a year, C is the cost of heat composed by the price of the energy source (in this case, C_{CO} or C_{elect}) and associated tax emissions (C_{CO_2} and C_{NO_x}), x is a vector containing all variables in the model, and x_{lb} and x_{ub} are the corresponding lower and upper bounds, respectively. The extra two terms in the objective function are regularization terms, which help the solver converge to a local solution, since the flow distribution within the DH system is not necessarily unique for some Q profiles and V_{TES}. Note that, here, C is also used as a weighting parameter for the waste-heat dump term.

Eq. (2) was discretized using implicit Euler with time step of one hour and implemented in Julia using JuMP as the mathematical modelling language (Dunning et al., 2017) and IPOPT as the nonlinear programming solver (Wächter and Biegler, 2006). Table 1 shows the values of parameters and variable bounds used for the calculation.

Simultaneous Optimal Operation and Design of a Thermal Energy
Storage Tank for District Heating Systems with Varying Energy Source
1955

Table 1. Parameters and variable bounds for Eq. (2) (The cost of CO-gas is confidential).

Parameter	Value	Bounds	Value
Electricity cost, C_{elect}	€ 0.087/kWh	T lower bound	40 °C
CO_2 emission tax, C_{CO_2}	€ 58.82/t CO_2	T upper bound	120 °C
NO_x emission tax, C_{NO_x}	€ 2,340.9/t NO_x	$Q_{whb,used}$ upper bound	22 MW
Annual interest rate, r	5 %	Q and q lower bound	0
Initial tank temp., $T_{TES}(0)$	95 °C	q_{whb} upper bound	333 kg/s

4. Results

The results for operational conditions considering electricity and CO-gas as peak heating were the same. In both cases, the optimal volume was 6323 m³ and Figure 3 shows some of the optimal operational conditions. The bottom plot shows peak heating and waste heat used, as well as the peak heating use without a TES tank. The total peak heating originally used during the period considered was 876.4 MWh. With the implementation of a TES tank of the optimal volume, this consumption is reduced in 48 % in total for the period. The top plot shows the TES tank, the supply temperatures to the DH system, and the corresponding return temperature. Initially during this the month, up to around 300 h, heat demand from the city is mostly greater than waste-heat supply, so the energy initially stored in the tank is consumed. Then, the TES tank temperature increases as excess waste-heat is available and reaches the maximum temperature at the end of the period.

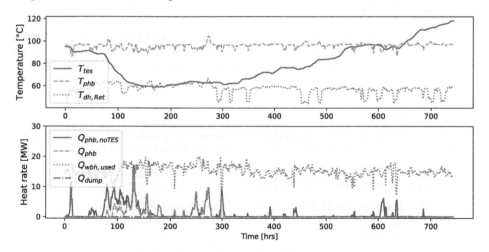

Figure 3. Optimal operation conditions for the optimal TES tank.

Regarding the economic aspect, if we consider that there are 3 months such as the representative period per year, and that the remaining months are not able to induce significant savings, the payback time for electricity as peak-heating source would be 13.7 years. Since in Norway, electricity is mainly from hydropower, the corresponding emission tax is lower. Consequently, for the case with CO-gas as peak-heating source the payback period is reduced to 12.2 years. Although these values imply large investment costs, uncertainties in the cost parameters, such as varying electricity price and emission taxes, the latter expected to increase in the next years (Klima- og Miljødepartementet, 2021), can reduce the payback time. Indeed, if the CO_2 tax is increased to the value expected by the Norwegian government in 2030, the payback time is reduced in about half for CO-gas as peak-heating source.

Since the investment cost of the TES tank is directly related to its volume, the payback period is also dependent on it. The bottom plot of Figure 3 shows that no waste heat is discarded, i.e., $Q_{dump} = 0$, and the large volume obtained for this TES tank is due to minimizing heat dump. The weighting parameter C can be seen as a cost for dumping heat and, in this case study, we used the actual cost of peak-heating. Decreasing its value could potentially allow for some excess waste-heat to be discarded, which, in turn, could reduce the tank volume. However, that would also increase peak heating and a balance should be found.

5. Conclusions and Future Work

The results show that using a single dynamic optimization model based on operation conditions can indeed be applied to design a TES tank and systematically investigate the influence of parameters subjected to uncertainties. The calculated TES tank volume for the case study is relatively large, which is a result from the selected input data (one month), and the available price parameters. For future work, we seek to apply this optimization model to a longer horizon that can comprehend an entire season and find a systematic approach to balance storing enough heat to obtain significant savings while keeping the tank as small as possible to reduce investment costs.

Acknowledgements

This work was supported by the Research Council of Norway (RCN) through FRIPRO Project SensPATH and HighEFF - Centre for an Energy Efficient and Competitive Industry for the Future under the FME-scheme (Centre for Environment-friendly Energy Research, 257632).

References

D. Connolly, H. Lund, B.V. Mathiesen, S. Werner, B. Möller, U. Persson, T. Boermans , D. Trier, P.A. Østergaard and S. Nielsen, 2014, Heat Roadmap Europe: Combining District Heating with Heat Savings to Decarbonise the EU Energy System. Energy policy, 65, 475-489

I. Dunning, J. Huchette and M. Lubin, 2017, JuMP: A modeling language for mathematical optimization, SIAM review, 59, 2, 295-320.

Klima- og Miljødepartementet. Heilskapeleg plan for å nå klimamålet. Regjeringen. 8 Jan. 2021: https://www.regjeringen.no/no/aktuelt/heilskapeleg-plan-for-a-na-klimamalet/id2827600/. Accessed 1 July 2021.

B. Knudsen, D. Rohde, and H. Kauko, 2021, Thermal Energy Storage Sizing for Industrial Waste-Heat Utilization in District Heating: A Model Predictive Control Approach, Energy, 234, 121200.

H. Li, J. Hou, T. Hong, Y. Ding and N. Nord, 2021, Energy, Economic, and Environmental Analysis of Integration of Thermal Energy Storage into District Heating Systems Using Waste Heat from Data Centres, Energy, 219, 119582.

A. Mirandola and E. Lorenzini, 2016, Energy, Environment and Climate: From the Past to the Future. International Journal of Heat and Technology, 34, 2, 159-164

A. Wächter and L.T. Biegler, 2006. On the implementation of an interior-point filter line-search algorithm for large-scale nonlinear programming, Mathematical programming, 106, 1, 25-57.

Proceedings of the 14th International Symposium on Process Systems Engineering – PSE 2021+
June 19–23, 2022, Kyoto, Japan © 2022 Elsevier B.V. All rights reserved.
http://dx.doi.org/10.1016/B978-0-323-85159-6.50326-2

A flexible energy storage dispatch strategy for day-ahead market trading

Jude O. Ejeh[a], Diarmid Roberts[a], Solomon F. Brown[a*]

[a]*Department of Chemical & Biological Engineering, The University of Sheffield, Mappin Street, Sheffield, S1 3JD, United Kingdom*

s.f.brown@sheffield.ac.uk

Abstract

In this work, we present a two-stage optimisation-based approach to obtain key metrics for use in a rules-based energy storage dispatch strategy. In electrical power systems, electrical energy storage (EES) devices have been shown to improve power reliability, quality and reduce electricity bills in behind-the-meter applications. However, owing to problems of a prolonged pay-back period, the scheduling of these EES devices play an important role in for asset owners. Existing optimisation-based approaches heavily rely on a rigid implementation of the obtained solutions, perfect foresight, and may not perform well even when uncertainties are considered. A flexible alternative, quite common in practice, involves the use of rules to guide battery actions. In our approach, we propose a two-stage approach to determine the value of key metrics which can be used in rules-based strategy. The first stage solves a 2-step optimisation model to determine the optimal charging and discharging electricity price from previous historical data, and the second stage simulates, in real-time, the battery actions based on the price rules initially created. This proposed method was applied to a microgrid with local load and PV power generation with access to the UK Day-Ahead energy market, with results showing an improvement in electricity cost savings across board when compared with the more popular time-based rules dispatch strategy.

Keywords: Energy storage dispatch, Energy arbitrage, Optimisation

1. Introduction

In electrical power systems, electrical energy storage (EES) devices have been shown to improve power reliability, flexibility, and quality, and reduce electricity bills in front-of-meter and/or behind-the-meter applications, especially with the increased penetration of intermittent renewable energy (RE) generators (Ma et al., 2018). Owing to problems of a prolonged pay-back period in large scale deployment of these devices, quite a number of research efforts have been focused on revenue stacking – where a collection of tasks are performed by the EES device in order to generate more revenue (Roberts and Brown, 2020). In such cases, the scheduling of the EES device plays an important role both in the total revenue generated as well as the lifetime of the device.

A great deal of research has thus focused on optimising the operation of such devices under differing conditions of energy demand, energy generation sources, electricity prices and/or accessible revenue streams (García Vera et al., 2019). Hannan et al. (2020) presented a review on existing optimisation methods/algorithms, amongst others, for EES sizing and scheduling in microgrid (MG) applications. Zia et al. (2018) also presented a critical review on methods and solutions for energy management systems - generation dispatch, frequency regulation, etc. Despite the accuracy these optimisation-based methodologies reviewed achieve given the underlying assumptions, results obtained are almost always a single set of time-dependent decision variables, which must be followed precisely to obtain the same objective function value considered in the model. In reality, these solutions require rigid implementation, and do not always perform well when conditions of energy demand, price, generation, or any other adopted data fall far from assumptions or predictions, even with uncertainty considerations. An alternative and more flexible approach towards energy dispatch scheduling involves using a rules-based approach. The rules-based approach uses a simple algorithm, iterating over each time step to determine the EES devices operating mode and action in real-time. Key metrics or system parameters, for example, the threshold prices or times during a day, which characterise the optimal operation of the EES device and thus specify periods for certain battery action (charging/discharging), are examples of rules-based strategies. Kanwar et al. (2015) compared optimisation and rules-based strategies for a MG on a time-of-use (ToU) tariff, with the former strategy obtaining just a 2% increase in savings with additional computational complexities. These rules-based methods also present an easy-to-implement strategy for EES device owners and are applicable over a wider range of system variability. They do however, have the drawback of lacking a guarantee on cost optimality. An ideal strategy will therefore comprise a combination of the cost optimality merits of the optimisation-based strategy and the ease-of-implementation and flexibility of the rules-based approach. Zhang et al. (2017) proposed rules-based strategies with some metrics determined via a linear programming (LP) problem to increase the accuracy of the rules generated. These LP problems were, however, solved at each period of operation.

To this end, we present a 2-step optimisation-based approach to obtain key metrics for use in a rules-based energy storage dispatch strategy. The key metrics to be identified in this work include the minimum/maximum prices, and thus the corresponding times, to charge/discharge an EES device in real-time. Kanwar et al. (2015) adopted a ToU tariff with known electricity price values per time which may not be applicable to wholesale market trading. Our proposed strategy further seeks to evaluate the threshold prices for rules-based battery operation. The first step solves an optimal energy dispatch optimisation model to obtain a set of distinct optimal solutions, and the second is a feature extraction stage which finds the optimal price range where charge and discharge actions are executed so as to minimise the total electricity cost of the MG. The approach is applied to a MG with an EES asset having access to the UK day-ahead energy market. Given a load demand which it must satisfy, this approach proffers the optimal price range (and times) for charging and discharging the EES device to minimise the electricity cost.

In the rest of the paper, the proposed strategy and associated mathematical formulation of the mixed integer linear programming (MILP) models are described in section 2. The strategy is then applied to a case study is section 3., with findings discussed and some conclusions drawn in section 4.

2.   Methodology

Figure 1 gives a flow diagram of the proposed strategy. The first stage involves two optimisation models. The first model (an MILP scheduling model - MILP_SH) minimises the total electricity cost of the MG. The second model (an MILP selection model - MILP_SL) determines similar battery actions from a set of distinct optimal solutions generated from MILP_SH, by maximising the number of similar battery actions.

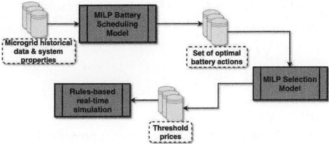

Figure 1: Proposed rules-based strategy

2.1.   Problem description & assumptions

The problems solved by these models are described as follows. For the MILP scheduling model (MILP_SH), *given* a microgrid (MG) with known local electricity demands over time (D_t^B), local photovoltaic (PV) power generation (G_t), and an EES device (battery) with a known initial state of charge (SOC, SOC^0), maximum power output (P^{max}), safety capacity ranges (SOC^{min}, SOC^{max}) and charge (η^C) and discharge (η^D) efficiencies; with direct access to the UK Day-Ahead (DA) markets with known historical buy (ρ_t^I) and sell (ρ_t^E) prices;
Determine the optimal schedule of the EES device - charging and discharging times, power (P_t^{s+}, P_t^{s-}) and SOC (SOC_t) - *so as to* minimise the total cost of electricity for the MG.
For the MILP selection model (MILP_SL), *given* a set of differing optimal solutions from the MILP_SH model, and their corresponding battery charging and discharging actions, *determine* the set of common battery charging and discharging actions per time amongst a subset of optimal solutions, *so as to* maximise the total number of similar battery actions.
The corresponding buy and sell electricity prices associated with the solution of MILP_SL model provides the threshold price ranges for charging and discharging the ESS which are used in a price-based rules strategy. It is assumed that the MG acts as a price taker, with its electricity demand not having an impact on the DA market prices.

2.2. Mathematical Formulation

The objective of the proposed MILP_SH model given by eq. (1) is to minimise the total electricity cost subject to eqs. (2) - (9).

$$min \quad \sum_t \Delta(\rho_t^I \cdot P_t^I - \rho_t^E \cdot P_t^E) \tag{1}$$

subject to:

$$SOC^{min} \leq SOC_t \leq SOC^{max} \qquad \forall\, t \tag{2}$$

$$SOC_t = SOC^0 \mid_{t=0} + SOC_{t-1} \mid_{t>0} + \Delta \cdot (\eta^C P_t^{s+} - \frac{P_t^{s-}}{\eta^D}) \qquad \forall\, t \tag{3}$$

$$P_t^I - P_t^E = \frac{D_t^B}{\Delta} + P_t^{s+} - P_t^{s-} - \frac{G_t}{\Delta} \qquad \forall\, t \tag{4}$$

$$P_t^I \leq M \cdot (1 - B_t^N) \qquad \forall\, t \tag{5}$$

$$P_t^E \leq M \cdot B_t^N \qquad \forall\, t \tag{6}$$

$$B_t^C + B_t^D \leq 1 \qquad \forall\, t \tag{7}$$

$$P_t^{s+} \leq P^{max} \cdot B_t^C \qquad \forall\, t \tag{8}$$

$$P_t^{s-} \leq P^{max} \cdot B_t^D \qquad \forall\, t \tag{9}$$

Eq. (2) ensures that the SOC of the battery is within predefined safety limits. The SOC at time t is evaluated using eq. (3) as the net charging/discharging action with respect to its initial SOC, where Δ represents the time step. The total power imported/exported is determined by eq. (4) as the local energy demand less the amount of solar and battery energy generated. At any given time, energy may only be imported or exported (eqs. (5) and (6); $B_t^N \in \{0,1\}$) to/from the MG. Finally, the battery cannot simultaneously charge and discharge (eq. (7); $B_t^C, B_t^D \in \{0,1\}$), and its output power must not exceed its rating (eqs.. (8) - (9)).

Distinct optimal (& near-optimal) solutions for MILP_SH are obtained by including the integer cut given by eq. (10) where τ is an index denoting the saved optimal solution in question, U_τ^C and U_τ^D represent the set of time periods in which B_t^C and B_t^D equal 1 respectively; L_τ^C and L_τ^D represent the set of time periods in which B_t^C and B_t^D equal 0 respectively; and σ is a positive integer which represents the degree of variation between generated optimal solutions. MILP_SH model with eq. (10) included is solved repeatedly, updating the sets $U_\tau^C, U_\tau^D, L_\tau^C, L_\tau^D$ in order to generate a set of distinct optimal solutions, $\mathcal{T}$.

$$\sum_{t\in U_\tau^C} B_t^C + \sum_{t\in U_\tau^D} B_t^D - \sum_{t\in L_\tau^C} B_t^C - \sum_{t\in L_\tau^D} B_t^D \leq |\, U_\tau^C \,| + |\, U_\tau^D \,| - \sigma \qquad \forall\, \tau \in \mathcal{T} \tag{10}$$

The objective of the proposed MILP_SL model given by eq. (11) is to maximise the total number of similar battery actions (charging, β_t^C, and discharging, β_t^D) amongst distinct optimal solutions subject to eqs. (12) - (14).

$$max \quad \sum_t \beta_t^C + \beta_t^D \tag{11}$$

subject to:

$$\sum_\tau \overline{B}_{t\tau}^C + |\, \mathcal{T} \,| (1 - \beta_t^C) \geq \psi \qquad \forall\, t \tag{12}$$

$$\sum_{\tau} \overline{B}_{t\tau}^{D} + \mid \mathcal{T} \mid (1 - \beta_t^D) \geq \psi \qquad \forall \, t \tag{13}$$

$$\beta_t^C + \beta_t^D \leq 1 \qquad \forall \, t \tag{14}$$

$\overline{B}_{t\tau}^{C}$ and $\overline{B}_{t\tau}^{D}$ denote the saved battery charging and discharging actions for each generated optimal solution τ at time t respectively. Eqs. (12) - (14) ensure that a battery action is evaluated as similar only if it occurs in some predefined minimum number of distinct optimal solutions ψ. Eq. (14) is a feasibility constraint that ensures that whatever similar battery actions are selected by the model, the battery still does not charge and discharge at the same time period. The threshold price range for charging and discharging is then obtain by matching the buy and sell electricity prices with the optimal values of β_t^C and β_t^D for each time period respectively.

3. Case study

For a case study, a MG with local load, a 3kW PV cell and a 25kWh, 16.7kW Li-ion battery with a round-trip efficiency of 81% was considered. The MG has direct access to the UK DA energy market to purchase/sell electricity. Local electricity demand was generated from ELEXON's 10-year average for 2018 and 2019. Each of the proposed optimisation models were solved using Pyomo 5.6.8 with Gurobi 9.0 to a 0% relative gap using an Intel Xeon E-2146G with 32GB and 4 threads running Windows 10.

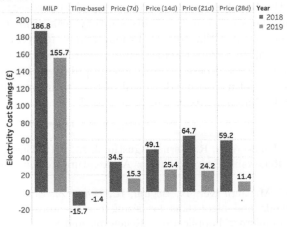

Figure 2: Annual electricity cost savings

Six scenarios were explored to demonstrate the impact of the proposed strategy. First, the MILP_SH model was solved as is using the historical data to provide a basis for the maximum electricity savings possible for the MG assuming perfect foresight (MILP). A rules-based approach based on set times of charging and discharging the battery in any day (Time-based) was simulated to reflect a popular approach used by asset owners. Finally, using the proposed 2-stage approach four additional scenarios were solved differing in how many historical days were used by models MILP_SH and MILP_SL to obtain the threshold prices for the price-based rules approach. "Price (7d)" thus consisted of a scenario where for each 7-day period simulated in a year using the price-based rules strategy, the threshold prices were generated using the previous 7-days load, PV and energy prices data. In each of the rules-based strategies, the algorithm was implemented over a 7-day period. Figure 2 shows the electricity savings for each of the scenarios solved for the years 2018 and 2019 respectively. In both years, results show that the time-based approach (charging at night and discharging at peak hours in the afternoon) made a loss when compared to scenarios without any battery installed. All cases using the proposed rules-based strategy obtained electricity savings in varying degrees. The maximum electricity cost savings were however obtained from generating threshold prices using 14 or 21 days of historical data.

4. Conclusions

In this work, a flexibile energy storage dispatch strategy was proposed for DA market trading by a MG. The strategy consisted of a two-stage approach. In the first stage, threshold buy and sell prices required for real-time simulation of battery actions were obtained by generating multiple optimal solutions for an MILP scheduling model which minimises the total electricity cost of the MG. The MG was assumed to have a local load, PV generation, and an installed battery with known capacity with access to the UK DA market. Next, an MILP selection model, is used to extract the threshold prices by maximising the total number of similar battery actions from the set of optimal solutions. The first stage thus uses historical data to generate the threshold prices which are used in a real-time simulation of battery actions in the second stage. This strategy was applied to a MG with a 3kWp PV system and 25kWh, 16.7kW Li-ion battery for years 2018 and 2019. Results showed that up to 34% of the maximum possible electricity cost saving was captured, exceeding the time-based simulations popularly adopted which made a loss in both years. As future work, additional metrics to improve the performance of the proposed rules-based strategy will be included.

References

García Vera, Y. E., Dufo-López, R., Bernal-Agustín, J. L., 2019. Energy Management in Microgrids with Renewable Energy Sources: A Literature Review. Applied Sciences 9, 3854.

Hannan, M., Faisal, M., Jern Ker, P., Begum, R., Dong, Z., Zhang, C., 2020. Review of optimal methods and algorithms for sizing energy storage systems to achieve decarbonization in microgrid applications. Renewable and Sustainable Energy Reviews 131, 110022.

Kanwar, A., Hidalgo Rodriguez, D. I., Von Appen, J., Braun, M., 2015. A comparative study of optimization and rule-based control for microgrid operation. Power and Energy Student Summit, 1–6.

Roberts, D., Brown, S. F., 2020. Identifying calendar-correlated day-ahead price profile clusters for enhanced energy storage scheduling. Energy Reports 6, 35–42.

Zhang, Y., Lundblad, A., Campana, P. E., Benavente, F., Yan, J., 2017. Battery sizing and rule-based operation of grid-connected photovoltaic-battery system: A case study in Sweden. Energy Conversion and Management 133, 249–263.

Zia, M. F., Elbouchikhi, E., Benbouzid, M., 2018. Microgrids energy management systems: A critical review on methods, solutions, and prospects. Applied Energy 222, 1033–1055.

Proceedings of the 14th International Symposium on Process Systems Engineering – PSE 2021+
June 19-23, 2022, Kyoto, Japan © 2022 Elsevier B.V. All rights reserved.
http://dx.doi.org/10.1016/B978-0-323-85159-6.50327-4

Monetizing Flexibility in Day-Ahead and Continuous Intraday Electricity Markets

Niklas Nolzen[ab], Alissa Ganter[ab], Nils Baumgärtner[d], Ludger Leenders[a], André Bardow[abc*]

[a]*Energy & Process Systems Engineering, Department of Mechanical and Process Engineering, ETH Zurich, Tannenstrasse 3, 8092 Zurich, Switzerland*
[b]*Institute of Technical Thermodynamics, RWTH Aachen University, Schinkelstraße 8, 52062 Aachen, Germany*
[c]*Institute of Energy and Climate Research - Energy Systems Engineering (IEK-10), Forschungszentrum Jülich GmbH, Wilhelm-Johnen-Straße, 52425 Jülich, Germany*
[d]*Energy Trading and Dispatching, Currenta GmbH & Co. OHG, Kaiser-Wilhelm-Allee 80, 51373 Leverkusen, Germany*
abardow@ethz.ch

Abstract

The rising share of renewable energies increases supply uncertainty in the energy system. To make short-term adjustments more cost-efficient, the continuous intraday market was introduced. The continuous intraday market allows flexible capacity to exploit the price volatilities by asset-backed trading. In asset-backed trading, flexible capacity is continuously traded depending on the real-time electricity price and the marginal cost for electricity production. However, the flexibility for the continuous intraday market needs already be considered during the commitment on the day-ahead market. Hence, this paper proposes an optimal joint bidding strategy for day-ahead and continuous intraday market participation. For this purpose, we employ option-price theory and stochastic optimization. A case study for a flexible multi-energy system shows savings of 11 % by participating in both markets compared to only the day-ahead market. Thus, the bidding strategy provides efficient decision support in short-term electricity markets.

Keywords: electricity markets, stochastic optimization, optimal bidding strategy, spot markets, continuous trading

1. Introduction

The expansion of renewable energies increases supply uncertainty in the electricity grid. To make short-term adjustments in the grid more economical, several European countries introduced the continuous intraday market that settles imbalances with continuous real-time trading (Koch and Hirth, 2019). The continuous intraday market is characterized by strong electricity price volatilities driven by updates in the renewables forecast (Kremer et al., 2020). These price volatilities can be monetized by asset-backed trading (Löhndorf and Wozabal, 2021). Asset-backed trading uses flexible capacity. Its electricity output is continuously traded depending on the current electricity price and the marginal cost for electricity production. When electricity prices rise above marginal cost, electricity from flexible capacity is sold. When prices fall below marginal cost, electricity is purchased on the market. Hence, asset-backed trading is particularly valuable when the electricity price fluctuates around the marginal costs. Eventually, the sum of all trades determines the actual electricity output of the flexible capacity.

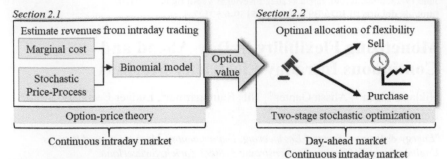

Figure 1: Method for an optimal bidding strategy in the day-ahead and continuous intraday market. First, the option value derives the revenue from trading flexible capacity in the continuous intraday market. Second, a two-stage stochastic optimization models the sequential decision-making process to optimally allocate flexibility in both markets.

Participation in the day-ahead and continuous intraday market is a sequential decision-making process. The day-ahead market clears first. Afterward, trading in the continuous intraday market is possible until shortly before delivery. An optimal bidding strategy considers both markets simultaneously to determine an optimal amount of flexible capacity for asset-backed trading. However, before day-ahead market clearing, only limited information is available on the continuous intraday market.

Due to its recent introduction, only a few studies investigate the continuous intraday market as a trading opportunity. Garnier and Madlener (2015) propose a bidding strategy for renewable energies participating in the continuous intraday market. Based on a multi-period lattice, intraday market participation is simulated using dynamic programming. Corinaldesi et al. (2020) analyze the flexibility of end-user technologies in the day-ahead and continuous intraday market. Intraday market trading is considered with an hourly updated rolling horizon with new price forecasts. Additional intraday market participation saves around 8 %. However, the rolling horizon limits the interaction between both markets as only the first optimization considers the day-ahead market participation.

This paper proposes a method that considers trading in the continuous intraday market in a multi-market optimization. The method determines an optimal bidding strategy for a flexible energy system in the day-ahead and continuous intraday market.

2. Method for optimal bidding strategies in day-ahead and intraday markets

The method allocates the flexible capacity of a market participant to the day-ahead and continuous intraday market to determine an optimal bidding strategy (Figure 1). In sequential decision making, the value of the intraday market opportunity needs to be considered while deciding on the day-ahead market participation. In Section 2.1, the value of trading in the continuous intraday market is derived based on forecast data using option-price theory (Björk, 2009). In Section 2.2, a two-stage stochastic optimization optimizes flexible capacity allocation to the day-ahead and continuous intraday market.

2.1. Deriving the option value for the continuous intraday market

Here, we propose to estimate the revenues from trading in the continuous intraday market with the option value. At the time of day-ahead market clearing, the joint market participation requires knowledge on the revenues in the continuous intraday market. The option-price theory allows estimating the revenues based on parameters available one day before delivery, i.e., at the time of commitment to the day-ahead market.

The option value is derived with the multiperiod binomial model based on Cox et al. (1979). In the multiperiod binomial model (Figure 2), the trader adjusts the

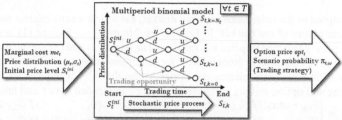

Figure 2: Calculation of revenues from trading in the continuous intraday market using the multiperiod binomial model from option-price theory. Therein, the risk-neutral trading strategy can be derived at each trading opportunity to replicate the option value for each traded hour $t \in T$.

purchased/sold share of flexible capacity at each trading opportunity to realize the estimated option value within a trading session. This adjustment is based on the stochastic price process and the marginal costs for electricity production, assuming a risk-neutral trader. Hence, we refer to this strategy as the risk-neutral asset-backed trading strategy. Using this strategy, option price theory assumes that the option value is realized independent of the price scenario. Therefore, the price volatility in the continuous intraday market is monetized without any financial risk. Therein, we assume that the stochastic price process sufficiently captures the price volatility.

For the sake of simplicity, we consider only trading hourly electricity contracts in the continuous intraday market, thus neglecting other trading opportunities such as half-hourly and quarter-hourly products. The trading session for each traded hour $t \in T$ starts after clearing the day-ahead market and ends shortly before delivery. In each trading session, we discretize continuous trading with N_t trading opportunities. At each trading opportunity, the adjustment of the trading position follows the risk-neutral asset-backed trading strategy. Furthermore, the marginal costs mc_t are known for the flexible capacity. Each trading session starts with the forecasted initial price-level S_t^{ini}. Then, the electricity price follows a stochastic price process modeled as arithmetic Brownian motion (Alexander et al., 2012). In contrast to the often used geometric Brownian motion, the arithmetic Brownian motion models the absolute price change (in €/MWh). Hence, the stochastic price process can also lead to negative electricity prices. Assuming the arithmetic Brownian motion throughout a trading session, the price moves up u_t or down d_t at each trading opportunity. The up-movement u_t and down-movement d_t are determined as follows:

$$u_t = \mu_t \cdot \frac{1}{N_t} + \sigma_t \cdot \sqrt{\frac{1}{N_t}} \quad \text{and} \quad d_t = \mu_t \cdot \frac{1}{N_t} - \sigma_t \cdot \sqrt{\frac{1}{N_t}} \ \forall t \in T, \tag{1}$$

by multiplying the price drift μ_t and the price volatility σ_t with the trading frequency $\frac{1}{N_t}$. Hence, the price drift μ_t and the price volatility σ_t are allocated over the trading session. At the end of a trading session, the last electricity price $S_{t,k}$ deviates from the initial price-level S_t^{ini}. In Eq. (1), we assume that the absolute price deviation from the initial price-level S_t^{ini} is normally distributed with $\mathcal{N}(\mu_t, \sigma_t^2)$. The arithmetic Brownian motion determines the last price as the summation of up-movements and down-movements. In the stochastic price process, $N_t + 1$ last prices $S_{t,k}$ are possible for each traded hour:

$$S_{t,k} = S_t^{ini} + k \cdot u_t + (N_t - k) \cdot d_t \ \forall t \in T, k \in \{0,1,\dots,N_t\}. \tag{2}$$

The option value $opt_t^{sell/pu}$ for positive (sell) and negative (pu) flexible capacity for each traded hour t is derived with Eq. (3) for the multiperiod binomial model:

$$opt_t^{sell/pu} = \sum_{k=0}^{N_t} \underbrace{\binom{N_t}{k}}_{(1)} \cdot \underbrace{\left(\frac{-d_t}{u_t - d_t}\right)^k}_{(2)} \cdot \underbrace{\left(\frac{u_t}{u_t - d_t}\right)^{N_t - k}}_{(3)} \cdot \underbrace{\Phi^{sell/pu}(S_{t,k})}_{(4)} \ \forall t \in T. \tag{3}$$

Eq. (3) is adapted to the continuous intraday market, assuming a zero interest rate due to the short-term nature of the market. In Eq. (3), the binomial coefficient (1) derives the absolute frequency that the last price k is reached. The terms (2) and (3) are the martingale measures, i.e., the risk-neutral probabilities, for an up-movement and down-movement, respectively. The term (4) evaluates the option value $\Phi^{sell/pu}$ at the end of a trading session for the sell option and the purchase option for each traded hour t and last price k:

$$\Phi^{sell}(S_{t,k}) = \begin{cases} S_{t,k} - mc_t, & if\ S_{t,k} > mc_t \\ 0, & if\ S_{t,k} \leq mc_t \end{cases} \text{ and } \Phi^{pu}(S_{t,k}) = \begin{cases} 0, & if\ S_{t,k} \geq mc_t \\ mc_t - S_{t,k}, & if\ S_{t,k} < mc_t \end{cases}.$$

In summary, the option value sums up and weights the revenues from all last prices. Therein, the scenario probabilities $\pi_{t,\omega}$ for utilizing positive and negative flexibility are obtained from the stochastic price process and the marginal cost. The derived option value $opt_t^{sell/pu}$ accounts for the revenues from asset-backed trading. Both parameters are the input parameters for the two-stage stochastic optimization presented in the following.

2.2. Two-stage stochastic optimization of day-ahead and intraday market participation

The two-stage stochastic optimization derives an optimal bidding strategy in the day-ahead and continuous intraday market. Hence, at the 1st stage, decisions are made to buy or sell electricity in the day-ahead market. Moreover, flexible positive capacity (sell option) and negative capacity (purchase option) are blocked for trading in the continuous intraday market. At the 2nd stage, trading in the continuous intraday market ends, and the energy system operation is adapted according to intraday market trading. The 2nd stage considers two scenarios $\omega \in \Omega = \{S < mc, S > mc\}$ that depend on the marginal costs for electricity production mc and the electricity price S at the end of a trading session. In the scenarios, we assume that trading flexibility in the continuous intraday market utilizes the blocked positive and negative flexibility.

The objective function of the two-stage stochastic optimization minimizes the expected costs consisting of operational costs $C_{t,\omega}^{op}$ subtracting the revenues from the day-ahead market R_t^{DA} and the continuous intraday market $R_{t,\omega}^{ID}$:

$$min \sum_{t \in T} \sum_{\omega \in \Omega} \pi_{t,\omega} \cdot \left(C_{t,\omega}^{op} - R_t^{DA} - R_{t,\omega}^{ID} \right) \tag{4}$$

$$s.t.\ energy\ balances\ \forall\ \omega \in \Omega, t \in T, e \in energy\ forms \tag{5}$$

$$market\ constraints\ \forall \omega \in \Omega, t \in T, market \in \{DA, ID\} \tag{6}$$

$$technical\ constraints\ \forall \omega \in \Omega, t \in T, u \in units \tag{7}$$

Therein, the costs and revenues are weighted with the scenario probability $\pi_{t,\omega}$. Furthermore, the energy system is operated to fulfill the product balances and to comply with the constraints set by market participation. In both scenarios $\omega \in \Omega$, trading flexible capacity $ID^{sell/pu}$ in the intraday market is reimbursed with the respective option value for positive and negative flexibility. Hence, the revenues $R_{t,\omega}^{ID}$ can be expressed with

$$R_{t,\omega}^{ID} = ID_{t,\omega}^{sell}(opt_t^{sell} + s_\omega^{sell} \cdot mc_t) + ID_{t,\omega}^{pu}(opt_t^{pu} - s_\omega^{pu} \cdot mc_t)\ \forall t \in T, \omega \in \Omega, \tag{8}$$

whereas s_ω^{sell} (s_ω^{pu}) is a binary parameter that is 1 if electricity is sold (purchased) in the continuous intraday market or 0 if no electricity is sold (purchased) in the continuous intraday market. By following the risk-neutral asset-backed trading strategy, the option value for positive and negative flexibility $opt_t^{sell/pu}$ is realized independently from the last price in the intraday market. Hence, the arising (saved) endogenous generation costs are compensated in Eq. (8) for the sell (purchase) option in scenario $S > mc$ ($S < mc$). Finally, the stochastic optimization models trading in the continuous intraday market as a series of sell and purchase options, whereas the sell option (purchase option) models the positive (negative) flexibility of the energy system.

3. Case study: market participation of a multi-energy system

The method is applied to a multi-energy system based on Baumgärtner et al. (2019). The multi-energy system consists of 4 combined-heat-and-power engines, 2 electrode boilers, 3 adsorption chillers, 3 compression chillers, and 4 gas boilers. The multi-energy system covers time-varying demands for electricity, heating, and cooling. The multi-energy system participates in the day-ahead and continuous intraday market in Germany.

The case study is conducted for Wednesday, October 9th, 2019. The day is chosen because the average day-ahead market price is in the same range as the marginal costs. Hence, asset-backed trading in the continuous intraday market is particularly interesting.

The day-ahead market is assumed to be deterministic with time-varying electricity prices taken from Bundesnetzagentur | SMARD.de (2021). For the calculation of the option value, the stochastic price process and the marginal costs are modeled as follows: the day-ahead market price approximates the initial price-level S_t^{ini}. The price drift μ_t and the price volatilities σ_t are derived using historical data from the years 2019 and 2020 from EPEX SPOT based on the ID3 price (EPEX SPOT, 2021). For each traded hour of 2019 and 2020, the price deviation between the continuous intraday market and the day-ahead market is calculated as the difference between ID3 price and day-ahead market price. All price deviations are clustered based on the hourly wind generation forecast, solar generation forecast, and residual load forecast available at ENTSO-E (2021). In each cluster, a normal distribution is fitted to obtain the price drift μ and the price volatility σ. Finally, each traded hour is matched with the respective cluster and u_t and d_t are derived based on Eq. (1). The marginal costs mc_t are derived from operational optimizations with varying electricity demands but without market participation for each traded hour. A linear regression is applied to determine the operational costs as a function of the electricity demands. The slopes are the marginal costs mc_t.

Two cases are compared for market participation: participation only in the day-ahead market (DA) and participation in the day-ahead and continuous intraday market (DA, ID). Both cases are solved to optimality with the solver Gurobi 9.1.1 in less than 5 minutes. Our method lowers expected operational expenditures by 11 % by intraday market participation (DA, ID) compared to only participating in the day-ahead market (DA).

Figure 3 shows the electricity exchange with the day-ahead and continuous intraday market. In case (DA), electricity is sold if the electricity price is higher than the marginal costs and purchased vice versa. In case (DA, ID), the commitment in the day-ahead market differs from the case (DA). For some traded hours, no electricity is purchased or sold in the day-ahead market. Thereby, the multi-energy system uses its positive and negative flexibility in the continuous intraday market. For the remaining traded hours, selling electricity on the day-ahead market maximizes negative flexibility. Overall, the flexibility is used for asset-backed trading in the continuous intraday market, while the

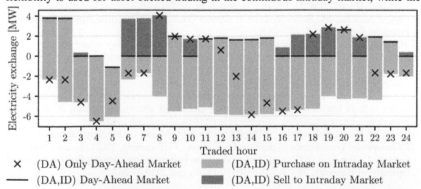

Figure 3: Electricity exchange with the day-ahead and continuous intraday market. In case (DA), electricity is only delivered to the day-ahead market. In case (DA, ID), two scenarios arise as the exchange of electricity in the continuous intraday market depends on the last price in the continuous intraday market S compared to the marginal costs mc.

last price in the continuous intraday market determines the overall physical delivery of electricity. Hence, utilizing the flexibility of the multi-energy system lowers operational expenditures. Finally, adapting the operation of the multi-energy system realizes the flexibility provision while still satisfying the internal heating, cooling, and electricity demand.

Hence, participation in the continuous intraday market allows marketing the flexibility of the multi-energy system. Overall, the method shows the ability to optimally deploy the flexibility of an energy system in the markets.

4. Conclusions

This paper presents a method based on option-price theory and two-stage stochastic optimization to derive an optimal bidding strategy for the day-ahead and continuous intraday market. The option-price theory derives the value of trading in the continuous intraday market based on forecast data. Afterward, the estimated option value serves as the input parameter for the two-stage stochastic optimization. The two-stage stochastic optimization models the sequential bidding process of joint day-ahead and continuous intraday market participation. The proposed method is applied to a case study of a multi-energy system. In this case study, savings of 11 % are expected by the proposed method for a selected day. Overall, the method is an efficient decision-making tool for operational optimization one day ahead of delivery, incorporating the complex market structure of the day-ahead and continuous intraday market.

Acknowledgements

This study is funded by the German Federal Ministry of Economic Affairs and Energy (ref. no.: 03EI1015A). The support is gratefully acknowledged.

References

D.R. Alexander, M. Mo, A.F. Stent, 2012, Arithmetic Brownian motion and real options, European Journal of Operational Research, 219, 1, 114–122, 10.1016/j.ejor.2011.12.023.

N. Baumgärtner, R. Delorme, M. Hennen, A. Bardow, 2019, Design of low-carbon utility systems: Exploiting time-dependent grid emissions for climate-friendly demand-side management, Applied Energy, 247, 2, 755–765, 10.1016/j.apenergy.2019.04.029.

T. Björk, 2009, Arbitrage theory in continuous time, 3rd ed., Oxford University Press, Oxford.

Bundesnetzagentur | SMARD.de, 2021, Marktdaten, https://www.smard.de/home/downloadcenter/download-marktdaten.

C. Corinaldesi, D. Schwabeneder, G. Lettner, H. Auer, 2020, A rolling horizon approach for real-time trading and portfolio optimization of end-user flexibilities, Sustainable Energy, Grids and Networks, 24, 01, 100392, 10.1016/j.segan.2020.100392.

J.C. Cox, S.A. Ross, M. Rubinstein, 1979, Option Pricing: A Simplified Approach, Journal of Financial Economics, 7, 229–263.

ENTSO-E, 2021, ENTSO-E: Transparency Platform, https://transparency.entsoe.eu/.

EPEX SPOT, 2021, Data of the ID3 index of the continuous intraday market.

E. Garnier, R. Madlener, 2015, Balancing forecast errors in continuous-trade intraday markets, Energy Syst, 6, 3, 361–388, 10.1007/s12667-015-0143-y.

C. Koch, L. Hirth, 2019, Short-term electricity trading for system balancing: An empirical analysis of the role of intraday trading in balancing Germany's electricity system, Renewable and Sustainable Energy Reviews, 113, 109275, 10.1016/j.rser.2019.109275.

M. Kremer, R. Kiesel, F. Paraschiv, 2020, An Econometric Model for Intraday Electricity Trading, Philosophical Transactions of the Royal Society A., Forthcoming, 10.2139/ssrn.3489214.

N. Löhndorf, D. Wozabal, 2021, Gas storage valuation in incomplete markets, European Journal of Operational Research, 288, 1, 318–330, 10.1016/j.ejor.2020.05.044.

Proceedings of the 14th International Symposium on Process Systems Engineering – PSE 2021+
June 19-23, 2022, Kyoto, Japan © 2022 Elsevier B.V. All rights reserved.
http://dx.doi.org/10.1016/B978-0-323-85159-6.50328-6

Planetary boundaries analysis of Fischer-Tropsch Diesel for decarbonizing heavy-duty transport

Margarita A. Charalambous[a], Juan D. Medrano-Garcia[a], Gonzalo Guillén-Gosálbez[a]*

[a]*Institute for Chemical and Bioengineering, Department of Chemistry and Applied Biosciences, ETH Zürich, Vladimir-Prelog-Weg 1, 8093 Zürich, Switzerland*
*gonzalo.guillen.gosalbez@chem.ethz.ch

Abstract

Here we evaluated Fischer Tropsch-diesel (FT-diesel) use in heavy-duty trucks based on various production pathways differing in the CO_2 and H_2 provenance. To better understand the global environmental implications of fuelling heavy-duty trucks (HD trucks) with FT-diesel, we quantified environmental impacts over the entire life cycle using seven Planetary Boundaries (PBs) regulating the Earth's resilience. Our environmental assessment follows a well-to-wheel scope with the functional unit based on the global annual freight demand. The baseline scenario corresponds to the conventional fossil fuel. Our results show that the fossil fuel alternative is unsustainable as it transgresses the climate change PBs. Using FT-diesel based on captured CO_2 could help operate within the safe operating space but it could induce critical burden-shifting if the CO_2 and H_2 sources are not adequately selected.
Keywords: Energy, Food and Environmental Systems

1. Introduction

In recent years, liquid fuels based on renewable carbon that can substitute conventional ones with minimal changes to current infrastructure have attracted increasing interest. Notably, fossil diesel can be replaced with "drop-in" fuels with similar or better characteristics. FT-diesel is a promising alternative to fossil diesel due to the high cetane number and improved properties with the potential to optimize combustion efficiency and decrease emissions.

So far, studies related to FT-diesel have focused mainly on the production process based on biomass as a raw material (Martín and Grossmann 2011), and very few on the CO_2-based production process (Al-Yaeeshi et al. 2019). Environmental assessments of FT-diesel often quantify impacts based on conventional life cycle assessment (LCA) metrics (Holmgren and Hagberg 2009; Wernet et al. 2016), which are hard to interpret due to the absence of thresholds that can classify the studied systems as environmentally unsustainable. Hence, the absolut environmental sustainability implications of this fuel remain unclear.

Here we evaluated FT-diesel use in HD trucks based on various production pathways differing in the provenance of the raw material using seven PBs. The PBs concepts developed initially by Rockström et al., 2009, provides a framework to carry out absolute environmental sustainability assessments considering the Earth's carrying capacity. The framework considers 11 control variables linked to nine Earths' biophysical subsystems or processes. These include climate change, stratospheric ozone depletion, ocean acidification, biogeochemical flows of nitrogen and phosphorus, land system change, freshwater use, biosphere integrity, atmosphere aerosol loading, and introduction of novel entities. All the PBs jointly establish the so-called safe operating space (SOS) for humanity. Consequently, for a scenario to be regarded as sustainable, none of the planetary boundaries should be transgressed. In essence, referring the LCA results to the safe operating space (SOS) delimited by these environmental guardrails facilitates the interpretation phase, particularly when evaluating systems that can be potentially deployed at a large scale. Based on this concept, we conducted an absolute environmental assessment over the global annual freight demand (33 trillion tkm), covering eight FT-diesel HD truck scenarios while benchmarking them against the fossil diesel HD truck counterpart –business as usual (BAU scenario)–. Hence, going well beyond standard LCAs, we analyze whether the FT-diesel trucks would help humanity operate safely within the PBs.

2. Methodology

2.1. Life cycle assessment and planetary boundaries

Following the general life cycle assessment (LCA) methodology, we carried out an environmental assessment based on the ISO 14040/44 framework using the SimaPro 9.0 software. The environmental assessment aims to assess the absolute sustainability of fuelling the global freight activities with FT-diesel from various sources. The functional unit corresponds to the global annual tkm demand for on-road HD truck activities, estimated by the International Energy Agency to be around $33 \cdot 10^{13}$ tkm.

For our analysis, we adopted a well-to-wheel scope using an attributional approach. The system boundaries cover all the upstream activities, from the production of H_2 and CO_2, through the FT-diesel synthesis, to the fuel combustion in HD trucks. We estimated the life cycle inventory for FT-diesel by assuming that most of the emissions can be attributed to CO_2 and H_2, as shown in Galán-Martín et al. (2021) for various bulk chemicals.

In order to construct the life cycle inventory, we first simulated the production of FT-diesel. Our calculations are based on the works of Shafer et al. (2019) for the FT-reactor, and Tomasek et al. (2020), for the wax hydrocracker. The Anderson-Shulz-Flory distribution (a) of choice for maximum diesel production is 0.88, and H_2/CO equals 2, based on which we calculated the product distribution of the FT-reactor. To calculate the total CO_2 and H_2 needed for the process, we assume that all the CO is converted into C_1-C_{22} in the FT-reactor. Hydrocracking of the waxes was modelled according to Tomasek et al. (2020). The total production of diesel, gasoline, kerosene, and C_1-C_4, coming from the FT-reactor and the wax hydrocracker are summed in order to get the final products. Furthermore, the light ends are combusted to produce CO_2, which is then recycled to the water-gas shift reactor (WGSR). Since gasoline and kerosene are co-produced, our LCA considers a system expansion approach with avoided burdens. The final inputs of the FT-diesel life cycle inventory are presented in Table 1.

Table 1: Life cycle inventory for the production of 1 kg FT-Diesel from CO_2 and H_2

FT-diesel	1	kg
Avoided products		
Gasoline	0.89	kg
Kerosene	0.72	kg
Inputs		
H_2	0.29	kg
CO_2	1.42	kg

The foreground system, i.e., truck and road constructions, etc., is based on the "Lorry 16-32 metric ton, EURO6" of the Ecoinvent v3.5 database. In essence, we consider the LCI for the BAU scenario, replacing the fossil diesel's inventory with that of FT-diesel, and adjusting the direct emissions based on Schemme et al. (2017). The scenarios differ in the origin of the educts (Figure 1). For this study, CO_2 is captured either from point sources at coal power plants, or directly from air (Coal and DAC). H_2 is produced through an electrolytic or thermochemical route. Polymeric water electrolysis is considered as the electrolytic route, powered by different energy sources, i.e., onshore wind, nuclear, and bioenergy with CCS (BECCS). Furthermore, for the thermochemical route, the conversion of biomass to hydrogen with carbon capture and storage (BTH CCS) is considered.

Data for the production of electrolytic H_2 were taken from Bareiß et al. (2019), considering for wind power a capacity factor of 0.34, respectively. For the different electricity sources, we used data from Ecoinvent v1.03, except for the BECCS scenario, based on Oreggioni et al. (2017). For the thermochemical route, the inventory was retrieved from Susmozas et al. (2016). Concerning the capture of CO_2, the Coal scenario is based on Iribarren, Petrakopoulou, and Dufour (2013), and the DAC scenario on Keith et al. (2018).

The life cycle impact assessment (LCIA) quantifies the absolute environmental sustainability level of FT-diesel by converting the LCI elementary flows into impacts on the control variables of seven PBs. Consequently, for a scenario to be regarded as sustainable, none of the planetary boundaries should be transgressed Our study follows the characterization factors proposed by Ryberg et al. (2018) to quantify the impact on six PBs, together with the ones introduced by Galán-Martín et al. (2021) to evaluate the impact on biosphere integrity.

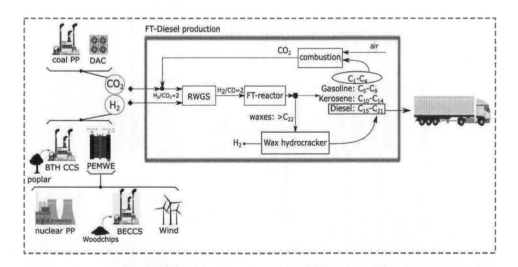

Figure 1: System boundaries of the different scenarios. From the production of CO_2 and H_2 from different sources to the production of FT-Diesel, and lastly, the end-use in HD trucks.

3. Results

3.1. Relative impact to the Safe Operating Space

Figure 2 shows that the current BAU scenario is unsustainable due to the transgressions of the climate change PB (CO_2, EI). Overall, all the FT-diesel scenarios have the potential to decrease the impacts of the BAU scenario. However, only two are sustainable, namely those based on DAC with electrolytic H_2 from nuclear, and H_2 from Biomass with CCS. These scenarios operate within the SOS for all th PBs. On the CO_2 boundary (75%, -140%, respectively), 1% and 2% in nitrogen flows (N-flows), and 7% and 25% in biosphere integrity (BII). All the remaining scenarios fail to be sustainable because they lead to burden-shifting to the N-flows and BII. Focusing on the scenarios with electrolysis routes powered with wind and nuclear electricity (DAC + Wind, DAC + Nuclear, coal + Wind, coal + Nuclear), undoubtedly, DAC scenarios would perform better than those relying on coal. The CO_2 coming from fossil resources was modelled as a positive emissions entry in contrast to the DAC scenario, where CO_2 is coming from the air, and hence is modeled as a negative emissions entry. Ultimately, these scenarios represent an interim solution as fossil fuels should be ultimately phased out. Scenarios that make use of CCS and biomass (DAC + BTH CCS, DAC + BECCS, coal + BTH CCS, coal + BECCS) show a great potential in the GHG-related PBs (CO_2, EI, OA), but lead to burden-shifting due to biomass growth. Notably, the N-flows are affected by the fertilisers, and the BII category by the use of land, e.g., DAC + BECCS and coal + BECCS take 30% of the SOS in the N flows.

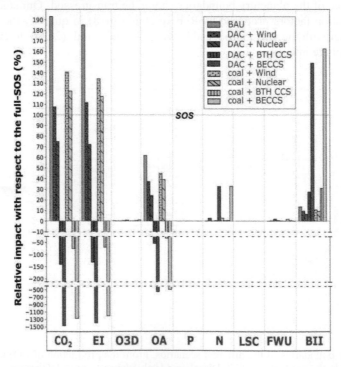

Figure 2: Relative impact in percentage of the safe operating space (SOS). The abbreviations of the PBs are: CO_2 (Climate change CO_2 concentration), EI (Energy imbalance), O3D (Stratospheric ozone depletion), P (Phosphorus flows), N (Nitrogen flows), LSC (Land system chage), FWU (Fresh water use), BII (Biosphere integrity).

3.2. Impact breakdown

The breakdown of impacts in Figure 3 (upper) shows that most carbon-positive impacts come from the combustion emissions, i.e., 45%-85% of the total positive contributions in all the scenarios. Carbon negative impacts come from the biomass based scenarios with CCS (BTH CCS, BECCS), and are linked to the H_2 production. Electrolysis powered with BECCS has the most negative impacts due to the carbon-negative nature of the electricity generated, which requires large amounts of biomass. Furthermore, H_2 from nuclear performs 1.2-fold better in the CO_2 PB compared to H_2 from Wind. Regarding the CO_2 capture technologies, DAC is the only technology that can provide negative impacts since CO_2 is modelled as a negative emission entry. With regard to the biosphere integrity (BII), Figure 3 (lower) shows that the biggest impacts come BTH CCS and BECCS, with the latter being the worst. These high impacts are linked to the extensive land use for biomass growth. All the other scenarios perform better comparing to the BAU, however, it is important to mention that 65% of the carbon positive impacts are coming from the combustion emissions and 25% from the construction of roads.

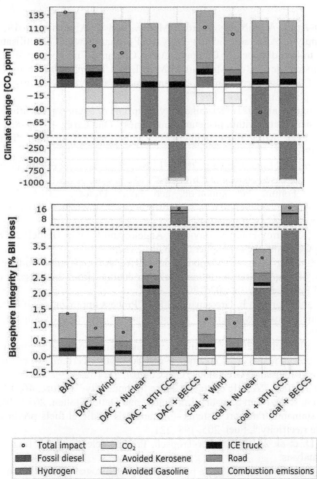

Figure 3: Breakdown of impacts in the CO_2 boundary (upper), nitrogen flows (middle), and biosphere integrity (lower).

4. Conclusions

Using CO_2-based fuels has attracted increasing attention, yet their broad sustainability implications remain unclear. Here we assessed the absolute environmental sustainability of FT-diesel from renewable carbon as an alternative fuel for HD trucks using seven planetary boundaries (PBs). We found that the current fossil-based fuel alternative transgresses the climate change-related PBs, while renewable-carbon fuels could help operate within these ecological limits. However, burden-shifting to BII and N-flows may occur. This collateral damage would be more critical in the biomass-related scenarios, which remove large amounts of CO_2 but need land and fertilisers for biomass growth. Hence, the CO_2 and H_2 sources for producing these fuels should be selected carefully to mitigate climate change without exacerbating the damage in other critical Earth-system processes, thereby preserving the planet's stability.

Acknowledgements: This work was created as part of the NCCR Catalysis, a National Centre of Competence in Research funded by the Swiss National Science Foundation.

References

Al-Yaeeshi, Ali Attiq, A. AlNouss, Gordon McKay, and T. Al-Ansari, 2019, "A model based analysis in applying Anderson–Schulz–Flory (ASF) equation with CO_2 utilisation on the Fischer Tropsch gas-to-liquid process.", Computer Aided Chemical Engineering, 46, 397–402.

Bareiß, Kay, Cristina de la Rua, Maximilian Möckl, and Thomas Hamacher, 2019, "Life cycle assessment of hydrogen from proton exchange membrane water electrolysis in future energy systems.", Applied Energy, 237, 862-72.

Galán-Martín, Ángel et al., 2021, "Sustainability footprints of a renewable carbon transition for the petrochemical sector within planetary boundaries." One Earth, 4, 565–83.

Goedkoop, Mark et al., 2016, "Introduction to LCA with SimaPro Colophon. "Introduction to LCA with SimaPro."

Holmgren, Kristina, and Linus Hagberg, 2009, "Life cycle assessment of climate impact of Fischer-Tropsch diesel based on peat and biomass."

Iribarren, Diego, Fontina Petrakopoulou, and Javier Dufour., 2013, "Environmental and thermodynamic evaluation of CO_2 capture, transport and sstorage with and without enhanced resource recovery." Energy, 50, 477–85.

Keith, David W., Geoffrey Holmes, David St. Angelo, and Kenton Heidel, 2018, "A process for capturing CO_2 from the atmosphere.", Joule, 2,1573–94.

Li, Pengcheng, Zhihong Yuan, and Mario R. Eden, 2016, "A comparative study of Fischer-Tropsch synthesis for liquid transportation fuels production from biomass.", Computer Aided Chemical Engineering, 38, 2025–30.

Martín, Mariano, and Ignacio E. Grossmann. 2011. "Process optimization of FT-diesel production from lignocellulosic switchgrass." Industrial and Engineering Chemistry Research 50(23): 13485–99.

Oreggioni, Gabriel D. et al. 2017. "Environmental assessment of biomass gasification combined heat and power plants with absorptive and adsorptive carbon capture units in Norway." International Journal of Greenhouse Gas Control, 57, 162–72.

Rockström, J. et al. 2009. "A Safe operation space for humanity." Nature, 46, 472–75.

Schemme, Steffen, Remzi Can Samsun, Ralf Peters, and Detlef Stolten. 2017. "Power-to-Fuel as a key to sustainable transport systems – An analysis of diesel fuels produced from CO_2 and renewable rlectricity.", Fuel, 205, 198–221.

Shafer, Wilson D. et al. 2019. "Fischer-Tropsch: Product selectivity-the fingerprint of synthetic fuels." Catalysts.

Susmozas, Ana et al., 2016, "Life-Cycle performance of hydrogen production via indirect biomass gasification with CO_2 Capture." International Journal of Hydrogen Energy, 41, 19484–91.

Szabina Tomasek, Ferenc Lonyi, Jozsef Valyon, Anett Wollmann, Jeno Hancsok., "Hydrocracking of Fischer–Tropsch paraffin mixtures over strong acid bifunctional catalysts to engine Fuels."

Proceedings of the 14th International Symposium on Process Systems Engineering – PSE 2021+
June 19-23, 2022, Kyoto, Japan © 2022 Elsevier B.V. All rights reserved.
http://dx.doi.org/10.1016/B978-0-323-85159-6.50329-8

Renewable Power Systems Transition Planning using a Bottom-Up Multi-Scale Optimization Framework

Ning Zhao[a*], Yanqiu Tao[a], Fengqi You[a]

aCornell University, Ithaca, New York, 14853, USA
nz225@cornell.edu

Abstract

In this work, we propose a novel multi-scale bottom-up optimization framework to address the decarbonization transition planning for power systems, which incorporates multiple types of information for each existing or new unit in the power systems, including its technology, capacity, and age. To reduce the computational challenge, a novel approach integrating Principal Component Analysis (PCA) with clustering techniques is proposed to obtain representative days. To illustrate the applicability of the proposed framework, a case study for New York State was presented. The proposed approach obtaining representative days using PCA coupled with K-means shows better performance than multiple state-of-the-art clustering approaches.

Keywords: decarbonization, renewable electricity transition, multi-scale optimization.

1. Introduction

Power systems decarbonization has been a priority topic for countries around the world (Gong et al., 2015). It facilitates the design of power systems decarbonization transition pathways to simultaneously optimize the systems' capacity changes and simulate the corresponding hourly operations, while considering each individual unit in the power systems (Zhao et al., 2020, 2021). Existing multi-scale energy transition optimization models typically include two time scales on yearly and hourly bases (Bennett et al., 2021). The yearly time scale accounts for the decisions of capacity changes to the power systems, while the operational decisions are made on an hourly basis in conjunction with the design decisions (Brown et al., 2018). To reduce the computational requirements associated with simultaneous planning for the energy transition pathways and simulating the hourly systems operations for the next multiple decades (Prina et al., 2020), the representative day approach has been widely applied in multi-scale energy transition optimization studies (Teichgraeber et al., 2019). Multiple approaches have been used to obtain the representative days, such as rule-based selection, agglomerative hierarchical clustering, and K-means clustering (Gabrielli et al., 2018). On the other hand, most of the existing multi-scale bottom-up energy transition models include only the capacity and technology information of a unit, while including the ages of both existing and future units in the framework is crucial for developing more reliable transition pathways, because existing units with large ages and new units with short facility lifetimes may retire during the transition period of decades owing to the lifespan limits. To the best of our knowledge, there is no existing research work on the multi-scale bottom-up renewable electricity transition optimization that incorporates multiple dimensions of information for each individual unit, including its technology, capacity, and age. To fill the knowledge gap, we propose a multi-scale bottom-up optimization framework that incorporates multiple dimensions of information for a unit as well as a machine learning-based approach to construct the reduced model.

2. Multi-scale bottom-up energy transition optimization framework

The overview of the proposed framework is presented in Figure 1. The proposed framework consists of three steps: (1) data processing based on machine learning, (2) multi-scale transition optimization, and (3) detailed hourly operational simulation.

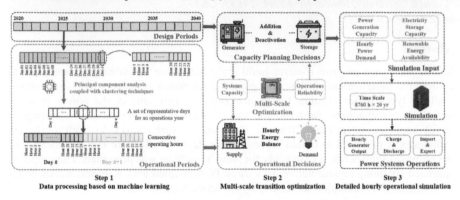

Figure 1. Overview of the proposed multi-scale bottom-up energy transition framework.

In the first step, machine learning techniques, which have been widely used in optimization (Shang et al., 2019), are applied to obtain the representative days based on the power load data from an entire year for developing a reduced optimization model, as it can be extremely computationally demanding for the multi-scale energy transition optimization to account for the yearly capacity planning and the hour-by-hour operational simulation simultaneously for the whole planning horizon of multiple coming decades. Specifically, a novel approach is proposed to obtain the representative days by coupling Principal Component Analysis (PCA) with clustering techniques that include agglomerative hierarchical clustering (AHC), Gaussian mixture model (GMM), Dirichlet process mixture model (DPMM), and K-means clustering. The data being clustered is the 24-dimension hourly power loads for all days in a year. We investigate the performances of using PCA coupled with each clustering approach. The clustering performances are evaluated by three metrics, namely intra-cluster variance, inter-cluster variance, and the Calinski-Harabasz index.

In the second step, multi-scale energy transition optimization is conducted based on a reduced model using representative days that are obtained from the first step. Two time scales are applied in the proposed optimization framework, namely the design periods and the operational periods. The planning horizon is equally partitioned on an annual basis, and the capacity planning decisions that include the additions and deactivations of generators and storage units should be determined for each year of the resulting design periods. On the other hand, to ensure the reliability and energy balance of the deep-decarbonized electric power systems with high penetration of renewable energy, hourly systems operations during the operational periods are incorporated in the proposed optimization model in conjunction with the changes to the electric power sector resulted from the capacity planning decisions.

In the third step, the hourly power systems operations, namely unit comment (Padhy, 2004), are simulated via optimization of each design period, based on the optimal energy transition results from the multi-scale optimization step, hourly power demand projections, and the hourly availability of renewable energy. Specifically, the generation and electricity storage capacities are fixed in the simulation according to the optimal capacity planning decisions in the second step, and the simulation aims to minimize the total operational cost by determining the outputs of each generator, the charging and discharging of energy storage units, and the importation and exportation of electricity on an hourly basis (Qiu et al., 2020), while ensuring the reliability, potential faults (Ajagekar and You, 2021), and balances of the electric power systems.

3. Case study for New York State

A case study on the renewable electricity transition for the New York State is presented to illustrate the applicability of the proposed multi-scale bottom-up optimization framework. The renewable electricity requirements and the climate targets for the New York State are set following the state legislation. The generation and storage capacity data, the annual electricity generation projections, and the scheduled power systems capacity changes for the New York State are obtained based on a report from the New York Independent System Operator (NYISO). In addition, the data on generation capacities for existing distributed solar PV in the state are collected following a study of the New York State Energy Research and Development Authority (NYSERDA). The technological and economic data projections for the power generation and electricity storage technologies are collected from a recent study (Tian and You, 2019). The hourly operations data for the state are obtained from the NYISO energy market and operation data, while the hourly availability of solar, on-land wind, and offshore wind are retrieved from the literature (Ning and You, 2019, 2022).

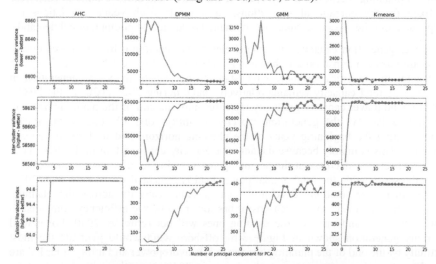

Figure 2. Intra-cluster variance, inter-cluster variance, and the Calinski-Harabasz index using PCA coupled with clustering approaches. Horizontal lines indicate performance without PCA.

To obtain the representative days, we investigate the performances of coupling PCA with multiple clustering approaches that include AHC, DPMM, GMM, and K-mean, as well as the performances using each particular clustering technique without coupling with PCA, as shown in Figure 2. From the performance evaluation results, using PCA coupled with clustering techniques could provide more effective or at least the same clustering results compared with using these techniques individually without PCA. For AHC, all three metrics are not as good as the other types of clustering techniques, regardless of whether it couples with PCA or not, and the best clustering result using AHC coupled with PCA is the same as using AHC alone. On the other hand, coupling PCA with other clustering techniques could improve the data grouping performances compared to using these techniques without PCA, as shown by lower intra-cluster variances, higher inter-cluster variances, and higher Calinski-Harabasz indices when PCA is involved. The improvement is owing to the effectiveness of PCA in capturing the correlations of the high-dimensional input data.

The optimization programs of the energy systems transition problem are coded in GAMS 27.3 on a PC with an Intel Core i7-8700 @ 3.20 GHz and 32.00 GB RAM, running on a Windows 10 Enterprise, 64-bit operating system. The energy transition planning is solved using CPLEX 12.9.0.0 with an optimality tolerance of 1%. The problem has 4,889 integer variable, 2,294,943 continuous variables, and 1,596,333 constraints. The optimal objective value is $ 96,343MM, and it takes 12,294 CPUs to solve the problem using the proposed optimization framework. Furthermore, to obtain detailed optimization results, power systems operations are simulated on an hourly basis for the entire planning horizon, based on the optimal transition pathway for electric power systems. The total transition cost under detailed operational simulation is $97,729MM, indicating that the difference between the optimal costs from energy transition planning and detailed hourly simulation is less than 1.5%. The simulation time of less than 70 CPUs is significantly less than the optimization time for energy transition planning, because the capacities of generators and storage units in each year are fixed for the operational simulations. As a result, each year's simulations are independent of other years, leading to a substantial reduction of computational demand.

The power generation capacity and annual electricity generation by the source during the decarbonization transition are shown in Figure 3(a) and Figure 3(b), respectively. As for generation capacities, offshore wind starts to participate in power generation in 2024, and its total capacity remains relatively stable during 2025-2030. In the 2030s, offshore wind power capacity gradually increases until the end of the planning horizon. Regarding solar PV, the generation capacity of utility solar PV has no significant changes during the beginning years of the planning horizon, and it starts to increase after 2027. This is mainly because the annual electricity consumption in the New York State is expected to decrease at the beginning years, owing to efficiency improvements across the state, while the total annual power load is projected to increase after 2027. On the other hand, distributed solar PV has a stable capacity across the planning horizon, which is owing to two reasons: it has lower economic efficiency compared to utility solar PVs, and most of the existing ones will not retire by 2040. For annual electricity generation shown in Figure 3(b), offshore wind would generate the most electricity by the end of the planning horizon, while hydropower and utility solar PV are the other two primary generation technologies in 2040. Note that although the total capacity of offshore wind is less than that of utility solar PV in 2040, offshore wind turbines tend to have much higher average capacity factors than utility solar PV, which enables them to generate more electricity on an annual basis. Hydropower currently accounts for the majority of renewable electricity generation in the state, and it continues to provide stable electricity on an annual basis across the planning horizon, owing to its relatively stable total generation capacity over the planning years.

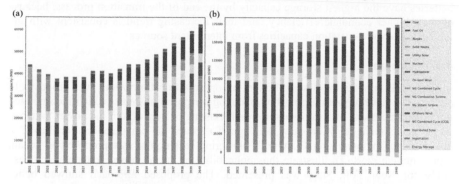

Figure 3. Electricity generation capacity and annual electricity generation by the source during the renewable electricity transition. (a) Electricity generation capacity according to the optimal transition pathway. (b) Annual electricity generation according to the operational simulations.

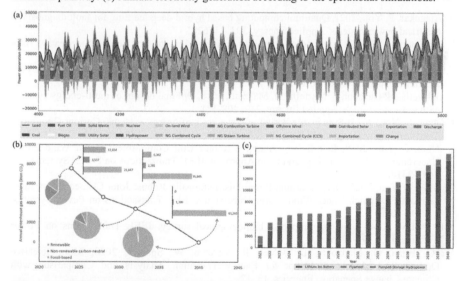

Figure 4. (a) Hourly power systems operations. (b) Annual greenhouse gas emissions. (c) Electricity storage capacities.

Figure 4 shows the hourly power systems operations, annual greenhouse gas emissions, and electricity storage capacities. Note that the fluctuation of electricity supply capacity is significant, owing to the high penetration level of variable renewable energy, such as solar and wind, and consequently, the charging and discharging activities are conducted frequently on a large scale. Solar PVs show clear periodic power outputs, because of the limited availability of solar energy during the evening, while the power outputs from offshore wind show no clear intra-day correlations. As for greenhouse gas emissions, they reduce decrease almost linearly across the planning horizon, while the reduction rate at the beginning years is slightly higher compared to the later periods, owing to more deactivated fossil-based power generation capacities in this period. Note that the greenhouse gas emissions reach zero in 2040, indicating that the goal of 100% decarbonized power systems is achieved. For electricity storage changes, lithium-ion batteries have the highest storage capacity by the end of the transition process, because of their higher economic efficiency, and their increasing trend is consistent with the increasing power generation capacities from intermittent sources.

4. Conclusion

In this work, a novel multi-scale bottom-up optimization framework was proposed to address the decarbonization transition planning for power systems, which incorporated multiple dimensions of information for each existing or new unit in the power systems, including its technology, capacity, and age. To reduce the computational challenge, a novel approach integrating PCA with clustering techniques was proposed to obtain representative days. To illustrate the applicability of the proposed framework, a case study for New York State was presented. The proposed approach showed better performance than multiple state-of-the-art clustering approaches.

References

A. Ajagekar, F. You, 2021, Quantum computing based hybrid deep learning for fault diagnosis in electrical power systems. Applied Energy, 303, 117628.

J. A. Bennett et al., 2021, Extending energy system modelling to include extreme weather risks and application to hurricane events in Puerto Rico. Nature Energy, 6, 3, 240-249.

T. Brown, J. Hörsch, D. Schlachtberger, 2018, PyPSA: Python for Power System Analysis. Journal of Open Research Software, 6, 1, 4.

P. Gabrielli, M. Gazzani, E. Martelli, M. Mazzotti, 2018, Optimal design of multi-energy systems with seasonal storage. Applied Energy, 219, 408-424.

J. Gong, F. You, 2015, Sustainable design and synthesis of energy systems. Current Opinion in Chemical Engineering, 10, 77-86.

C. Ning, F. You, 2019, Data-Driven Adaptive Robust Unit Commitment Under Wind Power Uncertainty: A Bayesian Nonparametric Approach. IEEE Transactions on Power Systems, 34, 2409-2418.

C. Ning, F. You, 2022, Deep Learning Based Distributionally Robust Joint Chance Constrained Economic Dispatch Under Wind Power Uncertainty. IEEE Transactions on Power Systems, 37, 191-203.

N.P. Padhy, 2004, Unit commitment-a bibliographical survey. IEEE Transactions on Power Systems, 19, 1196-1205.

M. G. Prina, G. Manzolini, D. Moser, R. Vaccaro, W. Sparber, 2020, Multi-Objective Optimization Model EPLANopt for Energy Transition Analysis and Comparison with Climate-Change Scenarios. Energies, 13, 12.

H. Qiu, F. You, 2020, Decentralized-distributed robust electric power scheduling for multi-microgrid systems. Applied Energy, 269, 115146.

H. Teichgraeber, A. R. Brandt, 2019, Clustering methods to find representative periods for the optimization of energy systems. Applied Energy, 239, 1283-1293.

C. Shang, F. You, 2019, Data Analytics and Machine Learning for Smart Process Manufacturing: Recent Advances and Perspectives in the Big Data Era. Engineering, 5, 1010-1016.

X. Tian, F. You, 2019, Carbon-neutral hybrid energy systems with deep water source cooling, biomass heating, and geothermal heat and power. Applied Energy, 250, 413-432.

N. Zhao, F. You, 2020, Can renewable generation, energy storage and energy efficient technologies enable carbon neutral energy transition? Applied Energy, 279, 115889.

N. Zhao, F. You, 2021, New York State's 100% renewable electricity transition planning under uncertainty using a data-driven multistage adaptive robust optimization approach with machine-learning. Advances in Applied Energy, 2, 100019.

Proceedings of the 14th International Symposium on Process Systems Engineering – PSE 2021+
June 19-23, 2022, Kyoto, Japan © 2022 Elsevier B.V. All rights reserved.
http://dx.doi.org/10.1016/B978-0-323-85159-6.50330-4

Design and Operation of Urban Energy Network: Integration of Civic, Industrial, and Transportation Sectors

Ruonan Li[a], Vladimir Mahalec[a*]

[a]*Department of Chemical Engineering McMaster University, 1280 Main St. West, Hamilton ON L8S 1A8, Canada*
mahalec@mcmaster.ca

Abstract

Integration of distributed energy systems for entities can further reduce greenhouse (GHG) emissions beyond the minimum emissions achieved by the individually operated energy systems. This work introduces an optimization approach with relative sizes of integrated entities, design and operation of the integrated energy system equipment, and production rates of plants as decision variables to maximize GHG emissions reduction brought by the integrated operation. The approach also differentiates temperature levels of heating demands to ensure feasible heat transfer by formulating heat balance for each process that requires heating. Results from case studies on an integrated system with a residential building with electric vehicles, a supermarket, a confectionery plant, a bakery plant, and a brewery show that, when optimizing the size of entities, the maximum GHG emissions reduction achieved by the integrated system is relatively constant under the various sizes of the residential building.

Keywords: Distributed energy network; GHG emissions reduction; Light industry; Energy system integration; Energy, Food and Environmental Systems.

1. Introduction

The integration of energy sectors reduces GHG emissions of urban areas by combining the heating, cooling, and electricity demands of civic structures, industrial plants, and transportation sectors (Fichera et al., 2017). The combined cooling, heating, and power (CCHP) system provides a solution for integrating energy sectors, where the power generation unit (PGU) is the critical equipment. It combusts fuel to generate electricity and uses the waste heat for the heating and cooling demand of an entity. Generally, the electricity and heat generated cannot be entirely consumed at the same time. Besides implementing additional equipment and energy resources, the unbalanced energy load and supply can also be solved by integrating individually operated CCHP systems of entities through heat and electricity transfer. Thus, an energy network forms, where each entity performs both as an energy supplier and consumer.

Knowledge gaps in existing studies on the integration of energy systems are identified as the following: 1. Existing studies assume the heating demands of all entities are at a uniform temperature. The assumption is not valid when the integrated system includes industrial plants, which require utilities at different temperatures for production. 2. Existing studies are based on fixed industrial energy demand profiles. It can lead to all entities having high energy demands at the same time. 3. There lacks an approach that identifies the optimal relative size of entities being involved.

Thus, this work introduces a novel optimization approach for the integration of energy systems in residential buildings with electric vehicles, commercial buildings, and industrial plants. The approach addresses the following areas not previously explored: 1. differentiates temperature levels of heating demands to ensure feasible heat transfer among entities; 2. sets production rates of integrated industries as decision variables to adjust industrial energy demands and further increase GHG emissions reduction of the integrated system; 3. provides optimal relative sizes of integrated entities, design, and operation of energy system equipment that maximize GHG emissions reduction. The formulation of the optimization problem is presented in Section 2. The approach has been tested by the case study presented in Section 3, where the results are shown in Section 4. Section 5 summarizes the main findings.

2. Optimization problem formulation

The optimization approach intends to maximize GHG emissions reduction brought by the integrated operation, compared to the non-integrated system. It is achieved by finding the optimal design and operation of energy system equipment, the relative size of entities in the integrated network, and the production rates of integrated industries. This work assumes the energy system of an entity has the structure shown in Figure 1.

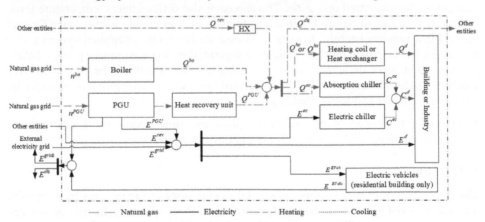

Figure 1 – A representation of the energy system in an entity.

2.1. Decision variables

Decision variables of the optimization approach can be classified as design decision variables, which are time-invariant, and operation decision variables, which can be manipulated during the operation.

Design decision variables include (i) size of energy system equipment: Cap_i^{eqp}; (ii) size of each entity: S_i. Operation decision variables include (i) amount of fuel used by PGU and boiler: $n_{i,t}^{PGU}$ and $n_{i,t}^{bo}$; (ii) amount of electricity an entity purchases and sells to the external grid: $E_{i,t}^{grid}$ and $E_{i,t}^{grids}$; (iii) whether there is heat and electricity transfer between two entities: $y_{i,i',t}$ and $z_{i,i',t}$ and the amount: $Q_{i,i',t}^{dis}$ and $E_{i,i',t}^{dis}$; (iv) production rate of the plants: $x_{i,t}$; (v) whether electric vehicles are charged or discharged: u_t and w_t and the corresponding amount: E_t^{EV-ch} and E_t^{EV-dis}.

2.2. Constraints

Constraints for the optimization problem can be further divided into four categories: energy generation of equipment, electric vehicles, energy transfer, and energy balances.

(1) Energy generation of equipment: Constraints under this category calculate the amount of heat ($Q_{i,t}^{eqp-out}$), electricity ($E_{i,t}^{eqp-out}$), or cooling ($C_{i,t}^{eqp-out}$) generated by a piece of equipment ($Egy_{i,t}^{eqp-out}$) based on its energy consumption.

$$Egy_{i,t}^{eqp-out} = \eta_i^{eqp} Egy_{i,t}^{eqp-in} \tag{1}$$

η_i^{eqp} is the efficiency of the equipment. Depending on the equipment, $Egy_{i,t}^{eqp-in}$ can represent fuel, heat, or electricity used by the equipment. Taking PGU and boiler as examples, $Egy_{i,t}^{eqp-in}$ represents the amount of fuel combusted ($n_{i,t}^{PGU-in}$ and $n_{i,t}^{bo-in}$) in the equipment. Besides energy consumption, heat, electricity, or cooling generated by an entity should be less than or equal to the equipment capacity (Cap_i^{eqp}).

(2) Electric vehicles: This work assumes all electric vehicles (EVs) can only be charged or discharged in the residential building and investigates all EVs as an aggregated subsystem to simplify the formulation. The amount of electricity in the battery (E_t^{EV}) equals the amount of electricity at the previous time (E_{t-1}^{EV}), plus the charged electricity (E_t^{EV-ch}), and minus the electricity discharged (E_t^{EV-dis}).

Since the charging and discharging behavior cannot occur at the same time, binary variables - u_t and w_t are used, as formulated in Eq. (2). ER^{EV} represents the maximum charging and discharging rate of EVs. Eq. (3) restricts EVs to be fully charged when leaving the building (t^l) and defines the amount of electricity in EVs when returning home (t^r). E^{EV-con} is the amount of electricity consumed by EVs outside.

$$E_t^{EV-ch} \leq ER^{EV} u_t, E_t^{EV-dis} \leq ER^{EV} w_t, u_t + w_t \leq 1, u_t, w_t \in \{0,1\} \tag{2}$$

$$E_{t=t^l}^{EV} = Cap^{EV}, E_{t=t^r}^{EV} = Cap^{EV} - E^{EV-con} \tag{3}$$

(3) Energy transfer: Constraints associated with energy transfer of the integrated system (i) ensure energy transfer between two entities at a period is unidirectional by using the binary variables $y_{i,i',t}$ and $z_{i,i',t}$, which indicate whether heat and electricity are being dispatched from entity i to i', respectively; (ii) restrict the amount of energy dispatched based on the amount of heat and electricity generated in the entity and the heat transfer pipe size; (iii) calculate the available heat and electricity an entity received by excluding energy loss during the transfer from the dispatched heat.

(4) Energy balance: Constraints under this category ensure the energy demand of an entity can be fully satisfied by all available energy resources. Since the proposed optimization approach differentiates temperatures of heating demands, the heat balance of each process is developed individually to reflect whether the process can use the transferred heat or not. For processes that cannot use the transferred heat, Eq. (4) is applied without the received heat item – $Q_{i,i',p,t}^{rev}$.

$$\frac{Q_{i,p,t}^d}{\eta_{i,p}^{hx}} = Q_{i,p,t}^{PGU-out} + Q_{i,p,t}^{bo-out} + \sum_i Q_{i,i',p,t}^{rev} \tag{4}$$

$Q_{i,p,t}^d$ represents the heating demand of a process p. For industrial plants, the energy demand of a process is calculated based on the production rate $- x_i$ (decision variable) and the energy used to produce a unit of product. $Q_{i,p,t}^{PGU-out}$ and $Q_{i,p,t}^{bo-out}$ are heat generated by the PGU and boiler. For residential buildings and commercial buildings whose heating demands are at a uniform temperature, p equals one.

2.3. Objective function

As mentioned, the objective function of the problem is maximizing GHG emissions reduction led by the integrated operation. It is based on the minimum GHG emissions of the integrated system and the non-integrated system, as shown in Eq. (5).

$$GHGD \% = \left(GHG_{non-integrated} - GHG_{integrated}\right)/GHG_{non-integrated} \tag{5}$$

The minimum GHG emissions of the integrated ($GHG_{integrated}$) and non-integrated system ($GHG_{non-integrated}$) are calculated based on the amount of fuel ($n_{i,t}^{PGU-in}$ and $n_{i,t}^{bo-in}$), electricity ($E_{i,t}^{grid}$) used by the system, and electricity sold to the external grid ($E_{i,t}^{grids}$). σ_{NG} and σ_E are GHG emissions coefficients associated with using fuel and grid electricity. Eq. (6) shows an example of GHG emissions for the integrated system.

$$GHG_{integrated} = \sum_i \sum_t \sigma_{NG}\left(n_{i,t}^{PGU-in} + n_{i,t}^{bo-in}\right) + \sigma_E E_{i,t}^{grid} - 0.5\sigma_E E_{i,t}^{grids} \tag{6}$$

The minimum GHG emissions of the non-integrated system can be simplified as a linear equation related to entity sizes. The linear relationship holds because there exist optimal operation patterns for the energy system of each non-integrated entity, which minimizes GHG emissions of the non-integrated system. When the sizes of the entities change, the optimal operation patterns of equipment do not change; however, the equipment sizes increase or decrease correspondingly to maintain the minimum GHG emissions.

3. Case study description

An integrated system with a residential building, a supermarket, a confectionery plant, a brewery, a bakery plant, and EVs has been used for case studies. The residential building and supermarket are assumed to have fixed energy profiles based on published information from Sullivan (2020) and Ghorab (2019), respectively. The sizes of the two entities are relative sizes, which represent the number of buildings having the base-case energy demands. Sizes of the confectionery plant, brewery, and bakery plant are the maximum production rates of the plants. Energy used to make a unit of product for the three plants is obtained based on the process studied by Singh (1986) and Therkelsen et al. (2014). Since the supermarket requires a large amount of energy for low-temperature refrigeration, the supermarket has been assumed does not have the PGU, heat recovery unit, or absorption chiller. For reducing the computation time, the relative size of the supermarket is set to be equal to the size of the residential building. Additionally, each unit of the relative size of residential building has been assumed to have 870 EVs.

4. Results and discussion

It has been found that the integrated operation can reduce GHG emissions of the system by a maximum of 17.5 %. The reduction is achieved by integrating a residential building, a supermarket, 870 EVs, a brewery with a capacity of 2,720 kg/hr, and a 5,000 kg/day bakery plant. There is no confectionery plant in the system. The reduction is due to, with energy transfer among entities, the integrated system purchasing 65.0 % less electricity from the external grid and operates boilers 79.6 % less.

As shown in Figure 2, although the maximum GHGD% (Case 1) is achieved when the relative size of the residential building is one, under larger relative sizes of the residential building, the highest GHGD % (Case 2, Case 3, and Case 4) at lower values can be achieved when optimizing sizes of the other entities. The values are 17.2 %, 16.7 %, and 15.9 % when relative sizes of the building are three, five, and ten, respectively. Table 1 presents the corresponding plant sizes. The case studies reflect the situation when there are specific requirements on entity sizes of the integrated system.

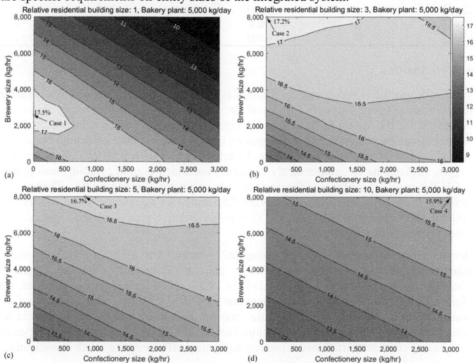

Figure 2 – GHGD % of the integrated system under various entity sizes.

According to the results shown in Figure 2 and Table 1, there is a small difference between the maximum GHGD % and the highest GHGD % when increasing the relative size of the residential building. The results indicate requirements on entity size do not significantly impact GHGD % of the overall integrated system. With requirements on entity sizes, the optimization problem can be formulated by adding additional constraints to define the desired entity sizes and find the optimal sizes of other entities.

Results also show that as the relative size of the residential building increases, the highest GHGD % decreases. It is due to upper bounds on industrial capacities force the integrated system to operate in a way, which deviates from its optimal relative entity sizes and optimal operation pattern. Thus, the maximum GHGD % cannot be held.

With an increase in the size of the residential building, the size of the bakery plant reaches its upper bound first (5,000 kg/day), followed by the brewery (8,000 kg/hr) and the confectionery plant (3,000 kg/hr). It is due to the bakery plant performing as a critical energy supplier in the integrated system, where all heat transfer and 20.2 % electricity transfer is dispatched by the bakery plant under the optimal entity sizes (Case 1). The brewery reaches its maximum size before the confectionery plant, which indicates the brewery is more suitable for the integrated operation. Compared to the confectionery plant, the brewery requires heating at lower temperatures. Thus, the brewery has a better ability to use the transferred heat.

Table 1 – Highest GHGD % under different relative sizes of the residential building.

	Relative size of residential building	Bakery plant (kg/day)	Brewery (kg/hr)	Confectionery plant (kg/hr)	GHGD %
Case 1	1	5,000	2,720	0	17.5 %
Case 2	3	5,000	8,000	0	17.2 %
Case 3	5	5,000	8,000	800	16.7 %
Case 4	10	5,000	8,000	3,000	15.9 %

5. Conclusions

This work quantifies reductions in GHG emissions that can be achieved by cross-sector integration of energy systems. Even if the energy systems within each sector are optimized for the lowest GHG emissions within that sector, further reduction in GHG emissions can be accomplished by integration between the sectors (residential buildings, commercial buildings, light industries, and electric vehicles). Each entity in the integrated system implements an independently operating combined cooling, heating, and power (CCHP) system, where there are heat and electricity transfers among entities. The optimal design and operation of energy systems are determined for equipment in each entity, the optimal production rate of plants, and the optimal relative size of entities, considering temperatures of heating demands.

Results from case studies on an integrated system with a residential building, a supermarket, a confectionery plant, a bakery plant, and a brewery show the integrated operation can lead to a maximum GHGD % of 17.5 %. If optimizing sizes of entities, the highest GHGD % can be maintained between 15.9 % and 17.5 %, even when there are requirements on sizes of specific entities and the integrated system deviates from its optimal relative entity sizes and operation. Future studies, which include more types of entities and consider partial load effects on equipment efficiency are worth investigating.

References

Fichera, Alberto, Mattia Frasca, and Rosaria Volpe. 2017. "Complex Networks for the Integration of Distributed Energy Systems in Urban Areas." *Applied Energy* 193: 336–45.

Ghorab, Mohamed. 2019. "Energy Hubs Optimization for Smart Energy Network System to Minimize Economic and Environmental Impact at Canadian Community." *Applied Thermal Engineering* 151(June 2018): 214–30.

Singh, R.P. 1986. "Energy Accounting of Food Processing Operations." In *Energy in Food Processing*, ed. R.P. Singh. New York: Elsevier Science Publishing Company Inc., 19–68.

Sullivan, Brendan. 2020. "A Comparison of Different Heating and Cooling Energy Delivery Systems and the Integrated Community Energy and Harvesting System in Heating Dominant Communities." McMaster.

Therkelsen, Peter, Eric Masanet, and Ernst Worrell. 2014. "Energy Efficiency Opportunities in the U.S. Commercial Baking Industry." *Journal of Food Engineering* 130: 14–22.

Proceedings of the 14th International Symposium on Process Systems Engineering – PSE 2021+
June 19-23, 2022, Kyoto, Japan © 2022 Elsevier B.V. All rights reserved.
http://dx.doi.org/10.1016/B978-0-323-85159-6.50331-6

Sustainable Design of Hybrid Energy Systems for Net Zero Carbon Emission

Xueyu Tian[a*], Fengqi You[a]

[a] *Cornell University, Ithaca, New York, 14853, USA*
xt93@cornell.edu

Abstract

This article addressed the sustainable design of carbon-neutral energy systems with earth source heat, lake source cooling, on-site electricity generation, and peak heating systems. A multi-period optimization model given time horizon and temporal resolution is built based on the proposed superstructure of carbon-neutral energy systems to minimize the total annualized cost. The aim is to determine the optimal design of the carbon-neutral energy systems in the target region, seasonal operations, energy mix, and corresponding capacity of each base-load and peak-load technology involved while fulfilling the seasonal demand for electricity, heat, and cooling. The applicability of the proposed modeling framework is illustrated through case studies using Cornell University as the living laboratory.

Keywords: carbon neutrality; energy systems; renewables; decarbonization.

1. Introduction

The Paris Agreement sets a goal to curb global greenhouse gas (GHG) emissions, driving vast penetration of renewable energy worldwide. Extensive research on deep decarbonization of energy systems is conducted at the city-level (Wiryadinata et al., 2019), state-level (Zhao and You, 2020), and country-level (Vaillancourt et al., 2017). Electrification of heat and cooling generation and decarbonization of electricity generation is identified as a promising lever to address the ambitious climate goals (de Chalendar et al., 2019). However, heat and cooling generation stand a chance to destroy the stability of the power system due to the surge in electric load involved if they are electrified in an uncontrolled way (Sánchez-Bautista et al., 2017). Therefore, it seems to be a reliable and promising decarbonization option by exploring renewable heat and cooling generation technologies rather than simply using electrified counterparts (Gong and You, 2015). Among the vast array of renewable heat and cooling generation technologies, geothermal energy and deep water source cooling system show great potentials for the decarbonization transition of energy systems (Lee et al., 2019). Recent research efforts have also identified the values of green hydrogen (Dodds et al., 2015), large-scale heat pumps (Bach et al., 2016), biomass and biogas (Kassem et al., 2020), and thermal energy storage (Ochs et al., 2020) for decarbonizing the heating system. There is a lack of studies addressing the sustainable design of energy systems toward carbon neutrality by simultaneously exploring renewable electricity, heat and cooling generation, and electrified heating and cooling options in the region with a humid continental climate, such as New York State (NYS) (Zhao and You, 2021). In this paper, a multi-period optimization model, given time horizon and temporal resolution for total annualized cost (TAC) minimization, is built. The aim is to determine the optimal design of the carbon-neutral energy systems in the investigated region, seasonal operations, energy mix, and corresponding capacity of each base-load and peak-load technology involved. The applicability of the proposed modeling framework is illustrated through case studies developed using the data from the Cornell campus.

2. Problem Statement and Model Summary

We are given a superstructure of carbon-neutral energy systems, including a set of renewable electricity generation technologies and a set of renewable and electrified heating and cooling options, as shown in Figure 1. To capture the optimal design, seasonal operations, energy mix, and corresponding capacity of each base-load and peak-load technology in the carbon-neutral energy systems, a time horizon and a set of time periods are specified to improve the temporal resolution of the model. The multi-period optimization problem of the proposed carbon-neutral energy system with earth source heat, lake source cooling (LSC), on-site electricity generation, and peak heating options for the total annualized cost minimization is formally defined in this section. The aim is to determine the optimal design of the carbon-neutral energy systems in the target region, seasonal operations, energy mix, and the corresponding capacity of each base-load and peak-load technology involved while fulfilling the seasonal demand for electricity, heat, and cooling.

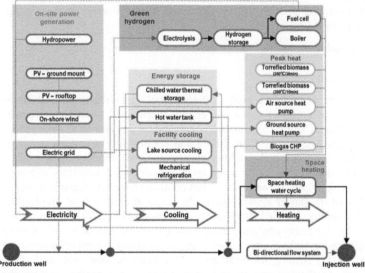

Figure 1. Superstructure of the proposed carbon-neutral energy systems.

The general multi-period optimization model is subjected to the mass balance and configuration constraints, energy balance constraints, logic constraints, and techno-economic evaluation constraints. The integer decision variables represent the selection of technologies. The number of geothermal well-pairs is an integer decision variable (Tian et al., 2019). Other essential decision variables such as the mass flow rates, energy flows, and capacities are continuous variables. The objective function, total annualized cost, includes integer variables such as the numbers of production wells and injection wells and thus is a mixed-integer function. The nonlinear terms mainly come from the separable concave terms induced by the economy of scale. Therefore, the resulting problem is a mixed-integer nonlinear programming (MINLP) problem. The general form of this MINLP problem is summarized as follows.

$$\min \quad TAC = AIC + AOC + RE \tag{1}$$

s.t. mass balance and configuration constraints;
    energy balance constraints;
    logic constraints;
    techno-economic evaluation constraints;

where *AIC*, *AOC*, and *RE* refer to the annualized investment cost, annual operating cost, and replacement cost, respectively.

3. Global Optimization Strategy

The resulting MINLP problem embraces both integer and continuous variables, along with nonlinear functions, so the global optimization of this problem is likely to be computationally challenging for general-purpose global optimization solvers. A global optimization strategy is utilized to solve the proposed MINLP problem efficiently (Gong and You, 2018). Specifically, we substitute the separable concave functions induced by the economy of scale for capital investment estimation with successive piecewise linear relaxations. The resulting MINLP problem is solved iteratively following the tenet of the branch-and-refine algorithm (You and Grossmann, 2011). The pseudocode of the global optimization algorithm is presented in Figure 2. *ub* and *lb* stand for the upper and lower bound, respectively.

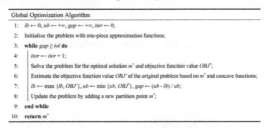

Figure 2. The pseudocode of the global optimization algorithm.

4. Application to Cornell University Campus Energy Systems

The proposed multi-period optimization modeling framework for energy systems decarbonization is applied to address the optimal design of the carbon-neutral energy systems using the main campus of Cornell University located in Ithaca as the living laboratory. Based on the optimization results, the optimal configuration of the carbon-neutral energy system in the target region, seasonal operations, energy mix, and corresponding capacity of each base-load and peak-load technology involved are determined while accommodating the seasonal demand for electricity, heat, and cooling across the main campus of Cornell University located in Ithaca, NYS. Three case studies are developed based on the real data from the main campus of Cornell University located in Ithaca with a consideration of different scopes of peak-load technologies (Tian et al., 2019). The first case study aims to obtain the global optimal solution of the multi-period optimization problem with a monthly model resolution for the proposed carbon-neutral energy system with earth source heat, LSC, on-site electricity generation, and peak heating options, including biomass or biogas heating, heat pumps, green hydrogen, and thermal storage. The second one explicitly excludes biomass or biogas as the peak heating options to evaluate the economic potential of electrified peak heating systems based on heat pumps.

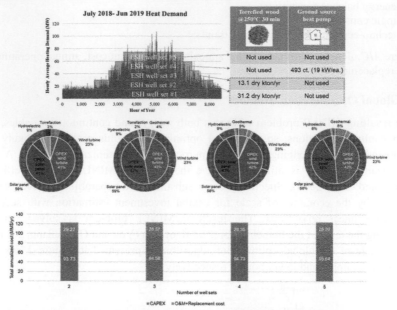

Figure 3. Selection of technologies and economic performance for case study 1.

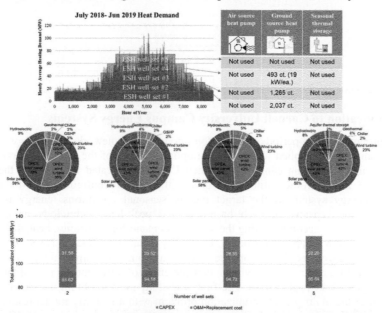

Figure 4. Selection of technologies and economic performance for case study 2.

Figure 3 shows the selection of technologies of the global optimal solution and the corresponding economic performance. When the capacity of base-load earth source heat is low, burning torrefied biomass is selected as the optimal peak heating technology through optimization. However, as the number of the geothermal well-pairs attains four, the ground source heat pump outperforms the torrefied biomass from the economic perspective. No peak heating technologies are needed when five geothermal well-pairs serve as the base-load heat supplier based on a monthly model resolution. Specifically, the annual consumption of torrefied biomass is 31.2 dry kton and 13.1 dry kton for the two-well-set case and three-well-set case, respectively. Overall, torrefied biomass outperforms the ground source heat pumps when the base-load capacity is low. We note that as the number of geothermal well-pairs increases from two to five, the annualized investment cost increases from $93.73 MM/yr to $95.64 MM/yr. The annual operating costs corresponding to the four cases are $29.27 MM/yr, $28.57 MM/yr, $28.35 MM/yr, and $28.20 MM/yr, respectively. When the number of base-load geothermal well-pairs equals two and three, torrefied biomass treated at 250 ℃ for 30 minutes is selected, accounting for 3% and 2% of the annualized investment cost, respectively. When the number of geothermal well-pairs attains four, 493 ground source heat pumps with a typical capacity of 19 kW in the North American region are employed to address the peak-load heat demand.

Figure 4 demonstrates the peak heating options with different base-load capacities, where only ground source heat pumps are employed with no thermal energy storage. Specifically, 2,037, 1,265, and 493 ground source heat pumps with a typical capacity of 19 kW for each in the North American region are deployed to handle the peak-load heat demand for the cases with two, three, and four base-load earth source heat pumps, respectively. The number of geothermal well-pairs is chosen as the investigated input parameter, ranging from two to five. We note that as the number of geothermal well-pairs increases from two to five, the annualized investment cost increases from $93.62 MM/yr to $95.64 MM/yr, while the annual operating cost decreases from $31.56 MM/yr to $28.20 MM/yr. In terms of capital investment, solar panels (58%-59%), wind turbines (23%), and hydroelectric power plant (8%-9%) are the major contributors. When the base-load earth source heat capacity is low, the operating cost associated with the ground source heat pump is more pronounced. The remaining annual operating cost is mainly sourced from the operations of solar and wind farms.

When the capacity of base-load earth source heat is low, i.e., the number of geothermal well-pairs equalling two, both green hydrogen and hot water tanks are needed through optimization with capacities of 53.5 GWh and 35.5 GWh, respectively. As the geothermal well-pairs attains three and four, green hydrogen is no longer needed to pursue the lowest total annualized cost. When five geothermal well-pairs serve as the base-load heat supplier based on a monthly model resolution, i.e., some short-term peak-load demands are neglected by considering a monthly average, no peak heating technologies are needed. Specifically, the capacity of hot water tanks for the three-well-set case and four-well-set cases are 41.4 GWh and 7.0 GWh, respectively. We note that as the number of geothermal well-pairs increases from two to five, the annualized investment cost decreases from $191.33 MM/yr to $95.64 MM/yr, while the annual operating costs decrease from $57.85 MM/yr to $28.20 MM/yr. When the number of base-load geothermal well-pairs equals two, both green hydrogen and hot water tank are in need to manage the peak-load heat demand, which accounts for 42% and 11% of the annualized investment cost, respectively. In addition to the operating cost of solar and wind farms, other operating costs are associated with green hydrogen and the replacement of hot water tanks. The total annualized cost corresponding to the two-well-set case is substantially higher than the other cases by a factor of 1.6 - 2.0 due to the high capital investment of hydrogen generation through electrolysis and storage.

5. Conclusion

A multi-period optimization model, given time horizon and temporal resolution for total annualized cost minimization, was built. The aim is to simultaneously determine the optimal design of the carbon-neutral energy systems in the investigated region, seasonal operations, energy mix, and corresponding capacity of each base-load and peak-load technology involved while fulfilling the seasonal demand for electricity, heat, and cooling. The applicability of the proposed modeling framework was illustrated through three case studies developed by leveraging the real-world data from the main campus of Cornell University, located in Ithaca, NYS.

References

B. Bach, J. Werling, T. Ommen, et al., 2016, Integration of large-scale heat pumps in the district heating systems of Greater Copenhagen. Energy, 107, 321-334.

J.A. de Chalendar, P.W. Glynn, S.M. Benson, 2019, City-scale decarbonization experiments with integrated energy systems. Energy & Environmental Science, 12(5), 1695-1707.

P.E. Dodds, I. Staffell, A.D. Hawkes, et al., 2015, Hydrogen and fuel cell technologies for heating: A review. International Journal of Hydrogen Energy, 40(5), 2065-2083.

J. Gong, F. You, 2015, Sustainable design and synthesis of energy systems. Current Opinion in Chemical Engineering, 10, 77-86.

J. Gong, F. You, 2018, A new superstructure optimization paradigm for process synthesis with product distribution optimization: Application to an integrated shale gas processing and chemical manufacturing process. AIChE Journal, 64, 123-143.

N. Kassem, J. Hockey, C. Lopez, et al., 2020, Integrating anaerobic digestion, hydrothermal liquefaction, and biomethanation within a power-to-gas framework for dairy waste management and grid decarbonization. Sustainable Energy & Fuels, 4(9), 4644-4661.

I. Lee, J.W. Tester, F. You, 2019, Systems analysis, design, and optimization of geothermal energy systems for power production and polygeneration: State-of-the-art and future challenges. Renewable and Sustainable Energy Reviews, 109, 551-577.

F. Ochs, A. Dahash, A. Tosatto, M.B. Janetti, 2020, Techno-economic planning and construction of cost-effective large-scale hot water thermal energy storage for Renewable District heating systems. Renewable Energy, 150, 1165-1177.

A.d.F. Sánchez-Bautista, J.E. Santibañez-Aguilar, et al., 2017, Optimal Design of Energy Systems Involving Pollution Trading through Forest Plantations. ACS Sustainable Chemistry & Engineering, 5, 2585-2604.

X. Tian, T. Meyer, H. Lee, et al., 2020, Sustainable design of geothermal energy systems for electric power generation using life cycle optimization. AIChE Journal, 66, e16898.

X. Tian, F. You, 2019, Carbon-neutral hybrid energy systems with deep water source cooling, biomass heating, and geothermal heat and power. Applied Energy, 250, 413-432.

K. Vaillancourt, O. Bahn, E. Frenette, et al., 2017, Exploring deep decarbonization pathways to 2050 for Canada using an optimization energy model framework. Applied Energy, 195, 774.

S. Wiryadinata, J. Morejohn, K. Kornbluth, 2019, Pathways to carbon neutral energy systems at the University of California, Davis. Renewable Energy, 130, 853-866.

F. You, I.E. Grossmann, 2011, Stochastic inventory management for tactical process planning under uncertainties: MINLP models and algorithms. AIChE Journal, 57(5), 1250-1277.

N. Zhao, F. You, 2020, Can renewable generation, energy storage and energy efficient technologies enable carbon neutral energy transition? Applied Energy, 279, 115889.

N. Zhao, F. You, 2021, New York State's 100% renewable electricity transition planning under uncertainty using a data-driven multistage adaptive robust optimization approach with machine-learning. Advances in Applied Energy, 2, 100019.

Proceedings of the 14th International Symposium on Process Systems Engineering – PSE 2021+
June 19-23, 2022, Kyoto, Japan © 2022 Elsevier B.V. All rights reserved.
http://dx.doi.org/10.1016/B978-0-323-85159-6.50332-8

Prediction of Charge / Discharge Behavior of Tri-Electrode Zinc-air Flow Battery Using Linear Parameter Varying Model

Woranunt Lao-atiman[a], Amornchai Arpornwichanop[a,b] and Soorathep Kheawhom[a,b*]

[a] *Department of Chemical Engineering, Faculty of Engineering, Chulalongkorn University, Thailand*
[b] *Center of Excellence in Process and Energy Systems Engineering, Chulalongkorn University, Bangkok, Thailand*
*soorathep.k@chula.ac.th

Abstract

Currently, energy storage systems are in the research spotlight as they can support the application of renewable energy. Owing to their high energy density and low cost, zinc-air flow batteries (ZAFBs) are seen to have great potential for use as renewable energy storage devices. However, the battery management system (BMS) for ZAFBs is still underdeveloped as a precise prediction of their nonlinear behavior is required. To overcome this drawback, a linear parameter varying (LPV) model is established via a multiple linear time-invariant (LTI) model with charge/discharge current and state of charge (SOC) as scheduling parameters. Validation of the developed model is carried out using various battery data from different experimental batches. According to the discharge current and SOC, results demonstrate that the nonlinear behavior of the ZAFBs can be predicted by the LPV model developed. Thus, the LPV model is found to be comparable with the linear model for local accuracy. In the case of global accuracy, it is seen that the LPV model outperformed the linear model. Such a result reveals the ability of the LPV model to predict the dynamics of the ZAFBs and their feasibility for use in the BMS.

Keywords: Zinc-air battery; battery modeling; Linear-parameter varying model.

1. Introduction

Nowadays, renewable energy technologies have attracted widespread interest due to the increase in energy consumption and the critical environmental crisis. Renewable energy sources such as solar and wind energy are intermittent by nature. Therefore, the power generated from these energy sources is found to be inconsistent. To address this issue, energy storage systems (ESSs) exhibit great potential because they can provide the stability for energy utilization. Zinc-air batteries (ZABs) are promising candidates for ESSs due to their high energy density and low cost (Lao-atiman et al., 2019, Radenahmad et al., 2021, Khezri et al., 2020).

In battery research, BMS is generally studied as it can enhance the safety and operability of the battery (Pop et al., 2008). One of the research aspects involving the development of BMS is modeling. Battery modeling can be done in a variety of ways. For example, simulation using a theoretical continuum model has been conducted to analyze the phenomena occurring inside the battery (Schröder and Krewer, 2014, Maia et al., 2017). The empirical model has also been studied for use in BMS in various types of batteries because this type of model is suitable for online prediction due to its speed and simplicity of calculation. Recently, a LPV model has been proposed to predict the discharge behavior of ZABs (Lao-atiman et al., 2020). Another example is the data-driven model developed for lithium-ion battery health monitoring (Sukanya et al., 2021). One important feature of BMS, which must be established, is battery state estimation, especially SOC estimation (Chang, 2013). As regards SOC estimation, an empirical model has also been used with various adaptive filters such as state observer (Hu and Yurkovich, 2012) or Kalman filter-based estimator (Wassiliadis et al., 2018). Nevertheless, studies of empirical models involving SOC changes are still inadequate in ZAB research.

This work aims to develop the LPV model and use it to predict the dynamic behavior of tri-electrode ZAFBs including the influence of charge/discharge current and SOC. Battery data used for dynamic identification and validation have been measured from the laboratory-made tri-electrode ZAFB. Linear state-space models identified from different conditions have been used to create the LPV model, which has discharge current level and SOC as scheduling parameters. After that, the LPV model was tested for validity using battery data from different experimental batches.

2. Description of tri-electrode ZAFB and experimental data

As shown in Figure 1A, the laboratory made ZAFB in this work was designed as a tubular cylinder cell. The cylinder support structure was made of poly vinyl chloride (PVC). The cell is circulated by an electrolyte, which is an 8 M solution of potassium hydroxide (KOH) having 0.5 M zinc oxide (ZnO). The anode active material is Zn electroplated onto the current collector, which is nickel (Ni) foam. The cathode active material is oxygen from the ambient air. The oxygen reduction reaction (ORR) occurred at the cathode current collector, which is Ni foam coated with a catalytic layer and gas diffusion layer. The charging electrode is made of Ni foam.

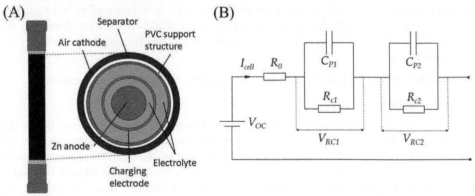

Figure 1 (A) Schema of laboratory-made tri-electrode ZAFB and (B) Electrical equivalent circuit diagram of battery: second-order RC model.

To obtain the experimental data, battery testing equipment (NEWARE, CT-4008-5V20mA, Neware Technology Ltd., Shenzhen, China) was used. Sampling time was 1 s. The used data are battery response data with current as input and voltage as output. For the dynamic identification, the pattern of the used profile was charge / discharge for 50 mAh alternating with a rest period of 5 min. The procedure was repeated until the voltage cutoff of 0.5 V was met for discharging or the capacity reached 500 mAh for charging. The applied current values were 500, 1000, 1500 and

2000 mA. As regards model validation, different sets of random charging and discharging data were used.

For the nomenclature of data, the data starting with DSOC is the discharge steps data and the data starting with CSOC is the charge steps data. The following numbers define the discharge current, for example, the data DSOC500A contain discharge steps with a current of 500 mA for 50 mAh alternating with a rest for 5 min. The data named MULTI A and MULTI B are discharge steps data with multiple current levels. Likewise, CMULTI refers to the charge steps data with multiple discharge currents.

3. LPV modeling

The LPV model is a time-variant model whereby parameters are varied as a function of scheduling parameters (p). The LPV model was constructed from a set of discrete state space model. The model contains 3 states: V_{RC1}, V_{RC2} and SOC. Both V_{RC1} and V_{RC2} represent the overpotential of the zinc electrode and counter electrode (air or charging), respectively. SOC represents the state of charge of the battery. SOC is included as a state because the model is used with a state estimator to estimate SOC in the next part. To make the LPV model more practical, the model was established having an equivalent circuit second-order RC model, as illustrated in Figure 1B. The LPV model can be written as follows:

$$\begin{bmatrix} V_{RC1}(k+1) \\ V_{RC2}(k+1) \\ SOC(k+1) \end{bmatrix} = \begin{bmatrix} A_1 & 0 & 0 \\ 0 & A_2 & 0 \\ 0 & 0 & 1 \end{bmatrix} \begin{bmatrix} V_{RC1}(k) \\ V_{RC2}(k) \\ SOC(k) \end{bmatrix} + \begin{bmatrix} B_1 \\ B_2 \\ \dfrac{\Delta t}{3600 \cdot C_n} \end{bmatrix} I_{cell}(k) \tag{1}$$

$$V_{cell}(k) = V_{oc} + \begin{bmatrix} 1 & 1 & 0 \end{bmatrix} \begin{bmatrix} V_{RC1}(k) \\ V_{RC2}(k) \\ SOC(k) \end{bmatrix} + D \cdot I_{cell}(k) \tag{2}$$

where A_1, A_2, B_1, B_2 and D are state space parameters. V_{cell}, V_{OC} and I_{cell} are cell voltage, open circuit voltage and cell current, respectively. Δt and C_n are sampling time and nominal capacity, respectively.

The calculation of SOC in Eq.(1) is based on the coulomb counting (CC) method. Eq.(2), which includes V_{OC}, ensures that the model is able to calculate cell voltage; V_{cell}. I_{cell} and SOC have been selected as scheduling parameters for A_1, A_2, B_1, B_2 and D. However, some arbitrary assumptions have been made in order to adapt the model with the scenarios of battery data. Firstly, for discharging, when SOC decreased, Zn at the electrode fully depleted. Thus, V_{RC1} is affected but V_{RC2} is not affected by the SOC change. This outcome enabled both A_2 and B_2 to become functions of only the current level for discharging. The next assumption is that the SOC effect on the overpotential is less significant for both electrodes when the battery is charged. Hence, B_1 and B_2 become functions of the current level for charging. The last assumption made infers that the internal resistance of the system is independent of current (Zhong et al., 2021). As seen in Figure 1B, D is equivalent to R_0 which is related to ohmic resistance. D is, therefore, assumed to be a function of SOC for both discharging, i.e., $A_2(I)$, $B_2(I)$ and $D(SOC)$ and for charging, i.e., $B_1(I)$, $B_2(I)$ and $D(SOC)$.

Regarding varying model parameters, the correlations between model parameters and scheduling parameters were constructed from the identified model parameters. For instance: model parameters as identified from the discharge step data with a discharge current of 1000 mA and SOC of 0.5 demonstrated scheduling current levels and SOCs of 1 A and 0.5, respectively. After the correlations were accounted for, the LPV model was validated according to the various dataset, including data obtained via a different batch of the experiment. For comparison purposes, the linear state space model was introduced. The expression of the linear model is the same as Eqs. 1 and 2 but the model parameters are not varied with scheduling parameters.

4. Results and discussion

As mentioned previously, the LPV model for the tri-electrode ZAFB was developed from the state space model as a based linear time-invariant (LTI) model. The state space parameters of the identified model were fitted to make the correlation with the scheduling parameters. From Eqs.(1) and (2), five parameters enabled the correlations to be set up viz. A_1, A_2, B_1, B_2 and D. There were also two scheduling parameters, including current level and SOC; the fitted correlations became surface functions. Nevertheless, some model parameters are functions of only one scheduling parameter according to the assumption made in the previous section. Additionally, V_{OC} has also varied with SOC. Therefore, a correlation with V_{OC} was also made. In Table 1, a list of the functions used to fit the correlations for both discharging and charging are tabulated.

Table 1 Functions used to fit the model parameter correlations.

Fitting function	parameter
Discharging	
$\alpha exp(\beta(SOC) + \gamma(I_{cell})) + \delta exp(\varepsilon(I_{cell}) + \theta(SOC)) + \vartheta$	A_1
$\alpha exp(\beta(SOC) + \gamma(I_{cell})) + \delta exp(\varepsilon(I_{cell}) + \theta(SOC))$	B_1
$\alpha exp(\beta(I_{cell})) + \gamma exp(\delta(I_{cell}))$	A_2, B_2
$\alpha exp(\beta(SOC)) + \gamma exp(\delta(SOC))$	D
Charging	
$\mu_{00} + \mu_{10}(SOC) + \mu_{01}(I_{cell}) + \mu_{11}(SOC)(I_{cell}) + \mu_{02}(I_{cell})^2$	A_1, A_2
$\alpha exp(\beta(I_{cell})) + \gamma exp(\delta(I_{cell}))$	B_1, B_2
$\alpha exp(\beta(SOC)) + \gamma exp(\delta(SOC))$	D
Open Circuit Voltage	
$\alpha exp(\beta(SOC)) + \gamma exp(\delta(SOC))$	V_{OC}

Regarding validation of the LPV model, the model was tested by predicting various response data. Testing data included the same data used to identify the LTI model and the data obtained from a different batch of the experiment. In Figure 2, the fit percentages of the prediction of the LPV model and linear model are shown. Results revealed that the LPV model was more accurate than the linear model in most cases for both discharging and charging. Such an outcome occurred due to the effect of SOC change in the LPV model. Thus, the LPV model proved to be more accurate than the linear model over a wider range of SOC. For further validation, the model was tested with data obtained from a different batch of the experiment including the data named MULTI A and MULTI B for discharging and CMULTI for charging. In Figure 2, it was observed that the fit percentage of LPV model prediction was acceptably high although the data were obtained from a different experimental batch.

In Figure 3, graphical prediction results for data MULTI A and CMULTI are displayed. In Figure 3A, for data MULTI A, prediction errors were still observed at some current levels. The highest error occurred upon discharging near the battery depletion zone. Besides, the LPV model performed adequately in predicting the response at the resting zone. As regards CMULTI data, a high error occurred at the beginning range of the predicted response. Such an error may have arisen from the mismatch between the V_{OC} correlation of the model and the resting voltage of this data. It is evident that the LPV model was quite accurate as it was able to address the effect of different currents and SOC changes. This outcome demonstrated the potential of the LPV model and its feasibility for use in SOC estimation with a model-based state estimation algorithm.

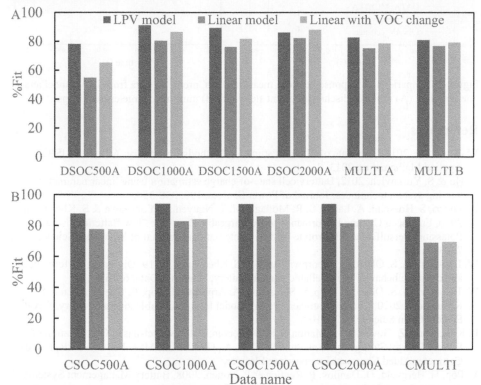

Figure 2 Comparison of fit percentage of model prediction between various models and data for (A) discharging and (B) charging. Fit % can be expressed as: $100 \times \left(1 - \frac{mean|y - \hat{y}|}{mean|y - mean(y)|}\right)$.

5. Conclusion

An LPV model for a tri-electrode ZAFB was proposed. Thus, parameters of the LTI model at various current and SOC values were used to construct the correlations for the LPV model. The experimental data used for dynamic identification and validation were obtained from the laboratory-made tri-electrode ZAFB. Results of the validation confirmed that the LPV model was able to predict the behavior of the ZAFB. Moreover, it was found that the LPV model was more accurate than the linear model based on the comparison using normalized mean absolute error. The proposed LPV model was also able to handle the effect of SOC changes as SOC was one of the scheduling parameters of the model. Overall, results show that the LPV model is a promising dynamic model for predicting the dynamic behavior of a tri-electrode ZAFB.

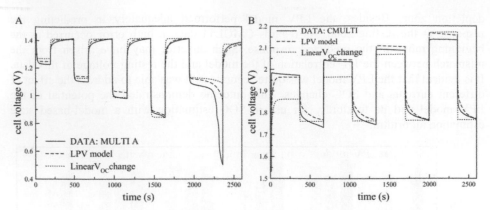

Figure 3 Comparison of response between measured data, predicted data from LPV model and linear model: (A) multiple discharge current steps and (B) multiple charge current steps.

References

W.-Y. Chang, 2013, The State of Charge Estimating Methods for Battery: A Review, ISRN Applied Mathematics, 2013, 7.

Y. Hu & S. Yurkovich, 2012, Battery cell state-of-charge estimation using linear parameter varying system techniques, Journal of Power Sources, 198, 338-350.

R. Khezri, S. Hosseini, A. Lahiri, S. R. Motlagh, M. T. Nguyen, T. Yonezawa & S. Kheawhom, 2020, Enhanced Cycling Performance of Rechargeable Zinc–Air Flow Batteries Using Potassium Persulfate as Electrolyte Additive, International Journal of Molecular Sciences, 21, 7303.

W. Lao-Atiman, S. Olaru, A. Arpornwichanop & S. Kheawhom, 2019, Discharge performance and dynamic behavior of refuellable zinc-air battery, Scientific Data, 6, 168.

W. Lao-Atiman, S. Olaru, S. Diop, S. Skogestad, A. Arpornwichanop, R. Cheacharoen & S. Kheawhom, 2020, Linear parameter-varying model for a refuellable zinc-air battery, Royal Society Open Science, 7, 201107.

L. K. K. Maia, Z. Güven, F. La Mantia & E. Zondervan, 2017, Model-based Optimization of Battery Energy Storage Systems, In: ESPUÑA, A., GRAELLS, M. & PUIGJANER, L. (eds.) Computer Aided Chemical Engineering: Elsevier.

V. Pop, H. Bergveld, D. Danilov, P. Regtien & P. Notten, 2008, Battery Management Systems: Accurate State-of-Charge Indication for Battery-Powered Applications.

N. Radenahmad, R. Khezri, A. A. Mohamad, M. T. Nguyen, T. Yonezawa, A. Somwangthanaroj & S. Kheawhom, 2021, A durable rechargeable zinc-air battery via self-supported MnOx-S air electrode, Journal of Alloys and Compounds, 883, 160935.

D. Schröder & U. Krewer, 2014, Model based quantification of air-composition impact on secondary zinc air batteries, Electrochimica Acta, 117, 541-553.

G. Sukanya, R. Suresh & R. Rengaswamy, 2021, Data-driven prognostics for Lithium-ion battery health monitoring, In: TÜRKAY, M. & GANI, R. (eds.) Computer Aided Chemical Engineering: Elsevier.

N. Wassiliadis, J. Adermann, A. Frericks, M. Pak, C. Reiter, B. Lohmann & M. Lienkamp, 2018, Revisiting the dual extended Kalman filter for battery state-of-charge and state-of-health estimation: A use-case life cycle analysis, Journal of Energy Storage, 19, 73-87.

Y. Zhong, B. Liu, Z. Zhao, Y. Shen, X. Liu & C. Zhong, 2021, Influencing Factors of Performance Degradation of Zinc–Air Batteries Exposed to Air, Energies, 14, 2607.

Proceedings of the 14th International Symposium on Process Systems Engineering – PSE 2021+
June 19-23, 2022, Kyoto, Japan © 2022 Elsevier B.V. All rights reserved.
http://dx.doi.org/10.1016/B978-0-323-85159-6.50333-X

An optimized resource supply network for sustainable agricultural greenhouses: A circular economy approach

Sarah Namany, Ikhlas Ghiat, Fatima-Zahra Lahlou, Tareq Al-Ansari[*]

aCollege of Science and Engineering, Hamad Bin Khalifa University, Qatar Foundation, Doha, Qatar
*talansari@hbku.edu.qa

Abstract

In light of the ever-increasing demand on food products and the associated intensification of agricultural activities, the environment and natural capital are witnessing unprecedented pressures. Alleviating these stresses will require the deployment of more sustainable and resilient food systems that offer a cost efficient and environmentally friendly alternatives to the conventional energy and water intensive food technologies. Greenhouses represent a promising solution to accommodate sustainable food supplies despite uncooperative external climate conditions. However, conventional greenhouses continue to rely on resource inflows originating from unstainable practices and systems such as groundwater abstraction and energy intensive production of fertilisers. Circular economy represents an opportunity to enhance resources utilisation and mitigate environmental burdens associated with their intensive exploitation at competitive costs through deploying novel solutions. In this study, waste integration involving CO_2 enrichment and sequestration along with wastewater reuse in irrigation are investigated for tomato cultivation in a greenhouse system in the State of Qatar. In this regard, a multi-objective optimisation model based on a mixed-integer linear program (MILP) is proposed to determine the optimal technology configuration and supply network that delivers resources for a greenhouse operation. The purpose of this framework is to minimise water and carbon footprints in addition to the total costs associated with resource provision to the greenhouse. Considering results from the pareto front, the optimal distribution requires 54,3579 \$/year and generates around 20,043 kg of CO_{2eq}/year. This option suggests a sustainable cultivation alternative attributed to its environmental efficiency with regards to water savings, CO_2 offsetting, and fertiliser use reduction.

Keywords: Circular economy, greenhouse, wastewater, CO_2 sequestration, optimisation, MILP.

1. Introduction

One of the most pressing global challenges faced nowadays is ensuring food security under restrained resource availability and a continuously growing population. Demand for food is prognosticated to increase by 60% by 2050, which will further increase the consumption of resources, mainly water and energy (Godfray et al., 2010). The agricultural sector already accounts for 70% of the total freshwater withdrawals, and 3.5% to 4.8% of the total energy consumption. As such, there is an impetus to shift agricultural systems to more sustainable practices that will not only produce more food, but also enable an efficient use of resources and reduce the associated environmental impact (Ghiat et al., 2021a, 2020). A key aspect of sustainable agriculture is water irrigation management. Irrigation water is traditionally supplied from unstainable sources, mainly groundwater or desalinated seawater which is an energy-intensive process. The reuse of wastewater for irrigation requirements represents an opportunity for the agricultural sector, whereby it can alleviate stresses on freshwater resources and curb environmental emissions related to desalination processes as well as the production of fertilisers. The use of treated wastewater for irrigation also provides a solution for fertilisation in which nutrient requirements such as nitrogen (N), phosphorus (P) and potassium (K) can be met. The levels of nutrients and contaminants present in the treated wastewater can vary depending on the sector or type of wastewater (e.g. municipal wastewater, industrial wastewater) and can also vary within the same sector (Lahlou et al., 2021). Several studies investigated the use of treated wastewater to meet plant nutrient requirements and concluded that it can be a valuable solution to substitute industrial fertilisers either partially or completely. The reuse of wastewater instead of conventional discharge can also save water resources and reduce the greywater footprint, because it eliminates the need of diluting wastewater with freshwater resources before sea discharge. In fact, the agricultural sector can tolerate higher levels of nutrients and contaminants, especially N, P, and K, as compared to wastewater discharge in the sea, which renders its use in agriculture not only meet plant water requirements but also their nutrient requirements (García-Delgado et al., 2012; Lahlou et al., 2020; Musazura et al., 2019; Raju and Byju, 2019). The need for sustainable intensification of food systems has also led researchers to investigate novel techniques to enhance production and reduce resource consumption such as CO_2 enrichment. CO_2 levels between 800-1200 ppm in the air have been linked to enhanced yields and reduced water consumption due to improved photosynthesis and lower transpiration rates. In this case, CO_2 is perceived as a commodity which can be supplied from industrial carbon capture systems, compressed, and transported either via pipeline or trucks to agricultural production sites (Ghiat et al., 2021b; Nederhoff, 1994). The expensive costs and complexities related to CO_2 capture and transportation are still a major challenge in the implementation of this practice. Many studies have investigated the supply chain of CO_2 enrichment in agricultural greenhouses in efforts to optimise the economic and environmental costs (Govindan and Al-Ansari, 2019). Similarly, this paper contributes to solving logistical complexities in the supply of novel and sustainable agricultural methods.

Previous studies have not considered the optimised allocation of both treated wastewater and CO_2 captured from wastewater treatment plants (WWTPs) to agricultural production systems. Thus, the objective of this study is to develop an optimisation model for the sustainable supply of water, fertilisers, and CO_2 to agricultural greenhouses all from WWTPs. The presented model can be applied to a network of multiple WWTPs and greenhouses. The optimisation model is applied to a case study in the State of Qatar, comprising of two existing WTTPs and an agricultural greenhouse producing tomatoes. The aim is to supply the greenhouse with the necessary water, nutrients, and CO_2 requirements while; 1) minimising the greywater footprint of wastewater, 2) minimising the carbon footprint, and 3) minimising the economic cost. A mixed integer linear program (MILP) model is presented to allocate the necessary resources to the greenhouse from the WWTPs at a reduced greywater footprint, environmental and economic costs.

2. Methodology

2.1. Model Formulation

The methodology proposed in this study consists of designing a multi-objective optimisation model that investigates the benefits of supplying an agricultural greenhouse with the required water, fertilisers, and CO_2 for an enhanced crop yield. The model fosters the concept of circular economy by introducing the reuse of treated wastewater for irrigation and fertilisation purposes, in addition to the enrichment of the greenhouse using CO_2 as means to improve the photosynthesis of the grown crops. In order to conduct this optimisation, a multi-objective mixed-integer linear program (MILP) is developed, wherein the first objective is to maximise the amount of greywater saved by the process of reusing wastewater instead of discharging it (equation 1). The second objective consists of minimising the environmental impact associated with the allocation of water and CO_2 to the greenhouse (equation 2). As for the third objective, it aims to minimise the economic costs associated with the reuse of treated wastewater and the CO_2 generated from different sources (equation 3). The mathematical formulation of the model is presented in the following section.

$$\text{Maximise: } GW = Q_w \sum_{i=1}^{m} \frac{n_i - N_{max}}{(N_{max} - N_w)} x_i^w 10^{-3} \tag{1}$$

$$\text{Minimise: } EC = \sum_{i=1}^{m} x_i^w c_i^w + \sum_{i=1}^{m} x_i^{CC} c_i^{CC} - Q_w \sum_{i}^{m} x_i^w n_i c^N 10^{-3} \tag{2}$$

$$\text{Minimise: } Env = \sum_{i=1}^{m} x_i^w e_i^w + \sum_{i=1}^{m} x_i^{CC} e_i^{CC} - Q_w \sum_{i}^{m} x_i^w n_i CF^N 10^{-3} \tag{3}$$

The decision variables are x_i^w and x_i^{CC} which represent the percentage contribution of different water sources and CO_2 origins to the total requirements of the greenhouse, respectively.

Subject to the following constraints:

$$x_i^w; \ x_i^{CC} > 0 \tag{4}$$

$$\sum_{i=1}^{m} x_i^w = 100\% \tag{5}$$

$$\sum_{i=1}^{m} x_i^{CC} = 100\% \tag{6}$$

Such that:

GW is the grey water footprint in m^3;

EC is the total cost in $;

Env is the total environmental impact represented by Global Warming Potential (GWP) emissions in kg of CO_{2eq};

m is the total number of treated wastewater sources;

i is the index of the treated wastewater source;

Q_w is the total quantity of water required for growing the crop inside the greenhouse in m^3;

n_i is the amount of nitrogen content in wastewater coming from the different sources in mg/L;

N_{max} is the maximum allowable concentration of nitrogen in the treated wastewater mg/L;

N_w is the concentration of nitrogen in the freshwater mg/L;

c^N is the cost of 1kg of nitrogen fertilizer in $/kg_N$;

CF^N is the carbon footprint associated with the production, packaging, and transportation of 1 kg of nitrogen fertilizer in kg-CO_{2-eq}/kg_N;

c_i^w and c_i^{CC} are the total economic costs associated with each source of treated wastewater and CO_2 capture, respectively. They include the cost of transportation of the treated water or CO_2 by means of pipelines, CO_2 compression in addition to the operational cost of the treatment or the capture.

They can be defined using the following equations:

$$c_i^w = (c_i^{wT} d_i + c_i^{wO}) Q_w \tag{7}$$

$$c_i^{CC} = (c_i^{CCT} d_i + c_i^{CCO}) Q_{CC} \tag{8}$$

where c_i^{wT} and c_i^{CCT} are the unit costs of transporting water and CO_2. d_i is the distance between the greenhouse and the wastewater treatment plant, which is also to location of the captured CO_2. c_i^{wo} and c_i^{CCO} are the unit environmental impacts associated with transporting and generating water and CO_2. As for Q_{CC} is the total quantity of CO_2 to be injected in the greenhouse.

2.1. Illustrative example and available data

Increasing water scarcity levels, limited land available for agricultural activities and harsh climatic conditions have rendered the satisfaction of local demands for food and water a challenging target. To overcome this problem, the State of Qatar has shifted the focus to some alternative water and food systems that can meet the local need while preserving the environment. In fact, greenhouses were adopted to support the local food production by providing an adequate environment for crops cultivation. As part of the strategies deployed to improve the efficiency of greenhouses and enhance their yields, CO_2 enrichment represents a sustainable technique to increase the productivity while offsetting the emissions generated from energy intensive technologies such as wastewater treatment. The reuse of wastewater also provides an opportunity to lift the burden on groundwater and represents a cleaner alternative in terms of the engendered emissions in comparison with desalination. In this study, the potential of utilising treated wastewater and CO_2 from two water treatment plants in a greenhouse producing tomatoes. The example is formulated as a planning problem where the allocation of resources from the sources (wastewater treatment plants (WWTP) to the sink (greenhouse) is optimised following the formulation presented in section 2.1. Tables 1 and 2 presents the input data used in the model.

Table 1. Water and carbon requirements of the greenhouse and the distance to the different sources.

Sink	Water Requirements	CO₂ Requirements	Distances from	
			Doha North WWTP	Shahaniyah
	Q_w (m³/year)	Q_{CC} (kg/year)	d_i (km)	
Greenhouse	1,664	14,300	23	8

Table 2. Economic and environmental costs input data.

		Operations (production or treatment)	Transportation and CO2 compression
Unit economic cost	Water treatment	0.11 \$/ m³	0.000671 \$/ m³/km
	CO₂ Capture	0.000241 \$/kg	0.0000172 \$/ kg of CO₂/km
Unit Environmental cost	Water treatment	0.49 kg of CO2eq/ m3	0.00162 kg of CO₂eq/ m³/km
	CO2 Capture	0 kg of CO2eq/ kg	0.036 kg of CO₂eq/ kg of CO₂/km

3. Results

Results of the multi objective optimisation are summarised in Figure's 1 and 2. The optimal solution suggests that 60% of the amount of water supplied to the greenhouse can be sourced from Doha North treatment plant, while the remaining 40% can be supplied by Shahaniyah. As for the CO_2 requirements, Shahaniyah contributes with 96% of total needed quantity while only 4% is taken from Doha North. Considering the costs and savings associated with the optimal solution, the Pareto Front demonstrates that an amount of 2,629 m³/year could be achieved in terms of greywater footprint. As for the economic and environmental costs, the optimal distribution requires 54,3579 $/year and generates around 20,043 kg of CO_{2eq}/year. As for the nutrient's intake, the wastewater used for irrigation supplies 20 kg/year of Nitrogen, which can substitute 15% of the nitrogen supplied from industrial fertilisers that are produced using energy-intensive technologies and processes.

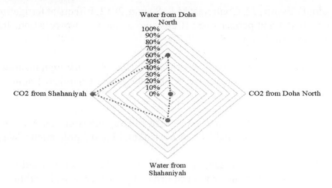

Figure 1: The optimal contribution of each resource source to the sink.

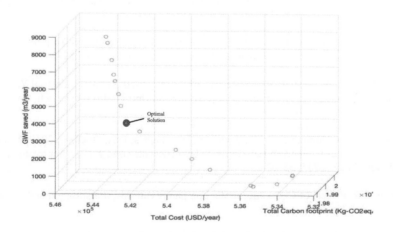

Figure 2: The Pareto Front.

4. Conclusion

Shifting to sustainable agricultural practices is essential in the light of the increasing pressures on the natural resource base and the environment. Adopting alternative water sources and deploying efficient cultivation techniques is deemed beneficial in alleviating the environmental burden that energy-intensive systems are inflicting. In this paper, the potential of using treated wastewater and CO_2 carbon sequestration in a greenhouse is investigated through a multi-objective optimisation model aiming to minimise environmental impacts and economic costs. Results of the study assert that requirements of the greenhouse can be fulfilled from two wastewater treatment plants that can deliver the necessary water, CO_2, and nitrogen. The optimal solution implies an investment of 54,3579 \$/year and produces 20,043 kg of CO_{2eq}/year with a significant reduction of greywater footprint amounting to 2,629 m^3/year. In addition, the use of wastewater instead of the conventional aquifer or desalinated water offsets almost 15% of the nitrogen required by the plant, reducing the dependency on industrial fertilisers.

5. References

C. García-Delgado; E. Eymar; J.I. Contreras; M.L. Segura, 2012, Effects of fertigation with purified urban wastewater on soil and pepper plant (Capsicum annuum L.) production, fruit quality and pollutant contents. Spanish J. Agric. Res., 10, 209.

I. Ghiat; R. Govindan; S. Namany; T.Al-Ansari, 2020, Network optimization model for a sustainable supply network for greenhouses, Computer Aided Chemical Engineering 48, 1885-1890.

I. Ghiat; H.R. Mackey; T. Al-Ansari, 2021, A Review of Evapotranspiration Measurement Models, Techniques and Methods for Open and Closed Agricultural Field Applications. Water, 13, 2523.

I. Ghiat; F. Mahmood; R. Govindan; T. Al-Ansari, 2021, CO2 utilisation in agricultural greenhouses: A novel 'plant to plant' approach driven by bioenergy with carbon capture systems within the energy, water and food Nexus. Energy Convers. Manag., 228, 113668.

H.C.J. Godfray; J.R. Beddington;I.R. Crute; L. Haddad;D. Lawrence; J.F. Muir; J. Pretty; S.Robinson;S.M. Thomas; C. Toulmin, 2010, Food Security: The Challenge of Feeding 9 Billion People. Science (80) 327, 812–818.

F.-Z. Lahlou; H.R. Mackey; T. Al-Ansari, 2021, Wastewater reuse for livestock feed irrigation as a sustainable practice: A socio-environmental-economic review. J. Clean. Prod., 294, 126331.

F. zahra. Lahlou; S. Namany; H.R. Mackey; T. Al-Ansari, 2020, Treated Industrial Wastewater as a Water and Nutrients Source for Tomatoes Cultivation: an Optimisation Approach. Comput. Aided Chem. Eng., 48, 1819–1824.

W. Musazura; A.O. Odindo; E.H. Tesfamariam; J.C. Hughes; C.A. Buckley, 2019, Nitrogen and phosphorus dynamics in plants and soil fertigated with decentralised wastewater treatment effluent. Agric. Water Manag. , 215, 55–62.

E.M. Nederhoff, Effects of CO2 concentration on photosynthesis, transpiration and production of greenhouse fruit vegetable crops.

R. Govindan; T. Al-Ansari, 2019, Simulation-based reinforcement learning for delivery fleet optimisation in CO2 fertilisation networks to enhance food production systems. In Proceedings of the Proceedings of the 29th European Symposium on Computer Aided Process Engineering.

J. Raju; G. Byju, 2019 Quantitative determination of NPK uptake requirements of taro (Colocasia esculenta (L.) Schott). J. Plant Nutr., 42, 203–217

Proceedings of the 14th International Symposium on Process Systems Engineering – PSE 2021+
June 19-23, 2022, Kyoto, Japan © 2022 Elsevier B.V. All rights reserved.
http://dx.doi.org/10.1016/B978-0-323-85159-6.50334-1

Exergoeconomic optimization of a double effect evaporation process of coffee extract

Tinoco-Caicedo D.L.,[a,b,*] Calle-Murillo J.,[a] Feijoó-Villa E.[a]

[a] *Facultad de Ciencias Naturales y Matemáticas, Escuela Superior Politécnica del Litoral, 090903 Guayaquil, Ecuador*
[b] *Department of Process Engineering, Universidad de las Palmas de Gran Canaria, 35017 Gran Canaria, Spain*
*dtinoco@espol.edu.ec

Abstract

The exergoeconomic optimization performed on different industrial processes is used to reduce investment costs, operational costs and the exergy destruction rate in order to increase at the same time the rentability and sustainability of a factory. This study focuses on exergoeconomic optimization of a double effect evaporation process of coffee extract in a factory located in Ecuador. The specific product cost was minimized, and the exergetic efficiency was maximized by a parametric study with an iterative methodology and the integration of a sub-optimized steam recompression system. The parametric study showed that the exergetic efficiency could be increased by 5% and the exergy destruction cost cost rate reduced by $1143/h by changing the concentration of the feed extract, the pressure of the 1st effect and the outlet pressure in the expansion valve to 26 w/w%, 35 kPa and 300 kPa, respectively. However, by integrating the mono-objective optimized steam recompression system in the process, the exergy destruction cost rate could be reduced up to 73%, the exergetic efficiency could be increased by 13% and the specific product cost could be reduced by $225/t.

Keywords: Exergoeconomic Optimization; Double Effect Evaporation; Exergetic efficiency

1. Introduction

The economic growth in the world requires an increase of the energy consumption. The World Council of Energy estimated that by 2030 there will be an increase of approximately 35% in the energy consumption (Tvaronavičienė & Ślusarczyk, 2019). The food factories, such as the plants that produces instant coffee, have one of the highest consumptions of energy (around 60.2 MJ/kg of product) using as energy source fossil fuels (95%), and electricity (5%) (Maroulis & Saravacos, 2007). An exergetic optimization is necessary to improve the sustainability of these industrial processes, by the reduction of the exergy destruction foot print (Romero & Linares, 2014). Also exergoeconomic optimization could be used as a tool to simultaneously improve the thermodynamic and economic performance of a system (Abusoglu & Kanoglu, 2009).

Many methods for the optimization of process have been developed, but they possess certain strengths and weaknesses. The optimization by GA (genetic algorithm) allows for a rapid search of solutions, however they could be lost in a sub-optimal solution (Ding et al., 2020). Utilizing a parametric study results in a better solution when the studied system is complex and modeled by a high resolution simulator or neural networks (Alirahmi & Assareh, 2020). The exergoeconomic optimization has mainly developed for power plants

(Alharbi et al., 2020), trigeneration systems (Gholizadeh et al., 2020), fuel cells (Feili et al., 2020) and similar systems, but for food factories few studies exist

This study proposes a re-engineering of a double effect evaporation process (DEEP) of coffee extract, by integrating steam recompression system. The aim of the present work is minimize the specific product cost by a mono-objective optimization of the recompressor in the evaporation process. Also a parametric study is done to the process in order to balance the investment cost and the exergy destruction cost rates of the system. This study is based on the results of an advanced exergoeconomic analysis of the DEEP of coffee extract presented in a previous study by the authors (Tinoco-Caicedo et al., 2021).

2. Materials and Methods

2.1. System Description

The DEEP of coffee extract integrated to a steam recompressor is presented on Figure 1. The coffee extract (stream 1) is an aqueous solution which is heated in the E-102 with steam (stream S1) generated by a boiler (B-101). The heated coffee extract (stream 3) is pumped to enter to a double effect evaporation system (D-101 and D-102) in order to increase the concentration of soluble solids by using steam as the heating medium. The concentrated coffee extract (stream 9) is cooled with chilled water (stream W3) to prevent development of microorganism, and then it (stream 10) is sent to a drying process in order to produce instant coffee powder. The evaporate water obtained in the evaporators (stream 11) is sent to a mechanical steam recompression (C-101) in order to increase its pressure (stream 12) so it can be reused for heat exchange. For the process simulation, the SRK thermodynamic model is used for high pressure steam and the ideal model for low pressure steam and flue gas because they are at low pressure. The liquid phase is assumed ideal because it is a dilute solution.

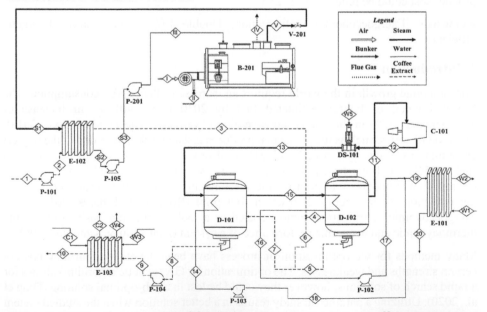

Figure 1. Process flow diagram of the double effect evaporation of coffee extract with a mechanical steam recompression.

2.2. Exergoeconomic Optimization

A mono-objective optimization was performed to the recompressor in order to minimize the specific product cost. The mathematical formulation of the optimization problem is shown in Eq (1)-(4) and based in the literature (Bejan et al., 1996).

$$Z_k^{CI} = \beta(TCI_k) \qquad (1) \qquad\qquad \min_{\varepsilon_k} \dot{C}_{P,k} = \dot{C}_{F,k} + \dot{Z}_k^{CI} + \dot{Z}_k^{OM} \qquad (2)$$

$$TCI_k = B_k \left(\frac{\varepsilon_k}{1-\varepsilon_k}\right)^{n_k} \dot{E}x_{p,k}^{m_k} \qquad (3) \qquad Z_k^{OM} = \gamma_k(TCI_k) + \omega_k \tau \dot{E}x_{p,k} + R_k \qquad (4)$$

Where the subscript k is the k-component, CI the Capital Investment, TCI the Total Capital Investment, β the Capital Recovery Factor, τ is the average annual plant operation hours, $c_{F,k}$ the specific fuel cost which is the electrical energy price, γ_k is a coefficient that takes into account the Operational and Maintenance Cost (OM) and is assumed as 1.06. The ω_k and R_k constants have not been taken into consideration because it is assumed that the variability in OM are negligible, as suggested by Liu & He, 2020. For determining the non-linear regression constants B_k, n_k, and m_k, the TCI was fitted as a function of the exergetic efficiency (ε_k) and product exergy ($\dot{E}x_{P,k}$) by using the software Wolfram Mathematica®. The isentropic efficiencies for the compressor were varied between 50% to 90%. With this assumption the Eqs (5)-(7) are used in order to find the optimal exergetic efficiency (ε_k^{OPT}), and exergoeconomic factor (f_k^{OPT}), for the compressor C-101.

$$\varepsilon_k^{OPT} = \frac{1}{1+F_k} \qquad (5) \qquad F_k = \left(\frac{(\beta+\gamma_k)B_k n_k}{\tau c_{F,k}\dot{E}x_{P,k}^{(1-m_k)}}\right)^{\frac{1}{n_k+1}} \qquad (6) \qquad f_k^{OPT} = \frac{1}{1+n_k} \qquad (6)$$

Furthermore an iterative methodology proposed by (Hamdy et al., 2019) was followed for the exergoeconomic optimization, the routine is presented on Figure 2. For the parametric study, the operational parameters were selected in function of the most critical components presented in the previous exergoeconomic analysis of the base case (Tinoco-Caicedo et al., 2021). In order to balance the overall exergetic destruction cost rate ($\dot{C}_{D,k}$) and investment cost rate ($\dot{Z}_k$) of the process, a value of 50% was considered as the optimum exergoeconomic factor (f_k).

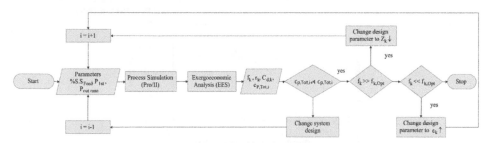

Figure 2. Algorithm used for the exergoeconomic optimization

3. Results and Discussion

The parametric study was performed first by analysing the effect of some important operational parameters presented on Table 1. These parameters were selected because they affect significantly the components which have the highest exergy destruction cost rate ($\dot{C}_{D,k}$) and investment cost rates ($\dot{Z}_k$).

Table 1. Recommended changes for the main components of the process.

Component	$f_k(\%)$	$\varepsilon_k(\%)$	Goal	$S.S_{Feed}$	P_{1st}	$P_{out\,man}$
D-101	85.8%	52%	$\dot{Z}_k \downarrow$	↑	↓	↑
D-102	54.8%	60%	$\dot{Z}_k \downarrow$	↑	↑	-
E-101	2.0%	19%	$\dot{C}_{D,k} \downarrow$	↑	-	-

The table shows some recommended changes in order to adjust the exergoeconomic factor to 50%. In order to achieve a reduction in the investment cost rate of the evaporators (D-101 y D-102), and the exergy destruction cost rate in the heat exchanger E-101, an increment in the concentration of soluble solids (S.S_{Feed}) in the feed extract is required.

On Figure 3. a) it is shown that an increment of the concentration of the feed extract to 26% allows the reduction of the overall exergy destruction cost rate of the process.

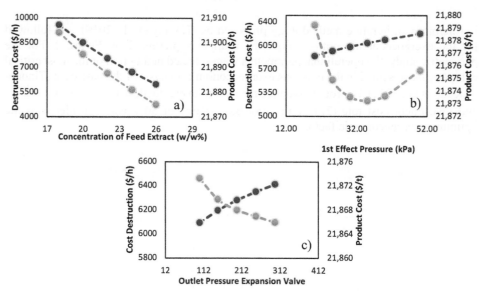

Figure 3. Effect of a) the initial concentration of soluble solids in coffee extract; b) the 1ˢᵗ effect pressure; c) the outlet pressure expansion valve; on the overall exergy destruction cost rate (•) and the product cost (•).

Furthermore, on Figure 3.b) it is shown that a reduction in the 1st effect pressure (P_{1st}) causes a reduction of the exergy destruction cost rate and a negligible increase of the product cost. An intermediate pressure of 33 kPa could minimize the product cost.

Additionally, an increment of the outlet pressure in the expansion valve ($P_{out\ man}$) causes a negligible reduction of the specific product cost as is shown in Figure 5. This is caused because the purchased equipment cost of the first effect evaporator is slightly reduced with the increment of pressure of the steam used for the heat exchanging.

For the mono-objective optimization of the recompression system, the constants for the non-linear regression model (B_k:275.834, n_k:0.85488 and m_k:0.9564) were estimated, with a R^2 of 0.9972. The optimum values of f_k y ε_k were 53,9% and 64% respectively. This result could be obtained when the compressor C-101 has an isentropic efficiency of 65%.

The results of the parametric study and the structural change are summarized on Table 2. It is shown that the steam recompression (structural change) allows for a more significant reduction of the specific product cost and an increase of 10% in the exergetic efficiency. This is caused because most of the exergy destruction rate is due to the condensation of the steam after the evaporation process. When this steam is recompressed and reused in the process, the exergy destruction rate is significantly reduced and therefore the specific product cost is also reduced.

Table 2. Results of the exergoeconomic optimization

Parameter	Units	Base Case	1st	2nd	3rd	Optimized Case
				Parametric Study		Structural Change
$S.S_{Feed}$	%	22	26	26	26	26
P_{1st}	kPa	25	25	35	35	35
$P_{out\ man}$	kPa	101.3	101.3	101.3	300	300
$\dot{C}_{D,Tot}$	$/h	7555	5973	6092	6412	2034
$c_{P,Tot}$	$/t	21888	21875	21873	21866	21663
$\dot{Z}_{Tot}$	$/h	199	182	179	163	203
f_{Tot}	%	2.57	2.96	2.85	2.48	9.09
ε_{Tot}	%	80	84	84	83	93

4. Conclusions

The purpose of this study was to perform an exergoeconomic optimization of the recompressor of the DEEP of coffee extract to minimize de product cost and a parametric study to balance the investment costs and the exergy destruction cost rate. The results of this research show that the optimized recompressor has an isentropic efficiency of 65% and an exergetic efficiency of 64%. Furthermore the structural change had a more significant effect because the specific product cost was reduced by $225/t and the exergy destruction cost rate was reduced by 73%, while the exergetic efficiency increase was 13%. The parametric study allows to reduce the exergy cost rate and increase the exergetic efficiency by 15% and 5% respectively, while the specific product cost was not affected significantly. The initial concentration of soluble solids proved to be a significant parameter for the process, given that an 8 w/w% increase of the initial concentration of soluble solids reduced the avoidable exergy destruction cost by 15%. Finally, the results suggest that a reduction of the exergy destruction rate in the system can be achieved and it is possible to have annual savings of $\$8.37x10^5$ in the overall operating costs. Further research is needed to optimize the solid-liquid extraction of coffee, in order to achieve a higher initial concentration of soluble solids in the coffee extract. Also, the recompression system has to be proved in a pilot scale in order to confirm the results of the simulation.

References

Abusoglu, A., & Kanoglu, M. (2009). Exergoeconomic analysis and optimization of combined heat and power production: A review. *Renewable and Sustainable Energy Reviews*, *13*(9), 2295–2308.

Alharbi, S., Elsayed, M. L., & Chow, L. C. (2020). Exergoeconomic analysis and optimization of an integrated system of supercritical CO2 Brayton cycle and multi-effect desalination. *Energy*, *197*, 117225.

Alirahmi, S. M., & Assareh, E. (2020). Energy, exergy, and exergoeconomics (3E) analysis and multi-objective optimization of a multi-generation energy system for day and night time power generation - Case study: Dezful city. *International Journal of Hydrogen Energy*.

Bejan, A., Tsatsaronis, G., & Moran, M. (1996). *Thermal Design and Optimization*. Wiley.

Ding, P., Yuan, Z., Shen, H., Qi, H., Yuan, Y., Wang, X., Jia, S., Xiao, Y., & Sobhani, B. (2020). Exergoeconomic analysis and optimization of a hybrid Kalina and humidification-dehumidification system for waste heat recovery of low-temperature Diesel engine. *Desalination*.

Feili, M., Ghaebi, H., Parikhani, T., & Rostamzadeh, H. (2020). Exergoeconomic analysis and optimization of a new combined power and freshwater system driven by waste heat of a marine diesel engine. *Thermal Science and Engineering Progress*, *18*, 100513.

Gholizadeh, T., Vajdi, M., & Rostamzadeh, H. (2020). Exergoeconomic optimization of a new trigeneration system driven by biogas for power, cooling, and freshwater production. *Energy Conversion and Management*, *205*(August 2019), 112417.

Hamdy, S., Morosuk, T., & Tsatsaronis, G. (2019). Exergoeconomic optimization of an adiabatic cryogenics-based energy storage system. *Energy*, *183*, 812–824.

Liu, Z., & He, T. (2020). Exergoeconomic analysis and optimization of a Gas Turbine-Modular Helium Reactor with new organic Rankine cycle for efficient design and operation. *Energy Conversion and Management*, *204*(November), 112311.

Maroulis, Z. B., & Saravacos, G. D. (2007). Food Plant Economics. In *Food Plant Economics*.

Romero, J. C., & Linares, P. (2014). Exergy as a global energy sustainability indicator. A review of the state of the art. *Renewable and Sustainable Energy Reviews*, *33*, 427–442.

Tinoco-Caicedo, Feijoó-Villa, E., Calle-Murillo, J., Lozano-Medina, A., & Blanco-Marigorta, A. M. (2021). Advanced exergoeconomic analysis of a double effect evaporation process in an instant coffee plant. In *Computer Aided Chemical Engineering*.

Tvaronavičienė, M., & Ślusarczyk, B. (2019). *Energy Transformation Towards Sustainability*. Elsevier Ltd.

Proceedings of the 14th International Symposium on Process Systems Engineering – PSE 2021+
June 19-23, 2022, Kyoto, Japan © 2022 Elsevier B.V. All rights reserved.
http://dx.doi.org/10.1016/B978-0-323-85159-6.50335-3

Ecohydrological modeling and dynamic optimization for water management in an integrated aquatic and agricultural livestock system

A. Siniscalchi [a,b], R. J. Lara [c], M. S. Diaz [a,b,*]

[a]*Planta Piloto de Ingeniería Química (PLAPIQUI CONICET-UNS), Camino La Carrindanga km. 7, Bahía Blanca, Argentina*
[b]*Departamento de Ingeniería Química, Universidad Nacional del Sur (UNS), Bahía Blanca, Argentina*
[c]*Instituto Argentino de Oceanografía (IADO-CONICET), Camino La Carrrindanga 7 Km7, Bahía Blanca, Argentina*
sdiaz@plapiqui.edu.ar

Abstract

In this work, we address the water-food-energy nexus by formulating an optimal control problem to mitigate extreme events, such as droughts and floods, by taking into account aquatic, agricultural livestock systems with associated carbon capture, as well as the required energy for irrigation.

The resulting dynamic optimization problem is subject to a differential algebraic equation system with four modules that include ecohydrological, agricultural and livestock models, as well as the associated carbon dioxide capture. Energy is calculated to pump water in aqueducts, if necessary, and for drip irrigation. Numerical results show that the integrated management of the aquatic and agricultural livestock system in a semiarid region can effectively mitigate, even avoid, the effects of droughts and floods, while improving the economic incomes in the region.

Keywords: optimal control, ecohydrological model, dynamic optimization, water-food-energy-health nexus.

1. Introduction

According to the UN Food and Agriculture Organization (FAO), water footprint associated to food production is around 70% of global water consumption. Water footprint is an indicator of the amount of water required for a production process and is used as a basis for more efficient management of this valuable resource (FAO, 1998). In arid and semi-arid regions, water is a resource of highly variable availability, where management becomes fundamental to reduce uncertainty in predicting mid- and long-term oscillations and thus supporting sustainable ecosystem services, particularly food safety food production. Droughts cause death of livestock, since animals cannot get access to drinking water, do not fatten and do not produce milk due to high temperatures associated to the lack of water and shade. Further, the uncontrolled variation of hydrochemical properties of water bodies can produce deleterious effects on reproduction and growth of valuable fish species. In Argentina, floods cause economic losses that can represent 1.1% of the Gross Domestic Product (World Bank, 2000), including crops, rural and urban inundation They increase uncertainty in investment strategies of touristic developments related to water bodies. The formation of transient, slow-moving, shallow water surfaces can promote the explosive proliferation of disease vectors such as mosquitoes, snails, parasites, etc. Similarly, migration of wild birds that use wetlands for breeding and resting can exchange viruses with domestic poultry. In the last years, the water-food-energy (WFE) nexus concept has been promoted as a tool for achieving sustainable development accounting the relationship among these three resources. From a nexus perspective, integrated aquatic and agricultural livestock systems approaches, can help to solve some water-food-energy issues. Therefore, to achieve sustainability of hydrographic basins, interactions among the WFE fluxes and the ecosystems fluxes. For this reason, mathematical models that allow simulation and optimization representing environmental extreme events, such as droughts and floods, become fundamental tools for decision-making in the socio-economic development of a region (Siniscalchi et al, 2018; 2019).

In this work, we propose the formulation of an optimal control problem to address mitigation of extreme environmental events such as floods and droughts. The preservation of a valuable fish species within a salt lake is also addressed, as well the use of water as a resource in a productive livestock-agricultural system located in a semi-arid region. The optimization problem is constrained with a system of differential and algebraic equations representing ecohydrological and agricultural livestock models of a salt lake and its basin. Within this framework, four main objectives are addressed: a) to prevent flooding of a nearby village and its touristic areas during a wet period by diverting part of the flow from a Chasicó Lake tributary into a constructed reservoir (the diversion flowrate is a control variable); b) to optimize management of the constructed reservoir to keep salinity in the lake within a desired value for fish species (silverside) during drought periods; c) to include restoration strategies for native species that comprise a xerophilic woodland currently existing within the salt lake basin, combining new plantations of Prosopis sp, with drought resistant crops (*Chenopodium quinoa*) and pasture (*Eragrostis curvula*) ,irrigated with freshwater taken from the proposed constructed artificial reservoir and d) to provide drinking water and shade to cattle. For the last three objectives, the outlet freshwater flowrate from the constructed reservoir is a control variable. Numerical results show that if water is accumulated in an artificial reservoir during wet periods (six-year period, with average annual precipitations of 650 mm), a subsequent ten-year drought period (average annual precipitations 250 mm) can be mitigated, while maintaining the salinity of Chasicó Lake for the conservation of silverside fishing. In this way, during the dry period, quinoa and pasture can be sown and Prosopis sp. can provide shade and fodder to cattle and long-term ecosystem benefits. The proposed dynamic optimization model has proven to be a powerful tool for water management in an integral way.

2. Methodology

We propose an optimal control problem for medium term planning of management strategies in integrated aquatic, agricultural livestock systems in semiarid regions under extreme events, such as droughts and floods. The main concept is to address this problem under the water-food-energy (WFE) nexus. The dynamic optimization problem is subject to a differential algebraic equations system that represents the integrated model. The objective function is an integral one that aims to keep a salt lake volume (and its associated salinity, as it is an endorheic basin), at a desired value, to avoid flooding of the nearby village and fields and to keep salinity at optimal values for reproduction of a valuable fish species, silverside (*Odontheses bonariensis*), which has been exported to several countries, as Japan. There are two control variables: the diverted stream flow to a nearby constructed reservoir to avoid flooding while keeping salinity at 23 kg/m^3 in a wet scenario and the stream flow that is fed to the salt lake from the constructed reservoir to keep lake salinity at the desired value in dry periods $Q_{Res}(t)$. When solving the dynamic optimization problem, the definition of a "wet" or "dry" period is not required, since this is determined by the input profiles provided for precipitations, temperature, wind and current conditions at daily time intervals.

The agroecohydrological model has four integrated submodels:

a) Ecohydrological model. It includes dynamic mass balances for water and salt within a salt lake and an artificial fresh water reservoir that is constructed to derive water during wet periods and to provide water for irrigation and to keep lake salinity within low values in dry periods.

$$\frac{dm}{dt} = \left[Q_{pp}(t)\left(\frac{V(t)}{h(t)}\right) + Q_{river}(t) + Q_{gw}(t) - \text{Evap}(t)\left(\frac{V(t)}{h(t)}\right) \right] \delta_w / 1000 \tag{1}$$

here m is the total water mass in the lake (kg); δ_w, is water density (kg.m^{-3}), which is assumed constant; V corresponds to salt lake volume (m^3) and h is average depth (m). Q_{pp} (L.day^{-1}.m^{-2}) corresponds to precipitations, Q_{gw} (L.day^{-1}) is groundwater flowrate. Q_{river} L.day^{-1}) is the tributary discharge to Chasicó Lake, which is calculated as

$$Q_{river}(t) = Q_{cr}(t) - Q_{divert}(t) + Q_{res}(t) \tag{2}$$

where Q_{divert} (L.day-1) is the stream flowrate that could be diverted when necessary from Chasicó River (Q_{cr}) (L.day-1) to the reservoir during wet periods, to keep Chasicó Lake salinity within desired values (it is a control variable). Q_{res} (L.day-1) is the daily water amount that could be diverted from the reservoir to Chasicó River and, subsequently to the salt lake for salinity and volume control (this is also a control variable) in drought periods, if required. Evaporation per unit area (*Evap*, L.day-1.m-2) is calculated taking into account energy and momentum balances (Penman, 1948; Siniscalchi et al., 2018a). As salt concentration in both groundwater and the tributary is negligible, we assume that salt mass is constant within the salt lake and salt concentration (Cs) is calculated as:

$$\frac{dCs}{dt} = -\frac{Cs(t)}{V(t)} \left[Q_{pp}(t)\left(\frac{V(t)}{h(t)}\right) + Q_{river}(t) + Q_{gw}(t) - Evap(t)\left(\frac{V(t)}{h(t)}\right) \right] \delta_w / (\delta_w 1000) \tag{3}$$

The water mass balance in the artificial reservoir is as follows

$$\frac{dVres}{dt} = Q_{divert}(t) + Q_{well}(t) + Q_{pp}(t)A(t)\delta_{H_2O} - Evap(t)A(t)\delta_{H_2O} - Q_{res}(t) - \sum_j Q_j(t) , \qquad j = Prosopis\ sp,\ \text{quinoa, } E.\ curvula.,\ \text{cattle} \tag{4}$$

Where $Q_{well}(t)$ corresponds to a stream from a flowing well; $Q_j(t)$ is the daily water requirement for *Prosopis sp* (once seedlings are transplanted to field), quinoa, *Eragrostis curvula* irrigation and water requirement for cattle, respectively. Water requirement for trees and crops is calculated as a function of crop evapotranspiration $ETq(t)$ and precipitations.

b) Agricultural model. It includes grow models and evapotranspiration calculations for *Prosopis sp.*, quinoa (a high value crop), and a native pasture species, *Eragrostis curvula*. Biomass for Prosopis is calculated as follows.

$$BP = e^{(-2.14+2.530*ln\,(DBH))} \tag{5}$$

$$DBH = 6.1\ 10^{-4}(t) + 2.43 \tag{6}$$

where DBH stands for diameter breast height of Prosopis stem
Actual evapotranspiration is $ETi(t)$ is calculated as:

$$ETi(t) = Ki(t) * ET0(t)\quad ,i=Prosopis,\ quinoa,\ E.\ curvula. \tag{7}$$

where $Ki(t)$ correspond to the cultivation "coefficient" for each species, which is represented with Fourier series, which were adjusted based on data on their water requirement through the different growth stages in a semiarid region, based on data from from the literature. $ET0$ is calculated as function of air temperature, solar radiation, wind, vapor pressure, etc (Siniscalchi et al., 2019).

c) Livestock model. It includes water requirement calculation for both breeding (270kg) and fattening (450kg) cattle, which is calculated at each time interval, as follows:

$$Q_{cattle270}(t) = [16.765 * e^{0.0295*T_{air}}] * N_1 \tag{8}$$

$$Q_{cattle455}(t) = [27.64 * e^{0.0288*T_{air}}] * N_2 \tag{9}$$

where N_1 and N_2 is the number of cows from each size. Correlations are based on data collected from the semiarid region that constitutes our case study.

d) Carbon capture model. Carbon capture (kg C/d) is calculated as function of tree and crop biomass as follows,

$$CC_{Eragrostis}(t) = 0.3 * Biomass_{Ec} \tag{10}$$

$$CC_{Prosopis}(t) = 0.5 * Biomass_{Psp} \tag{11}$$

Finally, energy calculations are carried out for pumps used in drip irrigation for crop and pasture.

3. Numerical results

In this work, the case study is Chasicó Lake and its endorheic basin, located in a semiarid region in Argentina. It is a salt lake. The optimal control problem described in Section 2 has been implemented and solved with a control vector parameterization methodology within gPROMS (Siemens Process Systems Engineering, 2020), for a time horizon of ten years of drought, considering an initial lake salinity of 22.9 kg/m^3 and an initial reservoir volume of 4.48E8 m^3. Freshwater flowrate from a 700 m depth flowing well is 348 m^3/h ($Q_{well}(t)$). Precipitation, temperature, wind profiles are represented by Fourier series whose parameters have been estimated based on historical data for dry periods (Siniscalchi et al., 2019). Average total annual precipitations are 256 mm and average calculated evaporation in both the salt lake (6000 ha surface area) and the artificial reservoir (2800 ha surface area) is 1104 and 770 mm, respectively. The objective is to keep the salt lake volume within 4.8E8 m^3 or, equivalently, lake salinity within 23 kg/m^3, which is optimal for silverside fish reproduction. The dynamic optimization problem has 13 differential equations and 42 algebraic ones. Total CPU time was 7168 s, considering time intervals of 15 days within a time horizon of 10 years.

Figure 1 shows salinity profiles without management; i.e., simulation results. It also shows profiles obtained with the proposed management strategy; i.e., solving the optimal control problem (by deriving freshwater from the artificial reservoir). It can be seen that, without management, salinity values increase to 73 kg/m^3, deeply affecting fish reproduction and survival.

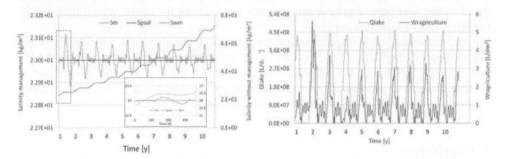

Figure 1: Salinity profiles with (green lines) and without management (blue line)

Figure 2: Water requirement for agriculture (blue line) and optimal profile for control variable (Q_{lake}) (orange line)

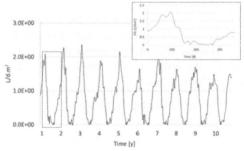

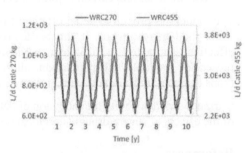

Figure 3: Evapotranspiration profile for *Eragrostis curvula*

Figure 4: Water requeriment for breeding (red line) and fattening (blue line) cattle

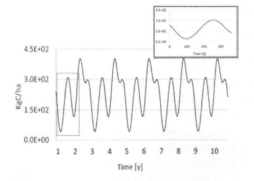

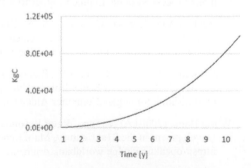

Figure 5: Carbon storage by *E. curvula*

Figure 6: Carbon storage in *Prosopis* wood

When the optimal control problem is solved, salinity is kept around 23 kg/m^3 throughout the 10-year dry period, by deriving freshwater, Q_{lake}, from the artificial reservoir. Q_{lake} profiles (orange lines) are shown in Fig. 2, together with other water stream flows taken from the artificial reservoir and used for trees, crop and pasture irrigation. Figures 3 to 6 show variable profiles in the optimal solution, i.e., obtained by applying the proposed management strategy. Figure 3 shows *Eragrostis c.* evapotranspiration profile, with a detail of the first year. Figure 4 shows water requirement for cattle, which includes 30 breeding and 70 fattening cattle.

Figure 5 shows carbon storage in *Eragrostis c.*, with a detail of the first year profile, while Fig. 6 shows carbon storage in *Prosopis sp.* (a native tree species) wood, during the 10-year period. The function is a sigmoidal one for a 30-year life cycle, but the studied period corresponds to the exponential growth phase. It can be noted that *Eragrostis c.* carbon capture is two orders of magnitude lower, as it is pasture for cattle.

4. Conclusions

In this work we have formulated an integrated model for planning management strategies under extreme weather events in aquatic, agricultural and livestock systems, by formulating an optimal control problem. Dynamic models are formulated for each system, rendering dynamics for water flowrates, salinity within the lake, volume, crop, pasture and trees biomass and carbon capture, among others.

The model has proven to be an efficient tool to plan management within the concept of water-food-energy nexus. It can be demonstrated that the effects of droughts can be effectively mitigated, even avoided, while improving the economic incomes in the region preserving valuable fish population, allowing for the cultivation of high value-added crops (quinoa), pasture and even growing cattle in dry periods.

Current work deals with the inclusion of detailed energy calculation, as well as the study of the effects of severe droughts on local population health.

References

FAO - Food and Agriculture Organization of the United Nations, Allen, R., Pereira, L., Raes, D., Smith, M. Crop evapotranspiration - Guidelines for computing crop water requirements (1998)

Siemens Process Systems Engineering, gPROMS User Manual (2020)

Siniscalchi A.G., Kopprio G., Raniolo L.A., Gomez E.A., Diaz M.S., Lara R.J. Mathematical modelling for ecohydrological management of an endangered endorheic salt lake in the semiarid Pampean region, Argentina. Journal of Hydrology(2018) 563: 778–789

Siniscalchi, A.G., García Prieto, VC., Raniolo, A., Gomez, E., Lara, R.J., Diaz, M.S. Ecosystem Services Valuation and Ecohydrological Management in Salt Lakes with Advanced Dynamic Optimisation Strategies.Computer Aided Chemical Engineering (2019) 46:1579-1584

World Bank. (2000) Argentina - Water Resources Management: Policy Elements for Sustainable Development in the 21st Century, Main Report. Washington, DC. https://openknowledge.worldbank.org/handle/10986/14980.

Proceedings of the 14th International Symposium on Process Systems Engineering – PSE 2021+
June 19-23, 2022, Kyoto, Japan © 2022 Elsevier B.V. All rights reserved.
http://dx.doi.org/10.1016/B978-0-323-85159-6.50336-5

Parametric Analysis of Ortho-to-Para Conversion in Hydrogen Liquefaction

Amjad Riaz[a], Muhammad Abdul Qyyum[b], Arif Hussain[c], Muhammad Islam[a], Hansol Choe[a], Moonyong Lee[a*]

[a]*Process Systems Design & Control Lab (PSDC), School of Chemical Engineering, Yeungnam University, Gyeongsan-si, Gyeongsangbuk-do 38541, Republic of KOREA*
[b]*Department of Petroleum & Chemical Engineering, Sultan Qaboos University, Muscat, OMAN*
[c]*Department of Chemical Engineering, COMSATS University Islamabad, Lahore Campus, 54000, Lahore, PAKISTAN*
mynlee@yu.ac.kr

Abstract

Hydrogen is an energy carrier and is produced just like electricity. Hydrogen is liquefied for storage and transportation purposes to overcome the shortcomings of its low molecular weight and energy density per unit volume. The liquefaction of hydrogen is different from that of other substances as it involves a reactive transformation of its isomers: ortho-hydrogen and para-hydrogen. As the temperature decreases, the equilibrium concentration shifts toward a higher para- content from the normal concentration of 25 % at 25 °C. Para-hydrogen is preferred because of its lower boil-off rate, which is a major challenge at cryogenic temperatures. Ortho-para conversion, heat leak, sloshing, and flashing are considered as the reasons for such losses. The self-conversion rate of hydrogen in a non-equilibrium state is extremely slow; however, at cryogenic temperatures, o-p conversion is an exothermic affair. From the liquefaction point of view, this exothermic heat of conversion is an added work, increasing he liquefaction energy requirement by about 15 %. Catalysts are used to achieve the equilibrium concentration of p-H_2 at a finite rate. Little work has been done from the process systems point of view regarding o-p H_2 conversion. Therefore, parametric analysis of this vital conversion reaction, the spatial distribution of intermediate heat exchangers, and impact on the energy efficiency of the liquefaction process have been studied and partially presented here.

Keywords: Hydrogen Economy; Hydrogen Liquefaction; Ortho-Para Conversion; Hydrogen Energy Network.

1. Introduction

Hydrogen (H_2) is considered a fuel that may revolutionize the future energy mix. It is the only zero-emission, sustainable, and flexible energy carrier with the potential to overcome multifaceted problems and serve as a comprehensive solution to the carbon footprint. High purity, high energy mass density, and high-power density are the hallmarks of hydrogen. Whereas, apart from its production woes, H_2 does face some storage and transportation issues. However, the low molecular weight of H_2, which results in a very low volumetric energy density, impedes its adoption as an energy vector, especially from the perspective of storage and bulk transport (Abdalla et al., 2018; Zheng et al., 2019). H_2 is usually stored and transported as compressed gas or cryogenic liquid. Cryogenic liquid hydrogen (LH$_2$) has three to four times higher volumetric energy density than compressed H$_2$ (International Energy Agency (IEA), 2019). LH$_2$ is usually preferred over the gaseous form for longer distances or high volumes. With the advent and commercial success of liquified natural gas (LNG), H$_2$ liquefaction is drawing more interest and attention ever.

LH$_2$ has a very low boiling point (-253 °C). Commercially, H$_2$ is liquefied in a three-stage process: precooling (up to -193 °C), cooling (up to -243 °C), and liquefaction (up to -253 °C). Liquid nitrogen is used as a refrigerant in the precooling phase, while a combination of JT valves or expanders and H$_2$ itself is used in the subsequent stages (Riaz et al., 2021). The specific energy consumption (SEC) is the benchmark used for comparing commercial and conceptual process designs; the ideal value for a feed at 25 bar is approximately 2.7 kWh/kg$_{LH2}$ (Aasadnia & Mehrpooya, 2018; Stolzenburg et al., 2013). However, the commercial plants operate at SEC in the range of 12.5–15.0 kWh/kg$_{LH2}$ (Yin & Ju, 2020).

Moreover, there are two variants of molecular H$_2$ depending upon the relative orientation of the nuclei of its constituent atoms: ortho-H$_2$ (nuclei spin in the same direction) and para-H$_2$ (nuclei spin in the opposite direction). The mixture composed of 75 % ortho-H$_2$ (o-H$_2$) and 25 % para-H$_2$ (p-H$_2$) at 25 °C is called Normal H$_2$ (n-H$_2$). The two isomers are not at the same energy level with o-H$_2$ being the excited state. The temperature dependent equilibrium between the two isomers shifts towards a more stable form with the decrease in temperature. This shift is called ortho-to-para conversion (OPC), and has a significant contribution to the SEC.

The OPC is critical from the storage and transportation point of view. If n-H$_2$ is liquefied without OPC, a portion of liquid will boil and trigger mass vaporization and energy losses. This happening is called boil-off and may vaporize 50% of the storage (Sherif et al., 2014) starting from 1% per hour (Baker & Shaner, 1978; Weitzel et al., 1958). The best way to mitigate these boil-off losses is to convert and store as p-H$_2$ (Ghorbani et al., 2019). As the initial concentration of o-H$_2$ decreases, there are lesser chances of boil-off losses, as depicted by Figure 1. Therefore, OPC is not neglected commercially, and a p-H$_2$ concentration >95 % is ensured (Stolten & Scherer, 2013). The best way is to conduct OPC simultaneously with the liquefaction. Therefore, the commercial processes use the approach of converting ortho- to para-hydrogen together with liquefaction.

2. Ortho–to–Para Conversion

The impact of this spin on the properties of matter is so weak that the spin isomers almost remain conserved (Zhang et al., 2021). The fast cooling and instant liquefaction does not alter the molecular composition of H$_2$ because of the slow OPC. It has been reported that the reaction is second-order (reaction rate = 0.0114 h^{-1}) and takes a long time to establish equilibrium in the absence of a catalyst. This mode of OPC is called self-conversion, which is also governed by temperature.

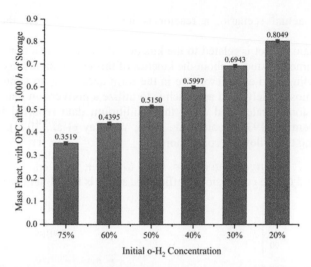

Figure 1 Effect of OPC on the liquid hydrogen boil-off rate as a function of storage time (McCarty et al., 1981)

The introduction of a suitable catalyst enhances the rate of reaction manifolds, magnetic materials, and radiation fields are often employed for the purpose. Generally, catalysts like iron oxide and nickel silica are pakced on the hydrogen side of the exchangers in an arrangement similar to a shell and tube heat exchanger (Jacob H. Stang et al., 2006). The following equation describes the chemical reaction inside the catalytic heat exchanger reactor:

$$o - H_2 \quad \rightarrow \quad p - H_2 \ + \ \Delta H \tag{1}$$

It is said that the catalysis increases the energy requirements by 15 %, but it is the OPC that increases the heat load by around 15–20 % (Stetson et al., 2016; Stolten & Scherer, 2013). The additional heat duty is the exothermic enthalpy of conversion, i.e., 527 kJ/kg, higher than the latent heat of H_2 (~447 kJ/kg), resulting in vaporizing the liquid or heating the already cooled part.

The equilibrium constants of the ortho-para conversion of H_2 in the ideal gas state are independent of pressure. Accordingly, pressure does not appreciably change the ortho-para ratio under equilibrium conditions. Although the lowest rotation levels of the ortho and para varieties differ, ΔE (internal energy change) for the reaction is zero.

3.   Research Methods

In recent times, most studies related to H_2 liquefaction are theoretical, i.e., modeling and simulation. The fact that the properties of spin isomers of H_2 differ, especially those related to temperature. Estimation of thermodynamic properties is at the core of process design, either modeling or simulation. Simulation is preferred because of the availability of more reliable components database and precise thermodynamic models. Peng-Robinson is the most widely used model for H_2 liquefaction processes; however, recent studies have preferred the modified version of the original Benedict-Webb-Rubin equation of state. At present, simulating such a reactive exchanger with the catalyst filled in the H_2 side of the exchanger is not possible in commercial simulation software.

To mimic the actual scenario, a reactor is installed right after the exchanger in simulation.

Another important aspect is related to the kinetics of the OPC reaction. Although there are many experimental studies about the kinetics of this conversion, very few are related to kinetic modeling, and even fewer are in the form acceptable to simulation software. Therefore, the most widely used approach is to utilize a conversion reactor. The overall percent conversion is calculated using the equilibrium data reported in the literature (Harkness & Deming, 1932; Scott et al., 1964; Woolley et al., 1948), as presented in Figure 2 by using the following correlation.

$$Conversion\ (\%) = C_o + C_1 \times T + C_2 \times T^2 \tag{2}$$

where C_o, C_1, and C_2 are conversion coefficients, and T is the H_2 temperature in Kelvin.

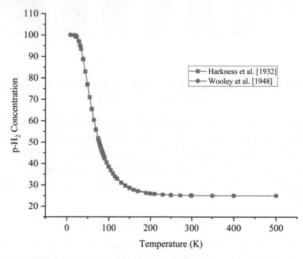

Figure 2 Change in equilibrium p-H_2 concentration with temperature (Harkness & Deming, 1932; Woolley et al., 1948)

4. Further Discussion

Aspen Hysys® has been used in this study which uses Eq. (2) to calculate the percent total conversion. Depending on the process configuration and the number of reactors used, the values reported in the literature vary. The values calculated for the conversion coefficients of each reactor in our recent contribution (Qyyum et al., 2021) are presented in Table 1 as an example. The conversion coefficients were obtained via fitting such that the p-H_2 concnetraion and temperature at each reactor's outlet coincided with the reported data. A similar approach was adopted by Ghorbani et al. (2019), Hammad & Dincer (2018), and Sadaghiani & Mehrpooya (2017) to name a few. The problem with this approach is that the operating conditions, especially temperature, have to be fixed at the reactor outlet; otherwise, the coefficients must be calculated time and again. Also, the heat of conversion is not entirely reflected in the reactor outlet stream leading to errors in the SEC and exergy efficiency calculations. A more appropriate way is to use the equilibrium H_2 data so that the temperature dependency and heat of conversion are accounted for.

Another aspect is the mode of operation for these reactors. The commercial processes use liquid nitrogen/hydrogen baths to carry out the OPC in an isothermal environment, while adiabatic reactors are also used in between. The current process schemes do not consider this very important factor. The conceptual studies' specific energy and exergy calculations err by approximately 20 %, to say the least. The temperature thresholds for each exchanger/reactor shall not remain the same once an adiabatic reactor is considered, whose impact on the overall energy scenario is challenging to comprehend in the simulation environment.

Table 1 Coefficients for percentage conversion to be used in the conversion reactor model in Aspen Hysys® (Qyyum et al., 2021)

Reactor ID	Conversion Reactor Coefficients		
	C_0	C_1	C_2
R-1	66.12	-0.4125	1.168×10^{-3}
R-2	85.35	-0.3325	1.118×10^{-2}

5. Conclusions

The present study has considered a brief analysis of OPC in the broader context of H_2 liquefaction. The limitations of the current simulation approaches have been highlighted, ranging from property estimation techniques to the use of conversion reactors. The mode of operation of these reactors results in incorrect energy consumption estimates. Therefore, a more thorough analytical analysis may pave the way to develop an understanding of the complex H_2 liquefaction process, which not only saves energy but also help solve the process design issues.

Acknowledgment

This work was supported by the National Research Foundation of Korea (NRF) grant funded by the Korean government (MSIT) (2021R1A2C1092152); Priority Research Centers Program through the National Research Foundation of Korea (NRF) funded by the Ministry of Education (2014R1A6A1031189); "Human Resources Program in Energy Technology" of the Korea Institute of Energy Technology Evaluation and Planning (KETEP), granted financial resource from the Ministry of Trade, Industry & Energy, Republic of Korea. (*No. 20204010600100*).

References

Aasadnia, M., & Mehrpooya, M. (2018). Large-scale liquid hydrogen production methods and approaches: A review. In *Applied Energy*. https://doi.org/10.1016/j.apenergy.2017.12.033

Abdalla, A. M., Hossain, S., Nisfindy, O. B., Azad, A. T., Dawood, M., & Azad, A. K. (2018). Hydrogen production, storage, transportation and key challenges with applications: A review. In *Energy Conversion and Management* (Vol. 165, pp. 602–627). Elsevier Ltd. https://doi.org/10.1016/j.enconman.2018.03.088

Asadnia, M., & Mehrpooya, M. (2017). A novel hydrogen liquefaction process configuration with combined mixed refrigerant systems. *International Journal of Hydrogen Energy*, *42*(23), 15564–15585. https://doi.org/10.1016/j.ijhydene.2017.04.260

Baker, C. R., & Shaner, R. L. (1978). A study of the efficiency of hydrogen liquefaction. *International Journal of Hydrogen Energy*. https://doi.org/10.1016/0360-3199(78)90037-X

Ghorbani, B., Mehrpooya, M., Aasadnia, M., & Niasar, M. S. (2019). Hydrogen liquefaction process using solar energy and organic Rankine cycle power system. *Journal of Cleaner Production*. https://doi.org/10.1016/j.jclepro.2019.06.227

Hammad, A., & Dincer, I. (2018). Analysis and assessment of an advanced hydrogen liquefaction system. *International Journal of Hydrogen Energy*. https://doi.org/10.1016/j.ijhydene.2017.10.158

Harkness, R. W., & Deming, W. E. (1932). The Equilibrium of Para and Ortho Hydrogen.

Journal of the American Chemical Society, 54(7), 2850–2852.
https://doi.org/10.1021/ja01346a503

International Energy Agency (IEA). (2019). The Future of Hydrogen - Seizing today's opportunities. In *The Future of Hydrogen - Seizing today's opportunities* (Issue June). https://doi.org/10.1787/1e0514c4-en

Jacob H. Stang, Nekså, P., & E. Brendeng. (2006). On the design of an effient hydrogen liquefaction process. *16th World Hydrogen Energy Conference 2006, WHEC 2006.*

McCarty, R. D., Hord, J., & Roder, H. M. (1981). *Selected Properties of Hydrogen (Engineering Design Data), National Bureau of Standards Monograph 168.* https://doi.org/Library of Congress Catalog Card Number: 80-600195

Qyyum, M. A., Riaz, A., Naquash, A., Haider, J., Qadeer, K., Nawaz, A., Lee, H., & Lee, M. (2021). 100% saturated liquid hydrogen production: Mixed-refrigerant cascaded process with two-stage ortho-to-para hydrogen conversion. *Energy Conversion and Management, 246*, 114659. https://doi.org/10.1016/J.ENCONMAN.2021.114659

Riaz, A., Qyyum, M. A., Min, S., Lee, S., & Lee, M. (2021). Performance improvement potential of harnessing LNG regasification for hydrogen liquefaction process: Energy and exergy perspectives. *Applied Energy, 301*(April), 117471. https://doi.org/10.1016/j.apenergy.2021.117471

Sadaghiani, M. S., & Mehrpooya, M. (2017). Introducing and energy analysis of a novel cryogenic hydrogen liquefaction process configuration. *International Journal of Hydrogen Energy.* https://doi.org/10.1016/j.ijhydene.2017.01.136

Scott, R. B., Denton, W. H., & Nicholls, C. M. (Eds.). (1964). *Technology and Uses of Liquid Hydrogen.* Pergamon Press.

Sherif, S. A., Goswami, D. Y., Stefanakos, E. K., & Steinfeld, A. (2014). *Handbook of Hydrogen Energy.* Taylor & Francis. https://books.google.co.kr/books?id=jGkLBAAAQBAJ

Stetson, N. T., McWhorter, S., & Ahn, C. C. (2016). Introduction to hydrogen storage. In *Compendium of Hydrogen Energy* (pp. 3–25). Woodhead Publishing. https://doi.org/10.1016/b978-1-78242-362-1.00001-8

Stolten, D., & Scherer, V. (2013). Transition to Renewable Energy Systems. In *Transition to Renewable Energy Systems.* https://doi.org/10.1002/9783527673872

Stolzenburg, K., Berstad, D., Decker, L., Elliott, A., Haberstroh, C., Hatto, C., Klaus, M., Mortimer, N. D., Mubbala, R., Mwabonje, O., Nekså, P., Quack, H., Rix, J. H. R., Seemann, I., & Walnum, H. T. (2013). *Efficient Liquefaction of Hydrogen: Results of the IDEALHY Project. November,* 7–9. https://www.idealhy.eu/uploads/documents/IDEALHY_XX_Energie-Symposium_2013_web.pdf

Weitzel, D. H., Loebenstein, W. V., Draper, J. W., & Park, O. E. (1958). Ortho-para catalysis in liquid-hydrogen production. *Journal of Research of the National Bureau of Standards, 60*(3), 221–227. https://doi.org/10.6028/jres.060.026

Woolley, H. W., Scott, R. B., & Brickwedde, F. G. (1948). Compilation of thermal properties of hydrogen in its various isotopic and ortho-para modifications. *Journal of Research of the National Bureau of Standards, 41*(5), 379–475. https://doi.org/10.6028/jres.041.037

Yin, L., & Ju, Y. (2020). Review on the design and optimization of hydrogen liquefaction processes. *Frontiers in Energy, 14*(3), 530–544. https://doi.org/10.1007/s11708-019-0657-4

Zhang, X., Karman, T., Groenenboom, G. C., & Avoird, A. van der. (2021). Para-ortho hydrogen conversion: Solving a 90-year old mystery. *Natural Sciences, 1*(1), e10002. https://doi.org/10.1002/NTLS.10002

Zheng, J., Chen, L., Wang, J., Xi, X., Zhu, H., Zhou, Y., & Wang, J. (2019). Thermodynamic analysis and comparison of four insulation schemes for liquid hydrogen storage tank. *Energy Conversion and Management, 186*, 526–534. https://doi.org/10.1016/j.enconman.2019.02.073

Proceedings of the 14th International Symposium on Process Systems Engineering – PSE 2021+
June 19-23, 2022, Kyoto, Japan © 2022 Elsevier B.V. All rights reserved.
http://dx.doi.org/10.1016/B978-0-323-85159-6.50337-7

Model agnostic framework for analyzing rainwater harvesting system behaviors

Qiao Yan Soh[a], Edward O'Dwyer[a], Salvador Acha[a], Nilay Shah[a]

[a]*Centre for Process Systems Engineering, Department of Chemical Engineering, Imperial College London, London SW7 2AZ, United Kingdom*
qiaoyan.soh13@imperial.ac.uk

Abstract

To evaluate risks and characterise the responses of a rainwater harvesting system under different rainfall types, this paper presents a model agnostic evaluation framework where a k-means clustering approach is supplemented with a statistical Partial Least Squares model. Four response modes were identified for a studied system. Using these response modes, a higher risk of system overflow was found in 4.5% of simulated scenarios with inadequate water supplies found in 48.2% scenarios. The rainfall distribution in time was found to be crucial in determining the response mode of the system, with sporadic high intensity events or consistent, high total volume events allowing the system to operate in a response mode corresponding to lower system stresses, but with reduced provision of rainwater.

Keywords: Rainwater harvesting and detention; Environmental Systems; Modelling, Analysis and Simulation

1. Introduction

Rainwater harvesting (RWH) systems have shown considerable promise as an alternative or addition to existing urban water management systems as a strategy for ensuring long-term sustainability. Acting as a temporary detention system and a secondary water source, an RWH system facilitates achieving long-term sustainability by reducing stresses in centralised wastewater management systems, improving efficiencies in local water reuse, and minimizing the need for potable water supply.

In the current climate of increasing pressures on urban water supply, however, the adoption rates for RWH systems have been relatively low. Cities require assurances of the long-term effectiveness and sustainability of new infrastructural investment, but the process of evidencing the efficiency of RWH systems can be complex and time consuming. Modelling these systems provides researchers and system managers a cost-effective method of analysis that would otherwise be capital intensive. Model accuracy is therefore a key challenge in the implementation of RWH systems.

In addressing these barriers, much of the existing research has focused on showcasing the effectiveness of RWH systems operating under an extensive list of climatic conditions (Jing et al., 2018), demand scenarios (Nnaji et al., 2017) and operational strategies (Soh et al., 2020). With the intention of ensuring model accuracy or improving the accessibility of modelling methods to non-researchers, a plethora of modelling and tank design approaches have been developed around the world. Quinn et al. (2021) studied how best to characterise and evaluate the performance of RWH systems in terms of meeting both the objectives of supplying water and in stormwater management. In comparing model outputs with empirical data, Ward et al. (2012) showed that detailed methods provided the best approximation of real RWH performance, but require detailed, long-term data to achieve their full potential. This is complicated by the fact that RWH systems are sensitive towards their local environments and in-depth modelling and evaluation is typically required for each implementation, which can become impractical for water managers around the world. In this paper, a model agnostic statistical framework is presented for evaluating system risk and performance as a practical strategy for evidencing the long-term sustainability and efficiency of RWH systems.

2. Methods

To map RWH system behaviors with rainfall types and characteristics, system response modes were derived from k-means clustering of performance indicators. This was coupled with a Partial Least Squares (PLS) model which relates the rainfall statistical markers with their corresponding system performance indicators in latent space as a fast and holistic method for evidencing RWH system performance and sustainability.

2.1. System Modelling

A digital twin which utilised mass balances for modelling storage tanks and orifice flow equations for flows between tanks within the system was previously developed for a three-tank urban RWH system (Soh et al., 2020). With historical rainfall data available at five-minute intervals, the rainfall volumes were uniformly disaggregated to match the one-second time resolution used in the digital twin model.

Using historical rainfall data divided into 24-hour windows, 1496 samples of simulated system behaviors were recorded with a focus on the system's ability in controlling water quantities under a passive operational model. The performance indicators were developed accordingly with the purposes of maximizing the water availability provided by the RWH system, protecting against surface inundation, and the prospects of doing so with minimal spatial costs. More specifically, these are:

- **Harvested Volumes:** Total volume of water collected at the end of the simulation, preferably maximized to establish the water availability levels achievable by the system.
- **Unused Capacities:** Tank capacities that have not been used throughout entire simulation period, if possible, maximized to determine the minimal tank sizes required to handle different rainfall patterns.
- **Overflow:** Maximum amount of water in a single one-second timestep that overflows from the system, which should be minimized to ensure that excess water is adequately removed from the urban catchment in the maintenance of the health and safety objectives of implementing a RWH system.
- **Maximum Discharge Rates:** Discharge rates out of the system into the larger wastewater network, measuring stresses the system may provide to its counterparts downstream and should be minimized.

2.2. K-means Clustering

The k-means clustering works to separate the system responses into groups of equal variances through minimizing the within-cluster-sum-of-squares, in turn identifying the main response modes of the RWH system. The number of significantly different response modes is determined through the optimal number of clusters for a k-means algorithm, established empirically using the Python Scikit-Learn library (Pedregosa et al., 2012) and selecting the number of clusters that demonstrated the highest average silhouette score.

Table 1: List of rainfall statistical markers used in PLS model.

Marker	Description	Representation
Maximum observed rain rate	Highest volume of water observed in a single timestep in the timeseries.	Volume, Intensity
Rain-only mean	Mean of all non-zero timesteps in the timeseries	Volume, Intensity
Rain-only standard deviation	Standard deviation of all non-zero timesteps in rainfall timeseries	Volume, Intensity
Non-zero timesteps	Number of wet timesteps in the timeseries	Duration, Distribution
Maximum continuous rainfall	Maximum number of continuous timesteps with non-zero rainfall in the timeseries.	Duration, Distribution
Adjusted precipitation concentration index (PCI)	Measures the spread of rainfall within the timeseries. Defined as a function of timeseries with N timesteps with mean μ and standard deviation σ: $$PCI = \frac{1}{N}\left[1 + \left(\frac{\sigma}{\mu}\right)^2\right] \times 100(\%)$$	Duration, Distribution

2.3. PLS Model

The PLS model works to relate model inputs with its outputs though their latent spaces and was built using the Python toolbox PyPhi (García-Muñoz, 2019). As the PLS is a projection-based method, the selection of input variables should be independent. From the set of rainfall timeseries data, a set of statistical markers, presented in Table 1, were extracted such that the volumes, intensities, duration, and distribution of rainfall in a single 24-hour timeseries is represented as much as possible.

2.4. Supplementing clustering results with PLS scores

Scores, or projections of the statistical markers into latent space, provided by the PLS model were coupled with the results derived from the k-means clustering. With each score labelled by their corresponding cluster label, clusters in the latent space, as well as their corresponding driving factors can be identified. The relative positions between input samples in latent space highlights the differences in the original input variables for these samples. This can be used to indicate the properties that change between two given sample points, and by extension, used to identify the rainfall properties that distinguish two given response modes.

3. Results and discussion

3.1. Clustering and the classification of response modes

Through the clustering of the performance indicators, four main response modes for the system were identified. Boxplots visualising the distribution of the performance indicators in each cluster were used to characterise each response mode, where:

- **Cluster 0** contains system responses that have low water availability, but a smaller tank would suffice and has minimal discharge rates for high integrability.
- **Cluster 1** represents system behaviors that demonstrate high water availability and discharge rates, but without utilizing much of the existing tank capacity.
- **Cluster 2** responses show low discharge rates out of the system, with high water availability and little need for a large tank capacity.
- **Cluster 3** includes responses that have the highest discharge rates and water availability out of all the response modes. However, this is also accompanied by the need for a larger tank capacity.

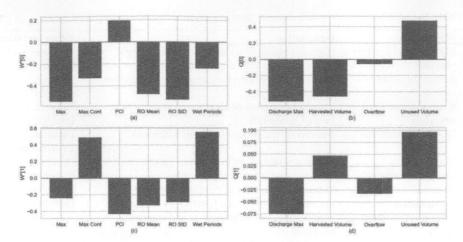

Figure 1: Projection coefficients used in PLS model for (a) input vectors onto latent variable 1, (b) output vectors onto latent variable 1, (c) input vectors for latent variable 2, and (d) output vectors for latent variable 2.

An analysis of the rainfall profiles contained in each cluster showed that rainfall volume is a significant driver of the response modes of the system. This follows with the rainfall profiles in Cluster 3 with the highest rainfall peaks, and profiles in Cluster 0 containing rainfall data with low intensities and volumes. Both clusters have sub-optimal performance, where Cluster 3 demonstrates a risk of overflow occurring, and Cluster 0 providing low water yields. The cluster sizes for these modes quantified the risk of undesirable system performances, with 4.5% and 48.2% of the total rainfall scenarios demonstrating possible overflow and low water availability respectively. The responses in Clusters 1 and 2 are much more desirable, and the rainfall characteristics distinguishing between these response modes and the sub-optimal modes would be identified alongside the statistical PLS model.

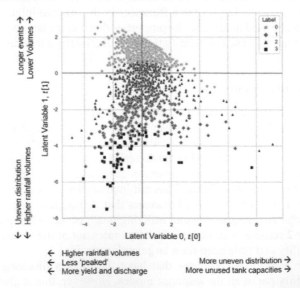

Figure 2: Score scatter of rainfall markers in latent space and main variable shifts along axes.

3.2. PLS model and the identification of main system driving forces

The PLS model identified two main driving forces in the RWH system, which are represented by the two latent variables found to best represent the input and output vectors. The latent variables were found to be able to explain the variation in 87.5% of the input, and 59.3% of the output. Whilst this study focuses on characterizing the impact of rainfall on the system performance, the behavior of a RWH system is expected to also be highly dependent on its internal design and is assumed, in this case, to account for the other 40.7% of variation observed in the outputs. Figure 1 shows the projection coefficients derived from the PLS model for both latent variables in the input and output vectors. The first input latent variable has coefficients for all inputs variables except the PCI marker positively correlated with each other and hence it is associated generally with the input rainfall volumes. The second input latent variable is associated with the rainfall duration and distribution, demonstrating that longer rainfall events are typically correlated with lower rainfall volumes and intensity.

In examining the relationship between the input and output variables in latent space, the model suggests that a higher rainfall volume, regardless of its distribution in time, is positively correlated with a higher discharge rate and improved water availability. Higher rainfall volumes also tend to require a larger tank, as observed in the first latent variable, the dimension associated with the overall rainfall volumes. Coefficient values in the second output latent variable are very low, hence correlations shown between the input and output vectors in this dimension are not significant.

3.3. Distinguishing rainfall properties between response modes

The input latent variables were plotted in latent space and coupled with response mode labels derived from the k-means clustering discussed in Section 3.1. Figure 2 shows the input variables represented in latent space and the associated changes in input variables in each axis direction. There are clear distinctions between each response mode in the input latent variable space. With Cluster 3 lying mostly within the high rainfall volume quarters and Cluster 0 in the low rainfall quarters, this confirms the initial analysis about rainfall volumes playing a key role in influencing the system's response mode.

For a given value for the second latent variable, the extreme values of the first latent variable typically belong to a response mode different from that of the central values, giving a parabolic structure to the clusters. In conjunction with the loadings plot shown in Figure 1(a), these extreme values in the first latent variables would correspond to either, in the negative direction, rainfall with more consistent and high rainfall volumes, or in the positive direction, more unevenly distributed rainfall patterns.

Rainfall days either with consistent but low intensity rainfall delivering a high total rainfall volume, or sporadic but high-intensity rainfall delivering lower total rainfall volumes would, therefore, drive the system towards a response mode with lower system stresses. This is characterised by smaller tank capacities, better integrability with its downstream systems, but accompanied by a reduced ability in ensuring water availability. This is evident in how the samples belonging to Cluster 2, which is associated with lower discharge rates and water availability than in Cluster 1, wraps the input latent space samples belonging that cluster in the second latent variable space.

4. Conclusions

The coupling of the k-means clustering method with a statistical PLS model successfully identified specific rainfall features that distinguishes between the response modes of the

RWH system in evaluation. For the given case study, the consistency in the delivery of the rainfall input is key in determining its response mode. Uniform rainfall rates that contribute to a high total rainfall volume, or sporadically intense rainfall would lead the system towards a response mode with lower water availability, but with lower capacity requirements and improved integrability with downstream systems.

This framework for evaluating the performance of a given RWH system also characterised the risks of the system responding sub-optimally. Under passive operation, the three-tank RWH system examined in this work has been found to underperform in 52.7% of the simulated scenarios. With the inclusion of design and or control parameters, their associated impacts can also be evaluated to ensure that the response modes for a RWH design is adequately within the desirable performance bounds. In understanding the rainfall characteristics that drive the system from one response to another, the framework can be used to assess future risks and hence the long-term sustainability of the system using predicted changes to rainfall runoff characteristics. With this information, necessary precautions that target the root changes can be easily designed into the system.

Acknowledgements

This research is supported by the Singapore Ministry of National Development and the National Research Foundation, Prime Minister's Office under the Land and Liveability National Innovation Challenge (L2 NIC) Research Programme (L2 NIC Award No. L2NICTDF1-2017-3). Any opinions, findings, and conclusions or recommendations expressed in this material are those of the author(s) and do not reflect the views of the Singapore Ministry of National Development and National Research Foundation, Prime Minister's Office, Singapore.

References

Jing, Xueer, Shouhong Zhang, Jianjun Zhang, Yujie Wang, Yunqi Wang, and Tongjia Yue. 2018. "Analysis and Modelling of Stormwater Volume Control Performance of Rainwater Harvesting Systems in Four Climatic Zones of China." Water Resources Management 32 (8): 2649–64.

García-Muñoz, Salvador. 2019. "PyPhi." https://github.com/salvadorgarciamunoz/pyphi.

Nnaji, Chidozie Charles, Praise God Chidozie Emenike, and Imokhai Theophilus Tenebe. 2017. "An Optimization Approach for Assessing the Reliability of Rainwater Harvesting." Water Resources Management 31 (6): 2011–24.

Pedregosa, Fabian, Gaël Varoquaux, Alexandre Gramfort, Vincent Michel, Bertrand Thirion, Olivier Grisel, Mathieu Blondel, et al. 2012. "Scikit-Learn: Machine Learning in Python." Environmental Health Perspectives 127 (9): 97008.

Quinn, Ruth, Charles Rougé, and Virginia Stovin. 2021. "Quantifying the Performance of Dual-Use Rainwater Harvesting Systems." Water Research X 10.

Soh, Qiao Yan, Edward O'Dwyer, Salvador Acha, and Nilay Shah. 2020. "Optimization and Control of a Rainwater Detention and Harvesting Tank." In Computer Aided Chemical Engineering, 48:547–52.

Ward, S., F. A. Memon, and D. Butler. 2012. "Performance of a Large Building Rainwater Harvesting System." Water Research 46 (16): 5127–34.

Proceedings of the 14th International Symposium on Process Systems Engineering – PSE 2021+
June 19-23, 2022, Kyoto, Japan © 2022 Elsevier B.V. All rights reserved.
http://dx.doi.org/10.1016/B978-0-323-85159-6.50338-9

Global assessment and optimization of renewable energy and negative emission technologies

Lanyu Li[a,b], Jiali Li[a] and Xiaonan Wang[a,c,*]

[a] Department of Chemical and Biomolecular Engineering, National University of Singapore, Singapore, 117585
[b] NUS Environmental Research Institute, National University of Singapore, Singapore, 138602
[c] Department of Chemical Engineering, Tsinghua University, Beijing, P.R. China, 100084
wangxiaonan@tsinghua.edu.cn

Abstract

Energy consumption can be a great environmental burden with heavy greenhouse gas emissions. Renewable energy, negative emission technologies, and waste-to-energy technologies are promising methods to assist in transitioning the energy- and carbon-intensive current energy systems towards low-carbon systems, mitigating emissions, contributing to carbon neutrality targets, and even achieving negative emissions. The availability of renewable resources is temporally uncertain and geographically different. Besides, the optimization and the conflicting economic and environmental trade-off of such systems have not yet been fully investigated in the literature. This study aims to provide a versatile framework for the assessment and sizing of renewable and negative emission technologies for global regions. The study also sheds light on the economic viability and carbon mitigation/reduction potential achievable by the combination of renewable energy and negative emission technologies regionally and globally.

Keywords: Hybrid renewable energy system, Negative emission, Design, Optimization, Analysis.

1. Introduction

Throughout history, we have gone through the energy transition from human and animal power to steam and from steam to electricity. The evolution of electricity has brought us tremendous economic prosperity and shocking changes in our daily lives, from the way we live to the way we think. However, while we enjoy an improved quality of life and convenience, we also pay a huge price for burning fossil fuels, rich reserves of carbon that can be stored well underground for thousands of years without human extraction. The world of fossil fuels has brought with it a dramatic increase in greenhouse gases in the atmosphere and the associated problem of climate change.

Figure 1 shows the emissions of the six major carbon emitters globally. Until the mid-20th century, the United States and Europe dominated the world's major carbon emissions. Emissions from the rest of the world began to rise significantly from the second half of the 20th century, with major increases in Asia, particularly China. Since 2006, China has surpassed the United States as the world's largest emitter of carbon dioxide. The main reasons can be attributed to the country's rapid industrialization, urbanization, and reliance on major fossil fuels such as coal. However, in terms of per capita greenhouse gas emissions, the United States., Russia, and Europe continue to be the highest per capita emitters. But since 2000, the per capita carbon emissions in the United States has started to fall, which is against the climb and stabilization of China's carbon emissions around the same period.

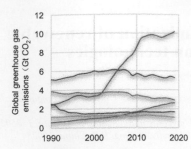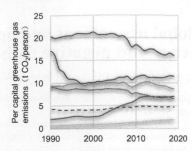

Fig. 1 Greenhouse gas emissions of the top six emitters around the world (Andrew, 2020; Hannah Ritchie; Max Roser, 2020).

Having experienced previous energy transitions, today, in the first half of the 21st century, we are about to witness another energy transition: from a fossil fuel-dependent energy system to a low-carbon, renewable, sustainable energy system. There is a growing awareness of the drawbacks and dilemmas of producing energy from fossil fuels, air pollution, greenhouse gas emissions and depletion, and the more difficult extraction that comes with it. Initiated by non-governmental organizations, the IPCC helped set the goal of carbon emissions, a goal of limiting global warming to well below 2°C, preferably 1.5°C, compared to pre-industrial levels. To achieve this goal, many countries set up their targets to peak their domestic greenhouse gas emissions in order to achieve a carbon-neutral world by the middle of the century.

Since 2015, 193 countries have submitted their climate commitments. At least 50 of these countries have set net-zero emission targets (*Members of the Carbon Neutrality Coalition*, 2021; Darby and Gerretsen, 2021; Wallach, 2021). Table 1 lists the carbon neutrality targets for several countries, including the major carbon-emitting economies such as China, the United States, the European Union, and Japan (ranked by annual carbon emissions). In addition to China's neutrality target, the United States has also proposed a statement of intent to achieve net-zero emissions by 2050 and 100% clean electricity by 2035. The governments of South Africa and Chile have also expressed their policy position of achieving net-zero emissions by 2050. The EU, Japan, South Korea, Canada, and Brazil have established explicit legislation or submitted written commitments to the United Nations to become carbon neutral in the coming decades.

Table 1 Net-zero carbon emissions goals set by different countries.

Country	Target date for carbon neutrality	Commitment
China	2060	Statement of Intent
United States	2050	Statement of Intent
EU	2050	Submitted to the United Nations
Japan	2050	Law
Korea	2050	Submitted to the United Nations
Canada	2050	Law
Brazil	2060	Submitted to the United Nations
South Africa	2050	Policy Position
Chile	2050	Policy Position

Under the common vision of humanity to reduce carbon emissions, renewable energy and negative emission technologies are promising technologies in the blueprint of low-carbon transition of energy systems (Li *et al.*, 2019). However, hybrid renewable energy systems combined with negative emission technologies, which have a synthetic effect and promising potential in building low-carbon energy systems, are still under-explored. Therefore, here we develop a versatile framework for the assessment and scaling of renewable and negative emission technologies in global regions, and to elucidate the economic feasibility and carbon reduction potential of these technologies.

2. Methodology

The study is carried out through data collection, modeling, and optimization based on the modeling method in Li *et al.* (2019), which is summarized in Figure 2. The inputs for the optimization model include meteorological, biomass, land area, economic, and social data for each of the countries. The coordinates for cities are obtained from *ArcGIS online* (2021). The models for renewable energy include solar PV, onshore wind, gasification, incineration, and pyrolysis. Biomass resources available for waste-to-energy conversion include horticultural wastes and wood wastes. The economic performance of the system is evaluated by the net present value. The environmental performance is quantified by the equivalent CO_2 emission. Either or both of them can be used as the objective function of the optimization. The constraints of technology placement include the resource limit for power generation, the requirement of demand satisfaction, and the design of operating constraints for the specific technologies. It is interesting to see what the technology planning would be like if the optimization is done to maximize each country's own benefit and the case maximizing for the global benefit as a whole. Therefore, we are carrying out the scenario analysis for three different scenarios: the case for maximized economic performance for each country, the case for maximized economic performance for the world as a whole, and the case for minimized greenhouse gas emissions globally.

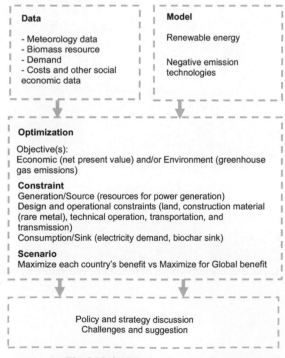

Fig. 2 Methodological framework.

3. Results and Discussion

We first calculate the maximum technology capacity achievable around the world without considering any constraints to have an overview of the resource potential by technology and by region. Table 2 summarizes the maximum potential of technology capacity sorted by power generation method.

Table 2. The maximum potential of technology capacity by power generation method (kW)

	Solar	Wind	Combustion	Gasification	Pyrolysis
Capacity	4.99×10^{13}	7.08×10^{12}	1.11×10^{7}	9.51×10^{6}	6.73×10^{6}
Percentage over total available potential	88%	12%	~ 0%	~ 0%	~ 0%

It can be found that the magnitudes of solar and wind availability are much higher than those for biomass conversion methods including combustion, gasification, and pyrolysis. Besides, the maximum total energy generation potential by the technologies by country is shown in Figure 3. It can be found that Russia, French Southern Territories, Chile, Bolivia, and the United States have the most abundant (mainly solar) energy resource for the technologies considered.

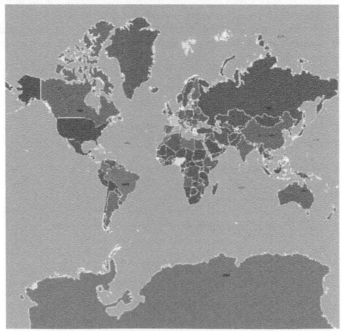

Figure 3. The maximum potential of renewable generation by country.

With this basis, we carried out the optimization by maximizing the case for maximized economic performance for each country (Scenario 1), maximizing the economic performance for the world as a whole (Scenario 2), and minimizing the greenhouse gas emissions globally (Scenario 3). The optimized results are provided in Table 3. It was found that maximizing the economic performance for countries and regions individually and optimizing globally as a whole result in almost the same economic and environmental performance. Moreover, minimizing the environmental performance as the objective function could result in an overall negative greenhouse gas emission, but it may lead to a dramatic increase in cost compared to the optimal NPV scenario. For the cases of maximizing the economic benefit, it was found that almost all regions, except 15 of them, are economically preferable for the utilization of the considered technologies no matter it is optimized for each country (Scenario 1) or globally (Scenario 2). Solar energy is the most selected technology for most regions in these cases. On the other hand, fourteen regions have no technology selected when the greenhouse gas emission is minimized as an environmental objective mainly due to the limited biomass resources.

Table 3. Result for different optimization scenarios.

	Scenario 1: Maximum economic performance for each country	Scenario 2: Maximum economic performance globally	Scenario 3: Minimum environmental performance globally
Net present value	2.15×10^{14}	2.18×10^{14}	-3.4×10^{10}
Greenhouse gas emissions	1.11×10^{12}	1.11×10^{12}	-3.7×10^{10}

4. Conclusion

This work presents the economic assessment and the potential for carbon reduction via renewable energy and negative emission technologies for countries around the world. The geographical diversification of renewable resources and system design was observed in global analysis. The result shows that except for 15 places, almost all countries and regions around the world were decided to be profitable locations for the proposed system when net present value is maximized. Negative emission was possible to be achieved globally if greenhouse gas emission was minimized, but it may lead to a dramatic increase in cost compared to the optimal NPV scenario. The methodology framework was demonstrated to be versatile and conveniently applicable to study the feasibility of the proposed renewable and negative emission technologies in multiple regions. In future studies, the impact of grid integration of renewables, a higher-precision analysis accounting for the temporal-spatial variation of various factors can be carried out to provide more insights.

Acknowledgment

This research is supported by the National Research Foundation, Prime Minister's Office, Singapore under its Campus for Research Excellence and Technological Enterprise (CREATE) program.

References

Andrew, R. (2020) *Figures from the Global Carbon Budget 2020*. Available at: https://folk.universitetetioslo.no/roberan/GCB2020.shtml (Accessed: 23 August 2021).

ArcGIS online (2021). Available at: https://www.arcgis.com/index.html.

Darby, M. and Gerretsen, I. (2021) *Which countries have a net zero carbon goal?* Available at: https://www.climatechangenews.com/2019/06/14/countries-net-zero-climate-goal/ (Accessed: 1 August 2021).

Hannah Ritchie; Max Roser (2020) *CO_2 and Greenhouse Gas Emissions, OurWorldInData.org*.

Li, L. *et al.* (2019) 'Optimal design of negative emission hybrid renewable energy systems with biochar production', *Applied Energy*. Elsevier, 243, pp. 233–249. doi: 10.1016/J.APENERGY.2019.03.183.

Members of the Carbon Neutrality Coalition (2021). Available at: https://carbon-neutrality.global/members/ (Accessed: 1 August 2021).

Wallach, O. (2021) *Race to Net Zero: Carbon Neutral Goals by Country*. Available at: https://www.visualcapitalist.com/race-to-net-zero-carbon-neutral-goals-by-country/ (Accessed: 1 August 2021).

Proceedings of the 14th International Symposium on Process Systems Engineering – PSE 2021+
June 19-23, 2022, Kyoto, Japan © 2022 Elsevier B.V. All rights reserved.
http://dx.doi.org/10.1016/B978-0-323-85159-6.50339-0

The Trade-Off between Spatial Resolution and Uncertainty in Energy System Modelling

Maria I. Yliruka[a*], Stefano Moret[b], Francisca Jalil-Vega[c], Adam D. Hawkes[a], Nilay Shah[a]

[a]*Department of Chemical Engineering, Imperial College London, United Kingdom*
[b]*Business School, Imperial College London, United Kingdom*
[c]*Facultad de Ingeniería y Ciencias, Universidad Adolfo Ibáñez, Chile*
m.yliruka@ic.ac.uk

Abstract

In energy system models, computational tractability is often maintained by adopting a simplified temporal and spatial representation in a deterministic model formulation i.e., neglecting uncertainty. However, such simplifications have been shown to impact the optimal result. To address the question of how to prioritize the limited computational resources, the trade-off between spatial resolution and uncertainty is assessed by applying a novel method based on global sensitivity analysis to a peer-reviewed heat decarbonization model. For all output variables apart from the total system and fuel cost, spatial resolution is ranks amongst the five most important model inputs. It is the most relevant factor for investment decisions on network capacities. For the total fuel consumption and emissions, spatial resolution turns out to be more relevant than the fuel prices themselves. Compared across all outputs, the analysis suggests the impact of spatial resolution is comparable the impact of heat demand levels and the discount rate.

Keywords: Spatial Resolution, Uncertainty, Mixed-integer Linear Program, Energy System Model, Global Sensitivity Analysis

1. Introduction

Energy system models help to explore different decarbonization pathways to reach net zero by 2050. Given the time horizon and scope, these models are often large and rely on long-term forecasts of input parameters. Computational tractability is maintained by adopting a simplified temporal and spatial representation of the system as well as a "deterministic" model formulation (i.e., neglecting uncertainty). However, the shortcomings of these approaches have been shown, notably in power systems applications: At low temporal resolutions, the dispatchable generation and storage capacities are underestimated whereas the renewable generation capacity is overestimated due to the smoothed production and demand profiles. This impact is especially pronounced for systems with high shares of intermittent and non-dispatchable renewables (Pfenninger, 2017). While the impact of temporal resolution and suitable aggregation methods have already been reviewed, the literature on the impact of spatial resolution is more limited and less conclusive. In power systems, the impact of spatial resolution on the total system cost is small compared to the changes in generation and flexibility technologies. Hörsch and Brown (2017) observe an increased investment in transmission capacity and a decrease in solar PV with increased spatial resolution for Europe, while Krishnan and Cole (2016) report the opposite effect for the US. In Jalil-Vega and Hawkes

(2018a), the averaging of heat demand densities at lower spatial resolution leads to an underestimation of district heating potential for local authorities (LA) with heterogenous demand levels in the UK. Global sensitivity analysis (GSA) has been applied to national energy system models to quantify the impact of uncertainty of model inputs (input parameters) (Pye et al., 2015; Moret et al., 2017). Economic parameters such as fuel prices and investment cost of technologies have commonly been identified as most relevant. Overall, these recent studies leave the modelling community with the question of how to prioritize computational resources: including uncertainty or increasing spatial/temporal resolution?

To our knowledge, this paper is the first to compare the impact of spatial resolution and uncertainty of input parameters. A novel methodology is applied to a case study in literature to assess this trade-off over a wide range of output variables. The fundamental novelty is that spatial resolution is considered as an uncertain input parameter in a GSA allowing to rank its importance relative to other uncertain input parameters.

2. Methodology

2.1. Novelty

GSA methods allow to quantify the impact of input parameters on one or more model outputs (Saltelli et al., 2008), where *impact* is defined as the ability of an input parameter to significantly alter a given output of interest when varied from its nominal value that corresponds to its most likely realization. Modelling choices such as spatial resolution are not commonly referred to as 'input parameters' and have therefore never been considered in GSA studies. In this work, we propose using GSA methods in a novel way that allows to assess the trade-off between spatial resolution and uncertainty in energy system models within the same methodological framework.

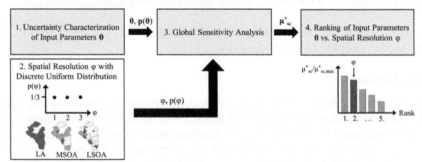

Figure 1: A summary of the three steps (1, 3, 4) involved in a conventional GSA (top) with the addition of spatial resolution (2) as an uncertain input parameter with discrete uniform distribution (bottom). In the case study, three distinct levels of spatial resolution (N=3) are modelled, having equal probability of 1/3.

Our proposed modified GSA method is illustrated in Figure 1. In conventional GSA studies, the uncertainty distributions $p(\theta)$ of the input parameters θ serve as inputs to the GSA. In our novel approach, the spatial resolution is included as an additional "uncertain" parameter (φ) characterized by a discrete uniform distribution $p(\varphi)$. The finite number of spatial resolution levels are characterized by a numerical value 1, 2, ..., N with N being total number of different resolutions considered in the analysis. Each level of resolution has equal probability $1/N$. This modification allows us to consider spatial resolution alongside the usual input parameters in the GSA, hence assessing its relative impact with respect to uncertainty in model inputs. In the following, spatial resolution and input parameters will collectively be referred to as *input factors*.

2.2. Global Sensitivity Analysis

The modified Morris Method is applied to obtain a qualitative ranking over a large set of input factors. To determine the elementary effect (EE) of each input factor on the outputs of interest, the input space is discretized into a *p*-level grid and systematically sampled using *r* trajectories. At each step of the trajectory, one of the *k* input factors is varied changing its value by Δ with $\Delta=p/(2(p-1))$ (Saltelli et al., 2008). To avoid the computational burden of oversampling, the enhanced Sampling for Uniformity (eSU) strategy is applied (Chitale et al., 2017). The EE of input factor θ_i on output variable Y_j in trajectory *m*, EE_{mij}, is subsequently calculated as the ratio between the change in the input factor and the consequent change in the output variable, as shown in Eq. (1).

$$EE_{mij} = \frac{\delta Y_j}{\Delta} \tag{1}$$

Campolongo et al. (2007) have shown that μ^*_{ij}, the mean of the distribution of the absolute values of EE_{mij}, is a good proxy to the total effect sensitivity index, that is normally determined using computationally expensive, variance-based methods. To compare the impact on multiple output variables of varying magnitudes and units, a similar approach to Sin and Gernaey (2009) is chosen by scaling μ^*_{ij} by σ_j, the standard deviation of the output variable Y_j:

$$\mu^*_{sc,ij} = \frac{1}{\sigma_j}\left(\frac{1}{r}\sum_{m=1}^{r}\left|EE_{mij}\right|\right) \tag{2}$$

2.3. Case Study

The method is applied to a spatially resolved, mixed-integer linear model for urban energy systems (Jalil-Vega and Hawkes, 2018b), which was previously used to study the impact of spatial resolution on the district heating uptake in six LAs with varying rural-urban character (Jalil-Vega and Hawkes, 2018a). The LA of Winchester is chosen and modelled at LA (1 node), middle layer super output area (MSOA, 10 nodes) and lower layer super output area (LSOA, 49 nodes) level as shown in Fig. 1. With increasing spatial resolution, the initial network capacity between and within cells, the heat and electricity demand and the distance between cells become more resolved. The domestic heat, gas and electricity supply infrastructure are explicitly modelled. The design and operation of the heat supply system minimizing the total system cost is determined from today until 2050. The same nomenclature as in Jalil-Vega and Hawkes (2018b) is used.

Apart from the lifetime of technology and network, all model parameters presented in Jalil-Vega and Hawkes (2018b) are considered as uncertain: Fuel prices ($Cost^E/Cost^G$), capital cost of technologies ($Cost^C_{tech}$), technological performance (η^{Th}_{tech} /COP_{tech} /η^E_{tech}), capital cost of intranodal network capacity ($Cost^{ND}_{E/G/H}$), capital cost of internodal network capacity ($Cost^{NT}_{E/G/H}$), operation and maintenance cost ($Cost^M_{tech}$), losses in heat networks ($Loss_T$), discount rate (r), electricity and heat demand (Dem^E/Dem^H). The uncertainty ranges, $R_\%$, summarized in Table 1 are determined based on UK specific data where possible (Yliruka et al., in preparation). For all input parameters, a uniform distribution is assumed.

Table 1: Uncertainty ranges, $R_\%$, for the input parameters in Jalil-Vega and Hawkes (2018b) (Yliruka et al., in preparation).

Parameter	$R_\%$ [%]	Parameter	$R_\%$ [%]
$Cost^C_{B/Erad/HXT}$	[−40, 42]	$Cost^E$	[−13, 15]
$Cost^C_{ASHP}$	[−31, 49]	$Cost^G$	[−38, 0]
$Cost^C_{GSHP}$	[−40, 44]	$Cost^{CO2}$	[−50, 50]
$Cost^C_{CHP}$	[−11, 9]	Dem^E	[−0.4, 15]
$Cost^C_{PV}$	[−57, 76]	Dem^H	[−39, 6]
$\eta^{Th}_{B/Erad/HXT}$	[−0.2, 0.4]	$Cost^{NT}_{E/G/H}$	[−39.3, 39.3]
COP_{ASHP}	[−14, 14]	$Cost^{ND}_{E/G/H}$	[−39.3, 39.3]
COP_{GSHP}	[−11, 14]	$Cost^M_{tech}$	[−49.2, 35.7]
η^E_{CHP}	[−10, 13]	$Loss_T$	[−2, 2]
η^E_{PV}	[−14, 17]	r	[−81, 4]

3. Results and Discussion

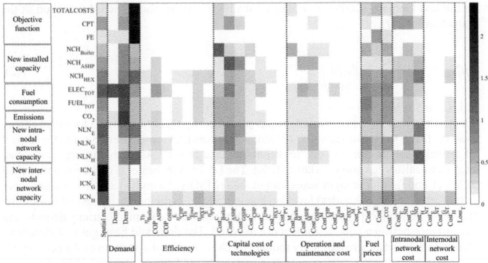

Figure 2: The impact of each input parameter on a selected subset of output variables is indicated by the shading in the heat map. The inputs and outputs are grouped by boxes and categorized by vertical lines. The horizontal line separates the network related outputs.

The results are based on $r=24$, $p=4$ for each uncertain input parameter and $p=3$ for the spatial resolution. The impact of the different input factors on a selection of output variables are summarized in Fig. 2. The fuel consumptions, emissions, technology and network capacities are summed over the multi-year time horizon. The scaling by σ_j in Eq. 2 also allows to compare the qualitative impact across all output variables. The darker the column, the more relevant is the parameter across multiple output variables.

The degree of spatial resolution is by far the most important factor on the installed capacity of all internodal networks ($ICN_{E/G/H}$). This ranking holds true even if the large change in $ICN_{E/G/H}$ between LA and MSOA level is excluded. Spatial resolution also remains the most important factor for the intranodal gas and electricity network capacity ($NLN_{E/G}$). However, in the case of heat networks (NLN_H), the investment cost of the pipelines and the discount rate are more relevant. While for gas and electricity networks the initial network needs to be reinforced, the heat network has to be built from scratch. Therefore, its deployment is limited by the high upfront cost that is sensitive to the capital cost of the network and the discount rate. The general observation that the network design is the most sensitive output variable to the choice of spatial resolution agrees with Hörsch and Brown's (2017) observations in power system models.

For the other output variables presented in Fig. 2, the impact of spatial resolution is less conclusive. The total system cost (TOTALCOSTS), capital cost (CPT) and fuel cost (FE) are dominated by the discount rate. Here, the level of spatial resolution becomes irrelevant. Garcia-Gusano et al. (2016) have previously discussed the decisive role of the discount rate on the system cost in long-term energy system models. Out of all heat technologies, the installed capacity of heat exchangers (NCH_{HEX}) is most sensitive to the choice of spatial resolution. Its impact is comparable to the heat demand and heat network investment cost. However, like NLN_H, NCH_{HEX} is the most sensitive to the discount rate. For the installed capacity of boilers (NCH_B) and ASHPs (NCH_{ASHP}), the investment cost of the technology itself is most important.

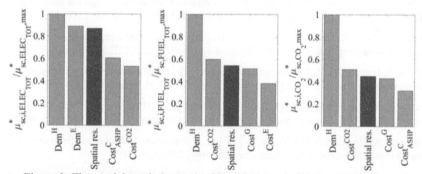

Figure 3: The spatial resolution ranks third (*dark grey*) with respect to the total emissions, electricity and natural gas consumption.

The impact of spatial resolution on the fuel consumption ($FUEL_{TOT}$, $ELEC_{TOT}$) and CO_2 emissions (CO_2) has not yet been discussed in the literature. As shown in Fig. 3, spatial resolution has a higher impact than the fuel prices. As the emissions are mostly caused by the natural gas consumption, both output variables are most sensitive to the heat demand and carbon tax. For $ELEC_{TOT}$, the demand levels for both heat and electricity are the most important. The high rank of spatial resolution indicates that the split in fuel consumption is governed by the installed network capacities that are a highly dependent on the level of spatial resolution. The choice in heat technology is secondary and subsequently determined by the relative capital cost. As the ranking can vary for different $R_\%$ (Moret et al., 2017), these findings should be confirmed for other $R_\%$ of the fuel prices. Spatial resolution ranks amongst the five most impactful input factors for all output variables apart from the total system and fuel cost. As shown previously, spatial resolution is the most relevant for the network capacities. Studying for the first time its impact on the total fuel consumption, spatial resolution turns out to be more relevant than the uncertainty in fuel prices. Overall, spatial resolution is found to be comparable to the discount rate and heat demand levels.

4. Conclusions

Neglecting uncertainties of input parameters and temporal/spatial aggregation have been shown to impact the optimal solution of energy system models, leaving the modelling community with the question of how to prioritize the limited computational resources. To guide this decision, this paper compares for the first time the impact of spatial resolution and uncertain input parameters across a range of design and operational variables of a peer-reviewed heat decarbonization model.

The choice of spatial resolution is the most relevant for the design of the networks whereas a single node system is well-suited to estimate the total system cost and capacity of heat technologies. The rankings for the total system cost and network variables agree with previous literature which serves as a verification of the method. For the first time, the impact of spatial resolution on the fuel consumption and emission levels is assessed and identified as relevant, ranking third after demand levels and carbon tax.

The novel, GSA-based framework can help prioritize the allocation of limited computational resources either on spatially detailed deterministic models or stochastic models with coarser spatial resolution in the early stages of the model development.

References

F. Campolongo, J. Cariboni, A. Saltelli, 2007, An effective screening design for sensitivity analysis of large models, Environmental Modelling & Software, 22, 10, 1509-1518

J. Chitale, Y. Khare, R. Munoz-Carpena, G. S. Dulikravich, C. Martinez, 2017, An effective parameter screening strategy for high dimensional models, IMECE2017, Tampa, USA

J. Hoersch, T. Brown, 2017, The role of spatial scale in joint optimisations of generation and transmission for European highly renewable scenarios, EEM, Dresden, Germany

D. Garcia-Gusano, K. Espegren, A. Lind and M. Kirkengen, 2016 The role of the discount rates in energy systems optimisation models, Renewable and Sustainable Energy Review, 59, 56–72

F. Jalil-Vega, A. Hawkes, 2018a, The effect of spatial resolution on outcomes from energy systems modelling of heat decarbonisation, Energy, 155, 339-350

F. Jalil-Vega, A. Hawkes, 2018b, Spatially resolved model for studying decarbonisation pathways for heat supply and infrastructure trade-offs, Applied Energy, 210, 1051-1072

V. Krishnan, W. Cole, 2016, Evaluating the Value of High Spatial Resolution in National Capacity Expansion Models using ReEDS, PESGM, Boston, USA

S. Moret, V. Codina Girones, M. Bierlaire, F. Marechal, 2017, Characterization of input uncertainties in strategic energy planning models, Applied Energy, 202, 597

G. Sin, K. V. Gernaey, 2009, Improving the Morris method for sensitivity analysis by scaling the elementary effects, Computer Aided Chemical Engineering, 26, 925

S. Pfenninger, 2017, Dealing with multiple decades of hourly wind and PV time series in energy models: A comparison of methods to reduce time resolution and the planning implications of inter-annual variability, Applied Energy, 197, 1-13

S. Pye, N. Sabio, N. Strachan, 2015, An integrated systematic analysis of uncertainties in UK energy transition pathways, Energy Policy, 87, 673-684

A. Saltelli, M. Ratto, T. Andres, F. Campolongo, J. Cariboni, D. Gatelli, M. Saisana, S. Tarantola, 2008, Global Sensitivity Analysis: The Primer, John Wiley & Sons, England

M. I. Yliruka, S. Moret, N. Shah, in preparation, Uncertainty characterisation for the UK energy system

Proceedings of the 14th International Symposium on Process Systems Engineering – PSE 2021+
June 19–23, 2022, Kyoto, Japan ©2022 Elsevier B. V. All rights reserved.
http://dx.doi.org/10.1016/B978-0-323-85159-6.50340-7

Designing a Resilient Biorefinery System under Uncertain Agricultural Land Allocation

Varun Punnathanam[a], Yogendra Shastri[a*]

[a]*Department of Chemical Engineering, Indian Institute of Technology Bombay, Mumbai*

yshastri@iitb.ac.in

Abstract

Agricultural residues are excellent feedstock for lignocellulosic biorefineries. However, the land allocated to various crops in a region can vary annually, thus impacting the feedstock availability for biorefineries. This work provides an optimization framework that considers uncertainty in land allocation for designing a biorefinery system that is resilient to such changes. A recently proposed decomposition-based approach is utilized to perform stochastic optimization, and the resulting design was compared to a deterministic design that considered mean land allocation. Lignocellulosic ethanol production for the state of Maharashtra, India, was taken as a case study, and the performances of both designs were evaluated on a set of 100 random land allocation instances. The resilient design had a smarter feedstock procurement strategy which resulted in a significant decrease in variation of feedstock procurement and transportation expenses. As a result, the variation in ethanol cost was 4% for the resilient design, as compared to 11% for the deterministic design.

Keywords: Stochastic optimization, large-scale optimization, uncertain feedstock availability, second generation biorefinery.

1. Introduction

The production of ethanol in India is expected to ramp up significantly to meet the target of 20% blending with gasoline by the year 2025, with significant contribution from second generation biorefineries. The feedstock for these biorefineries are to be primarily lignocellulosic biomass in the form of agricultural residues. A systems based approach is needed to design such production systems to address challenges of distributed biomass availability, multiple feedstock types, seasonality of feedstock, and the low maturity of the processing technologies (Daoutidis et al., 2013; Ng and Maravelias, 2017). Additionally, the uncertainty in demand, raw material price, and conversions pose further challenges to decision makers (Gong et al., 2016; Guo et al., 2022).

For an agricultural residue based biorefinery system, the residue availability is directly affected by the agricultural land allocation of the region, which in turn is governed by various external factors. This work presents a stochastic optimization based approach for designing a system of biorefineries under uncertain agricultural land allocation. The model is applied to a case study of Maharashtra, India, and the performance of the resilient design obtained from stochastic optimization is compared to a deterministic design.

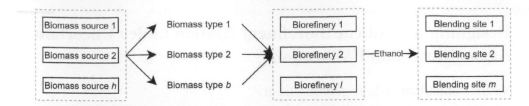

Figure 1: Schematic representation of the biorefinery system.

2. Problem formulation

The supply chain and process synthesis model for the biorefinery system from Punnathanam and Shastri (2021) is adopted in this work and schematically shown in Figure 2. This model considers multiple types of agricultural residue as feedstock, seasonal feedstock availability, biomass storage at biorefineries, multiple feedstock source locations, and multiple ethanol-gasoline blending sites. Additionally, the model considers multiple biorefinery location options, with the allowance of selection of one or more locations for setting up biorefineries. The transportation of feedstock from source to biorefineries and ethanol from biorefineries to blending sites is via trucks and tankers, respectively. The biorefinery configuration is fixed based on previous studies and is reported in Section 4.. The model is formulated as a large mixed integer linear programming problem (MILP) and is given as follows:

$$\min \quad z = \sum_{l \in L} \left(\mathbf{c}_l^{\text{TC}} + \mathbf{c}_l^{\text{VO}} + \mathbf{c}_l^{\text{FO}} \right) \tag{1}$$

subject to Biorefinery processing constraints at location l $\hspace{2cm} \forall l \in L$ $\quad$ (2)

$$\sum_{l \in L} \mathbf{f}_{h,b,l,t}^{proc} \leq \mathbf{F}_{h,b,t} \hspace{2cm} \forall h \in H, b \in B, t \in T \quad (3)$$

$$\sum_{l \in L} \mathbf{e}_{l,m,t}^{\text{prod}} \geq \mathbf{D}_m \hspace{2cm} \forall m \in M, t \in T \quad (4)$$

The objective is to minimize to total annualized cost (TAC) of the system, which is the sum of the capital and operating expenses. The two primary sets of constraints reflect the availability constraints on the feedstock and the minimum demand for ethanol that needs to be met at each blending site by all the biorefineries. The constraints reflecting the conversion of biomass to ethanol at biorefinery l are grouped together under "Biorefinery processing constraints at location l", and are taken from Vikash and Shastri (2019). The binary variables for the model are the selection of biorefinery locations and the selection of biomass to be processed at each biorefinery at each time period. The key continuous variables are biomass procurement and ethanol distribution at each biorefinery, the capacities of the equipment within each biorefinery, and the biomass storage at each biorefinery. The mean value of the random variable can be utilized in this formulation to obtain a deterministic design for the biorefinery system. Here, design refers to the selection of biorefinery locations, capacities of equipment within each biorefinery, and the feedstock procurement plan. For the deterministic design case, the feedstock procurement plan is identical to the actual feedstock procurement in the solution to the optimization problem.

3. Stochastic optimization

While the deterministic design is generated considering the mean value of the random variable, the resilient design is generated by considering a set of random instances. The goal of the resulting stochastic optimization problem is to obtain the optimal biorefinery design that minimizes the expected TAC considering all the random instances. This optimization is formulated as follows:

$$\min \ \bar{z} = E(z) = \frac{\sum_{s \in S} z'_s}{N_S} \tag{5}$$

subject to Biorefinery processing constraints at location l

$$\text{for instance } s \qquad\qquad \forall l \in L, s \in S \tag{6}$$

$$z'_s = \sum_{l \in L} \left(\mathbf{c}'^{\text{TC}}_{l,s} + \mathbf{c}'^{\text{VO}}_{l,s} + \mathbf{c}'^{\text{FO}}_{l,s} \right) \qquad\qquad \forall s \in S \tag{7}$$

$$\sum_{l \in L} \mathbf{f}'^{\text{proc}}_{h,b,l,t,s} \leq \mathbf{F}_{h,b,t} \qquad\qquad \forall h \in H, b \in B, t \in T, s \in S \tag{8}$$

$$\sum_{l \in L} \mathbf{e}'^{\text{prod}}_{l,m,t,s} \geq \mathbf{D}_m \qquad\qquad \forall m \in M, t \in T, s \in S \tag{9}$$

The objective function of TAC from the model presented in Section 2. has been replaced with the expected TAC, which is the average TAC across all instances assuming all instances have identical probabilities. All constraints have an additional index $s \in S$ to reflect the set of random instances. Similarly, all variables except for the design variables have an additional index $s \in S$. The design variables, being common for all instances, do not have this additional index. Each instance considered within the stochastic optimization framework has a corresponding feedstock procurement variable. The feedstock procurement plan, however, is a design variable and common for all instances. The positive difference between the instance specific feedstock procurement and the feedstock procurement plan is the quantity of feedstock that would be procured on short notice. Feedstock procured in this manner is termed as short-term procurement, and is more expensive. This is taken into account as part of the stochastic optimization formulation.

Depending on the number of random instances considered, the stochastic problem can potentially be very big. For the case study in consideration, the stochastic problem considering 10 random instances has 140,297 constraints, 420,354 variables, and 19,899 binary variables, and was not solvable in a reasonable time on an INTEL® i7-4770 3.40GHz CPU with 4GB RAM using the CPLEX® 12 MILP solver. Hence, the decomposition based approach presented in Punnathanam and Shastri (2020) was employed in this work. This method employs Dantzig-Wolfe decomposition (DWD) to simplify the original MILP such that it can be solved using the CPLEX® solver. The solution method involves the following steps: first, the deterministic optimization problem was decomposed to be solved within a DWD framework, where each sub-problem represents the constraints corresponding to a single biorefinery location. Next, the mean value of the random variable was used to solve the problem up to a specified termination criteria using DWD. Here, sub-problems and the master problem are iteratively solved; the sub-problem solutions are integer feasible solutions for each biorefinery and the mater problem assigns weights to the solutions obtained

Table 1: Types of feedstock in the form of agricultural residues and their prices ($/ton).

Cotton stalk	Sugarcane bagasse	Rice straw	Wheat stalk	Sorghum Stalk
6.7	53.3	10.4	9.5	7.2

from the sub-problems. DWD stage is terminated when the change in objective value of the master problem over iterations is below a specified threshold. On termination, based on the weights assigned to the sub-problem solutions by the master problem at the final iteration, certain binary selections corresponding to biorefinery location and biomass to be processed are rejected. This heuristic drastically simplifies the original MILP. The resulting simplified problem can be solved in the stochastic optimization framework to obtain the resilient design for the biorefinery system. As a result of the DWD-assisted simplification step, the final simplified problem had half the original number of constraints and variables, and 92% fewer binary variables, and could be solved in 12 hours. However, due to the heuristic which was employed for simplification, the optimality of the final solution is not guaranteed.

4. Case study details

The presented model is applied for the case study of ethanol production for Maharashtra, India. Maharashtra is an agriculturally intensive state with a large variety of crops under cultivation. The feedstock for biorefineries are assumed to be residues from five prominent crops; cotton, wheat, sugarcane, sorghum, and rice. The residues and their costs are provided in Table 1. These feedstocks are available only in specific months each year, and only 10% of the residue generated in farms are available for biorefineries as the rest are consumed within the farms for various purposes. Feedstock procured in short term is assumed to be twice as expensive as its nominal price. The major towns of 33 districts in the state act as collection sites from which feedstock can be distributed to biorefineries. These 33 locations also act as potential biorefinery locations. The processing configuration within each biorefinery is as follows: hammer milling for size reduction followed by dilute acid pretreatment, washing and detoxification, bioconversion via SSCF, and purification by conventional and extractive distillation. This configuration was found to be optimal for this case study Punnathanam and Shastri (2021). Ethanol blending sites are assumed to be located in 35 districts in the state. The demand for ethanol is calculated as the quantity of ethanol required to meet a 10% blending target at each district. The resulting total ethanol demand for the state was 42.47 million litres per month. Additional details on the case study is available in Punnathanam and Shastri (2021).

The land allocated towards the cultivation of a specific crop directly impacts the quantity of the corresponding crop residue available in that region. The allocation of agricultural land to various crops in the different districts of Maharashtra changes every year based on numerous external factors and hence is considered as an uncertain parameter in this work. The mean land allocated towards each crop in each district between 2010 to 2018 was utilized for the deterministic optimization problem presented in Section 2. and the DWD stage in the solution method presented in Section 3.. Random land allocation instances are generated by identifying the lower and upper bounds on land allocated towards each crop

Table 2: Comparison of the feedstock procurement plan recommended by the deterministic and resilient designs ('00 million kg).

Design	Cotton stalk	Sugarcane bagasse	Rice straw	Wheat stalk	Sorghum stalk	Total
Deterministic	634	210	190	446	507	**1,988**
Resilient	772	147	251	585	578	**2,334**

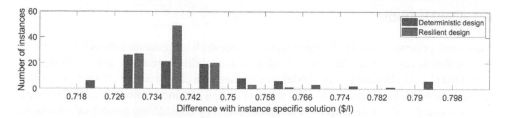

Figure 2: Frequency chart representing the performance of the deterministic and resilient designs over 100 random land allocation instances.

in each district between 2010 and 2018, and sampling from uniform distributions between these bounds. Note that each district and crop has a unique lower and upper bound. Additionally, a lower and upper bound on the total agricultural land allocated at each district was determined from historical data as well, and instances where the total land allocated was outside these bounds were scaled appropriately. For stochastic optimization, increasing the number of random instances can improve the design but significantly increase the computational time required for optimization. Based on this computational restriction, 10 random instances were considered for stochastic optimization. Similarly, a set of 100 random instances were generated for evaluating the performance of the deterministic and resilient biorefinery system designs. For a fair comparison, the total expense allocated for the feedstock procurement plan is enforced to be identical for both designs.

5. Results and discussion

The deterministic and resilient designs recommended similar locations for setting up biorefineries; with the deterministic design recommending 11 and the resilient design recommending 10 biorefineries. The feedstock procurement plans for both designs are presented in Table 2. The resilient design planned for the procurement of significantly higher quantities of cotton, sorghum, rice, and wheat based biomass, while the deterministic design planned for a higher quantity of sugarcane bagasse. Note that sugarcane bagasse is a much more expensive feedstock as compared to the others and that both designs had identical feedstock procurement plan expenses. Hence, the total quantity of feedstock planned by the resilient design was 17.4% higher than the deterministic design. As a result, the resilient design was better equipped to handle the uncertainty in feedstock availability.

Figure 2 presents the performance of both designs on the set of 100 random land allocation instances in the form of a frequency chart. At the lower end, the deterministic design was observed to perform marginally better; the lowest ethanol costs obtained by both designs varied by 0.5%. However, at the higher end, the resilient design performed significantly better than the mean design; the highest ethanol cost obtained by the resilient design was 6% lower than the deterministic design. Moreover, the variation in cost over 100 instances was 4.2% for the resilient design as compared to 10.7% for the deterministic design. The difference in costs between the designs was primarily due to the differences in operating expenses, in particular, the feedstock procurement and transportation expenses.

6. Conclusions

This work presents a stochastic optimization framework to generate a biorefinery system design that is resilient to changes in feedstock availability caused by annual changes in agricultural land allocation. The model was applied to a case study of ethanol production in the state of Maharashtra, India. The resilient design obtained from stochastic optimization was compared to a deterministic design obtained by considering mean land allocation. The performances of both designs were evaluated on a set of 100 random land allocation instances. The feedstock procurement plan for the resilient design took into consideration the uncertainty in feedstock availability. Hence, the variation in ethanol cost for the resilient design was 61% lesser as compared to the deterministic design. This work can be extended to consider uncertainties in other parameters, such as the composition of feedstock and the yield at various processing stages.

References

Daoutidis, P., Kelloway, A., Marvin, W. A., Rangarajan, S., and Torres, A. I. (2013). Process systems engineering for biorefineries: new research vistas. *Current Opinion in Chemical Engineering*, 2(4):442–447.

Gong, J., Garcia, D. J., and You, F. (2016). Unraveling optimal biomass processing routes from bioconversion product and process networks under uncertainty: an adaptive robust optimization approach. *ACS Sustainable Chemistry & Engineering*, 4(6):3160–3173.

Guo, C., Hu, H., Wang, S., Rodriguez, L. F., Ting, K., and Lin, T. (2022). Multiperiod stochastic programming for biomass supply chain design under spatiotemporal variability of feedstock supply. *Renewable Energy*.

Ng, R. T. and Maravelias, C. T. (2017). Design of biofuel supply chains with variable regional depot and biorefinery locations. *Renewable Energy*, 100:90–102.

Punnathanam, V. and Shastri, Y. (2020). Efficient optimization of a large-scale biorefinery system using a novel decomposition based approach. *Chemical Engineering Research and Design*, 160:175–189.

Punnathanam, V. and Shastri, Y. (2021). Optimization based design of biomass-based energy systems: A case study of the state of maharashtra, india. *Clean Technologies and Environmental Policy*.

Vikash, P. V. and Shastri, Y. (2019). Conceptual design of a lignocellulosic biorefinery and its supply chain for ethanol production in India. *Computers & Chemical Engineering*, 121:696–721.

Proceedings of the 14th International Symposium on Process Systems Engineering – PSE 2021+
June 19-23, 2022, Kyoto, Japan © 2022 Elsevier B.V. All rights reserved.
http://dx.doi.org/10.1016/B978-0-323-85159-6.50341-9

LCA modelling as a decision-tool for experimental design: the case of extraction of astaxanthin from crab waste

Carina L. Gargalo[a*], Liliana A. Rodrigues[b,c], Alexandre Paiva[c], Krist V. Gernaey[a], Ana Carvalho[c]

[a] *Process and Systems Engineering Centre (PROSYS), Department of Chemical and Biochemical Engineering, Technical University of Denmark, Kgs. Lyngby, Denmark*
[b]*iBET - Instituto de Biologia Experimental e Tecnológica, Oeiras, Portugal*
[c]*NOVA School of Science and Technology, Universidade NOVA de Lisboa, Caparica, Portugal*
[c]*CEG-IST, Instituto Superior Técnico, Universidade de Lisboa, Av. Rovisco Pais, 1, 1049-001 Lisboa, Portugal*
carlour@kt.dtu.dk

Abstract

The worldwide consumption of crustaceans, mainly crabs and shrimp, has increased significantly over the last decades. Noteworthy is that this waste inherently contains high-value-added compounds, such as astaxanthin. Hence, it is economic- and environmentally beneficial to extract astaxanthin effectively and sustainably from seafood production wastes and thrive towards a circular economy.

Therefore, the objective of this work is three-fold: (i) to propose a novel integrated process for the extraction of astaxanthin-oleoresin (AXT-oleoresin) from crab shell wastes; (ii) to assess the environmental performance of this new process at an industrial scale in Portugal; and, (iii) to identify latent production and environmental bottlenecks as well as provide suggestions for process and/or model re-design.

A process environmental and human health impact must become an essential consideration when designing new processes. Henceforth, aiming at a more sustainable process development and production, by applying LCA at the early stage of process development, we aim to identify critical issues early on, then to re-design, and reassess. Hence, this study concludes that the side-by-side application of LCA modelling and experimental process development is the best proactive approach for designing new processes and strategies as early as at the experimental level.

Keywords: LCA, astaxanthin, oleoresin, circular economy

1. Introduction

More than 6 million tons of crustacean shell waste is produced worldwide per year (Rodrigues et al., 2020). Astaxanthin (AXT), a lipid-soluble carotenoid from the xanthophyll family, is among other high-value compounds present in this waste. It has the leading antioxidant activity compared to other antioxidants (e.g., lycopene, vitamins E and A). Other natural sources of AXT are *Haematococcus pluvialis* (highest yield) and the yeast *Xanthophyllomyces dendrorhous* (Rodrigues et al., 2020; Sajna et al., 2015). Although there are many challenges when competing with synthetic AXT production, natural AXT has been approved as food coloring agent and to be used in the formulation of cosmetics and nutraceuticals (Rodrigues et al., 2020; Sajna et al., 2015).

Therefore, to transition to a circular economy emphasizing the conversion of waste into value, in this work, we propose a novel process for the extraction and production of astaxanthin-oleoresin (AXT-oleoresin) from crab shell wastes. Astaxanthin (AXT) is isolated from the shells using a menthol:myristic acid-based deep eutectic solvent as an alternative to conventional approaches. This leads to a bioactive extract, AXT-oleoresin, with combined solute and solvent properties that can find potential applications as a functional ingredient and as a natural preservative in the pharmaceutical, nutraceutical, or cosmetic/personal care industries (Rodrigues et al., 2020). Furthermore, to ensure sustainable progress, LCA is applied at the early stage of process development to identify critical issues early on, then to re-design, and reassess.

2. Methodology

The objective of this work is three-fold: (i) to propose a novel integrated process for the extraction and production of astaxanthin-oleoresin (AXT-oleoresin) from crab shell wastes; (ii) to assess the environmental performance of this new process at an industrial scale in Portugal; and, (iii) to identify latent production and environmental bottlenecks as well as provide recommendations for process and/or model re-design. To this end, the methodology illustrated in Figure 1 is implemented, which implies applying the LCA framework in a stepwise manner.

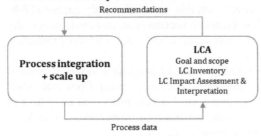

Figure 1: Methodology outline.

The LCA was performed following the ISO 14040 and 14044 guidelines (*ISO* 2006a,b). SimaPro vs.9.1 (Pré Consultants 2020) was the software used. Consequential modelling was used in the inventory analysis as described in (Weidema et al., 2009).

2.1. Goal and scope

As previously stated, the goal of the LCA study is two-fold; (i) to assess the environmental performance of the novel AXT-oleoresin production process in Portugal at industrial scale; and, (ii) to identify latent production/environmental bottlenecks and provide suggestions for process re-design. Figure 2 shows the product system for the overall production of AXT-oleoresin. LCA is performed with cradle-to-gate boundaries. The functional unit requires an additional 1 kg of AXT-oleoresin at the production facility's gate. It is important to note that the cut-off criteria are applied to the crab shell wastes according to the ILCD handbook (JRC-IES, 2010): the environmental impacts related to the production of crab shells are not included in the model. It is assumed that the crab fishery industry is responsible and accounts for all impacts related to the production and transport of crab shell wastes.

Furthermore, as presented in Figure 2, two critical aspects of consequential modelling have been employed: (a) the marginal use of crab shell wastes is identified as being composting (Muñoz et al., 2018) (identification of marginal suppliers), and (b) the process by-product rich in minerals is assumed to substitute mineral fertilizer in the market (product substitution).

The life cycle impact assessment method chosen is the one identified by the International Life Cycle Data (ILCD) handbook, at the midpoint level.

2.1.1. Production process: production of astaxanthin-oleoresin (AXT-oleoresin)

The first step involves the drying and milling of the raw materials since the efficiency of the subsequent extraction process may be influenced by the moisture content and/or particle size. The dried and milled residue is then directed to the extraction step, in which AXT is extracted from the shells using a menthol:myristic acid (ME:MA,8:1) deep eutectic solvent (DES). The AXT-rich extract is separated from the solid residue by filtration, which leads to the final product and a by-product stream. The final product is the AXT-oleoresin, which is composed of DES and AXT-rich extract. The by-product stream is rich in minerals and proteins.

Due to the novelty of the proposed production process, it has only been developed and tested at the lab scale. Therefore, in this work, the experimental results have been appropriately scaled up (Piccinno et al., 2016) in order to reflect a credible industrial scale (see Section 3).

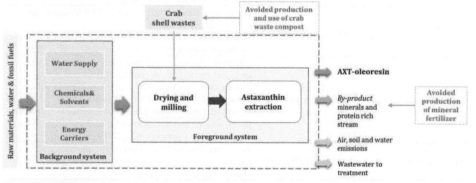

Figure 2: Product system for the production of AXT-oleoresin. System boundaries: cradle-to-gate.

2.2. Life Cycle Inventory

We believe a credible scenario to be the production of AXT-oleoresin that fulfills 50% of the forecasted amount of natural AXT to be consumed in 2021 in Europe (corresponding to approx. 6.9 tons) ("Global Natural Astaxanthin Market insights"). We assume that the natural AXT is produced/consumed in the form of AXT-oleoresin. Therefore, the hypothetical plant's production capacity is approximately 6.9 tons of AXT-oleoresin. The foreground data was obtained by applying appropriate scale-up factors (Piccinno et al., 2016) to the experimental results of the process described in Section 2.1. As previously stated, the tasks in the background system were modelled with the consequential version of the ecoinvent database v.3.1 (ecoinvent Centre 2016). The life cycle inventory and details concerning primary data sources and assumptions are presented in Table 1.

Table 1: Life cycle inventory of the industrial production of AXT-oleoresin from crab shell waste.

Exchanges	Unit	Amount	LCI data
Output of products/services:			
AXT-oleoresin	ton	6.9	Reference flow. Yearly production of AXT-oleoresin (AXT + ME:MA)
Avoided products:			
Mineral Fertilizer	ton	1.85	Displaced production of mineral fertilizer: disrupted production of mineral fertilizer through traditional routes. Composition approx. Minerals (65%), Protein (21%), Chitin (11%). Ecoinvent dataset: Lime fertilizer, from sugar production, at plant/ES Mass Assumption: it is assumed that the by-product stream is directly used and replaces mineral fertilizer.
Brown crab waste composting and soil application	ton	3.73	Avoided composting of crab shell wastes and use of compost. Amount of crab shell waste used. Model developed in this work based on brown crab shell composition and the modelling strategy proposed in (Muñoz et al., 2018) and references within.
Input of products/services:			
Crab shell waste	ton	3.73	Inflow of waste. Applied cut-off criterion, no environmental impacts associated.
Menthol (ME)	ton	6.25	Ecoinvent dataset: Cyclohexanol {RER}| market for cyclohexanol | Conseq, U Assumption: production of cyclohexanol used as proxy for the production of Menthol. (Kendall et al., 2011a)
Myristic acid (MA)	ton	1.14	Ecoinvent dataset: Crude palm kernel oil (incl. LUC incl. peat emissions), at producer/GLO Assumption: production of myristic acid from palm kernel oil. 1.14 ton of MA result in 8.90 ton of palm kernel oil. Based on the content of fatty acids in the oil (82%) and the content of myristic acid within the fatty acids (15.7%). (Tambun et al., 2019)
Heating energy	MJ	6.76E+03	Ecoinvent dataset: Heat, from steam, in chemical industry {RER}| market for heat, from steam in chemical industry | Conseq, U
Electricity	MWh	3.49E-02	Ecoinvent dataset: Electricity, medium voltage {PT}| market for | Conseq, U Assumption: production plant to be located in Portugal.
Emissions to air: direct emissions from composting and use of the compost.			
Wastewater treatment:	ton	1.88	Ecoinvent dataset: Wastewater, average (waste treatment) {RoW}| treatment of, capacity 5E9l/year | Conseq, U

3. Results & Discussion

Figure 3 shows the life cycle impact assessment results (LCIA) and the relative contribution of the different activities in producing AXT-oleoresin in Portugal, from cradle-to-gate. To simplify the interpretation of results, only nine categories are selected out of sixteen reported in ILCD. Due to space constraints, the numerical values are omitted here. The LCIA shows credits (savings, negatives values in Figure 3) for several impact categories. This is especially noticeable in the categories: Acidification (AC), Terrestrial eutrophication (EUTT), Marine eutrophication (EUTM), and water use/depletion (WD). The water savings are associated with the avoided traditional mineral fertilizer production (water-intensive) since mineral fertilizer (minerals-rich side stream) is a by-product of AXT-oleoresin production. The savings in AC, EUTT, and EUTM are mostly related to the fact that, as previously mentioned, the crab shell wastes are being diverted from the composting activity and subsequent use of compost. The use of compost usually leaves behind excessive amounts of nutrients in the soil (e.g., nitrogen and phosphorus), which can be leached into lakes, streams, and coastal waters and thus lead to a gradual increase in nutrient concentration (terrestrial and marine eutrophication). Additionally, commonly the same nutrients also cause soil acidification by changing the soil pH levels. Moreover, there are also some minor credits in the Climate Change (CC), Particulate Matter (PM), and Photochemical ozone formation (POF) categories of impact due to avoided air emissions of small amounts of, for example, nitrous oxide gas.

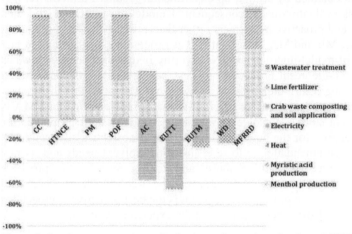

Figure 3: Impact assessment results for extraction and production of AXT-oleoresin from crab shell wastes in Portugal by activity (characterization values, per kg of AXT-oleoresin). Climate change (CC), Human toxicity, non-cancer effects (HTNCE), Particulate matter (PM), Photochemical ozone formation (POF), Acidification (AC), Terrestrial eutrophication (EUTT), Marine eutrophication (EUTM), Water resource depletion (WD) and Mineral, Fossil & Ren resource depletion (MFRRD).

Figure 3 also illustrates the effect of other activities in the LCIA results. The production and use of menthol (ME) and myristic acid (MA) play a crucial role; they are the most environmentally damaging activities in all impact categories. Thus, they are critical process hotspots. In this study, MA is produced from palm kernel oil, which affects/disrupts the palm oil kernel market, thus leading to increased extraction and associated environmental impacts. This is further aggravated by the low MA to oil yield. The production of cyclohexanol was taken as a proxy for ME production (Kendall et al., 2011b) due to the lack of information. This also implies a lack of accuracy and the propagation of uncertainty to the results. The production of cyclohexanol is based on the hydrogenation of benzene (resource intensive). Henceforth, this is in line with, for example, the results obtained for the Mineral, Fossil & Renewable resource depletion (MFRRD) category. However, it is important to note that the AXT-oleoresin production process's energy needs are somewhat irrelevant among the other contributors. Besides, the process does not consume freshwater. Therefore, this is a positive indication that if improving the critical process points (ME and MA models), the process has a good probability of improving its environmental performance. For example, the production of MA from nutmeg butter and coconut oil will be investigated (model improvement). Although both ME and MA have been identified as the process' critical points, replacing these solvents is not in question at this point. Both solvents were chosen due to their properties for producing a bioactive extract for the potential formulation of cosmetics and nutraceuticals. Henceforth, to overcome the described process limitations and hotspots, the overall recommendations are to (i) test if smaller quantities of solvents will have the same properties for the formulation of the bioactive extract, (ii) have a dialogue with the solvent producer in order to find a greener source, and/or (iii) attempt to synthesize the solvents in-house and reassess the process's environmental performance.

4. Conclusions

To the best of our knowledge, we have presented the first model of the cradle-to-gate LCA for the novel production of natural AXT-oleoresin from crab shell wastes using DES. Primary data was obtained by scaling up experimental results. Furthermore, this study addresses process evaluation using consequential modelling principles and thus performs a prospective and proactive assessment rather than retroactive. The production of the DES components, ME and MA, was identified as process hotspots, which in fact translates into targets for model improvement. Noteworthy is that the energy and water consumption of the AXT-oleoresin production process seems to be immaterial when comparing to the remaining contributors. This is a reasonable indication that the process has a good environmental performance. As previously stated, replacing ME and MA is not an option at this point since both solvents were chosen due to their properties for producing a bioactive extract for the potential formulation of cosmetics/nutraceuticals. Therefore, the overall recommendation is to re-evaluate the solvent needs and its raw materials, potentially synthesize them in-house, and finally reassess the process's environmental performance. This study concludes that the side-by-side application of LCA modelling and experimental process development is the best proactive approach for designing new processes and strategies as early as at the experimental level.

References

Global Astaxanthin Market insights, forecast to 2025, 2019.

ISO 14040—environmental management—life cycle assessment—principles and framework, 2006. . Genève.

ISO 14040: Environmental management – Life cycle assessment – principles and framework, 2006. . Gnève.

JRC-IES, 2010. ILCD handbook—general guideline for life cycle assessment—detailed guidance. First edition. Ispra.

Kendall, A., Yuan, J., Brodt, S., Jan Kramer, K., 2011a. Carbon Footprint of U.S. Honey Production and Packing Report to the National Honey Board.

Kendall, A., Yuan, J., Brodt, S., Jan Kramer, K., 2011b. Carbon Footprint of U.S. Honey Production and Packing Report to the National Honey Board.

Muñoz, I., Rodríguez, C., Gillet, D., M. Moerschbacher, B., 2018. Life cycle assessment of chitosan production in India and Europe. Int. J. Life Cycle Assess. 23, 1151–1160.

Piccinno, F., Hischier, R., Seeger, S., Som, C., 2016. From laboratory to industrial scale: a scale-up framework for chemical processes in life cycle assessment studies. J. Clean. Prod. 135, 1085–1097.

Rodrigues, L.A., Pereira, C. V., Leonardo, I.C., Fernández, N., Gaspar, F.B., Silva, J.M., Reis, R.L., Duarte, A.R.C., Paiva, A., Matias, A.A., 2020. Terpene-Based Natural Deep Eutectic Systems as Efficient Solvents To Recover Astaxanthin from Brown Crab Shell Residues. ACS Sustain. Chem. Eng. 8, 2246–2259.

Sajna, K.V., Gottumukkala, L.D., Sukumaran, R.K., Pandey, A., 2015. White Biotechnology in Cosmetics. Ind. Biorefineries White Biotechnol. 607–652.

Tambun, R., Ferani, D.G., Afrina, A., Tambun, J.A.A., Tarigan, I.A.A., 2019. Fatty Acid Direct Production from Palm Kernel Oil. IOP Conf. Ser. Mater. Sci. Eng. 505.

Weidema, B.P., Ekvall, T., Heijungs, R., 2009. Guidelines for applications of deepened and broadened LCA. Deliverable D18 of work package 5 of the CALCAS project.

Proceedings of the 14th International Symposium on Process Systems Engineering – PSE 2021+
June 19-23, 2022, Kyoto, Japan © 2022 Elsevier B.V. All rights reserved.
http://dx.doi.org/10.1016/B978-0-323-85159-6.50342-0

Decomposition of Organic Compounds in Water from Oil Refineries

Shoma Kato[a], Yasuki Kansha[a*]

[a]*Organization for Programs on Environmental Sciences, Graduate School of Arts and Sciences, The University of Tokyo, 3-8-1 Komaba, Meguro-ku, Tokyo 153-8902 JAPAN*

kansha@global.c.u-tokyo.ac.jp

Abstract

In oil refineries, water is used for many different purposes and a substantial amount of wastewater is generated. The traditional wastewater treatment process in oil refineries is comprised of primary and secondary treatment, and the treated water is discharged to water bodies. By recycling the treated water, water withdrawal from the environment can be reduced. Additionally, the organic sulfur, organic nitrogen, and other organic compounds may not be fully removed by biological treatment. For the treatment of these organics and for the recycling of wastewater from oil refineries, a wastewater treatment system for oil refineries has been proposed. The proposed process consists of 1) primary and secondary treatment 2) the removal of remaining organic sulfur and nitrogen using hydrogen, and 3) the removal of remaining organics by photocatalysis. To realize the proposed system, the removal of organic sulfur using hydrogen by simulation and photochemical reactions by experiments were studied. The removal of organic sulfur using hydrogen was analyzed using oil refinery plant data. The results showed that organic sulfur and nitrogen were removed in the process, and the conversion ratio highly depended on the chemical structure of the components. For the experiment, as a representative of the remaining organic compound in the wastewater from the oil refinery, phenol was chosen. In the experiment, the decomposition of phenol using UV and TiO_2 was investigated. The results showed that phenol was successfully decomposed. From the results of the investigations, this process shows promise to improve the treatment of industrial wastewater and contribute to the conservation of water resources.

Keywords: Water Treatment, Photocatalyst, Environmental Systems, Process Design

1. Introduction

Globally, there is a rise in the demand for freshwater resulting from the growing world population and industrialization. In fact, the global demand for water has been increasing by 1 % per year (UNESCO, 2020). For sustainable development and the well-being of the global ecosystem, it is necessary that water usage is reduced, that water is properly treated, and that water is recycled. Wastewater is generated from the agricultural sector, the industrial sector, and the domestic sector.

The contaminants present in industrial wastewater vary from industry to industry. In the case of the petroleum industry, most processes in oil refineries use water, so large amounts of wastewater are generated. For example, 0.60 – 0.71 L of water is used to produce 1 L of gasoline (Sun et al., 2018). To reduce the environmental impact of the discharge of wastewater and to reduce the water withdrawal of the petroleum industry, the treated wastewater can be recycled to use in the oil refinery or other industries. For example, the Kalundborg Industrial Symbiosis in Denmark was developed to recycle the by-products in an industrial community, and the initial reason for the development was to reduce the usage of groundwater in oil refineries (WWAP, 2017).

The traditional treatment of wastewater from oil refineries includes primary treatment to remove the oils and sediments physically and secondary treatment to remove the organics in water biologically. Then, the water is discharged to water bodies. For the water to be recycled, additional treatment of water is necessary to remove the remaining contaminants after primary and secondary treatment of wastewater as the contaminants left in the treated water may affect the industrial processes such as fouling and degradation of catalysts. Additionally, the treatment process should be non-toxic and safe for the environment.

Advanced Oxidation Processes (AOPs) that generate strong oxidants such as OH radicals are effective to treat the refractory pollutants remaining in the water (Miklos et al., 2018). However, the chemical oxidation using AOPs is expensive, so they may be used in combination with biological treatment to reduce operation costs and for the effective treatment of contaminants in the wastewater (Oller et al., 2011).

In this research, a process was proposed to enhance the recycling of the water generated from oil refineries. In the proposed process, the remaining organic sulfur and nitrogen are removed after the biological treatment of water. Then, the remaining organic is removed by photocatalysis which is an AOP with the generation of OH radicals. The validity of the process was verified through simulation and experiments.

2. Proposed process for the treatment of water from oil refineries

Figure 1 shows the proposed process to recycle the water generated from the oil refineries. The proposed process aims to recycle the water in the oil refinery to reduce water withdrawal from the environment.

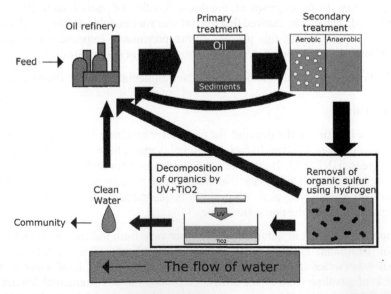

Figure 1: The proposed wastewater treatment system for the recycling of water from oil refineries is shown.

After primary treatment to remove the oils and sediments and secondary treatment to remove some of the organic compounds, organic sulfur, organic nitrogen, and other organic compounds may remain in the water. This is because wastewater from oil refineries contains a high concentration of refractory organics which are difficult to remove by biological treatment, and it may show higher COD values even after primary and secondary treatment (Diya'uddeen et al., 2011).

To treat the remaining contaminants in the water for recycling, two additional processes were added. The first process is to treat the remaining organic sulfur and organic nitrogen using hydrogen. In an oil refinery, desulfurization processes such as residue desulfurization (RDS) use hydrogen to upgrade residual fuel by removing contaminants, by converting them into lighter products, and by promoting hydrogenation (Marafi et al., 2006). Like the RDS process, the remaining organic sulfur and organic nitrogen may be removed using hydrogen. After the removal of organic sulfur and organic nitrogen, the remaining contaminants are further removed through photocatalysis with TiO_2 as the catalyst. TiO_2 was chosen as the catalyst for its ability to oxidize organic pollutants, chemical stability, and nontoxicity (Nakata and Fujishima, 2012). Depending on the necessary quality of the water needed for the industrial processes, the water can be recycled back to the oil refinery instead of going through the full cycle. Additionally, the process is safe for the environment as it does not require the use of flocculants or other chemical agents. The energy usage of the process can be further reduced by using sunlight as the light source for the photocatalyst process.

3. Methods

The removal of organic sulfur, organic nitrogen, and other organic compounds was explored through simulation. The treatment of the remaining organic contaminants was evaluated through experiments using phenol as a representative and TiO_2 as the photocatalyst.

3.1. The removal of organic sulfur, organic nitrogen, and other organic compounds using hydrogen

The data of the components of the feed and product data from the RDS process was used to calculate the overall conversion ratio of the organic sulfur and nitrogen. The conversion ratio was calculated as shown in Eq.(1).

$$Conversion\ ratio = \frac{Product\ concentration}{Feed\ concentration} \tag{1}$$

By understanding the reaction in the process using hydrogen, the effectiveness of the process can be evaluated, and the potential products from the reaction can be known.

3.2. The treatment of phenol through photocatalysis

Phenol (purity 99 %, FUJIFILM Wako Pure Chemical Corporation) was used as the representative for the organic compound. It was dissolved in Milli-Q water so that the concentration was 50 mg/L. Aeroxide TiO_2 P 25 (specific surface area 53 m^2/g, NIPPON AEROSIL CO., LTD.) was used as the photocatalyst, and it was added to the phenol solution so that the TiO_2 dosage was 1 g/L. A 300 mL Pyrex beaker (diameter 77 mm and height 110 mm) was used as the vessel. The vessel was filled with 200 mL of the mixture of TiO_2 and phenol solution. After filling the vessel, it was covered with a quartz lid (thickness 3 mm). A 4 W UV lamp with the main wavelength at 365 nm (model LUV-4,

AS ONE Corporation) was placed directly on top of the quartz lid. The experiments were conducted at room temperature.

The samples were taken from the vessel using syringes, and they were filtered using PES syringe filters of pore size 0.1 μm to remove TiO_2 particles. The absorbance of the samples was measured with a UV-vis spectrophotometer (model ASUV-1100, AS ONE Corporation) at 270 nm which is the absorbance peak of phenol. The phenol removal was calculated by the following equation:

$$Phenol\ removal\ (\%) = \frac{C_0 - C_t}{C_0} \times 100 \tag{2}$$

where C_0 and C_t are the concentrations of phenol at time 0 and at time t respectively.

4. Results and Discussion

4.1. The overall conversion ratio of the organic sulfur and organic nitrogen

Table 1 shows the conversion ratio by using hydrogen for organic sulfur and organic nitrogen present in the feed and product data. Depending on the structure of the component, the conversion ratio varies. Oil B was a heavier oil compared to Oil A, and the conversion ratio for Oil B was less than that of Oil A. In both oils, there is remaining organic sulfur and organic nitrogen which needs further treatment. Additionally, the process consumes a large amount of hydrogen. Further investigation is needed to increase the efficiency of this process to reduce hydrogen consumption.

Table 1: The conversion ratio by hydrogen treatment is shown.

Name	Conversion ratio of molecule with at least one sulfur atom	Conversion ratio of molecule with at least one nitrogen atom
Oil A	0.655	0.141
Oil B	0.575	0.084

4.2. The removal of phenol by photocatalysis

Figure 2 shows the comparison of the removal of using UV only versus using UV and TiO_2 as the photocatalyst. For UV photolysis alone, only 0.14 % of phenol was removed after 180 min. The result that phenol is difficult to degrade through direct photolysis is consistent with that of Lin et al. (2011). For UV+TiO_2 photocatalysis, 8.4 % of the phenol was removed after 180 min. With the presence of the TiO_2 photocatalyst, phenol which is the representative organic can be removed. However, the rate of removal is slow under the experimental conditions. The removal rate may be improved by increasing the light intensity, by optimizing TiO_2 dosage.

For the photocatalytic process, it is necessary to consider the recovery of the TiO_2 catalyst. One way is to use a TiO_2 photocatalytic membrane reactor which combines photocatalytic oxidation and membrane filtration (Leong et al., 2014). In this way, the TiO_2 can be recovered. Additionally, sunlight may be used as the light source to reduce energy consumption.

Figure 3 shows the possible mechanism of the removal of the organic compounds using the TiO_2 membrane. By irradiating the TiO_2 membrane with UV light, water may be converted into H^+ ions and OH radicals (Sobczyński et al., 2004). Additionally, through a series of reactions, the organic compounds are completely mineralized into water and CO_2.

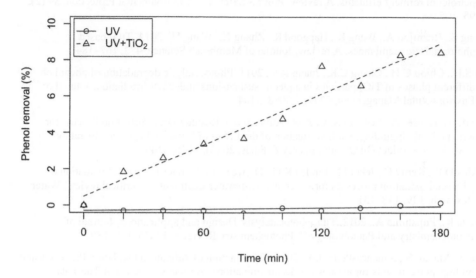

Figure 2: Comparison of the phenol removal using UV photolysis only versus UV+TiO_2 photocatalysis.

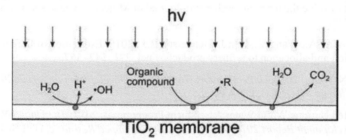

Figure 3: The schematic of photocatalytic treatment of organic contaminants in wastewater.

5. Conclusion

A system was proposed to recycle the wastewater from an oil refinery. The proposed process consists of 1) primary treatment, 2) secondary treatment, 3) treatment of remaining organic nitrogen, organic sulfur, and other organics by hydrogen, and 4) treatment of the remaining organics by photocatalysis with TiO_2 as the catalyst. To evaluate the treatment using hydrogen, simulation was conducted. The results showed that although the process can treat some of the contaminants, further treatment is necessary to remove the remaining contaminants. To evaluate the treatment using photocatalysis, experiments were conducted. The results showed that phenol, a representative of a refractory organic can be removed. Depending on the demanded quality of the water, the water may be recycled back to the oil refinery and the industrial community to reduce the water withdrawal from the environment. The proposed process is promising to reduce the amount of water withdrawal, and it is a green process that does not require the use of toxic chemicals.

References

Diya'uddeen B.H., Daud W.M.A.W., Abdul Aziz A.R., 2011, Treatment technologies for petroleum refinery effluents: A review, Process Safety and Environmental Protection, 89 (2), 95–105.

Leong S., Razmjou A., Wang K., Hapgood K., Zhang X., Wang H., 2014, TiO_2 based photocatalytic membranes: A review, Journal of Membrane Science, 472, 167–184.

Lin S.H., Chiou C.H., Chang C.K., Juang R.S., 2011, Photocatalytic degradation of phenol on different phases of TiO_2 particles in aqueous suspensions under UV irradiation, Journal of Environmental Management, 92 (12), 3098–3104.

Marafi A., Hauser A., Stanislaus A., 2006, Atmospheric Residue Desulfurization Process for Residual Oil Upgrading: An Investigation of the Effect of Catalyst Type and Operating Severity on Product Oil Quality, Energy & Fuels, 20 (3), 1145–1149.

Miklos D.B., Remy C., Jekel M., Linden K.G., Drewes J.E., Hübner U., 2018, Evaluation of advanced oxidation processes for water and wastewater treatment – A critical review, Water Research, 139, 118–131.

Nakata K., Fujishima A., 2012, TiO_2 photocatalysis: Design and applications, Journal of Photochemistry and Photobiology C: Photochemistry Reviews, 13 (3), 169–189.

Oller I., Malato S., Sánchez-Pérez J.A., 2011, Combination of Advanced Oxidation Processes and biological treatments for wastewater decontamination—A review, Science of The Total Environment, 409 (20), 4141–4166.

Sobczyński A., Duczmal Ł., Zmudziński W., 2004, Phenol destruction by photocatalysis on TiO_2: an attempt to solve the reaction mechanism, Journal of Molecular Catalysis A: Chemical, 213 (2), 225–230.

Sun P., Elgowainy A., Wang M., Han J., Henderson R.J., 2018, Estimation of U.S. refinery water consumption and allocation to refinery products, Fuel, 221, 542–557.

UNESCO, UN-Water, 2020, *United Nations World Water Development Report 2020: Water and Climate Change*, UNESCO, Paris, France.

WWAP (United Nations World Water Assessment Programme), 2017, *The United Nations World Water Development Report 2017. Wastewater: The Untapped Resource*, UNESCO, Paris France.

Proceedings of the 14th International Symposium on Process Systems Engineering – PSE 2021+
June 19-23, 2022, Kyoto, Japan © 2022 Elsevier B.V. All rights reserved.
http://dx.doi.org/10.1016/B978-0-323-85159-6.50343-2

Energy Harvesting Wireless Sensors Using Magnetic Phase Transition

Yasuki Kansha[a], Masanori Ishizuka[b]

[a]Oraganization for Programs on Environmental Sciences, Graduate School of Arts and Sciences, The University of Tokyo, 3-8-1 Komaba, Meguro-ku Tokyo 153-8902, Japan
[b]Collaborative Research Center for Energy Engineering, Institute of Industrial Science, The University of Tokyo, 4-6-1 Komaba, Meguro-ku, Tokyo 153-8505, Japan
kansha@global.c.u-tokyo.ac.jp

Abstract

In this research we proposed energy harvesting data acquisition sensors using the magnetic phase transition resulting from changes in temperature, and electromagnetic induction resulting from changes in magnetic flux. The proposed system can provide wireless temperature or velocity sensors that directly measure electromotive forces generated by a solenoid following Faraday's law without any additional energy input. Our proposed energy harvesting sensors have the potential to contribute significantly to the development of CPS in the near future.

Keywords: Wireless sensors; Energy harvesting; Cyber-physical systems

1. Introduction

The Japanese government has proposed 'Society 5.0', under which, not only industry, but society itself will be changed by information and communication technology (ICT) and the internet of things (IoT) to allow sustainable development. In the energy related field, dig data such as climate and environmental information and energy usage in communities will be acquired and analysed by Artificial Intelligence (AI). Furthermore, energy will supply following the analysed data for providing a stable energy supply etc. (Cabinet Office in Japan).

To deploy cyber-physical systems (CPS) and the IoT in society, it is necessary to develop overall security systems (Alguliyev et al., 2018), efficient information and communication technologies (ICT) including data transfer systems in the network, and acquisition systems, such as sensors and actuators (Patil and Fiems, 2018). It is also important to find new energy sources to connect and transfer digital data to the cloud and to develop intelligent decision/control systems (Shen et al., 2019) and learning algorithms (Zhang et al., 2018) for rational operation of the overall networks, incorporating the use of artificial intelligence.

In meeting the requirements of energy for CPS, the term 'energy harvesting' is commonly used. Energy harvesting involves electric power generation for online operation of sensor and electric devices from currently unused low level energies such as vibrations, radio frequencies, light and low temperature heat, into electricity. Piezoelectric elements and antennae are common devices used for converting vibrations or radio frequencies into electricity (Wang et al., 2018). Photovoltaics is also a familiar way of converting light into electricity to supply CPS (Gunduz and Jayaweera, 2018). Thermoelectric elements based on the Seebeck effect are also commonly used to convert heat into elcctricity (Ando

Junior et al., 2018). However, the energy efficiency of many of these energy harvesting technologies is still low, and much research is focused on increasing their efficiency. In fact, the thermos-electric device generate the electric power by p- and n-type semiconductors. However, it is well-known that the efficiency of this device is so low due to \small entropy change by electron or hole transfer (Chen and William, 1996)

On the contrary, much recent research has focused on the possibility of combining energy harvesting and sensors, including wearable sensors for human information (Myers et al., 2017) and wireless sensors for environmental information (Babayo et al., 2017). These sensors called energy harvesting sensors directly sense the measured target without any additional energy conversion, leading to increase overall energy efficiency of CPS. Thus, the energy harvesting sensors are expected as a key technology for propagation of a CPS (Kausar et al., 2014).

2. Energy harvesting using magnetic phase transition and sensors

Kansha et al. (2018) proposed an energy harvesting system from sub-ambient heat sources, such as exhaust heat from refrigerators or coolers, that uses magnetic phase transition integrated with electromagnetic induction. In this system, a magnetic material such as gadolinium is cooled by a sub-ambient temperature heat to below the Curie temperature (292 K for gadolinium). The material becomes ferromagnetic. It is placed near a solenoid as shown in Figure 1 a) and heated by a high temperature heat to its Curie temperature. The material is demagnetized, becoming paramagnetic, and the magnetic flux from the material that passes through the solenoid decreases as the temperature increases. An electromotive force is generated in the solenoid following Faraday's law of induction;

$$\varepsilon = -N\frac{d\Phi}{dt} \qquad (1)$$

where ε is the electromotive force, N is the number of turns on the solenoid, Φ represents the magnetic flux, and t is time.

Figure 1 b) shows a schematic image of the thermodynamic cycle in temperature-entropy diagram of the series of action. The theoretical thermodynamic cycle of this system transitions to a closed trilateral cycle suitable for sensible heat recovery to generate electric power.

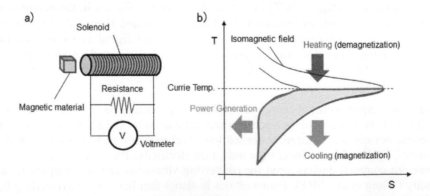

Figure 1: Thermodynamic cycle of the proposed power generation system.

A study of the adiabatic temperature change due to the magnetic phase transition during magnetization of gadolinium from 0 T to 1 T indicates that the change is almost linear from 270 K to the Curie temperature (292 K) and from the Curie temperature to 320 K at a peak of 292 K as shown in Figure 2. Furthermore, the adiabatic efficiency of magnetization/demagnetization changes of gadolinium was examined about 0.92 (Kotani et al., 2013). From these aspects, the change in entropy at the magnetic phase transition around the Curie temperature may have a linear relationship with the temperature.

Thus, observing the flux changes for magnetic materials at different temperatures could allow it to be used as a temperature sensor integrated with energy harvesting around the Curie temperature, without requiring any additional energy instead of thermocouples.

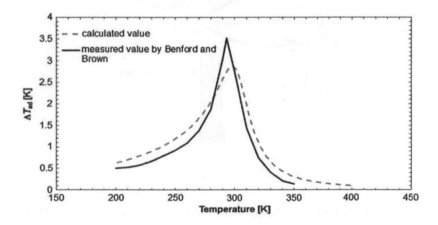

Figure 2: Adiabatic temperature change when gadolinium is magnetized from 0 T to 1 T (Kotani et al., 2013).

3. Experimental procedure and set-up

To examine the possibility of energy harvesting sensors, the relationship between the temperature of a magnetic material and the change in its magnetic flux was investigated by measuring the electromotive force. Gadolinium was selected as the magnetic material because of its Curie temperature.

Before measuring the electromotive forces, a gadolinium nugget (2.73 g) was left at each temperature for more than 30 mins until the nugget temperature became constant. The maximum magnetic flux, which the gadolinium nugget creates, was about 25 mT measured by a gauss meter (HMMT-6J04-VF, Lake Shore Cryotronics Inc.).

Using the following experimental set-up shown in Figure 3, the electromotive forces generated by Faraday's law of induction were monitored at four different temperatures (256, 280, 292, 296 K) by an oscilloscope (InfiniiVision DSO-X 2002A, Agilent Technologies Inc.). A permanent magnet (275 mT) was positioned at the edge of a solenoid with 500 coils of iron wire.

Using the gauss meter, the magnetic flux density at the other end of the solenoid, the nearest point to the gadolinium, was 19 mT. The gadolinium nugget, which had a magnetic flux density of 0 mT without the magnetic field, was fixed to the end of a 40-mm arm, which was rotated horizontally by a motor (rotation speed: 545 degree/s). The gadolinium was passed over the solenoid. The minimum distance from the gadolinium to the solenoid was 3 mm. Thus, the magnetic flux through the solenoid changed with the position of the gadolinium. It is noted that the magnetic flux was changed by heat transfer to the magnetic materials to use the proposed system as an ambient temperature sensor. However, this change might be too sensitive to examine in the experiments. Therefore, the gadolinium was forced to move by motor for changing the distance to the solenoid at the constant temperature in order to sense the targeted temperature in this experiment. The electromotive force produced by electromagnetic induction and the current were measured by the oscilloscope with a shunt resistance of 4.7 Ω.

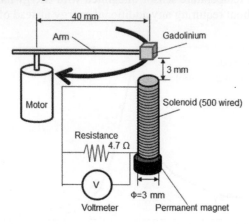

Figure 3: Adiabatic temperature change when gadolinium is magnetized from 0 T to 1 T.

4. Experimental results

The large electromotive forces generated are shown in Figure 4 at the Curie temperature (292 K). The grey trace shows the raw data and the black line is the 10-ms moving average. The figures show positive and negative peaks. A positive peak was created when the gadolinium came close to the solenoid and the negative peak was created when the gadolinium passed. To determine the relationship between temperature and the generated electromotive force, the amplitude of the electromotive force was measured. Table 1 lists the amplitude of the electromotive force generated as a moving average at each temperature. From this table, it can be understood that the maximum power output is distributed following the temperature. Thus, the proposed system has a possibility to use it as a temperature sensor with calibration.

Table 1. Relationship between temperature and amplitude of electromotive forces.

Temperature [K]	256	280	292	296
minimum value of electromotive force [mV]	-0.182	-0.466	-0.505	-0.228
maximum value of electromotive force [mV]	0.023	0.239	0.267	-0.018
amplitude of electromotive force[1] [mV]	0.102	0.352	0.386	0.105
maximum power ouput[2] [nW]	1.8	28.2	34.7	2.4

[1] amptitude = (maximum - minimum)/2

[2] power output = V^2/R

Figure 5 shows a comparison of the 10-ms moving averages of the electromotive forces generated for several motor rotation speeds (182, 363, and 545 degree/s) at 292 K; the amplitudes were 0.188, 0.309, and 0.386 and the intervals between the peaks (53.5, 40.5 and 31.5 ms) changed following the rotational speed of the gadolinium. The amplitudes are linearly increase with rotation speeds. Therefore, it can be worked as a velocity sensor.

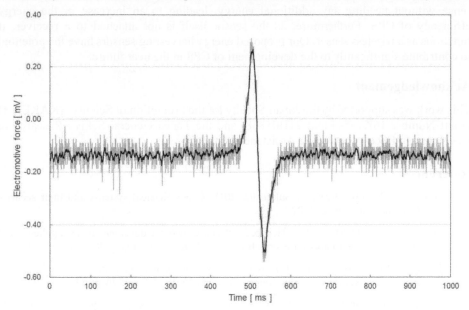

Figure 4: Generated electromotive force at 292 K during one cycle.

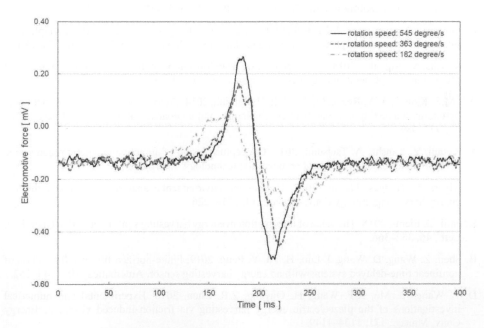

Figure 5: Moving average of generated electromotive force at 292 K by different rotation speeds.

5. Conclusion

This paper proposes designs for energy harvesting temperature and velocity sensors. By integrating magnetic phase transition with electromagnetic induction, these sensors can sense without needing any additional energy, leading to an increase in the energy efficiency of CPS. Furthermore, as the sensor itself is not attached to a receiver, it functions as a wireless sensor. Our proposed energy harvesting sensors have the potential to contribute significantly to the development of CPS in the near future.

Acknowledgement

This work was supported by the Japan Society for the Promotion of Science, (KAKENHI Grant Number 16K14544 and 21H01868) and the Tonen General Sekiyu Research & Development Encouragement & Assistance Foundation.

References

R. Alguliyev, Y. Imamverdiyev, L. Sukhostat, 2018, Cyber-physical systems and their security issues, Comput. Ind., 100, 212–223.

O.H. Ando Junior, A.L.O. Maran, N.C. Henao, 2018, A review of the development and application of thermoelectric microgenerators for energy harvesting, Renew. Sust. Energ. Rev., 91, 376–393.

A.A. Babayo, M.H. Anisi, I. Sli, 2017, A review on energy management schemes in energy harvesting wireless sensor networks, Renew. Sust. Energ. Rev., 76, 1176–1184.

Cabnet office in Japan https://www8.cao.go.jp/cstp/english/society5_0/index.html (access 10/Nov./2021)

K. Chen, S.B. William, 1996, An analysis of the heat transfer rate and efficiency of TE (thermoelectric) cooling systems. Int. J. Energy Res., 20, 399–417.

H. Gunduz, D. Jayaweera, 2018, Reliability assessment of a power system with cyber-physical interactive operation of photovoltaic systems, Int. Electr. Power Energy Syst., 101, 371–384.

Y. Kansha, M. Ishizuka, 2018, Power generation from low temperature waste heat using magnetic phase transition, Chemical Engineering Transactions, 70, 43–48.

A.S.M.Z. Kausar, A.W. Rezan, M.U. Saleh, H. Ramiah, 2014, Energizing wireless sensor networks by energy harvesting systems: Scopes, challenges and approaches, Renew. Sust. Energ. Rev., 38, 973–989.

Y. Kotani, Y. Kansha, A. Tsutsumi, 2013, Conceptual design of an active magnetic regenerative heat circulator based on self-heat recuperation technology, Energy, 55, 127–133.

A. Myers, R. Hodges, J.S. Jur, 2017, Human and environmental analysis of wearable thermal energy harvesting, Energy Conv. Manag., 143, 218–226.

K. Patil, D. Fiems, 2018, The value of information in energy harvesting sensor networks, Oper. Res. Lett., 46, 362–366.

B. Shen, Z. Wang, D. Wang, J. Luo, H. Pu, Y. Peng, 2019, Finite-horizon filtering for a class of nonlinear time-delayed systems with an energy harvesting sensor, Automatica, 100, 144–152.

D.W. Wang J.I. Mo, X.F. Wang, H. Ouyang, Z.R. Zhou, 2018, Experimental and numerical investigations of the piezoelectric energy harvesting via friction-induced vibration, Energy Conv. Manag., 171, 1134–1149.

X. Zhang, T. Yu, Z. Xu, Z. Fan, 2018, A cyber-physical system with parallel learning for distributed energy management of a microgrid, Energy, 165, 205–221.

Proceedings of the 14th International Symposium on Process Systems Engineering – PSE 2021+
June 19-23, 2022, Kyoto, Japan © 2022 Elsevier B.V. All rights reserved.
http://dx.doi.org/10.1016/B978-0-323-85159-6.50344-4

Competitive Adsorption of Copper, Nickel, and Chromium Ions onto Amine Functionalized SBA-15

Bawornpong Pornchuti[a*], Yuttana Phoochahan[a], Prarana Padma[a],
Suchada Ruengrit[a], Pravit Singtothong[b]

[a]*Chemical Engineering Department, Mahanakorn University of Technology, Bangkok 10530, Thailand*
[b]*Chemistry Department, Mahanakorn University of Technology, Bangkok 10530, Thailand*
bawornpong@yahoo.com

Abstract

Wastewater from electroplating industry is composed of heavy metals. Adsorption is an effective method treating this type of hazardous wastewater. Many works studied adsorption of single metal ion however industrial wastewater is always multicomponent system. As a consequence, this work aimed to study the competitive adsorption of copper, nickel, and chromium ions onto amine functionalized SBA-15. It was found that adsorption capacity increased with an increase of pH. Moreover, removal efficiency of metal ion decreased when initial metal concentration increased. To represent the adsorption data, Langmuir, Freundlich, extended Langmuir, and modified competitive Langmuir models were selected. Their SSEs (Sum of squared errors) were 98.58, 118.40, 165.84, and 156.18, respectively.

Keywords: competitive adsorption, heavy metals, multicomponent isotherms, SBA-15.

1. Introduction

Electroplating is one of finishing steps used to improve mechanical and chemical properties of products. It also gives an attractive surface. However, wastewater from electroplating process is always comprised of heavy metals such as copper, nickel, and chromium. These heavy metals are harmful to human and can accumulate in ecosystems. Adsorption is one of the promising techniques treating wastewater containing heavy metals. As a consequence, many researches aimed to develop novel adsorbents having high adsorption capacities.

SBA-15 was chosen in this work because it has large surface area and easy to modify its surface. If any functional groups containing N, O, S, or P atom is incorporated onto adsorbent surface, the removal efficiency of metal ions is improved (Maleki, 2016). Therefore adsorption of copper, nickel, and chromium ions onto amine functionalized SBA-15 was studied in this work.

Most works in literature reported the adsorption of single heavy metal ion while industrial wastewater is generally composed of several heavy metal ions. The adsorption of a component may be affected by the other component (Girish, 2017). Therefore, the study of multicomponent adsorption is necessary. Isotherm models are used to predict the adsorption behaviour. Langmuir, Freundlich, extended Langmuir, and modified competitive Langmuir isotherm models were chosen in this study.

2. Materials and methods

2.1. Materials

TEOS (Tetraethyl orthosilicate), Pluronic P123, toluene, ethanol, APTES (Aminopropyltriethoxysilane), and hydrochloric acid were purchased from Sigma Aldrich or Merck. They were used as received.

2.2. Synthesis of SBA-15

SBA-15 was synthesized by using the method of Naik et al. (2011). Briefly, 6.64 g of Pluronic P123, 13.5 mL of HCl solution, and 202 mL of deionized water was mixed together. Next, 13.86 g of TEOS was dropped into the mixture. After aging at 90 °C for 24 h, the solid obtained was filtered and washed with deionized water. Then, the solid was dried and calcined at 550 °C for 3 h.

2.3. Preparation of amine functionalized SBA-15

The method of Parida and Rath (2009) was used to graft amino functional group onto the surface of SBA-15. In brief, 2.0 g of SBA-15 was mixed with 60 mL of toluene. Next, 1.2 mL of APTES was dropped into the mixture. After reflux for 8 h, the solid was washed with ethanol and deionized water, respectively. Then it was dried at room temperature for 12 h.

2.4. Batch adsorption studies

In general, 100 mg of adsorbent was added into 50 mL of heavy metal solution. The mixture was shaken at 105 rpm for 48 h. After vacuum filtration, the metal concentration was determined by atomic absorption spectroscopy (AAS). The adsorption capacity was calculated by Eq. (1).

$$q_e = \frac{(C_i - C_e)V}{m} \tag{1}$$

q_e is the adsorption capacity at equilibrium; C_i is the initial metal concentration; C_e is the metal concentration at equilibrium; V is the volume of metal solution; m is the mass of adsorbent. For the calculation of removal efficiency, Eq. (2) was used.

$$\text{Removal efficiency} = \frac{(C_i - C_e)}{C_i} \times 100\% \tag{2}$$

3. Results and discussion

3.1. Effect of pH

To study the effect of pH, the experiments were conducted at pH 2 – 5. It was found that the amount of nickel ion adsorbed by amine functionalized SBA-15 was very low as shown in Figure 1. The nickel adsorption was suppressed by copper and chromium ions, suggesting antagonistic interaction. This effect was also found in the adsorption of nickel in the presence of copper when using olive stones as adsorbent (Girish, 2018). Moreover, adsorption of copper and chromium ions increased with an increase of pH. This may be caused by competition with hydrogen ion at low pH (Sertsing et al., 2018).

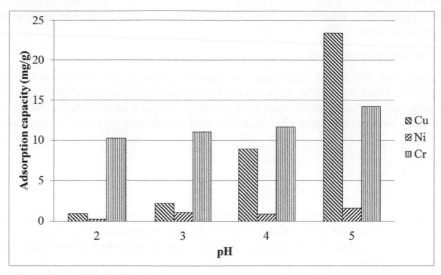

Figure 1. Effect of pH on adsorption capacity of copper, nickel, and chromium ions in multicomponent adsorption.

3.2. Effect of initial metal concentration

The effect of initial metal concentration was studied at pH 5. The results were illustrated as Figure 2. When initial metal concentration increased, the removal efficiency of metal ion decreased. This phenomenon was also found in single metal adsorption (Pornchuti et al., 2020). By fixing the amount of adsorbent, the number of adsorption site was limited. That was why the reduction of the removal efficiency occurred.

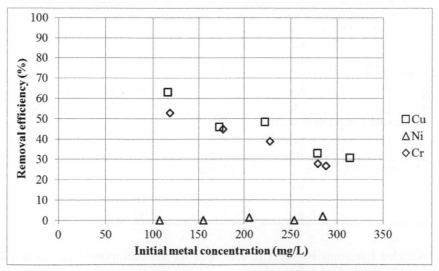

Figure 2. Effect of initial metal concentration on removal efficiency of metal ion in multicomponent adsorption.

To treat wastewater containing copper, nickel, and chromium ions by amine functionalized SBA-15, at least two adsorption columns are required. The first one is used to adsorb copper and chromium ions. The other one is used to remove nickel ion.

When copper and chromium ions are removed, nickel ion can be adsorbed in the second adsorption column.

3.3. Mathematical modeling

Langmuir and Freundlich isotherms were used widespread for adsorption of single component. There were some works using these models representing experimental data of multicomponent adsorption such as the work of Qi and Pichler (2017). Due to their simplicity, both models were chosen in this work. Langmuir isotherm is expressed as Eq. (3).

$$q_{e,i} = \frac{q_{m,i} b_{L,i} C_{e,i}}{1 + b_{L,i} C_{e,i}} \tag{3}$$

$q_{e,i}$ is the adsorption capacity at equilibrium for component i; $q_{m,i}$ is the maximum adsorption capacity for component i; $b_{L,i}$ is the Langmuir constant for component i; $C_{e,i}$ is the metal concentration at equilibrium for component i. Freundlich isotherm is expressed as Eq. (4).

$$q_{e,i} = K_{F,i} C_{e,i}^{1/n_i} \tag{4}$$

$K_{F,i}$ is the Freundlich constant for component i; n_i is the adsorption intensity for component i. Isotherm parameters of both models can be determined from the linearized forms of each model and listed in Table 1. Since the amount of adsorbed nickel was very low, the experimental data of nickel was excluded from the models.

Table 1. Adsorption isotherm parameters.

Adsorption isotherm parameters	Cu	Cr
Langmuir isotherm		
$q_{m,i}$	50.2513	43.8597
$b_{L,i}$	0.0511	0.0407
Freundlich isotherm		
$K_{F,i}$	18.4204	16.7687
n_i	5.7438	6.2972
Extended Langmuir isotherm		
$q_{m,i}$	111.7629	76.2243
$b_{L,i}$	0.0249	0.0284
Modified competitive Langmuir isotherm		
$q_{m,i}$	78.1250	163.9344
$b_{L,i}$	0.0230	0.0268
$\eta_{L,i}$	1.0174	3.2165

The extended Langmuir isotherm is developed for multicomponent adsorption. This model can be represented as Eq. (5).

$$q_{e,i} = \frac{q_{m,i} b_{L,i} C_{e,i}}{1 + \sum_{j=1}^{N} (b_{L,j} C_{e,j})} \tag{5}$$

N is the total number of ions in the solution. The evaluation of parameters can be achieved by minimization of the error in non-linear regression analysis. The parameters of this model were listed in Table 1. In case of Modified competitive Langmuir isotherm, the interaction factor ($\eta_{L,i}$) is included to explain the competitive effect between the individual components. This model can be expressed as Eq. (6).

$$q_{e,i} = \frac{q_{m,i} b_{L,i} \left(C_{e,i} / \eta_{L,i} \right)}{1 + \sum_{j=1}^{N} \left(b_{L,j} \left(C_{e,j} / \eta_{L,j} \right) \right)} \tag{6}$$

All parameters in this model except the interaction factor, was determined from experimental data of single component adsorption. The interaction factor was evaluated from experimental data of multicomponent system by minimization of the error in non-linear regression analysis. The parameters were summarized in Table 1.

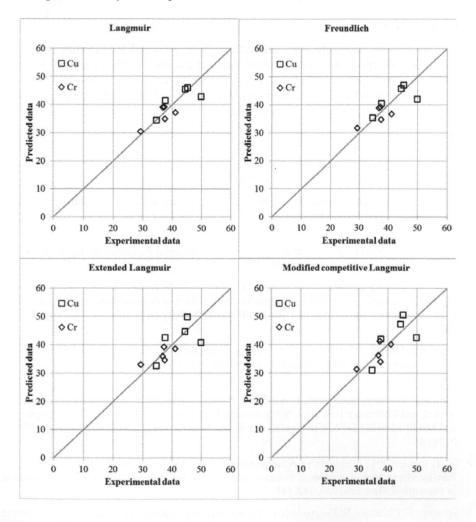

Figure 3. Comparison of experimental and predicted adsorption capacities.

The predicted data of adsorption capacity from each model were compared with the experimental data in Figure 3. The SSEs of Langmuir, Freundlich, extended Langmuir, and modified competitive Langmuir models were 98.58, 118.40, 165.84, and 156.18, respectively. Langmuir model was better than Freundlich model. Anyway, application of both models is limit. In addition, they do not include the effect of other components. When other metal concentrations were changed, the adsorption capacity was not altered as shown in Eq. (3) and (4). As a result, the adsorption behavior could not be described.

Since wastewater is always composed of various metal ions, the multicomponent models are useful for the design of adsorption column. The effect of other components is included in multicomponent adsorption models. Extended Langmuir isotherm and modified competitive Langmuir isotherm were developed for multicomponent adsorption. Modified competitive Langmuir model gave better result because it includes interaction factor which showed the competitive effect of metal ions (Girish, 2017).

4. Conclusions

Our study showed that there were interactions between metal ions during adsorption process. The adsorption affinities of metal ions depend on pH. The efficiency of metal removal decreased when initial metal concentration increased. Although SSEs of Langmuir and Freundlich models were better than those of multicomponent adsorption models, the application of single component model was limit.

References

C. R. Girish, 2017, Various Isotherm Models for Multicomponent Adsorption: A Review, Journal of Civil Engineering and Technology, 8, 80-86

C. R. Girish, 2018, Multicomponent Adsorption and the Interaction between the Adsorbent and the Adsorbate: A Review, International Journal of Mechanical Engineering and Technology, 9, 177-188

H. Maleki, 2016, Recent Advances in Aerogels for Environmental Remediation Applications: A Review, Chemical Engineering Journal, 300, 98-118

B. Naik, V. Desai, M. Kowshik, V. S. Prasad, G. F. Fernando, N. N. Ghosh, 2011, Synthesis of Ag/AgCl-Mesoporous Silica Nanocomposites using a Simple Aqueous Solution-Based Chemical Method and a Study of Their Antibacterial Activity on E. coli, Particuology, 9, 243-247

K. M. Parida, D. Rath, 2009, Amine Functionalized MCM-41: An Active and Reusable Catalyst for Knovenagel Condensation Reaction, Journal of Molecular Catalysis A: Chemical, 310, 93-100

B. Pornchuti, B. Pongpattananurak, D. Sutthiard, P. Singtothong, 2020, Adsorption of Copper, Nickel, and Chromium Ions using Silica Aerogel Synthesized by Ambient-Pressure Drying and Modified with EDTA, IOP Conference Series: Materials Science and Engineering, 778, 012133

P. Qi, T. Pichler, 2017, Competitive Adsorption of As(III), As(V), Sb(III), and Sb(V) onto Ferrihydrite in Multi-Component Systems: Implications for Mobility and Distribution, Journal of Hazardous Materials, 330, 142-148

S. Sertsing, T. Chukeaw, S. Pengpanich, B. Pornchuti, 2018, Adsorption of Nickel and Chromium Ions by Amine-Functionalized Silica Aerogel, MATEC Web of Conferences, 156, 03014

Proceedings of the 14th International Symposium on Process Systems Engineering – PSE 2021+
June 19-23, 2022, Kyoto, Japan © 2022 Elsevier B.V. All rights reserved.
http://dx.doi.org/10.1016/B978-0-323-85159-6.50345-6

Use of Environmental Assessment and Techno Economic Analysis (TEA) to Evaluate the Impact and Feasibility of Coatings for Manufacturing Processes

Antoine Merlo[a]*, Grégoire Léonard[a]

aDepartment of Chemical Engineering, University of Liège, Quartier Agora B6a Sart-Tilman, 4000, Belgium
*antoine.merlo@uliege.be

Abstract

Coatings have become ubiquitous in modern manufacturing processes as the mechanical improvements or the new properties that they offer bring a lot of added value in the processes in which they are used. Coatings offer interesting challenges in terms of environmental and economic assessment as the deposition process and the use phase of the coating are often decoupled, although both need to be considered for a proper evaluation.

In this paper, the impact of the choices made during the deposition process will be demonstrated through a case study about TiAlN coatings for machining applications. Two deposition techniques are evaluated: Magnetron Sputtering in Direct Current (DC-MS) and in Hi-Powered Impulse (HiPIMS) regime. While coatings deposited with the HiPIMS technology are costlier and have a higher carbon impact, their increased coating life compensates that higher cost. Inclusion of the impact and costs of other aspects such as the steel substrate production would further increase the benefits of using HiPIMS.

Keywords: Environmental Assessment, Economic Assessment, Machining, TiAlN, HiPIMS

1. Introduction

Machining is an important part of many modern supply chains, for the automotive or electronic industries for example. Global market for machining is estimated to have reached a worth of $341.91 bn in 2019 in which China is the biggest shareholder [1]. In just the machining sector, the global market for cutting tools reached a size of $34.42 bn [2]. Due to the impact of that sector, finding ways to increase the performance of the machining process would be of great importance. One of the ways the machining process can be improved is, for example, by using coatings to improve the mechanical properties and the durability of the cutting tools. In the present paper, a method for economic and environmental joint assessment for different coating technologies will be presented to evaluate feasibility of the HiPIMS technology for TiAlN coatings.

2. Goal and scope

The goal of the present work is is to compare Ti0.5Al0.5N coatings deposited by HiPIMS (Hi-Powered Impulse Magnetron Sputtering) and by DC-MS (Direct Current Magnetron Sputtering) on cutting tools. The motivations are that previous works have shown that using HiPIMS to deposit TiAlN has the capacity to greatly extend the tool life time under cutting conditions (75 vs 50 minutes) [3]. That result combined with the fact that HiPIMS, while having interesting properties, tends to be costlier due to lower deposition rates and higher investment costs [4], make up an interesting trade-off between cost and performance to study in the use phase. The target public for this work is mostly the scientific community interested in either joint economic and ecological assessment, HiPIMS and cutting tools technology. The functional unit (FU) is an 8 mm X 8 cm high-speed steel (HSS) tool coated by 4µm Ti0.5Al0.5N coatings sputtered by DC-MS or HiPIMS.

This study will take into account the extraction phase for all materials and the coating phase. Instead of a full Life Cycle Analysis, the study will focus on CO_2 emissions only because they are the main environmental impact linked to cutting and to the present coating technologies as their main difference is electricity consumption. This is also a way to streamline data acquisition.

3. Data collection and inventory

Several subprocesses are necessary to assess the cost and impacts of the coating process. Those are: production of the TiAlN target, gas production and the deposition process itself.

Target production itself is comprised of several different phases: Metal extraction, melting, powderization, annealing, compaction and sintering. To assess the energy consumption of those phases, the calculation of Kruzhanov is used [5]. The generic data provided in [5] is used for all phases except for metal extraction where emission factors (EF) from [6] are used for the CO_2 emitted for aluminium and titanium production. The energy requirements are shown in Table 1.

Table 1: Energy requirements of TiAlN target production

	Powder production	Compaction	Sintering	Total
Energy requirement (kWh per kg)	1.32	1	4	6.32

Because the target's production country is unspecified, the EF of the energy production is taken as the global average: 475 gCO_2/kWh [7]. Cost of target is assumed to be of 400 €/kg based on prices of partners.

The deposition process itself must also be evaluated, which takes place in a vacuum chamber. A representation of the chamber is given in Figure 1. The chamber is assumed to be octagonal and to be able to accommodate 5 rows of 5 tools on each of its sides. The substrate holder is able to rotate and to heat the tools to 450°C. A thickness of 4 µm and a deposition rate of 1.8 µm/h for HiPIMS are assumed [4]. The only difference for the two technologies is that a deposition rate of twice the one for HiPIMS, i.e. 3.6 µm/h, is assumed for DC-MS [3]. No material losses are assumed as a first approximation as the impact of TiAlN is negligible in the results.

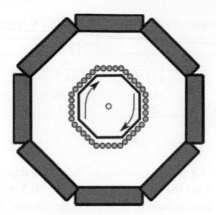

Figure 1: Representation of the deposition chamber for DC-MS and HiPIMS

The details for chamber operation are given in Table 2 for calculation of energy and gas consumption through deposition time.

Table 2: Energy consumption and time requirements of the deposition phase subprocesses of HiPIMS and DC-MS

	Power (W)	Time (Min)	Source
Heating	2000	30	[3]
Etching	120000	30	[3]
Deposition	120000	HiPIMS: 133.3 DC-MS: 66.7	[3]
Cooling	2000	20	[8]
Grinding	1000	10	[8]
Loading/Unloading	-	20	[8]
Pumping	8000	continuous	[8]

Coating is assumed to take place in Belgium, the EF for the energy production is then of 174 gCO_2/kWh [9]. Based on invoices, cost of nitrogen is assumed to be 0.332 $/kg. Cost of argon is assumed to be 0.976 $/kg. A dollar-euro conversion rate of 1.18 $/€ is used for all of the present work. The EF for argon is considered to be 0.385 $kgCO_2/m^3$ [10]. As for nitrogen, the energy needed for air separation is 243 kWh/t of nitrogen [11] and global average EF for the energy used in the separation process is considered. A consumption of 400 sccm of Ar is assumed during the heating, etching and deposition phases as well as a consumption 400 sccm of N_2 during the deposition phase.

4. Cost and environmental assessment

The first goal is to differentiate the costs and impacts of DC-MS and HiPIMS for the different steps of coatings production.

4.1 Gas production

Following the assumptions made in section 3.2, consumption, cost and emissions of the gases used in the deposition process for a batch of 200 tools are summarized in Table 3.

Table 3: Gas consumption, costs and related CO_2 emissions per batch

	HiPIMS		DC-MS	
	Ar	N	Ar	N
Gas consumption (g)	131.5	64.2	86.1	32.1
Total Gas cost (€)	0.11	<0.02	0.08	<0.01
Gas CO_2 emissions (g CO_2)	29.8	7.7	19.5	3.8

4.2 Target production

Using a TiAlN density of 4.8 g/cm³, a coating of 38.6 mg is deposited on every tool, from which 28.1 mg comes from the sputtered target and 10.5 mg comes from nitrogen. Following assumptions from section 3, 62.4 gCO_2 are emitted for target production per batch for both DC-MS and HiPIMS. Every batch induces a cost of 2.24 € of target consumption.

4.3 Electricity consumption

The electricity consumption for the coating process is summarized in Table 4.

Table 4: Electricity consumption of the deposition subprocesses

Subprocess	HIPIMS	MS
Heating (kWh/**batch**)	5.00	5.00
Etching (kWh/**batch**)	65.00	65.00
Deposition (kWh/**batch**)	297.78	148.89
Cooling (kWh/**batch**)	0.67	0.67
Grinding (kWh/**batch**)	0.17	0.17
Total (kWh/batch)	368.61	219.72
Total (kWh/tool)	1.84	1.10

Most of the energy is used in the deposition and the etching phase due to the area of targets in those types of MS installations. Using the Belgian EF, 64 138.3 gCO_2 and 38 231.7 gCO_2 are emitted per batch for HiPIMS and DC-MS respectively. Using a cost of 0.0807 €/kWh, electricity costs per batch are 29.7 € and 17.7 € for HiPIMS and DC-MS respectively.

4.4 Labor, maintenance and annuities

The final metrics to evaluate the costs of each technology are labor, maintenance and annuities costs. In order to assess those aspects per coating produced, it is necessary to establish working parameters.

For maintenance, the annual costs are assumed to be 2% of the equipment costs each year. For annuities, a time of return on investment of 5 years is assumed, while inflation is neglected as a first approximation. For labor, each installation is assumed to work 300 days a year with 2 shifts of 8 hours every day. A single operator per shift with a hourly salary cost of 40 € is assumed to be assigned to the installation. Due to the longer time per batch for the HiPIMS process, the annuity cost per FU will be higher.

4.5 Summary

Using the previously stated costs and assumptions, costs and CO_2 emissions per FU can be worked out. The total costs per FU are shown in Figure 2. They amount to 2.36 € per HiPIMS coating and 1.67 € per DC-MS coating.

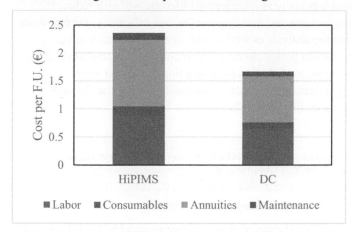

Figure 2: Cost breakdown of the coatings for DC-MS and HiPIMS

Most of the costs are related to labor and annuities. This is due to the relatively high investment cost of the installations. For HiPIMS, that investment cost is higher, and the cost per coating is exacerbated by HiPIMS' lower productivity.

CO_2 emissions for each technology are presented in Figure 3. Electricity production is overwhelmingly responsible for CO_2 emissions for both technologies. This is mainly due to the large amount of energy required for coating compared to the amount of material deposited. Due to the lower deposition rate of the HiPIMS technology, the amount of CO_2 emitted by this technology compared to DC-MS is noticeably higher (320.7 gCO_2 vs 191.2 gCO_2).

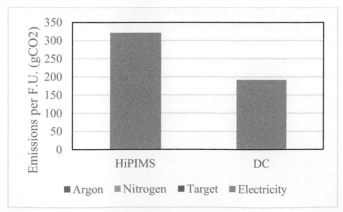

Figure 3: Total CO2 emissions for DC-MS and HiPIMS technologies

5. Conclusions

In summary, HiPIMS coatings have a 41% increased cost and a 68% increase in CO2 emissions compared to their DC-MS counterpart. However, with tool lifetimes of 75 minutes and 50 minutes for HiPIMS and DC respectively, one can reasonably justify the choice of HiPIMS coatings, at least in terms of costs. Indeed, the tool lifetime is increased by 50% for HiPIMS, while its costs only increases by 41%, making this choice a priori advantageous especially as a longer tool lifetime will also reduce the downtimes of the machining process and thus improve its cost efficiency. Further work will include the use phase in the analysis as well as technical data in order to verify these assumptions. Finally, further perspectives will also consider the cost of the HSS tool substrate and the CO_2 emissions linked to its production. Despite a lower productivity of HiPIMS, these elements will also presumably benefit the HiPIMS technology due to the longer tool lifetime it allows.

6.   References

[1] Beroe Inc., 2019, Machining Market to Reach $414.17 Billion by 2022, Says Beroe Inc[Online]. Last accessed: 19/10/2021 [Available at : https://www.prnewswire.com/news-releases/machining-market-to-reach-414-17-billion-by-2022--says-beroe-inc-300939464.html]

[2] IndustryArc, 2021, Cutting Tools Market - Forecast(2021 - 2026)

[3] Weichart, J., Lechthaler, M., 2012, Titanium aluminum nitride sputtered by HIPIMS, M. IOP Conf. Series: Mater. Sci. Eng., 39, 012001.

[4] Anders, A.,2010, Deposition Rates of High Power Impulse Magnetron Sputtering: Physics and Economics. Lawrence Berkeley National Laboratory. Retrieved from https://escholarship.org/uc/item/7cz4g23b

[5] Kruzhanov, V., Arnhold, V., 2012, Energy consumption in powder metallurgical manufacturing. Powder Metallurgy, 55(1). DOI 10.1179/174329012X13318077875722

[6] Nuss, P. and Eckelman, M. J., 2014, Life Cycle Assessment of Metals: A Scientific Synthesis. PLoS ONE 9(7): e101298. https://doi.org/10.1371/journal.pone.0101298

[7] IEA, 2019, Global Energy & CO2 Status Report 2019

[8] Centre de Recherche Métallurgique (CRM)'s installations

[9] European Environment Agency, 2021, Indicator assessment: Greenhouse gas emission intensity of electricity generation in Europe. https://www.eea.europa.eu/data-and-maps/indicators/overview-of-the-electricity-production-3/assessment-1

[10] Nakhla, H., Shen, J.Y. and Bethea, M., 2012, Environmental Impacts of Using Welding Gas. The Journal of Technology, Management, and Applied Engineering, 28(3).

[11] European Industrial Gases Association, 2010, Position Paper, Indirect CO2 emissions compensation: Benchmark proposal for Air Separation Plants

Proceedings of the 14th International Symposium on Process Systems Engineering – PSE 2021+
June 19-23, 2022, Kyoto, Japan © 2022 Elsevier B.V. All rights reserved.
http://dx.doi.org/10.1016/B978-0-323-85159-6.50346-8

Forecasting Operational Conditions: A case-study from dewatering of biomass at an industrial wastewater treatment plant

Sebastian Olivier Nymann Topalian[a], Pedram Ramin[a], Kasper Kjellberg[b], Murat Kulahci[c], Xavier Flores Alsina[a], Damien J. Batstone[d], Krist V. Gernaey[a]

[a]*Technical University of Denmark, Department of Chemical and Biochemical Engineering, Søltofts Plads 228A, Kgs. Lyngby, 2800, Denmark*
[b]*Novozymes A/S, Hallas Alle 1, Kalundborg, 4400, Denmark*
[c]*Technical University of Denmark, Department of Applied Mathematics and Computer Science, Richard Petersens Plads 324, Kgs. Lyngby, 2800, Denmark*
[d]*Advanced Water Management Centre, The University of Queensland, Brisbane, QLD 4067, Australia*

Corresponding Author's E-mail: kvg@kt.dtu.dk

Abstract

In this paper, we present a data-driven approach to predicting polymer dosages for industrial decanters based on upstream production data. First, a data extraction algorithm using on-line sensors is developed to identify when the operational mode is changed with a 99 % accuracy. Next, an investigation of process delays in the collected data is carried out by analysing partial autocorrelation matrix eigenvalues upon which is it concluded to transform the data by summarising the data by batch and including lagged summaries to account for a time delay of 2 hours. Finally, a random forest forecasting model is trained capable of learning structured information from the lagged summaries producing decent predictions for both low and high polymer dosages (RMSE 14.89). The proposed approach could potentially save operators 3-6 hours a day.

Keywords: Control; Operation; Forecasting; Environmental Systems;

1. Introduction

Decanters are widely used in the biotech industry to carry out solid-liquid separations. Achieving adequate separation in the decanters reduces the amount of energy, and therefore money, required to treat the reject water in the waterline, however this is at the cost of adding chemicals to the decanters. At the current point in time, it is deemed unfeasible to control the settings of the decanters based on a cost-benefit analysis due to the complexity of accounting for savings in the waterline, so the problem is reduced to achieving decent separation with the necessary operational conditions as determined by the operators. Running decanters with a highly variable feed composition in a satisfactory manner is a task for operators who manually carry out flocculation tests to estimate the proper chemical dosage at a given point in time. This procedure is labour intensive and cumbersome. It also has a strong time component i.e delay between the result of the flocculation essay and the need of deciding for a dosage strategy. In order to circumvent this limitation, we developed a mathematical model based on on-line sensor data to estimate dosing of polymer based on upstream process information. The proposed methodology is tested at case study largest industrial wastewater treatment plant in Northern Europe.

2. Methods

2.1. Plant description

The plant has a hall with decanters that treat approximately 200 m³/h of biomass and waste activated sludge with operation around the clock. The stream that goes into the decanter hall comes from one of two storage tanks that are operated in a semi-continuous manner, and the biomass in the storage tanks consists of inactivated biomass from the upstream biotechnological production as well as waste activated sludge from the waterline at the plant. Apart from storing material, the storage tanks also buffer the flow to the decanter hall allowing for smoother operation, and a regular emptying pattern, whereas the filling pattern is dictated by upstream production. There are approximately 4-6 batches going through the decanter hall every day. The decanters are controlled by operators who adjust the polymer dosage, among other settings, depending on the biomass and sludge that is being treated for each batch. The operators judge the quality of the separation by looking at the reject water from the decanters with a visual test, and then decide to change the polymer dosage, among other settings, accordingly. A conceptual illustration of the plant is provided in Figure 1.

Conceptual Plant Illustration

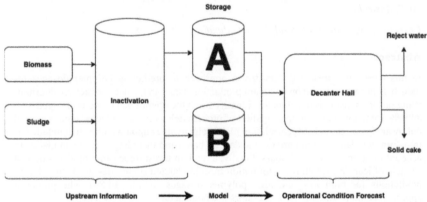

Figure 1 – Conceptual illustration of the plant. Biomass and waste activated sludge are inactivated and stored in two storage tanks that semi-continuously feed into a hall with centrifugal decanters. The upstream information is used to produce operational condition forecasts to assist operators in controlling the decanters.

2.2. Data availability

The plant is equipped with online sensors of pH, temperature, tank volumes, flowrate among others, and historic data from the sensors is available from an online database. Historic time-series data from the 1st of July to 31st of December 2020 were extracted from the online database with a resolution of 15-minute averages. Categorical information such as production codes were 15-minute medians. In total 302 variables from the plant were selected for analysis. The data retrieved from the online database is ordered by time of observation, however the plant contains several unit operations with residence times larger than the 15-minute averages for which data is obtained, and hence comparing upstream variables with downstream variables at the same time should exhibit poor correlation unless the variables being compared are severely autocorrelated.

2.4. Batch data-extraction algorithm

A custom algorithm is developed to automatically identify when a batch begins when the other tank starts to empty (see Figure 2). In this way, a batch change is identified by checking that the tank volume is above approx. 50 %, and that the gradient of that volume is negative, and that the gradient is larger than 0.5 % as to prevent identifying a batch change time when operation is stopped due to sensor noise (approx. 0.1 %).

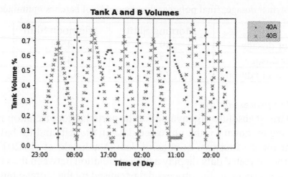

Figure 2 – Algorithm for detection of batch changes for the semi-continuous tank system. Identified batch changes are marked with a vertical red line.

2.3. Partial autocorrelation matrix (PACM)

The partial autocorrelation matrix (PACM) is used to study process delays due to unit operations. Due to the semi continuous operation of the tanks, a large delay is expected that correlates the upstream data with the downstream data at a lag corresponding to the residence time in the storage tanks. Summarising The PACM is constructed by calculating the partial correlation matrix for lagged versions of the dataset, and the eigenvalues for each matrix. Next, the obtained PACM is compared to the eigenvalues of the matrix at prior and future lags where large and small eigenvalues indicate overall good and correlation between the variables and their lagged states respectively (Vanhatalo et al., 2017).

2.5. Predictive method

Random forests are widely regarded as a good off-the-shelf regression model for structured data, such as the data presented herein (Hastie et al., 2017). The random forest (RF) algorithm is used to construct a forecast model that predicts the operator decided polymer dosage for a batch based on the upstream information obtained while the batch is in storage. The first 80 % of the batch observations are used as training data to decide which hyper-parameters to utilize, and the final 20 % of the batch observations are used to evaluate the model performance by recursively refitting the model and predicting the next batch polymer dosage.

3. Results

3.1. Analysis of batch data

The proposed algorithm identified batch changes with a 99% accuracy. In total 869 batches were analyzed for the 6-month period. The target objective of the forecast model is the polymer dosage after the operators have evaluated and changed the operational conditions, implying that the mean for a batch averages out the conditions before and after the operators have done their job, so to have a more accurate estimate of the polymer dosage a zero-excluded median is calculated for the decanters in the hall, since this removes idle machines or machines that exhibit irregular behaviour. The zero-excluded median is calculated after rounding the dosages to nearest base 5 as this is the typical magnitude of change that the operators will apply and the final value for a batch is considered as the suggested dosage from the side of the operators. In table 1 the initial polymer dosage, final polymer dosage and percentage of batches with new conditions is shown for all the identified batches. A batch is deemed to have new conditions if the calculated starting median differs from the final one. The statistics are shown for all three shifts in Table 1. The afternoon shift appears to utilize less polymer and change their operational conditions more frequently than the morning and night shift.

Table 1- Initial polymer dosage, final polymer dosage and % of batches optimized.

	Morning Shift	Afternoon Shift	Night Shift
Initial polymer dosage	49.6	45.8	46.3
Final polymer dosage	47.4	45.5	50.2
Batches w. new conditions	79.5 %	92.4 %	81.2 %

4.2. Assessment of process delays

In Figure 3 the largest absolute eigenvalue for the partial autocorrelation matrix (PACM) as a function of lag is shown for the untreated data on the left, and for batch summarised data on the right for the first two months of data (Vanhatalo et al., 2017). The PACM eigenvalues for the untreated data display a high correlation between the variables at delay 20, which could correspond to the process delay caused by the storage tanks that feed the decanter hall.

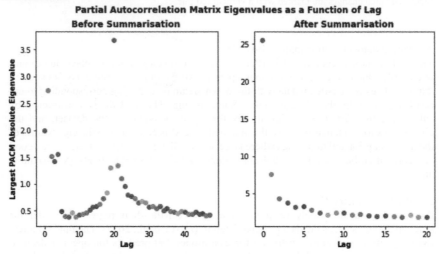

Figure 3 – Largest absolute eigenvalues for the PACM at different lags before and after batch summarization left and right respectively.

The summarised data is calculated by averaging all values for a given batch time, and the lag corresponds to moving the starting and end point of the batch by one timestep, to check if different variables should be summarised by different time indices to account for delays occurring before the semicontinuous storage tanks. An illustration is provided in Figure 4. The eigenvalues past lag 8 appear to stabilize, so for the forecasting model these lagged summaries will be included when developing the forecasting model, effectively accounting for the past 2 hours.

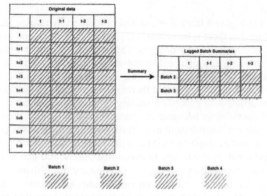

Figure 4 – Batch summarization illustration. For each identified batch the mean is calculated for each variable, and for each lagged version of the variables.

4.3. Random forest
4.3.1. Model training
The sci-kit learn library is used to carry out the hyper-parameter optimization and train the models (Pedregosa et al., 2011). The hyper-parameter optimization is carried out by building 500 random forests with 1024 trees where the complexity parameter, maximum number of features and maximum depth are drawn from the uniform distributions U(0.2, 0.8), U{1, 2700} and U{1,10} for each random forest respectively. The random forest with the best performance has complexity parameter, maximum number of features and maximum depth values of 0.23, 105 and 9 respectively.

4.3.2. Model predictions
For each observation in the test set the model is refit on all prior data to produce the forecasts. The root mean square error (RMSE) on the test set is 14.89 whereas using a naïve guess such as the mean of polymer dosage results in an RMSE of 18.44 and the RF algorithm thus successfully learns structured information, however considering that the average polymer dosage is between 45-50 the RMSE is still considered as high. In Figure 5 two residual plots are shown. The left one shows the residuals of the model on the test set as a function of time, and the right one shows the residuals as a function of the model prediction. Each residual is shown with a dot, if the batch conditions are new compared to the previous conditions, and a cross, if the batch conditions are old corresponding to unchanged conditions from the last batch.

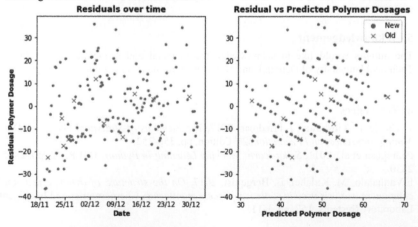

Figure 5 – Model residuals a function of date on the left, and as a function of predicted polymer dosage on the right. A dot represents a batch with new operational conditions, and a cross old operational conditions compared to that of the previous batch.

4.3.3. Discussion

From the left plot in Figure 5 there does not appear to be an increasing trend in the model residuals indicating that there is no systematic error in the model, and that the model can predict both low and high polymer dosages. The batches that utilize old conditions have a lower residual on average, which could correspond to batches where the operators deem the performance of the decanters sufficient, and therefor keep the old operational conditions. The forecasting model could therefore be applied in situations where the old conditions are insufficient to generate a starting point for the operators. For each batch where the model provides an adequate estimate 1-2 hours of time could potentially be saved for the operators allowing them to perform other crucial tasks. Upon investigating the random forests feature importance it becomes evident that the tree utilizes primarily flow rates, and information about time to make predictions, and that information related to upstream product information, pH and temperature are not utilized as much. Models that inherently utilize time as information can be impractical since they often do not convey implications of physical phenomena, however for systems with temporal patterns they can provide a significant boost in forecasting accuracy. As a means of comparison, a naïve forecast where the prediction for each value is the prior value yields a RMSE of 17.43 which is 17.06 % larger than that of the RF model. Random forests are considered as a good off-the-shelf data mining procedure, and here they also achieve moderate success encouraging further data collection and investigation of more sophisticated and time-consuming forecasting methods such as neural network approaches. One inherent drawback of predicting the actions of operators is that the operation could change depending on the operators working a shift and including the operator schedule could lead to a better forecasting model and provide a method for evaluating operator performance through data-engineered key performance indicators. The framework presented herein can also be utilized for predicting other operational conditions, and transferred to other plants with similar plant layouts, or for predicting decanter failure.

4. Conclusion

This study demonstrates performance analysis of an industrial-scale process using plant-wide operational data. With proper data treatment, data analysis, feature selection and data-driven modelling it was possible to make an automated algorithm to handle complex datasets for prediction purposes and later optimization and control. We propose that it could save operators between 3-6 hours of work every day, leaving room to carry out other important tasks at the plant, however further model development and following verification is required to increase the accuracy of the forecasts.

5. Acknowledgement

The authors would like to acknowledge the financial and academic support of the Technical University of Denmark and Novozymes for their contributions to the project.

6. References

T. Hastie, R. Tibshirani, J. Friedman, 2017, *The Elements of Statistical Learning: Data Mining, Inference, and Prediction*, 2nd Edition, pp. 350-352.

Pedregosa et al., 2011, *Scikit-learn: Machine Learning in Python*, JMLR, 12, pp. 2825-2830.

E.Vanhatalo, M. Kulahci, B. Bergquist, 2017, *On the structure of dynamic principal component analysis used in statistical process monitoring*, Chemometrics and Intelligent Laboratory Systems, 167, pp. 1-11.

Proceedings of the 14th International Symposium on Process Systems Engineering – PSE 2021+
June 19-23, 2022, Kyoto, Japan © 2022 Elsevier B.V. All rights reserved.
http://dx.doi.org/10.1016/B978-0-323-85159-6.50347-X

Plant wide modelling of a full-scale industrial water treatment system

Vicente T. Monje[a], Helena Junicke[a], Kasper Kjellberg[b], Krist V. Gernaey[a], Xavier Flores Alsina[a]

[a]*Technical University of Denmark, Department of Chemical and Biochemical Engineering, Søltofts Plads, Building 228A, 2800 Kgs. Lyngby, Denmark*
[b]*Novozymes A/S, Hallas Alle 1, 4400 Kalundborg, Denmark*
Corresponding Author's E-mail: xfa@kt.dtu.dk

Abstract

In this study a set of mathematical tools are developed and assembled together to assess and predict mass and volumetric flows in industrial water treatment systems (iWTS). The proposed approach is constructed upon a set of data reconciliation methods, influent fractionation routines and process simulations models (and model interfaces) to balance, analyse, reproduce and forecast the behaviour of different compounds within treatment facilities. The proposed approach is tested on full-scale data collected after a five week measuring campaign at the largest iWTS in Northern Europe. Results show that the proposed approach is capable to predict the occurrence, transformation and fate of COD, N, P, S and multiple metals (Na, K, Ca, Mg and Al).

Keywords: Data reconciliation, Mass balancing, Model simulation, Process systems engineering, Scenario analysis, Wastewater

1. Introduction

Industrial wastewaters have very diverse dynamics (compared to urban wastewater), which is a result of different production schemes/schedules within the factory. Variable pH, influent biodegradability, non-standard COD/N and COD/P ratios might challenge traditional biological processes. In some cases, high loads decrease methane/biogas production (and potential energy recovery). This reduction is attributed to two factors: 1) loss of electron equivalents due to the presence of sulfate reducing bacteria; and, 2) decrease of acetoclastic and hydrogenotrophic methanogenesis due to sulfide inhibition. Metals and some inorganic/organic compounds can inhibit microbial growth and/or have severe toxicity effects. The high content of cations and anions promotes the formation of precipitates at different locations in the reactor (granules, pipes), which can have detrimental (decrease of methanogenic activity) or catastrophic (cementation) effects on reactor performance (Feldman et al., 2017). Hence, mathematical models describing iWTS should include all these (hostile) phenomena in order to produce reliable predictions.

The objective of this study is to present a set of mathematical tools to assess and predict mass and volumetric flows in iWTS. The study presents the following novelties: 1) The results of a 5 week sampling campaign at the largest iWTS Norther Europe; 2) A reconciled mass balance analysis showing the occurrence, transformation and fate of traditional (COD, N, P & S) but also non-traditional (Na, K, Ca, Mg, Al) compounds; and, 3) A customized/calibrated model library describing multiple technologies treating different types of waste streams (liquid, solid).

This work goes beyond state of the art by presenting a modelling approach: 1) dealing with extremely concentrated streams (2,5 M PE in 10,000 m³/day); 2) modifying the existing mathematical model structures to adapt to the harsh industrial conditions; and, 3) extending the quantity of monitored compounds up to 10 (Q, traditional and non-traditional compounds) + pH + VSS/TSS ratio; 4) presenting for a first time an integrated plant-wide model dealing with a large industrial iWTS at this level of detail.

2. Methods

2.1 Plant description and measuring campaign

Figure 1 shows a schematics + detailed description of the case study. Influent flow may be treated anaerobically (PAT & AGSR) or aerobically (ASR, SEC, FLOT) or both. Biogas goes through a cleaning process (SCRUBB, REAC & SET) before being introduced to a gas motor for energy (electricity / heat) recovery purposes. Reject water from biomass dewatering can be sent to either the PAT & AGSR or the ASR, SEC, FLOT section (both is also an option). The output of PAT & AGSR is sent to the ASR.

A five week measuring campaign was conducted. Samples were taken from 11 locations within the plant: 3 influent streams ($PWW_{2, 3 \& 4}$), after PRIM ($PRIM_{over}$), after PA (PAT_{liq}), after AGSR ($AGSR_{liq}$), after ASR (ASR_{liq}), after SEC (SEC_{eff}, SEC_{RASS} & SEC_{WAS}) and after dewatering (DEW_{under}, DEW_{over}). Additional one day samples were taken to characterize: primary underflow ($PRIM_{under}$) and the output of the inactivation tank (IT_{liq1}, IT_{liq2}). Measurements involved the determination of: TSS, VSS, COD, TN, TP and TS and multiple metals (Al, Ca, Mg, Na and K), in both unfiltered and filtered samples. The analysis also includes the quantification of nitrates (NO_x).

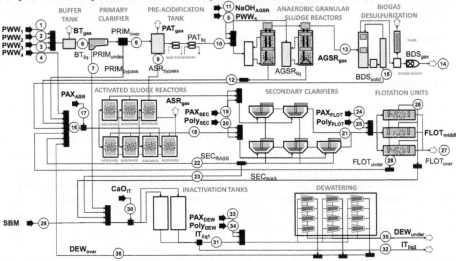

Figure 1. Flow diagram, measuring points, defined sub-systems and mass balances of the iWTS under study: **1-5** process waste water (PWW_{1-5}), **6** - effluent buffer tank (BT_{liq}), **7-8** –primary clarifier (PRIM) overflow and underflow ($PRIM_{under}$, $PRIM_{over}$) **9**- by pass activated sludge reactor (ASR) to pre-acidification tank (PAT) (ASR_{bypass}), **10** – PAT effluent (PAT_{liq}), **11** – NaOH addition to the anaerobic granular sludge reactor (AGSR)($NaOH_{AGSR}$), 12 – AGSR effluent ($AGSR_{liq}$), **13** – AGSR biogas (untreated) ($AGSR_{gas}$), **14** – AGSR biogas (treated) (BDS_{gas}), **15** – S recovered (BDS_{Solid}), **16** – input AER, **17** – Poly-aluminum chloride (PAX) addition to ASR (PAX_{ASR}), **18** – effluent ASR (ASR_{liq}), **19-20** – PAX and Polymer addition to secondary settler (SEC) (PAX_{SEC}, $Poly_{SEC}$), **21-23** – SEC effluent, recirculation and waste flow (SEC_{eff}, SEC_{RASS}, SEC_{WAS}), **24-26** – PAX and polymer addition to flotation (FLOT) (PAX_{FLOT}, $Poly_{FLOT}$), **26-28** – flotation streams ($FLOT_{over}$, $iWTS_{eff}$, $FLOT_{under}$), **29** – spent biomass stream (SBM), **30** – lime addition to inactivation tanks (IT) (CaO_{IT}), **31-32** – IT outputs (IT_{liq1}, IT_{liq2}), 33-34 – PAX and Polymer addition to dewatering (DEW) (PAX_{DEW}, $PolyD_{EW}$), **35-36** – dewatering under (cake) and overflow (reject water) (DEW_{under}, DEW_{over})

2.3 Mass balancing and data reconciliation

A five step methodology is used to reconcile the data obtained during the measuring campaign (Puig et al., 2008, Behami et al., 2019): 1) definition of the identity matrix; 2) curation, processing, cleansing and data analysis; 3) estimation of the missing fluxes; 4) calculation of optimal flows using Lagrange multipliers; and, 5) new data set quality verification.

2.4. Influent fractionation

Influent fractionation is based on earlier work (Feldman et al., 2017; Monje et al., 2021). Essentially, ADM states were estimated by assuming: 1) degree of COD biodegradability (D_{BIO}) in both COD_{sol} and COD_{part}; 2) degree of acidification (D_{ACID}) in CODs; 3) fraction of ethanol (D_{ETOH}). Hence, it is possible to determine S_I, X_I, S_{ac} and S_{ETOH}. Once this is established, X_{li}, X_{prot} and S_{aa} are quantified using P and N content of the aforementioned compounds. Since the influent wastewater originates from the fermentation industry, sugars ($S_{su} = 0$) and carbohydrates ($X_{ch} = 0$) should not be present (consumed upstream). The quantity of P not associated with organics is assumed to be inorganic and precipitated (mainly calcium phosphate) (X_{Ca-P}). Any remaining particulate calcium and magnesium will be assumed to be calcium carbonate (X_{CaCO3}). The influent inorganics (X_{ISS0}) are determined using the TSS/VSS ratio minus the quantity of precipitates.

2.5. Main mathematical models

The main model is based on: (1) a biological model; (2) a physico-chemical model; and, (3) model interfaces. The biological models (BM) comprise an anaerobic digestion model (ADM) and an activated sludge model (ASM). The ADM is used to describe influent conditions, BT, PRIM, PA, AGSR, IT and DW, while the ASM describes the ASR, SEC, and FLOT units. The physico-chemical model (PCM) includes an aqueous phase + precipitation model and a gas transfer model. Finally, the model interfaces include an ADM/ASM/ADM interface and PCM/ADM/ASM interface. The ADM/ASM is incorporated before the AS unit and the ASM/ADM after the SEC. The outputs of the ASM/ADM at each integration step are used as inputs for the PCM module to estimate pH, ion speciation/pairing, precipitation potential and stripping. A comprehensive description of these models can be found elsewhere (Feldman et al., 2017; Flores-Alsina et al., 2019).

3. Results

3.1. Reconciled mass balances

Result of reconciled mass balances revealed that the effluent of the iWTS under study (iWTS$_{eff}$) contains the largest Q fraction (95 %). The dewatered cake (DEW$_{under}$) and IT liquor (IT$_{liq2}$) only account for a marginal contribution (<5 %). About 32 % of the incoming COD is captured in the anaerobic granular sludge reactor (AGSR$_{gas}$) and then potentially converted into electricity and heat. The remaining COD is lost in intermediate operations (BT$_{gas}$, PAT$_{gas}$) (12 %), burned aerobically/anoxically in the activated sludge section (ASR$_{gas}$) (23 %), part of the effluent (iWTS$_{eff}$) (1 %) or trucked to an external biogas facility as sludge cake (32 %) (IT$_{liq2}$, DEW$_{under}$). Regarding N, 65 % is removed in the activated sludge section via nitrification/denitrification (ASR$_{gas}$). A significant part of N ends up being part of the bio-solids to be disposed (IT$_{liq2}$, DEW$_{under}$) (33 %) while a small fraction leaves with the effluent (iWTS$_{eff}$) (2 %). The analysis of P, Ca, Mg and Al reveals that these compounds are basically accumulated in the bio-solids (IT$_{liq2}$, DEW$_{under}$) (> 70 %) as precipitates. Contrary to that, Na, K and S remain soluble and leave via the liquid stream (iWTS$_{eff}$) (> 70 %). It is important to mention that 11 % of S is converted to H_2S in the anaerobic granular sludge reactor (AGSR$_{gas}$,) and then captured in a desulfurization tower. Next, it is re-oxidized again to S_0 and introduced (BSD) in the biological reactor.

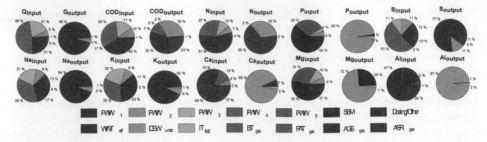

Figure 2. Contribution to influent and effluent loads of the different streams. For each reconciled magnitude (Q, COD, N, P, S, Na, K, Ca, Mg, Al), pie charts represent influent and effluent compositions, respectively.

3.2. Influent fractionation

The results of the influent fractionation are illustrated in **Figure 3.** Four examples are presented: 1) process wastewater 2, 3 & 4 (PWW $_{2, 3, 4}$); and, 2) the spent biomass stream (SBM). Indeed, **Figure 3** depicts the proportions of the different model states (both soluble and particulate) in form of a stack bar adding up to 100 % of the particulate or soluble fraction of each reconciled major component (COD, N, P & Ca). In all PWW streams, COD_{sol} and COD_{part} is mainly composed of fatty acids (S_{fa}) and lipids (X_{li}), whereas SBM mainly contains amino acids (S_{aa}) and proteins (X_{pro}). The remaining fraction is allocated to non-biodegradable organics (S_i, X_i) and VFAs (S_{ac}). With respect to N (soluble), PWW_3 and PWW_4 have a large contribution of nitrate, which comes from the use of nitric acid for cleaning equipment. The remaining N is linked to ammonia (S_{NHx}) and amino acids (S_{aa}). Particulate N is (almost) entirely allocated to X_{prot}. All the soluble P is assumed to be phosphate (S_{PO4}). The particulate fraction is allocated to lipids (X_{li}) and calcium precipitates (X_{Ca-P}). $PWW_{2, 4}$ and SBM have an important contribution of precipitates, which is also confirmed by their lower VSS/TSS ratios. In contrast, P in PWW_3 is mainly linked to organics (X_{li}). In both, N and P cases, a small fraction is associated with organic inert material (S_I and X_I). Ca has a soluble (S_{Ca+2}) and a particulate fraction in the form of carbonates (X_{CaCO3}) and phosphates (X_{Ca-P}). In the vast majority of cases, calcium carbonate is assumed to be the major component, but a small fraction is linked to phosphates depending on the availability of inorganic (particulate) P (see $PWW_{2,4}$ and SBM). A special case is PWW_3 where the low content of inorganic material is in the form of phosphates.

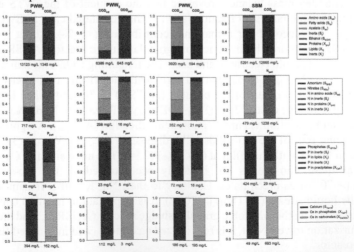

Figure 3. Fractionation of $PWW_{2, 3, 4}$ and SBM streams: COD, N, P and Ca.

A similar approach is used for Mg i.e. the soluble part is S_{Mg+2} and the particulate form is magnesium carbonate (X_{MgCO3}). For the sake of simplicity, the soluble fraction of S is associated to sulfate (S_{SOx}), while the particulate fraction is assumed to be sulfur mineral (X_{S0}). With respect to the remaining compounds (see **Fig 4**): 1) K and N are assumed to be only soluble and present in ionic form (S_{Na+} and S_{K+}); 2) Al is assumed to be only in particulate form (X_{AlOH}). The pH values were adjusted by modifying the Cl concentration. A pool of undefined particulates is used to calibrate the VSS/TSS ratio.

3.3. Computer simulations

3.3.1. Buffer tank (BT) + primary clarifier (PRIM)

COD_{sol}/COD_{part} values are the result of hydrolysis, acidogenesis and H_2 stripping. N and P transformations include hydrolysis and uptake/release during biomass growth/decay. Specifically, for N the model considers dissimilatory nitrate reduction to ammonium (see TN, NH_X and NO_x). Precipitation of P is assumed to happen (justified with the high alkalinity of PWW_2). The other studied compounds are assumed to be non-reactive (S, Mg, K, Al). In PRIM, a fraction of all compounds with a particulate fraction (COD, N, P, Ca, Mg, Al) is removed. The model also reflects well the hydraulic balance (see Q values), weak acid chemistry (see pH values) and the differences in settling velocities between organics and inorganics (see VSS/TSS values) (see **Figure 4 & 5**)

3.3.2. Pre-acidification tank (PAT)

The model predicts a reduction of COD_{sol} and an increase of the COD_{part} as result of the acidogenic activity. The first is due to hydrogen formation (and subsequently stripping). The second is the result of biomass production/growth. The models also suggest a change in the composition of COD_{sol} (~60 % is acidified) which would explain the lower pH experimental values (well predicted by the model too). Both, plant data and simulation values, show total denitrification NO_x (~ 0 g/m^3) (see **Figure 5**). Since TN values do not seem to change, DNRA was again the main mechanism. No further reactions were assumed for the other compounds (see **Figure 4 & 5**).

3.3.3. Anaerobic granular sludge reactor (AGSR)

The proposed approach is capable to reproduce COD values (soluble/particulate) resulting from the methanogenic microorganisms (see **Figure 4**). In addition, it also predicts the S removal by sulfate reducing bacteria (see **Figure 4**). No further reactions e.g. precipitation were assumed to occur. The pH and VSS/TSS predictions match well with plant data (see **Figure 4 & 5**)

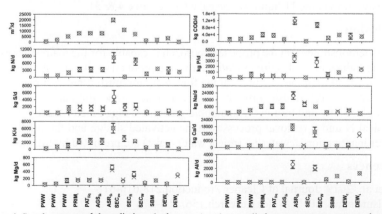

Figure 4. Steady state model predictions (red crosses) and reconciled measurements mean and standard deviation (black circles and whiskers) for several plant locations (see labels in X-axis). A: volumetric flow rate; B: TCOD; C: TN; D: TP; E: TS; F: TNa; G: TK; H: TCa; I: TMg; J: TAl. Deviation between measurements and simulations = 10 %

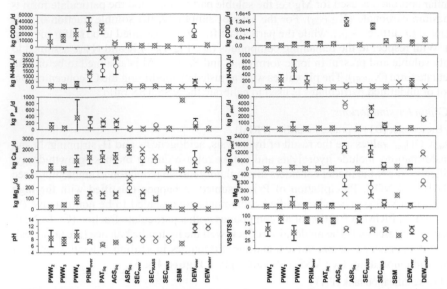

Figure 5. Steady state model predictions (red crosses) and reconciled measurements mean and standard deviation (black circles and whiskers) for several plant locations (see labels in X-axis). A, CODpart; B, CODsol; C, NH4+; D, NO3-; E, TPpart; F, TPsol; G, TCapart; H, TCasol; I, TMpart; J, TMgsol/TSS ratio. K, pH; L, VSS/TSS ratio. Deviation between measurements and simulations = 12 %

3.3.4. Activated sludge reactor (ASR) + secondary clarifier (SEC) + flotation (FLOT)

Simulations results show that both biological reactor + secondary settler model describe COD, N and P removal (the latter due to precipitation with Al). Another important aspect predicted by the model is the VSS/TSS ratio, including the effect of the particulate Ca compound arriving via reject water on the VSS/TSS ratio. The latter leads to an increase of the biomass operational mixed liquor suspended solids (MLSS) concentration in the reactor to achieve enough nitrification capacity (see **Figure 4** & **5**)

3.3.5. Inactivation tank (IT) + dewatering unit (DEW)

The model predicts three types of behaviour: 1) heavily hydrolysed; 2) precipitated; and, 3) unaltered. The dewatering module can reproduce the measuring data, i.e. soluble compounds were concentrated in the reject water stream, while particulate compounds exited the system via the IT liquor or the cake (see **Figure 4** & **5**).

4. Conclusion

This study demonstrates that the proposed approach is capable for the main streams in the iWTS to reproduce neutralization, volatile fatty acid production, particulate removal and nitrate denitrification in the first units of the flow diagram (buffer tank, primary clarifier, pre-acidification tank). It also correctly predicts biogas composition and COD recovery in form of electricity and heat in the anaerobic granular sludge reactor. Lastly, biological and chemical N and P removal processes in the activated sludge and the quality of bio-solids after inactivation/dewatering (reject water /cake) are predicted by the model.

5. References

Behami et al., 2019. Journal of Cleaner Production.218.616-618

Flores-Alsina et al., 2019. Water Research.156. 264-276

Feldman et al., 2017. Water Research. 126,488-500

Puig et al., 2008. Water Research.42,18,4645-4655

Monje et al., 2021. Journal of Environmental Managament. 293, 112806

Proceedings of the 14th International Symposium on Process Systems Engineering – PSE 2021+
June 19-23, 2022, Kyoto, Japan © 2022 Elsevier B.V. All rights reserved.
http://dx.doi.org/10.1016/B978-0-323-85159-6.50348-1

A Systematic Framework for the Integration of Carbon Capture, Renewables and Energy Storage Systems for Sustainable Energy

Manali S. Zantye, Akhilesh Gandhi, Mengdi Li, Akhil Arora, M. M. Faruque Hasan*

Artie McFerrin Department of Chemical Engineering, Texas A&M University, College Station, TX 77843-3122, USA
Email: hasan@tamu.edu

Abstract

In this work, we address the challenges associated with decarbonizing electricity grids through a decentralized integration scheme of individual fossil power plants with energy storage and flexible CO_2 capture. To this end, we develop a technology design and downselection framework and demonstrate a prototype tool called THESEUS (TecHno-Economic framework for Systematic Energy storage Utilization and downSelection) which enables an extensive techno-economic analysis of integrating fossil power plants with several candidate energy storage technologies which include cryogenic, molten-salt, compressed air, and batteries. The core of THESEUS is a large-scale mixed integer nonlinear programming (MINLP)-based optimization that determines an optimal selection and combination of energy storage technologies for minimizing the cost of meeting the grid electricity demand. We demonstrate THESEUS through a case study for reducing the fossil power plant cycling and minimizing carbon emissions while satisfying sharp spikes in energy demands.

Keywords: Energy Storage, Simultaneous Design and Operation, Downselection.

1. Introduction

As the global energy demand increases, there is a push to adopt sustainable renewable energy sources. However, the seamless integration of clean renewable energy with electricity grids requires measures to address the challenges arising from its inherent intermittency. Conventional measures include the cycling of the fossil-based generating units and the installation of large-scale energy storage. While power plant cycling reduces its efficiency and leads to thermal and mechanical stresses in critical plant components, grid-scale energy storage is cost-intensive with only a limited number of suitable technologies. These integration challenges make it difficult for renewables to completely replace the dispatchable fossil generators from electricity grids. CO_2 capture presents a promising solution to decarbonize fossil-based power generation, but its large-scale deployment is limited by its high energy requirement and cost (Hasan et al., 2012).

To address the challenges associated with the decarbonization technologies, we propose that the operational synergies between them can be leveraged through their localized integration with individual fossil power plants to achieve low-cost as well as reliable clean energy systems. To this end, we study a decentralized power generation system comprising of a natural gas combined cycle (NGCC) power plant integrated with co-located CO_2 capture and energy storage facilities. The benefits include the ability to meet demand spikes without increased cycling of the power plant. Furthermore, along with reducing plant emissions by up to 90%, CO_2 capture also acts as a form of 'indirect storage' to counter renewable intermittency (Zantye et al., 2021).

The integrated system is depicted in Figure 1 where it is connected to the electricity grid and is required to satisfy the time-varying electricity demand. The net demand is considered to incorporate the variability of renewable energy and is given by the total grid demand less the renewable generation. Four candidate storage technologies are considered for the storage block: mechanical energy storage through compressed

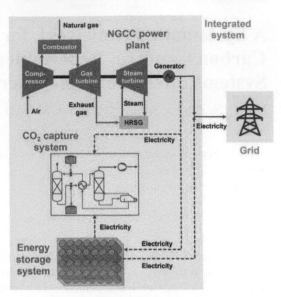

Figure 1: Integrated system schematic.

air energy storage (CAES), thermal energy storage using cryogenic and molten salt-based high temperature storage (CES and HTTS, respectively), and electrochemical energy storage in the form of sodium sulfur (NaS) battery systems. For the CO_2 capture system, storage of the solvent enables time-varying operation of the energy-intensive solvent regeneration step and enables the flexible operation of the capture system. The power generated by the NGCC plant is used to meet the grid demand. A portion of this can also be used for CO_2 capture or stored in the energy storage for cases when the net demand is low from excess renewable availability. On the other hand, if the power plant output is insufficient to meet the grid demand, the energy stored in the storage system can be discharged to provide the required electricity.

The economic viability of this integration is influenced by the high investment cost of the energy storage and CO_2 capture systems. The candidate storage technologies also exhibit trade-offs between the various factors such as lifetime, efficiency and cost. It is crucial to determine the overall dynamic operation of the different systems to cost-effectively ensure that the time-varying grid demand is met while accommodating variable renewable energy. To consider these trade-offs and determine if it is profitable to invest in the integrated system under the spatio-temporal variability of electricity markets and renewable availability, we develop a mathematical programming-based simultaneous design and scheduling framework. This framework forms the back-end of our user-friendly software program: THESEUS (TecHno-Economic framework for Systematic Energy storage Utilization and downSelection). THESEUS enables the user to evaluate and compare the different energy storage alternatives for various demand profiles, region specific factors, and power plant types and conditions. Section 2 in this article presents the THESEUS program and the back-end optimization formulation. Section 3 depicts the framework demonstration to determine system design and operation for a case incorporating a sharp demand spike.

2. THESEUS Framework

The overall framework is depicted in Figure 2. From the user interface, the user can input the power plant parameters such as the cost parameters, ramping limits, nominal capacity and minimum load factor. The user can also input several region-specific parameters including the cost of electricity, ambient conditions, as well as specify a time-varying electricity demand profile to be met by the system. These inputs are combined with the power plant models and the technology models for the various direct and indirect storage

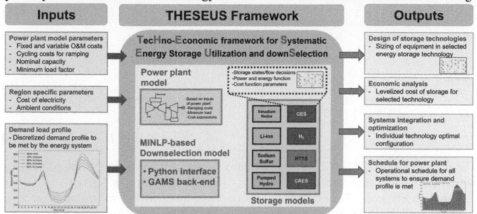

Figure 2: THESEUS software framework for energy storage and CO$_2$ capture (indirect storage) technology downselection.

systems in the back-end. Our mathematical programming-based optimization formulation which accounts for the trade-offs between system costs and flexibility comprises the unifying element connecting the different modules in an overall decision framework. Here, the problem statement is as follows: given a time-varying net electricity demand profile, determine the optimal storage technology, size and operation to integrate decentralized energy storage with existing fossil power plants for minimizing the overall system cost of meeting the demand. The general formulation of the optimization framework is given below:

$$
\min TC = \sum_{i=1}^{NI}\left(C_i^{S,iv} + C_i^{S,of} + \sum_{t=1}^{NT}(C_{i,t}^{S,ov} + C_{i,t}^{FP,ov}) + C_t^{FP,rc} + C^{os} + C^{us} \right. \tag{1}
$$
$$
\left. + C^{co2} \right)
$$

$$
P_t^{dem} = \sum_i P_{i,t}^S + P_t^{FP} - P_t^{os} + P_t^{us} - P_t^{co2} \tag{2}
$$

$$
|P_{t+1}^{FP} - P_t^{FP}| \le ro^{FP} P^{FP,nom}\Delta t \qquad\qquad \forall t \in T \setminus \{NT+1\} \tag{3}
$$

$$
P^{FP,nom} l f^{FP,min} \le P_t^{FP} \le P^{FP,nom} \qquad\qquad \forall t \in T \tag{4}
$$

$$
C_t^{ov,FP} = c4\, P_t^{FP}\eta^{FP,nom}/\eta_t^{FP} \qquad\qquad \forall t \in T \tag{5}
$$

$$
C_t^{FP,rc} = c5|P_{t+1}^{FP} - P_t^{FP}| \qquad\qquad \forall t \in T \setminus \{NT+1\} \tag{6}
$$

$$
P_{i,t}^S = f1_i\big(s_{i,t},\, l_{i,t},\, z_{i,t}^b,\, x_i\big) \qquad\qquad \forall t \in T, \forall i \in I \tag{7}
$$

$$E_{i,t} = f2_i(s_{i,t}) \qquad\qquad \forall t \in T, \forall i \in I \qquad (8)$$

$$-z_{i,t}^C P_{i,t}^S \leq y_i P_i^{max,C} \qquad\qquad \forall t \in T, \forall i \in I \qquad (9)$$

$$z_{i,t}^D P_{i,t}^S \leq y_i P_i^{max,D} \qquad\qquad \forall t \in T, \forall i \in I \qquad (10)$$

$$E_{i,t+1} = E_{i,t} - (\eta_i^S z_{i,t}^C + z_{i,t}^D) P_{i,t}^S \Delta t \qquad\qquad \forall t \in T \setminus \{NT+1\}, \forall i \in I \qquad (11)$$

$$E_{i,t=NT+1} = E_{i,t=1} \qquad\qquad \forall i \in I \qquad (12)$$

$$z_{i,t}^{idle} + z_{i,t}^C + z_{i,t}^D = 1 \qquad\qquad \forall t \in T, \forall i \in I \qquad (13)$$

$$0 \leq x_i \leq E_i^{max} y_i \qquad\qquad \forall i \in I \qquad (14)$$

$$0 \leq E_{i,t} \leq x_i \qquad\qquad \forall t \in T, \forall i \in I \qquad (15)$$

$$C_i^{S,iv} = c1_i(x_i) CRF_i\, T/8760 \qquad\qquad \forall i \in I \qquad (16)$$

$$C_i^{S,of} = c2_i(P_i^{max,D})\, T/8760 \qquad\qquad \forall i \in I \qquad (17)$$

$$C_{i,t}^{S,ov} = c3_i(P_{i,t}^S) \qquad\qquad \forall t \in T, \forall i \in I \qquad (18)$$

The set $t \in T = \{1, 2, \ldots, NT, NT+1\}$ denotes the set of time periods in the scheduling horizon, while the candidate storage technology set is given by: $I = \{ces, htts, caes, nas\}$. The time resolution is denoted by Δt, with T representing the time horizon length (in hrs). The optimization design decisions comprise of the selection y_i, energy capacity x_i, design discharging and charging power $P_i^{max,D}$ and $P_i^{max,C}$ respectively of storage technology i. The operational decisions for technology i includes the manipulating/flow variable $l_{i,t}$. In addition, the state of operation $z_{i,t}^{idle}$, $z_{i,t}^C$, $z_{i,t}^D$ i.e. if the technology is in the idle, charging or discharging state comprises the operational storage decisions. The power output from the power plant at time t, P_t^{FP}, is the plant-level operational decision. Here, the cost minimization objective given by Eq.(1) represents the sum of the storage investment cost $C_i^{S,iv}$, fixed operating cost $C_i^{S,of}$, variable operating cost $C_{i,t}^{S,ov}$, the fossil power plant operating cost $C_{i,t}^{FP,ov}$, plant cycling cost $C_t^{FP,rc}$, the electricity oversupply penalty C^{os}, the undersupply penalty C^{us}, and the cost of CO_2 capture C^{co2}. The optimization constraints consist of the following 3 categories: the grid-level constraints, the power plant model and energy storage model.

Eq.(2) represents the grid-level constraints and denotes the overall energy balance. Here, P_t^{dem}, $P_{i,t}^S$, P_t^{os} and P_t^{us} denote the grid power demand, power output from the storage, electricity oversupply and undersupply at time t, respectively. $P_{i,t}^S$ is positive when the storage is discharging and negative when charging. Eqs.(3)-(6) represent the power plant model. Here, ro^{FP}, $P^{FP,nom}$, $lf^{FP,min}$, $\eta^{FP,nom}$, $c4$ and $c5$ are constants representing the unit ramping rate of the power plant (%/MW.hr), nominal plant capacity (MW), minimum load factor (%), base-load efficiency (%), unit operational cost ($/MW.hr) and the unit cycling cost ($/MW.hr) respectively. Eqs.(7)-(8) represent the technology-specific models for the storage power $P_{i,t}^S$ and the storage energy capacity $E_{i,t}$ in terms of the storage state variable $s_{i,t}$, storage flow variable $l_{i,t}$, storage operational state $z_{i,t}^b$, and storage design x_i. Eqs.(9)-(15) represent the general operational model of each storage technology, while Eqs.(16)-(18) denote the general cost models. For these equations, η_i^S and E_i^{max} denote the storage efficiency and maximum possible energy capacity, while $c1_i$, $c2_i$ and $c3_i$ are constants denoting the unit storage investment, fixed operating and variable operating costs respectively. The operational and cost models of the CO_2 capture system are adapted from Zantye et al., 2021.

3. Results and Discussion

We demonstrate THESEUS for an NGCC power plant of 641 MW nominal capacity to study if it is beneficial for the plant to invest in a co-located energy storage facility and/or a CO_2 capture system. We consider a scheduling time discretization of 5 minutes over a time horizon of one day and a net demand profile with a sharp spike in the evening hours. Furthermore, we consider a futuristic carbon pricing scenario with a CO_2 tax of \$80/ton and a selling price of \$35/ton. This scenario is found to be one of the economically viable cases to provide enough incentive for investment in carbon reduction technologies (Zantye et al., 2021).

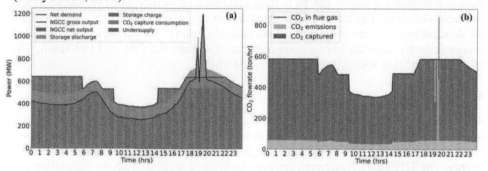

Figure 3: Optimal operational profiles for (a) the integrated system including power plant, CO_2 capture and energy storage, (b) CO_2 emission and capture.

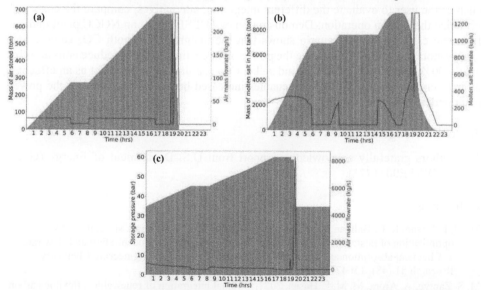

Figure 4: Optimal storage operational profiles for (a) CES technology, (b) HTTS technology, (c) CAES storage system.

THESEUS reports the optimal solution for the NGCC plant to invest in energy storage comprised of CES, HTTS and CAES technologies and a CO_2 capture system to minimize the overall cost. The optimal storage integration size for CES is 100 MW/71 MWh, HTTS is 217 MW/386 MWh, and CAES is 242 MW/66 MWh. The optimal operational profile of the integrated system is demonstrated in Figure 3a. The operation of the CO_2 capture system is shown in Figure 3b. We observe that during periods of troughs in the net demand curve, the power plant produces additional power than the grid demand to charge energy in storage systems and in CO_2 capture. The stored energy is then discharged by the storage system to meet the demand spike during hours 17-22. From Figure 3b, we can see that the high CO_2 price ensures that the capture system is operating almost throughout the day, with the cumulative capture of 89%. The capture system also enables the decreased cycling and increased base-load operation of the power plant and reduces the overall cycling by 24% compared to a case without CO_2 capture. The optimal operational profiles for the selected CES, HTTS and CAES are shown in Figure 4. We find that the storage technologies slowly charge to their maximum capacities when excess energy is available during the day and are discharged almost instantaneously in the evening when the demand spike occurs. Among the three technologies, HTTS has the highest installed energy capacity with the lowest levelized cost of storage (LCOS) of \$149/MWh. The LCOS for CAES is \$302/MWh and CES is \$345/MWh.

4. Conclusions

A decentralized integration schematic of energy storage and CO_2 capture systems with individual fossil power plants is proposed to address the challenges associated with power plant cycling and large-scale energy storage while accommodating variable renewable energy. The THESEUS framework can systematically determine both the integration and dynamic operational decisions. By extensively modelling the interactions between the system components in an overall optimization-based decision framework, THESEUS enables the user to evaluate the different integration alternatives, compare the costs and visualize the system operation. Demonstration of THESEUS for an NGCC plant under a futuristic carbon pricing scenario shows that the integration of both CO_2 capture and energy storage is optimal to reduce the power plant cycling by 24%, reduce emissions by nearly 90%, and meet a sharp demand spike. CO_2 capture is shown to act as an effective indirect energy storage system and enables increased base-load operation of the power plant under renewable integration.

5. Acknowledgements

The authors gratefully acknowledge support from U.S. Department of Energy (Grant number DE-FE0031771).

References

M. M. F. Hasan, R. C. Baliban, J. A. Elia, C. A. Floudas, 2012. Modeling, simulation, and optimization of postcombustion CO_2 capture for variable feed concentration and flow rate. 1. Chemical absorption and membrane processes. Industrial & Engineering Chemistry Research 51 (48), 15642–15664.

M. S. Zantye, A. Arora, M. M. F. Hasan, 2021. Optimal integration of renewables, flexible carbon capture, and energy storage for reducing CO_2 emissions from fossil power plants. In: Computer Aided Chemical Engineering. Vol. 50. Elsevier, pp. 1535-1540.

M. S. Zantye, A. Arora, M. M. F. Hasan, 2021. Renewable-integrated flexible carbon capture: A synergistic path forward to clean energy future", Energy & Environmental Science, 14, 3986 - 4008.

Proceedings of the 14th International Symposium on Process Systems Engineering – PSE 2021+
June 19-23, 2022, Kyoto, Japan © 2022 Elsevier B.V. All rights reserved.
http://dx.doi.org/10.1016/B978-0-323-85159-6.50349-3

Integration of experimental study and computer-aided design: A case study in thermal energy storage

Shoma Fujii[a*], Yuichiro Kanematsu[b], Yasunori Kikuchi[abc]

[a]*Institute for Future Initiatives, The University of Tokyo, 7-3-1 Hongo, Bunkyo-ku Tokyo, 113-8654, Japan*
[b]*Presidential Endowed Chair for "Platinum Society", The University of Tokyo, 7-3-1 Hongo, Bunkyo-ku Tokyo, 113-8656, Japan*
[c]*Department of Chemical System Engineering, The University of Tokyo, 7-3-1 Hongo, Bunkyo-ku Tokyo, 113-8656, Japan*
shoma.fujii@ifi.u-tokyo.ac.jp

Abstract

The integration of experimental studies and computer-aided design strengths the coupling of the interfaces between different research scales. Using the mobile thermal energy storage system as a case study, we demonstrated the connection from the small scale of the material-level to the system-level researches. A full-scale conceptual design was carried out by numerical analysis including material properties that was validated by experiments, and the relationship between design parameters and performance indicators was summarized. By utilizing sensitivity and regression analysis using design parameters, inventory data that can be easily used in process flow models and system evaluations was generated. In addition, system hot spots were extracted from the system evaluation, and the summarized relationship between the design parameters and performance indicators provided feedback on the requirements for proceeding to the real site demonstration.

Keywords: Interdisciplinary approach, Life cycle assessment, Techno-economic analysis.

1. Introduction

To achieve a sustainable society, the time to adoption of emerging technologies must be minimized. To facilitate the adoption of technologies, the connectivity of interfaces between different research areas, and between scales from the material to the system level needs to be further enhanced. For example, integrated simulations of agricultural and industrial processes have made it possible to study the generation of by-products from different cultivars and the introduction of technologies to exploit them (Ohara et al., 2019; Ouchida et al., 2017). As for differences in scale, for example, in thermal energy storage technology, there were cases where bench-scale experiments (Nonnen et al., 2016) and numerical analysis (Mette et al., 2014a) were conducted based on material-level analysis (Mette et al., 2014b). However, now that lifecycle thinking is strongly required, it is necessary to consider new technological developments, combinations of mature technologies, and changes in socioeconomic and environmental conditions in design. This requires not only modeling of material properties, design of equipment incorporating material-level research, and process flow of the plant, but also seamless integration of future system-level analysis such as life cycle assessment and techno-economic analysis in computer-aided design.

In this study, the integration of experimental studies and computer-aided design is demonstrated through a case study of a thermal energy storage system (Fujii et al., 2019) that can charge unused heat from the industrial waste heat and renewable resources, shift the heat in time or space, and release the stored heat according to the heat demand.

2. Material and Methodology

A mobile thermal energy storage (m-TES) system based on the water vapor ad/desorption cycle of zeolites can eliminate the spatial and temporal mismatch between unused heat generation and heat demand. Zeolites generate heat (heat of adsorption) when they capture water vapor (= heat discharging). The adsorbed water can be released by adding dry and heated air at the location of unused heat generation (=heat charging). The heat-charged zeolite can be packed in containers and transported by truck to surrounding heat demand areas. We studied the implementation of a m-TES system between a sugar mill and a food processing factory in Japan. Sugar mills usually burn sugarcane bagasse as fuel to generate steam for their demand. However, sugarcane bagasse is generated more than what is needed, therefore, more bagasse than necessary is burned in the bagasse boiler, emitting unused heat with a temperature of around 200 °C. On the other hand, the surrounding food processing factories consume fossil fuels for process steam generation. Industrial symbiosis of heat through m-TES systems has the potential for effective utilization of unused heat derived from biomass.

Scenario analysis, which considers various socioeconomic and environmental conditions and their combination with other technologies, plays an important role in the design of system. However, when conducting system-level analysis, inventory data extracted from lab-scale experimental results have large errors with real data. For example, in the case of a cylindrical packed bed reactor, the proportion of heat leakage from outside the walls to the total heat balance of the reactor is larger in a laboratory system with a smaller diameter than in a full-scale reactor. In this example, the inventory data extracted from the experimental results will underestimate the heat recovery efficiency. Therefore, it is necessary to predict the full-scale performance utilizing an experimentally validated numerical model and use the inventory data calculated from that simulation. However, it is difficult for researchers in system-level analysis and planners to build numerical simulation models and calculate inventories because it requires specialized skills in the target technology. In addition, when estimating the effects of horizontal deployment, combination with other technologies, and introduction under various socioeconomic and environment conditions, it takes an enormous amount of computational load and time to perform detailed numerical simulations at the reactor level for each condition. Using the development of the m-TES system as a case study, we extracted the essentials for the integration of experimental study and computer-aided system design, including experiments of material properties and their integration into numerical analysis, conceptual design of devices, demonstration experiments, and full-scale performance prediction by numerical analysis.

3. Case study in thermal energy storage

3.1. Experimentally validated computer-aided design

The requirements for the m-TES system were to be able to retrofit a heat charging system into the sugar mill and to be able to charge a large amount of heat continuously at the heat charging site, and to be able to generate process steam continuously at the heat discharging site, which had not been achieved so far. To meet these requirements, a zeolite moving bed reactor was adopted for both the heat charging (named heat charger) and discharging devices (named zeolite boiler). For the zeolite boiler, it was experimentally demonstrated that pressurized steam could be supplied continuously by adopting an indirect heat exchange process (Fujii et al., 2021). For the heat charger, a counter-current contact type moving bed was adopted, and it was demonstrated that continuous heat charging was possible (Arimoto et al., 2019). To incorporate the adsorption equilibrium, kinetics, and heat transfer model of the zeolite into the numerical analysis, a separate series of packed-bed tests were conducted. The developed numerical analysis was found to be able to simulate the temperature distribution in the packed bed, which can be used to predict the full-scale performance and optimize the design parameters. However, these numerical analyses require a lot of computational loads, and we also found that it is difficult to implement numerical analysis in both the process flow model of the plant, and all of the calculations for each case of life cycle assessment and techno-economic analysis as the system-level analysis.

3.2. Seamless connection to system-level analysis utilizing experimental data

The above simulation of the device for each condition of the scenario analysis takes a lot of computational load and time. Therefore, it is necessary for researchers of individual technologies to clarify the relationship between input parameters and output results (performance and required auxiliary power) in advance by sensitivity analysis of input parameters using the constructed numerical model. In addition, regression analysis of the input parameters and output results can reduce the computational load and time for system-level analysis. When researchers of individual technologies are studying and demonstrating concepts of new materials and devices, they can correctly reflect the information of the technology in the system-level analysis by conducting even regression analysis on a full-scale basis, instead of taking the laboratory-level results as the final result. In this way, researchers and system planners working on future system-level analysis can predict system performance based on experimentally validated numerical simulations while utilizing appropriate inventory data, seamlessly linking laboratory and future system predictions.

Figure 1 shows the schematic of the zeolite boiler and the relationship between design parameters and performance as an example. The zeolite boiler uses a moving bed and indirect heat exchange system. The heat charged zeolite is supplied from the top of the zeolite boiler, and the steam generated by the existing boiler is injected into the zeolite boiler. The zeolite adsorbs the injected steam and generates adsorption heat. The adsorption heat raises the temperature of the zeolite bed, and this heat is transferred to the feed water that is introduced from the bottom to the heat exchanger to produce pressurized steam for the process. The full-scale performance was predicted using numerical analysis, and the relationship between the design parameters and the performance was summarized by conducting sensitivity analysis with the design parameters. As an example, Figure 2 shows the sensitivity analysis for the initial water uptake on zeolite for the zeolite boiler. The lower the water uptake at input, the higher

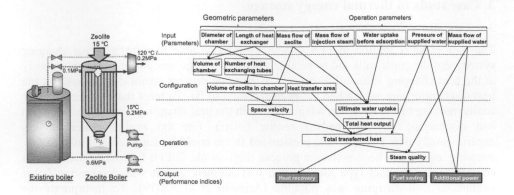

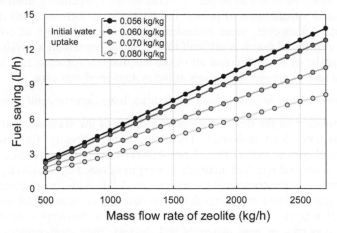

Figure 1 Schematic and linkage flow diagram of relationship among each parameter of zeolite boiler.

Figure 2 Effect of initial water uptake for zeolite boiler on fuel savings

the heat storage density, and thus the higher the fuel saving effect at the same scale. All point data are the result of optimizing other parameters to maximize the fuel saving based on the established design methodology using numerical analysis. The results show that the performance can be linearized with respect to scale at each initial water uptake, and this linear relationship was used in the system evaluation without going through complex numerical analysis. The same method was applied to other design parameters of the zeolite boiler and the heat charger, and the results were reflected in the process flow of the sugar mill and the heat demand for life cycle assessment and techno-economic analysis.

3.3. Research flow of integrating experimental study and system-level analysis

From the case of m-TES system, the integration of experimental study and computer-aided system design can be achieved by following procedure.

Material-level
-Testing the material properties and heat transfer characteristics

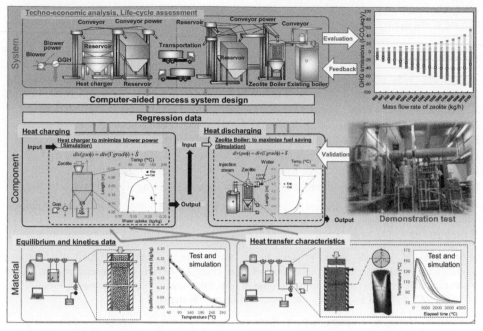

Figure 3 Research flow of seamless analysis from maerial to system level

Component-level

-Developing a numerical model that incorporates the material level information, and validate the model by demonstration tests

-Designing a full-scale equipment utilizing the validated numerical model, and predicting the performance, summarizing the relationship between design parameters and performance, and creating the regression data by sensitivity analysis

System-level

-Reflecting regression data in the process flow model of the plant

-Conducting system-level analysis such as life cycle assessment and techno-economic analysis utilizing inventory data obtained from the regression data.

-Identifying hot spots based on the results of the system-level analysis and feed them back to the target of technology in the reverse order above.

These are summarized in Figure 3. By creating this research flow, when a new material or device design is devised, the effect can be immediately evaluated from the system perspective by computer-aided system design, and it can also be used for scenario analysis in combinations of multiple technologies and various socioeconomic and environmental conditions.

4. Conclusions

In this study, a m-TES system with a zeolite water vapor ad/desorption cycle was used as a case study to extract the essentials for the integration of experimental studies and computer-aided system design. Using experimentally validated numerical analysis, sensitivity analysis, and regression analysis with design parameters of the equipment, computer-aided design (numerical analysis, process flow modelling, and future system-level analysis including environmental and socioeconomic parameters) can be integrated with experimental studies, and each scale from material-level to system-level can be consistently linked.

Acknowledgement

This work was supported by Shinko Sugar Mill Co., Ltd., JSPS Grant-in-Aid for Research Activity Start-up (Grant Number 20K23360), JST COI-NEXT (Grant Number JPMJPF2003), and the Environment Research and Technology Development Fund (Grant Number JPMEERF20213R01 and JPMEERF20192010) of the Environmental Restoration and Conservation Agency of Japan. The activities of the Presidential Endowed Chair for "Platinum Society" at the University of Tokyo are supported by the KAITEKI Institute Incorporated, Mitsui Fudosan Corporation, Shin-Etsu Chemical Co., ORIX Corporation, Sekisui House, Ltd., East Japan Railway Company, and Toyota Tsusho Corporation.

References

N. Arimoto, Y. Abe, S. Fujii, Y. Kikuchi, Y. Kanematsu, T. Nakagaki, 2019, Performance prediction and experimental validation of heat charging device in thermochemical energy storage and transport system utilizing unused heat from sugar mill (in Japanese), Proceedings of the 24th National Symposium on Power and Energy Systems.

S. Fujii, N. Horie, K. Nakaibayashi, Y. Kanematsu, Y. Kikuchi, T. Nakagaki, 2019, Design of zeolite boiler in thermochemical energy storage and transport system utilizing unused heat from sugar mill, Applied Energy, 238, 561-571.

S. Fujii, Y. Kanematsu, Y. Kikuchi, T. Nakagaki, 2021, Effect of residual heat recovery of zeolite boiler in thermochemical energy storage and transport system, Proceedings of the 15th International Conference on Energy Storage ENERSTOCK

B. Mette, H. Kerskes, H. Druck, 2014a, Experimental and numerical investigations of different reactor concepts for thermochemical energy storage, Energy Procedia, 57, 2380-2389.

B. Mette, H. Kerskes, H. Druck, H. Muller-Steinhagen, 2014b, Experimental and numerical investigations on the water vapor adsorption isotherms and kinetics of binderless zeolite 13X, International Journal of Heat and Mass Transfer, 71, 555-561.

T. Nonnen, S. Beckert, K. Gleichmann, A. Brandt, B. Unger, H. Kerskes, B. Mette, S. Bonk, T. Badenhop, F. Salg, R. Glaser, 2016, A thermochemical long-term heat storage system based on a salt/zeolite composite, Chemical Engineering & Technology, 39, 2427-2434.

S. Ohara, Y. Kikuchi, K. Ouchida, A, Sugimoto, T. Hattori, T. Yasuhara, Y. Fukushima, 2019, Reduction of greenhouse gas emissions in the introduction of inversion system to produce sugar and ethanol from sugarcane (in Japanese), Journal of Life Cycle Assessment, Japan, 15, 86-100.

K. Ouchida, Y. Fukushima, S. Ohara, A. Sugimoto, M. Hirao, Y. Kikuchi, 2017, Integrated design of agricultural and industrial processes: A case study of combined sugar and ethanol production, AIChE J., 63, 560-581.

Proceedings of the 14th International Symposium on Process Systems Engineering – PSE 2021+
June 19-23, 2022, Kyoto, Japan © 2022 Elsevier B.V. All rights reserved.
http://dx.doi.org/10.1016/B978-0-323-85159-6.50350-X

Design support toolbox for renewable-based regional energy systems; The concept, data integration, and simulator development

Yuichiro Kanematsu[a*], Shoma Fujii[b], and Yasunori Kikuchi[a,b,c]

[a] Presidential Endowed Chair for "Platinum Society", The University of Tokyo, Tokyo 113-8656, Japan
[b] Institute for Future Initiatives, The University of Tokyo, Tokyo 113-8654, Japan
[c] Department of Chemical System Engineering, The University of Tokyo, Tokyo 113-8656, Japan
kanematsu@platinum.u-tokyo.ac.jp

Abstract

The design and implementation of renewable-based energy systems in various regions are increasingly necessitated. However, the data required are diverse and distributed across multiple ministries and local governments, and it often takes a considerable amount of time and efforts just to collect the basic data. Many studies regarding simulations of system designs for various resources and regions have been conducted, but they are not always reusable. In this study, we propose a design support toolbox integrated with databases for renewable-based regional energy systems, and the status of the development is introduced.

Keywords: Life cycle assessment, Open data, Decarbonization.

1. Introduction

Under the strong global demand for decarbonisation, there is an urgent need for action plans at national and local level. In Japan, though many municipalities have declared their carbon neutral by 2050, most of them do not have the clear plans of technology implementation. Renewable energy is the key technology for the decarbonization. The introduction of solar and wind power has been progressing, but because of their variability, energy storage and energy carriers are becoming increasingly important (Sinsel *et al.*, 2020). Biomass can be stored as the state of fuel, but there are difficulties in design of supply chain and whole energy system (Zahraee *et al.*, 2020). Toward the achievement of carbon neutral, complicated design for the combination of these technologies will be required. Many energy system analysis tools have been developed for various spatial and temporal targets (Ringkjøb et al, 2018), but very few of them can consider such emerging technologies.

In this study, a design support toolbox for renewable-based regional energy systems is proposed. Case studies of designing regional energy system were carried out to clarify the problems and barriers in design procedure. The requirement of the support toolbox is defined through building activity model and the data model of the system design. The status of tool development is also introduced.

2. Case studies of designing renewable-based regional energy systems

To clarify the problems and barriers in designing renewable-based regional energy systems, we had been carried out some case studies of design and evaluation of them. A case of designing combined cooling, heating, and power system using local woody biomass toward sustainable forestry (Kanematsu *et al.*, 2017a) was analysed. As shown in Figure 1, the design and evaluation procedure mainly consist of "Data collection", "Alternative generation", "Simulation", and "Evaluation". Data collection was required for identify and quantify the local resources and energy demand. In the alternative generation, the system flow diagrams are drawn, similar to the process flow diagrams for chemical processes, for multiple alternatives. To enable the simulation of material and energy balances, simulation modules were developed for the elemental technologies, e.g., power plant, district heating and cooling system, wood chipping machine, transport, and so on. Estimations for hourly variations in energy demand of each consumer based on a published knowledge was also required for executing simulation considering supply-demand balance. As the system evaluation, life cycle assessment (LCA) was carried out using the simulation results as inventory data. As another case study, industrial symbiosis around sugar mill using the excess heat and bagasse (Kikuchi et al., 2016) was also carried out, and the case had the similar structure.

We identified the problems that such design and evaluation could be carried out as research activity, but it was time consuming, labour intensive and requires specialized knowledge and its application. In order to achieve rapid social implementation, it needs to be possible to implement it in different multiple areas in a shorter time. Just as process simulators and chemical databases have contributed to speeding up and saving labour in chemical process design, tools such as simulators and integrated databases will be essential for rapid design of regional energy system.

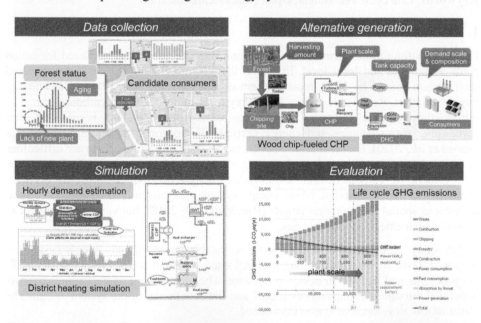

Figure 1 Overview of the case study of designing combined cooling, heating, and power system using regional woody biomass (compiled from Kanematsu et al., 2017a)

3. Requirement definition of the design support toolbox

Activity model and data model were built to define the requirements of the design support toolbox for renewable-based regional energy systems by re-analysing the abovementioned case studies.

3.1. Activity model

The designing procedure was visualized by activity model in order to clarify how we should support what activities. IDEF0 functional modelling method was employed for expressing the activity model. In IDEF0 method, an activity is represented by a box and information is represented by four types of concepts (Input, Output, Control and Mechanism), each of which is represented by an arrow with a different direction toward the activity box. The activities were expressed from the viewpoint of the designing working group as in Figure 2. The main designing tasks consists of [Examine present system], [Generate alternatives], [Simulate flows in alternatives] and [Evaluate alternatives]. Draft of proposals are created from these designing tasks, and it will be reviewed among the decision makers in the [Review proposals] activity.

In this activity modelling, we especially focused on the Mechanisms, which enable the execution of the activities including tools, human resources, and datasets. We identified the tools that should be developed in order to carry out activities with less time and effort. The required functions of these tools were identified by carefully examining the Input, Output and Control of each activity.

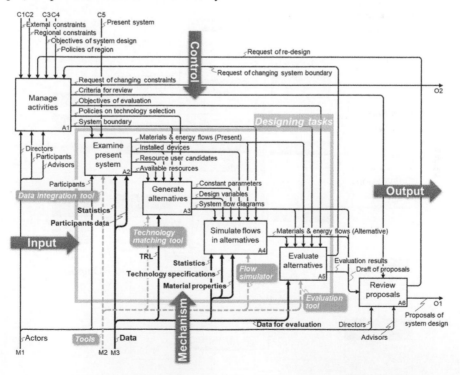

Figure 2 IDEF0 activity model: the activity of "Propose the design of renewable-based regional energy systems" (modified from Kanematsu *et al.*, 2017b)

3.2. Data model for flow simulation

A data model was also built to clarify the types of data required and the linkages between them in executing the simulation. The step of simulation was focused because it was revealed to be the central activity in the design procedure from the viewpoint of data processing. Data model can also be used as the conceptual design diagram of required database for the simulator. UML class diagram was applied for expressing the data model as shown in Figure 3. The flame of "Participant" means the entities which will participate in the regional energy system to be developed, e.g., resource providers, energy suppliers, local industries, consumers, and so on. The participant is described as "Prosumer" because it can be both of consumer and producer of resources. Participant can be simply expressed which produces output from input via conversion devices. For example, sugar mill produces raw sugar, bagasse, molasses, heat, power, and other by-products from sugarcane via sugar milling processes includes in-house cogeneration system. In actual simulation, multiple modules that functionalize the participants are connected to each other, and the energy and material balance can be calculated.

3.3. The function of design support toolbox

The required functions of design support toolbox were identified and defined through the re-analysis of the activity and data models. We have developed the concept of the toolbox as in Figure 4. The toolbox consists of the tools that support the respective activities and the databases linked with these tools.

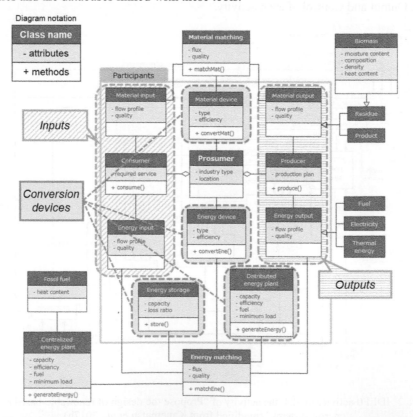

Figure 3 Data model for the simulation of renewable-based regional energy system

Design support toolbox for renewable-based regional energy systems;
The concept, data integration, and simulator development

2105

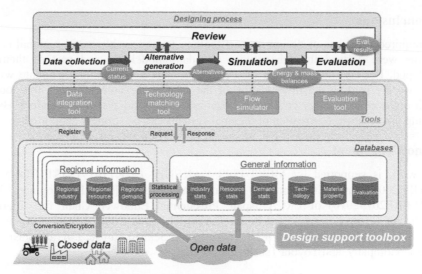

Figure 4   The concept of the design support toolbox for renewable-based regional energy systems

4. Development of modules in the toolbox

4.1. Data integration and visualization of resources and demands

A web-based application that enables the semi-automation of collection, analysis, and visualization of scattered data sources for regional resources was developed as a prototype. Open data originally organized by different ministries were integrated, such as renewable energy potential and the installation status of renewables under the Feed-in-Tariff scheme. An algorithm to estimate the energy demand for each municipality is also developed and will be combined to this application.

4.2. Matching between resources and technologies

A matching system integrated with technology database is conceptually designed under the cooperation with an association of engineering enterprises. The system can search conversion technologies that can produce value-added products from the local resources that found in the stage of data collection. This system can connect the novel technologies and local players that have not been known each other well.

4.3. Simulation of the material and energy balances

By applying the developed woody biomass simulator to other area, the time required was reduced by a factor of 20 compared to the first case. Additionally, we developed and are combining the simulators for thermal energy storage (TES) and transportation with zeolite (Fujii *et al.*, 2019) and the technologies for massive installation of variable renewable energy such as battery-assisted solar-derived hydrogen production (Kikuchi *et al.*, 2019, Sako *et al.*, 2020) and wind-TES (Yamaki *et al.*, 2020).

4.4. Environmental and socio-economic evaluations

Once the material and energy flows in the region with future installation of the new technologies could be calculated by the simulator, evaluations become executable by combining them with the data for evaluations such as life cycle inventory databases or input-output table available from national or local government. The integration mechanism of these data is under development.

5. Conclusions

The required functions of design support toolbox for renewable-based regional energy systems were defined through activity and data modelling, and part of them are developed. Because the workload of the development and continuous update will be huge, we are starting the investigation of the scheme for co-creation of this toolbox which involves technology developers, local industries, and system integrators. By co-creating this toolbox, co-creation of regional systems can be strongly supported.

Acknowledgement

This work is supported by MEXT/JSPS KAKENHI (21K17919), JST COI-NEXT (JPMJPF2003), and Environment Research and Technology Development Fund (JPMEERF20192010) of ERCA Japan. Presidential Endowed Chair for "Platinum Society" of UTokyo is supported by the KAITEKI Institute Inc., Mitsui Fudosan Co., Ltd. , Shin-Etsu Chemical Co., ORIX Corporation., Sekisui House, Ltd., the East Japan Railway Company, and Toyota Tsusho Corporation.

References

S. Fujii, N. Horie, K. Nakaibayashi, Y. Kanematsu, Y. Kikuchi, T. Nakagaki, 2019, Design of Zeolite Boiler in Thermochemical Energy Storage and Transport System Utilizing Unused Heat from Sugar Mill, Applied Energy, 238, 561-571

Y. Kanematsu, K. Oosawa, T. Okubo, Y. Kikuchi, 2017a, Designing the Scale of a Woody Biomass CHP Considering Local Forestry Reformation: A Case Study of Tanegashima, Japan. Applied Energy, 198, 160–172

Y. Kanematsu, T. Okubo, Y. Kikuchi, 2017b, Acitivity and Data Models of Planning Processes for Industrial Symbiosis in Rural Areas, Kagaku Kogaku Ronbunshu, 43 (5), 347-357

Y. Kikuchi, T. Ichikawa, M. Sugiyama, M. Koyama, 2019, Battery-assisted low-cost hydrogen production from solar energy: Rational target setting for future technology systems, International Journal of Hydrogen Energy, 44, 1451–1465

Y. Kikuchi, Y. Kanematsu, M. Ugo, Y. Hamada, T. Okubo, 2016, Industrial Symbiosis Centered on a Regional Cogeneration Power Plant Utilizing Available Local Resources: A Case Study of Tanegashima. Journal of Industrial Ecology. 20 (2), 276–288

H.-K. Ringkjøb, P. M. Haugan, I. M. Solbrekke, 2018, A review of modelling tools for energy and electricity systems with large shares of variable renewables, Renewable and Sustainable Energy Reviews, 96, 440-459,

N. Sako, M. Koyama, T. Okubo, Y. Kikuchi, 2021, Techno-Economic and Life Cycle Analyses of Battery-Assisted Hydrogen Production Systems from Photovoltaic Power, Journal of Cleaner Production, 298, 126809

S. R. Sinsel, R. L. Riemke, and V. H. Hoffmann, 2020, Challenges and solution technologies for the integration of variable renewable energy sources—a review, Renewable Energy, 145, 2271–2285

A. Yamaki, Y. Kanematsu, Y. Kikuchi, 2020, Lifecycle Greenhouse Gas Emissions of Thermal Energy Storage Implemented in a Paper Mill for Wind Energy Utilization. Energy, 205, 118056

S. M. Zahraee, N. Shiwakoti, P. Stasinopoulos, 2020, Biomass Supply Chain Environmental and Socio-Economic Analysis: 40-Years Comprehensive Review of Methods, Decision Issues, Sustainability Challenges, and the Way Forward. Biomass and Bioenergy, 142, 105777

Proceedings of the 14th International Symposium on Process Systems Engineering – PSE 2021+
June 19-23, 2022, Kyoto, Japan © 2022 Elsevier B.V. All rights reserved.
http://dx.doi.org/10.1016/B978-0-323-85159-6.50351-1

Circular Economy Integration into Carbon Accounting Framework for Comprehensive Sustainability Assessment

Nasyitah Husniyah Mahbob[a], Haslenda Hashim[a,b*]

a School of Chemical and Energy Engineering, Faculty of Engineering, Universiti Teknologi Malaysia, 81310 Skudai, Johor, Malaysia
b Process Systems Engineering Malaysia (PROSPECT), School of Chemical and Energy Engineering, Faculty of Engineering, Universiti Teknologi Malaysia, 81310 Skudai, Johor, Malaysia
haslenda@utm.my

Abstract

Circular economy is an approach to develop economy without giving harms to the social and environment. It uses 'cradle-to-cradle' concept which design out pollution and waste while keeping materials and products in loop for as long as possible. This economy concept which is also in line with several SDG goals such as ecosystem restoration, responsible consumption, as well as climate action makes it more preferrable than the old 'cradle-to-grave' concept. There are a lot of circular economy actions that can be adopted by the industry as an effort in shifting towards circular economy. Therefore, a systematic framework to aids the industry in selecting appropriate circular economy actions is needed. In this paper, a framework for selection of circular economy actions is proposed by considering its environmental and economic impact. This framework integrates Total Circularity Index (TCI) and integrated carbon accounting and mitigation framework (INCAM) to assess the environmental impact of the actions. The practicability of the framework is illustrated through a relevant palm oil mill case study. The results show that the application of this framework enables the industry to select circular economy actions that is appropriate to its environmental and economic status.

Keywords: circular economy, carbon mitigation, sustainability, decision-making framework

1. Introduction

The world is currently facing climate change where the temperature rises leading to other consequences such as melting of glaciers and more frequent occurrence of natural disaster. Based on The Sustainable Development Goals Report 2020 released by United Nations (2020), 2019 was the second warmest year on record and the global temperature are projected to rises to 3.2 °C by 2100. The rises in Earth's temperature are due to the increasing amount of greenhouse gases emission and according to data by Climate Watch (2021), global greenhouse gases emission gradually increases each year, and it reaches 48.94 Gt CO2e in 2018.

Numerous efforts have been made to tackle climate change at various level. Paris Agreement which is a legally binding agreement that was adopted by many countries in 2015 is one of the efforts done to fight climate change at global level. This agreement aims to limit the global warming at 1.5 °C and many countries have pledged to reduce their greenhouse gases emission through their nationally determined contributions (NDCs) to achieve this target. Apart from that, utilization of renewable energy and energy efficiency program are also part of the efforts done to tackle climate change since more than 50 % of the global greenhouse gases emission comes from energy sector (Climate Watch, 2021).

In addition, shifting towards circular economy can also be one of the ways to tackle climate change. Circular economy is an approach that uses 'cradle-to-cradle' concept which aims to decouple economic growth from consumption of finite resources. It is built to design out waste and pollution and elongate the life cycle of products and materials while regenerating natural system. Therefore, circular economy can help reduce the greenhouse gases emission by eliminating its sources (waste, pollution, and manufacturing process) and providing more carbon sinks. This statement can be proven through a study by Ramboll et al. (2020) that shows 61% of GHG emissions across European Union can be reduced by implementing circular economy actions (CE actions). Research done by Ellen MacArthur Foundation (2019) also discussed the role of circular economy in tackling climate change.

There are numerous CE actions that can be applied in an economic system ranging from utilization of renewable resources to reuse or recycling of materials and products. Therefore, a methodology to select suitable circular economy actions is required to assist the industry in decision-making process. A study conducted by Ramboll et al. (2020) has proposed a method to select and assess circular economy actions and its impact on climate change mitigation. A generic methodology to quantify the potential CO_2 emission reduction was developed by integrating the life cycle analysis (LCA) and material flow modelling. The methodology is designed to be applicable to all economic sectors. However, this methodology can be challenging for some users as it requires knowledge and experiences in LCA and material flow modelling. Moreover, uncertainty and biodiversity in LCA also makes it more challenging as LCA studies depend on assumptions and scenarios.

The aim of this study is to propose an easy and systematic framework for the industry to select suitable circular economy actions by considering its economic and environmental impacts. This paper integrates the Total Circularity Index (TCI) and integrated carbon accounting and mitigation framework (INCAM) to assess the environmental impacts of the circular economy actions. The methodology proposed enables the industry to find the hot spots for carbon emissions in their plant or organization as well as selecting befitting circular economy actions based on its economic and environmental impacts.

2. Circular Economy Actions Selection Framework

According to the framework for selection of CE actions shown in Figure 1, the selection of CE actions is made based on three criteria which are the emission reduction, its circularity as well as the payback period. There are three main steps in this framework which are: (i) identifying the hot spots for GHG emissions, (ii) proposing possible CE actions, (iii) evaluating and ranking of CE actions. To identify hot spots for GHG emissions, an integrated carbon accounting and mitigation (INCAM) framework by Hashim et al. (2015) will be applied. Based on INCAM, the methodology to identify hot spots include defining Carbon Accounting Centre (CAC), developing carbon checklist, and calculating carbon emission index (CEI) and carbon emission profile (%) of each CAC. CAC with the highest carbon emission profile will be the hot spots.

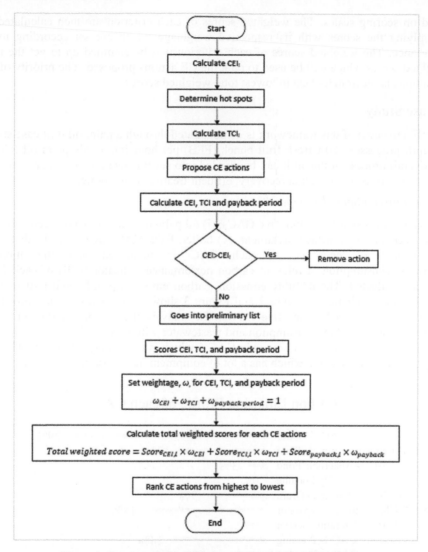

Figure 1: Framework for Selection of Circular Economy Actions

After the hot spots have been identified, total circularity index (TCI) of the hot spots will be calculated. TCI is an indicator to measure the circularity of a process route which consider both material and energy aspects. Circular Material Use (CMU) is used to calculate the circularity of the material by dividing the product produced through secondary materials with the total demand of the products. On the other hand, energy aspects will be measured through Circular Exergy Use (CEU) which is the amount of secondary exergy produced over the exergy demand. CMU and CEU will be multiplied by its respective weightage before being added up to get TCI. The TCI value will be between 0 and 1 with the latter represents full circularity of the process route. CE actions is then proposed accordingly.

CEI, TCI and payback period of each CE actions proposed will be estimated. Then, CE actions with higher CEI than CEI_i, which is the baseline, will be eliminated from the options. The CEI, TCI, and payback period of the remaining CE actions will be scored based on scoring scales. The weighted scores of each criterion are then calculated by multiplying the scores with its respective weightage which are set according to its importance. The weighted scores of each criterion will be summed up to get the total weighted scores which will be used to rank the CE actions proposed. The priority of the CE actions is set from highest to lowest total weighted scores.

3. Case Study

The effectiveness of this framework is demonstrated through a palm oil mill case study. The mill processes 120 t fresh fruit bunch (FFB) per hour into crude palm oil (CPO). The overall process of this mill can be divided into several sections which are palm oil extraction, oil recovery, kernel recovery, effluent treatment, and boiler.

3.1. Identify Hot Spots Through INCAM

The whole process is break into five CACs, (i) oil palm extraction, (ii) oil recovery, (iii) kernel recovery, (iv) effluent treatment, (v) boiler. Each CAC have several sub-CACs. The source of emission of each CAC is identified and the monthly generation/consumption of relevant carbon performance indicator (CPI) of each CAC has been collected. The monthly emission, carbon emission profile, and CEI of each CAC has been calculated. Figure 2 and Figure 3 show the carbon emission profile of each CAC and each CPI respectively. From the analysis, the top three carbon emission are from electricity, fuel consumption and wastewater with CPI of 68 %, 28 % and 3 % respectively. The highest emission is in CAC 5 which is the boiler followed by CAC 2, palm oil extraction section, which has a lot of equipment that consumes electricity.

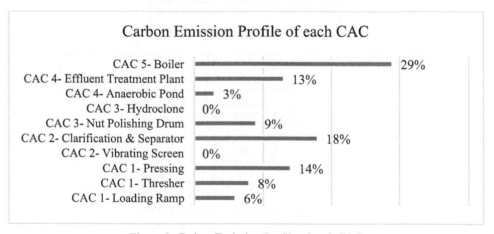

Figure 2: Carbon Emission Profile of each CAC

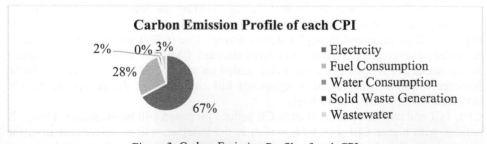

Figure 3: Carbon Emission Profile of each CPI

3.2. Propose CE Actions and Priority Ranking

Before proposing CE actions, TCI of the palm oil is calculated in order to identify any circularity actions that has been implemented in the mill. In this case study, the demand for product and exergy used in CMU and CEU calculation is the amount of product and exergy demand by the palm oil mill and its plantation. The total exergy demand is 3086 GJ/month. From the calculation, the CMU, CEU and TCI of the palm oil mill is all 0. This means that there is no circular action has been implemented in the mill. After measuring the circularity of the mill, CE actions to reduce carbon emissions are proposed according to the hot spots identified earlier. The CE actions proposed will focus on reducing electricity and fuel consumption as well as reducing emissions from effluent and boiler. There are three CE actions proposed which are electricity generation using biogas from POME, change boiler fuel from diesel to biogas, and production of fertilizer from POME. The contributions - CEI, TCI and payback period of all CE actions proposed has been estimated and summarized in Table 1. Since all actions proposed has lower CEI than the baseline CEI, all actions proposed will be considered for implementation.

The contributions of each CE actions are scored based on scoring scales. For this case study, the weightage for emission reduction, TCI and payback period has been set at 0.4, 0.2, and 0.4 respectively. This is because the mill wants to prioritize reducing emission with low costs. Table 2 shows the priority rank of the CE actions according to the total weighted scores. From the analysis, it shows that production of fertilizer from POME should be implemented first since it has the highest total weighted scores compared to other options. This option can reduce 13,325 kgCO2e/month and improves the TCI of the mill by 50 %. It also has the lowest payback period which is 0.84 years. The second highest ranking is to change boiler fuel from diesel to biogas which has average scores in all criteria. The third priority is to generate electricity using biogas from POME. Although this option can reduce the most carbon emissions, 309,725 kgCO2e/month, it requires high payback period which is 7.64 years and has the lowest TCI improvement which makes it less favorable.

Table 1: CEI, TCI and payback period of CE actions proposed

CE Actions	Emission (kgCO2e/month)			Total Circularity Index		Payback Period (year)
	Baseline	After	Reduction	Before	After	
Biogas from POME to generate energy		127,737.43	309,725.03		0.23	7.64
Change boiler fuel to biogas (POME)	437,462.46	313,850.36	123,612.10	0	0.27	1.71
POME to fertilizer		424,137.43	13,325.03		0.50	0.84

Table 2: Scores and priority ranking for CE actions proposed

CE Actions	Scores							Priority Ranking
	Emission Reduction	Weighted Score (0.4)	TCI	Weighted Score (0.2)	Payback Period	Weighted Score (0.4)	**Total Weighted Scores**	
Biogas from POME to generate energy	4	1.6	2	0.4	2	0.8	**2.8**	2
Change boiler fuel to biogas (POME)	2	0.8	2	0.4	4	1.6	**2.8**	2
POME to fertilizer	1	0.4	3	0.6	5	2.0	**3**	1

4. Conclusions

In this paper, a framework to select circular economy actions has been proposed. This framework aims to ease the decision maker to choose which CE actions to be prioritized and implemented by considering the emission reduction, circularity, and the payback period. The practicality of this framework has been demonstrated through a palm oil mill case study and the result shows that this framework is able to guide decision-maker in choosing mitigation actions that can improve carbon performance, circularity, and economy. Improvement that can be made for this framework is by considering other constraints in the selection process such as the reduction target and financial budget.

5. Acknowledgement

The authors would like to gratefully acknowledge UTM University Grant Vot. No. Q.J13000.3009.02M81 and Q.J130000.2409.08G96 for the financial in completing this project.

References

Climate Watch. (2021). Historical GHG Emissions. Retrieved from Climate Watch: https://www.climatewatchdata.org

Ellen MacArthur Foundation. (2019). Completing the Picture: How the Circular Economy Tackles Climate Change www.ellenmacarthurfoundation.org/publications

H., Hashim, M. R., Ramlan, L. J., Shiun, H. C., Siong, H., Kamyab, M. Z., Majid, & C. T., Lee (2015). An Integrated Carbon Accounting and Mitigation Framework for Greening the Industry. Energy Procedia, 2993-2998.

Ramboll, Fraunhofer, & Ecological Institute. (2020). The Decorbonisation Benefits of Sectoral Circular Economy Actions.

United Nations. (2020). The Sustainable Development Goals Report 2020. New York: United Nations Publications.

Proceedings of the 14th International Symposium on Process Systems Engineering – PSE 2021+
June 19-23, 2022, Kyoto, Japan © 2022 Elsevier B.V. All rights reserved.
http://dx.doi.org/10.1016/B978-0-323-85159-6.50352-3

Design and analysis of fuel-assisted solid oxide electrolysis cell combined with biomass gasifier for hydrogen production

Shih-Chieh Chen, Jyh-Cheng Jeng*

Department of Chemical Engineering and Biotechnology, National Taipei University of Technology, Taipei 10608 Taiwan
jcjeng@ntut.edu.tw

Abstract

To solve environmental problems, the development of alternative energy sources is becoming more and more important. Hydrogen is a very high potential energy carrier because it has a higher energy density compared to other energy carriers and clean in use. This study proposes a novel hydrogen production process that combines a biomass gasifier and a fuel-assisted solid oxide electrolysis cell (SOFEC). The syngas fed to the anode of SOFEC for fuel-assisted electrolysis is formed by gasification of biomass. The advantages of combining these two systems are that the operating temperatures are similar to each other, which reduces energy waste, and that part of the electrical energy used in electrolysis is replaced by the chemical energy of the fuel, which significantly reduces the demand for external power supply during electrolysis. In addition, the thermal energy of the high-temperature flow from the exhaust of SOFEC can be used to preheat the feed. In this study, we consider the effect of gasifier operating parameters, including gasification temperature, dried biomass moisture, and equivalence ratio on system. The results show that there is a positive effect on the system when the gasification temperature and the moisture content of the biomass are higher; With different gasifier operating conditions, the efficiency of the system is highest when the equivalence ratio is between 0.15 and 0.25, mainly depending on the amount of syngas production from the gasifier.

Keywords: Hydrogen production, Gasifier, Fuel-assisted solid oxide electrolysis cell, System simulation.

1. Introduction

Since the industrial revolution, fossil fuel combustion has been the main method of energy generation, contributing the largest proportion of global energy demand. However, fossil fuels are non-renewable resources, making them more and more scarce in the future and bound to face problems such as energy depletion, and greenhouse gas emissions from fossil fuel use already have a serious impact on global warming and air pollution. Therefore, governments are actively looking for alternative energy sources. Considering the economic growth and energy demand, as well as the environmental issues, the research on renewable energy and its storage has been increasing rapidly in recent years.

Although producing hydrogen through fossil fuels is the most efficient and economical way to produce hydrogen, the production results in greenhouse gas emissions that cause environmental pollution. Therefore, the generated hydrogen cannot be regarded as green hydrogen. Hydrogen production through renewable energy is the way to produce green hydrogen, and in the medium term, the solid oxide electrolysis cell is the most potential solution for the development of green hydrogen production technology. Therefore, this study proposes a hydrogen production system using syngas generated from biomass gasifier to assist the solid oxide electrolysis cell, replacing part of the electrical energy in electrolysis with chemical energy from syngas and thus reducing the electrical energy demand in the electrolysis of hydrogen. In this study, the impact of the operating parameters of the gasifier on the system efficiency is investigated.

2. Models of Gasifier and Fuel-Assisted Solid Oxide Electrolysis Cell

2.1. Gasifier model

When fresh biomass enters the gasifier for gasification reaction, the process is divided into drying zone, decomposition zone, gasification zone, and sepearation zone. These sections are realized through different modules in Aspen Plus®, as the model shown in Figure 1. First, the biomass source will enter the RStoic reactor and dry the biomass with the temperature set at 150°C as the drying temperature. After drying, the moisture content of biomass will decrease and become dry biomass and then enter the RYield reactor to simulate the pyrolysis of biomass in which the reaction temperature is set at 550°C. After the pyrolysis reaction, biomass will enter the gasification reactor (Rgibbs) in the form of elements, and additional air will be used as the gasification agent. The reacted syngas will be separated from ash and carbon through SEPARATE unit to simulate the ash and syngas separation section of the gasifier, and H2O-SEP is used to separate the water from the biomass to calculate the dry basis of biomass syngas (Tavares et al., 2020).

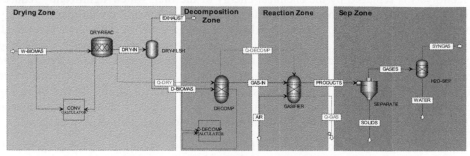

Figure 1 Gasifier Model in Aspen Plus®.

2.2. SOFEC model

The actual voltage of an electrolysis cell during electrolysis is expressed by the equivalent voltage added to its irreversible loss. Voltage is the most important parameter that affects the performance of electrolytic cells. Therefore, the amount of polarization will affect the power consumption of the entire electrolytic cell :

$$V_{SOFEC} = V_{SOFEC}^{OCV} + \eta_{act} + \eta_{ohm} + \eta_{conc} \tag{1}$$

Because the electrochemical reactions in the SOFEC cathode and anode are opposite reactions, there is no standard potentials of electrochemical reactions in the calculation of open-circuit potential, but the quotient of the component concentrations is retained according to the Nernst equation. The equilibrium voltage is as shown in Eq.(2) where R is the ideal gas constant (8.314 J/mole K), T is the operating temperature (K), and $P_{H2,ca}$, $P_{H2O,ca}$, $P_{H2,an}$, $P_{H2O,an}$ are the surface partial pressure of each component at the inlet of the cathode and anode sides, respectively (Salzano et al., 1985).

$$V_{SOFEC,Nernst} = \frac{RT}{2F} \ln\left(\frac{P_{H_2,ca} P_{H_2O,an}}{P_{H_2O,ca} P_{H_2,an}} \right) \tag{2}$$

For the calculation of the actual voltage in the electrolysis, the actual voltage is greater than the equilibrium voltage due to the irreversible loss. The irreversible losses in electrolysis are activation polarization, ohmic polarization, and concentration polarization. The activation polarization is given by Butler-Volmer equation as: (Yahya et al., 2018)

$$\eta_{act,i} = \frac{RT}{2F} \ln\left[\frac{J}{2J_{0,i}} + \sqrt{\left(\frac{J}{2J_{0,i}}\right)^2 + 1} \right] \tag{3}$$

$$J_{0,i} = \frac{RT}{2F}\left(\frac{P_{H_2O}}{P_{ref}}\right)^{1.0433} \left(\frac{P_{H_2}}{P_{ref}}\right)^{0.9653} k_i \exp\left(-\frac{E_i}{RT}\right) \tag{4}$$

Here, i can be cathode or anode, and J and $J_{0,i}$ represent the current density and exchange current density, respectively. The ohmic polarization is given by: (Ferguson et al., 1996)

$$\eta_{ohm} = J\left(\frac{d_{an}}{\sigma_{an}} + \frac{d_{ele}}{\sigma_{ele}} + \frac{d_{ca}}{\sigma_{ca}} \right) \tag{6}$$

where d and σ are the thickness and conductivity, respectively. The concentration polarizations in the SOFEC anode and cathode can be calculated by

$$\eta_{conc,SOFEC} = \frac{RT}{2F} \ln\left(\frac{P_{H_2,ca}^{TPB} P_{H_2O,ca}}{P_{H_2,ca} P_{H_2O,ca}^{TPB}} \right) + \frac{RT}{2F} \ln\left(\frac{P_{H_2O,an}^{TPB} P_{H_2,an}}{P_{H_2O,an} P_{H_2,an}^{TPB}} \right) \tag{7}$$

The power input or output to the SOFEC stack is given in Eq.(8), and the current can be calculated according to Faraday's law, which means that for every mole of water vapor electrolyzed, two moles of electrons are generated

$$P_{SOFEC} = IV_{SOFEC} \tag{8}$$

$$I = 2F \times \dot{n}_{steam,consumed} \tag{9}$$

3. Integrated System process description

The process system created by Aspen Plus consists of two main subsystems, the biomass gasifier and the SOFEC. The SOFEC system can be divided into a cathode and an anode side. At the cathode side of the SOFEC, the electrolysis reaction of steam is carried out. Fresh feed water is pressurized and heated to generate steam through an evaporator. The electrolysis temperature of the SOFEC is set to 800°C. In order to avoid the breakage of the SOFEC structure and the formation of oxidizing conditions, the steam is heated to a higher temperature through the heat exchanger and mixed with the partial recirculation of cathode exit stream before entering the cathode. The cathode outlet stream contains hydrogen and unreacted steam, so the water is removed from the stream by condensation and the pure hydrogen is separated as the hydrogen product of the system.

The anode side is mainly used for fuel-assisted electrolysis reaction, where the fuel source is the syngas generated by the gasifier. Firstly, the solid biomass is converted into syngas through a gasifier. In order to increase the amount of hydrogen in the syngas and to avoid carbon deposition on the surface of the SOFEC anode, the syngas will be passed through the reformer for the steam reforming reaction before entering the SOFEC anode side. The reformed fuel is heated through heat exchange and then entering the anode for the O^{2-} reduction reaction, while the unreacted fuel at the anode exit will enter the afterburner for combustion. The burned streams contain a large amount of heat enaegy that can be used to preheat the inlet streams of the SOFEC system, and at the end the excess heat is used for heat recovery applications.

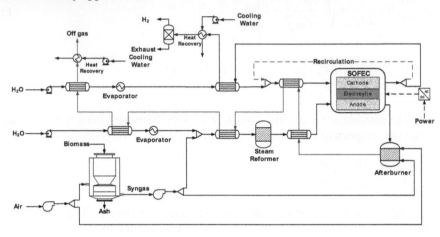

Figure 2 Gasifier-SOFEC integration system.

4. Results and Discussions

4.1. Effect of equivalence ratio on the system

Figure 3 shows the effect of changing the gasification equivalence ratio on the system efficiency. When the equivalence ratio is at 0.2, the total system efficiency is the highest at each operating steam utilization, and when the biomass gasifier equivalence ratio exceeds 0.2, the system efficiency decreases as the equivalence ratio increases because the flow rate of the syngas produced at an equivalence ratio of 0.2 is the highest, thus allowing the system to electrolyze more steam on the cathode side.

(a)   (b)   (c)

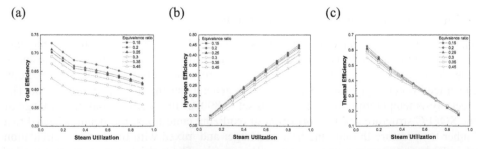

Figure 3 Effect of gasifier equivalent ratio on system efficiency (a)Total efficiency (b)Hydrogen production efficiency (c)Thermal efficiency.

The total efficiency decreases as the SOFEC steam utilization increases. The highest overall efficiency is 72.70% at a SOFEC steam utilization of 0.1 and a gasifier equivalence ratio of 0.2. As the steam utilization gradually increases, the hydrogen production efficiency gradually increases, while the thermal efficiency gradually decreases. At an equivalence ratio of 0.2 and an SOFEC steam utilization rate of 0.9, the maximum hydrogen production efficiency of the system is about 44.99%. Because the amount of syngas consumed in the electrochemical reaction is less when the SOFEC steam utilization is small, the amount of unused fuel at the anode outlet is higher and more fuel is burned in the afterburner, resulting in a large amount of heat that can be recovered at the outlet.

4.2. Effect of gasification temperature on the system

Figure 4 show shows the effect of gasification temperature on system efficiency, the maximum total efficiency occurs at the gasifier temperature of 900°C and SOFEC electrolysis rate of 0.1, which is about 68.67%, while the total efficiency is much lower than other operating temperatures when the gasifier is operated at 600°C. In the hydrogen production efficiency, it can be seen that when the temperature of the biomass gasifier is operated at 600°C, the efficiency decreases by about 10% compared to other temperatures. At other temperatures, the hydrogen production efficiencies of the three almost overlap. The main difference is in thermal efficiency because the carbon conversion rate of the gasifier is low at the gasification temperature of 600°C, resulting in less syngas production in the gasifier, which reduces the amount of fuel that can be assisted, making the amount of steam that can be electrolyzed in the cathode lower.

(a)　　　　　　　　　　　　(b)　　　　　　　　　　　　(c)

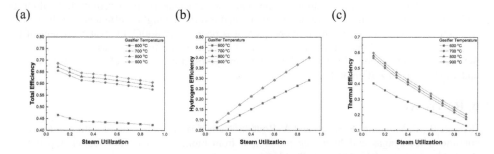

Figure 4 Effect of gasification temperature on system efficiency: (a) Total efficiency (b) Hydrogen production efficiency (c) Thermal efficiency.

4.3. Effect of different gasifier operating conditions on system efficiency

According to the previous analysis, changes in the operating parameters of the gasifier will affect the overall efficiency of the system. The higher the production of syngas in the gasifier, the higher the overall efficiency of the integrated system. By varying the moisture content of the biomass after drying, the highest efficiency of the gasifier occurs at moisture contents of 5%, 9% and 15% with equivalent ratios of 0.25, 0.2 and 0.15, respectively. From Figure 5, the maximum efficiency of the system at 800°C occurred at 15% biomass moisture content and 0.15 equivalents, with a maximum total efficiency of 75.42% for SOFEC operated at 0.1 steam utilization. The maximum hydrogen production efficiency was 46.97% at a steam utilization of 0.9. From the results, it can be confirmed that when the biomass gasifier operates at maximum efficiency, the integrated system also has the highest efficiency.

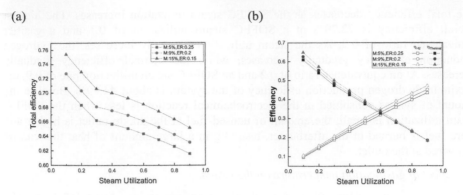

Figure 5 System efficiency under different gasification conditions: (a) Total efficiency (b) Hydrogen production and Thermal efficiency.

5. Conclusions

In the study, the effects of operating parameters of the gasifier on the overall efficiency and specific energy consumption of the gasifier-SOFEC integrated system were analyzed, and it can be learned that the key parameters of the gasifier are the equivalence ratio of the gasification agent and the gasification temperature, while the impact of the moisture content of the biomass drying on the system efficiency and specific energy consumption is not obvious. When the gasification temperature is higher than 700°C, the effect on system efficiency is reduced. When the syngas production from the gasifier is higher, the efficiency of the integrated system increases, mainly because the higher syngas production enables more water vapor to be electrolyzed, resulting in higher hydrogen production. Therefore, from the final results, the highest efficiency of the system was achieved when the gasification temperature was operated at 800C, the water content of the biomass was 15%, and the equivalence ratio was 0.15. The total efficiency was 75.42% under the SOFEC steam utilization of 0.1, and the maximum hydrogen production efficiency was 46.97% under the steam utilziation of 0.9.

6. Reference

Tavares, R., E. Monteiro, F. Tabet, and A. Rouboa, Numerical investigation of optimum operating conditions for syngas and hydrogen production from biomass gasification using Aspen Plus. Renewable Energy, 2020. **146**: p. 1309-1314.

Salzano, F.J., G. Skaperdas, and A. Mezzina, Water vapor electrolysis at high temperature: Systems considerations and benefits. International Journal of Hydrogen Energy, 1985. **10**(12): p. 801-809.

Yahya, A., D. Ferrero, H. Dhahri, P. Leone, K. Slimi, and M. Santarelli, Electrochemical performance of solid oxide fuel cell: Experimental study and calibrated model. Energy, 2018. **142**: p. 932-943.

Ferguson, J.R., J.M. Fiard, and R. Herbin, Three-dimensional numerical simulation for various geometries of solid oxide fuel cells. Journal of Power Sources, 1996. **58**(2): p. 109-122.

Proceedings of the 14th International Symposium on Process Systems Engineering -PSE 2021+
June 19-23, 2022, Kyoto, Japan ©2022, Elsevier B. V. All right reserved.
http://dx.doi.org/10.1016/B978-0-323-85159-6.50353-5

Plasma-Based Pyrolysis of Municipal Solid Plastic Waste for a Robust WTE Process

Hossam A.Gabbar*, Emmanuel Galiwango, Mustafa A. Aldeeb, Sharif Abu Darda, Kiran Mohammed

Faculty of Energy Systems and Nuclear Science, University of Ontario Institute of Technology, Oshawa, ON L1G0C5, Canada.
* *Hossam.Gaber@ontariotechu.ca*

ABSTRACT

The process design and material charactrization for a futuristic plasma-based pyrolysis/gasification for waste to energy is presented. The direct current as the source of thermal energy and the plasma-based reactor were designed. The plastic solid waste was analyze for pyrolytic, functional groups and morphological properties. Pyrolysis results using thermogravimetric analysis showed increase in thermal energy in the system with increasing heat rate and one major degradation peak was observed between 400-550 °C. There was no observed difference in functional groups between raw plastic solid waste and its ash. The structural morphologies of the plastic ash sample showed uniform rough surfaces. The physicochemical results of high volatile matter, low moisture and oxygen content revealved the ease of pyrolysis and postullate for quality products out of the futuristic plasma-based pyrolysis/gasification process.

Keywords: Municipal solid waste; Control; Operation; Characterization

1. Introduction

Solid wastes constitute a significant fraction of waste in the environment. The world generates 2.01 billion tonnes of municipal solid waste annually and this amount is expected to increase to 3.40 billion tonnes by 2050, and is double the population growth rate [1]. In addition, over at least 33 percent of that is extremely conservatively not managed in an environmentally safe manner [2]. Proper management and recycling of large amounts of solid waste are necessary to reduce its environment burdens and to minimize risks to human health [3, 4]. Sustainable and safe solid waste management is an under tapped field with great potential for energy production [5, 6]. Proper utilization of solid waste for conversion into energy and valuable products currently face various challenges such as heterogeneous nature of the waste, large moisture content, low calorific value, making it industrially undesirable [7-9]. Although appreciable amount of research in the field has been carried out on the conversion of solid waste to energy, there is still lack of comprehensive and efficient conversion methods in literature. Conventional conversion technologies of waste-to-energy (WTE) include biological (such as fermentation, anaerobic digestion and etc), thermochemical processes (such as incineration, pyrolysis, hydrothermal oxidation, and gasification) [10, 11]. These technologies enjoy several advantages such as biogas from biological process produce huge amounts of CH_4 and CO_2 with high energy value (1 m^3 of biogas was reported to be equivalent to 21 MJ of energy, and it could generate 2.04 kW h of electricity considering the 35% of generation efficiency [12]. Furthermore, Incinerators can reduce the volume of solid wastes up to 80–85%, and thus, significantly reduce the necessary volume for disposal. Also, biofuels from processes such as pyrolysis/gasification are positive indicators of good approach for WTE [13]. However, there setbacks in these technological advances such as pollution challenges from components like nitrogen, hydrogen sulfide and oxygen embedded in the waste [14, 15]. In addition, the heterogeneous nature of most solid wastes makes energy recovery, product yields and quality difficult [16]. With the improvement in air emission control systems and commitment from policy makes to enforce more strict environmental regulatory rules that significantly decreasing potential human health effects, more research on efficient processes for solid waste conversion to WTE and oil is gaining more attention.

The overall goal was to contribute to the solution of this challenge, we proposed the integration of DC/Microwave driven Inductively Coupled Atmospheric pressure thermal Plasma (RF/M-ICAP) to provide energy from municipal solid waste for transportation systems such as: trains, ships, garbage trucks etc. The proposed system includes new MICAP torch and reactor chamber design to produce highly efficient plasma jet suitable to process any waste type for different applications. MICAP, not only applied to reduce the volume of solid waste, but also to produce fuel from plastic and electricity through a process called pyrolysis and gasification. The suggested process has a potential to provide solution for waste-to-energy within transportation infrastructures, with reduced waste volume and increase clean energy generation while reducing emissions of trips resulting from extra transportation trips of waste transfer. Engineering designs are proposed for the target mobile waste-to-energy to be integrated with transportation infrastructures and develop discharge mechanisms of produced clean fuel and electricity in distributed stations based on demand profiles and generation capacities. This paper introduces novel design of modular chamber torches and waste characteristics with innovative features to maximize energy efficiency and minimize losses while improving the waste conversion to energy with minimum losses. Intelligent control systems will be developed to control plasma torch, chamber, and the associated energy systems.

2. Materials and Methods
2.1. Materials and chemicals
The shredded plastic MSW (average particle size <1mm) were obtained from industrial partner (Pro-flange, ON, Canada). The samples were air dried, separated according to their uniform compositions prior to the physicochemical characterization. The ultimate analysis was analyzed in external laboratory (Biotron Experimental Climate Change Research Centre at Western University in London, Ontario, Canada). The proximate analysis were determined on site in triplicates according to ASTM D3173 (inherent moisture content), ASTM D3174 (ash), ASTM D3175 (volatile matter), ASTM D3172-07a (Fixed carbon) and bulky density according to the reference method [17].

The torches and reactor designs were designed using solid works and COMSOL simualtor.
2.2. Waste characterization
2.2.1 Thermogravimetric analysis
The samples were subjected to thermogravimetric analysis (TGA) to investigate their pyrolytic properties of the waste material. In all experimental runs, 8±2 mg of the sample was placed on a platinum thermobalance crucible and loaded to the TGA analyser (Q50 series, TA instrument). The samples were equilibrated at 30 °C for 5 minutes before heating to 800 °C at heating rates of 10, 20, 30 and 40 °C/min under continuous inert N_2 flow at 20 mL/min. The results of thermal decomposition were continuously recorded as a change in weight as a function of temperature and time.

2.2.2 Fourier Transform Infra-red spectroscopy (FTIR) analysis
The ATR-FTIR spectroscopic analysis was performed to investigate the possible structural alteration between the samples before and after pyrolysis. Plastic samples before and after pyrolysis were pressed uniformly against a diamond surface by a fixed sample holder anvil, spectra were observed using a Bruker optics vertex system with inbuilt diamond-germanium ATR single reflection crystal. Spectra were obtained over a range of 400 and 4000 cm^{-1} with 34 average numbers of scans and spectral resolution of 4 cm^{-1}.

2.2.3 Scanning electron microscope (SEM) analysis

A FlexSEM1000 Scanning Electron Microscope (SEM) operated at 5 kV; spot size of 40 was used to image the char sample. To improve conductivity and quality of image, samples were coated with Au/C using a vacuum sputter coater.

2.3 Thermal source design

The torch design featured different major parts that aid in simulation to provide optimum conditions for the pyrolysis and gasification of the waste.

3. Results and Discussion

3.1 Waste characterization

The material characteristics are presented in Table 1. The low moisture content is important in lowering the energy requirements often needed to offset the inherent high moisture content in many MSW. This makes the waste under investigation a economically a good candidate for production of valuable products and electricity. In addition, the high volatile matter and low fixed carbon is an indication of the ease for combustion of this type of plastic waste, thus economical viability of the process. The ultimate analysis results show high hydrogen to carbon ratio (0.17) with low oxygen content, hence an anticipation of good quality products with less oxygenated fractionates.

Table 1: Material physicochemical charateristics

Proximate analysis (wt.%)				
Moisture content	Ash content	Volatile matter	Fixed carbon	Bulky density (Kg/m3)
0.13 ± 0.02	0.79 ± 0.01	86.8 ± 0.4	12.28 ± 0.1	84.43 ± 1.5
Ultimate analysis (wt.%)				
C	H	N	S	O
85.44	14.34	0.02	<MRL	0.22

MRL: method reporting limit

The TGA results in Figure 1 revealed the possible isothermal conversion (α) of the waste to be > 90 wt.% at all tested heating rates. Morever, the increase in the heating rate shifted the TA curves towards higher temeperature side, an indication of increase in thermal energy to the system with increasing heating rate. The differential thermograms showed one major degradation peak, a postullation of a single component under thermal pyrolysis. Pyrolysis/gasification of this category of waste is poised to be successful across different heating rates and product distribution will be easy to understand given the DTG degradation that suggests similarity in degradation chemistry of the material under pyrolysis.

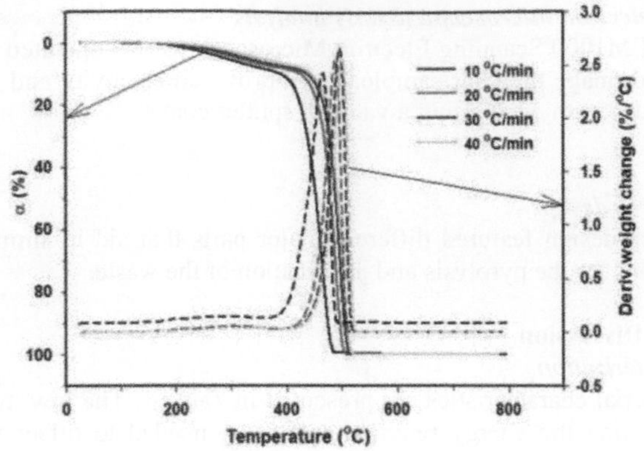

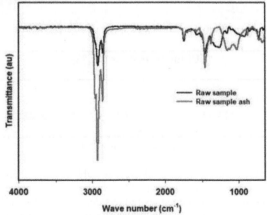

Figure 1: TGA and DTG curves for mixture of plastic waste at heating rates of 10, 20, 30 and 40 °C/min.

The study of the functional groups in Figure 2 shows characteristic similarity between the raw waste and its ash with minimal changes in the functional groups but rather more spectral intensity for the ash. These results prove the known phenomena about the material being a thermoplastic. The major vibrations between 2900-2800 cm^{-1} were assigned to the C-H and CH$_2$ bond vibrations characteristic to hydrocarbons present in most plastic waste. The absorption bands below 2000 cm^{-1} were associated with the stretching of the C=C, C-H, C-O, symetric and asymetric CH$_2$ bonds inherent in the aromatic ring of plastic polymer and the aliphatic structural makeup of the plastic [18, 19]. It is worthy to note that the absence of absorption at wave number around 3500 cm-1 a region characterstic of O-H bond vibration proves low existance of water molecule bonds as already seen from the low moisture content measurements.

Figure 2: FTIR measurements of raw plastic sample and its ash after pyroysis at 800 °C.

The morphological studies following pyrolysis of the plastic MSW samples in Figure 3 revealed agglomerates with majorly rough surfaces and a few smooth surfaces observed X5 and X10 magnification. The micrographs depicts a good thermal interaction with the plastic waste's polymer matrix.

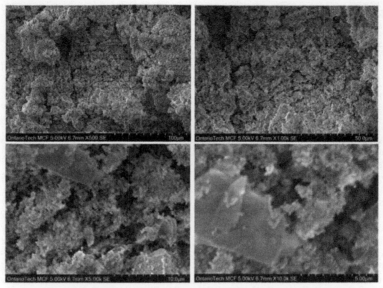

Figure 3: Plastic ash SEM images at different magnifications.

3.1 Thermal source design

Figure 4 represents the design for the DC torch and the plasma-based reactor for the futuristic process for production of renewable environmentally safe energy from waste.

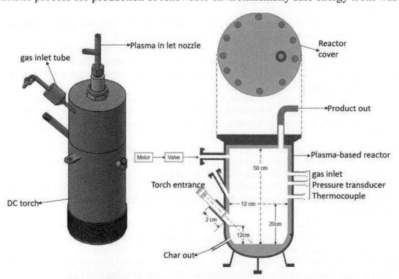

Figure 4: DC torch and plasma-based reactor drawings.

4. CONCLUSION

The conceptual design and charaterization of the plastic municipal solid waste provides a deeper understanding and new approach to waste management in an environmentally and economically friendly manner. The low moisture content and ash content combined with high voltility of the waste and H-C ratio signifies a potential ease of the futuristic plasma-based pyrolysis/gasification process both in terms of energy requirements and reactor configuration.

Acknowledgements: Thanks to lab members of Advanced Plasma Engineering Lab (APEL). Authors would like to thank Proflange Ltd. for their funding and continuous support. The research is also supported by NSERC, OCI, and Mitacs.

References

[1] T.W. Bank, Trends in Solid Waste Management, in, 2016.

[2] M.A. Nwachukwu, M. Ronald, H. Feng, Global capacity, potentials and trends of solid waste research and management, Waste Management & Research, 35 (2017) 923-934.

[3] H. Yang, M. Ma, J.R. Thompson, R.J. Flower, Waste management, informal recycling, environmental pollution and public health, J Epidemiol Community Health, 72 (2018) 237-243.

[4] P. Alam, K. Ahmade, Impact of solid waste on health and the environment, International Journal of Sustainable Development and Green Economics (IJSDGE), 2 (2013) 165-168.

[5] S.A.T. Muawad, A.A.M. Omara, Waste to energy as an alternative energy source and waste management solution, in, IEEE, pp. 1-6.

[6] P. Prajapati, S. Varjani, R.R. Singhania, A.K. Patel, M.K. Awasthi, R. Sindhu, Z. Zhang, P. Binod, S.K. Awasthi, P. Chaturvedi, Critical review on technological advancements for effective waste management of municipal solid waste-Updates and way forward, Environmental Technology & Innovation, (2021) 101749.

[7] S. Das, S.H. Lee, P. Kumar, K.-H. Kim, S.S. Lee, S.S. Bhattacharya, Solid waste management: Scope and the challenge of sustainability, Journal of cleaner production, 228 (2019) 658-678.

[8] J. Ali, T. Rasheed, M. Afreen, M.T. Anwar, Z. Nawaz, H. Anwar, K. Rizwan, Modalities for conversion of waste to energy—challenges and perspectives, Science of The Total Environment, 727 (2020) 138610.

[9] K. Moustakas, M. Loizidou, M. Rehan, A.S. Nizami, A review of recent developments in renewable and sustainable energy systems: Key challenges and future perspective, in, Elsevier, 2020.

[10] N. Tulebayeva, D. Yergobek, G. Pestunova, A. Mottaeva, Z. Sapakova, Green economy: waste management and recycling methods, in, EDP Sciences, pp. 01012.

[11] D. Ghosal, R. Kaur, D.A. Jadhav, Utilization and Management of Waste Derived Material for Sustainable Energy Production: A mini review, Academia Letters, (2021).

[12] J.D. Murphy, E. McKeogh, G. Kiely, Technical/economic/environmental analysis of biogas utilisation, Applied Energy, 77 (2004) 407-427.

[13] T. Rasheed, M.T. Anwar, N. Ahmad, F. Sher, S.U.-D. Khan, A. Ahmad, R. Khan, I. Wazeer, Valorisation and emerging perspective of biomass based waste-to-energy technologies and their socio-environmental impact: A review, Journal of Environmental Management, 287 (2021) 112257.

[14] M.S. Qureshi, A. Oasmaa, H. Pihkola, I. Deviatkin, A. Tenhunen, J. Mannila, H. Minkkinen, M. Pohjakallio, J. Laine-Ylijoki, Pyrolysis of plastic waste: Opportunities and challenges, Journal of Analytical and Applied Pyrolysis, 152 (2020) 104804.

[15] J. Lui, W.-H. Chen, D.C.W. Tsang, S. You, A critical review on the principles, applications, and challenges of waste-to-hydrogen technologies, Renewable and Sustainable Energy Reviews, 134 (2020) 110365.

[16] M. Singh, S.A. Salaudeen, B.H. Gilroyed, S.M. Al-Salem, A. Dutta, A review on co-pyrolysis of biomass with plastics and tires: recent progress, catalyst development, and scaling up potential, Biomass Conversion and Biorefinery, (2021) 1-25.

[17] I. Obernberger, G. Thek, Physical characterisation and chemical composition of densified biomass fuels with regard to their combustion behaviour, Biomass and bioenergy, 27 (2004) 653-669.

[18] R.C. Asensio, M.S.A. Moya, J.M. de la Roja, M. Gómez, Analytical characterization of polymers used in conservation and restoration by ATR-FTIR spectroscopy, Analytical and bioanalytical chemistry, 395 (2009) 2081-2096.

[19] I. Noda, A.E. Dowrey, J.L. Haynes, C. Marcott, Group frequency assignments for major infrared bands observed in common synthetic polymers, in: Physical properties of polymers handbook, Springer, 2007, pp. 395-406.

Proceedings of the 14th International Symposium on Process Systems Engineering – PSE 2021+
June 19-23, 2022, Kyoto, Japan © 2022 Elsevier B.V. All rights reserved.
http://dx.doi.org/10.1016/B978-0-323-85159-6.50354-7

Hybrid Modelling Strategies for Continuous Pharmaceutical Manufacturing within Digital Twin Framework

Pooja Bhalode[a#], Yingjie Chen[b#], Marianthi Ierapetritou[b,*]

[a]*Department of Chemical and Biochemical Engineering, Rutgers University, 98 Brett Road, Piscataway, NJ 08854, USA*
[b]*Department of Chemical and Biomolecular Engineering, 150 Academy Street, Newark, DE 19716, USA*
[*]*Corresponding author – mgi@udel.edu*
[#]*These authors contributed equally to this work*

Abstract

The application of Industry 4.0 related technologies has enabled the development of digital twins to model and mirror the physical systems in a virtual construct. Despite the successful adoption of digital twins in other industries, a complete digital twin framework is yet to be developed for continuous pharmaceutical manufacturing processes. Challenges related to the specific industry include the integration of multi-scale information ranging from powder properties to process flowsheets, and the need of model adaptability to capture changes in operation. In this work, hybrid multizonal compartment models and hybrid adaptive models are developed to address such challenges. The computationally efficient and self-adaptive hybrid modelling strategies can aid the developing of digital twin for continuous pharmaceutical manufacturing.

Keywords: Hybrid modelling; Digital twin; Pharma 4.0; Continuous pharmaceutical manufacturing; multi-zonal compartment modelling.

1. Introduction

Industry 4.0 revolution has been catalysed by the advances in digitization, cyber-infrastructure, artificial intelligence, and Internet of Things. In pharmaceutical industry, this digitalization move is popularly named as Pharma 4.0, and it is incentivized by increasing market competition, along with the encouragement from regulatory agencies to develop agile, flexible, and robust manufacturing lines (O'Connor et al., 2016, Chen et al., 2020). Pharma 4.0 technologies enable the development of a digital twin, which can be defined as an integrated digitized framework consisting of virtual and physical components, with a seamless connection between the two (Chen et al., 2020). In continuous pharmaceutical manufacturing, although some efforts have been made to create data integration framework and various types of models, a fully integrated digital twin that allows for real-time process monitoring, control, and optimization is still in its infancy. From a modelling perspective, multi-scale models that can characterize powder properties and describe process flows need to be integrated, which is a challenging task. Furthermore, models developed are required to adapt to process changes, instead of being static. These challenges are bottlenecks in development of a digital twin for continuous pharmaceutical manufacturing.

We propose the use of hybrid modelling strategies to address some of these issues. In general, hybrid models can be considered as a combination of data-driven and white-box (mechanistic) models (Zendehboudi et al., 2018). The modelling components can be arranged in a serial, parallel, or combined manner (Chen and Ierapetritou, 2020). This modelling strategy provides a unique way to combine known mechanistic knowledge with data. In this work, a novel two-stage solution approach based on hybrid modelling strategies is proposed. The first stage addresses the challenge of multi-scale models by utilizing hybrid multi-zonal compartmentalization method, which is discussed in Section 2. The second stage focuses on the hybrid adaptive model framework, as discussed in Section 3. The two stages, when integrated, would be a central part of a computationally efficient and self-adaptive digital twin for continuous pharmaceutical manufacturing.

2. Hybrid Multi-zonal Compartmentalization

Multi-zonal compartmentalization methodology is a computationally efficient approach used to model complex systems (Jourdan et al., 2019). Compartmentalization methodology aims at combining systemic (unit operation) and local (particle) level information using multi-scale models and has been applied for complex process operations, such as biologics (Delafosse et al., 2014) Here, the systemic information obtained from mechanistic models is combined with local process information obtained from high-fidelity simulations such as computational fluid dynamics (CFD) or discrete element modelling (DEM). The detailed local information is compressed by dividing the simulation domain into compartments based on the process variables of interest, such that each compartment has process variables within user-defined limits based on the degree of scrutiny required for modelling the system.

In this paper, compartmentalization methodology is used to develop a predictive model for continuous powder blender (Bhalode and Ierapetritou, 2021). The process models developed in literature for blenders lack understanding and prediction of powder mixing within the systems and systemic models do not capture this level of detail (Vanarase and Muzzio, 2011). Important for assuring drug product quality, it is crucial to develop predictive models that incorporate local information concerning powder mixing (Lee et al., 2015). The details of the developed compartment model for continuous powder blender are described in Figure 1 where, the blender is broken down into periodic sections and a section is simulated using DEM. To demonstrate the proposed strategy, two types of spherical particles with 1 mm particle size are added to the section, to ensure reasonable computational times of DEM simulation, and can be easily extended for more particle types. The simulation is performed with an Intel Xeon E5-2650 v4 2.2GHz processor and 128 GB RAM using EDEM 2021 (Altair Solutions, Michigan, USA), leading to a total computation time of 8 hrs with 4 CPU cores. The simulated section is then divided into 10 equal slices along the X axis for ease of computation and each slice is further divided into 20x20 grids along Y and Z axes for post-processing. Average particle velocities along all directions (v) are extracted for all grids, and the grids are compartmentalized into zones where v is positive, negative, zero, and zone with no particles. Using these zones, radial compartment maps are developed for all slices, and these maps are overlaid on top of each other (Bhalode and Ierapetritou 2021). Using the overlaid radial map, interconnection flowrates are determined based on the interfacial area and the velocities of the respective compartments, converting the compartment map into an inter-connected network of compartments. A similar approach is adopted along the axial direction to develop network of compartments based on interconnection flow rates between the compartments. The axial network is eventually combined with the radial to obtain the overall the multizonal compartment model for the periodic section. This model thus incorporates systemic as well as local level information, allowing detailed assessment of powder flow.

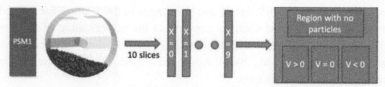

Figure 1: Zones developed along axial and radial direction for the periodic section

Following the model development, validation is performed along both radial and axial directions. For validation along the radial direction, the degree of powder mixing is compared between the compartment model and the DEM simulation. For comparison, a mixing index – relative standard deviation (RSD) is used to obtain the time at which the system has reached 95% of the total mixing in the system (T_{95}). T_{95} is evaluated for different RPMs and shown in Table 1 along with the absolute relative error. Validation along the axial direction is performed by comparing the axial flow using residence time distribution (RTD) profiles, as shown in Figure 2, where similar RTD profiles are obtained using DEM and compartment models. Thus, compartment model provides accurate prediction of powder flow compared to the computationally intensive DEM simulation, along with significant time savings (in order of 1-2 mins) Lastly, the compartment model can be extended to model the entire length of continuous blender by connecting periodic sections, as shown in Figure 3.

Table 1. T_{95} mixing times for radial validation of DEM compared to Compartment model

Blade speed	T_{95} (DEM)	T_{95} (Compartment model)	Absolute relative error
50 RPM	14.812	14.548	0.018
75 RPM	12.366	12.129	0.019
100 RPM	8.911	9.268	0.040

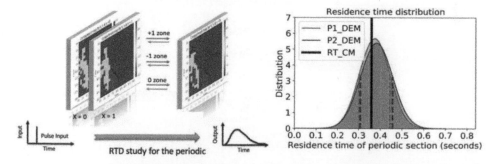

Figure 2. Axial validation performed using residence time distribution profiles.

3. Hybrid Adaptive Modelling

As mentioned in Section 1, the development of hybrid models involves the combination of data-driven and mechanistic models. Given the structure of hybrid models, some parts are data-driven and typically trained on defined sets of historical data. Therefore, the resulting hybrid models are often time-invariant and can only reflect the system for the range of operational conditions, environmental variables, equipment status, and material properties covered in the training set (Gama et al., 2014), leading to a challenge related to model updates. Although the hybrid models may accurately describe the current state of a system, the process can shift slowly due to changes in the features mentioned above. These informed or uninformed changes can impact the performance of a process and, eventually, the critical quality attributes of a product leading to decline of model prediction accuracy (Gama et al., 2014). Appropriate model update strategies thus need to be established to resolve this issue.

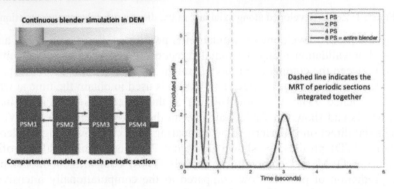

Figure 3. Compartment model extended for the entire continuous powder blender

The proposed framework for hybrid adaptive modelling is shown in Figure 4. Using historical data with adequate model selection criteria, an initial data-driven sub-model can be built and integrated into an appropriate hybrid model structure. Prediction results from the hybrid model are compared with plant outputs in a continuous manner. Based on defined criteria, the adaptive algorithm determines if an update on the hybrid model is required, and carries out the necessary steps to retrain the model. The overall computational time depends on the selection of model training and adaptive algorithms.

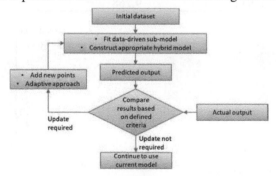

Figure 4. Hybrid adaptive modelling framework.

Two adaptive modelling algorithms are applied. The moving window method with fixed window size, as depicted in Figure 5(a), refers to a blind adaptation technique where the model is adapted using a fixed number of most recent data points, regardless if a change is observed (Gama et al., 2014). As new samples are streamed in, the window slides to include the newest samples and forgets the older ones. This strategy also implies that model update is always required. The adaptive windowing method, as shown in Figure 5(b), is an informed adaptation technique (Gama et al., 2014), which includes a change detection component and an adaptive algorithm. The number of data points included in a window grows until a significant change is detected. The detection algorithm splits the window into two sub-windows in all possible permutations and compares the means. When the difference of two means is larger than a defined tolerance, a process change is detected, and the points before such change (i.e., the ones in the first sub-window) are discarded. Both approaches can be integrated into different data-driven modelling methodologies to enable model updates, but the adaptive windowing can be more computationally expensive because of the change detection algorithm.

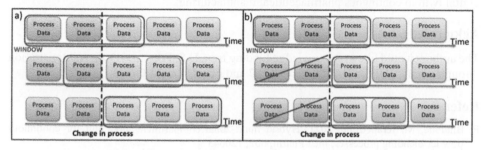

Figure 5. Schematic of (a) moving window with fixed window size, and (b) the adaptive windowing method.

Both algorithms are tested on a continuous pharmaceutical manufacturing line via direct compaction route. Such process is explained in (Wang et al., 2017). The baseline models are neural networks with the combined structure, detailed in (Chen and Ierapetritou 2020).

A case study of sudden and patterned process change is developed, for which the hybrid adaptive modelling framework experiences an uninformed change of excipient bulk density from the nominal 400 kg/m^3 to 520 kg/m^3 (and back for the patterned change), with other variables remaining unaffected. The impacts of the change onto tablet weight are monitored, and the results are shown in Figure 6.

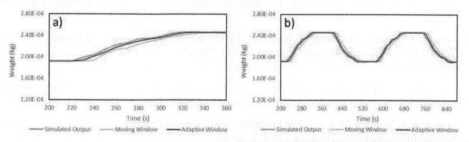

Figure 6. Comparison of results for between simulated output, hybrid adaptive model predictions based on moving window and adaptive windowing for (a) sudden, and (b) patterned change.

The hybrid adaptive models can follow the changes in process, and the adaptive windowing technique outperforms moving window. This observation is expected as the change detection algorithm in adaptive windowing makes it very sensitive to this type of sudden changes, resulting in a faster response.

4. Conclusions

The proposed article focuses on utilizing hybrid modelling strategies to address the challenges associated with developing a digital twin for pharmaceutical manufacturing. Hybrid multi-zonal compartment models are developed to integrate multi-scale information concerning powder flow properties and operation mechanics within the developed process model, demonstrated for continuous powder blender. Hybrid adaptive modelling framework is developed to facilitate model updates with process changes or newly acquired data, showcased with a case of mid-process change. The resulting hybrid and adaptive model can be used as an online predictive model in a fully integrated digital twin. The two strategies can be further integrated together to move towards an integrated digital twin in continuous pharmaceutical manufacturing.

Acknowledgement

The authors would like to acknowledge funding from U.S. Food and Drug Administration through grants DHHS-FDA-U01FD006487 and FDABAA-20-00123.

References

P. Bhalode, M. Ierapetritou, 2021, Hybrid multi-zonal compartment modeling for continuous powder blending processes, Int J Pharm, 602, 120643

Y. Chen, M. Ierapetritou, 2020, A framework of hybrid model development with identification of plant-model mismatch, AIChE Jounal, 66, 10, e16996.

Y. Chen, O. Yang, C. Sampat, P. Bhalode, R. Ramachandran, M. Ierapetritou, 2020, Digital Twins in Pharmaceutical and Biopharmaceutical Manufacturing: A Literature Review, Process, 8, 9.

A. Delafosse, M. Collignon, S. Calvo, F. Delvigne, M. Crine, P. Thonart, D. Toye, 2014, CFD-based compartment model for description of mixing in bioreactors, Chemical Engineering Science, 106, 76-85.

J. Gama, I. Zliobaite, A. Bifet, M. Pechenizkiy, A. Bouchachia, 2014, A survey on concept drift adaptation, ACM Computing Surveys, 46, 4, 1-37.

N. Jourdan, T. Neveux, O. Potier, M. Kanniche, J. Wicks, I. Nopens, U. Rehman, Y. Le Moullec, 2019, Compartmental Modelling in chemical engineering: A critical review, Chemical Engineering Science, 210, 115196.

S. Lee, T. O'Connor, X. Yang, C. Cruz, S. Chatterjee, R. Madurawe, C. Moore, L. Yu, J. Woodcock, 2015, Modernizing Pharmaceutical Manufacturing: from Batch to Continuous Production, Journal of Pharmaceutical Innovation, 10, 3, 191-199.

T. O'Connor, L. Yu, S. Lee, 2016, Emerging technology: A key enabler for modernizing pharmaceutical manufacturing and advancing product quality, Int J Pharm, 509, 1-2, 492-498.

A. Vanarase, F. Muzzio, 2011, Effect of operating conditions and design parameters in a continuous powder mixer, Powder Technology, 208, 1, 26-36.

Z. Wang, E. Sebastian, M. Ierapetritou, 2017, Process analysis and optimization of continuous pharmaceutical manufacturing using flowsheet models, Copmuters & Chemical Engineering, 107, 77-91.

S. Zendehboudi, N. Rezaei, A. Lohi, 2018, Applications of hybrid models in chemical, petroleum, and energy systems: A systematic review, Applied Energy, 228, 2539-2566.

Proceedings of the 14th International Symposium on Process Systems Engineering – PSE 2021+
June 19-23, 2022, Kyoto, Japan © 2022 Elsevier B.V. All rights reserved.
http://dx.doi.org/10.1016/B978-0-323-85159-6.50355-9

Determination of probabilistic design spaces in the hybrid manufacture of an active pharmaceutical ingredient using PharmaPy

Daniel Laky[a], Daniel Casas-Orozco[a], Francesco Rossi[a], Jaron S. Mackey[a], Gintaras V. Reklaitis[a], Zoltan K. Nagy[a]*

aDavidson School of Chemical Engineering, Purdue University, West Lafayette, IN, 47907, USA
Corresponding author's e-mail: zknagy@purdue.edu

Abstract

The pharmaceutical industry has seen more interest in shifting from traditional fully batch operation to continuous manufacturing. With these shifts in mind and the adoption of industry 4.0 standards, digital tools are required to ensure critical medicines can be manufactured with quality guarantees. PharmaPy is one such tool that can create a digital twin of a pharmaceutical manufacturing process and enable the digital design of optimized manufacturing routes. This new tool has particular strengths in modelling batch, continuous and hybrid manufacturing systems. PharmaPy is shown throughout this work to be capable of digitally addressing quality-by-design (QbD) through design space identification for processes containing a variety of operating modes.

Keywords: design space, pharmaceutical manufacturing, process systems engineering, hybrid processing

1. Introduction

The development and transition of pharmaceutical manufacturing from predominantly batch processing to end-to-end continuous or hybrid operation mode has been prompted by benefits such as operational robustness, consistency of product quality and, potentially, lower environmental impact (Içten et al. 2020). However, these benefits typically are not trivial to identify during the process design phase, especially for hybrid, dynamic process systems.

Traditionally, a process operating region must be identified in which process inputs correspond to outputs that meet product quality constraints, also known as the design space. In pharmaceutical manufacturing, the Food and Drug Administration (FDA) launched the now well-known quality-by-design (QbD) paradigm (FDA, 2004), promoting process development of optimal manufacturing pathways through design space analysis with process robustness and quality assurance in mind.

With the widespread adoption of the QbD approach, a numerical tool that handles the simulation of end-to-end batch, hybrid, or end-to-end continuous systems, as well as laying the groundwork for in silico design space identification and other process system analysis, is desirable. Recently, we have developed a simulation platform, PharmaPy (Casas-Orozco et al. 2021a), structured as an object-oriented tool based on a set of robust declarative representations (Marquardt 1992), which offers a rich suite of first-principles, dynamic models for drug substance manufacturing. In addition to the process model library and unit operation simulation capabilities, PharmaPy allows the creation of digital twins by offering both parameter estimation and statistical analysis capabilities for uncertainty quantification and propagation.

The numerical capabilities of PharmaPy regarding process simulation, parameter estimation and uncertainty quantification make the tool useful in determining the design space for a fixed pharmaceutical flowsheet or set of candidate flowsheets, thus facilitating the analysis of the combined effect of operational variables (flowrates, inlet compositions, etc.) and parametric uncertainty (e.g., kinetics and transport phenomena) (Bano et al. 2019; Laky et al. 2019). The evaluation of the design space under uncertainty makes the mapping of feasible operating regions more comprehensive and provides a valuable tool for process validation from a regulatory perspective.

In this work, a two-reactor process flowsheet for the synthesis of a low volume/high value active pharmaceutical ingredient (API) is presented as a case study to demonstrate the simulation/design-space generation framework. Different flowsheets resulting from the combination of reactor operating modes (batch and continuous) for a pilot scale production plant are evaluated in terms of waste generation (unreacted reagents) and API productivity. To account for uncertainty, parameter estimation was performed on one reaction system using PharmaPy to identify variance/covariance information regarding model parameters for reaction kinetics. Using this model uncertainty, probabilistic design spaces, representing feasible regions for process operation, are generated via adaptive sampling combined with appropriate uncertainty propagation strategies. Specifically, the probabilistic design space for a continuous process, a plug flow reactor (PFR) followed by a continuous stirred tank reactor (CSTR), is compared with that from a semibatch reactor to demonstrate the benefits in terms of probability of adhering to critical quality attributes of an API reaction product mixture.

2. Methodology

To perform probabilistic design space analysis, model parameters must have a tangible uncertainty representation to infer confidence that some chosen operating point will guarantee that all critical quality attributes (CQAs) are maintained. For this purpose, PharmaPy was used to fit experimental spectral data in a sequential manner. First, a multivariate, Principal Component Regression (PCR) calibration model was gathered to compute species concentrations during experiments. Real-time IR spectra were then gathered during several experiments at varying temperatures and molar ratios between reactants, which allowed to capture temperature dependence of the reaction parameters for a set of representative molar concentrations. Finally, experimental data were bootstrapped (Chernick, 2008) to generate the nominal values and parametric uncertainty of those reaction kinetics parameters. The parameter estimation results were first shown in Casas-Orozco et. al. (2021b).

Uncertainty is a key component of generating probabilistic design spaces. Generation of the probabilistic design space is as follows. First, the desired or proposed continuous operating region is transformed into a discrete counterpart by dividing each operating parameter range into equally-spaced increments, generating a finite number of operating points. Then, for each operating point, the model parameters are sampled 100 times. For each of these random samples, the PharmaPy model of the process is simulated.

Once the simulation has finished, the resulting state variables can be used to evaluate the set of CQAs. If all CQAs are met, (i.e. all critical quality constraints are less than zero), this sample is marked as successful. The number of those samples that are successful at a given operating point are tallied and once those 100 samples finish, the probability that the current operating point provides output at acceptable quality is the number of successful samples divided by the number of samples run. Performing this analysis for the complete set of discretized operating conditions and interpolating the results provides a digital representation of the probabilistic design space of the given process. After analysis is complete, review of the probabilistic design space can lead to operating conditions that guarantee quality with a quantitative confidence.

3. Case Study

For this analysis, in-house data on the synthesis of Lomustine was used. Lomustine is an API used to treat brain cancer and has been a compound of interest while developing an end-to-end optimal pharmaceutical modelling framework using PharmaPy. Lomustine synthesis is generated through two synthesis steps:

$$CHA + ISOCN \xrightarrow{k_1} INT \tag{1}$$

$$INT + TBN \xrightarrow{k_2} Lom + TBOH \tag{2}$$

Where CHA is cyclohexylamine, ISOCN is 1-Chloro-2-isocyanate, INT is the lomustine intermediate 1-(2-Chloroethyl)-3-cyclohexylurea, TBN is tert-Butyl nitrite, Lom is Lomustine, and TBOH is tert-Butyl alcohol. For this work, the form of reaction kinetics that were fit followed the standard Arrhenius rate law shown in Eq. (3).

$$k_i = A_i exp\left(\frac{-E_{a,i}}{R} * \left(\frac{1}{T}\right)\right) \tag{3}$$

Experiments have shown that the relative speed which reaction 1 occurs is sufficiently fast and thus can be modelled accordingly by setting low activation energy $E_{a,1}$ and high preexponential factor A_1. Parameters for the second reaction were estimated for varying temperatures as previously mentioned. The nominal parameter values for preexponential factors A_i and activation energies $E_{a,i}$ are shown below in Eq. (4).

$$\{A_1, A_2\} = \left\{2.1 \frac{L}{mol*s}, 1.877 * 10^7 \frac{L}{mol*s}\right\} ; \{E_{a,1}, E_{a,2}\} = \tag{4}$$
$$\left\{2.0 \frac{kJ}{mol}, 52.52 \frac{kJ}{mol}\right\}$$

The reaction 2 parameters were transformed to exhibit improved numerical behaviour from the standard Arrhenius rate law such that the log of a reference preexponential factor at temperature T_{ref} and activation energy were fit from experimental data. This transformation is shown below in Eq. (5).

$$ln(k_2) = ln(A_{2,ref}) - \frac{E_{a,2}}{R}\left(\frac{1}{T} - \frac{1}{T_{ref}}\right) \tag{5}$$

Using this transformation, uncertainty in parameters $A_{2,ref}$ and $E_{a,2}$ were evaluated through bootstrapping and are shown as covariance matrix Σ below in Eq. (6).

$$\Sigma = \begin{bmatrix} 9.797 * 10^{-5} & -1.176 * 10^{-3} \\ -1.176 * 10^{-3} & 4.839 * 10^{-1} \end{bmatrix} \tag{6}$$

Using these reaction kinetics, two processing routes were explored, described below in Figure 1: Schematic of both process alternatives for the productions of Lomustine. On the left, the continuous flowsheet with two unit operations. On the right, the one-pot semibatch reaction. The first is a continuous reaction train using a PFR followed by a CSTR. Under normal operation, optimal design for the continuous reactors is a PFR volume of approximately 695 mL with a flowrate of 16.66 L/h (Laky et al., 2021). The CSTR volume is set at 20 L to ensure high residence time to achieve adequate conversion of intermediate to Lomustine. A schematic of the continuous process is shown below.

Also shown is the second processing route, a semibatch reactor. Here an initial loading of CHA and ISOCN are set to react for a small amount of time (approximately 30 minutes). Then, pure TBN is added over the course of 15 minutes. The system continues reacting until a total batch time of 4 hours is reached. It is known that Lomustine is unstable above 40 °C. Thus, all reactors begin at 25 °C and are heated to 35 °C to increase reaction rate while mitigating thermal degradation of the API. Throughout this work, both processes were modelled using PharmaPy.

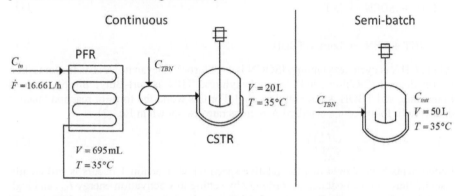

Figure 1: Schematic of both process alternatives for the productions of Lomustine. On the left, the continuous flowsheet with two unit operations. On the right, the one-pot semibatch reaction.

There were two operating parameters explored in this work. First is the inlet concentration or initial concentration of CHA and ISOCN in the first reactor, shown as C_{in} in Figure 1: Schematic of both process alternatives for the productions of Lomustine. On the left, the continuous flowsheet with two unit operations. On the right, the one-pot semibatch reaction. Second is the concentration of TBN for the second reaction step, shown as C_{TBN}. For this study, the concentration of CHA and ISOCN were fed or initialized in an equimolar ratio. Inlet concentration C_{in} has bounds from 0.04 M to 0.08 M and C_{TBN} from 0.06 M to 0.12 M. Both operating parameters were discretized with 11 points for a total of 121 total discrete operating points over the design space.

Purification of Lomustine using crystallization is the next processing step. With this in mind, CQAs can be defined as below:

$$C_{Lom} - \alpha C_{Lom,Sat} \leq 0 \tag{7}$$

$$C_{CHA} + C_{ISOCN} + C_{INT} + C_{TBN} - C_{Lom} \leq 0 \tag{8}$$

$$4.0\, C_{INT} - C_{Lom} \leq 0 \tag{9}$$

Eq. (7), ensures that no premature crystallization of Lomustine occurs. Eq. (8), ensures that all reagents are sufficiently converted to Lomustine. Eq. (9), ensures that the intermediate, which is known to crystallize, adequately converts to Lomustine.

4. Results

Using the processes and discretization schemes described above, 100 samples at each operating point were taken in the model parameter space, and the probabilistic design spaces were generated. These resulting design spaces are shown below in Figure 2: Contour plot of the probabilistic design space generated for both operating modes. Contours represented by confidence that an operating point adheres to CQAs as a fraction of 1.0 where lighter tones represent higher confidence of feasibility..

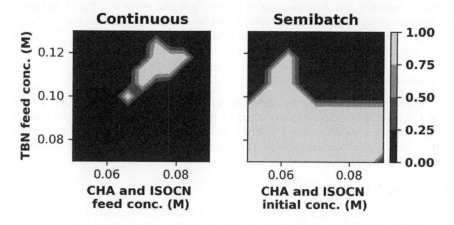

Figure 2: Contour plot of the probabilistic design space generated for both operating modes. Contours represented by confidence that an operating point adheres to CQAs as a fraction of 1.0 where lighter tones represent higher confidence of feasibility.

Generation of the design space for the continuous process took 243 minutes, whereas the semibatch design space took about 46 minutes (Macbook Pro 2016, 2.6 GHz Quad-Core i7 processor, 16GB 2133 MHz LPDDR3 RAM). The difference in simulation time for the continuous case results from the numerical solution of PDE models versus ODE models in the semibatch case. From the figure, the semibatch operating mode provides a much wider range of feasible operation than the continuous case. Nearly half of the explored operating space results in feasible operation, whereas only a small region is feasible in the continuous case.

The discrepancy in probabilistic design space between the two designs highlights the advantage of the high conversion of batch systems. In the continuous case, limitations on the residence time and volume of the CSTR lead to lower overall conversion. The integration of two unit operations with their individual design spaces, also leads to a smaller overall feasible design space for the integrated process. If lower flowrates were possible, the residence time of the CSTR could increase and lead a larger feasible operating region. The results indicate that while continuous integrated processes present numerous advantages compared to batch or semibatch systems, the overall robust design space for these tend to be smaller, and the implementation of suitable control systems can be critical to keep the process within the robust operating space.

5. Conclusions

In this work, PharmaPy was used to generate probabilistic design spaces of API synthesis routes using statistical sampling of experimentally determined uncertainty in the relevant model parameters. PharmaPy has strengths in easily modelling varying operational modes, from fully batch, to hybrid, to fully continuous operation. Using these strengths, the synthesis of a cancer drug API, Lomustine, was analysed for both continuous and semibatch operation. It was found that in the case of batch processing, when reaction time is long enough, conversion of raw materials to the API allows for a larger feasible operating space than in the continuous case. Also, these design spaces consider no control action, and likewise the feasibility region of the continuous manufacturing process will necessarily increase when control action is implemented.

Overall, PharmaPy is an emerging tool in leveraging digital twins and digital analysis in the pharmaceutical manufacturing space. It can be seamlessly called in many relevant digital process analysis techniques, such as those facilitating QbD and QbC. This is just one example highlighting these capabilities.

6. Acknowledgement

This project was supported by the United States Food and Drug Administration through grant U01FD006738. The views expressed by the authors do not necessarily reflect the official policies of the Department of Health and Human Services; nor does any mention of trade names, commercial practices, or organization imply endorsement by the United States Government.

References

G. Bano, P. Facco, M. Ierapetritou, F. Bezzo, and M. Barolo. 2019, Design Space Maintenance by Online Model Adaptation in Pharmaceutical Manufacturing, Computers and Chemical Engineering, 127, 254–71.

D. Casas-Orozco, D. Laky, V. Wang, M. Abdi, X. Feng, E. Wood, C. Laird, G. V. Reklaitis, and Z. K. Nagy, 2021a, PharmaPy: An object-oriented tool for the development of hybrid pharmaceutical flowsheets, Computers and Chemical Engineering, 153

D. Casas-Orozco, D. Laky, J. Mackey, A. Mufti, G. V. Reklaitis, and Z. Nagy, 2021b, Chemometric techniques for the combined calibration/parameter estimation of pharmaceutical drug substance manufacture, AIChE Annual Meeting 2021, Boston, MA

M.R. Chernick, 2008. Bootstrap Methods: A Guide for Practitioners and Researchers, 2nd ed. Wiley - Interscience, New Jersey.

FDA 2004, Pharmaceutical CGMPs for the 21s Century-A Risk-Based Approach, Rockville, MD

E. Içten, A. J. Maloney, M. G. Beaver, D. E. Shen, X. Zhu, L. R. Graham, J. A. Robinson, S. Huggins, A. Allian, R. Hart, S. D. Walker, P. Rolandi, and R. D. Braatz, 2020, A Virtual Plant for Integrated Continuous Manufacturing of a Carfilzomib Drug Substance Intermediate, Part 1: CDI-Promoted Amide Bond Formation, Organic Process Research & Development, 24, 10, 1861-1875

D. Laky, S. Xu, J. S. Rodriguez, S. Vaidyaraman, S. García Muñoz, and C. Laird. 2019. An Optimization-Based Framework to Define the Probabilistic Design Space of Pharmaceutical Processes with Model Uncertainty, Processes, 7 (2)

D. Laky, D. Casas-Orozco, G. V. Reklaitis, C. Laird, and Z. Nagy. 2021, Simulation-optimization framework for grey-box optimization using PharmaPy, AIChE Annual Meeting – Boston 2021

W. 1992. An Object-Oriented Representation of Structured Process Models, Computers and Chemical Engineering 16: S329–36

Proceedings of the 14th International Symposium on Process Systems Engineering – PSE 2021+
June 19-23, 2022, Kyoto, Japan © 2022 Elsevier B.V. All rights reserved.
http://dx.doi.org/10.1016/B978-0-323-85159-6.50356-0

Hybrid Modelling of CHO-MK Cell Cultivation in Monoclonal Antibody Production

Kozue Okamura[a], Sara Badr[a*], Sei Murakami[b], Hirokazu Sugiyama[a]

[a]*Department of Chemical System Engineering, The University of Tokyo, 7-3-1, Hongo, Bunkyo-ku, 113-8656, Tokyo, Japan*
[b]*Manufacturing Technology Association of Biologics, 2-6-16, Shinkawa, Chuo-ku, 104-0033, Tokyo, Japan*
badr@pse.t.u-tokyo.ac.jp

Abstract

Monoclonal antibodies (mAbs) are essential as drug substances and are getting increased attention from industry and academia. In order to improve mAb production efficiency, optimising the process is necessary. Optimising the cell cultivation step where mAbs are produced by host cells is especially important due to its influence on cost and time. Adequate process models are essential to perform the needed simulations. This work aims to propose a new modelling approach that avoids the failures of previously developed kinetic models with new cell lines. A new hybrid modelling approach is introduced, which combines a mechanistic module with a data-driven one to account for cell phases and varying environmental conditions. The developed hybrid approach was tailored for a newly developed Chinese Hamster Ovary (CHO) cell line (CHO-MK 9E-1). The hybrid approach gave a higher accuracy in the depiction of lactate and glucose concentrations compared to the mechanistic model alone. Insights are gained through analysing the results of the data-driven module. Such insights can be used as feedback in an opportunity to develop more versatile mechanistic models.

Keywords: Cell culture; Chinese hamster ovary (CHO) cells; Hybrid modelling; Principal component analysis; Biopharmaceutical production

1. Introduction

Monoclonal antibodies (mAbs) are essential as an active pharmaceutical ingredient. They are produced using Chinese Hamster Ovary (CHO) cells through the process shown in Figure 1. Process models are necessary to perform simulations for process optimisation and reduce the number of required experiments. In particular, models for the cell cultivations are required due to the high contribution of this step to the overall production cost and time (Yang et al., 2019).

Many efforts have been made to develop a kinetic cultivation model that represents the cultivation phenomena well. The recent works are well summarised by Tang et al. (2020). One of the latest works is done by Badr et al. (2021), and the proposed Monod-type-based model is validated with both the most common cell line (CHO-K1) and a newly established cell line (CHO-MK CL1001) cultivation.

However, despite of the efforts of many researchers in this field, developing a versatile kinetic model has been a challenge due to the complexity of the cell cultivation phenomena and variations in the cell lines. For example, previous models were not successful in depicting lactate concentrations for a new cell line (CHO-MK 9E-1).

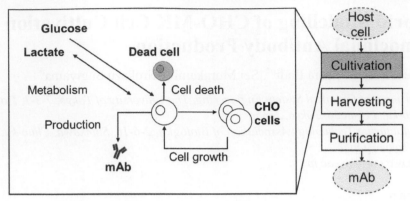

Figure 1: A monoclonal antibody (mAb) production process flow and cultivation phenomena.

This cell line is known to have a faster metabolism and is accordingly sensitive to lactate concentrations, which affects other aspects of cell viability and growth.

To overcome the challenge of needing to describe the impact of behaviour at different cell phases and environmental conditions, a hybrid modelling approach is introduced in this work. In this hybrid approach, a mechanistic module involving kinetic models of cell metabolism is combined with a data-driven one. Mechanistic models are easier to interpret but require a comprehensive understanding of the underlying mechanisms. On the other hand, data-driven models are easy to develop and apply even when the underlying mechanisms are not fully understood. However, they require a high volume of experimental data for model development and are more difficult to interpret or extrapolate. Hybrid modelling approaches aim to combine the advantages of both modelling techniques (Hong et al., 2018). The applications of hybrid modelling in pharmaceutical bioprocesses are well summarised by Narayanan et al. (2019).

This work proposes a new hybrid modelling approach that drastically improved the lactate modelling accuracy of CHO-MK 9E-1. A discussion is provided for the insights gained from the data-driven module to help improve the mechanistic understanding of the underlying phenomena. The developed model can subsequently be further exploited to achieve a better description of cell death and subsequent impurity generation, such as host cell proteins (HCP) and DNA.

2. Methodology

2.1. Experimental Data

Experimental data for model validation was obtained from the Kobe GMP consolidated lab of Manufacturing Technology Association of Biologics. Two cultivation modes were represented: the fed-batch mode, where nutrients continue to be fed to the cells throughout the operation, and the perfusion mode, where nutrients are continuously fed to the reactor and outlet streams are constantly removed to achieve continuous operations. Data from four experiments are used in this work: two experiments of CHO-MK 9E-1 cell in fed-batch mode (a 50 L stirred tank and a 2 L glass vessel, respectively); one experiment of CHO-K1 cell in fed-batch mode (a 50 L stirred tank); one experiment of CHO-MK CL1001 in perfusion mode (a 2 L glass vessel).

2.2. Cultivation Model

In this study, a hybrid modelling approach was introduced to improve the accuracy of lactate concentration modelling. The mechanistic module involved a kinetic model for fundamental cell metabolism. The model developed by Badr et al. (2021) was used in this work (Eqs.(1)-(7)). Badr et al. (2021) validated the model with the same experimental data of CHO-K1 cell in fed-batch mode and CHO-MK CL1001 in perfusion mode.

$$\frac{dVX_V}{dt} = (\mu - \mu_d)VX_V - F_{bleed}X_V \tag{1}$$

$$\frac{dVP}{dt} = Q_P VX_V - (F_{harvest} + F_{bleed})P \tag{2}$$

$$\frac{dV[GLC]}{dt} = -\left(\frac{\mu - \mu_d}{Y_{X_V/glc}} + m_{glc}\right)VX_V + F_{in}c_{in} + F_{suppl}c_{suppl} \tag{3}$$

$$- (F_{harvest} + F_{bleed})[GLC]$$

$$\frac{dV[LAC]}{dt} = Y_{lac/glc}VX_V - (F_{harvest} + F_{bleed})[LAC] \tag{4}$$

$$\frac{dV}{dt} = F_{in} + F_{suppl} - F_{harvest} + F_{bleed} \tag{5}$$

$$\mu = \mu_{max}\left(\frac{[GLC]}{K_{glc} + [GLC]}\right)\left(\frac{KI_{lac}}{KI_{lac} + [LAC]}\right) \tag{6}$$

$$\mu_d = k_d\left(\frac{[LAC]}{KD_{lac} + [LAC]}\right)\left(\frac{KD_{glc}}{KD_{glc} + [GLC]}\right) \tag{7}$$

where, X_V is the viable cell density. V is the solution volume inside bioreactor. F_{bleed}, $F_{harvest}$, F_{in}, and F_{suppl} are flow rates of bleeding, harvesting, feeding, and supplementary glucose solution feeding respectively. P, $[GLC]$, and $[LAC]$ represent concentrations of mAb, glucose (GLC), and lactate (LAC), respectively. c_{in} and c_{suppl} are glucose concentrations in the feeding media and supplementary glucose solution. respectively. μ and μ_d are cell growth and death rates. μ_{max} and k_d are their maximum values. Q_P is the specific mAb production rate. Y is the yield coefficient. m_{glc} is the glucose consumption coefficient for cell maintenance. K is Monod parameter.

The mechanistic module is followed by a data-driven one based on principal component analysis followed by linear regression (PCR), where LAC and GLC concentrations are calculated according to Eq. (8).

$$[Y_j] = \sum_{i=1}^{N} a_i PC_i + C \tag{8}$$

where $[Y_j]$ represents the concentration of component j representing in this work (LCA or GLC). PC_i represents the ith principal component derived from the principal

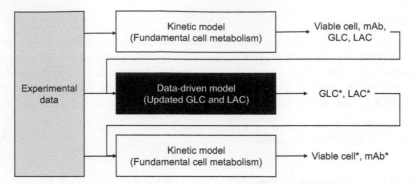

Figure 2: Outline of the hybrid modelling flow. * represents updated calculation results.

component analysis, a_i represents the coefficient of i_{th} principal component, N represents the appropriate number of principal components, and C is a constant.

Figure 2 shows how the two modules are combined in the hybrid approach. After fitting the kinetic model to the experimental data, the model parameters are obtained with initial predictions of viable cell density, and mAb, GLC, and LAC concentrations. These concentrations along with the estimated Monod parameters are fed as input to the data-driven module. Experimental conditions not included in the mechanistic model are also added to the data-driven module (e.g. dissolved oxygen (DO), pH, Temperature). The model input also includes time and information about the cell line and operating mode. The data-driven module then gives updated LAC and GLC concentrations as an output. These concentrations are fed back to the kinetic model to update the values of fitted parameters and obtain updated VCD and mAb values.

3. Results and Discussion

3.1. Cell Cultivation Modelling Results

Figure 3 shows part of the results of applying the model to predict concentrations in one experiment (CHO-MK 9E-1 in fed-batch mode with a 50L stirred tank). Lactate concentrations were better predicted by the hybrid model compared to the kinetic model alone. Other concentration profiles (GLC, viable cell, and mAb) were also well predicted. Similar results were obtained when using the model with each of the other experiments in the training set, with the hybrid model outperforming the kinetic model. A leave-one-out-cross-validation is also carried out to confirm model stability.

There is still a gap in viable cell toward the end of cultivation even though LAC modelling was improved. It might be interpreted as a need to further modify Eqs.(6) and (7), which define the effect of GLC and LAC on cell growth rate and death rate.

3.2. Insights from PCA loading trends

Figure 4 shows the cumulative explained variance ratio by each principal component (PC) and the categorised variable composition of each PC. With 9 PCs, the cumulative explained ratio reached 0.96. In order to make it easier to see the overall trend, the input variables were categorised into four groups: cell line and measurement factors; cell metabolism factors; environmental factors; operating factors.

In the case of PC_2 and PC_3, cell metabolism factors were the dominant ones. This indicates that such factors have a relatively high influence on the variance of lactate profiles. They also indicate different cell behaviour at different operating conditions of each experiment despite using similar cell lines. For example, incorporating an additional mechanistic model describing lactate consumption by the cells, it might be possible to further describe the complex behaviour of lactate metabolism. The variables in this category are the outputs of the kinetic model. The data-driven module thus uses both experimental data and such initial kinetic parameters to close the gap between the experimental data and calculated results of the kinetic model.

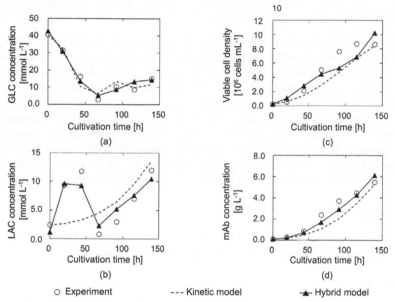

○ Experiment   --- Kinetic model   ▲ Hybrid model

Figure 3: Modelling results of CHO-MK 9E-1 in fed-batch mode (a 50L stirred tank), showing (a) GLC concentration, (b) LAC concentration, (c) viable cell density, and (d) mAb concentration. The accuracy of the hybrid model are as follows: (a) GLC (R^2=0.961), (b) LAC (R^2=0.863), (c) viable cell (R^2=0.851), and (d) mAb (R^2=0.942)

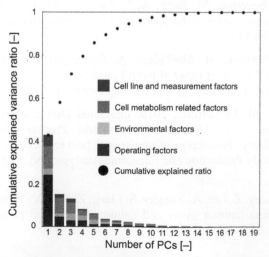

Figure 4: Cumulative explained variance ratio and categorised variable composition of each PC.

On the other hand, the ratio of environmental factors toward latter PCs becomes relatively larger along with cell metabolism factors. Here, by looking into the details of PC loadings, some possible relationships could be observed, such as between cell growth or death rate and DO or pH. This kind of insight could tell us the critical factors which have not been considered in the kinetic model yet. Utilising such feedback in kinetic model development might contribute to deeper mechanistic understanding towards obtaining more versatile kinetic models.

4. Conclusions

This work presents a new hybrid modelling approach for cell metabolism. The approach worked well with experimental data of different tested cell lines, including CHO-MK 9E-1, especially to achieve higher LAC modelling accuracy (R^2=0.863). Also, opportunities to utilise the insights obtained from the data-driven modelling part for further development of kinetic models were presented. Analysing the biological meaning of PC loading compositions and providing feedback to kinetic model development might contribute to a deeper understanding of cell cultivation phenomena. Experiments to further confirm the validity of the modelling approach with further cell lines and experimental conditions are still required. Additionally, exploration of different data-driven approaches is still required to enhance the modelling versatility and performance.

Acknowledgements

This work was supported by the Japan Agency for Medical Research and Development (AMED) [grant No. JP21ae0121015, JP20ae0101064, JP20ae0101058].

References

S. Badr, K. Okamura, N. Takahashi, V. Ubbenjans, H. Shirahata, H. Sugiyama, 2021, Integrated design of biopharmaceutical manufacturing processes: Operation modes and process configurations for monoclonal antibody production, Comput Chem Eng, 153, 107422

M. S. Hong, K. A. Severson, M. Jiang, A. E. Lu, J. C. Love, R. D. Braatz, 2018, Challenges and opportunities in biopharmaceutical manufacturing control, Comput Chem Eng, 110, 106-114

H. Narayanan, M. Sokolov, M. Morbidelli, A. Butté, 2019, A new generation of predictive models: The added value of hybrid models for manufacturing processes of therapeutic proteins, Biotechnology and Bioengineering, 116, 10, 2540-2549

O. Yang, S. Prabhu, M. Ierapetritou, 2019, Integrated Design of Biopharmaceutical Manufacturing Processes: Operation Modes and Process Configurations for Monoclonal Antibody Production, Comparison between Batch and Continuous Monoclonal Antibody Production and Economic Analysis, Ind. Eng. Chem. Res., 58, 15, 5851-5863

P. Tang, J. Xu, A. Louey, Z. Tan, A. Yongky, S. Liang, Z. J. Li, Y. Weng, S. Liu, Kinetic modeling of Chinese hamster ovary cell culture: factors and principles, Crit. Rev. Biotechnol., 40, 2, 265-281

Proceedings of the 14th International Symposium on Process Systems Engineering – PSE 2021+
June 19-23, 2022, Kyoto, Japan © 2022 Elsevier B.V. All rights reserved.
http://dx.doi.org/10.1016/B978-0-323-85159-6.50357-2

Multimodal modelling of uneven batch data

Atli Freyr Magnússon[a,b], Jari Pajander[b], Gürkan Sin[b], Stuart M. Stocks[b]

[a]*Technical University of Denmark, Søltofts Plads 227, Kgs. Lyngby 2800, Denmark*
[b]*Leo Pharma A/S, Industriparken 55, Ballerup 2750, Denmark*
flydkl@leo-pharma.com

Abstract

This work explores the application of a novel tri-linear regression methodology known as Shifted Covariates REgression Analysis for Multi-way data (SCREAM) to predict the quality of a fed-batch process. The SCREAM model shows promise as it is the only known multilinear regression tool that can directly handle three-way data arrays of different lengths. Thus, it provides an alternative modelling tool that does not require complicated time warping methods as a preprocessing step. The model was tested on a simulated fed-batch dataset based on industrial simulation of penicillin production. Variations were intentionally included in the simulations to create uneven data arrays. The SCREAM model outperforms traditional staples of multivariate models like NPLS and UPLS when warping is not considered and thus shows promise for application in fed-batch processes.

Keywords: Fed-Batch, Multimodal Modelling, PLS, Multivariate analysis

1. Introduction

The biochemical industry readily uses fed-batch processes for the production of various chemicals and pharmaceuticals. It is common in the industry to utilize a recipe driven approach, where operations are not adjusted to accommodate the variations in feed or initial blending of the batch. Biochemical processes by their very nature will introduce batch-to-batch variations if the system is not very tightly controlled. Knowledge of how the system variations may affect the final batch yield or quality can be vital for proper recovery of the final product. Most industries monitor key process parameters throughout the process, resulting in large amount of data. This has given rise to assumption free modelling (Westad et al, 2015). Data driven approaches can be utilized to establish correlations between captured data and the process quality.

For most data-driven regression methods, it is required that the data is available as two-dimensional matrix. However, batch data consisting of K batches is a three-way dataset $\underline{X}(I \times J \times K)$ where each batch is measured on I process variables at J time points, while quality variables may be measured only after the conclusion of the batch. For a regression type problem, the most common method of modelling three-way array is to reshape the 3-way array into a 2-D matrix in a process known as unfolding. If preserving the batch mode, the resulting matrix will be $X(K \times IJ)$. The most common multivariate analysis is to use Partial Least Squares (PLS) on the unfolded matrix, the overall process is referred to as UPLS (Wold et al, 1987). There also exist tri-linear models that work directly with 3-way arrays. They are useful if one expects the multilinear structure to affect the overall system variation. For regression purposes the easiest one to use would be the multilinear PLS or NPLS (Bro, 1996). Direct trilinear models are more strict than working with unfolded data and thus usually do not necessarily perform better when looking at prediction errors. They are still useful as they preserve the multi-linear structure and have more interpretative properties.

However, unfolding the matrix or utilizing NPLS requires that the lengths of the batch data matrix $(I \times J)$ be consistent across all batches. Variations of batch operation and initial blend will create uneven data with shifts and shape changes, where different events or peaks in data take place at different times. The simplest way to handle uneven lengths of batches is to identify the shortest runtime of a batch and remove excess time points from all others, a process known as cut-to-shortest. In case of severe shifts or variation in batch length, it is possible to synchronize the data structure in a process known as time warping (Gonzáles-Martinéz et al, 2018).

Alternatively, the Shifted Covariates REgression Analysis for Multi-way data (SCREAM) method (Marini and Bro, 2013) can be used. This method is the multi-modal version of the Principal Covariates Regression (PovR), but it utilizes a Parallel Factor Analysis 2 (PARAFAC2) decomposition method (Kiers 1991) instead of the two-way equivalent Principal Component Analysis (PCA). PARAFAC2 is a decomposition method that allows for shifts and uneven lengths in a single mode. This makes SCREAM an interesting candidate for modelling batches as it can directly model the data without unfolding the three-way structure and requires no time warping as a preprocessing step.

2. Materials and Methods

2.1. SCREAM Model

The SCREAM model utilizes a PARAFAC2 fitting algorithm based on an Alternating Least Squares approach. PARAFAC2 models are expressed as

$$X_k = AD_kB_k^T + E_k \qquad k = 1, \dots K \qquad (1)$$

Here X_k is a single slab of the entire three-way structure $\underline{X}$, or in this case the data from a single batch. For a PARAFAC2 model with F components, the matrix A is a matrix $(I \times F)$ of loadings in the I direction. For batch data, this is usually the variable loadings. D_k is a diagonal matrix $(I \times F)$ containing the k'th row of the matrix $C(K \times F)$ which contains the loadings in the K or batch direction. C is similar to a score matrix in ordinary 2-way PCA. Finally, B_k is the loadings in the J direction or the time point direction. Generally, B_k hold the loadings where the shifts happen. Finally, E_k contains the residuals. PARAFAC2 models are made unique by the constraint that the cross-product of each B_k is the same i.e. $B_kB_k^T = H$ for all $k = 1, \dots K$. The standard PARAFAC does not have a unique loading matrix in the J direction for each slab X_k but rather uses a single B for the entire three-way structure. Different B_k loadings allow PARAFAC2 to directly model three-way arrays of batch data of different lengths but also makes it more flexible when handling shifts in batch data.

Fitting a PARAFAC2 model is the least squares minimization of the following loss function

$$\sum_{k=1}^{K} ||(X_k - AD_kB_k^T)||^2 \qquad (2)$$

Note that C is a 2D matrix and a direct multi-linear regression onto Y is possible. This would be the multimodal equivalent to Principal Component Regression (PCR). However, there is no guarantee that the score matrix C is predictive of Y as it is attempting to summarizing the entire $\underline{X}$ array. Thus, changes in $\underline{X}$ that may have no significance on the output Y, will still affect the C matrix.

For prediction purposes it is sought to seek a score matrix **C** that is relevant for predicting **Y**. For a single dependent variable **y**, it is achieved by minimizing the following

$$||y - Cr||^2 \tag{3}$$

where **r** is a vector of regression coefficients. Making a predictive model that is relevant for both $\underline{X}$ and **y** requires the minimization of both equations (2) and (3). This is the same setup as in the two-way Principal Covariate Regression (PCovR) where a weighing parameter α $0 \leq \alpha \leq 1$ is introduced. This parameter controls to what degree the fitting should summarize $\underline{X}$ or predict **y**. The SCREAM model is then fitted by minimizing the following

$$\alpha \sum_{k=1}^{K} ||(X_k - AD_k B_k^T)||^2 + (1 - \alpha)||y - Cr||^2 \tag{4}$$

The PARAFAC2 direct fitting algorithm (Kiers et al 1999) is utilized to solve this minimization problem while maintaining the uniqueness constraint.

This modelling techniques has two hyperparameters that must be selected, the number of factors F and the value of the weighing parameter α. Improper selection of these parameters leads to models that do not predict or overfit on **y**. For most practical applications, optimization of the hyperparameters is done via cross-validation. A common method is minimizing the Root Mean Square Error of Cross-Validation.

2.2. Preprocessing

Preprocessing is important for any multivariate or multi-modal methods. Centering removes constant offsets in the data, while the obvious reason for scaling is to adjust for scale differences between variables measured in different units. Since the goal is for the model to capture the variations between different batches, the centering will be across the batch mode. This is done by computing a separate mean for the measured variable for each time point across batches and then subtracting the mean from each measurement.

Scaling is performed on the centered data. There are three types of scaling for three-way data. Column scaling, single-slab scaling, and double-slab scaling. This study will utilize the single-slab scaling technique as it has shown better performance in regression modeling of batch data. (Mears et al, 2016). For single-slab scaling all time points for a single variable are scaled to unit root-mean-square.

3. Case Study: Industrial Simulation of Penicillin Fermentation

Exploring the applicability of the SCREAM method for modelling batch fermentations a dataset is needed. For this purpose, the industrial simulation model (Birol et al, 2002s) is used for simulating an industrial fed-batch fermentation for the production of penicillin. 100 batches are simulated with a recipe driven approach given in Table 1, Batch-to-batch variations are created by varying the initial conditions of biomass and glucose in each batch, the total runtime of each batch is also varied, as well as the time to start the feed. For simplification variables that are controlled such as pH and Temperature are not simulated but a random Gaussian noise is added around the set point for each time index.

In the generated dataset, a total of 11 process variables are assumed to be monitored continuously on each batch which will be utilized for multi-modal modelling. The collection of the 11 process variables at equal timepoint distances of one hour will create the three-way array $\underline{X}$.

Table 1 Recipe used in the fed-batch simulation for penicillin production

Recipe Setup	Value	Potential variations
Initial Biomass	0.5 g/L	±0.05 g/L
Initial Glucose	40 g/L	±2 g/L
Initial Volume	100 m^3	±100 L
Feed Rate Set Point	200 L/h	±1 L/h
Feed Glucose Concentration	600 g/L	±10 g/L
Agitation	Maintain kla of 400 h^{-1}	±1 on kla
Aeration	1200 m^3/h	±20 m^3/h
Temperature Set Point	298 K	±0.05 K
pH Set Point	5	±0.01
Fermentation Duration	400 h	±10 h
Feed Start Time	50 h	±4 h

The quality variable y of interest is the total harvested Penicillin at the end of the batch. Figure 1 shows the batch-to-batch variation of penicillin harvest. Because of the variation in runtime of up to 20 hours, the three-way array $\underline{X}$ is uneven which is common in industrial batch processes.

4. Results and Discussions

The SCREAM model was built to predict the penicillin harvest based on monitored variables. Because the data is simulated the data there are observable effects of tank geometry or start time of batches, thus the partition into modelling and test set is done randomly. 30 Batches are held over for testing while the model is built on 70 batches. Utilizing a 10-fold venetian blinds cross validation it is found that setting F=5 and α = 0.3 provided a decent regression without overfitting.

Model predictions are shown in Figure 2 where predicted harvest from SCREAM model and measured harvest from the simulation are reported. The Root Mean Square Error of the test set was calculated to be 13.01 which results in a less than 5% absolute error when predicting the harvest of new batches.

For comparison the traditional three-way NPLS model is built as well as the most common type of multivariate model UPLS for batch data. To make the batches even for these types of models the cut-to-shortest time warping method is used. The performance of these models is shown in Table 3. Both NPLS and UPLS started overfitting the data with 3 components but with only 1 component it failed to establish any correlation between process data and penicillin harvest, thus 2 components were used for building the models.

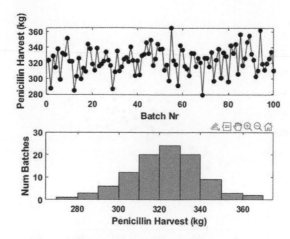

Figure 1 Batch-To-Batch variations in the penicillin harvest of the simulated dataset

However, SCREAM outperforms both when looking at the test set. This may be due to NPLS being too strict to handle the shifts, but unlikely as the UPLS is more flexible but has the same problems when predicting the penicillin harvest. The most likely issue is that the data at the end of the batch contains crucial information for establishing the correlations which is missing with a cut-to-shortest time warping. Dynamic Time Warping (DTW) or Correlation Optimized Warping (COW) may improve the performance of both UPLS or NPLS models but are difficult and time-consuming methods to employ. However, SCREAM models do not require these treatments and thus are able to predict the test-batches with more accuracy.

Table 2: Comparison of different regression models on the simulated dataset.

Model Type	RMSE (Calibration Set)	RMSE (Test Set)
SCREAM	11.23	13.01
NPLS	6.98	17.61
UPLS	7.82	18.00

5. Conclusions

A regression method that can directly model uneven batch data was tested on a simulated industrial fed-batch dataset. The SCREAM model was developed as a modification to Multivariate covariate regression, which allows it to handle shifts in dataset and uneven three-way matrices. The fed-batch simulation is based on an industrial model and disturbances were introduced in the recipe to create the variations in runtime and shifts the data, which is common in industry. The SCREAM model outperforms both NPLS and UPLS when used on the data when complex warping techniques are not utilized with no overfit and little bias. There is even room for improvement by utilizing non-derivative optimization method for the model hyperparameters, it may well be possible to find a set of hyperparameters that outperforms the model reported here.

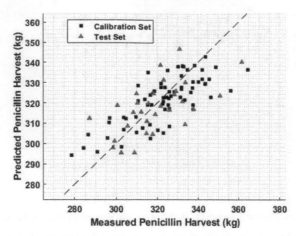

Figure 2: Predicted Penicillin harvest from the SCREAM model vs the actual simulated value.

Further work will be analysing the loadings and diagnostics of the model to see if the model can identify the key variables and time points of an industrial fermentation to extract information without access to the original simulation. Overall, the SCREAM model technique is a promising alternative in the modelling fed-batch systems.

References

G. Birol, C. Ündey, A. Çinar, 2002, A modular simulation package for fed-batch fermentation: penicillin production, Computers and Chemical Engineering, 26, 11 1553-1565

R. Bro, 1996, Multiway calibration. Multilinear PLS, Journal of Chemometrics, 10, 1, 47-61

J.M. Gonzáles-Martínez, J. Camarcho, A. Ferrer, 2018, MVBatch: A matlab toolbox for batch process modeling and monitoring, Chemometrics and Intelligent Laboratory Systems, 183, 122-133

H. Kiers, 1989, An alternating least squares algorithm for fitting the two- and three-way dedicom model and the idoscal model, Psychometrika, 54, 3, 515-521

H. Kiers, J. Ten Berge, R. Bro, 1999, PARAFAC2 – Part I. A direct fitting algorithm for the PARAFAC2 model, Journal of Chemometrics, 13, 3-4, 275-294

F. Marini, R. Bro, 2013, SCREAM: A novel method for multi-way regression problems with shifts and shape changes in one mode, Chemometrics and Intelligent Laboratory Systems, 129, 64-75

L. Mears, R. Nørregård, G. Sin, K. Gernaey, S. Stocks, M. Albaek, K. Villez, 2016, Functional Unfold Principal Component Regression Methodology for Analysis of Industrial Batch Process Data, AiChE Journal, 62, 6, 1986-1994

F. Westad, L. Gidskehaug, B. Swarbrick, G. Flåten, 2015, Assumption free modeling and monitoring of batch processes, Chemometrics and Intelligent Laboratory Systems, 149, 66-72

S. Wold, P. Geladi, K. Esbensen, J. Öhman, 1987, Multi-way principal components-and PLS-analysis, Journal of Chemometrics, 1, 1, 41-56

Proceedings of the 14th International Symposium on Process Systems Engineering – PSE 2021+
June 19-23, 2022, Kyoto, Japan © 2022 Elsevier B.V. All rights reserved.
http://dx.doi.org/10.1016/B978-0-323-85159-6.50358-4

Application of MHE-based NMPC on a Rotary Tablet Press under Plant-Model Mismatch

Yan-Shu Huang[a*], M. Ziyan Sheriff[a], Sunidhi Bachawala[b], Marcial Gonzalez[b,c], Zoltan K. Nagy[a], Gintaras V. Reklaitis[a]

[a]Davidson School of Chemial Engineering, Purdue University, West Lafayette, IN 47907, USA
[b]School of Mechanical Engineering, Purdue University, West Lafayette, IN 47907, USA
[c]Ray W. Herrick Laboratories, Purdue University, West Lafayette, IN 47907, USA
huan1289@purdue.edu

Abstract

Active control strategies play a vital role in modern pharmaceutical manufacturing. Automation and digitalization are revolutionizing the pharmaceutical industry and are particularly important in the shift from batch operations to continuous operation. Active control strategies provide real-time corrective actions when departures from quality targets are detected or even predicted. Under the concept of Quality-by-Control (QbC), a three-level hierarchical control structure can be applied to achieve effective setpoint tracking and disturbance rejection in the tablet manufacturing process through the development and implementation of a moving horizon estimation-based nonlinear model predictive control (MHE-NMPC) framework. When MHE is coupled with NMPC, historical data in the past time window together with real-time data from the sensor network enable model parameter updating and control. The adaptive model in the NMPC strategy compensates for process uncertainties, further reducing plant-model mismatch effects. The frequency and constraints of parameter updating in the MHE window should be determined cautiously to maintain control robustness when sensor measurements are degraded or unavailable. The practical applicability of the proposed MHE-NMPC framework is demonstrated via using a commercial scale tablet press, Natoli NP-400, to control tablet properties, where the nonlinear mechanistic models used in the framework can predict the essential powder properties and provide physical interpretations.

Keywords: pharmaceutical manufacturing; continuous manufacturing; process control; nonlinear model predictive control; moving horizon estimation.

1. Introduction

Several factors currently drive the transition of the pharmaceutical manufacturing industry from batch to continuous process operation. These include potential improvement in both product quality homogeneity and process controllability. Quality control traditionally followed a Quality-by-Testing (QbT) approach, wherein product quality was tested at the end of each batch processing step. However, with improved product and process understanding, a Quality-by-Design (QbD) approach was adopted to enable systematic design of the operating space using mechanistic models. More recently, there has been a desire to adopt a Quality-by-Control (QbC) approach, wherein quantitative and predictive understanding can be leveraged for active process control and aid robust process design and operation, thereby enabling smart manufacturing (Su et al., 2019).

An important part of any real-time process monitoring and control strategy is the ability to identify and manage the impact of plant-model mismatch (PMM). PMM can arise in the continuous manufacture of oral solid dosage for numerous reasons, e.g., disturbances that affect critical material attributes (CMAs) such as the bulk density can be introduced during the feeder refill step (Destro et al., 2021). As this can result in a deviation in the critical quality attributes (CQAs), PMM needs to be identified and handled appropriately. Several approaches have been developed in order to identify and assess the impact of PMM, e.g., based on mutual-information (Chen et al., 2013) or autocovariance (Wang et al., 2017). Stringent regulations placed by regulatory bodies make it essential to track CQAs and CMAs online, but they may be unmeasurable in practice as process analytical technology (PAT) sensing methods may not be available to track these states or parameters, e.g., bulk density. Therefore, this work proposes the use of an on-line, real-time parameter estimation approaches to accurately track model parameters online, to guide operating decisions. It is important to note that most work in the continuous manufacturing domain utilize linear model predictive control (MPC) strategies, that are derived from the linearization of the nonlinear system and may not be adequate for nonlinear process models and unit operations such as the rotary tablet press (Ierapetritou et al., 2016).

A recent in-silico study by (Huang et al., 2021) demonstrated that a combined MHE-NMPC framework could satisfy the dual requirement of efficient estimation and control. Unfortunately, there are no case studies in the literature that demonstrate the application of the proposed framework to real data from a continuous pharmaceutical manufacturing process. Therefore, the primary objective of this work is to validate the practical applicability of the proposed framework using a Natoli NP-400 rotary tablet press.

2. Methodology

The moving horizon estimation-based nonlinear model predictive framework (MHE-NMPC) aims to satisfy the dual requirement of estimation and control, by combining the effective estimation capabilities of MHE with the control performance provided by NMPC. Given a nonlinear state-space model:

$$\dot{x} = g(x, u, \theta, w) \tag{1}$$

$$y = l(x, u, \theta, v) \tag{2}$$

where x, u, θ, and y are vectors that represent the state variables, input variables, model parameters, and measurements, respectively. Process and measurement noise are denoted by w and v, respectively. In this work, the model is described by a set of explicit algebraic equations with no differential states, and f and h will represent these algebraic equations. MHE can then be formulated as follows (López-Negrete and Biegler, 2012):

$$\min_{\hat{\theta}_k} J = \sum_{t=k-N_{past}}^{k} (\epsilon_t)^T W_E \, \epsilon_t + \left(\hat{\theta}_k - \hat{\theta}_{k-1}\right)^T W_\theta \left(\hat{\theta}_k - \hat{\theta}_{k-1}\right) \tag{3a}$$

subject to

$$\hat{x}_{k-N_{past}+j+1} = f\left(\hat{x}_{k-N_{past}+j}, u_{k-N_{past}+j}, \hat{\theta}_k\right) \tag{3b}$$

$$\hat{y}_{k-N_{past}+j} = h(\hat{x}_{k-N_{past}+j}) \tag{3c}$$

$$\epsilon_{k-N_{past}+j} = y_{k-N_{past}+j} - \hat{y}_{k-N_{past}+j} \tag{3d}$$

$$\hat{x}_{k-N_{past}+j+1} \in \mathbb{X}, \qquad \epsilon_{k-N_{past}+j} \in \Omega_\epsilon, \qquad \hat{\theta}_k \in \Omega_\theta \qquad (3e)$$

$$j = 0, 1, \dots, N_{past} \qquad (3f)$$

where $\hat{\theta}_k$ are estimated uncertain parameters, bounded in compact set Ω_θ. y_t and u_t are measurements of output and input variables at time t, respectively; $\hat{y}_t$ and $\hat{x}_t$ are estimated output and state values, respectively; ϵ_t are output disturbances, bounded in compact set Ω_ϵ; and W_E and W_θ are weighting matrices. Once the MHE optimization problem is solved at time $t = k$, the estimated state $\hat{x}_{k-N_{past}+1|t=k}$ is chosen as the initial state value for the next time step $t = k + 1$, i.e., $\hat{x}_{k-N_{past}+1|t=k+1} = \hat{x}_{k-N_{past}+1|t=k}$.

This study utilizes the median of the error distribution in the past time window to represent output disturbances ζ_k at time $t = k$, i.e.,

$$\zeta_k = median\left\{\epsilon_{k-N_{past}+j}\right\}, \quad \text{for } j = 0, 1, \dots, N_{past} \qquad (4)$$

The NMPC framework at time $t = k$ is defined as follows:

$$\min_{\Delta u_t} J = \sum_{t=k}^{k+N_p}(\hat{y}_t - y_{sp})^T W_y(\hat{y}_t - y_{sp}) + \sum_{t=k}^{k+N_c-1}(\Delta u_t^T W_{\Delta u}\Delta u_t) \qquad (5a)$$

subject to

$$\hat{x}_{k+j+1} = f(\hat{x}_{k+j}, \hat{u}_{k+j}, \hat{\theta}_k) \qquad (5b)$$

$$\hat{y}_{k+j} = h(\hat{x}_{k+j}) + \zeta_k \qquad (5c)$$

$$\Delta u_{k+j} = \hat{u}_{k+j+1} - \hat{u}_{k+j} \qquad (5d)$$

$$\hat{x}_{k+j} \in \mathbb{X}, \qquad \hat{u}_{k+j} \in \mathbb{U}, \qquad \Delta u_{k+j} \in \Omega_{\Delta u} \qquad (5e)$$

$$j = 0, 1, \dots, N_p - 1 \qquad (5f)$$

where N_c is the length of the control time window, and y_{sp} are the setpoints of the output variables. W_y and $W_{\Delta u}$ are weighting matrices. Control movements Δu are constrained in compact set $\Omega_{\Delta u}$. A detailed discussion of the MHE-NMPC framework including its computational feasibility is provided in (Huang et al., 2021).

3. Case Study

3.1. Tablet press model

The tablet press is responsible for the formation of solid tablets via mechanical compression. The weight of a convex tablet W and the tablet production rate $\dot{m}_{tablet}$ are given by the following relationships (Huang et al., 2021):

$$W = \rho_b V_{fill}\left(1 - \xi_1\frac{n_T}{n_F} + \xi_2\frac{H_{fill}}{D}\right) \qquad (6)$$

$$\dot{m}_{tablet} = W n_T N_{station} \qquad (7)$$

where D, V_{fill}, H_{fill}, ρ_b, n_T, and n_F, are the diameter of the die, volume of the die cavity, dosing position, powder bulk density, turret speed, and feed frame speed, respectively. $N_{station}$ refers to the number of stations in the tablet press. ξ_1 and ξ_2 are empirical model parameters that are estimated from experimental data. The volume of the die cavity for the D-type tooling is provided by (Huang et al., 2021).

The pre-compression force F_{pc} and the main compression force F_{punch} can be computed as follows:

$$F_{pc} = \frac{\pi D^2}{4b} \left[\frac{\rho^{pc} - \rho_c}{\rho^{pc}(a-1) + \rho_c} \right] \tag{8}$$

$$\rho^{pc} = \frac{W}{V^{pc}\rho_t} \tag{9}$$

$$F_{punch} = \frac{\pi D^2}{4b} \left[\frac{\rho^{in-die} - \rho_c}{\rho^{in-die}(a-1) + \rho_c} \right] \tag{10}$$

$$\rho^{in-die} = \frac{W}{V^{in-die}\rho_t} \tag{11}$$

where parameters a and b are Kawakita constants, which represent the maximum degree of compression and the reciprocal of the pressure applied to attain this degree of compression, respectively. ρ^{pc} and ρ^{in-die} are the pre-compression and in-die relative densities, respectively. ρ_t refers to the true density of the powder. The pre-compression volume, V^{pc}, and in-die volume V^{in-die} are provided by (Huang et al., 2021).

3.2. Applying MHE to tablet press: Comparison of fixed model and adaptive model

To investigate state estimation and parameter updating, an experiment was performed via open-loop control. Setpoint changes of input variables were introduced to the tablet press, and corresponding measurements of output variables and model predictions are recorded and shown in Figure 1 (a) with fixed model parameters and Figure 1 (b) with adaptive model parameters.

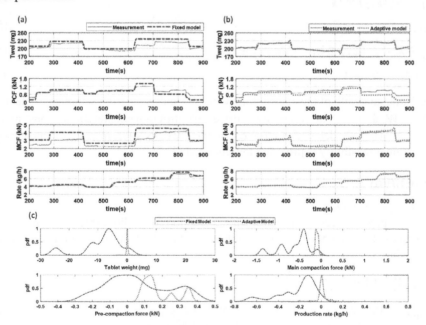

Figure 1. Real-time monitoring of PMM when uncertain parameters are (a) fixed or (b) adaptive with (c) error distribution of estimated output variables.

In this study, since the mathematical model is represented by a set of explicit algebraic equations the MHE only updates two uncertain parameters: (1) the bulk density (ρ_{bulk}), to compensate for the effects of disturbance on the estimated value of the tablet weight, which further affects estimated values of pre-compression force, main compression force, and the production rate, and (2) the critical relative density (ρ_c) to provide flexibility to estimate the pre-compression and main compression forces more accurately in the adaptive model compared to the case of only bulk density being updated. The adaptive model predicts output variables more accurately compared to the fixed model.

To quantify model accuracy and precision, error distributions of the estimated output variables are provided in Figure 1 (c), where the probability density functions (pdf) of the error distributions are rescaled (meaning that the area under the density curve is not 1). Once MHE is applied to update the bulk and critical relative densities, the absolute values of the median error and error spans of all output estimations are significantly reduced, as shown in Figure 1 (c). However, an exception is found in pre-compression force, whose error span is reduced from 0.75 kN to 0.37 kN, while median error is increased from 0.07 kN to 0.22 kN. Since pre-compression force and main compression force share the same model parameters as shown in Equation (8-11), there exists the need to establish a compromise between the accuracy of these two output variables.

3.3. Experimental verification of MHE-NMPC

Control profiles for a representative experimental run of the 4 input variables, 4 output variables, and 2 uncertain model parameters are shown in Figure 2 (a), (b), and (c), respectively. Offsets in the output variables are observed as open-loop control is applied at the start of operation (highlighted in red). When the MHE-NMPC algorithm is implemented from t = 200 s, offset free control is achieved. Additional setpoint changes are introduced for the tablet weight at t = 600 s, 800 s, 1500 s, the main compression force at t = 1100 s, and the production rate at t = 1000 s.

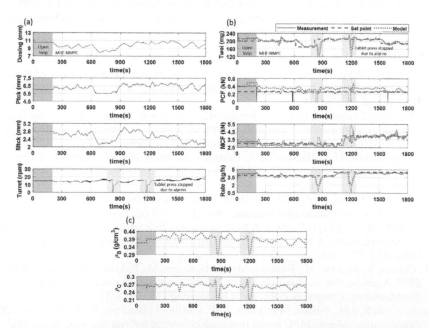

Figure 2. MHE-NMPC control performance of the tablet press with (a) input variables, (b) output variables, and (c) uncertain parameters.

All results of setpoint tracking are satisfactory as shown in Figure 2 (b). During the experimental run, an internal alarm stopped the tablet press twice at $t = 820$ s and $t = 1160$ s (highlighted in yellow) forcing the turret speed to drop to 0 rpm. The machine stop prevents the distributed control system (DCS) from collecting in-house tablet weight and production rate measurements, where the time delay can be attributed to the 10 s moving average window. The strength of the MHE-NMPC algorithm can once again be noted, as offset-free control is quickly achieved once the tablet press resumes operation. As the uncertain parameters are updated in real-time as shown in Figure 2 (c), the mismatch can be mitigated as presented in Figure 2 (b). While mismatch for pre-compression force cannot be completely mitigated, as it shares the same parameters as the main compression force, the disturbance term used in controller model described in Equation 5c still guarantees offset-free control of the pre-compression force.

4. Conclusions

Real-time process monitoring and control are essential to enable continuous operation of modern pharmaceutical manufacturing processes. The MHE-NMPC framework demonstrates satisfactory control performance and parameter updating in the rotary tablet press to handle plant-model mismatch (PMM). Future work will include sensor fusion studies to incorporate at-line measurements with long sampling time to the framework. Accurate estimation is required to enable the control of critical quality attributes such as tensile strength, which need to be predicted from soft sensors due to limited availability of real-time measurements because of the destructive nature of the testing methods used.

Acknowledgement

This work was supported by the United States Food and Drug Administration under grant 1U01FD006487-01. The authors would like to thank Natoli Engineering Company for the availability of the tablet press and Carmelo Hernandez-Vega for his technical support.

References

Chen, G., Xie, L., Zeng, J., Chu, J., Gu, Y., 2013. Detecting Model–Plant Mismatch of Nonlinear Multivariate Systems Using Mutual Information. Industrial & Engineering Chemistry Research 52, 1927–1938.

Destro, F., García Muñoz, S., Bezzo, F., Barolo, M., 2021. Powder composition monitoring in continuous pharmaceutical solid-dosage form manufacturing using state estimation – Proof of concept. International Journal of Pharmaceutics 605, 120808.

Huang, Y.-S., Sheriff, M.Z., Bachawala, S., Gonzalez, M., Nagy, Z.K., Reklaitis, G. v., 2021. Evaluation of a Combined MHE-NMPC Approach to Handle Plant-Model Mismatch in a Rotary Tablet Press. Processes 9, 1612.

Ierapetritou, M., Muzzio, F., Reklaitis, G., 2016. Perspectives on the continuous manufacturing of powder-based pharmaceutical processes. AIChE Journal 62, 1846–1862.

López-Negrete, R., Biegler, L.T., 2012. A Moving Horizon Estimator for processes with multi-rate measurements: A Nonlinear Programming sensitivity approach. Journal of Process Control 22, 677–688.

Su, Q., Ganesh, S., Moreno, M., Bommireddy, Y., Gonzalez, M., Reklaitis, G. v, Nagy, Z.K., 2019. A perspective on Quality-by-Control (QbC) in pharmaceutical continuous manufacturing. Computers & Chemical Engineering 125, 216–231.

Wang, S., Simkoff, J.M., Baldea, M., Chiang, L.H., Castillo, I., Bindlish, R., Stanley, D.B., 2017. Autocovariance-based MPC model mismatch estimation for systems with measurable disturbances. Journal of Process Control 55, 42–54.

Proceedings of the 14th International Symposium on Process Systems Engineering – PSE 2021+
June 19-23, 2022, Kyoto, Japan © 2022 Elsevier B.V. All rights reserved.
http://dx.doi.org/10.1016/B978-0-323-85159-6.50359-6

Gray-box modelling of pharmaceutical roller compaction process

Shuichi TANABE[a*], Shubhangini AWASTHI[b], Daiki KAKO[a], and Srikanth R. GOPIREDDY[b]

[a]*Formulation Technology Research Laboratories, Daiichi Sankyo Co., Ltd., Kanagawa 254-0014, Japan*
[b]*Formulation Technology, Daiichi Sankyo Europe GmbH, Pfaffenhofen 85276, Germany*
Corresponding author's email: tanabe.shuichi.h3@daiichisankyo.co.jp

Abstract

A novel method was developed that allows the rolling theory of granular solids (Johanson, 1965) to predict the ribbon density accurately. In this study, a gray-box model of roller compaction process was developed based on the rolling theory of granular solids to demonstrate a practical application of the accurate and descriptive roller compaction model for process development. A placebo formulation composed of mannitol, microcrystalline cellulose, and magnesium stearate was used to generate 26 samples of roller compaction experiment. Compressibility factor and elastic recovery rate were predicted using regression models to consider the dependence on material attribute and process parameter (PP). The gray-box model composed of the modified rolling theory proposed by Reynolds et al. (2010) and the complemental regression models showed a better prediction performance compared to the white-box model reported by Reynolds et al. The root mean square error of cross validation of the ribbon density in white-box and gray-box models were 0.07 g/cc and 0.04 g/cc, respectively. With the gray-box model the effect of PP on the ribbon density and mass throughput was visualized, which is beneficial to identify the target and the acceptable ranges of PP for manufacturing.

Keywords: Dry granulation, Modeling and Simulation, Ribbon density, Control strategy.

1. Introduction

The granulation process is a critical process that impacts the quality of pharmaceutical products such as the dissolution and the uniformity of active ingredient content of tablets, capsules, powder filled bottles, etc. Roller compaction is a dry granulation processes for producing granules from powder blends. In roller compaction powder blends are continuously compressed by the two counter-rotating rolls and subsequently the generated ribbons are milled to obtain the granules. In general, ribbon density is considered as a critical material attribute (CMA), i.e., a factor of drug product quality and therefore it needs to be controlled via process parameter (PP) such as roll force, roll speed, and roll gap. The pressure on the powder blends in the roller compactor directly correlates to the ribbon density. Due to the difficulty in direct pressure measurement, process models that use the measurable PP to predict process outputs are needed to control the drug product quality.

Johanson (1965) developed the rolling theory of granular solids, which correlates the powder compaction in roller compaction process with the raw material properties and the PP. In the rolling theory, the space in between the rolls are divided into three different regions, i.e., the slip region, the nip region, and the release region. A sketch of the three different regions are provided in Figure 1a. In the slip region, powder blends slip along with the rotation of rolls whereas in the nip region, the powders are trapped by the rolls and move at the same speed as the roll surface, which results in the compaction of the powders to form roller compacted ribbons. In the release region, the ribbons show elastic recovery due to the release from compaction force by the rolls. The nip angle α is the transition angle from slip to nip region.

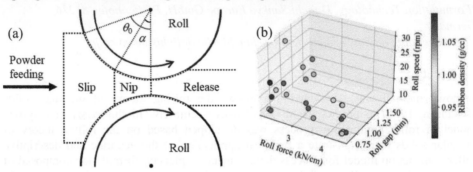

Figure 1 (a) Sketch of the three different regions in roller compaction. (b) Scatter plot of experimental runs with different PP setting.

The maximum pressure at the minimum separation of rolls, which is defined as roll gap, is estimated from the pressure distribution equation as a function of roll angle. The output parameters such as ribbon density and mass throughput are calculated based on the estimated maximum pressure. Due to the difficulty in measuring the required pre-consolidation pressure, limited works have been reported with focus on the practical application of the rolling theory for process understanding and control. Reynolds et al. (2010) proposed a modified rolling theory as a practical compromise with an assumption that the pre-consolidation pressure is 1 MPa. Based on this assumption, the pre-consolidation density and the compressibility factor are considered constant regardless of the PP, and were estimated from uniaxial experiments to predict ribbon density and mass throughput. While their approach provided accurate prediction of ribbon density, it has a limitation in predicting ribbon density accurately because their approach does not consider the effect of the PP on the pre-consolidation pressure. As presented by Reynolds et al., the pre-consolidation density and compressibility factor are different between the uniaxial measurement and the roller compaction process, which indicates that the pre-consolidation density and compressibility factor depend on the PP. Besides, the rolling theory and the variations do not consider the elastic recovery of the ribbons in release region even though the ribbon thickness is larger than the roll gap as reported in previous studies (Shi and Sprockel, 2016; Souihi et al., 2013). Since the off-line ribbon density measurement uses samples collected from release region, the relationship between the PP and the elastic recovery rate should be considered for higher prediction accuracy.

Gray-box model is one of the practical solutions to complement prediction performance of theoretical models with the use of data while keeping the theoretical structure

(Ahmad et al., 2020). The gray-box model is expected to present a higher interpretability compared to the black-box model, i.e., data-driven model. The gray-box models have been successfully applied to chemical and pharmaceutical processes (Van sprang et al., 2005; Ahmad et al., 2020). However, application of the gray-box model to the roller compaction process is not reported so far. In this study, a gray-box model was demonstrated on the rolling theory for the first time to show its practical applicability to the process development and control strategy setting.

2. Materials and Methods

A placebo formulation composed of Pearlitol 100SD as mannitol (Roquette, France), Ceolus UF-711 as microcrystalline cellulose (Asahi Kasei, Japan), and HyQual 5712 as magnesium stearate (Mallinckrodt, USA) in a weight ratio percentage of 79:20:1 was used for this study. A 5-L V-blender was used to manufacture the powder blends for roller compaction in 2-kg/batch scale. A roller compactor FP90 (Freund Turbo corporation, Japan) that have a roll diameter of 90 mm and a roll width of 30 mm with the textured roll surface was used for producing ribbons. 26 runs of roller compaction experiments were performed using two different lots of powder blends. Figure 1b shows the scatter plot of experimental runs with different PP setting.

2.1. Physical testing of powder blends and ribbons

The bulk and tapped density of powder blends were measured using a graduated cylinder. A quantity of the powder blends was poured to a graduated cylinder. The net weight of the material and the bulk volume were recorded. The tapped volumes were recorded using a tapping apparatus SZ-02 (Rinkan Kogyo Co., Ltd., Japan). The flow properties of the powder blends were measured using FT4 powder rheometer (Freeman Technology, UK). A 25 mm vessel in diameter was used to assess effective angle of internal friction (EAIF) and wall friction angle (WFA). The ribbon density was measured using GeoPyc 1365 and AccuPyc 1340 (Micrometrics, USA). A sample chamber with an internal diameter of 25.4 mm was used for GeoPyc 1365. The ribbon thickness was measured immediately after the sampling of ribbons during experiments using thickness gauge.

2.2. The modified rolling theory of granular solids and the gray-box model

The rolling theory of granular solids modelled the stress gradient in the slip and the nip region along with the roll angle θ as shown in Eq.(1) and Eq.(2).

$$\left(\frac{d\sigma}{d\theta}\right)_{slip} = \frac{4\sigma((\pi/2) - \theta - v)\tan\delta}{D/2\left(1 + S/D - \cos\theta\right)\left(\cot\left(((\theta + v + (\pi/2))/2) - u\right) - \cot\left(((\theta + v + \pi/2)/2) + u\right)\right)} \tag{1}$$

$$\left(\frac{d\sigma}{d\theta}\right)_{nip} = \frac{K\sigma_\theta(2\cos\theta - 1 - S/D)\tan\theta}{(D/2)((1 + S/D - \cos\theta)\cos\theta)} \tag{2}$$

where

$$v = \left(\pi - \sin^{-1}\left(\frac{\sin\varphi}{\sin\delta}\right) - \varphi\right)/2 \tag{3}$$

$$u = (\pi/4) - (\delta/2) \tag{4}$$

K is compressibility factor, δ is EAIF, $\emptyset$ is WFA, S is roll gap, and D is roll diameter. The nip angle α at which the pressure gradients for the slip and nip regions were equal is determined by equating Eq.(1) and Eq.(2). In the Reynolds' modified rolling theory, the pressure distribution between the rolls was used to relate the roll force R_f, which is a common and a measurable PP with the peak pressure P_{max} applied at the minimum separation of rolls as given in Eq.(5) and Eq.(6).

$$R_f = \frac{P_{max}WDF}{2} \tag{5}$$

where

$$F = \int_{\theta=0}^{\theta=\alpha(\delta,\emptyset,K)} \left[\frac{S/D}{(1 + S/D - \cos\theta)\cos\theta} \right]^K \cos\theta \, d\theta \tag{6}$$

The Eq.(5) and Eq.(6) represent a relationship between PP (R_f, S), geometric parameters (D, roll width W), and material properties of powder blends (δ, $\emptyset$, and K). Furthermore, the relationship between the material density and P_{max} was defined in Eq.(7) based on the assumption that the pre-consolidation pressure P_0 equals 1 MPa.

$$\gamma_R = \gamma_0 \left(\frac{P_{max}}{P_0}\right)^{\frac{1}{K}} \tag{7}$$

where γ_R is relative ribbon density, γ_0 is relative pre-consolidation blend density.

The elastic recovery rate β is defined as a ratio of ribbon thickness to roll gap. In the proposed gray-box model, K and β are assumed to be PP and material attribute dependent to capture unconsidered relationship in the modified rolling theory. The pre-consolidation density was assumed equal to bulk tapped density ρ_0 to provide further practicability. Utilizing regression models for K and β the ribbon density ρ_R and mass throughput M_R were calculated based on the Eq.(8) and Eq.(9).

$$\rho_R = \rho_0 (P_{max})^{1/K} / \beta \tag{8}$$

$$M_R = \pi DNWS\rho_0 (P_{max})^{1/K} \tag{9}$$

where N is roll speed. Gaussian process regression (GPR) was used as a regression method to cope with nonlinearity. In the regression analysis, material attributes of powder blends (bulk loose density, bulk tapped density, EAIF, WFA) and PP (roll force, roll gap, roll speed) were used as input parameters to predict K and β. The input and output parameters were centered by subtracting mean values and scaled by dividing by sample standard deviation (SD), which is a so-called auto-scaling. In the regression analysis, root mean squared errors of cross validation (RMSECV) and coefficient of determination (R2) for the ribbon density in leave-one-out cross validation (LOOCV) was used to show the validity of the gray-box model. The RMSECV and R2 for the white-box model according to the Reynolds' modified rolling theory were also evaluated using the mean K and β in the calibration set in LOOCV to compare the prediction performance.

3. Results

Figure 2 shows the prediction performances of the white-box and gray-box models in LOOCV. The gray-box model improved prediction accuracy of the white-box model in terms of the RMSECV (0.07 g/cc for white-box model and 0.04 g/cc for gray-box model) and R2 values (0.19 for white-box model and 0.54 for gray-box model). With the consideration of the sampling error in normal process variation, the prediction performance of the gray-box model for the ribbon density was practically sufficient for process development. The lower prediction accuracy in the white-box model could be derived from variations of the K and β in the calibration set. The RSD% of the K and β in the calibration set were 11.1% and 12.8%, respectively.

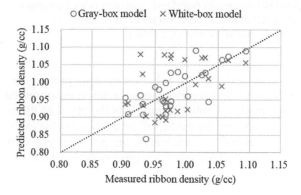

Figure 2 Scatter plot of measured and predicted ribbon density

To understand the relationship between the input and output parameters captured in the gray-box model, the effect of PP on the ribbon density and mass throughput was visualized into the contour plot as a function of roll gap and roll speed, see Figure 3a. The contour plot suggested that roll gap had a positive impact on both ribbon density and mass throughput. On the other hand, roll speed affected mass throughput only and had little impact on ribbon density. Same analysis can be performed to observe the effect of roll force on ribbon density and mass throughput. The reliability of the estimated impacts was also visualized using the normalized SD of the GPR prediction as shown in Figure 3b. The color progression from white to black represents the SD of the expected values in GPR, i.e., the predicted values of K and β. The highest SD is represented with black and color tones become closer to white as the SD decreases. In principle, the expected values with high SD in GPR prediction suggested that there are few data points in the calibration dataset around the predicting points and therefore the prediction is less reliable. Therefore, the PP with the higher SD of predicted K and β suggested that the predicted impacts are less reliable and will need additional experiments to clarify its actual responses. The threshold of SD to adopt the expected values and the gray-box model outputs would be determined considering actual variance of the observed data. These estimations provide a clear view on the control of PP to achieve target outputs and will be a justification for the future experiments.

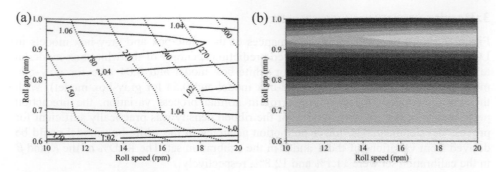

Figure 3 (a) Contour plots of the predicted ribbon density (solid line, g/cc) and mass throughput (dotted line, g/min). (b) Contour plot of the SD of the expected values in GPR. The color progression from white to black represents the SD of the expected values.

Overall, the applicability of the gray-box model for the roller compaction process was demonstrated. The gray-box model showed a higher prediction accuracy at the observed data points compared to the white-box model. A robust estimate of the effect of PP on the process outputs was presented by accounting for the SD of the expected value in the GPR model introduced in the gray-box model. In conclusion, the proposed gray-box model provides excellent applicability to the roller compaction process development.

4. Conclusions

The gray-box model composed of the rolling theory of granular solids for roller compaction process and the complemental regression models was presented to achieve high prediction accuracy and interpretability. The approach was found to provide reliable predictions for the process outputs such as ribbon density and mass throughput based on the material attributes, PP, and the geometric parameters. With the gray-box model the criticality of the PP were successfully visualized together with the reliability of prediction by means of the SD of the expected values, which would contribute to the process development activities such as risk assessment and design space setting.

References

I. Ahmad, A. Ayub, M. Kano, I.I. Cheema, 2020, Gray-box soft sensors in process industry: current practice, and fugure prospects in era of big data, Processes, 8, 243.

J.R. Johanson, 1965, A rolling theory for granular solids, Journal of Applied Mechanics, 32, 4, 842–848.

G. Reynolds, R. Ingale, R. Robert, S. Kothari, B. Gururajan, Practical application of roller compaction process modeling, Computers and Chemical Engineering 34, 1049-1057.

W. Shi, O.L. Sprockel, 2016, A practical approach for the scale-up of roller compaction process, 106, 15-19.

N. Souihi, M. Josefson, P. Tajarobi, B. Gururajan, J. Trygg, 2013, Design space estimation of the roller compaction process, Industrial & Engineering Chemistry Research, 52, 12408-12419.

E. N. M. Van sprang, H.-J. Ramaker, J.A. Westerhuis, A.K. Smilde, 2005, Statistical batch process monitoring using gray models, AIChE Journal, 51, 3 931-945.

Proceedings of the 14th International Symposium on Process Systems Engineering – PSE 2021+
June 19-23, 2022, Kyoto, Japan © 2022 Elsevier B.V. All rights reserved.
http://dx.doi.org/10.1016/B978-0-323-85159-6.50360-2

Multi-objective optimisation for early-stage pharmaceutical process development

Mohamad H. Muhieddine[a], Shekhar K. Viswanath[b], Alan Armstrong[c], Amparo Galindo[a], Claire S. Adjiman[a*]

[a]*Department of Chemical Engineering, The Sargent Centre for Process Systems Engineering, and Institute for Molecular Science and Engineering, Imperial College London, South Kensington Campus, London SW7 2AZ, United Kingdom*
[b]*Lilly Research Laboratories, Indianapolis, IN 46082, USA*
[c]*Department of Chemistry and Institute for Molecular Science and Engineering, Imperial College London, Molecular Sciences Research Hub, White City Campus, Wood Lane, London W12 0BZ, United Kingdom*
c.adjiman@imperial.ac.uk

Abstract

The pharmaceutical industry is under constant pressure to deliver its products quickly and effectively while minimising development costs and pursuing green pharmaceutical manufacturing methods. Given the many considerations in process development, a model-based method that takes multiple performance metrics into account is proposed for early process development. Several key performance indicators are identified, namely environmental footprint, cost, and conversion, selectivity, and yield. We employ multi-objective optimisation to assess the trade-offs between capital cost as one objective, and selectivity or conversion as a second objective, while exploring the interdependencies between all performance indicators. The approach is applied to two multiphasic reactions, each occurring in a 6-stage cascade CSTR: the hydrogenation of 4-Isobutylacetophenone (4-IBAP) to 1-(4-Isobutylphenyl)ethanol (4-IBPE) and the carbonylation of 4-IBPE to Ibuprofen (IBP).

Keywords: pharmaceutical process development, continuous manufacturing, multi-objective optimisation, Ibuprofen.

1. Introduction

The pharmaceutical industry faces increasing pressure to reduce development costs and time, and to meet increasingly stringent environmental regulations (Montes et al. 2017). Process Systems Engineering methods can help to improve process design/performance, enhance process understanding, and select optimal processing materials and operating conditions while exploring trade-offs between conflicting performance objectives. However, they have not yet been fully deployed for process development within the pharmaceutical and fine chemicals industries (Papadakis et al. 2018).

The complexity of the relationships between key process performance indicators (KPIs) such as feasibility, productivity, economics and environmental impact make it essential to adopt a holistic approach to process design. These KPIs are strongly linked to process-wide decisions such as process structure, unit size, operating conditions and even to molecular-level decisions (Adjiman et al. 2014). Pharmaceutical process development is a complex activity involving multiple pharmaceutically relevant objectives that need to be satisfied (Nicolaou and Brown, 2013). The coupling of process modelling with multi-objective optimisation allows for the quantification of process metrics that are of interest to pharmaceutical manufacturing, and the evaluation of trade-offs between conflicting KPIs. The use of model-based approaches to process development is especially relevant in continuous manufacturing, an increasingly important area, given the potential technical and economic benefits of implementing flow technology (McWilliams et al. 2018).

In this work, several relevant KPIs are identified and their use within a multi-objective design framework that can be employed in early-stage process development is explored. The approach is illustrated on the modelling and multi-objective optimisation of two multiphase reactors involved in the production of Ibuprofen via the Hoechst pathway (Elango et al. 1991), with a focus on moving to continuous production. Case Study 1 consists of a catalytic hydrogenation reactor to convert 4-Isobutylacetophenone (4-IBAP) to 1-(4-Isobutylphenyl)ethanol (4-IBPE), while Case Study 2 is focused on a homogeneous carbonylation reactor for the subsequent production of Ibuprofen (IBP).

2. Methodology

The methodology can be summarized as follows. First, a set of KPIs is defined and a reactor configuration is chosen. Next, a conceptual process model is developed using kinetic rate and gas solubility equations obtained from the literature or experimental investigations. Finally, a multi-objective optimisation problem involving KPIs as objectives and constraints is formulated and solved.

2.1. Selected KPIs and process configuration

For the optimal design of pharmaceutical processes, it is important to identify, quantify, and assess the interdependencies between pharmaceutically relevant KPIs. In this work, the KPIs of interest include the capital cost C_R ($) of the reactor system, the reaction selectivity S, the overall conversion X of raw materials, the yield Y of the pharmaceutical compound, and the Environmental factor (E-factor), defined as the mass ratio of waste product to desired product, and used to quantify environmental footprint. Mathematical expressions of these KPIs for Case Study 1 are shown in Table 1, where the expression of C_R is obtained from Douglas (1988). All symbols are defined in Table 2, where species j includes IBAP, IBPE, IBEB, H_2O and oligomers. Furthermore, the successful implementation of continuous flow technology for multiphase reactions requires efficient phase mixing and long residence times. While tubular reactors equipped with static mixers offer these features, they necessitate high volumetric flow rates which may be incompatible with slow reactions. Chapman et al. (2017) developed a multistage continuous-stirred tank reactor (CSTR) suitable for multiphasic reaction systems. Additionally, employing multiple CSTRs in series enhances system performance, minimising the total reactor volume required to achieve a specific conversion. Accordingly, a multi-stage cascade CSTR configuration is chosen for the studied multiphasic reactions. A schematic of such cascade reactor system with N reactors in series is shown in Figure 3.

2.2. Conceptual process model

A model of the cascade reactor system is developed to give a best case assessment of its performance. At this early stage of design, mass transfer limitations are neglected and thermodynamic equilibrium is assumed. The reactors are treated as isothermal.

2.2.1. Obtaining kinetic rate and gas solubility equations

Kinetic data for the hydrogenation of 4-IBAP over Pd/SiO$_2$ catalyst in *n*-decane as the solvent, following the scheme in Figure 1, are taken from Thakar et al. (2007). The solubility of hydrogen in the liquid phase is estimated using the Henry's law constant of Trinh et al. (2015). Kinetic and thermodynamic data for the carbonylation step in butanone (Figure 2) are taken from Seayad et al. (2003).

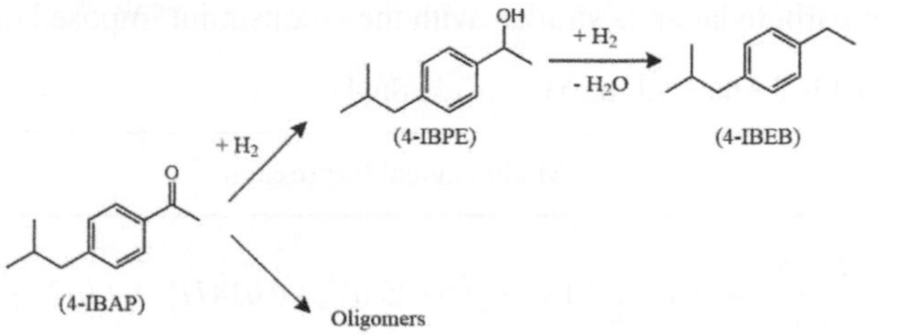

Figure 1 Reaction Scheme of the Catalytic Hydrogenation of 4-IBAP

The carbonylation of 4-IBPE is achieved using a homogeneous palladium complex catalyst dissolved in a water/butanone solvent mixture, as follows:

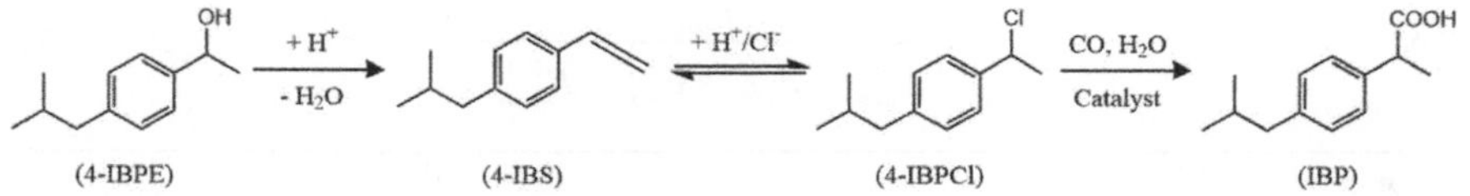

Figure 2 Reaction Scheme of the Homogeneous Carbonylation of 4-IBPE

2.2.2. Deriving flow reactor material balance equations of all reaction species

The hydrogenation system is modelled as a series of 6 triphasic CSTRs with side-stream addition of pure hydrogen. Similarly, the carbonylation reactor is modelled as a series of 6 biphasic CSTRs with side-stream addition of pure carbon monoxide. The steady-state material balance equations of all reaction species are then derived for each case study.

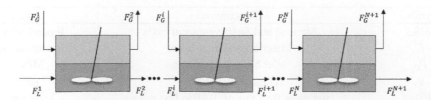

Figure 3 Gas-Liquid Cascade CSTR System

2.3. Formulating and solving the multi-objective reactor optimisation problems

For Case Study 1, the objectives of the optimisation problem are minimising cascade reactor capital cost and maximising overall conversion. For Case Study 2, the objectives are minimising reactor cost and maximising selectivity for all $X \geq 90\%$. The optimisation problems are solved using the ε-constraint method of Haimes et al. (1971). The second objective of Case Study 1 is transformed into a constraint, lower-bounding conversion by a series of ε values ranging between 50% and 98%. Similarly, the second objective of Case Study 2 is transformed into a constraint, lower-bounding selectivity by a series of ε values ranging between 70% and 98%. The identified KPIs are reported for every solution. Bounds on decision variables are imposed based on the experimental conditions of the corresponding kinetic study. The formulated problem is nonlinear and is solved using the CONOPT solver in the GAMS software, version 25.0.3. The formulation for carbonylation is similar, with the ε-constraint imposed on selectivity.

Table 1 Reactor KPIs for the Hydrogenation Case Study

KPI	Mathematical Expression	
C_R	$\dfrac{M\&S}{280} \times 482.37 \times \left[2.18 + F_m \left(0.0255 P_{H_2}^2 + 0.0387 P_{H_2} + 1.0136 \right) \right] \times V_T^{0.6287}$	(1)
X	$\dfrac{F_{IBAP}^1 - F_{IBAP}^{N+1}}{F_{IBAP}^1}$	(2)
S	$\dfrac{F_{IBPE}^{N+1}}{F_{IBPE}^{N+1} + F_{IBEB}^{N+1} + F_{H_2O}^{N+1} + F_{oligomers}^{N+1}}$	(3)
Y	$\dfrac{F_{IBPE}^{N+1}}{F_{IBAP}^1}$	(4)
$E-factor$	$\dfrac{\dot{m}_{n-decane} + \dot{m}_{H_2,L}^{N+1} + \dot{m}_{IBAP}^{N+1} + \dot{m}_{IBEB}^{N+1} + \dot{m}_{H_2O}^{N+1} + \dot{m}_{oligomers}^{N+1}}{\dot{m}_{IBPE}^{N+1}}$	(5)

Table 2 Nomenclature for the hydrogenation case study

Symbol	Description	Units
$M\&S$	Marshall and Swift equipment index	dimensionless
F_m	Material correction factor	dimensionless
P_{H_2}	Inlet hydrogen pressure	MPa
V_T	Total reactor volume	m^3
F_{IBAP}^1	Inlet molar flow rate of IBAP to first reactor	kmol/sec
F_j^{N+1}	Outlet molar flow rate of species j from reactor N	kmol/sec

$\dot{m}_{n-decane}$	Mass flowrate of n-decane	kg/s
$\dot{m}_{H_2,L}^{N+1}$	Outlet mass flowrate of dissolved H_2 from reactor N	kg/s
$\dot{m}_j^{N+1}$	Outlet mass flowrate of species j from reactor N	kg/s
F_G^i	Inlet molar flow rate of gas stream to reactor i	kmol/sec
F_L^i	Inlet molar flow rate of liquid stream to reactor i	kmol/sec

3. Results and discussion

The variation of normalised cascade reactor capital cost and other reactor performance metrics is shown for the two case studies (Figure 4) for different values of ε. The hydrogenation reaction mechanism involves three reactions, with 4-IBPE being hydrogenated upon its formation to produce 4-isobutylethylbenzene (4-IBEB) and water side products. Furthermore, the condensation of 4-IBAP produces oligomers as additional side products. This explains the decrease in 4-IBPE selectivity with increasing 4-IBAP conversion. The reaction yield increases across a wide range of conversions, but this trend is reversed at higher conversions where the concentration of side products exceeds that of 4-IBPE. On the other hand, the E-factor generally decreases with conversion due to the consumption of reacting materials, but increases at higher conversions due to side product formation and the reduction in the concentration of 4-IBPE. The carbonylation mechanism also involves three reactions, with IBP being produced in the third reaction. Once again non-monotonic behaviour is observed for the E-factor as a function of conversion, whereas selectivity and conversion are now found to follow the same trends. Generating performance metric plots such as those in Figure 4 can assist pharmaceutical manufacturers in understanding the trade-off and synergies that exist for a given reaction route, and make it possible to consider alternatives across a range of criteria.

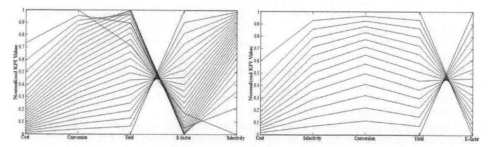

Figure 4 Trade-offs between KPIs for the hydrogenation (left) and carbonylation (right) case studies, shown as parallel coordinate plots. The two objectives are shown as the first two (leftmost) coordinates in both cases.

4. Conclusions

We have presented a set of KPIs for the assessment of process designs at the early stage, namely reactor capital cost, selectivity, E-factor, conversion, and yield. These can be used within a multi-objective optimisation framework to assess the trade-offs between different performance metrics, hence enabling more informed decisions on possible routes for pharmaceutical process design. This approach was illustrated by investigating two continuous multiphase reactors involved in Ibuprofen synthesis. The approach is currently being extended to include separation processes.

Acknowledgments

Funding from Eli Lilly and Company and the UK EPSRC, through the PharmaSEL-Prosperity Programme (EP/T005556/1), is gratefully acknowledged. AG acknowledges funding from the Royal Academy of Engineering and Eli Lilly and Company for support of a Research Chair (Grant RCSRF18193).

References

C.S. Adjiman, A. Galindo, and G. Jackson, 2014, Molecules matter: the expanding envelope of process design, Computer Aided Chemical Engineering, 34, pp. 55-64

M.R. Chapman, M.H. Kwan, G. King, K.E. Jolley, M. Hussain, S. Hussain, I.E. Salama, C. González Niño, L.A. Thompson, M.E. Bayana, A.D. Clayton, B.N. Nguyen, N.J. Turner, N. Kapur, and A.J. Blacker, 2017, Simple and versatile laboratory scale CSTR for multiphasic continuous-flow chemistry and long residence times, Organic Process Research & Development, 21(9), pp.1294-1301

J.M. Douglas, 1988, Conceptual design of chemical processes, 1110, McGraw-Hill New York

V. Elango, M. Murphy, B.L. Smith, K.G. Davenport, G.N. Mott, E.G. Zey, and G.L. Moss, 1991, Method for producing ibuprofen

Y. Y. Haimes, L. Lasdon, and D. Wismer, 1971, On a Bicriteria Formulation of the Problems of the Integrated System Identification and System Optimization, IEEE Transaction on Systems, Man, and Cybernetics, SMC-1, pp. 296-297

J.C. McWilliams, A.D. Allian, S.M. Opalka, S.A. May, M. Journet, and T.M. Braden, 2018, The evolving state of continuous processing in pharmaceutical API manufacturing: a survey of pharmaceutical companies and contract manufacturing organizations, Organic Process Research & Development, 22(9), pp.1143-1166

F. Montes, K.V. Gernaey, and G. Sin, 2017, Uncertainty and Sensitivity Analysis for an Ibuprofen Synthesis Model Based on Hoechst Path, Computer Aided Chemical Engineering, 40, pp. 163-168

C.A. Nicolaou and N. Brown, 2013, Multi-objective optimization methods in drug design, Drug Discovery Today: Technologies, 10(3), pp. 427-435

E. Papadakis, J.M. Woodley, and R. Gani, 2018, Perspective on PSE in pharmaceutical process development and innovation, Computer Aided Chemical Engineering, 41, pp. 597-656

A.M. Seayad, J. Seayad, P.L. Mills, and R.V. Chaudhari, 2003, Kinetic modeling of carbonylation of 1-(4-isobutylphenyl)ethanol using a homogeneous $PdCl_2(PPh_3)_2$/TsOH/LiCl catalyst system, Industrial & engineering chemistry research, 42(12), pp. 2496-2506

N. Thakar, R.J. Berger, F. Kapteijn, and J.A. Moulijn, 2007, Modelling kinetics and deactivation for the selective hydrogenation of an aromatic ketone over Pd/SiO_2, Chemical Engineering Science, 62(18-20), pp.5322-5329

T.K.H Trinh, J.C. de Hemptinne, R. Lugo, N. Ferrando, and J.P. Passarello, 2015, Hydrogen solubility in hydrocarbon and oxygenated organic compounds, Journal of Chemical & Engineering Data, 61(1), pp.19-34

Proceedings of the 14th International Symposium on Process Systems Engineering – PSE 2021+
June 19-23, 2022, Kyoto, Japan © 2022 Elsevier B.V. All rights reserved.
http://dx.doi.org/10.1016/B978-0-323-85159-6.50361-4

Quality by design and techno-economic modelling of RNA vaccine production for pandemic-response

Zoltán Kis[a,b*], Kyungjae Tak[a], Dauda Ibrahim[a], Simon Daniel[a], Damien van de Berg[a], Maria M Papathanasiou[a], Benoît Chachuat[a], Cleo Kontoravdi[a], and Nilay Shah[a]

[a] *Centre for Process Systems Engineering, Department of Chemical Engineering, Imperial College London, South Kensington Campus, London SW7 2AZ, United Kingdom*
[b] *Department of Chemical and Biological Engineering, The University of Sheffield, Mappin St, Sheffield S1 3JD, UK*
Corresponding Author's E-mail: z.kis@sheffield.ac.uk

Abstract

Vaccine production platform technologies have played a crucial role in rapidly developing and manufacturing vaccines during the COVID-19 pandemic. The role of disease agnostic platform technologies, such as the adenovirus-vectored (AVV), messenger RNA (mRNA), and the newer self-amplifying RNA (saRNA) vaccine platforms is expected to further increase in the future. Here we present modelling tools that can be used to aid the rapid development and mass-production of vaccines produced with these platform technologies. The impact of key design and operational uncertainties on the productivity and cost performance of these vaccine platforms is evaluated using techno-economic modelling and variance-based global sensitivity analysis. Furthermore, the use of the quality by digital design framework and techno-economic modelling for supporting the rapid development and improving the performance of these vaccine production technologies is also illustrated.

Keywords: techno-economic modelling; Quality by Design (QbD) modelling; process development; RNA vaccine production.

1. Introduction

Process Systems Engineering tools have a lot to offer and are not applied to their full potential in vaccine and biopharmaceutical product-process development, and during production process operation. Over the past decades, substantial progress has been made in the field of computational modelling and mechanistic, dynamic, machine learning and hybrid models have been successfully implemented in various manufacturing fields, outside of vaccine and biopharmaceutical manufacturing. These digital tools have been used to create a digital replica (or digital twins) of the manufacturing process. Vaccine and biopharmaceutical production are lagging behind in digitalisation, because vaccines and biopharmaceuticals are conventionally produced using cell-based processes that, due to their inherent complexity and variability, have been challenging to model. In addition, vaccine manufacturing is highly regulated, and improvements are not implemented rapidly to avoid the risk of negatively impacting product quality, safety, and efficacy. Modelling of complex biological systems, digitalisation, real-time monitoring, process control, automation, and knowledge-rich regulatory submissions are hindered by the lack of real-time or near-real-time hardware sensors for measuring vaccine and biopharmaceutical quality attributes. This is because several critical quality attributes (CQAs) and parameters are difficult, time-consuming, or expensive to measure or estimate in real-time, drastically limiting the information available for developing computational models. To overcome these limitations, software sensors are being developed.

To our knowledge, the quality of vaccines is currently assured without taking advantage of digitalisation and is currently tested after every production batch. Batches that fail to yield the product quality specifications are discarded, wasting valuable resources. Quality assurance could be improved by real-time product quality monitoring and by using model-predictive control. Assuring product quality with such digital tools would fit perfectly into the QbD framework (CMC-Vaccines Working Group, 2012). The use of the QbD framework is supported by regulatory authorities for systematic co-development of the vaccine product, vaccine production processes and of the process control strategies, based on sound science and quality risk management (ICH Expert Working Group, 2009). As far as we know, a full QbD framework has not been implemented for this purpose. The QbD framework combined with digital tools is also referred to as the Quality by Digital Design (QbDD) framework and this has the potential to replace quality by testing with assuring product quality by the design and operation of the production process.

Besides QbD modelling, techno-economic modelling also offers a valuable tool for assessing the productivity and cost profile of the holistic production process (Ferreira and Petrides, 2021; Ferreira et al., 2021; Kis et al., 2021a, 2021b, 2020; Pereira Chilima et al., 2020). Moreover, this process-cost modelling approach also helps to identify the production bottlenecks, and then de-bottlenecking approaches are evaluated to increase process performance. Techno-economic modelling is also used to evaluate various scenarios, for example different downstream configurations, at different production scales to identify the process configuration that leads to maximum productivity and lowest cost (Kis et al., 2021a, 2021b, 2020). Additionally, uncertainty and sensitivity analysis is performed combined with techno-economic modelling, to identify how the co-variation of many uncertainties would impact production throughputs and resource requirements (Kis et al., 2021b).

In this work, we showcase the use of the QbDD framework together with techno-economic modelling to guide the development and operation of new vaccine production platform technologies, such as the messenger RNA (mRNA), self-amplifying RNA (saRNA) and adenoviral vectored (AVV) vaccine platforms.

2. The AVV, mRNA and saRNA vaccine production platform technologies

The AVV production process was modelled based on the manufacturing of the replication-deficient chimpanzee adenovirus-vectored (ChAdOx1) vaccine which was co-developed by Oxford University and AstraZeneca plc (Kis et al., 2021b). The ChAdOx1 production process starts with preparing the HEK293 cell seed train and the adenovirus inoculum seed train. For this, the HEK293 cells are cultured at increasing volumes until the culture amounts required for the production bioreactor (commonly at 2000 L working volume) scale are obtained. These cells are then infected with the adenovirus which was genetically modified to express the SARS-CoV-2 spike protein. Following virus replication in HEK293 cells in the bioreactor, the virus culture and cell culture enter the downstream purification, whereby cells are initially lysed then the larger impurities are removed using microfiltration. Next, tangential flow ultrafiltration/diafiltration is carried out, followed by an ion-exchange chromatography step. After this, the adenoviral vector solution is sterile filtered, and the buffer can be exchanged for the formulation buffer, using tangential flow ultrafiltration/diafiltration. Subsequently, the adenovirus vaccine drug substance (active ingredient) is formulated and filled into vials or other containers, often at a different facility / location (Kis et al., 2021b).

The mRNA and saRNA vaccines (collectively referred to as RNA vaccines) are synthesised using the T7 RNA polymerase based on a DNA template in the *in vitro* transcription reaction, which is usually completed in 2 hours, substantially faster than AVV production (Kis et al., 2021b). Following RNA synthesis, the plasmid DNA is digested using the DNAse I endonuclease enzyme. Next, the RNA is purified out of the reaction mix using a series of conventional filtration- and chromatography-based unit operations (Kis et al., 2021a, 2020). These can include tangential flow ultrafiltration and diafiltration combined with one or two of the following chromatography techniques: ion-exchange, reverse-phase, oligo dT affinity, hydroxyapatite, hydrophobic interaction, multimodal hydrogen bonding and anion exchange, cellulose-based, and multimodal core-beads. After the RNA is purified out of the enzymatic reaction mix, the RNA is encapsulated in lipid nanoparticles. For this, the four lipid components contained in an ethanol stream are mixed with the RNA contained in an aqueous stream (e.g. in citric acid buffer). The mixing of the lipids with RNA can be achieved using a mixing device based on: microfluidics, T-junction, impingement jet, vortex, or pressurised stainless-steel tanks. Following formulation, the solution is sterile filtered and shipped to the fill-to-finish site for filling into glass vials or other containers (Kis et al., 2021b).

3. Techno-economic modelling of the AVV, mRNA and saRNA platforms

Rapid and global response to pandemics by mass vaccination is currently limited by the rate at which vaccine doses can be manufactured on a global scale. Here techno-economic modelling is presented for the AVV, mRNA and saRNA vaccine production platform technologies that were deployed during the COVID-19 pandemic. Unlike AVV and mRNA vaccines, several of which were approved by the regulatory authorities, the saRNA platform is not yet deployed at commercial scale for vaccines, with saRNA vaccines still undergoing clinical development. Herein, a combination of techno-economic modeling and variance-based global sensitivity analysis (GSA) is applied. This quantifies the performance of each platform in terms of their productivity and resource requirements, subject to key design and operational uncertainties, cf. Figure 1. GSA was carried out by interfacing SobolGSA with SuperPro Designer via MatLab and Excel Visual Basic for Applications (VBA). For GSA, 10,000 simulations were performed for each of the three platform technologies. For these simulations, model inputs were quasi-randomly sampled from a seven dimensional input space using Sobol sequences, as previously described (Kis et al., 2021b). These seven model inputs are: the scale of the production process, batch failure rate, titre/yield in the production bioreactor, cost of raw materials, cost of labour, drug substance amount per dose and cost of quality control (Kis et al., 2021b).

Cost and productivity results from the techno-economic modelling and GSA are shown below in Figure 2. The ranges and probability distributions of the number of drug substances and finished drug product doses that can be produced based on a one billion USD investment in operating expenses (OpEx) for the three platform technologies are shown in the violin plots in Figure 2A. A one billion USD investment in OpEx will produce a median of 2.66 (IQR=2.44-2.83) billion AVV drug product doses, a median of 0.95 (IQR=0.74-123) billion mRNA drug product doses and a median of 2.48 (IQR=2.36-2.58) billion saRNA drug product doses. OpEx includes the annualised capital costs, however it is worth noting that investment in facilities must be made upfront, because constructing, equipping, validating and starting up production can take several years. The ranges and probability distribution of the cost per dose for the drug substance and finished drug product for the three platform technologies is shown in Figure 2B. The drug product manufacturing cost per dose is 0.38 (IQR=0.35-0.41), 1.05 (IQR=0.81-1.35), 0.4 (IQR=0.38-0.42) for AVV, mRNA and saRNA vaccines, respectively.

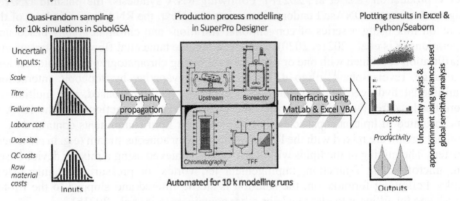

Figure 1. A computational framework for uncertainty quantification for AVV, mRNA and saRNA production. The uncertainty is propagated from the inputs via the model to the outputs. In addition, the sensitivity of the model output key performance indicators (KPIs) is attributed to the individual inputs, to determine the degree to which individual inputs impact the output KPIs. Modified from (Kis et al., 2021b).

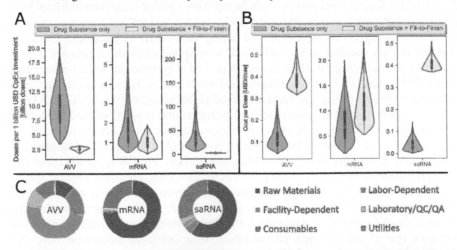

Figure 2. Cost distributions associated with AVV, mRNA and saRNA vaccine production. **A.** Violin plots showing the distribution of the estimated number of doses produced based on a 1 billion USD investment in operating expenses (OpEx). The OpEx contains the annualised facility costs. **B.** Violin plots showing the distribution of cost per dose values for AVV, mRNA and saRNA vaccine production. **C.** Doughnut charts showing the distribution of OpEx, the annualised capital costs are included in the facility-dependent costs.

In the centre of all the violin plots, box and whisker plots are shown with the median values indicated by the white dots; the 25th and 75th percentiles with the top and bottom of the boxes; and minimum and maximum values, excluding outliers, with the ends of the whiskers. The width of the violin plots represents the probability distributions. Figure 2C shows the breakdown of the annual production costs for the baseline scenarios (c.f. (Kis et al., 2021b)) for these three vaccine platform production technologies. Fixed costs dominate the AVV production costs, whereas mRNA and saRNA vaccines production is driven by variable costs. This implies that maintaining surge capacity based on the RNA platform will be more cost effective than based on the AVV platform. Fill-and-finish was modelled with 10-dose vials for AVV vaccines and 5-dose vials for mRNA and saRNA vaccines.

4. Integration of QbD and techno-economic modeling with the RNA platform

The mRNA vaccine production platform technology has been proven clinically successful during the COVID-19 pandemic. The cell-free nature and consequently the relative simplicity (compared to cell-based vaccine production) makes the RNA platform technology ideal for digitalisation and advanced automation with the QbDD framework. The integration of the QbDD framework with the RNA platform will accelerate product-process development, enhance production rates and production volumes, reduce costs, and assure high product quality. Moreover, the RNA vaccine production platform and the QbDD framework will form a powerful synergy as both tools are disease agnostic. This synergy will use prior platform knowledge, experimental data, clinical data, quality risk management and digital tools to accelerate product and process development. This will also accelerate and streamline the regulatory approval process based on knowledge-rich regulatory submissions and demonstrated product knowledge and process understanding.

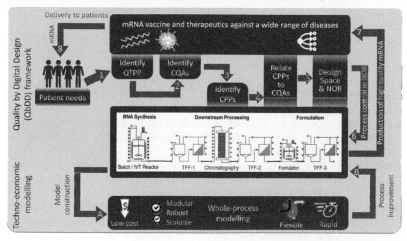

Figure 3. Integration of techno-economic and QbDD modelling with the RNA vaccine production process. Abbreviations: QbDD - Quality by Digital Design, QTPP - Quality Target Product Profile, CQAs – product Critical Quality Attributes for safety and efficacy, CPP - Critical Process Parameters, NOR - Normal Operating Range, within the design space.

The product specifications will be based on product performance instead of batch history, and the focus will shift from reproducibility to process-product robustness (Kis et al., 2020; van de Berg et al., 2021). In addition, the QbDD framework and the digital tools will be used to automate RNA vaccine manufacturing, building quality into the design and operation of the process. These features of the QbDD framework will also support scale-up and technology transfer. In addition, techno-economic modelling will guide cost-reduction, de-bottlenecking and improved process performance. The interplay between the RNA vaccine platform technology, the QbDD framework and techno-economic modelling is shown above in Figure 3.

5. Conclusions

In conclusion, on top of having surge vaccine manufacturing capacity available for future outbreaks, modelling tools such as those presented here can further accelerate vaccine development and improve the performance of the vaccine manufacturing processes. The combination of vaccine platform technologies and these disease-agnostic modelling tools provides a powerful approach for rapid-response vaccine deployment against currently known and unknown diseases.

Acknowledgements: This research was funded by the Department of Health and Social Care using UK Aid funding and is managed by the Engineering and Physical Sciences Research Council (EPSRC, grant number: EP/R013764/1). The views expressed in this publication are those of the author(s) and not necessarily those of the Department of Health and Social Care. Funding from UK Research and Innovation (UKRI) via EPSRC grant number EP/V01479X/1 on COVID-19/SARS-CoV-2 vaccine manufacturing and supply chain optimisation is thankfully acknowledged.

References

CMC-Vaccines Working Group, 2012. A-Vax: applying quality by design to vaccines.

Ferreira, R., Petrides, D., 2021. Messenger RNA (mRNA) Vaccine Large Scale Manufacturing – Process Modeling and Techno-Economic Assessment (TEA) using SuperPro Designer. https://doi.org/10.13140/RG.2.2.12780.28800

Ferreira, R.G., Gordon, N.F., Stock, R., Petrides, D., 2021. Adenoviral Vector COVID-19 Vaccines: Process and Cost Analysis. Processes. https://doi.org/10.3390/pr9081430

ICH Expert Working Group, 2009. ICH harmonised tripartite guideline on pharmaceutical development Q8 (R2) (No. August), ICH guideline Q8 (R2) on pharmaceutical development. International Council for Harmonisation of Technical Requirements for Pharmaceuticals for Human Use (ICH), Geneva, Switzerland.

Kis, Z., Kontoravdi, C., Dey, A.K., Shattock, R., Shah, N., 2020. Rapid development and deployment of high-volume vaccines for pandemic response. J. Adv. Manuf. Process. 2, e10060. https://doi.org/10.1002/amp2.10060

Kis, Z., Kontoravdi, C., Shattock, R., Shah, N., 2021a. Resources, Production Scales and Time Required for Producing RNA Vaccines for the Global Pandemic Demand. Vaccines. https://doi.org/10.3390/vaccines9010003

Kis, Z., Tak, K., Ibrahim, D., Papathanasiou, M.M., Chachuat, B., Shah, N., Kontoravdi, C., 2021b. Pandemic-response adenoviral vector and RNA vaccine manufacturing. medRxiv. https://doi.org/10.1101/2021.08.20.21262370

Pereira Chilima, T.D., Moncaubeig, F., Farid, S.S., 2020. Estimating capital investment and facility footprint in cell therapy facilities. Biochem. Eng. J. 155, 107439. https://doi.org/10.1016/j.bej.2019.107439

van de Berg, D., Kis, Z., Behmer, C., Samnuan, K., Blakney, A., Kontoravdi, C., Shattock, R., Shah, N., 2021. Quality by Design modelling to support rapid RNA vaccine production against emerging infectious diseases. NPJ Vaccines 6, 1–10. https://doi.org/10.1038/s41541-021-00322-7

Proceedings of the 14th International Symposium on Process Systems Engineering – PSE 2021+
June 19-23, 2022, Kyoto, Japan © 2022 Elsevier B.V. All rights reserved.
http://dx.doi.org/10.1016/B978-0-323-85159-6.50362-6

Design of Value Function Trajectory for State of Control in Continuous Manufacturing System

Tomoyuki TAGUCHI [a*], Toshiyuki WATANABE [b*], Shigeru KADO [b*] and Yoshiyuki YAMASHITA [c]

[a]Chiyoda Corporation, Pharmaceutical Engineering Department, 4-6-2, Minatomirai, Yokohama, Kanagawa 220-8765, JAPAN
[b]Chiyoda Corporation, Research and Development Center, 3-13, Moriya-cho, Yokohama, Kanagawa 221-0022, JAPAN
[c] Department of Chemical Engineering, Tokyo University of Agriculture and Technology, Tokyo 184-8588, JAPAN
taguchi.tomoyuki@chiyodacorp.com

Abstract

Recently, pharmaceutical manufacturing has been aimed at incorporating more efficient production systems for easy scale-up, higher quality and lesser usage of solvent, and observation of the operation state. A continuous manufacturing system can enable a system design that maintains the desired process conditions, with suitable devices and measurements. The process systems engineering (PSE) approach is very helpful for the visualization of the operation state, and the compensation of system features of process dynamics and disturbances. The investigated design space provides valuable information regarding the process and control strategy. The combination of steady-state data sets among control, process, and objective variables can visualize the range of the allowable operation zone, and desired objective control state. The proposed value function trajectory is an attractive method for designing tracking control as a multivariable function of the Hamiltonian, which connects the consistent approach of the design of process equipment, structure, and quality control.

Keywords: Process Design and Control, Continuous, Design Space, Optimization.

1. Introduction

Lee (2015) stated the kinds of features of continuous manufacturing, such as small footprints, short supply chain, stable operation with good monitoring, combination of processes with no-stop handling, easy scale-up, new synthetic routes, safe operation, and efficient high-throughput production. The understanding of the process features in the early design stage is a way to achieve agility, flexibility, and robustness of continuous manufacturing. Myerson (2014) listed the needs of technologies for innovating manufacturing systems: comprehension of process steady-state and dynamics, design of monitoring and control systems, development of systems integration, and data analysis to understand the design space, considering disturbances, nonlinearities, constraints, and uncertainties. It was an effective approach for reducing various operational risks. Diab (2020) described examples of visualization for certain active pharmaceutical ingredients and unit operations, under the consideration of structuring a continuous process. This method was highly attractive for both process design and operation philosophy.

To realize a continuous pharmaceutical manufacturing system, the comprehension of material and reaction properties, a priori knowledge of transport phenomena, and the innovation of devices and instruments must be closely integrated based on proven experimental data. Such basic data should be transferred to the control design and operational support through system design.

In this study, a novel design concept of using the steady-state calculation was proposed. To understand process characteristics, the design space was determined according to the process design and operational philosophy. In this process, the objective control state and possible operational zone could be defined by introducing a value function as the potential energy of the system characteristics. Through the implementation of this approach, The Hamiltonian expression combining the potential energy as the value function, and the kinetic energy as the process dynamics, could be derived from the case study on steady-state calculation, which consequently resulted in the structuring of a feedback control system.

2. Process control with a priori knowledge

The issues of control and process optimization are strongly influenced by the chemistry, process, control, and operation. Plutschack et al. (2018) illustrated the types of heat and mass transport measurements, and the numerous examples of instruments related to various chemistries. The comprehension of system features in the wide range of materials and process attributes was achieved, a process design was conducted using mathematical modelling and simulation. Larsson (2000) reviewed the plantwide control that imposed structural design: selection of controlled variables, manipulated variables, measurements, control configuration, and type of controller were the issues of determination. They strongly depended on the process structure, followed by the degrees of freedom; therefore, the focus was on the process-oriented approach. Shalifzadeh (2013) summarized the integration approach for process design and control. The traditional approach for process design and control is sequential, that is, the control design is only conducted for rigid manufacturing systems. The decision-making of system design between process and control, and the competition between efficiency and controllability are the main issues from the perspective of the designer and planner. The proposed method of trajectory tracking control was aimed at combining the acquisition of process a priori knowledge, and the design of control structure, like the simultaneous approach. The objective control state, and surrounding allowable operation zone were achieved as the dimensional reduction information through the constitution of the design space. To incorporate such a priori knowledge into a feedback control system, a data bank that included the value function and process state relationship, was implemented by the Hamiltonian. They were extracted from the steady-state simulation model for using a process model in an environment of low computational cost, and an easy-to-handle mathematical model.

3. Reactor Simulation Model

The reaction process plays a major role in the pharmaceutical and chemical industries. Over-reaction and by-product generation are the conventional problems of organic synthesis reaction. The following first-principle model is provided to illustrate the proposed approach:

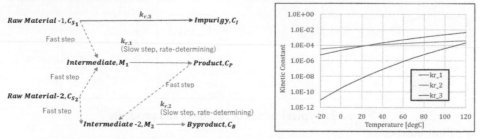

Figure 1: Synthesis route and temperature dependence of the three kinetic constants

3.1. Constraints of the reaction system

1) Reaction paths

   The following three reaction paths were provided to consider the typical non-linearity of the organic synthesis route, as shown in Figure 1.

   A) The second-order reaction of two raw materials for generating the main product.
   $$r_P = k_{r_1} C_{S_1} C_{S_2} \qquad (1)$$

   B) The second-order reaction of the main product and one of the raw materials for generating the by-product of over-reaction.
   $$r_B = k_{r_2} C_{S_2} C_P \qquad (2)$$

   C) The first-order reaction of one of the raw materials for generating the impurity.
   $$r_I = k_{r_3} C_{S_1} \qquad (3)$$

2) Kinetic constants

   Three kinetic constants whose characteristics were dependent on the temperature change, are shown in Figure 1.

3.2. Reactor configuration, size and flow condition

1) Reactor configuration

   A typical plug flow reactor, with a jacket to handle the liquid-phase reactants, was considered as the test bed for this study (Figure 2). The temperature of the jacket colorant was assumed to be 40 °C, and UA = 200 W/m²-K.

2) Size

   Internal diameter = 30 mm, Length = 500 mm were assumed.

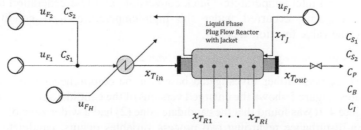

Figure 2: Typical plug flow reactor with inlet heat exchanger and colorant jacket for temperature control

Table 1: Study range of volume flow, molar concentration and temperature

Volume Flow Rate [mL/min.]		Molar Concentration [mol/m³]		Temp. [°C]
u_{F_1}	u_{F_2}	C_{S_1}	C_{S_2}	$x_{T_{in}}$
50~500	50~500	100~500	100~500	20~100

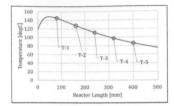 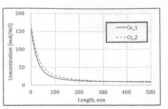 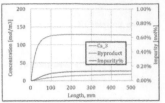

Figure 3: Typical trends of reactants and products

3) Flow balance

The following ranges (Table 1) of volume flow, molar concentration, and temperature were assumed.

3.3. Typical trends of reactants and products

Figure 3 shows an example of the simulation results. The inlet flow condition and temperature set-point strongly affected the resultant by-product and impurities.

4. Design Space

4.1. Determination of value function

To consider the non-linear constrains of reactants, the following value function, V was employed to design the objective control state. Therein, ψ_i is the weight factor, ξ_c is the conversion of substrates, ξ_y is the product yield, ξ_i is the impurity in mol%, with an upper limit of 0.01 mol%, and z_p is the assumed target production rate, in kg/h.

$$V = \psi_1(\xi_c - 1.0)^2 + \psi_2(\xi_y - 1.0)^2 + \psi_3\xi_i + \psi_4(z_p - 2.0)^2 \qquad (4)$$

In case, ξ_i is higher than the assumed upper limit of 0.01 mol%, ψ_3 becomes zero. Then, the optimized value function becomes zero under the corresponding control variables, $u_{F_{opt}}$, and state variables, $x_{T_{opt}}$.

4.2. Comprehension of wide-range process attributes

The overall comprehension of the possible process variable changes during plant operation was investigated. By changing the control variables listed in Table 1, followed by the state variables, the corresponding reactant results could be achieved through the simulation case studies. Figure 4 shows the results of the value function with respect to residence time, and inlet temperature. Quick conversion, and less generation of the by-product and impurity occurred under the high-temperature feed condition, and subsequently, the value function was low.

4.3. Robust design space against process disturbances

The design space was further investigated based on the comprehension of the overall process features. Figure 5 shows the enlarged versions of the candidate zones (1), and (2), shown in Figure 4. It was found that the candidate zone (2) had a wider zone of low-value function. If a disturbance regarding two process variables occurs, candidate zone (2) would be expected in the robust control state.

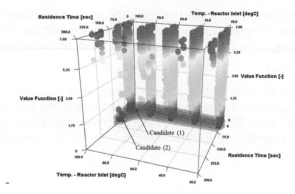

Figure 4: Wide-range three-dimensional plot of the value function

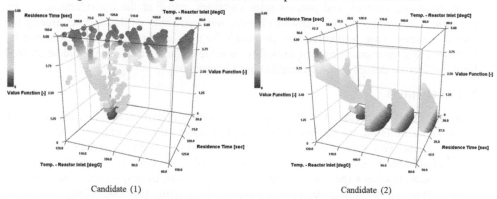

Candidate (1)   Candidate (2)

Figure 5: Narrow-range three-dimensional plot of the value function

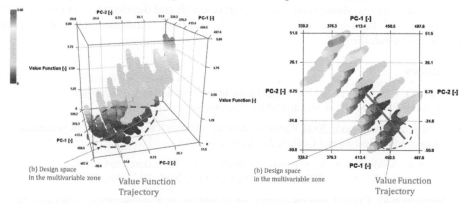

Figure 6: Trajectory of the value function and the allowable operation state

5. Trajectory Tracking Control

5.1. Determination of value function trajectory

The value function was affected not only by the residence time and inlet temperature, but also the feed flow and concentration change. To consider such a multivariable effect on the value function, the x-y axes of the candidate zone (2) in Figure 5 were transformed to the principle components, PC-1 and PC-2; and the results are shown in Figure 6.

The operation range of low-value function is clearly expressed in Figure 6, with the group of minimum values as optimized line in the design space, and the allowable operation range could be designed as the multivariable zone around the value function trajectory.

5.2. Proposed control

The trajectory and the surrounding allowable zone were considered as the potential function, and its features were quadratic. Ortega (2002) utilized passivity-based control to design robust controllers for the Euler-Lagrange equations of motion. This consideration were applied to the Hamiltonian function, described by the following equation:

$$H = V^\circ + V + F \tag{5}$$

where, V° is the value function at the optimal condition, V is the relationship between the value function, control, and process state variables, and F is the consideration of the dynamic state equation. The stepwise change, and sensitivity of the data set among the control, process state, and value function, could be translated to each coefficient matrix. It showed that the consequence of work flow to achieve the design space could be utilized for the effective design of multivariable control structures.

6. Conclusions

Using the reactor simulation model, the design space was considered, and the trajectory of the value function and the allowable operation zone were determined. The designed control space with the trajectory of the value function could be transformed into a multivariable feedback controller based on the Hamiltonian concept. To design a control system incorporating experimental and a priori knowledge of engineering, the proposed design, that is, a design-directed approach, is useful for combining the good corroborations of the process and the control design. The different functions of the cross-sectional departments, throughout the stages of research and development of commercial operation, could be interactively connected by utilizing fundamental knowledge of mathematical modelling.

References

S. Diab and D. I. Gerogiorgis, 2020, Design Space Identification and Visualization for Continuous Pharmaceutical Manufacturing, Pharmaceutics, 12, 3, 235.

T. Larsson and S. Skogestad, 2000, Plantwide control—A review and a new design procedure, Modeling, Identification and Control, 21, 4, 209-240.

S. L. Lee et al., Modernizing Pharmaceutical Manufacturing: from Batch to Continuous Production, 2015, Journal of Pharmaceutical Innovation, 2015, 10, 3, 191-199.

A. S. Myerson et al., Control systems engineering in continuous pharmaceutical manufacturing May 20–21, 2014 continuous manufacturing symposium, Journal of pharmaceutical sciences, 2015, 104, 3, 832-839.

R. Ortega et al., Interconnection and damping assignment passivity-based control of port-controlled Hamiltonian systems, Automatica, 2002, 38, 4, 585-596.

M. B. Plutschack et al., The Hitchhiker's Guide to Flow Chemistry, Chemical reviews, 2017, 117, 18, 11796-11893.

M. Sharifzadeh, Integration of process design and control: A review, Chemical Engineering Research and Design, 2013, 91, 12, 2515-2549.

Proceedings of the 14th International Symposium on Process Systems Engineering – PSE 2021+
June 19-23, 2022, Kyoto, Japan © 2022 Elsevier B.V. All rights reserved.
http://dx.doi.org/10.1016/B978-0-323-85159-6.50363-8

A Thermodynamic Approach for Simultaneous Solvent, Coformer, and Process Optimization of Continuous Cocrystallization Processes

Nethrue Pramuditha Mendis, Richard Lakerveld*

Department of Chemical and Biological Engineering, The Hong Kong University of Science and Technology, Clear Water Bay, Hong Kong, China
r.lakerveld@ust.hk

Abstract

Model-based optimization of cocrystallization processes involves the simultaneous identification of the optimal coformer and solvent types and the process operating conditions, which should suppress the formation of undesirable solid-state forms. New methods are needed for such optimization tasks. This work presents a computational framework for the optimal selection of coformers, solvents, and operating conditions for a cocrystallization process of a drug with low aqueous solubility. The method considers a cocrystal product that meets a specified target for solubility enhancement, which enhances product functionality. The proposed framework is demonstrated for the model drug carbamazepine. An optimization problem is formulated and solved with the proposed strategy, which illustrates its effectiveness for the optimization of cocrystallization processes with constraints on the pharmaceutical product performance.

Keywords: Cocrystallization, solvent selection, integrated product and process design, process optimization, PC-SAFT.

1. Introduction

Many active pharmaceutical ingredients (APIs) are orally administered as crystalline solids due to the various advantages they possess over other types of dosage forms. However, a low aqueous solubility can seriously limit the bioavailability of such an API. Pharmaceutical cocrystals (CCs) are solid-state forms that can improve the dissolution rate of APIs. They are formed when a neutrally charged API is crystallized with a neutrally charged coformer (CF). Cocrystals have a well-defined API:CF stoichiometric ratio in the crystal lattice and they can quickly dissolve into their molecular forms due to the non-covalent bonds.

Solution crystallization is a common strategy for industrial cocrystal synthesis (Lange & Sadowski, 2015). Solvents typically play a crucial role in any crystallization process. Thus, the solvent selection is a key decision. In general, systematic approaches for integrated solvent and process optimization have been reported for crystallization-based processes, for example, for the case of anti-solvent crystallization processes with recycles (Wang & Lakerveld, 2018). These problems are challenged by the strong interdependence between solvent selection and process operating conditions. Optimization of cocrystallization processes is uniquely challenging because the CF type plays a pivotal role in dictating the process and product performance. As cocrystallizing systems have the potential to form multiple solid-state forms, i.e., at least the pure API, pure CF, and the desired cocrystal, the choice of CFs, solvents, and operating conditions should facilitate the formation of the desired cocrystal, while suppressing the formation of other solid forms to obtain a product with a high purity. Furthermore, in the case of a pharmaceutical cocrystal, the degree to which the aqueous solubility may be enhanced for a given API is determined by the CF type. Therefore, the CF selection may not only affect the process performance but also the product efficacy. Although substantial work has been reported on related topics such as cocrystal discovery (e.g., (ter Horst et al., 2009)), optimization approaches for cocrystallization processes involving simultaneous CF, solvent, and operating condition identification have not been reported.

The objective of this work is to develop a thermodynamics-based optimization framework to select CFs, solvents, and operating conditions for cocrystallization processes to minimize the process operating costs and maximize the solubility enhancement of the final cocrystal form. Our optimization framework is based on the perturbed-chain statistical associating fluid theory (PC-SAFT) (Gross & Sadowski, 2001), which can model cocrystallizing mixtures well (Lange & Sadowski, 2015). Carbamazepine (CBZ) is selected as the model API along with eight of its CFs (glutaric acid, nicotinamide, saccharin, salicylic acid, oxalic acid, malonic acid, succinic acid, 4-aminobenzoic acid). CBZ is a solubility/dissolution rate-limited API.

2. Approach

2.1. Process Model Development

The model of the process (see Figure 1) consists of material balances (omitted here for brevity) and equilibrium relations with activity coefficients obtained from the PC-SAFT. The PC-SAFT provides a means of calculating the residual Helmholtz free energy of a mixture based on hard-chain, dispersion, and association interactions, which can serve as the basis to determine other thermodynamic properties like activity coefficients (Gross & Sadowski, 2001, 2002). In this work, all the compounds involved in the process, i.e., the API, the CF, and the solvent, are characterized by five parameters as in Gross & Sadowski, 2002: segment number (m), segment diameter (σ), dispersion energy parameter (ε), association energy (ε^{AB}), and association volume (κ^{AB}). The list of 48 solvents and their PC-SAFT pure component parameters considered in this work are obtained from our previous work (Wang & Lakerveld, 2018). Pure component parameters of the API and the eight CFs are estimated from solubility data reported in the literature, which is a standard procedure for estimating PC-SAFT parameters of solid compounds and omitted here for brevity.

The solubility of a pure compound is calculated according to:

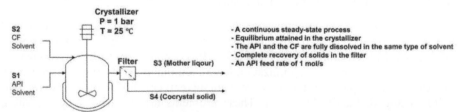

Figure 1: The proposed single-stage crystallization process configuration. The CF type, CF feed flow rate, solvent type, and the solvent flow rates in stream S1 and stream S2 are the free variables to be optimized.

$$x^{sat}\gamma^{sat} = \exp\left[\frac{\Delta H_f^m}{R}\left(\frac{1}{T_m}-\frac{1}{T}\right)\right],\tag{1}$$

where x^{sat}, γ^{sat}, ΔH_f^m, and T_m are the mole fraction, activity coefficient, enthalpy of fusion, and the melting temperature, respectively. R and T are the universal gas constant and the temperature, respectively.

The following inequality constraints are applied to ensure that both the API and the CF are fed to the process in fully dissolved form and do not crystallize in their pure forms:

$$x_{API,S1} \le x_{API}^{sat}, \; x_{CF,S2} \le x_{CF}^{sat}, \; x_{API,S3} \le x_{API}^{sat}, \; x_{CF,S3} \le x_{CF}^{sat}\tag{2}$$

The solubility of a cocrystal is expressed as a solubility product (Good & Rodríguez-Hornedo, 2009) as follows:

$$\left(x_{API,S3}\gamma_{API,S3}\right)^a \left(x_{CF,S3}\gamma_{CF,S3}\right)^b = K_{sp}\tag{3}$$

where a and b are the stoichiometric coefficients of the API and the CF in the cocrystal lattice, and K_{sp} is the solubility product.

2.2. Cocrystal Solubility Advantage

The potential solubility enhancement of cocrystals is often quantified by the cocrystal solubility advantage (SA), which can be defined as the ratio of the maximum API concentration due to cocrystal dissolution over the solubility of the pure API (Good & Rodríguez-Hornedo, 2009). The SA for a given API is primarily determined by the CF type. The pH of the dissolution medium is critical when calculating the SA for ionizable API/CFs. In this work, a pH of 2.5 is adopted for SA calculations, which represents the stomach pH. The extents of ionization of the CFs are determined using their k_a values reported in the literature. The ionization of the API (CBZ) is neglected (Good & Rodríguez-Hornedo, 2009).

The following expressions are aided to calculate the SA for a monoprotic weak acid-CF (*HA*),

$$x_{tot,CF} = x_{HA} + x_{A^-}, \; bx_{API} = ax_{tot,CF}, \; k_a = \frac{[A^-][H_3O^+]}{[HA]}, \; \left(x_{API}\gamma_{API}\right)^a \left(x_{HA}\gamma_{HA}\right)^b = K_{sp}\tag{4}$$

Note that only the nonionized portion of the CF contributes to the solubility product. In case the CF is a base, a similar approach is followed with the k_a value of the conjugate acid.

2.3. Optimization

In the optimization problem, CF type is characterized by the 10-dimensional vector $\mathbf{p}_{CF}$ whose elements include the five PC-SAFT pure component parameters, and additionally ΔH_f^m, T_m, and k_a of the CF, b/a ratio, and K_{sp} of the cocrystal. Solvent type is characterized by the 5-dimensional vector $\mathbf{p}_{sol}$, which consists of the five PC-SAFT pure component parameters of the solvent. The CF and solvent selection is formulated as:

$$\mathbf{p}_{CF} = \sum_i y_{CF,i} \mathbf{p}_{CF,i}, \ \sum_i y_{CF,i} = 1, \ i \in \{1,2,3,...,n_{CF}\}, \ y_{CF,i} \in \{0,1\} \tag{5}$$

$$\mathbf{p}_{sol} = \sum_j y_{sol,j} \mathbf{p}_{sol,j}, \ \sum_j y_{sol,j} = 1, \ j \in \{1,2,3,...,n_{sol}\}, \ y_{sol,j} \in \{0,1\} \tag{6}$$

where i and j are the existing CF and solvent candidates, whose choice is represented by binary variables $y_{CF,i}$ and $y_{sol,j}$. n_{CF} and n_{sol} are the total numbers of CF and solvent candidates, respectively. The above formulation ensures only one CF and solvent candidate each is selected. The normalized production cost (NPC) and the SA at pH 2.5 represent the process efficiency and the product performance, respectively, where the former needs to be minimized and the latter to be maximized. We formulate a multiobjective optimization problem as follows:

$$\min_{z_p, z_{sol}, z_{CF}} J_1 = \frac{c_{sol}\left[F_{S1,sol}\left(z_p, z_{sol}, z_{cf}\right) + F_{S2,sol}\left(z_p, z_{sol}, z_{cf}\right)\right]}{F_{S4,API}\left(z_p, z_{sol}, z_{cf}\right)}, \ \max_{z_{CF}} J_2 = SA(z_{CF}) \quad \text{(P1)}$$

s.t., process and thermodynamic models, inequality constraints in Eq.(2), Eq.(4), Eq.(5), Eq.(6)

z_p, z_{sol}, and z_{CF} stand for the operating conditions (CF feed flow rate and the solvent flow rates in stream S1 and S2), solvent type, and CF type, respectively. $F_{S1,sol}$ and $F_{S2,sol}$ are the solvent flow rates in streams S1 and S2, respectively. $F_{S4,API}$ is the API flow rate in stream S4 (product stream). The solvent cost is accounted for by a solvent cost parameter (c_{sol}) of 7.94 USD/L, which is estimated from data from Lab Alley (www.laballey.com). The CF cost is neglected as it is consumed in much smaller quantities compared to the solvent, and CFs are usually inexpensive chemicals. The optimization problem is solved by employing the epsilon-constraint method, where the lower bound for the SA (J_2) is increased from 0 to 40 to generate cases with different product performances, and the NPC (J_1) is minimized for each case.

The resulting optimization problem is an MINLP due to the discrete nature of CF/solvent selection and continuous operating conditions, which demands a relaxation strategy. The continuous mapping method (Bardow et al., 2010) is a commonly used MINLP relaxation strategy for problems involving simultaneous optimization of solvent types and operating conditions supported by the PC-SAFT. However, the optimization problem in this work is complicated by the simultaneous optimization of the CF type, which involves ten CF-related variables. A continuous mapping method may not be the most efficient solution strategy when the number of relaxation variables is large compared to the number of candidate compounds in the database. An approach similar to the traditional branch-and-bound approach is likely more efficient for such a case. Therefore, we propose a hybrid algorithm comprising the continuous mapping method for optimization of the solvent type and operating conditions and a branch-and-bound-like strategy for the CF selection. The algorithm involves two steps: 1) the optimization problem P1 is solved as an NLP by relaxing all integer variables, i.e., $y_{CF,i}$ in Eq.(5) and $y_{sol,j}$ in Eq.(6), to continuous variables, 2) P1 is solved separately for a selected set of CF candidates identified from step 1).

The solution for the relaxed problem in step 1 may not correspond to an existing CF candidate, which would be reflected by multiple nonzero values for the relaxed $y_{CF,i}$. In step 2, the optimization problem P1 is solved separately for each CF candidate i for which a nonzero $y_{CF,i}$ was found from the relaxed problem solution in step 1. As the CF type is fixed in P1 now, the solvent type and operating conditions can be optimized with an existing continuous mapping method from the literature (Wang & Lakerveld, 2018). Multiple integer solutions, i.e., solutions with an existing CF and solvent candidate, are obtained at the end of step 2, from which the optimal integer solution is identified. To verify that the integer solution obtained is the true optimal, step 1 is repeated, but without those CF candidates that have already been identified previously, i.e., n_{CF} is smaller now. If the objective function value of the new relaxed problem has not improved compared to the already identified optimal integer solution, the algorithm stops, and the identified optimal integer solution is the final solution. Otherwise, step 2 is repeated to identify more integer solutions, the optimal integer solution is updated, and step 1 is repeated again after excluding all the CF candidates already been identified. The algorithm stops when the relaxed problem objective function value does not improve compared to the already identified optimal integer solution. To solve the resulting NLPs, the CONOPT solver (Drud, 1985) was implemented in The General Algebraic Modeling System (GAMS Development Corporation).

3. Results and Discussion

When no lower bound is set for the SA, the optimal CF-solvent pair is oxalic acid (OA) and benzyl alcohol, which allows for the lowest NPC (Integer Solution 1 in Figure 2). Both the relaxed and integer solutions remain unchanged as the lower bound on the SA is increased from 0 to 6.78, which is consistent with the SA=6.78 for CBZ-OA cocrystal under the given conditions. When the lower bound of the SA is increased beyond 6.78, the NPC of the relaxed solution gradually increases. All these relaxed solutions correspond to the same integer solution (Integer Solution 2 in Figure 2), where nicotinamide (NA) and benzyl alcohol are the optimal CF and the optimal solvent type, respectively. Even though CBZ-NA cocrystal has a substantially higher SA of 41.72, the increase in NPC is substantial for the integer solution, likely because only eight CFs are considered in this work, i.e., it appears no CFs with intermediate properties are available. The optimal solvent type and required quantity for dissolving the API are the same for both optimal solutions. However, the higher total solvent consumption and the lower yield from Solution 2 cause an increase in the NPC. Finally, the optimal API:CF ratios are 1:1.46 and 1:3.16 for Solution 1 and Solution 2, respectively. This result shows that the determination of the optimal ratio at which the API and the CF have to be mixed to minimize the NPC is not trivial as it differs from the API:CF stoichiometric ratio (1:0.5 for CBZ-OA and 1:1 for CBZ-NA).

4. Conclusions

The proposed modeling and optimization framework for cocrystallization processes can simultaneously identify optimal CFs, solvent types, and operating conditions while considering both process design and product performance. The method can be used to balance the complex trade-offs that often exist in crystallization-based processes for the

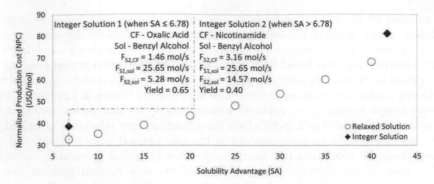

Figure 2: The Pareto fronts of the relaxed and integer solutions of the optimization problem (Eq.(P1)). The yield is defined as the fraction of the API in the feed that ends up in the cocrystal phase.

production of pharmaceutical cocrystals. This framework is particularly suitable for the early stages of process and product design, where the available experimental data are limited, and the identified set of CFs, solvents, and operating conditions can serve as the basis to launch an experimental program or more detailed simulations. Additionally, the method has the capability to recommend cocrystal property targets for efficient process and product design, which can guide future experimental efforts related to cocrystal discovery.

Acknowledgment

The work described in this paper was supported by a grant from the Research Grants Council of the Hong Kong Special Administrative Region, People's Republic of China (project no. 16214418).

References

Bardow, A., Steur, K., & Gross, J. (2010). Continuous-Molecular Targeting for Integrated Solvent and Process Design. Industrial & Engineering Chemistry Research, 49(6), 2834–2840.

Drud, A. (1985). CONOPT: A GRG code for large sparse dynamic nonlinear optimization problems. Mathematical Programming 1985 31:2, 31(2), 153–191.

Good, D. J., & Rodríguez-Hornedo, N. (2009). Solubility advantage of pharmaceutical cocrystals. Crystal Growth and Design, 9(5).

Gross, J., & Sadowski, G. (2001). Perturbed-Chain SAFT: An Equation of State Based on a Perturbation Theory for Chain Molecules. Industrial & Engineering Chemistry Research, 40(4), 1244–1260.

Gross, J., & Sadowski, G. (2002). Application of the Perturbed-Chain SAFT Equation of State to Associating Systems. Industrial & Engineering Chemistry Research, 41(22), 5510–5515.

Lange, L., & Sadowski, G. (2015). Thermodynamic Modeling for Efficient Cocrystal Formation. Crystal Growth & Design, 15(9), 4406–4416.

ter Horst, J. H., Deij, M. A., & Cains, P. W. (2009). Discovering New Co-Crystals. Crystal Growth & Design, 9(3), 1531–1537.

Wang, J., & Lakerveld, R. (2018). Integrated solvent and process design for continuous crystallization and solvent recycling using PC-SAFT. AIChE Journal, 64(4), 1205–1216.

Proceedings of the 14th International Symposium on Process Systems Engineering – PSE 2021+
June 19-23, 2022, Kyoto, Japan © 2022 Elsevier B.V. All rights reserved.
http://dx.doi.org/10.1016/B978-0-323-85159-6.50364-X

Optimizing the selection of drug-polymer-water formulations for spray-dried solid dispersions in pharmaceutical manufacturing

Suela Jonuzaj[a], Christopher L. Burcham[b], Amparo Galindo[a], George Jackson[a] and Claire S. Adjiman[a]

[a] *Department of Chemical Engineering, The Sargent Centre for Process Systems Engineering, Institute for Molecular Science and Engineering, Imperial College London, London SW7 2AZ, UK.*
[b]*Eli Lilly and Company, Indianapolis, IN, USA.*
c.adjiman@imperial.ac.uk

Abstract

In this work we present a systematic computer-aided design methodology for identifying optimal drug-polymer-water formulations with desired physical and chemical properties that are used in the spray drying of drug products. Within the proposed method, the UNIFAC model is employed to predict the solubility and miscibility of binary and ternary mixtures, whereas the Gordon-Taylor equation is used to estimate the glass transition temperature of a wide range of chemical blends. The design methodology is applied to the selection of optimal drug-polymer blends that maximize the loading of naproxen, while ensuring that stable formulations are designed. Finally, we explore the trade-offs between two competing objectives through multiobjective optimization, where the drug loading and water-content of API-polymer-water blends are maximized simultaneously. A ranked list of optimal solutions (mixtures with different chemicals and compositions) that can be used to guide experimental work is obtained by introducing integer cut inequalities into the model.

Keywords: Spray drying dispersion; optimal formulation; solubility; phase stability.

1. Introduction

Most of the new chemical entities in the drug discovery pipeline present unfavorable solubility properties, which often translates into low intrinsic bioavailability and makes the development of solid oral dosage forms challenging (Duarte et al., 2015). Spray-dried dispersion (SDD) is an effective technique to address these limitations; it has been successfully employed to improve the solubility and bioavailability of poorly soluble drugs in pharmaceutical manufacturing. SDD consists of an active pharmaceutical ingredient (API), preferably in its amorphous state to increase solubility, dispersed in a hydrophilic polymer matrix that stabilizes the amorphous form of the drug. Despite the benefits of this strategy, solid dispersions are often thermodynamically metastable, so there is a risk that the API may crystallize, leading to low-solubility behavior. To avoid such risks, it is important to select suitable API-polymer formulations that meet desired properties and ensure high solubility and bioavailability of the drug (Davis and Walker, 2018). In current industrial practice, heuristic approaches and experimental-based workflows are typically employed for pre-screening a small set of commonly used polymers for SDD. Such time-consuming and costly procedures can lead to longer development timelines and limited innovation. The development of computational methodologies that improve the selection strategies in spray drying formulations and reduce the experimental efforts and required resources is therefore highly desirable.

In recent years, several researchers have developed model-based approaches that focus on predicting the phase behavior of given API-polymer systems using common thermodynamic tools. Among others, Bansal et al. (2016) and Tian et al. (2013) employed Flory-Huggins (F-H) theory to predict the solubility and miscibility of drug-polymer blends. Sadowski and co-authors have studied the long-term thermodynamic stability of amorphous solid dispersions (ASD) and constructed phase diagrams of binary and ternary systems using the perturbed-chain statistical associating fluid theory (PC-SAFT). In particular, Prudic et al. (2014) have investigated the impact of copolymer composition on the phase behavior of solid dispersions of different drug-polymer blends. Lehmkemper et al. (2017a & 2017b) used experimental data and thermodynamic modeling to construct binary phase diagrams and explore ASD physical stability for different API-polymer systems at various humidity levels. Dohrn et al. (2020 & 2021) studied the impact of solvents on the phase separation of ASDs during drying, computing the solubility and liquid-liquid phase diagrams of API-polymer-solvent systems using PC-SAFT.

Despite these recent modeling advances, a limited set of pre-defined blends, with fixed ingredients, have been investigated to date. Thus, the selection of suitable polymers that can be more effective for stabilizing existing and/or new drugs has not been explored systematically or via optimization methods. In this work, we present a systematic approach for identifying optimal (i) API-polymer and (ii) API-polymer-water formulations that meet desired physicochemical properties and can lead to drug products with improved bioavailability. Within the proposed methodology, property-prediction models are employed to estimate the solubility, miscibility and glass transition temperatures of a wide range of binary and ternary blends. In addition, we exploit advanced optimization techniques (Jonuzaj et al., 2016, 2019) to design improved formulations that yield high solubility and stability of the drug. As a case study, the design approach is applied to the selection of optimal polymers that maximize the drug loading of naproxen in different API-polymer and API-polymer-water mixtures, while ensuring phase stability and a sufficiently high glass transition temperature of the final blend. The proposed optimization model yields a ranked list of diverse high-performing solutions, where solid dispersion blends of the API with different polymers, polymer proportions and sorbed water content are identified.

2. Design Methodology

2.1. Problem definition

The proposed mathematical model involves the design of optimal API-polymer and API-polymer-water formulations used in spray drying dispersions. The solid dispersion blends to be designed consist of a predefined API, an optimal polymer (p_1), and sorbed water (when ternary systems are considered). The following set $I = \{$API, $p_1, H_2O\}$ represents all components in the designed blend, where the optimal polymer is selected from a predefined set $P = \{1, ..., N_p\}$. Due to limited space, the mixed-integer programming problem (MINLP) presented in the next section considers the design of optimal binary API-polymer blends; it can be extended to formulating ternary API-polymer-water systems as shown in the supplementary information.

Optimizing the selection of drug-polymer-water formulations for spray-dried
solid dispersions in pharmaceutical manufacturing
2187

2.2. MINLP formulation for designing drug-polymer blends

The following mathematical formulation is derived to identify optimal polymers that maximize the drug loading of an API and satisfy pure and mixture property constraints:

$$\max_{w,x,y} \quad \frac{w_{API}}{w_{p_1}} \tag{1}$$

$$s.t. \quad \breve{T}_{g,mix} = \frac{\breve{w}_{API}T_{g,API}+K\breve{w}_{p_1}T_{g,p_1}}{\breve{w}_{API}+K\breve{w}_{p_1}} \geq T_g^L \tag{2}$$

$$K = \frac{\rho_{API}T_{g,API}}{\rho_{p_1}T_{g,p_1}} \tag{3}$$

$$w_{API} \leq 0.95\breve{w}_{API} \tag{4}$$

$$\frac{\partial \ln\gamma_{API}(T,x)}{\partial x_{API}} + \frac{1}{x_{API}} \geq 0 \tag{5}$$

$$\ln\tilde{x}_{API}^{eq} + \ln\gamma_{API}(T,\tilde{x}^{eq}) = \frac{\Delta H_{fus,API}}{R}\left[\frac{1}{T_{m,API}} - \frac{1}{T}\right] \tag{6}$$

$$w_i = \frac{x_iMW_i}{\sum_{j\in I}x_jMW_j}; \quad \widetilde{w}_i^{eq} = \frac{\tilde{x}_i^{eq}MW_i}{\sum_{j\in I}\tilde{x}_j^{eq}MW_j}; \quad i = API, p_1 \tag{7}$$

$$\sum_{p=1}^{N_p}y_{p_1,p} = 1 \tag{8}$$

$$x \in [x^L,x^U] \subset \mathbb{R}^I; \quad w \in [w^L,w^U] \subset \mathbb{R}^I; \quad y \in \{0,1\}^q$$

The vectors x and w represent the mole and mass fractions in the designed blend, respectively; $\widetilde{w}^{eq}$ and $\widetilde{w}$ represent mass fractions on the solid-liquid and glass transition phase boundaries, respectively; $y_{p_1,p}$ is a binary variable for assigning a polymer p from the set P to the polymer component p_1 in the designed blend. Eq. (1) is the objective function, where the ratio of API/polymer is maximized. Eqs. (2)-(8) are property constraints for estimating the glass transition temperature, the solid-liquid equilibrium and the miscibility of API-polymer blends. A schematic phase diagram of a drug-polymer blend is shown in Figure 1, where the blend is designed to be in areas A and B at system temperature T through eqs. (2)-(7). The glass transition temperature $\breve{T}_{g,mix}$ of a binary blend (green curve) is calculated using the Gordon-Taylor equation (Gordon and Taylor, 1952) given in eqs. (2) & (3), where ρ_i and $T_{g,i}$ are the density and glass transition temperature of component i, respectively. $\breve{T}_{g,mix}$ is set to be higher than a user-specified value T_g^L. To ensure that the API has low molecular mobility and does not undergo rapid crystallization in the designed blend, an offset in mass fraction is imposed in eq. (4), so that the glass transition temperature of the blend is well below $\breve{T}_{g,mix}$. Through eq. (5), the API-polymer blend is miscible at the chosen composition x. The solubility $\tilde{x}_{API}^{eq}$ is calculated in eq. (6), where the API heat of fusion $\Delta H_{fus,API}$ and melting point $T_{m,API}$ are taken from experimental data and R is the gas constant. Activity coefficients are

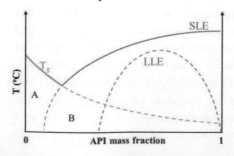

Figure 1: Schematic phase behavior of an API-polymer blend (adapted from Lehmkemper et al., 2017a). The blue curve represents solid-liquid equilibrium (SLE), the gray curve is liquid-liquid equilibrium (LLE), which may or may not be present, and the green curve the glass-transition temperature of the blend. In order to ensure the stability of the blend at room temperature, the desired operating region includes areas A & B.

calculated using the UNIFAC model (Fredenslund et al., 1975). The mass fractions of the API in the designed blend and at SLE are calculated in eq. (7), where MW_i is the molar mass of component i. Note that the binary blend can be above or below the SLE curve, so $\tilde{x}_i^{eq}$ and $\tilde{w}_i^{eq}$ could be calculated after the solution of the optimization problem. The logic relation presented in eq. (8) ensures only one polymer is selected from the set P.

3. Case study: optimal binary and ternary blends for the SSD of naproxen

3.1. Problem description

Naproxen (NPX) is a nonsteroidal anti-inflammatory drug commonly used to treat pain in joints and muscles. We selected naproxen to demonstrate the design methodology as it is a well-studied drug for SDD, and it can be modeled with UNIFAC. The proposed systematic methodology is applied to the design of optimal binary (NPX-polymer) and ternary (NPX-polymer-water) systems, where problems of increasing complexity are formulated and solved. First, a binary NPX-polymer system is considered, where optimal polymers are selected from a set of 25 candidates to maximize the drug loading of naproxen. The number of monomers (repeated units) in each polymer is allowed to vary between 10 and 200 in order to explore a wide range of polymers with different sizes and structures. Next, a larger design problem that takes into account a ternary NPX-polymer-water solid dispersion blend is formulated via multiobjective optimization (MOO), where the drug loading and water-content of the ternary mixture are optimized simultaneously. Relevant measured naproxen data are included in Table 1 and the optimal solutions obtained with the binary and ternary models are discussed in the next section.

Table 1: Measured naproxen property data and system temperature used in this case study.

API	Mw_{API} (g/mol)	$\Delta H_{fus,API}$ (J/mol)	$T_{m,API}$ (K)	$T_{g,API}$ (K)	T (K)	T_g^L (K)
Naproxen	230.26	32673.75	427.32	277.15	298.15	338.15

3.2. Results and discussion

The design formulations are implemented and solved in GAMS version 36.2.0, using local and global algorithms. The two MINLP models for designing binary and ternary blends for the SDD of naproxen can be found at *doi.org/10.5281/zenodo.5637599* and the results are given in Table 2 and Figure 2. Some of the solutions reported are global solutions obtained with SCIP. Where global optimality is not reached with either SCIP or BARON within 1000 s, local solutions obtained with SBB are presented. For cases where both SBB and SCIP terminate successfully, the two solvers converge to the same solution (triangles in Figure 2), giving confidence in the performance of the local solver.

The ranked list of optimal solutions presented in Table 2 shows that (hydroxypropyl)methyl cellulose p55, methacrylic acid-methyl methacrylate, and polyvinylpyrrolidone K30 are promising polymer candidates that yield high loading of naproxen in the polymeric carrier. The mass fraction of naproxen (w_{API}) in each binary mixture is higher than the mass fraction of the API at solid-liquid equilibrium ($\tilde{w}_{API}^{eq}$), and lower than the mass fraction at the glass transition boundary ($\tilde{w}_{API}$). In addition, the optimization model ensures the designed binary blends are miscible at optimal compositions of naproxen and polymer in the mixture, so that phase separation is prevented. Thus, the designed API-polymer blends are in the desired phase region, i.e., below the solubility and glass transition curve, and outside the immiscible area.

Table 2: Top 3 optimal solutions of the API-polymer model, including the optimal API/polymer ratios; the mass fraction of API in the designed blend (w_{API}); the SLE mass fraction ($\widetilde{w}_{API}^{eq}$); the mass fraction ($\breve{w}_{API}$) on the reference glass transition curve ($\breve{T}_{g,mix}$); the identity of the optimal polymers and the number of the repeated units (N_m) in each polymer.

w_{API}/w_{p1}	w_{API}	$\widetilde{w}_{API}^{eq}$	$\breve{w}_{API}$	$\breve{T}_{g,mix}(\breve{w}_{API})$	Polymers	N_m
1.601	0.616	0.001	0.648	338.15	HPMC p55	10
1.181	0.541	0.001	0.570	338.15	Eudragit L100	11
1.052	0.513	0.003	0.540	338.15	PVP K30	10

In the ternary model, the drug/polymer ratio and water content are optimized simultaneously via MOO in order to investigate the maximum amount of water that can absorbed in the designed blends while maintaining high drug loading. The set of Pareto optimum solutions given in Figure 2 is obtained using the ε-constraint method, in which the water mass fraction is maximized, and the API/polymer ratio is constrained by a given lower bound, ε. The value of ε is increased from 0.2 to 2 with a step-size of 0.2. All Pareto points share the same monomer structure, HPMC p55, but differ in the polymer size and water-polymer composition (cf. Table S1 in supplementary information). We note that the miscibility constraints were not included in the formulation of the ternary problem. Further investigation is required in order to ensure there is no phase separation of the optimal ternary blends designed.

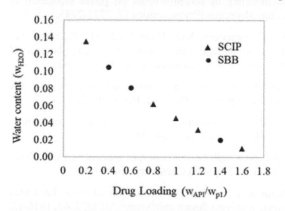

Figure 2: Optimal drug loading and water mass fraction when solving the multiobjective optimization problem for the ternary API-polymer-water system.

4. Conclusions

In this work we have developed a systematic computer-aided methodology for identifying optimal formulations for the spray drying of drug products. Within the proposed approach, two optimization models were formulated and solved. First, a two-component formulation was derived, where optimal API-polymer mixtures that meet desired property targets and satisfy given phase boundary conditions, were identified. A ranked list of optimal solutions (different binary blends with optimized polymer structure, size and compositions) was obtained by introducing integer cut inequalities into the model. Next, a ternary API-polymer-water system was formulated via multiobjective optimization, where the drug loading and water fraction were optimized simultaneously. Future work will focus on validating the modeling results with experimental data (where possible), and constructing binary and ternary phase diagrams of the optimal mixtures obtained. In addition, the models will be extended to designing API-polymer-solvent blends, taking into account key process constraints used in spray drying solvent selection, spray solution viscosity, spray atomization and final particle size of the solid spray dried particles. Finally, the proposed models will be incorporated in a polymer and solvent selection workflow for spray drying that can be used to identify better-performing designs in a more efficient and systematic way.

Acknowledgments: The authors gratefully acknowledge financial support from Eli Lilly and Company, and the EPSRC (EP/T005556/1).
Data statement: Data underlying this article can be accessed and used under the Creative Commons Attribution license on Zenodo at *doi.org/10.5281/zenodo.5637599*.

References

Bansal, K., Baghel, U.S., Thakral, S., 2016, Construction and validation of binary phase diagram for amorphous solid dispersion using flory-huggins theory, AAPS PharmSciTech 17, 318-327.

Davis, M., Walker, G., 2018, Recent strategies in spray drying for the enhanced bioavailability of poorly water-soluble drugs, Journal of Controlled Release 269, 110-127.

Dohrn, S., Reimer, P., Luebbert, C., Lehmkemper, K., Kyeremateng, S.O., Degenhardt, M., Sadowski, G., 2020, Thermodynamic modeling of solvent-impact on phase separation in amorphous solid dispersions during drying, Molecular Pharmaceutics 17, 2721-2733.

Dohrn, S., Luebbert, C., Lehmkemper, K., Kyeremateng, S.O., Degenhardt, M., Sadowski, G., 2021, Solvent influence on the phase behavior and glass transition of amorphous solid dispersions, European Journal of Pharmaceutics and Biopharmaceutics 158, 132-142.

Duarte, I., Santos, J.L., Pinto, J.F., Temtem, M., 2015, Screening methodologies for the development of spray-dried amorphous solid dispersions, Pharmac. Research 32, 222-237.

Fredenslund, A., Jones, R.L., Prausnitz, J.M., 1975, Group-contribution estimation of activity coefficients in nonideal liquid mixtures, AIChE Journal 21, 1086-1099.

Gordon, M., Taylor, J.S., 1952, Ideal copolymers and the second-order transitions of synthetic rubbers. i. non-crystalline copolymers, Journal of Applied Chemistry 2, 493-500.

Jonuzaj, S., Akula, P.T., Kleniati, P.M., Adjiman, C.S., 2016, The formulation of optimal mixtures with generalized disjunctive programming: A solvent design case study, AIChE J. 62, 1616-33.

Jonuzaj, S., Cui, J., Adjiman, C.S., 2019, Computer-aided design of optimal environmentally benign solvent-based adhesive products, Computers & Chemical Engineering 130, 106518.

Lehmkemper, K., Kyeremateng, S.O., Heinzerling, O., Degenhardt, M., Sadowski, G., 2017a, Long-term physical stability of PVP- and PVPVA-amorphous solid dispersions, Molecular Pharmaceutics 14, 157-171.

Lehmkemper, K., Kyeremateng, S.O., Heinzerling, O., Degenhardt, M., Sadowski, G., 2017b, Impact of polymer type and relative humidity on the long-term physical stability of amorphous solid dispersions, Molecular Pharmaceutics 14, 4374-4386.

Prudic, A., Kleetz, T., Korf, M., Ji, Y., Sadowski, G., 2014, Influence of copolymer composition on the phase behavior of solid dispersions, Molecular Pharmeceutics 11, 4189-4198.

Tian, Y., Booth, J., Meehan, E., Jones, D. S., Li, S., Andrews, G.P., 2013, Construction of drug–polymer thermodynamic phase diagrams using Flory–Huggins interaction theory: identifying the relevance of temperature and drug weight fraction to phase separation within solid dispersions, Molecular Pharmaceutics 10, 236–248.

Proceedings of the 14th International Symposium on Process Systems Engineering – PSE 2021+
June 19-23, 2022, Kyoto, Japan © 2022 Elsevier B.V. All rights reserved.
http://dx.doi.org/10.1016/B978-0-323-85159-6.50365-1

Integrated design of injectable manufacturing processes considering characteristics of process- and discrete-manufacturing systems

Masahiro Yamada[a], Isuru A. Udugama[a], Sara Badr[a], Kenichi Zenitani[b], Kokichi Kubota[b], Hayao Nakanishi[b], Hirokazu Sugiyama[a*]

[a]*Department of Chemical System Engineering, The University of Tokyo, Tokyo 113-8656, JAPAN*
[b]*Settsu Plant, Shionogi Pharma Co Ltd., Osaka 566-0022, JAPAN*
*sugiyama@chemsys.t.u-tokyo.ac.jp

Abstract

This work proposes a superstructure-based design approach to optimize an end-to-end injectable manufacturing process. At the core of this approach are unit operation models for batch and continuous operations and a plant-wide scheduling model that can explicitly model the semi-continuous operations of injectable manufacturing to a high level of precision. An integrated evaluation model consisting of a Net Present Value (NPV) and technology readiness modules is used to identify the optimal flowsheet. This approach can simultaneously optimize the design of a given process flowsheet alternative consisting of both batch and continuous unit operations considering product parameters, process conditions, and market characteristics. The overall approach developed was then demonstrated on an end-to-end injectable manufacturing case study with four process flowsheet alternatives (batch and continuous operations of compounding and lyophilization). Based on the techno-economic analysis, it was shown that the alternative with batch compounding and lyophilization was preferred.

Keywords: Process design; injectable manufacturing; optimization; continuous manufacturing; superstructure

1. Introduction

Injectables are an effective dosage form for many pharmaceutical products, including COVID-19 vaccines (Alharbi, 2021). Injectables are typically manufactured in a production line consisting of solution compounding, sterile filtration, filling, lyophilization, and inspection unit operations. Currently, these units are operated as batch processes. While continuous technologies are being developed for injectable manufacturing, they are still in the early stage of conceptualization and research. Examples include continuous compounding (Casola et al., 2015) and lyophilization (Bockstal et al., 2017). The main driving force for introducing these continuous technologies is the promise of reduced manufacturing cost, environmental load, and improved process flexibility.

As a result, in the future, there is the possibility to design an injectable manufacturing process where each unit can be operated in either batch or continuous modes. From a design perspective, this requires evaluating multiple potential process flowsheet alternatives considering multiple operational, design, and economic factors. As such, there is a need to develop decision-support tools that can enable informed decision-making without entirely relying on empirical knowledge.

Superstructure optimization is such a decision-support concept. In short, the superstructure-based approach evaluates the manufacturing system's overall performance rather than focusing on a single unit operation. As a result, this approach can find a globally optimal process design (Quaglia et al., 2015). In the chemical industry, superstructure based optimization is a commonly used method for determining process sequence (Tian et al., 2020). While in pharmaceutical production processes, superstructure-based approaches have been used for choosing between batch and continuous unit operations in tablet manufacturing (Matsunami et al., 2020). Figure 1 illustrates a tentative superstructure of the injectable manufacturing process, where operation mode and equipment materials choices are listed. The set of process alternatives that need to be evaluated are all possible choice combinations.

To accurately evaluate process alternatives generated from a superstructure, there is a need for an approach that can explicitly optimize their process design and operations. In the context of injectable manufacturing, this requires the development and integration of two types of models. One model that can accurately capture the continuous operations which are "process-manufacturing systems" where the bulk solution behavior is essential. Another model that can describe batch operations which are "discrete-manufacturing systems" where process operation behavior is essential. In addition, the scheduling of the overall manufacturing process is needed as the same production line needs to process multiple products while guaranteeing sterility levels. The output of these combined models then must be techno-economically evaluated. The rest of this work is organized as follows. In section 2, the overall superstructure-based approach is introduced. This is followed by section 3, where the proposed approach is demonstrated on a case study. Finally, in section 4, the conclusions are given.

2. Methodology

For process optimization, a model-based approach using superstructure was suggested. First, a superstructure covering all possible candidates was created based on literature data and discussions with production experts. Figure 1 illustrates the overall superstructure and the multiple decision variables that can be considered, such as the operation mode (batch vs continuous), equipment material (stainless vs resin), and tasks (machine vs operator inspection). In the future, further choices can be added to the superstructure.

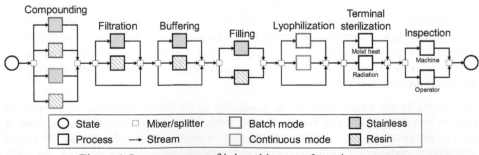

Figure 1 Superstructure of injectable manufacturing process

Integrated design of injectable manufacturing processes considering *characteristics of process-and discrete-manufacturing systems*

2193

Figure 2 displays the overall modeling framework followed to evaluate each process flowsheet candidate systematically. The entire model can be divided into three parts: i) process flowsheet model, ii) scheduling model, and iii) evaluation model. Through these models, the optimal candidate is obtained based on NPV and Technology Readiness Level (TRL) evaluation. Due to the level of fidelity of the process flowsheet model, this framework can optimize the economic performance of each process alternative identified considering multiple design choices, market conditions and process parameters. Hence, applying the approach developed with a superstructure-based flowsheet generation enables the optimal process alternative and the corresponding process design to be found simultaneously.

2.1. Process flowsheet model
The objective of the process flowsheet model is to calculate the required equipment size, the input and output material flows, and the expected unit operation processing time for a one-lot production. To carry out this evaluation, the process flowsheet model uses an integrated process and discrete production module. The process module explicitly considers Active Pharmaceutical Ingredient (API) crystal dissolution and Residence Time Distribution (RTD) factors. On the other hand, the discrete module considers lot residue, lot disposal, and operational defects which are critical operational factors for discrete process operations. The scheduling model receives the expected processing time for each unit operation for single-lot production from the process flowsheet model.

2.2. Scheduling model
The objective of the scheduling model is to evaluate if the production line can meet product demand and inventory requirements with the given operational constraints and changeover tasks. Suppose a single product is manufactured in a production line. In that case, the repetition of the lot production can meet the demand and inventory margin if there is sufficient time allocated for production. However, when multiple products are manufactured in a production line, the production sequence must be optimized to meet the constraints and minimize the non-production time. If there is no solution to satisfy the constraints, the design alternative is evaluated as unfeasible.

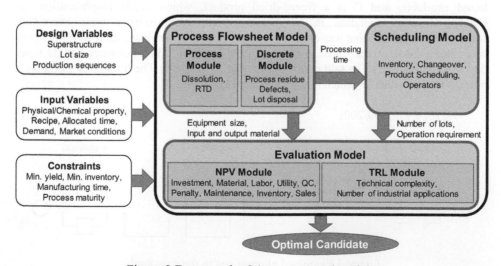

Figure 2 Framework of the constructed models

2.3. Evaluation model

The evaluation model consists of the NPV module and TRL module. The overall objective of these two evaluation modules is to calculate the economic performance of the process flowsheet (considering the intricacies of process- and discrete-manufacturing systems) and to calculate the overall technology readiness level.

In the NPV module, NPV is calculated as an economic indicator, as shown in Eq. (1).

$$\max NPV = -investment + \sum \frac{CF}{(1+r)^n} \qquad (1)$$

Investment cost *investment* [JPY] and annual cash flow *CF* are calculated by considering the following factors: the annual cash flow, material cost, utility cost, labor cost, quality control cost, penalty cost, maintenance cost, and inventory cost as well as the product sales. The process flowsheet model generates information related to equipment sizing and material mass per lot. The operational requirement information, e.g., the sum of production duration, required space for inventory management, and the number of lots for each product, is generated from the scheduling model. In the TRL module, the overall process flowsheet is given a score between a TRL of 1 and 9. The TRL score is based on the state of progress in research, experiment, and implementation of each unit operation that is selected. This evaluation is carried out according to the guidelines set in Li et al. (2019) and Silk et al. (2020).

3. Case study

The proposed approach was then applied to a case study examining four process alternatives, as illustrated in figure 3. Each alternative consisted of end-to-end injectable manufacturing operations where the operations of the compounding and lyophilization units were changed between batch and continuous modes. Following assumptions were made during the subsequent analysis: 1) Case study represents a grass-root design, where resin equipment and final sterilization were not considered; 2) Three products A, B, and C, with fixed demand, were manufactured in one production line. Product A and B is liquid products, and C is a freeze-dried product, which needs lyophilization; 3) Batch/continuous operation in compounding and lyophilization were examined. Inspection with a machine was chosen for liquid products, and inspection with operators was chosen for the freeze-dried product; 4) Production of 5 days per week and 24 h per day was assumed; 5) The sale period was set as 20 years; 6) The lot size is varied as a process design variable, which is shown in Eq. (2).

$$V_{\text{lot}} \, [\text{L}] \in \{50, 100, 200\} \qquad (2)$$

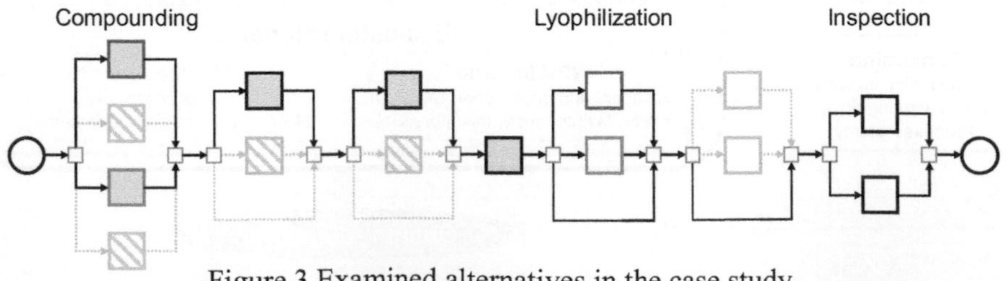

Figure 3 Examined alternatives in the case study

Over a half million design alternatives were analyzed for each of the four process alternatives. Figure 4 (a) illustrates a violin plot of production times observed for each process alternative. Analyzing Figure 4 (a), continuous operations in the lyophilization unit resulted in a noticeable reduction in the overall production time, which is in line with previous results (Pisano et al., 2019). In comparison, the mode of operation in the compounding unit had a minor effect. The design condition of each process alternative has a significant influence on the observed production duration. For instance, a good process alternative (e.g continuous compounding and lyophilization) can have a longer production time than a bad process alternative (e.g batch compounding and lyophilization) purely due to design choices. This shows the importance of simultaneous process alternatives selection and process design optimization during the decision-making processes.

Figure 4 (b) illustrates the NPV of selected design conditions for the four process alternatives where the lot size and production sequences were fixed. The production time of four alternatives is marked in red in Figure 4 (a). It can be seen that continuous lyophilization and batch compounding results in the best outcome. It can also be seen that continuous lyophilization choice has a notable positive impact on the NPV. This is mainly because of the reduction in labor costs due to the shortened production time. In comparison, the introduction of continuous compounding results in a decrease in NPV. This is because the cost of investment and initial start-up requirements outweigh other benefits. It should be noted that these designs have not been economically optimized by considering design factors such as lot size, production sequence, and other market factors. Instead, designs with a shorter overall production time were selected.

From a TRL point of view, batch compounding and lyophilization are tried and tested technologies implemented in the industry. Hence, batch compounding and lyophilization operations can be awarded a TRL of nine. In contrast, continuous operations are only implemented for slurry and liquid, and not for API crystal and water for injection. Hence continuous compounding and lyophilization can be awarded a TRL score of three, as only the fundamentals are established.

Overall the fully batch alternative is recommended over other alternatives as 1) the NPV of the fully batch alternative is only 10% lower than the most optimal process alternative and 2) the significant difference in TRL between batch and continuous operations.

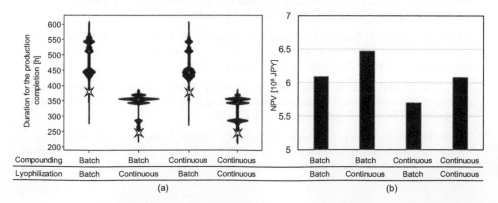

Figure 4 Evaluation result for (a) total duration for the production and (b) NPV

4. Conclusion

A model-based design approach for the injectable manufacturing process was presented. A superstructure was constructed to consider multiple process candidates while process and discrete models were developed for each unit operation to optimize the process and design conditions for each process candidate. An evaluation model consisting of an NPV and TRL analysis was developed to rate each candidate's performance. The approach was applied to a case study where batch and continuous compounding/lyophilization alternatives were examined. The case study illustrated the importance of simultaneous determination of process alternatives and design considering the product, process, market characteristics. In this instance, the alternative with batch compounding and continuous lyophilization was shown to be economically optimal. However, when considering TRL, the fully batch operations was a better alternative. As future work, the economic evaluation should be done for all alternatives, including the process parameter and production sequences. Model updates are also necessary to evaluate other process alternatives and different equipment materials.

5. References

N. Alharbi, 2021. Overview of SARS-CoV-2 and COVID-19 Vaccine, Bahrain Medical Bulletin, 43 (2), 511–515

G. Casola, S. Yoshikawa, H. Nakanishi, M. Hirao, H. Sugiyama, 2015. Systematic retrofitting methodology for pharmaceutical drug purification processes, Computers and Chemical Engineering, 80 (2), 177–188

B. Li, I. A. Udugama, S. S. Mansouri, W. Yu, S. Baroutian, K. V. Gernaey, B. R. Young, 2019. An exploration of barriers for commercializing phosphorus recovery technologies, Journalk of Cleaner Production 229 (20), 1342–1354

K., Matsunami, F. Sternal, K. Yaginuma, S. Tanabe, H. Nakagawa, H. Sugiyama, 2020. Superstructure-based process synthesis and economic assessment under uncertainty for solid drug product manufacturing, BMC Chemical Engineering, 2 (1), 1–16

R. Pisano, A. Arsiccio, L. C. Capozzi, B. L. Trout, 2019. Achieving continuous manufacturing in lyophilization: Technologies and approaches, European Journal of Pharmaceutics and Biopharmaceutics, 142, 265–279

A. Quaglia, C. L. Gargalo, S. Chairakwongsa, G. Sin, R. Gani, 2015. Systematic network synthesis and design: Problem formulation, superstructure generation, data management and solution, Computers and Chemical Engineering, 72 (2), 68–86

D. Silk, B. Mazzali, C. L. Gargalo, M. Pinelo, I. A. Udugama, S. S. Mansouri, 2020. A decision-support framework for techno-economic-sustainability assessment of resource recovery alternatives, Journal of Cleaner Production, 266 (1), 121854

G. Tian, A. Koolivand, N. S. Arden, S. Lee, T. F. O'Connor, 2019. Quality risk assessment and mitigation of pharmaceutical continuous manufacturing using flowsheet modeling approach, Computers & Chemical Engineering, 129 (4), 106508

P. Van Bockstal, S. Mortier, L. De Meyer, J. Corver, C. Vervaet, I. Nopens, T. De Beer, 2017. Mechanistic modelling of infrared mediated energy transfer during the primary drying step of a continuous freeze-drying process, European Journal of Pharmaceutics and Biopharmaceutics, 114, 11–21

Proceedings of the 14th International Symposium on Process Systems Engineering – PSE 2021+
June 19-23, 2022, Kyoto, Japan © 2022 Elsevier B.V. All rights reserved.
http://dx.doi.org/10.1016/B978-0-323-85159-6.50366-3

A bi-level decomposition approach for CAR-T cell therapies supply chain optimisation

Niki Triantafyllou[a], Andrea Bernardi[a], Matthew Lakelin[b], Nilay Shah[a], Maria M. Papathanasiou[a*]

[a]*Sargent Centre for Process System Engineering, Imperial College London, SW72AZ, London, United Kingdom*
[b]*TrakCel Limited, 10/11 Raleigh Walk, Cardiff, CF10 4LN UK*
maria.papathanasiou11@imperial.ac.uk

Abstract

Autologous cell therapies are based on bespoke, patient-specific manufacturing lines and distribution channels. They present a novel category of therapies with unique features that impose scale out approaches. Chimeric Antigen Receptor (CAR) T cells are an example of such products, the manufacturing of which is based on the patient's own cells. This automatically: (a) creates dependencies between the patient and the supply chain schedules and (b) increases the associated costs, as manufacturing lines and distribution nodes are exclusive to the production and delivery of a single therapy. The lack of scale up opportunities and the tight return times required, dictate the design of agile and responsive distribution networks that are eco-efficient. From a modelling perspective, such networks are described by a large number of variables and equations, rendering the problem intractable. In this work, we present a bi-level decomposition algorithm as means to reduce the computational complexity of the original Mixed Integer Linear Programming (MILP) model. Optimal solutions for the structure and operation of the supply chain network are obtained for demands of up to 5000 therapies per year, in which case the original model contains 68 million constraints and 16 million discrete variables.

Keywords: CAR T cell therapy; supply chain optimisation; MILP; personalised medicine; bi-level decomposition.

1. Introduction

Chimeric Antigen Receptor (CAR) T cell therapy is a type of immunotherapy, where the patient's own immune system is utilised to recognise and kill cancer cells (Sadelain *et al.* 2015). The patient's T cells are removed from the bloodstream and are genetically engineered to express the CAR, rendering them capable of recognising and attacking the target tumour cells. CAR T cells can be obtained from the patient's own blood (autologous) or the lymphocytes of another healthy donor (allogeneic). Following the success in Phase 1 of clinical trials, the US Food and Drug Administration (FDA) and the European Medicines Agency (EMA) have approved 5 of these therapies so far (UPMC, 2021). Currently, there are 6,581 active and ongoing clinical trials regarding CAR T cell treatments, with most of them being autologous, while their allogeneic counterpart is progressing as well (Caldwell *et al.* 2021). The relatively high prices of these therapies can be partially attributed to the high manufacturing, distribution and product administration costs (Spink *et al.* 2018). Time-intensive manufacturing processes, in-time delivery under hospital admission and daily monitoring of the patient for side effects are among the factors that increase the cost (Han *et al.* 2021). Another key challenge of the

CAR T cell therapy lifecycle is the minimisation of the turnaround time, which varies between 15 and 24 days for the commercially available treatments (Nucleus Biologics, 2021). The in-time delivery is of utmost importance for the patients as late administration may negatively impact the response to treatment. To address these challenges, digital tools such as mathematical models and optimisation policies are used to assist the decision-making process by coordinating the different tasks and identifying the optimal supply chain network structures (Sarkis *et al.* 2021a). The complexity of the CAR T cell supply chain can be easily observed by the product's lifecycle and autologous nature that challenge the identification of an optimal supply chain network (Sarkis *et al.* 2021b). The main steps of a typical CAR T cell therapy lifecycle are: (a) patient identification, (b) leukapheresis, (c) manufacturing, (d) Quality Control, (e) therapy administration.

There have been works in the literature focusing on the optimisation of CAR T cell therapies via Mixed Integer Linear Programming (MILP) models (Bernardi *et al.* 2021; Karakostas *et al.* 2020). The autologous nature of these therapies often results in novel supply chain formulations (Papathanasiou *et al.* 2020), where the MILP problem comprises a significantly high number of integer variables. The latter can range from 600,000 for clinical trial applications to over 16 million for commercial scales of average demand. This can render the convergence to global or sometimes even local optimality infeasible. Hence, methodologies to enable the solution of large-scale instances are required (Erdirik-Dogan and Grossmann, 2008; Terazzas-Moreno and Grossmann, 2011). In this work, we present a bi-level decomposition algorithm capable of providing candidate solutions with respect to the location, number and capacity of manufacturing sites, and the most suitable mode of transport. Optimal solutions for the structure and operation of the supply chain network are obtained for demands of up to 5000 therapies per year under three different time constraint scenarios (17, 18 and 19 days) total return time. The latter refers to the total duration of the therapy life cycle, starting from the leukapheresis procedure and ending with the delivery of the therapy at the hospital.

2. Materials and methods

The model examined in this work is an in-house MILP model that describes the CAR T cell supply chain, used for the identification of the optimal supply chain network structure for in-time delivery of the therapies (Bernardi *et al.* 2021). An overview of the model formulation is presented in Table 1. The supply chain network includes 4 nodes; namely, leukapheresis site, manufacturing site, quality control (QC), and hospital (Figure 1). More specifically, a patient is allocated to a specialised leukapheresis site, where T cells are isolated from the bloodstream. Subsequently, the leukapheresis material undergoes freezing (cryopreservation) and is shipped to the manufacturing site. After the completion of the manufacturing process, the final product is tested in the QC site, which is co-located with the manufacturing facilities. Finally, the cryopreserved CAR T cell therapy is transported to the hospital for administration.

The objective is to minimize the total cost of the therapies over a long-term planning horizon (year quarter), whilst operating the supply chain in the short-term (daily) and fulfilling several constraints. The supply chain network's performance (full space model) is assessed for different demand scenarios (200, 500, 1000, 2000, 3000 patients per year) generated by an in-house algorithm and different return times (17, 18, and 19 days). The model parameters, such as the demand profiles and cost coefficients are assumed to be deterministic. The study considers 4 leukapheresis sites and 4 hospitals in the UK and 6 manufacturing sites located in the UK and Europe. The manufacturing facilities have a capacity of 4, 10, or 31 parallel lines, and a forward-looking scenario of a manufacturing time of 7 days is considered.

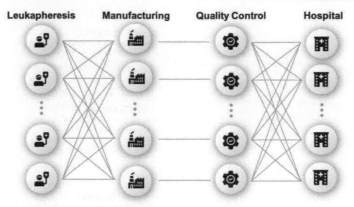

Figure 1. CAR T cell supply chain network with 4 nodes: (a) leukapheresis site, (b) manufacturing site, (c) Quality Control, and (d) hospital. The hospital and the leukapheresis centre are collocated in this case.

Table 1: Overview of the model formulation.

Index	Mathematical Formulation	Description
Objective function		
(1)	$min\ C_{total\ cost} = C_{manufacturing} + C_{transport} + C_{quality\ control}$	Total cost of therapies
Constraints		
(2)	$TRT_p = t_{delivery} - t_{start} \leq U^t$	Return time of therapy
(3)	$CAP_{m,t} = FCAP_m - \sum_p INM_{p,m,t}$	Capacity constraint
(4)	$X1_{c,m} \leq E1_m, \forall c,m, \quad X2_{m,h} \leq E1_m, \forall c,h, \quad \sum_m E1_m \leq U^M$	Network constraints
(5)	$INC_{p,c,t} = OUTC_{p,c,t+TLS}$, $INM_{p,m,t} = OUTM_{p,m,t+TMFE+TQC}$	Sample balances at each node (leukapheresis, manufacturing)
(6)	$LSR_{p,c,m,j,t} = LSA_{p,c,m,j,t+TT1_j}, \quad \forall p,c,m,j,t$, $FTD_{p,m,h,j,t} = MSO_{p,m,h,j,t+TT2_j}, \quad \forall p,m,h,j,t$	Transport constraints

An assessment of the model's complexity was conducted in order to evaluate the capabilities of the full space MILP problem. In this case, CPLEX results into global optimum solutions in less than an hour for small-scale problems. However, an increase in the number of patients and the planning horizon, makes the problem computationally intractable and leads to CPU times over 10,000s. It is shown that the CAR T cell supply chain problem is a problem with complicating constraints, and it is more sensitive to an increase in the number of therapies rather than an increase in the time horizon. The bottleneck constraints were identified via a model complexity analysis. These are the two transport constraints (Table 1) based on which the therapy is shipped from the leukapheresis site to the manufacturing site and from the manufacturing site to the hospital. Hence, the search space of the detailed model becomes very large for commercial scale problem instances, mainly because of the penta-dimensional transportation variables. The identified complicating constraints are key elements in the development of the bi-level decomposition algorithm. To keep the problem computationally tractable for a higher number of therapies, a bi-level decomposition approach is proposed. The original detailed model is decomposed into an upper-level planning and a lower- level scheduling problem in order to decrease the computational complexity and solve for larger instances of the problem.

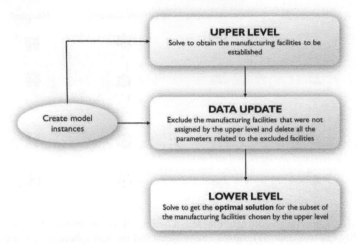

Figure 2. Flowchart for the proposed algorithm.

Figure 2 presents a flowchart of the proposed bi-level decomposition algorithm. The upper-level model is a relaxation of the original full space model, and it is responsible for strategic planning. More specifically, it chooses the number and the location of the manufacturing facilities to be established. The upper-level model is lower-dimensional compared to the original and thus the candidate transport modes and hospitals are considered fixed. The supply chain model for the upper level considers the network up until the end of manufacturing and all the constraints regarding material flows post manufacturing are ignored here. Consequently, the complicating constraint about the shipping of the therapy from the manufacturing site to the hospital is eliminated from the upper level. Almost all the variables in the upper level are identical to the ones in the original detailed model, apart from some that are redefined to fit the dimensional changes in the upper level. Given that the upper level is a relaxation of the original model, its solution provides a lower bound to the lower level subproblem. The lower-level model is a subproblem of the original detailed scheduling model, as it is solved for the subset of the manufacturing facilities chosen by the upper level. Specifically, manufacturing sites that were predicted not to be established by the upper level, are excluded from the lower level. Hence, the search space of the lower-level model is significantly reduced. The lower level can choose the same or a subset of the manufacturing sites that were predicted in the upper level. The optimal solution of the lower-level model becomes the final solution of the problem.

3. Results and discussion

All the models have been implemented in Python 3.7.1 and Pyomo 5.6.1 and solved with CPLEX 12.9. All computational experiments were performed in a 24-core Xeon E5-2697 machine with 96GB. Here we present the results of seven problem cases of increasing size. All cases were solved using both the bi-level decomposition algorithm, as well as the full-space model, aiming to assess the capabilities of the former. The bi-level algorithm was tested for cases of up to 5000 therapies per year, two different demand profiles (A and B) for each case, and different turnaround times (17, 18, and 19 days). It should be noted that with the proposed algorithm both subproblems reach global optimality. In addition, most of the solutions obtained by the bi-level decomposition algorithm are identical to the global optimum solutions of the full space model. Specifically, the solutions of the four first cases (200, 500, 1000, and 2000 patients per year) are identical to the global optimum solutions of the full space model. In the fifth case of 3000 patients per year, the proposed algorithm arrives at a significantly improved solution and both subproblems arrive at global optimality. In the last two cases, the full space model was unable to provide a solution in contrast to the bi-level algorithm. For these cases, both the upper and lower levels arrive at global optimality. Based on the above, the first four cases are used to guarantee that the algorithm arrives at global optimality, which has been proven from the full space model. Finally, the matches between the manufacturing facilities are the same across all solutions. Both the full space model and the bi-level algorithm result in the same supply chain network structures.

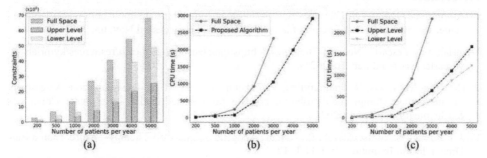

Figure 3. Comparison of the full space model and the bi-level decomposition algorithm for an increasing number of therapies in (a) the number of constraints, (b) the solution time in CPU seconds and (c) the solution time in CPU seconds of the full space model and the upper- and lower-level of the proposed algorithm.

As illustrated in Figure 3a, the proposed approach decreased the total number of constraints in the model. Additionally, the computational time reduced by more than 50% compared to the full space model (Figure 3b). The advantage of the bi-level algorithm is that the two separate subproblems are solved sequentially and thus the computational complexity of the one does not impact the other. This is clearly illustrated in Figure 3c, where the computational time for the upper- and the lower- level is displayed. For example, in the 3000 patients/year case the CPU time of the full space model is 2335 s, while the CPU time of the upper- and lower- level is 644 s and 409 s respectively. The algorithm will stop being efficient only when the two subproblems become computationally intractable. This leads to the conclusion that the bi-level algorithm can provide feasible and optimal solutions for instances higher than 5000 patients/year.

4. Conclusions

In this work, we have addressed the increasing demand in CAR T cell therapies by developing a novel bi-level decomposition algorithm. The proposed algorithm is rigorous and can provide global optimum solutions in reasonable CPU times, even when the full space model provides local or no solutions at all. The original detailed model is decomposed into an upper-level planning problem responsible for strategic planning decisions and a lower-level scheduling problem. As a result, the computational complexity is decreased and solutions for large-scale instances of the problem are obtained. The computational analysis showed that the proposed algorithm has made significant improvements over the full space model. The bi-level algorithm can provide feasible and optimal solutions for instances of up to 5000 patients/year. Nevertheless, the efficiency of the proposed algorithm creates new possibilities to explore, such as an even higher number of therapies per year or an increased time horizon.

5. Acknowledgements

Funding from the UK Engineering & Physical Sciences Research Council (EPSRC) for the Future Targeted Healthcare Manufacturing Hub hosted at University College London with UK university partners is gratefully acknowledged (Grant Reference: EP/P006485/1). Financial and in-kind support from the consortium of industrial users and sector organisations is also acknowledged.

References

FDA-approved CAR T-cell Therapies | UPMC Hillman. https://hillman.upmc.com/mario-lemieux-center/treatment/car-t-cell-therapy/fda-approved-therapies (accessed Nov. 05, 2021).

Kymriah vs. Yescarta Nucleus Biologics." https://nucleusbiologics.com/resources/kymriah-vs-yescarta/ (accessed Feb. 10, 2021).

A. Bernardi, M. Sarkis, N. Triantafyllou, M. Lakelin, N. Shah, M. M. Papathanasiou. Assessment of intermediate storage and distribution nodes in personalised medicine (in press). Computers & Chemical Engineering. 2021

K. J. Caldwell, S. Gottschalk, A. C. Talleur. Allogeneic CAR Cell Therapy—More Than a Pipe Dream. Front. Immunol. 2021;11:1–12.

M. Erdirik-Dogan, I. E. Grossmann. Simultaneous planning and scheduling of single-stage multi-product continuous plants with parallel lines. Comput. Chem. Eng. 2008;32(11): 2664–2683.

D. Han, Z. Xu, Y. Zhuang, Q Qian. Current progress in CAR-T cell therapy for hematological malignancies. J Cancer. 2021;12(2):326–334.

P. Karakostas, N. Panoskaltsis, A. Mantalaris, and M. C. Georgiadis. Optimization of CAR T-cell therapies supply chains. Comput. Chem. Eng. 2020;139:106913.

M. M. Papathanasiou, C. Stamatis, M. Lakelin, S. Farid, N. Titchener-Hooker, N. Shah. Autologous CAR T-cell therapies supply chain: challenges and opportunities?. Cancer Gene Ther. 2020;27:799–809.

M. Sadelain, R. Brentjens, I. Rivière, J. Park. CD19 CAR Therapy for Acute Lymphoblastic Leukemia. Am. Soc. Clin. Oncol. Educ. B. 2015;35:e360–e363.

M. Sarkis, A. Bernardi, N. Shah, M. M. Papathanasiou. Decision support tools for next-generation vaccines and advanced therapy medicinal products: present and future. Curr. Opin. Chem. Eng. 2021a;32(1):100689.

M. Sarkis, A. Bernardi, N. Shah, M. M. Papathanasiou. Emerging Challenges and Opportunities in Pharmaceutical Manufacturing and Distribution. Processes. 2021b;9(457):267–277.

K. Spink, A. Steinsapir. The long road to affordability: a cost of goods analysis for an autologous CAR-T process. Cell Gene Ther. Insights. 2018;4(11):1105–1116.

S. Terrazas-Moreno, I. E. Grossmann. A multiscale decomposition method for the optimal planning and scheduling of multi-site continuous multiproduct plants. Chem. Eng. Sci. 2011;66(19):4307–4318.

Proceedings of the 14th International Symposium on Process Systems Engineering – PSE 2021+
June 19-23, 2022, Kyoto, Japan © 2022 Elsevier B.V. All rights reserved.
http://dx.doi.org/10.1016/B978-0-323-85159-6.50367-5

An agent-based model for cost-effectiveness analysis in the manufacture of allogeneic human induced pluripotent cells in Japan

Yusuke Hayashi[a]*, Kota Oishi[a], Hirokazu Sugiyama[a]

[a]*Department of Chemical system Engineering, The University of Tokyo, 7-3-1, Hongo, Bunkyo-ku, Tokyo 113-8656, Japan*
y-hayashi@pse.t.u-tokyo.ac.jp

Abstract

This work proposes an agent-based model for cost-effectiveness analysis in the manufacture of allogeneic human induced pluripotent (hiPS) cells in Japan. The agent-based model was developed that can estimate the disability-adjusted life years (DALYs) of each patient and the total required manufacturing cost for allogeneic hiPS cells. The DALYs were defined as the effectiveness indicator, while the total required cost for manufacturing was applied as the cost indicator. Given the disease, the annual number of treated patients, and the treatment mode, the agent-based model can calculate these two indicators. The model was applied to analyze allogeneic hiPS cell therapy for two diseases that are under clinical studies in Japan. A case study demonstrated that the treatment mode would have a significant impact on the cost-effectiveness.

Keywords: Regenerative medicine, Healthcare, Disability-adjusted life year, Therapeutic effect, Kernel density estimation.

1. Introduction

Japan is one of the most advanced countries of research and development on allogeneic human induced pluripotent stem (hiPS) cells. The cells were first produced at Kyoto University (Takahashi et al., 2007), and are one of the most promising sources of regenerative medicine products. Along with successful clinical studies, e.g., Parkinson's disease, implementation of allogeneic hiPS cell therapy is in progress.

Recently in Japan, approval of several products by Pharmaceuticals and Medical Devices Agency (PMDA) received public attention, e.g., nivolumab and tisagenlecleucel. This was because both the high therapeutic effect and the price set by PDMA (e.g., JPY ca. 700,000 for nivolumab in 2014, the year of new introduction). In the national health insurance system of Japan, generally 70 % of the treatment cost is paid by public health system. Further approval of expensive products could lead concerns regarding budget deficits. The products based on allogeneic hiPS cells, once marketed, would also be given ultra-high prices considering the high therapeutic effects. Therefore, the balance between the therapeutic effect and the financial impacts should be considered, in order to pursue sustainability of the healthcare system in Japan. However, detailed investigations considering the individual patient's conditions, e.g., age, have yet to be performed.

Agent-based modeling is useful to deal with complex systems based upon agents, which enables consideration of the individual person's condition. Various topics have been analyzed by agent-based models, e.g., emissions trading considering exchange rates by

Peng et al. (2019), integrated energy systems planning and operation by Zhang et al. (2020), and market acceptance of electric vehicles in China by Huang et al. (2021).

This work proposes an agent-based model for cost-effectiveness analysis in the manufacture of allogeneic hiPS cells in Japan. The agent-based model was developed that can quantify the disability-adjusted life years (DALYs) of each patient and the required manufacturing cost for allogeneic hiPS cells in Japanese society. The DALYs were applied as the effectiveness indicator, while the total required cost was used as the cost indicator. Given the disease, the annual number of treated patients, and the treatment mode, the agent-based model can evaluate these two indicators. The model was applied to analyze allogeneic hiPS cell therapy for two diseases that are in clinical studies in Japan.

2. Methods

2.1. Effectiveness indicators of medical treatments

Generally, there are two indicators used for cost-effectiveness analyses of medical treatments: DALYs (Murray and Lopez, 1997) and quality-adjusted life years (QALYs; Zeckhauser and Shepard, 1976). DALYs indicate overall disease burden expressed as the number of life years lost. The coefficients used for calculating DALYs were provided by World Health Organization (World Health Organization, 2004). On the other hand, QALYs represent overall health condition expressed as the number of life years considering quality of life. The coefficients used for calculating QALYs require to be defined by questionnaires for patients. Allogeneic hiPS cell therapy is still in the middle of clinical studies, and it is quite difficult to perform questionnaires to patients. Hence, in this work DALYs is adopted as the effectiveness indicator.

2.2. Model overview

Figure 1 shows an overview of the developed agent-based model. The overall inputs are defined as the disease, θ [–] (e.g., Parkinson's disease), the annual number of treated patients, $N_{\text{patient}}^{\text{treat}}$ [person year^{-1}], the treatment mode, γ [–] (e.g., from youngest to oldest), and the treatment period, $t_{\text{treat}}^{\text{final}}$ [year]. The outputs are defined as the total DALYs of all patients at the end of the treatment period, *DALY* [year], and the total required manufacturing cost for allogeneic hiPS cells, C_{total} [JPY]. The output variables of *DALY* and C_{total} are used as the effectiveness indicator and the cost indicator, respectively.

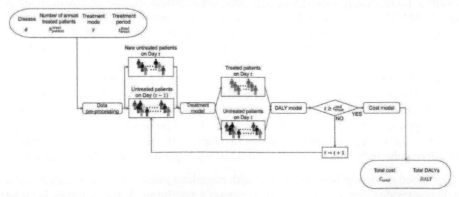

Figure 1. Model overview.

2.3. Data pre-processing

Generally, the data of patient's age is collected as discreate data. For example, the patient's age data in Japan is collected in five-year increments. However, continuous data of patient's age is needed for DALY estimation. Thus, the discrete data of patient's age is converted into continuous data as follows:

$$D_{\text{age}}(Y) = \frac{1}{N_{\text{patient}}^{\text{initial}} h} \sum_{i=1}^{N_{\text{patient}}^{\text{initial}}} K\left(\frac{Y - Y^i}{h}\right) \tag{1}$$

$$K(x) = \frac{1}{\sqrt{2\pi}} \exp\left(-\frac{x^2}{2}\right) \tag{2}$$

where D [–] is the distribution, Y [year] is the patient age, $N_{\text{patient}}^{\text{initial}}$ [person] is the initial number of patients, h [–] is the smoothing parameter, and K [–] is the kernel function.

2.4. Treatment model

Figure 2 shows the details of the treatment model. The inputs of the model are defined as the untreated patients on Day $(t - 1)$ and the new untreated patients on Day t. The treatment model is classified into reorder and cure models. In the reorder model, all untreated patients on Day t are sorted in the order of treatment. In the cure model, a fixed number of untreated patients based on the annual number of treated patients, $N_{\text{patient}}^{\text{treat}}$, is treated. Moreover, the treatment model is defined, based on the following assumptions.

- For diseases causing sudden death, it occurs with a probability of β [–].
- The effect of treatment is 100%.
- The required time to manufacture allogeneic hiPS cells can be ignored.

2.5. DALY model

The DALYs of a patient i, $DALY^i$ [year], and the total DALYs are estimated as follows (Murray and Lopez, 1997):

$$DALY^i = YLL^i + YLD^i \tag{3}$$

$$YLL^i = \int_{Y_{\text{dead}}^i}^{Y_{\text{life}}} \alpha \, dY \tag{4}$$

$$YLD^i = \int_{Y_{\text{sick}}^i}^{Y_{\text{cure}}^i} \alpha DW \, dY \tag{5}$$

$$DALY = \sum_{i=1}^{N_{\text{patient}}^{\text{final}}} DALY^i \tag{6}$$

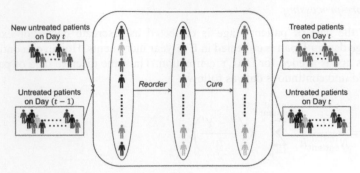

Figure 2. Details of the treatment model.

where YYL [year] is the years of life lost, Y_{life} [year] is the average life span, Y_{dead} [year] is the age of death, α [–] is the disability adjusted life year weight, YLD [year] is the years lost due to disability, Y_{cure} [year] is the age of complete recovery, Y_{sick} [year] is the age of onset, W [–] is the disability weight, and $N_{patient}^{final}$ [person] is the final number of all patients including completely cured patients.

2.6. Cost model

The total required manufacturing cost of allogeneic hiPS cells, C_{total}, is calculated using the following equations:

$$C_{total} = C_{const} + C_{man} + C_{trans} \tag{7}$$

where C_{const} [JPY] is the cost for construction, C_{man} [JPY] is the cost for manufacturing, and C_{trans} [JPY] is the cost for transportation. The value used in the calculation of C_{man} were defined with reference to Sugiyama et al. (2020).

3. Results and discussion

3.1. Data pre-processing

Figure 3 shows the relationship between the patient age, Y, and the probability density, φ [–], for the two diseases in Japan. For both diseases, the parks of the probability density are around 80 years old.

3.2. Case study: Investigation of treatment mode

The impact of the treatment mode on the cost-effectiveness of allogenic hiPS cell therapy for the two diseases was investigated in this section. For the cost-effectiveness analysis, the total required manufacturing cost of allogenic hiPS cells per one DALY, C_{total}^{DALY} [JPY year^{-1}], was defined as the evaluation indicator of cost-effectiveness, as shown in the following equations:

$$C_{total}^{DALY} = \frac{C_{total}}{\Delta DALY} \quad \left(N_{patient}^{treat} \geq 1\right) \tag{8}$$

An agent-based model for cost-effectiveness analysis in the manufacture of allogeneic human induced pluripotent cells in Japan

2207

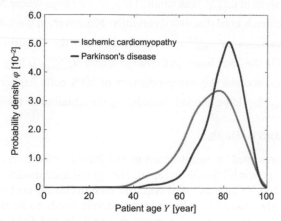

Figure 3. Distribution of the patient age.

$$\Delta DALY = -\{DALY(N_{patient}^{treat}) - DALY(0)\} \tag{9}$$

where $\Delta DALY$ [year] is the disability-adjusted life years that can be reduced by the treatment. The optimization problem was formulated as the following equation:

$$\min C_{total}^{DALY}(\gamma, N_{patient}^{treat}) \tag{10}$$

subject to

$$\gamma \in \{\text{From youngest to oldest, From oldest to youngest}\}$$
$$N_{patient}^{treat} \leq 6.0 \times 10^3$$
$$\theta \in \{\text{Ischemic cardiomyopathy, Parkinson's disease}\}$$

Figure 4 shows the relationship between $N_{patient}^{treat}$, C_{total}^{DALY}, γ, and θ. For both diseases, in the range where $N_{patient}^{treat}$ was small, the values of C_{total}^{DALY} differed greatly depending on the treatment mode. On the other hand, in the range where $N_{patient}^{treat}$ was large, the

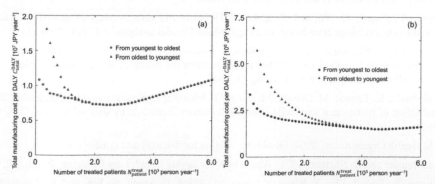

Figure 4. Impact of the treatment mode and the number of treated patients on the cost-effectiveness for (a) ischemic cardiomyopathy and (b) Parkinson's disease.

difference in the values of C_{total}^{DALY} was small. Thus, in the range where $N_{patient}^{treat}$ was small, the treatment mode needs to be discussed carefully. Moreover, in the range where $N_{patient}^{treat}$ is small, C_{total}^{DALY} decreases as $N_{patient}^{treat}$ increases because the construction cost per one patient decreases. On the other hand, in the range where $N_{patient}^{treat}$ is large, C_{total}^{DALY} increases as $N_{patient}^{treat}$ increases because the overproduction of hiPS cells happens. Hence, careful selection of $N_{patient}^{treat}$ should be needed considering the situation of the disease.

4. Conclusions and outlook

In this work, we presented an agent-based model for cost-effectiveness analysis in the manufacture of allogeneic hiPS cells in Japan. The agent-based model was developed that can quantify the DALYs and the total required manufacturing cost of allogeneic hiPS cells. The case study showed that the treatment mode needs to be discussed carefully when the number of annual treated patients is small. In the field of computer-aided process engineering, cell therapy related studies are becoming relevant, e.g., Moschou et al. (2020) and Hayashi et al. (2020). Further model-based studies in this area are encouraged.

Acknowledgements

Discussions with Prof. Masahiro Kino-oka at Osaka University are appreciated. H. S. is thankful for financial support by a Grant-in-Aid for Challenging Research (Exploratory) No. 20K21102 from the Japan Society for the Promotion of Science.

References

Y. Hayashi, I. Horiguchi, M. Kino-oka, and H. Sugiyama, 2020, Slow freezing process design for human induced pluripotent stem cells by modeling intracontainer variation, Comput. Chem. Eng., 132, 106597.

X. Huang, Y. Lin, F. Zhou, M. Lim, and S. Chen, 2021, Agent-based modelling for market acceptance of electric vehicles: Evidence from China, Sustain. Prod. Consum., 28, 206.

D. Moschou, M. Papathanasiou, M. Lakelin, and N. Shah, 2020, Investment planning in personalised medicine, Comput. Aided Chem. Eng., 48, 49.

C. Murray and A. Lopez, 1997, Global mortality, disability, and the contribution of risk factors: global burden of disease study, Lancet, 349, 1436.

Z. Peng, Y. Zhang, G. Shi, and X. Chen, 2019, Cost and effectiveness of emissions trading considering exchange rates based on an agent-based model analysis, J. Clean. Prod., 219, 75.

H. Sugiyama, M. Shiokaramatsu, M. Kagihiro, K. Fukumori, I. Horiguchi, and M. Kino-oka, 2020, Apoptosis-based method for determining lot sizes in the filling of human-induced pluripotent stem cells, J. Tissue Eng. Regen. Med., 14, 1641.

K. Takahashi, K. Tanabe, M. Ohnuki, M. Narita, T. Ichisaka, K. Tomoda, and S. Yamanaka, 2007, Induction of pluripotent stem cells from adult human fibroblasts by defined factors, Cell, 131, 861.

World Health Organization, 2004, Disability weights for diseases and conditions.

R. Zeckhauser and D. Shepard, 1976, Where now for saving lives?, Law Contemp. Probl., 40, 5.

Z. Zhang, R. Jing, J. Lin, X. Wang, K. Dam, M. Wang, C. Meng, S. Xie, and Y. Zhao, 2020, Combining agent-based residential demand modeling with design optimization for integrated energy systems planning and operation, Appl. Energy, 263, 114623.

Proceedings of the 14th International Symposium on Process Systems Engineering – PSE 2021+
June 19-23, 2022, Kyoto, Japan © 2022 Elsevier B.V. All rights reserved.
http://dx.doi.org/10.1016/B978-0-323-85159-6.50368-7

Design and operation of healthcare facilities using batch-lines: the COVID-19 case in Qatar

Brenno C. Menezes[a*], Mohamed Sawaly[a,b], Mohammed Yaqot[a], Robert E. Franzoi[a], Jeffrey D. Kelly[c]

[a]*Division of Engineering Management and Decision Sciences, College of Science and Engineering, Hamad Bin Khalifa University, Qatar Foundation, Doha, Qatar*
[b]*Department of Supply Chain, Hamad Medical Corporation, Doha, Qatar*
[c]*Industrial Algorithms Ltd., 15 St. Andrews Road, Toronto M1P 4C3, Canada*
bmenezes@hbku.edu.qa

Abstract

In the wake of the COVID-19 pandemic, hospitals worldwide have been overwhelmed and deprived of valuable resources such as bed capacities, medical equipment, personal protection equipment (PPE) stocks, and personnel. These factors imposed unforeseen challenges in the healthcare treatment systems. Mitigating inefficiencies by learning from COVID-19 is necessary to be better prepared to save lives and conserve resources. The main goal of this study is the development of an optimized healthcare treatment network by using predicted epidemiology curves to determine influxes of patients and bed capacities in a hospital facility for both in-patient (IP) wards (oxygen outlets) and intensive care units (ICU). Our model considers flows of patients by distinguishing them in terms of medical severity for their optimal allocation in an existing or installed healthcare facility treated as batch-lines (batch-processes in lines) with time-varying yields of a number of patients per day of treatment. Considering the hospital's admission and discharge of patients from 2020's 1st wave of COVID-19 in Qatar, we determine the bed space availability at any given future date for a hospital facility. This enables the prescription of engineered solutions to increase the capacity, responsiveness, and preparedness of healthcare systems infrastructure and management.

Keywords: Healthcare systems, supply chain resilience, optimization, COVID-19.

1. Introduction

The rapid spread of COVID-19 cases demonstrated the challenges of containing a pandemic whilst providing adequate care (Murthy et al., 2020). Design and operational inefficiencies are among the biggest reasons healthcare systems fail to minimize death rates and spreads of pandemics. Given the inevitable occurrence of future pandemics, healthcare systems must predict the growth and spread of the virus, implement strategies to contain it, and prepare their facilities and resources accordingly.

Several works address predictions on epidemiology curves (Santosh, 2020; Jewell et al., 2020) and their respective effects on resources such as personal protective equipment (PPE) (Tosh et al., 2014). Stübinger and Schneider (2020) propose a forecast of the future COVID-19 spread by addressing identified lead-lag effects using dynamic time warping from batch process monitoring and analysis. Garbey et al. (2020) use data obtained from the French Government during COVID-19 in a computational model to anticipate the patient load of each care unit, and the amount of PPE required by these units, as well as other key parameters that measure the performance of a healthcare system. Goodarzian et al. (2021) introduce a sustainable-resilience healthcare network for handling COVID-19 pandemic using meta-heuristics for allocation of medicine, resources, and staff throughout the supply chain elements considering capacities and flows among warehouses, distribution centers, pharmacies, hospitals, etc.

The proposition of this work is to develop prescriptive analytics for the optimal healthcare treatment systems in the planning, scheduling, and coordination of the disease treatment networks. With the utilization of the epidemiology curves, decisions can be made to determine optimal bed capacities needed during the COVID-19 pandemic, enabling the design and operation for a resilient medical supply chain to the COVID-19 pandemic.

2. Problem statement

The epidemiology data obtained for this study provide a daily prediction of positive cases from February 1st to May 31st, 2021 in Qatar. From the total number of suspected cases, it is assumed that 1% ends up in to the national healthcare systems' triage facility. We develop a mixed-integer linear programming (MILP) model for 120 days as time-horizon with 1-day time-step, in which 30% of the suspected patients at the triage result in negative diagnostic, and the remaining 70% result in positive. Among the admitted in hospital, 70% of the patients went as an in-patient (IP) and 30% in an intensive care unit (ICU). The model considers actual distributions (in terms of medical severity) of approximately 7,800 patients admitted into a hospital from March 2020 for one year, including the daily inflows and outflows of patients among the facility networks. It also considers the hospital's capacity as 335 IP and 230 ICU beds, with initial occupancies of 30% for each. The field hospital to be opened has a capacity of 160 IP and 80 ICU beds.

For the design and operation optimization of healthcare treatment systems, the network in Figure 1 shows a flowsheet of existing and future facilities and connections constructed in the unit-operation-port-state superstructure (UOPSS) from Kelly (2005) built-in in the Industrial Modeling and Programming Language (IMPL) (Kelly and Menezes, 2019). The shapes are considered as: a) unit-operations m for sources and sinks ($\diamond$), tanks or inventories ($\triangle$), batch-processes ($\square$) and b) the connectivity involving arrows ($\rightarrow$), inlet-port i ($\bigcirc$) and outlet-port j. Unit-operations and arrows are modeled by binary y and continuous x variables and the ports as yields of patients.

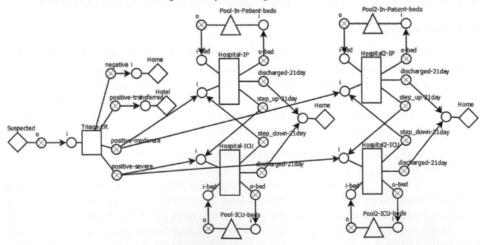

Figure 1. Base flowsheet of the healthcare system.

The model includes predicted yields of patient step-downs (from ICU to IP – meaning the patients' medical status improved), patient step-ups (from IP to ICU – meaning the patients' medical status deteriorated), patient admission yields, patient discharge yields, death yields, and transfer yields. The maximum batch-time is 21 days (which captures approximately 85 to 90% of hospitalized patients' lengths of stay), although there are time-varying yields from the actual distribution of patients in- and out-fluxes both IP and ICU wards from the observed cases. The 1- to 20-days yields and their connections are not represented in Figure 1 for simplicity.

3. Mathematical modeling

The objective function in Eq.(1) maximizes the pre-treatment of the suspected cases in the triage emergency room (ER), where $x_{j,i,t}$ represent number of cases for flows from the outlet port set j to inlet set i at time t. The variable $xh_{m,t}$ defines batch-processes' or hospital-units' holdups and pools of bed capacity in the model. All flows and holdups are governed by semi-continuous constraints of the shapes to themselves, such as $\bar{x}_{j,i,t}^L \, y_{j,i,t} \leq x_{j,i,t} \leq \bar{x}_{j,i,t}^U \, y_{j,i,t} \, \forall \, (j,i) \in JI, t$. The sets I and J represent in- and out-ports, respectively, while the set JI defines connecting patient flows between out- and in-ports. For the batch-processes (triage and hospitals), the holdup $xh_{m,t}$ is taken when they are starting up ($zu_{m,t}=1$) constrained by the respective bounds of the hospital facility capacities. The UOPSS formulation in Eq.(2) establishes that the holdup or inventory level bounds ($\overline{xh}_{m,t}^L$ and $\overline{xh}_{m,t}^U$) of the hospital facilities respect the sum of the flows arriving in and leaving from ports (in- and out-ports) whenever the respective startup variable $zsu_{m,t}$ is active. The sets M_{BATCH} include triage-ER, IP, and ICU facilities and M_{POOL} the IP/ICU bed's pools. In the indices in the summations from Eq.(1) to (5), the subsets of the I, J, and JI follow the flowsheet in Figure 1. For $x_{j,i,t}, xh_{m,t} \geq 0$; $y_{j,i,t}, y_{m,t} = \{0,1\}$; $zsu_{m,t} = (0,1)$:

$$Max \, Z = \sum_{t} \sum_{JI_{Suspected}} x_{j,i,t} \tag{1}$$

$$\overline{xh}_{m,t}^L \, zsu_{m,t} \leq \sum_{i \in I} x_{j,i,t} \leq \overline{xh}_{m,t}^U \, zsu_{m,t} \quad \forall \, (m,j) \in M_{BATCH}, t \tag{2}$$

$$\sum_{j \in J_{up}} x_{j,i,t} = xh_{m,t} \quad \forall \, (i,m) \in M_{BATCH}, t \tag{3}$$

$$x_{j,i \in I_{do},t+delay} = \bar{r}_{j,t+delay} \, xh_{m,t} \quad \forall \, (m,j) \in M_{BATCH}, t \tag{4}$$

$$xh_{m,t} = xh_{m,t-1} + \sum_{j_{up} \in J} x_{j_{up},i,t} - \sum_{i_{do} \in I} x_{j,i_{do},t} \quad \forall \, (i,m,j) \in M_{POOL}, t \tag{5}$$

$$y_{m_{up},t} + y_{m,t} \geq 2y_{j_{up},i,t} \quad \forall \, (m_{up}, j_{up}, i, m), t \tag{6}$$

$$\sum_{tt<t} zsu_{m',tt} + y_{m,t} \leq y_{j,i,t} \quad \forall \, (m', j, i, m), t \tag{7}$$

$$y_{m,t} - y_{m,t-1} - zsu_{m,t} + zsd_{m,t} = 0 \quad \forall \, m \in M_{BATCH}, t \tag{8}$$

$$y_{m,t} + y_{m,t-1} - zsu_{m,t} - zsd_{m,t} - 2zsw_{m,t} = 0 \quad \forall \, m \in M_{BATCH}, t \tag{9}$$

$$zsu_{m,t} + zsd_{m,t} + zsw_{m,t} \leq 1 \quad \forall \, m \in M_{BATCH}, t \tag{10}$$

Equations (3) and (4) are related to the modeling of hospitals as batch-processes and in a special case called batch-lines. This is applied in Menezes et al. (2020) for livestock planning to determine the initial procreation of the animal batches, in which there is no accumulation of amounts of batches at each time step, as in Eq.(3). Instead, balances of batch amounts at a single time-window and the delaying and yield of amounts leaving the facility are modeled, as in Eq.(4). In the hospital facilities, new batches of patients arrive in the Triage-ER, Hospital IP, and Hospital ICU every day. The unit-operations' inventory or holdup quantity balance of pools are determined in Eq.(5) for both IP and ICU bed capacities. These constraints manage the availability of beds (holdup) to be utilized in the hospitals by controlling the a) inlet flow, when patients are dispatched outside the system or change their status to step-up or step-down; and b) the outlet flow, when the beds are needed in the hospital facilities.

Equations (5) and (6) represent the constraints for the structural transitions that allow the setup $y_{m,t}$ or startup $zsu_{m,t}$ of connected out-port-states j and in-port-states i unit-operations. When the setup of unit-operations m and m' is equal to the unitary in Eq.(6), by implication, the setup variable the arrow stream $y_{j,i,t}$ between the neighbor unit-operations must be true. In Eq.(7), addressing the hospital facilities as batch-processes, as the setup variable of m' is changed by the summation of the startups. These logic valid cuts reduce the tree search in branch-and-bound methods. The temporal transition in Equations (8) and (9) control the operations for semi-continuous blenders from Kelly and Zyngier (2007). The binary variable $y_{m,t}$ manages the start-up ($zsu_{m,t}$) switch-over ($zsw_{m,t}$) and shut-down variables ($zsd_{m,t}$), which are relaxed in the interval [0,1]. Equation (10) guarantees the integrality of the relaxed variables.

4. Results

The optimization for the proposed MILP in Figure 1 for 120 days as time-horizon with 1-day time-step is solved in 63 seconds with GUROBI 9.1.1 and 256 seconds with CPLEX 20.1.0 both at 1.0% of MILP relaxation gap using an Intel Core i7 machine at 3.4 GHz (8 threads) with 64 GB of RAM. There are 68,721 constraints for 23,765 continuous variables and 15,248 binary variables in the problem. The results in Figure 2 show that the existing facility's IP capacity could not sustain the surge of patients caused by the new strain of COVID-19 (initiated on day 63), which triggered the opening of the new field hospital by day 85.

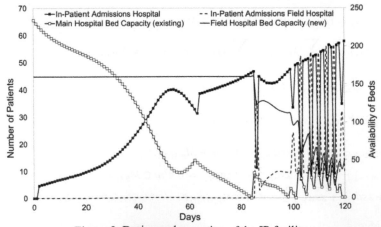

Figure 2. Design and operation of the IP facility.

Based on Figure 2, the existing hospital was only able to sustain the increase in patients for exactly 21 days (day 84), where the number of In-Patient admissions was 47 patients per day. From day 85 onwards, both facilities simultaneously received patients, which relieved the pressure on the existing facility and made the design feasible. Days 98, 101, and 119 demonstrate how the new field hospital's capacity helped sustain the operation when the main existing facility's capacity depleted. By the end of the time horizon, there were still 51 vacant IP beds in the field hospital (i.e., new facility). The lines in the plot are symmetrical as each facility worked hand in hand to handle the influx of patients.

Figure 3 demonstrates how the opening of 80 more ICU beds aided in sustaining the hospitalization of patients on days 83, 92, and onwards. The phenomenon observed from days 92 to 107, where the capacity of the field hospital remains zero, is illustrated in Figure 4.

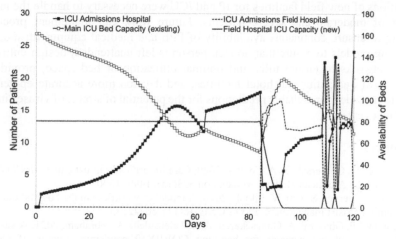

Figure 3. Design and operation of the ICU facility.

Figure 4 explains why the Field Hospital ICU capacity curve remains flat at zero from day 92 to 107, in which the capacity pool freeing up (by people being discharged) is occupied by other patients at the same rate. This phenomenon shows extreme efficiency since utilizes the limit opened capacity available, although it does not consider real-world factors such as the disinfection or preparation of ICU space.

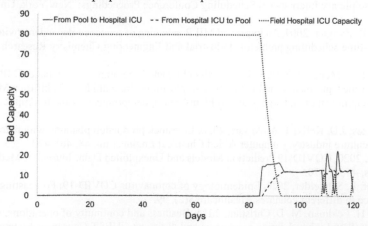

Figure 4. Pool of patients in and out causing the capacity to remain flat at zero.

5. Conclusions

It was evident from the events of COVID-19 that our world is interconnected in a way that virus outbreaks in a region can easily spread and cause impacts in a global sphere. Protecting the lives of humans entails that we must have a proper number of resources allocated in a timely and efficient manner. This work demonstrates how to be better prepared by designing and operating a healthcare treatment network with different facilities as the triage-ER, in-patient (IP), and intensive care unit (ICU) wards. The predicted epidemiology curves and the time-varying yields of the distribution of patients throughout the network served as inputs for the modeling and solving of batch-lines of patients interconnected among the facilities and their outlets. With such proposition, availability of bed capacities has been determined along the time-horizon, and installations of new field facilities for IP and ICU were necessary to handle the increased number of moderate and severe patients. Future work can implement procurement planning to ensure continuous availability of PPE and medical equipment; model staff scheduling models to ensure that no sick person is left unattended; design entire health system networks to ensure fully and optimal utilization of bed space; provide better utilization of quarantine and hotel facilities; and develop more accurate epidemiology curves to provide more reliable predictions for the potential of strains of viruses in further pandemic events.

References

S. Murthy, C. D. Gomersall, R. A. Fowler, 2020, Care for critically ill patients with COVID-19. Journal of the American Medical Association 323(15), 1499–1500.

M. Garbey, G. Joerger, S. Furr, V. Fikfak, 2020, A model of workflow in the hospital during a pandemic to assist management. PLOS ONE, 15(11), e0242183.

F. Goodarzian, P. Ghasemi, A. Gunasekaren, A.A. Taleizadeh, A. Abraham, 2021, A sustainable-resilience healthcare network for handling COVID-19 pandemic, Annals of Operations Research.

N. Jewell, J. Lewnard, B. Jewell, 2020, Caution warranted: using the Institute for Health Metrics and Evaluation model for predicting the course of the COVID-19 pandemic. Ann Intern Med, Epub ahead of print.

J.D. Kelly, 2005, The Unit-Operation-Stock Superstructure (UOSS) and the Quantity-Logic-Quality Paradigm (QLQP) for Production Scheduling in The Process Industries, In Multidisciplinary International Scheduling Conference Proceedings: New York, United States, 327-333.

J.D. Kelly, D. Zyngier, 2007, An improved MILP modeling of sequence-dependent switchovers for discrete-time scheduling problems, Industrial and Engineering Chemistry Research, 46, 4964-4973.

J.D. Kelly, B.C. Menezes, 2019, Industrial Modeling and Programming Language (IMPL) for Off- and On-Line Optimization and Estimation Applications. In: Fathi M., Khakifirooz M., Pardalos P. (eds) Optimization in Large Scale Problems. Springer Optimization and Its Applications, 152, 75-96.

B.C. Menezes, J.D. Kelly, T. Al-Ansari, 2020, Livestock production planning with batch-lines in the agriculture industry, Computer Aided Chemical Engineering, 48, 465-470.

K. Santosh, 2020, COVID-19 Prediction Models and Unexploited Data. Journal of Medical Systems, 44(9), 170.

J. Stubinger, L. Schneider, 2020, Epidemiology of coronavirus COVID-19: Forecasting the future incidence in different countries. Healthcare 8(2), 99.

P.K. Tosh, H. Feldman, M. D. Christian, 2014, Business and continuity of operations: Care of the critically ill and injured during pandemics and disasters: CHEST Consensus Statement. Chest, 146(4).

Proceedings of the 14th International Symposium on Process Systems Engineering – PSE 2021+
June 19-23, 2022, Kyoto, Japan © 2022 Elsevier B.V. All rights reserved.
http://dx.doi.org/10.1016/B978-0-323-85159-6.50369-9

Application of PSE Methods on Monoclonal Antibody Productivity Improvement and Quality Control

Ou Yang[a], and Marianthi Ierapetritou[b*]

[a] Department of Chemical and Biochemical Engineering, Rutger, the State University of New Jersey, Piscataway, NJ, 08854, USA
[b] Department of Chemical Engineering and Biomolecular Engineering, University of Delaware, Newark, DE, 19716, USA
mgi@udel.edu

Abstract

With a high demand for monoclonal Antibodies (mAbs) in the current biopharmaceutical market, there is a need to improve overall process productivity while maintaining product quality. This work introduces three case studies that apply process systems engineering (PSE) methods including flowsheet modeling, mechanistic modeling, and process optimization to provide strategies on production mode selection, process prediction, and operating design space determination. These approaches can be utilized in process development and ultimately improve the productivity during biopharmaceutical manufacturing.

Keywords: Process simulation; Dynamic modeling; Biopharmaceutical manufacturing

1. Introduction

MAbs is one of the most promising therapeutic products with its wide applications in cancers, infections, and autoimmune disorders treatment. It has been reported that the growth rate of mAb market ranges between 7.2 and 18.3% since 2016 and would reach $130-200 billion in the year 2022 (Grilo and Mantalaris, 2019). Most of mAbs belong to IgG class which contains two regions: antigen-binding (Fab) and crystallizable regions (Fc). Protein glycosylation in Fc region affects protein stability and efficacy (Xu et al. 2011, Zheng et al., 2011). Thus, maintaining operating conditions within a feasible region to ensure the required level of glycosylation is also critical and challenging in mAb production. Process systems engineering (PSE) methods such as process simulation, sensitivity analysis, and process optimization allow early-stage process design, system analysis, and process improvement, which has great potential to address the above challenges.

In this work, flowsheet modeling is used as a process decision-making tool to evaluate cost-effectiveness of fully integrated continuous operation over the conventional batch operation. Then mechanistic and surrogate models are used to capture nonlinear bioreactor dynamics under different operating conditions. The models are able to correlate operating parameters with productivity and quality. Furthermore, feasibility analysis is applied to process model to determine the design space for desired protein production and drug quality.

2. Flowsheet modeling and techno-economic analysis

In this case study, fed-batch and continuous mAb production lines are designed to evaluate the benefits of continuous operations. Different analysis approaches including deterministic cost analysis, and sensitivity analysis are used to discover the possibilities and challenges of continuous applications in biopharmaceutical manufacturing.

2.1. Process Description

Biopharmaceutical manufacturing for mAbs production includes inoculation, cell culture protein production, clarification, primary capture, polishing, and final formulation steps. The inoculation is a series of cell culture passages that contain test tubes, T flasks, shake flasks, and seed culture bioreactor. Fed-batch bioreactor and perfusion bioreactor are used as production bioreactors for batch process and continuous process, respectively. After production in batch operating mode, proteins are purified by centrifuge followed by microfilters, dead-end filters, Protein A chromatography, acid-based virus inactivation tanks, AEX chromatography, nano filter, ultrafiltration, and diafiltration. For continuous process, tangential filtration is used to harvest protein and send it to filters and periodic counter-current (PCC) protein A column for primary capture. Two virus inactivation tanks and AEX chromatography work alternatively to achieve continuous virus inactivation. For the final formulation staged single-pass tangential flow filtration and counter-current staged diafiltration are used for buffer exchange and protein concentration.

2.2. Simulation software and techno-economic analysis

SuperPro Designer (Intelligen, Scotch Plains, NJ) is a recipe-driven simulator that can be used to simulate both batch and continuous biopharmaceutical manufacturing. Material balances and process scheduling can be captured through the whole process. Process analysis including economic analysis with breakdown cost categories, throughput analysis, and sensitivity analysis are used to evaluate the two operating modes under different scales and process operating parameters.

Economic analysis is performed using SuperPro Designer with customer inputted equipment and material costs. The costs are referenced from literature and online vendor resources. The calculation includes capital investment and operating costs. The capital investment is based on equipment expenditure. The installation, piping, insulation costs are calculated based on the equipment cost. The operating cost contains material, consumables, utilities, labor-depended, quality control and quality assurance, and facility cost. Cost of goods per gram (COG/g) is used to represent cost per unit of production.

2.3. Results

Two integrated lines, one with batch operating mode and another with continuous operating mode are simulated. The base case scenario is adjusted to 620 kg/yr production rate and the selling price of mAb is assumed at \$20/mg. The result shows that upstream takes the highest percent of the overall cost in both operating modes. Comparing batch and continuous processes, the overall capital cost in the fed-batch process is \$165 million, which is 2 times higher than that of the continuous process. The operating cost of fed-batch and continuous processes are \$61 million/yr, and \$32 million/yr, respectively.

Figure 1 clearly shows that the main benefit of the continuous process stems from the capital cost investment and facility dependent cost, mainly due to smaller footprint of the continuous process. The consumables cost is also reduced due to resin cost savings in the PCC process.

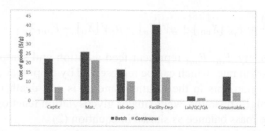

Figure 1 Cost of goods analysis between fed-batch and continuous

Sensitivity analysis is used to further investigate the impact of throughput and upstream parameters on the cost-effectiveness of the two operating modes. The manufacturing scale changes from 50 kg/yr to 1200 kg/yr as shown in Figure 2. Results show a decreased trend of unit operating cost with plant capacity changes, which is mainly attributed to the cost savings of labor dependent cost. The continuous process is more cost-effective than the batch process through the whole range of capacity. Similar results are also found whe titer changes as shown in Figure 3, where titer varies from 1.5 g/L to 5.5 g/L.

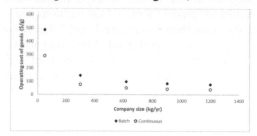

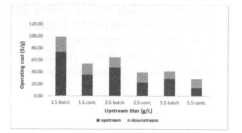

Figure 2 Manufacturing scales change vs. COG/g　　Figure 3 Upstream titer vs operating cost

3. Predictive modeling of cell culture and protein glycosylation processes

From the previous case study, it has been found that the upstream bioreactor takes the highest percentage of the overall cost and is usually identified as a bottleneck of the overall process production. In this section experimental data are used to develop a mechanistic model to understand cell culture and protein glycosylation process in order to improve production and achieve target product quality.

3.1. Background

In mAbs production, the protein glycosylation process is a post-translation modification process that affects product potency and efficacy. Operating conditions such as temperature, pH, and metabolites concentrations all affect glycosylation. To understand the effect of temperature on the glycosylation process, Chinese hamster ovary (CHO)-K1 cell is first cultured under 37 °C and switched to 35 °C, 37 °C and 39 °C on day 4. Different metabolite concentrations, protein titer and glycan fractions are measured at different time points and used to train a mechanistic model.

3.2. Equations

The mechanistic model includes two parts, the cell unstructured model and the structured glycosylation model. The unstructured model considers cells as a black box and only captures the mass balance of critical components in a bioreactor. Equation (1) represents the mass balance of cell density.

$$\frac{d(V[X_v])}{dt} = F_{in}\,[X_{v0}] + \mu V[X_v] - \mu_d V[X_v] - F_{out}X_v \tag{1}$$

where X_v is cell density, F_{in}, F_{out} represent feed addition and sampling during the cell culture; μ is cell growth rate, which can be represented by empirical equations correlated to other metabolites/nutrients in the solution; and μ_d is cell death rate. The single cell model considers Golgi Apparatus, where major glycosylation reaction happens, as a plug flow reactor and the mass balance as shown in equation (2).

$$\frac{\partial[G_m]}{\partial t} = -V_1\frac{\partial[G_m]}{\partial z} + \sum_n^{Enzyme} v_{m,n} r_n \tag{2}$$

where G_m is glycan fractions; r_n represents kinetic rate for enzyme reaction n; $v_{m,n}$ is the reaction coefficient of glycan m that catalyzed by enzyme n; V_1 represents the linear velocity that protein glycan transfers through the Golgi apparatus.

3.3. Results

Least-squares parameter estimation is used to fit the model to experimental data. Experimental data and simulation fitting results for product titer and G0 fraction (one of the glycan fractions) under different temperatures are shown in Figure 4, and 5. The result shows that the mechanistic model is able to capture the general trend of both titer and G0 fraction and the relative trends under different temperature can also be obtained. Cell density, glucose, ammonia concentrations and other glycan fractions can also be captured.

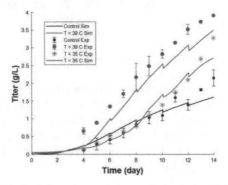
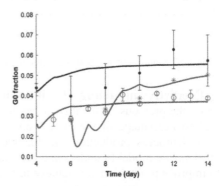

Figure 4 Experimental data and simulation fitting results for titer

Figure 5 Experimental data and simulation fitting results for G0 fraction

4. Fed-batch bioreactor modeling and design space identification

In this case study, dynamic kriging together with a surrogate-based adaptive sampling approach is used to capture the effects of temperature and pH on productivity (product titer) and product quality (glycan fraction in glycosylation process) and furthermore determine the design space for upstream bioreactor operation (Yang and Ierapetritou 2021).

4.1. Kriging and dynamic kriging

As an interpolation method, kriging equation is shown in Equation (3).

$$\hat{f}(x^i) = \beta f(x^i) + \varepsilon(x^i) \tag{3}$$

$\beta f(x^i)$ indicates a known regression model that defines the global trend of the data $f(x^i)$, β is unknown parameter; $\varepsilon(x^i)$ is a residual term that represents the error at location x^i

which is usually normally distributed with zero mean and variance σ^2. Dynamic kriging is a modification of kriging model as shown in equation (4).

$$\hat{f}(x_k^i) = \beta f(x_k^i, \hat{f}(x_{k-1}^i)) + \varepsilon(x^i, \hat{f}(x_{k-1}^i)) \tag{4}$$

The dynamic system is first discretized into different time points k, and the kriging model is used as an autoregressive model that collects the predicted results $\hat{f}(x_{k-1}^i)$ from the previous time point (k-1) and combines with the state variables or control input x_k^i to estimate the future time point $\hat{f}(x_k^i)$(Hernandez and Grover 2010). In this work, both models are built using DACE toolbox in MATLAB.

4.2. Feasibility analysis
Feasibility function is defined in equation (5).

$$\varphi(x) = \max_{j \in J} g_i(x) \tag{5}$$

where $g_i(x)$ represent different constraints including productivity and product quality, x includes temperature, pH and other operating parameters which are total operating time, initial conditions of cell density, glucose and mAb concentrations. Initial sample points can be generated by space filing sampling and a feasible region ($\varphi(x)<0$) can be obtained by calculating the feasibility function as shown in Equation (5). Adaptive sampling method is used to improve the accuracy of the feasible boundary by maximizing the modified EI function, shown in equation (6). The new sample points that are close to the boundary of the feasible region are used to update the kriging model.

$$EI_{feas}(x) = \hat{s}(x)\phi\left(-\frac{\hat{y}(x)}{\hat{s}(x)}\right) = \hat{s}(x)\frac{1}{\sqrt{2\pi}}e^{-0.5(\frac{\hat{y}(x)^2}{\hat{s}(x)^2})} \tag{6}$$

Standard error $\hat{s}(x)$ can be obtained from kriging prediction at location x, and $\hat{y}(x)$ is the predicted value. The detailed explanation of the modified EI function can be found in (Boukouvala and Ierapetritou 2014).

4.3. Results
The mechanistic model for mAbs production outlined in Section 3 is first used to generate training datasets to build kriging models. A two-level full factorial design is applied to generate data under different temperature and pH. Viable cell density, glucose concentration, protein titer and glycan fractions are obtained. The prediction from dynamic kriging and regular kriging are compared in Figure 6. Dynamic kriging provides higher prediction accuracy than regular kriging, because dynamic kriging considers more sample points (the previous time points) and correlations during the model prediction.

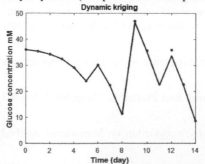
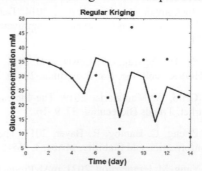

Figure 6 Comparison between dynamic kriging and regular kriging

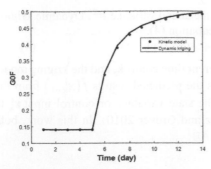

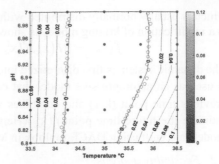

Figure 7 Prediction of glycan fraction using dynamic kriging

Figure 8 Design space obtained from adaptive sampling.

Input parameters (temperature and pH) together with viable cell concentration, protein titers, and glycan fractions at (t−1) time points are used to predict the glycan fractions at time t. Figure 7 provides the prediction of one of the glycan fractions from dynamic kriging which shows that dynamic kriging is able to predict the dynamic trend of glycan fractions with high accuracy. Feasibility analysis is used to ensure high product titer while maintaining glycan fractions within the required range. Figure 8 shows the contour plot for feasibility function values under different pH and temperatures. The feasible operating region is demonstrated inside the zero line.

5. Conclusions

This work shows the application of process modeling and system analysis methods on the improvement of biopharmaceutical manufacturing process. Three approaches are provided, including flowsheet simulation and single unit operation modeling and optimization. Flowsheet modeling is applied to process design and integrated process evaluation. Mechanistic modeling is used to correlate process parameters with cell growth and critical product quality attributes. A framework is built to improve process understanding as well as finding an optimum design space to satisfy required productivity and quality. For the future work, the mechanic model can be integrated to the flowsheet model to obtain an end-to-end biopharmaceutical manufacturing simulation system. Operation of single unit and integrated line can be optimized by adaptive sampling and surrogate based optimization. Validating and training the model with real data would further improve the robustness and reliability of the model.

References

F. Boukouvala, M. Ierapetritou, 2014. Derivative-free optimization for expensive constrained problems using a novel expected improvement objective function. AIChE Journal **60**(7): 2462-2474.

X. Xu, H. Nagarajan, S. Lewis, et al. 2011, The genomic sequence of the Chinese hamster ovary (CHO)-K1 cell line. Nat Biotechnol. 29 (8), 735-741

A. Grilo, A. Mantalaris, 2019, The Increasingly Human and Profitable Monoclonal Antibody Market. Trends Biotechnol. 37, 9–16

K. Zheng, C. Bantog, R. Bayer, 2011, The Impact of Glycoyslation on Monoclonal Antibody Conformation and Stability, MAbs, 3, 568-576

O. Yang, M. Ierapetritou, 2021, mAb Production Modeling and Design Space Evaluation Including Glycosylation Process. Processes

Proceedings of the 14th International Symposium on Process Systems Engineering – PSE 2021+
June 19-23, 2022, Kyoto, Japan © 2022 Elsevier B.V. All rights reserved.
http://dx.doi.org/10.1016/B978-0-323-85159-6.50370-5

Image classification of experimental yields for cardiomyocyte cells differentiated from human induced pluripotent stem cells

Samira Mohammadi, Ferdous Finklea, Mohammadjafar Hashemi, Elizabeth Lipke, and Selen Cremaschi*

Department of Chemical Engineering, Auburn University, Auburn, Alabama, USA
selen-cremaschi@auburn.edu

Abstract

Cardiovascular diseases (CVDs) are the number one cause of death worldwide. Mass production of engineered heart tissue using differentiation of human-induced pluripotent stem cells (hiPSCs) can substitute a large number of the lost heart muscle cells in patients with CVDs. However, the scale-up of the differentiation systems for heart tissue, i.e., cardiomyocyte (CM), production is challenging because many parameters affect the process. Machine learning (ML) techniques can be employed to identify critical process parameters for differentiation systems and build models to elucidate the impact of these parameters on process outcomes. Here, we present a ML model to predict CM content on day 10 of the differentiation. Phase-contrast images of microspheroid tissues on differentiation day 5 are the inputs of the ML model, and the output is CM content on 10 of differentiation, classified as either sufficient and insufficient. Support vector machines are used as the classifier models. We utilized feature extraction and selection methods. The best classifier had an accuracy of 77% in predicting the sufficient CM content class.

Keywords: cardiac differentiation, machine learning, support vector machines

1. Introduction

Heart muscle cells (cardiomyocytes (CMs)) are one of the least regenerative cells in the body. Cardiovascular diseases (CVDs) can lead to heart failure and loss of in the order of billion CMs (Kempf et al., 2016). Few viable treatments are present for patients with CVD and post-heart attack problems. Production of CMs via differentiation from human-induced pluripotent stem cells (hiPSCs) may contribute to developing and testing therapeutics for CVDs, e.g., in fields such as drug monitoring and cell therapy (Denning et al., 2016). Mass production of CMs and their implementation in cell therapy of CVD patients is another potential application of hiPSC-derived CMs (hiPSC-CMs).

The production of CMs by differentiation of hiPSCs in a 3D platform is a complex, expensive process, and a high number of parameters impact the system performance (Gaspari et al., 2018). The 3D platforms are promising for the scale-up of CM production, and identifying critical process parameters and their optimal ranges for 3D platforms is the first step towards scale-up. More specifically, distinguishing an unsuccessful batch from a successful one at an earlier time point of the differentiation would significantly reduce the expense and time required for CM production.

In recent years, machine learning (ML) techniques have been successfully used to study complex systems where fundamental understanding is limited. These techniques use information from data sets to infer the relationships between process parameters (inputs) and outcomes (outputs). With the progress in ML algorithms and computational power, many studies exploited the information contained in images to build models to study different systems, such as quantification of CM contraction using image correlation analysis (Kamgoué et al., 2009) and plant disease detection (Vishnoi et al., 2021).

This study investigates the ability to classify CM content on day 10 of hiPSC-laden microspheroid differentiation using images taken on day 5. The CM content is defined as the percentage of the cells which are CMs on the specific differentiation day. We hypothesize that the phase-contrast images of the cells taken during differentiation include information regarding differentiation progress and that a classifier model can capture this information to distinguish batches with sufficient CM content from those with insufficient. Support vector machines are trained using different extracted feature sets of the phase-contrast images to predict the CM content class. The best model had an accuracy of 77% and an MCC of 0.53.

2. Methods and Materials

2.1. Experiments

HiPSCs were encapsulated within PEG-fibrinogen (PF) by using a novel microfluidic system (Tian and Lipke, 2020) in microspheroids with different sizes and axial ratios (AR). After culturing the hiPSC-laden microspheroids in E8 or mTeSR-1 media for 3 days, the CM differentiation is carried out by supplemented CDM3 or RPMI/B27 minus insulin with CHIR on day 0 and IWP2 on days 1 and 3, respectively. Fresh CDM3 was added on days 3, 5, 7, and fresh RPMI/B27 minus insulin media was added on days 1 and 5. Following day 7 or 10, the microspheroids were cultured with RPMI/B27 (Gibco), and the media was exchanged every 3–4 days. (Figure 1). Phase-contrast images were taken throughout the differentiation timeline on days 0, 1, 3, and 5, shown in Figure 1.

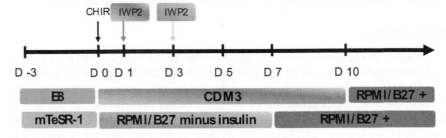

Figure 1. Differentiation protocol of hiPSC-laden microspheroids

2.2. Data Used to Build the Classifier Model

The initial training data set included 301 phase-contrast images, from day 5 of differentiation, with their corresponding CM content on day 10. Images on day 5 were used because day 5 is the earliest time point without any external stimuli or changes to the system with image availability. Each image contained 496 × 658 pixels. Figure 2 shows two representative images. The images were augmented to increase the number of training data points to improve the model's generalization. Each image was flipped and rotated (180°), increasing the number to 903.

CM content above 70% on the 10th differentiation day was defined as the *Sufficient* class, and batched with CM content below 70% belonged to the *Insufficient* class. The data was

Image classification of experimental yields for cardiomyocyte cells differentiated from human induced pluripotent stem cells

2223

split into test and train sets using 20% and 80% of data ratios, respectively. Different classifier models were compared based on their performance on the test set.

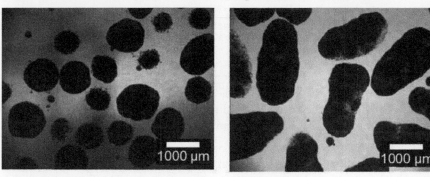

Figure 2. Representative phase-contrast images of microspheroids on day 5

2.3. Feature Extraction

The RGB color space features (color features) of the image pixels formed the initial input feature set. We used two techniques to extract additional features from the images, Histogram of Oriented Gradients (HOGs) (Freeman and Roth, 1994) and texture transformations (Haralick et al., 1973). The HOG feature descriptor is used for object detection and utilizes the local intensity gradient distributions to identify object edges in the images. In the texture transformation method, the grey level co-occurrence matrix (GLCM) is used to calculate six different statistical attributes to explain the image texture patterns. Four different directions, 0°, 90°, 45°, and 135°, were used to calculate the GLCM matrices. The six attributes derived from the co-occurrence matrix (Aborisade et al., 2014; Haralick et al., 1973) includes

1) **Contrast**, which is a measure of the local intensity variations,

$$Contrast = \sum_i \sum_j |i-j|^2 p(i,j) \qquad \text{Eq. 1}$$

2) **Dissimilarity**, which is a localized measure of distance for a pair of pixels,

$$Dissimilarity = \sum_i \sum_j |i-j| \, p(i,j) \qquad \text{Eq. 2}$$

3) **Angular Second Moment (ASM)**, which represents the orderliness of each window of the image,

$$ASM = \sum_{i,j} p(i,j)^2 \qquad \text{Eq. 3}$$

4) **Energy**, which is the square root of the ASM,

$$Energy = \sqrt{ASM} \qquad \text{Eq. 4}$$

5) **Homogeneity**, which represents the local homogeneity within the image by comparing the elements to the diagonal value of the GLCM matrix, and

$$Homogenity = \sum_i \sum_j \frac{1}{1 + |i - j|^2} \, p(i,j) \qquad\qquad \text{Eq. 5}$$

6) **Correlation**, which is a measure of the linear correlation between the grey-level values of neighbouring pixels.

$$Correlation = \sum_i \sum_j \frac{(i - \mu_i)\,(j - \mu_j)\,p(i,j)}{\sigma_i \sigma_j} \qquad\qquad \text{Eq. 6}$$

In Eqs. (1) – (6), $p(i,j)$ is the normalized value of the GLCM matrix element at row i and column j, and μ_i and σ_i are mean and variance for each row of the GLCM matrix components.

We constructed five feature sets as potential inputs for the classifier model using color features, HOG features, and texture transformation features. The first set includes all features (color+HOG+texture), the second color and HOG features (color+HOG), the third color and texture features (color+texture), the fourth HOG and texture features (HOG+texture), and the last one only texture (texture) features. Principal Component Analysis (PCA) (Hotelling, 1933) was used to reduce feature set dimensions. PCA uses orthogonal transformations to build components with a linear combination of the original input features to convert a set of possibly correlated features into uncorrelated ones. The principal components (PCs) explaining 95% of the variance in the input data were considered as classifier inputs.

2.4. Classifier Model Construction and Evaluation

Support Vector Machines (SVMs) (Drucker et al., 2002) were used as the classification models. Linear, radial basis function, and second and third-order polynomials, were evaluated as potential kernels for the SVMs. Kernel selection and regularization parameter tuning were carried out using five-fold cross-validation. Accuracy (Guyon and Elisseeff, 2003), recall (Sokolova and Lapalme, 2009), precision (Sokolova and Lapalme, 2009), and Mathew's correlation coefficient (MCC) (Matthews, 1975) were the metrics used for comparing the performance of the classifiers.

3. Results and Discussion

The performance of classification models in predicting the CM content class for the test points is shown in Figure 3. Figure 3 includes a plot of the performance metrics of the classifiers trained using each feature set. The classifiers were trained using the original data set and the augmented data set, and the performance metrics are plotted separately for these classifiers. The plots only include performance metrics calculated using the test data. Figure 3 reveals that the SVM employing the texture transformation features yielded the best performance with an accuracy of 77%, a recall of 92%, a precision of 75%, and an MCC of 0.53. The data augmentation improved the classifier model performance for the ones employing features other than textures transformations. Because texture features, except for those in which the GLCM matrix was calculated in 45° and 135° directions, are obtained using global transformations, their values are both rotation and flip invariant. As a result, the models that employ texture transformation features perform similarly when trained using the original data set or the augmented one.

The performance of classification models trained using PCs is given in Figure 4. The classifiers that employ the texture features had the best performance with an accuracy of

74% and an MCC of 0.51. The classifier model trained only using HOG and texture features for constructing the PCs, eliminating all color features, had the worst performance with recall, precision, and MCC of zero. Data augmentation, in general, improved the performance of the classifiers that used PCs as input sets. However, the performance metrics of the classifier models using PCs as inputs were lower (worse) than those of classifier models built using raw texture, color, and HOG features.

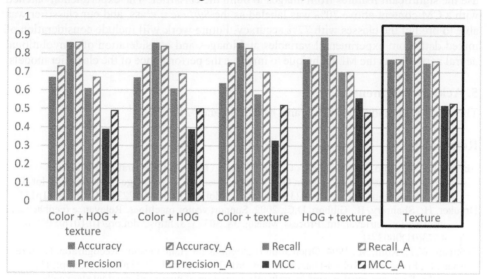

Figure 3. Bar plots of SVM classifier performance metrics trained using different feature sets for the original data set (solid bars) and augmented data set (dashed bars).

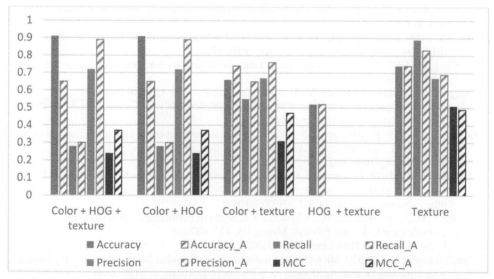

Figure 4. Bar plots of SVM classifier performance metrics trained using PC and different feature sets for the original data set (solid bars) and augmented data set (dashed bars).

4. Conclusions

Imaging is commonly used for tracking human induced pluripotent stem cell (hiPSC) differentiation. Using image-based classification, we built binary classification models to predict *Sufficient/Insufficient* classes of cardiomyocyte (CM) content in cells differentiated from hiPSCs. Feature extraction methods were implemented to identify and use the significant features from images to build the classifier. The experimental batched with a CM content above 70% was labeled as the *Sufficient* class, and our classifier was able to predict the classes with 77% accuracy. Future work will include consideration of mixed data from experimental variables and images and consideration of convolutional neural networks as the ML technique to improve the performance of the classifier models.

5. Acknowledgments

This work was funded by NSF grants #1743445 and #2135059.

References

Aborisade, D.O., Ojo, J.A., Amole, A.O., Durodola, A.O., 2014. Comparative Analysis of Textural Features Derived from GLCM for Ultrasound Liver Image Classification. Int. J. Comput. Trends Technol. 11, 239–244. https://doi.org/10.14445/22312803/ijctt-v11p151

Drucker, H., Shahrary, B., Gibbon, D.C., 2002. Support vector machines: Relevance feedback and information retrieval. Inf. Process. Manag. 38, 305–323. https://doi.org/10.1016/S0306-4573(01)00037-1

Freeman, W.T., Roth, M., 1994. Orientation Histograms for Hand Gesture Recognition. Gesture.

Gaspari, E., Franke, A., Robles-Diaz, D., Zweigerdt, R., Roeder, I., Zerjatke, T., Kempf, H., 2018. Paracrine mechanisms in early differentiation of human pluripotent stem cells: Insights from a mathematical model. Stem Cell Res. 32, 1–7. https://doi.org/10.1016/j.scr.2018.07.025

Guyon, I., Elisseeff, A., 2003. An introduction to variable and feature selection. J. Mach. Learn. Res. 3, 1157–1182.

Haralick, R.M., Shanmugam, K., Dinstein, I., 1973. Textural Features for Image Classification. IEEE Trans. Syst. Man. Cybern. SMC-3, 610–621.

Hotelling, H., 1933. Analysis of a complex of statistical variables into principal components. J. Educ. Psychol. 24, 417.

Kamgoué, A., Ohayon, J., Usson, Y., Riou, L., Tracqui, P., 2009. Quantification of cardiomyocyte contraction based on image correlation analysis. Cytom. Part A 75, 298–308. https://doi.org/10.1002/cyto.a.20700

Kempf, H., Andree, B., Zweigerdt, R., 2016. Large-scale production of human pluripotent stem cell derived cardiomyocytes. Adv. Drug Deliv. Rev. 96, 18–30. https://doi.org/10.1016/j.addr.2015.11.016

Matthews, B.W., 1975. Comparison of the predicted and observed secondary structure of T4 phage lysozyme. Biochim. Biophys. Acta - Protein Struct. 405, 442–451. https://doi.org/10.1016/0005-2795(75)90109-9

Sokolova, M., Lapalme, G., 2009. A systematic analysis of performance measures for classification tasks. Inf. Process. Manag. 45, 427–437. https://doi.org/10.1016/j.ipm.2009.03.002

Tian, Y., Lipke, E.A., 2020. Microfluidic Production of Cell-Laden Microspheroidal Hydrogels with Different Geometric Shapes. ACS Biomater. Sci. Eng. 6, 6435–6444. https://doi.org/10.1021/acsbiomaterials.0c00980

Vishnoi, V.K., Kumar, K., Kumar, B., 2021. Plant disease detection using computational intelligence and image processing, Journal of Plant Diseases and Protection. Springer Berlin Heidelberg. https://doi.org/10.1007/s41348-020-00368-0

Proceedings of the 14th International Symposium on Process Systems Engineering – PSE 2021+
June 19–23, 2022, Kyoto, Japan ©2022 Elsevier B. V. All rights reserved.
http://dx.doi.org/10.1016/B978-0-323-85159-6.50371-7

Prediction of API concentration using NIRS measured off-line and in-line instruments

Norihiko Fukuoka[a], Sanghong Kim[b*], Takuya Oishi[c], Ken-Ichiro Sotowa[a]

[a]*Department of Chemical Engineering, Kyoto University, Kyoto 615-8510, Japan*
[b]*Department of Applied Physics and Chemical Engineering, Tokyo University of Agriculture and Technology, Tokyo 184-8588, Japan*
[c]*Powrex, Hyogo 664-0837, Japan*

sanghong@go.tuat.ac.jp

Abstract

In the powder mixing process, it is important to uniformly mix components. Since it is difficult to directly measure the concentration in the equipment, a statistical model to predict concentration using NIRS (near infrared spectrum) has been studied in this work. Statistical models can be divided into two types: those constructed from data obtained by the in-line sensor (in-line models), and those constructed from data obtained by the off-line sensor (off-line models). In the off-line data acquisition, the amount of powder used in the experiment can be reduced, and mixing experiments using actual equipment are not required. Thus it would be better if data could be collected off-line. In this study, the prediction accuracies of the two models were compared, and it was found the prediction accuracy of the model tends to be higher when the data was measured in-line. Off-line model cannot predict in-line data because the measurement environment of off-line data is different from that of in-line data. And variable selection was used to improve the in-line prediction accuracy of the off-line model.

Keywords: NIR, Solid dosage forms, Mixing, PLS, PAT, Off-line, In-line

1. Introduction

In the powder mixing process, it is important to ensure the active pharmaceutical ingredients (API) and excipients are sufficiently mixed. However, it is difficult to directly measure API concentration in the mixing equipment. One of the ordinaly method to solve the problem is to construct a statistical model to predict the API concentration from the near infrared spectrum (NIRS). Statistical models can be divided into two types: those constructed from data obtained by the in-line sensor (in-line models), and those constructed from data obtained by the off-line sensor (off-line models). In the off-line data acquisition, the amount of powder and the scale of equipment are smaller than the in-line data acquisition. However, since off-line and in-line data are measured under different conditions, off-line model cannot accurately predict API concentration. There are many researches to construct a prediction model for API concentration in the mixing process as shown in Table 1. However, in previous studies, few studies have compared off-line models with in-line models.

Table 1: Prior studies

Number	Author	Year	Sensor type	Reference
1	Wee Beng Lee et al.	2019	In-line	[W. B. Lee et al. , 2019]
2	Barbara Bakri et al.	2015	In-line	[B. Bakri et al. , 2015]
3	Leonel Quinones et al.	2014	In-line	[L. Quinones et al. 2014]
4	Yleana M. Colon et al.	2014	In-line	[Y. M.Colon et al. , 2014]
5	Aditya U.Vanarase et al.	2013	In-line	[A. U.Vanarase et al. , 2010]
6	Otto Scheibelhofer et al.	2013	Off-line	[O.Scheibelhofer et al., 2013]
7	Sanghong Kim et al.	2011	Off-line	[S. Kim et al. , 2011]
8	Brian M. Zacour et al.	2011	In-line	[B. M. Zacour et al.2011]
9	Aditya U.Vanarase et al.	2010	In-line	[A. U. Vanarase et al., 2013]
10	Otto Berntsson et al.	2002	Off-line	[O. Berntsson et al., 2013]

The prediction accuracy of in-line data tends to be worse because off-line data was taken using an apparatus different from that of in-line data. In order to quantify the effect of the data acquisition method on the prediction accuracyand improve the in-line prediction accuracy of the off-line model, the prediction accuracy of the in-line and the off-lline models are calcurated and wavelength selection are used in this work.

2.   Experimental

Mixing experiments were conducted to mix acetaminophen(normal grade; Iwaki Seiyaku Co., Ltd) with lactose(Pharmatose 200M; DFE pharma). In the experiment, API and the excipient were put into the mixer and NIRS were measured through the sapphire window on a side of the equipment while the mixer was in operation. The equipment used in the experiment is shown in Figure 1(Powrex; MG-200 mixer). API concentration was changed from 1% to 40% as shown in Table 2. In the mixing experiment, the center blade rotation speed r_1 and scraper rotation speed r_2 (Figure 2) were set to the values in Table 3. The NIRS are combined with the API concentration calculated from the mass of feed powders. This data is called in-line data. Since mixing is not completed immediately, the spectra

Table 2: API concentration

Experiment number	API concentration [%]
1	1
2	10
3	15
4	20
5	30
6	40

Table 3: Operation condition of the mixer in each experiment

Number	r_1 [rpm]	r_2 [rpm]
1	0	0
2	500	20
3	0	0
4	500	100
5	0	0
6	1500	20
7	0	0
8	1500	100
9	0	0

measured after 150 seconds from the start of mixing operation were used to construct and validate the model. Therefore, the data measured in operation condition 1 and the first half of operation condition 2 were excluded from the analysis.

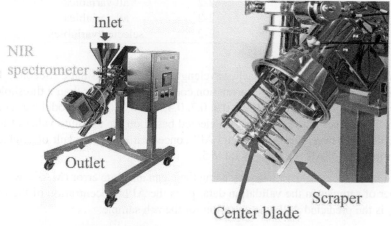

Figure 1: Powder mixer

Figure 2: Inside the mixer

After the powder was removed from the mixer and NIRS of the powder was measured off-line, the API concentration of the powder was measured by high performance liquid chromatography (HPLC). The data which consists of NIRS measured off-line and API concentration measured by HPLC is called off-line data.

3.　Model development and validation

The in-line and off-line data were split into two halves respectively, and one half was used for model construction and the other for validation. The measurement condition and usage of the data are shown in Table 4. The prediction models were constructed using partial

Table 4: Data overview

Data number	Sensor type	Usage
1-1	Off-line	Model construction
1-2	Off-line	Model validation
2-1	In-line	Model construction
2-2	In-line	Model validation

least squares (PLS). Standard normal variate was applied to NIRS, and the number of latent variables in PLS was set from 1 to 5. The models were validated with data 1-2 and 2-2 as shown in Table 5.

Correlation coefficients between absorbance and API concentration were calculated using data 1-1 and data 2-1 in order to select variables which have similar relationship between absorbance and API concentration in off-line and in-line. The procedure of variable selection can be divided into two steps. First, To calculate correlation coefficients between

Table 5: Condition of model construction and validation

Case	Model construction	Model validation	Variable selection
1	1-1	1-2	all variables
2	1-1	2-2	all variables
3	2-1	2-2	all variables
4	1-1	2-2	selected variables

absorbance and API concentration on each wavelength using off-line data. Second, To select variables whose absolute value of correlation coefficient is greater than a threshold. The threshold value is 0.98, 0.97, 0.96, 0.93, 0.9, 0.7, 0.5, 0.3, 0.1, 0. The model was constructed using a absorbance at the wavelength selected based on the absolute value of the correlation coefficient between absorbance and API concentration. The result of selected variables that minimizes MAE is case 4 in Table 5.

The accuracy of the models was evaluated by using the mean absolute error (MAE), where N is the number of samples in the validation data, y_n is the API concentration of the nth sample, and $\hat{y}_n$ is the predicted API concentration for the nth sample.

$$\text{MAE} = \frac{1}{N} \sum_{n=1}^{N} |\hat{y}_n - y_n| \tag{1}$$

4. Results

Figure 3 shows MAE obtained through validation for a model constructed using all variables. When the number of latent variables is 3, the MAE is 0.28%, 2.64% and 0.67% for the case 1~3, respectively. The minimum MAE was obtained for case1, in which

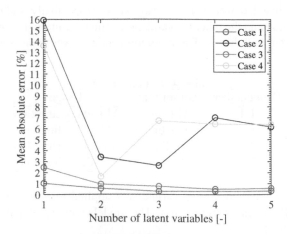

Figure 3: Validation result

off-line data was predicted with a model built with offline data. The MAE of case 2 are significantly larger than those in case 2 and 3.

In this study, the model construction with variable selection was also conducted. Figure 4 shows the relationship between the threshold used for variable selection and the minimum MAE of the model built with the chosen variables. The vertical axis is the MAE computed with the data 2-2, and the horizontal axis is the threshold value of variable selection. In the variable selection, two threshold values of 0.93 and 0.5 are taken as minima.

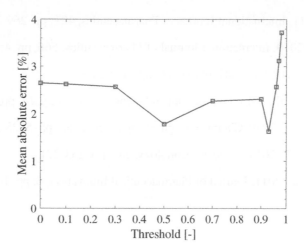

Figure 4: Variable selection result

MAE for the threshold value of 0.93 is smaller than that for the threshold value of 0.5. By comparing this result with case2, the prediction error reduces from 2.64% to 1.63% by using the threshold value of 0.93 for limiting the variables used for model building.

In Figure 4, the MAE is smaller when the thershold is between 0.5-0.93. When it is less than 0.5 the MAE becomes worse because useless input variables are inculded in the model. When the threshold is larger than 0.93 the MAE becomes worse because important input variables are not used for the prediction.

5. Conclusions

In this study, the prediction accuracy of the in-line and the off-line models was discussed. The results showed that there was a large difference in their prediction accuracy. Variable selection successfully reduced improving MAE of the off-line model from 2.64% to 1.63%. There could be some improvement in prediction accuracy since the best MAE of 1.63% for the offline model is slightly larger than the MAE of the inline model.

The reason for the difference in prediction accuracy between the two models is that there are some variables that are noisy in the inline measurements and these variables are adopted in the off-line model construction.

References

B. Bakri et al., 2015, European Journal of Pharmaceutical Biopharmaceutical, 97, pp. 78-89

O. Berntsson et al., 2002, Powder Technology, 123, pp. 185-193

Y. M.Colon et al., 2014, International Journal of Pharmaceutics, 417, pp. 32-47

S. Kim et al., 2011, International Journal of Pharmaceutics, 421, pp. 269-274

W. B. Lee et al., 2019, International Journal of Pharmaceutics, 566, pp. 454-462

L. Quinones et al., 2014, AIChE Journal, 60, pp.3123-3131

O. Scheibelhofer et al., 2013, Pharmaceutical Science Technology, 14, pp. 234-243

A. U. Vanarase et al., 2010, Chemical Engineering Science, 65, pp. 5728-5733

A. U. Vanarase et al., 2013, Powder Technology, 241, pp. 263-271

B. M. Zacour, et al., 2011, Journal of Pharmaceutical Innovation, 6, pp. 10-23

AUTHOR INDEX

Printed and bound by CPI Group (UK) Ltd, Croydon, CR0 4YY

03/10/2024

01040326-0013